FOURTH EDITION

Probability and Statistics for Engineers

RICHARD L. SCHEAFFER

University of Florida

JAMES T. McCLAVE

Infotech, Inc.

An Alexander Kugushev Book

Duxbury Press
An Imprint of Wadsworth Publishing Company
Belmont, California

Duxbury Press
An Imprint of Wadsworth Publishing Company
A division of Wadsworth, Inc.

Assistant Editor   *Jennifer Burger*
Editorial Assistant   *Michelle O'Donnell*
Production Editor   *Sandra Craig*
Cover and Text Designer   *Cloyce Wall*
Print Buyer   *Randy Hurst*
Art Editor   *Donna Kalal*
Permissions Editor   *Peggy Meehan*
Copy Editor   *Mary Roybal*
Cover Photograph   *FPG International*
Technical Illustrator and Compositor   *Interactive Composition Corporation*
Printer   *Banta Company*

 This book is printed on acid-free recycled paper.

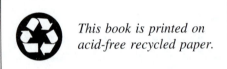 TM
International Thomson Publishing
The trademark ITP is used under license

Printed in the United States of America
7 8 9 10 — 03 02 01 00

Library of Congress Cataloging-in-Publication Data

Scheaffer, Richard L.
    Probability and statistics for engineers / Richard L. Scheaffer,
James T. McClave. — 4th ed.
        p.   cm.
    Includes bibliographical references and index.
    ISBN 0-534-20964-5
    1. Statistics. 2. Probabilities. I. McClave, James T.
II. Title.
TA340.S33   1995   93-43597
519.2'02462—dc20

T his book is a calculus-based introduction to probability and statistics, with emphasis on techniques and applications that are most useful in engineering. The examples and exercises strongly emphasize engineering applications. The problems presented, which deal with such topics as energy, pollution, weather, chemical processes, size and abundance of animals, global warming, earthquakes, and physical fitness, show the breadth of societal issues that engineers and scientists are called upon to solve.

Emphasis on Timely Topics

In addition to the general theme of quality improvement, certain topics suggest the applicability of this book to many areas of scientific investigation. Odds, odds ratios, and relative risk as tools for analyzing frequency data are presented in Chapter 2. The coverage of the normal distribution in Chapter 4 includes the use of quantile-quantile plots for assessing normality of data. Reliability concepts are also covered in Chapter 4. Control charts and process capability are introduced in Chapter 6 as applications of sampling distributions; all the major control charts used in industry are presented. Confidence intervals, prediction intervals, and tolerance intervals are covered in Chapter 7. The topic of model fitting through regression is covered in Chapters 9 and 10, which stress a modern data analysis point of view, with emphasis on regression diagnostics such as residual plots. The design aspect of problem solving is presented in Chapter 11, along with a modern view of analyzing data from designed experiments.

Quality and Process Control

The motivating theme of almost any successful business or industry today is improving the quality of processes. Whether the product is the food that nourishes us, the cars that transport us, the medicine that heals us, or the movies that entertain us, high quality at a reasonable price is in demand. Engineers, who are key personnel in many of these industries, must become proficient in the use of statistical techniques of process improvement. The theme of quality improvement is central to the development of statistical ideas throughout this book, beginning in Chapter 1 with an introduction to current process improvement ideas and techniques.

Modern Features

This book emphasizes problem solving through the collection and analysis of data. Examples and exercises contain real scenarios, often with the actual data presented.

"Activities for Students" sections, found near the end of most chapters, contain real data or outline methods for simulating data collection to illustrate practical points of statistical reasoning.

Appropriately, in its modern approach to data analysis, the book places much emphasis on graphical techniques, including stemplots, boxplots, and scatterplots to summarize data and look for patterns, quantile-quantile plots to assess distributional assumptions in data, and residual plots to check the underlying assumptions of regression.

To make these plots efficiently and to allow for easy handling of data in the context of making inferences and building models, each student should have access to a statistical software package on a computer. Many of the calculations in the text are made on Minitab, but any statistical software package will suffice. To ease the transition from text to computer, a data disk is provided with each copy of this textbook. The disk includes all medium to large data sets from the exercises and some from the examples.

Organization

The book is designed to be compact yet cover effectively the essentials of an introduction to statistics for those who must use statistical techniques in their work. Following a discussion in Chapter 1 of the use of data analysis techniques in quality improvement, Chapters 2 through 5 develop the basic ideas of probability and probability distributions, beginning with applications to frequency data in one- and two-way tables. Chapter 6, as a transition between probability and statistics, introduces sampling distributions for certain statistics in common use. Chapters 7 and 8 present the key elements of statistical inference through a careful look at estimation and hypothesis testing. Chapters 9 and 10 deal with the process of fitting models to data and then using these models in making decisions. Chapter 11 discusses the importance of designing experiments to answer specific questions and makes use of the models from earlier chapters in analyzing data from designed experiments.

Suggested Coverage

This book organizes the most important ideas in introductory statistics so all can be covered in a one-semester course. One possibility is to cover all of Chapters 1 and 2, followed by a more cursory coverage of Chapters 3, 4, and 5. Selections from Chapter 3 should include the binomial and Poisson distributions, while those from Chapter 4 should include the exponential and normal distributions.

For a more applications-oriented course, Chapter 5 can be skipped. The ideas of expectation of linear functions of random variables and correlation are presented again in later chapters, within the context of their application to data analysis and inference. Chapters 6, 7, and 8 present the heart of statistical inference and its more formal connections to quality assurance, and should be covered rather thoroughly. The main ideas of Chapters 9 and 10 should receive much attention in an introductory course, even at the expense of skipping some of the probability material in Chapters 3–5, because building statistical models by regression techniques is an essential topic for scientific investigations. Design and analysis of experiments, the main ideas of Chapter 11, are key to statistical process improvement and should receive considerable attention. The use of computer software for statistics will enhance coverage of Chapters 9, 10, and 11.

Some sections are marked as optional and can be skipped with no loss of continuity. The material on moment-generating functions is concentrated in Sections 3.9, 4.10, and 5.6. A method for approximating probability distributions is presented in Section 6.8. Maximum likelihood (Section 7.7) and Bayesian methods (Section 7.8) are not essential for the development of statistical methods presented in later chapters.

Acknowledgments

The authors hope that students in engineering and the sciences find the book both enjoyable and informative as they seek to learn how statistics is applied to their field of interest. We had much assistance in making the book a reality and wish to thank all of those who helped along the way. In particular, we want to thank the reviewers for their many insightful comments and suggestions for improvement: Robert L. Armacost, University of Central Florida; Paul I. Nelson, Kansas State University; and Harry O. Posten, University of Connecticut. A special thank you goes to Chris Franklin for her excellent work on the Solutions Manual. After many years, it was a real pleasure to work once again with Alex Kugushev as editor. He and his fine staff at Wadsworth Publishing Company made the work on this textbook much more pleasurable than it might have been; we thank them for their dedication to quality with a personal touch.

Richard L. Scheaffer
James T. McClave

Data and Decisions

About This Chapter

In today's information society, decisions are made on the basis of data. A student checks the calorie chart before selecting a fast-food lunch. A homeowner checks the efficiency rating before purchasing a new refrigerator. A physician checks the outcomes of recent clinical studies before prescribing a medication. An engineer tests the tensile strength of wire before it is wound into a cable. The decisions made from data—correctly or incorrectly—affect each of us every day of our lives. This chapter presents some basic ideas on how to systematically study data for the purpose of making decisions. These basic ideas will form the building blocks for the remainder of the book.

Contents

1.1   Quality Really Is Job One

"Quality is job one." This slogan is now synonymous with an American automobile manufacturer. But it is far more than a slogan! A commitment to quality has allowed this manufacturer to recapture a sizable share of the automobile market once lost to foreign competition. Similar stories can be told about many other firms whose products have risen in customer satisfaction since the firm began emphasizing quality within the production process, or continuous process improvement. Producing a product or service of high quality is a complex matter, but virtually all the success stories have one thing in common. Decisions were and are made on the basis of objective, quantitative information—data!

To see how Ford Motor Company emphasizes quality, data, and statistics, consider the following statement from its manual *Continuous Process Control*: "To prosper in today's economic climate, we—Ford, our suppliers and our dealer organizations— must be dedicated to never-ending improvement in quality and productivity. We must constantly seek more efficient ways to produce products and services that consistently meet customers' needs. To accomplish this, everyone in our organization must be committed to improvement and use effective methods. . . . The basic concept of using statistical signals to improve performance can be applied to any area where work is done, the output exhibits variation, and there is desire for improvement. Examples range from component dimensions to bookkeeping error rates, performance characteristics of a computer information system, or transit times for incoming materials."

Ford is not alone in its emphasis on effective use of statistics. The chairman/CEO of the Aluminum Company of America, Paul H. O'Neill, has stated: "As world competition intensifies, understanding and applying statistical concepts and tools is becoming a requirement for all employees. Those individuals who get these skills in school will have a real advantage when they apply for their first job."

Arno Penzias, Vice President for Research at AT&T Bell Laboratories, has written: "The competitive position of industry in the United States demands that we greatly increase the knowledge of statistics among our engineering graduates. Too many of today's manufacturers still rely on antiquated 'quality control' methods, but economic survival in today's world of complex technology cannot be ensured without access to modern productivity tools, notably applications of statistical methods" (*Science*, 1989, vol. 244, p. 1,025).

Even in our daily lives, all of us are concerned about quality. We want to purchase high-quality goods and services, from automobiles to television sets, from medical treatment to the sound at the local movie theater. Not only do we want maximum value for our dollar when we purchase goods and services, but we also want to live high-quality lives. Consequently, we think seriously about the food we eat, the amount of exercise we get, and the stresses that build up from our daily activities. How do we make the many decisions that confront us in our effort? We compare prices and value, we read food labels to determine calories and cholesterol, we ask our physician about possible side effects of a prescribed medication, and we make mental notes about our weight gains or losses from day to day. In short, we, too, are making daily decisions on the basis of objective, quantitative information—data!

Formally or informally, then, data provide the basis for many of the decisions made in our world. Use of data allows decisions to be made on the basis of factual information rather than subjective judgment, and the use of facts should produce better results. Whether or not the data lead to good results depends upon how the data were produced and how they were analyzed. For many of the simpler problems confronting us, data are readily available. Such is the case, for example, when we evaluate the caloric content of food or decide on the most efficient appliance. So techniques for analyzing data will be presented first, with ideas on how to produce good data coming later.

1.2   A Model for Problem Solving

Whether we are discussing products, services, or lives, improving quality is the goal. Along the pathway to improved quality, numerous decisions must be made—and made on the basis of objective data. It is possible, however, that various decisions could affect one another, so the wise course of action is to view the set of decisions together in a wider problem-solving context.

A civil engineer is to solve the problem of hampered traffic flow through a small town with only one main street. After the collection of data on the volume of traffic, the immediate decision seems easy—widen the main street from two to four lanes. Careful thought might suggest, though, that more traffic lights will be needed for the wider street so that traffic can cross it. In addition, traffic now using the side streets will begin using the main street once it is improved. A "simple" problem has become more complex once all the possible factors have been brought into the picture. The whole process of traffic flow can be improved only by taking a more detailed and careful approach to the problem.

The main theme of this chapter, and of this book, is that data analysis is to be used to improve the quality of a process, whether the "process" is traffic flow, production of automobiles, purchase of a VCR, or studying for an exam. All aspects of the process must be examined, as the goal is to improve the process throughout. Because this procedure may involve the study of many variables and how they interrelate, a systematic approach to problem solving will be essential to our making good decisions efficiently. In recent years, numerous models for solving problems have evolved within business and industry; the model outlined below contains the essential steps present in all of them.

1.2.1   State the Problem or Question

This sounds like an obvious and, perhaps, easy step. But going from a loose idea or two about a problem to a careful statement of the real problem requires careful assessment of the current situation and a clear idea of the goals of the study. [I am pressed to get my homework assignments done on time and I do not seem to have adequate time to complete all my reading assignments. I still want to work out each day and spend some time with friends. Upon review of my study habits, it seems that I study rather haphazardly and tend to procrastinate. The real problem, then, is not to find more hours for study but to develop a study plan that is efficient.]

1.2.2 Collect and Analyze Data

Now, a listing of all factors is made and data are collected on all factors thought to be important. A plan is directed toward solving the specific problem addressed in the first step. Appropriate data analysis techniques are used according to the data collected. [The data show that I study 5 hours a day and work late into the evening. They also show that the gym is crowded when I arrive, and this may slow down my workout.]

1.2.3 Interpret the Data and Make Decisions

After the data have been analyzed and the analysis carefully studied, potential solutions to the original problem or question may be posed. [I observe that the studying late into the evening often occurs because I watch television and visit with friends before studying and am tired when I begin studying. I will set a schedule that puts my study time early in the evening and my visiting later in the evening. Also, I will work out in the mornings rather than the afternoons to save time.]

1.2.4 Implement and Verify the Decisions

Once a solution is posed, it should be put into practice on a trial basis (if that is feasible). New data should be collected on the revised process to see if improvement is actually realized. [I tried the earlier study time for two weeks, and it worked fine. I seemed to have more time to complete my assignments, even though the data show that I was not devoting any more hours to study. The gym is just as busy in the morning, and so I realized no saving of time by the new strategy.]

1.2.5 Plan Next Actions

The trial period of step 4 may show that the earlier decision solved the problem. More likely, though, the decision led to results that were only partially satisfactory. In any case, there is always another problem to tackle, and a response to additional problems should be planned now, while the whole process is still firmly in mind and the data are still fresh. [I would still like to find more time for pleasure reading. I will see how I could fit that into my revised schedule.]

1.3 A Real Application of the Problem-Solving Model

Office personnel responsible for the processing of customer payments to a large utility company noticed that they were receiving complaints both from customers ("Why hasn't my check cleared?") and from the firm's accounting division ("Why did it take so long to get these receipts deposited?"). The office staff decided it was time for their quality-improvement training to swing into action. The following is a brief summary of their quality-improvement story.

1.3.1   State the Problem or Question

After brainstorming on the general problem, the team collected background data on elapsed time for processing payments and found that about 1.98% of the payments (representing 51,000 customers) took more than 72 hours to process. They suspected that many of these were multis (payments containing more than one check or bill) or verifys (payments in which the bill and the payment do not agree). Figure 1.1 is a *Pareto chart* that demonstrates the correctness of their intuition; multis account for 63% of the batches requiring more than 72 hours to process but comprise only 3.6% of the total payments processed. Verifys are not nearly so serious as first suspected but are still the second leading contributor to the problem. The problem can now be made specific: Concentrating first on the multis, reduce payments requiring more than 72 hours of processing time to 1% of the total.

FIGURE **1.1**
A Pareto Chart

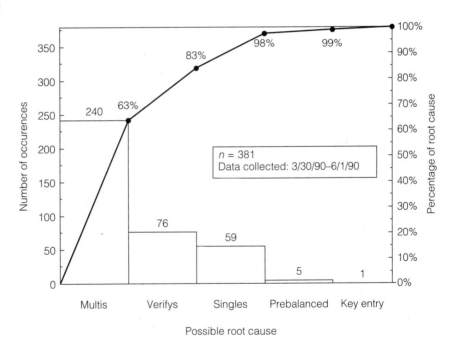

Batches Processed in Over 72 Hours

1.3.2   Collect and Analyze Data

What factors affect the processing of multis? A *cause and effect* diagram, shown in Figure 1.2, is used to list all important factors and to demonstrate how they might relate to one another. A missing check digit on a bill leads to an incorrect account number, which in turn causes a nonscannable bill. Sometimes a customer keeps the bill as a receipt, and the check cannot be processed. The largest single factor that could be corrected, in this case, was the fact that the computerized auto-balancing process was unable to handle multiple checks.

F I G U R E **1.2**   A Cause and Effect Diagram

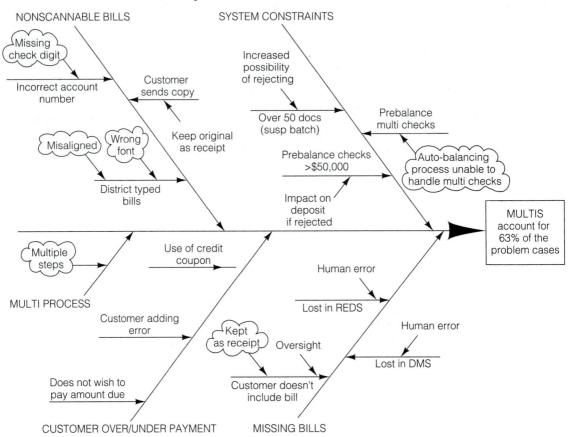

Another factor, cash carryover from the previous day, was shown to have an effect on the entire payment processing system. The effect on elapsed time of processing payments can be seen in the *scatterplot* of Figure 1.3. An efficient system for handling each day's bills must reduce the carryover to the next day.

F I G U R E **1.3**

A Scatterplot

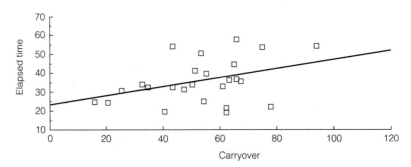

1.3.3   Interpret the Data and Make Decisions

With data in hand that clearly showed the computerized check processing system to be a major factor in the processing delays, a plan was developed to reprogram the machine so that it would accept multiple checks and coupons as well as both multis and verifys that were slightly out of balance.

1.3.4   Implement and Verify the Decisions

The reprogramming was completed and, along with a few additional changes, produced results that were close to the goal set forth in the problem statement. The record of processing times during a trial period is shown in the *time series plot* of Figure 1.4.

F I G U R E **1.4**   A Time Series Plot

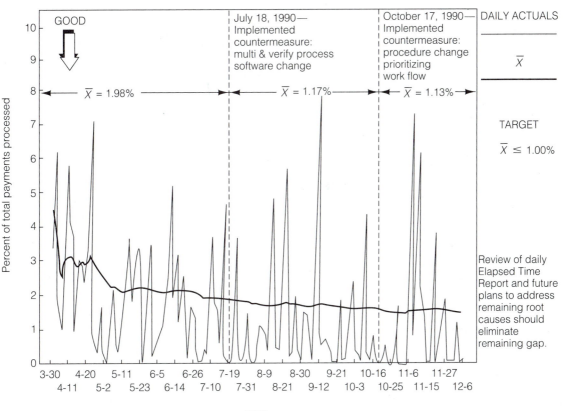

Payments with Elapsed Time Greater Than 72 Hours
Tuesday–Friday Elapsed Time Reported

1.3.5   Plan Next Actions

The carryover problem remains, and, along the way, it was discovered that the mail-opening machine had an excessive amount of downtime. Resolution of these two problems led to an improvement that exceeded the goal set in the problem statement.

1.4   Basic Tools for Analyzing Data

The tools implicitly introduced in the utility company example will now be explained in more detail, and more tools will be added to the list. These tools will be used throughout the book as basic components of data analysis problems. They will serve you well as keys to solving any problem that involves data collection and analysis.

What policies should be recommended in order to improve air quality in the United States? That overarching question motivates the data analysis examples of this section. On a question of this magnitude, we obviously will not be able to carry out a policy to see if it is effective and plan further action from that point. We can, however, take a serious look at the first three steps of the problem-solving model, all of which involve careful and constructive use of data.

Background work on assessing the quality of today's air and the improvements that may have come about in recent years can begin with data on the *pollutant standards index* (*PSI*). The pollutants that contribute most to the PSI are carbon monoxide (CO), ozone (O_3), particulate matter (PM), and sulfur dioxide (SO_2). Table 1.1 gives emission estimates for the past ten years for both carbon monoxide and ozone (which comes mainly from volatile organic compounds). The source categories and changes over the years can be seen from the table, but the important features of the data can be seen more clearly through graphic displays.

T A B L E **1.1**   Carbon Monoxide and Ozone Sources, 1981–1990 (million metric tons/year)

Source Category	1981	1982	1983	1984	1985	1986	1987	1988	1989	1990
Carbon Monoxide:										
Transportation	55.4	52.9	52.4	50.6	47.9	44.6	43.3	41.2	40.0	37.6
Fuel combustion	7.7	8.2	8.2	8.3	7.5	7.5	7.6	7.6	7.8	7.5
Industrial processes	5.9	4.4	4.3	4.7	4.4	4.2	4.3	4.6	4.6	4.7
Solid waste	2.1	2.0	1.9	1.9	1.9	1.8	1.8	1.7	1.7	1.7
Miscellaneous	6.4	4.9	7.8	6.4	7.1	5.1	6.4	9.5	6.3	8.6
Total[†]	77.5	72.5	74.5	71.9	68.7	63.2	63.4	64.7	60.4	60.1
Volatile Organic Compounds:										
Transportation	8.9	8.3	8.2	8.1	7.6	7.2	7.1	6.9	6.4	6.4
Fuel combustion	1.0	1.0	1.0	1.0	0.9	0.9	0.9	0.9	0.9	0.9
Industrial processes	8.3	7.5	7.9	8.9	8.5	8.0	8.3	8.1	8.1	8.1
Solid waste	0.7	0.6	0.6	0.6	0.6	0.6	0.6	0.6	0.6	0.6
Miscellaneous	2.4	2.1	2.7	2.6	2.5	2.2	2.4	2.9	2.5	2.7
Total[†]	21.3	19.6	20.4	21.2	20.1	19.0	19.3	19.4	18.5	18.7

Source: U.S. Environmental Protection Agency, Office of Air Quality Planning and Standards.

[†]The sums of subcategories may not equal total due to rounding.

Pareto Chart   The relative importance of the source categories for the various pollutants must be known before appropriate policy decisions can be made. A device for showing these relative contributions is the *Pareto chart*, a simple bar graph with the height of the bars proportional to the contributions from each source and the bars ordered from tallest to shortest. Figure 1.5 provides the Pareto charts for carbon monoxide and ozone, comparing 1981 to 1990. Clearly, transportation was the largest contributor to carbon monoxide in both 1981 and 1990, although its share was reduced. Transportation was the largest contributor to ozone in 1981, but improvements have been so dramatic that it fell into second place (behind industrial processes) by 1990. The cumulative relative frequency plot on the Pareto chart shows the contribution of combined categories; it is easily seen that, in 1990, industrial processes and transportation together were the source of over 77% of the ozone problem. The question on policy decisions might now be refined to specific goals related to improving the carbon monoxide pollution from transportation and the ozone pollution from industrial processes.

You might be interested in a little background on the Pareto chart. In the late nineteenth century, the Italian economist V. Pareto showed that the distribution of income is very uneven; most of the world's wealth is controlled by a small percentage of the people. The American economist M. C. Lorenz picked up this same theme and drew up a diagram to represent the way in which much wealth is controlled by very few people. An early expert in quality control, J. M. Juran, applied Lorenz's diagram to problems in industrial processes to show that most defects arise from a relatively small number of causes. He named the diagram a Pareto chart and referred to the process of sorting out the vital few causes from the trivial many as Pareto analysis. The Pareto chart is a simple device, but underlying it is a much deeper theory of process improvement: Sort out the few causes of most of your problems, and solve them first.

Cause and Effect Diagram   Many of the factors that affect air quality are now coming into the picture, and it is time to list them, think about how they are interrelated, and begin to weight them as to their relative importance. All of this can be shown succinctly on a *cause and effect diagram* (or *fishbone*). Figure 1.6 provides some of the essentials of a cause and effect diagram for the problem of improving air quality. Important factors such as transportation and industrial processes have already been singled out, but careful construction and study of a cause and effect diagram can show finer breakdowns. Here, for example, agriculture is seen as an important contributor to particulate matter, while chemical manufacturing is a major contributor to ozone. Even factors beyond our control, such as the weather, should be listed and their importance assessed.

Stem-and-Leaf Plot (Stemplot)   We have studied some of the components of the PSI; let us now turn to this measure itself and see what it tells us about air quality around the United States. Table 1.2 shows the number of days for which PSI was greater than 100 for each of ten years in each of 15 metropolitan areas. (A PSI greater than 100 indicates unhealthy air.) It is not easy to see patterns in data arrayed like this, so we look for simple graphical devices that will help us see patterns more clearly. One of the simplest, but most useful, is the stem-and-leaf plot or stemplot, as we will call it.

F I G U R E **1.5**
Carbon Monoxide and Ozone
Emissions

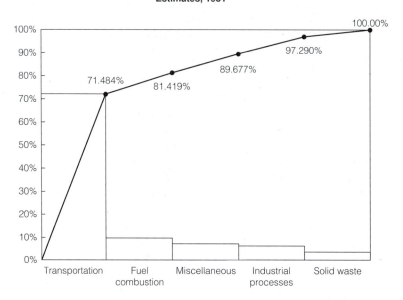

**National Carbon Monoxide Emission
Estimates, 1981**

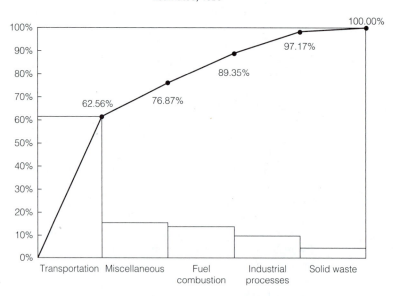

**National Carbon Monoxide Emission
Estimates, 1990**

F I G U R E **1.5**
(Continued)

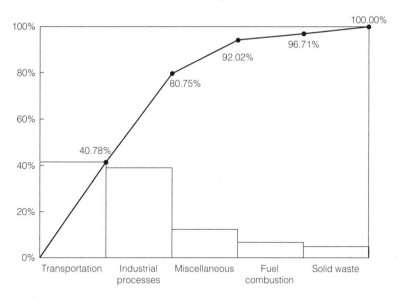

**National Volatile Organic Compound
Emission Estimates, 1981**

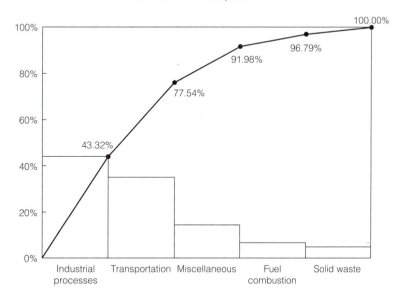

**National Volatile Organic Compound
Emission Estimates, 1990**

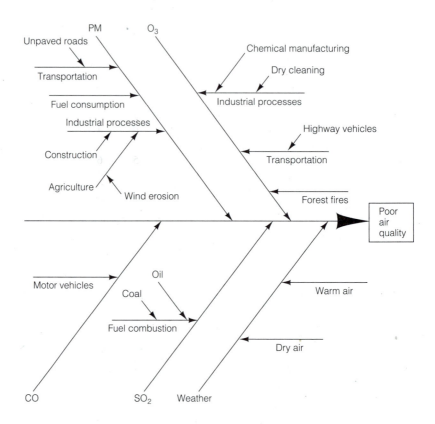

Stemplots of the annual data for Atlanta, Chicago, Houston, and New York are given in Figure 1.7. Looking at the Atlanta data in Table 1.2, observe that the numbers range from 3 to 23. The integer in the tens position gets repeated quite often, so we will use its values to form bins in which to cast the data. But this would produce only three bins, since only three integers (0, 1, 2) appear in the tens position. A more interesting plot can be made by splitting these bins in half so that unit values between 0 and 4 go into the first bin, those between 5 and 9 go into the second bin, and so on. The result of such plotting is a stemplot with the tens digit forming the stem and the units digit forming the leaves. The column to the extreme left gives the cumulative frequencies, counting in from either extreme.

Notice that other stemplots in Figure 1.7 have different patterns and break out the stems in different ways. The data for Chicago have so little variability that it is convenient to construct the stem by dividing each tens digits into five bins. The Houston data have much greater variability but are shifted toward higher numerical values.

In order to discuss patterns in data, we must agree upon a common language. The Atlanta stemplot is somewhat *mound-shaped* with two *clusters* of points and a *gap* between them. The Chicago stemplot is also somewhat mound-shaped but is *skewed* in the positive direction. That is, the distribution has a long tail in the direction of the larger values. The Houston stemplot is mound-shaped and almost *symmetric*, whereas the New York stemplot is somewhat symmetric and *U-shaped*.

T A B L E **1.2**   Pollution Standards Index, 1981–1990: Number of PSI Days Greater Than 100 at All Ozone Trend Sites

O_3 Trend Sites		Year									
		1981	1982	1983	1984	1985	1986	1987	1988	1989	1990
Atlanta	2	9	3	23	8	9	17	19	15	3	16
Boston	1	0	3	10	6	2	0	4	11	1	1
Chicago	5	3	3	12	6	6	2	9	15	1	0
Dallas	2	12	11	17	10	11	5	6	3	3	5
Denver	2	4	4	11	1	0	1	4	3	0	0
Detroit	8	12	17	16	4	1	3	6	16	10	3
Houston	4	33	22	43	30	30	26	29	31	19	35
Kansas City	5	4	0	4	11	3	3	2	3	1	2
Los Angeles	13	152	133	142	154	153	159	146	165	137	116
New York	5	21	20	33	16	13	6	14	30	5	8
Philadelphia	10	22	33	52	22	25	19	32	34	17	11
Pittsburgh	5	9	4	15	0	2	2	7	21	4	0
San Francisco	2	1	0	2	0	1	0	0	0	0	0
Seattle	1	1	0	0	0	0	1	0	1	0	2
Washington	11	12	19	38	12	12	9	18	33	4	5
Total	76	295	272	418	280	268	253	296	381	205	204

Source: U.S. Environmental Protection Agency, Office of Air Quality and Standards.

Figure 1.8 shows the stemplots for the 1981 and 1990 PSI data, across all cities. Notice the different selections of leaf unit. How would you explain the difference in patterns between these two years?

Histogram   Stemplots can become cumbersome to construct and study if the data sets are large. Histograms preserve some of the essential features of stemplots with regard to displaying the shape of data distributions, but they are not limited by the size of the data set. Since histograms are most useful for large data sets, the height of each bar usually represents the proportion or percentage (rather than the number) of data points that fall inside the interval covered by the bar. Such *relative frequency histograms* are shown in Figure 1.9 for each of the four main contributors to the PSI. The data are peak daily readings for 1990 from various sites around the country.

Particulate matter readings (Figure 1.9a) show a high percentage of values between 25 and 30, and a very high concentration of values between 20 and 35. The distribution is severely skewed toward the high values and shows two gaps toward the higher end. (Does it seem logical that peak readings should be skewed in this way?) Sulfur dioxide readings (Figure 1.9b) are similarly skewed toward the high values, but less severely. Carbon monoxide peak values (Figure 1.9c) appear to be distributed in a rather symmetric, mound-shaped fashion, while the ozone readings (Figure 1.9d) appear to have two mounds, a large one and a small one (often called a bimodal distribution).

F I G U R E **1.7**
PSI Days for Selected Cities

Stem-and-leaf of Atlanta $N = 10$
Leaf unit $= 1.0$

```
2          0 | 33
5          0 | 899
5          1 |
5          1 | 5679
1          2 | 3
```

Stem-and-leaf of Chicago $N = 10$
Leaf unit $= 1.0$

```
2          0 | 01
5          0 | 233
5          0 |
5          0 | 66
3          0 | 9
2          1 |
2          1 | 2
1          1 | 5
```

Stem-and-leaf of Houston $N = 10$
Leaf unit $= 1.0$

```
1          1 | 9
2          2 | 2
4          2 | 69
(4)        3 | 0013
2          3 | 5
1          4 | 3
```

Stem-and-leaf of New York $N = 10$
Leaf unit $= 1.0$

```
3          0 | 568
5          1 | 34
5          1 | 6
4          2 | 01
2          2 |
2          3 | 03
```

F I G U R E **1.8**
PSI Days for 1981 and 1990

Stem-and-leaf of 1981 $N = 10$
Leaf unit $= 10.0$

(11)	0	00000000111
4	0	223
1	0	
1	0	
1	0	
1	1	
1	1	
1	1	5

Stem-and-leaf of 1990 $N = 10$
Leaf unit $= 1.0$

(11)	0	00001223558
4	1	16
2	2	
2	3	5
1	4	
1	5	
1	6	
1	7	
1	8	
1	9	
1	10	
1	11	6

One note of caution is in order: When you are constructing histograms, the intervals that form the width of the bars should be of equal length. That allows the area of a bar, relative to the total area of all bars, to reflect the proportion of the data set that is actually inside the interval covered by the bar. A correct visual image of the data distribution is portrayed in this way.

Boxplot Stemplots and histograms are excellent graphic displays for focusing attention on key aspects of the shape of a distribution of data (symmetry, skewness, clustering), but they are not good tools for making comparisons among data sets. In Figure 1.8, for example, it is difficult to pick out the key differences in regard to PSI days between 1981 and 1990. A simple graphical display that is ideal for making comparisons is the *boxplot*. To construct a boxplot, we must look at some *numerical* summaries of data to accompany the graphical summaries we have been using.

Features of a set of data are conveniently divided between those that measure the *center* of the data distribution and those that measure the *variability* in the distribution. A simple measure of center is the *median*, the value that falls in the middle when the data values are ordered. A simple measure of variability is the *range*, the difference between the largest and smallest data values. A more sensitive measure of variability involves finding the *quartiles* and observing where they fall relative to the median. The *lower quartile* is the middle value among the data points below the median, and the *upper quartile* is the middle value among the data points above the median.

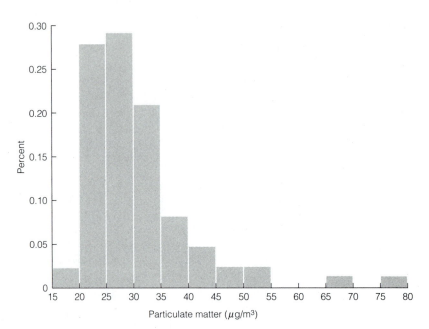

a

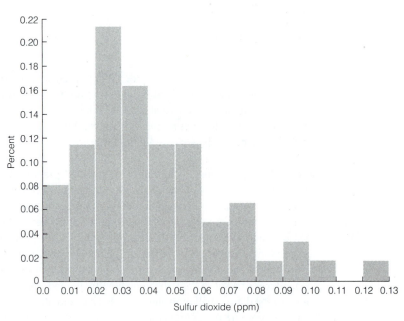

b

F I G U R E **1.9**
(Continued)

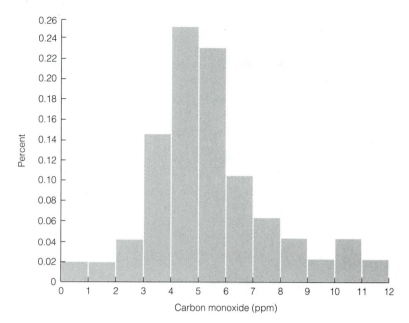

c

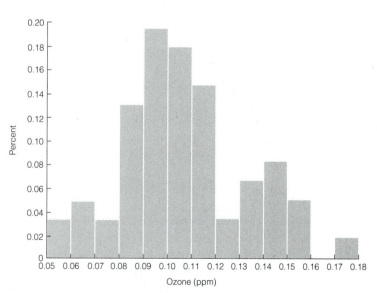

d

It is time for an example. Look at the Atlanta PSI data as displayed on the stemplot in Figure 1.7. There are ten data points here, so the median must fall between the two middle values (the fifth and sixth ones). The fifth value is 9, and the sixth value is 15; splitting the difference (in the spirit of medians) gives a median of 12. The middle value among the five data points below the median is 8, the lower quartile. The middle value among the five data points above the median is 17, the upper quartile. Note that the extreme values for the Atlanta data are 3 and 23.

A five-number summary that includes the extremes, the quartiles, and the median tells a great deal about key features of a data set and forms the basis of the boxplot. To construct a boxplot, we plot these five numbers on a real-number line, connect the quartiles by a narrow box, and mark the median with a line drawn across the box. Figure 1.10 shows such a boxplot for Atlanta, followed by boxplots for four other cities.

FIGURE **1.10**
Comparisons of PSI Days Among
Cities

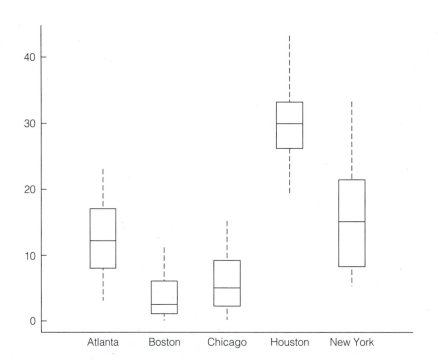

What features stand out in the parallel boxplots? First, we see some of the same features that were apparent in the stemplots and histograms. The Atlanta and Houston data sets are fairly symmetric, while Chicago's data are positively skewed. However, the clusters and gaps cannot be seen in the boxplots. Second, we see some new comparative features that could not be spotted easily in the stemplots. Houston has the highest values. Atlanta and New York have similar data values, but New York has greater variability than Atlanta. (What feature of the plot shows the greater variability?) Boston has smaller values with less variability than Chicago, but the two cities are similar. More could be said, but the main point is that boxplots summarize key features of center and variability and are ideal for comparing these features across data sets.

Data analysis at this exploratory level can be thought of as detective work with data, and one important bit of detection is to find data values that appear to be uniquely different from the rest of the data set. These different values are called *outliers*. An outlier might indicate that something went wrong with the data collection process, as you would likely think if you saw a human height recorded as 9.5 feet, or that something important caused an unusual value, as you might think if you saw a daily rainfall recording of 4.7 inches. In any case, a method for drawing attention to outliers is essential in data analysis. The method to be used here depends on the quartiles and the difference between the quartiles, called the *interquartile range (IQR)*. An outlier is defined as any data point more than a distance of 1.5(IQR) from either end of the box in a boxplot.

A summary of the information on medians, quartiles, and outliers follows.

Summary of Medians and Quartiles

For any data set, the **median** is the value in the middle of the ordered array, the **lower quartile** is the middle value of the half of the data below the median, and the **upper quartile** is the middle value of the half of the data above the median. Notationally,

$$Q_L = \text{lower quartile}$$
$$Q_M = \text{median (middle quartile)}$$
$$Q_U = \text{upper quartile}$$

Then an **outlier** will be any data point that lies below

$$Q_L - 1.5(\text{IQR})$$

or above

$$Q_U + 1.5(\text{IQR})$$

For the Atlanta data used above, IQR $= 17 - 8 = 9$ and $1.5(9) = 13.5$. The lower quartile is 8, and $8 - 13.5$ will be negative (no count of days could be negative). The upper quartile is 17, and $17 + 13.5 = 30.5$, which exceeds all the data points. Thus, the Atlanta PSI data have no outliers, but we will soon encounter outliers in another data set.

In addition to the comparisons among cities made in Figure 1.10, we might also want to compare years. Figure 1.11a shows the boxplots for 1981, 1985, and 1990. The data points not connected by dotted line to the box itself are the outliers. Los Angeles is an outlier all three years, and Houston becomes an outlier in 1990. Extreme outliers like Los Angeles tend to dominate the display and make it difficult to see other patterns. Figure 1.11b shows what happens when Los Angeles is removed from the data set. Houston now becomes an outlier in all three years (Why did this change?),

F I G U R E **1.11**
Comparisons of PSI Days Across
Years

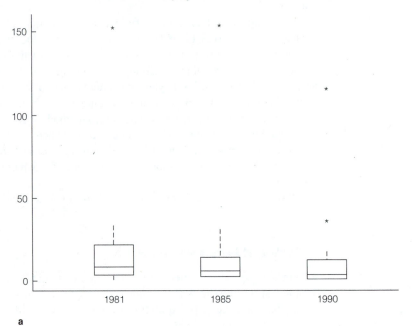

Number of PSI days greater than 100 at ozone trend sites

a

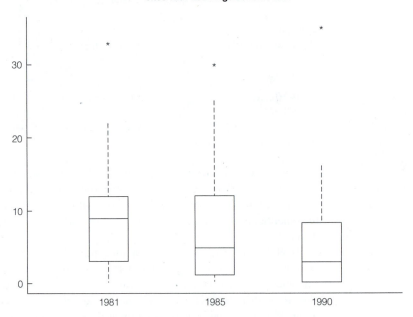

**Number of PSI days greater than 100 at ozone trend
sites with Los Angeles removed**

b

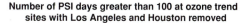

**Number of PSI days greater than 100 at ozone trend
sites with Los Angeles and Houston removed**

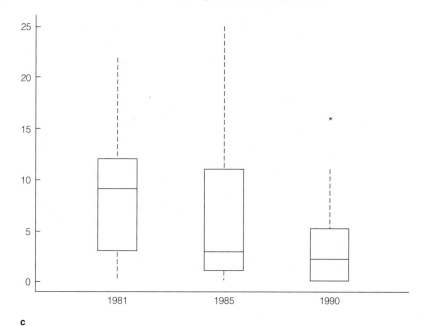

c

but a better comparative picture of the years emerges. Removing Houston from the data set (Figure 1.11c) produces other changes. (What city now becomes an outlier?) Clearly, the PSI days above 100 are decreasing in number as we move from 1981 through 1985 and on to 1990. In addition, the variability in scores among cities is decreasing, which is a further indication of improvement in air quality.

It should be noted at this point that most of the graphical displays shown above are somewhat time-consuming to draw by hand, especially with large data sets. As a user of statistics, you should become familiar with at least one computer software package that will make standard plots and perform routine statistical calculations. There are many on the market, including MINITAB, SYSTAT (or MYSTAT), STATGRAPHICS (or EXECUSTAT), DATA DESK, NCSS, SAS, and SPSS. In addition, many of the popular spreadsheet packages will produce standard statistical plots and calculations. One caution is in order: The software packages vary in the way they define quartiles. The lower quartile must be a point at approximately the middle of the lower half of the data set, as defined above, but some refinements can be made to the definition we used. So do not be surprised if boxplots produced by two different computer packages look slightly different; they will convey essentially the same features of the data.

Scatterplot So far in our study of data that might shed light on policies necessary for air-quality improvement, we have concentrated on plotting one variable at a time. Some aspects of the policies we formulate could, however, take advantage of associations between variables. If, for example, two cities seem to have similar records, perhaps the same policies would be good for both. If two of the pollutants seem to be associated with each other, perhaps one change in, say, manufacturing operations will improve both. The simplest tool available for looking for associations between two variables is the *scatterplot*.

A scatterplot is simply a plot of paired data points on a rectangular coordinate system. In Table 1.2, look at the two lines of PSI days across the years for Boston and Chicago. In 1981, Boston had a 0 and Chicago had a 3, which produces the ordered pair (0,3). Each of the other years produces a similar ordered pair; the plotted ordered pairs are shown in Figure 1.12a. Notice, in this plot, that Boston and Chicago tend to vary together; when one has a high value, the other tends to have a high value as well. This pattern is called *positive association* (or *positive correlation*). For Boston and Chicago, the association appears to be fairly strong, so the same policies might be appropriate for each city.

Looking at Figure 1.12b, we see that Atlanta and Boston may be positively associated (with regard to PSI days), but the association is much weaker than that between Boston and Chicago. The solution to Atlanta's problems may not have much in common with the solutions to Boston's problems (which, incidentally, are not great).

How are the 1990 PSI values, across cities, associated with those of 1981? Can we graphically display the changes across the decade of this study? Pairing on cities, Figure 1.13a shows the scatterplot of 1990 against 1981. It looks as if there is a positive association, as we would expect, but it is difficult to see what is happening for most of the cities because of the strong influence of Los Angeles. Putting the Los Angeles values on the graph with the others causes a scaling problem; all the other cities cluster together in one corner. To alleviate this difficulty, Los Angeles was removed from the data set; the new scatterplot is shown in Figure 1.13b. Now Houston has the greatest influence, but it is possible to see the pattern for the other cities, and the association is still positive.

To keep track of the cities, the points in the scatterplot can be labeled, as in Figure 1.13c. Which cities are making the largest gains? (Remember, the small numbers are good in this case.) Suppose we draw a $y = x$ line diagonally across the scatterplot. A point below that line would indicate that the 1981 value exceeds the 1990 value, which is a good sign. Fortunately, most of these data points do lie below the $y = x$ line. What can we say about a city that lies close to this diagonal line? What can we say about a city that lies above the $y = x$ line? Looking at the scatterplot in this way, can we identify those cities on which we should concentrate our efforts at improving air quality?

F I G U R E **1.12**
PSI Associations Between Cities
(number of PSI days greater
than 100)

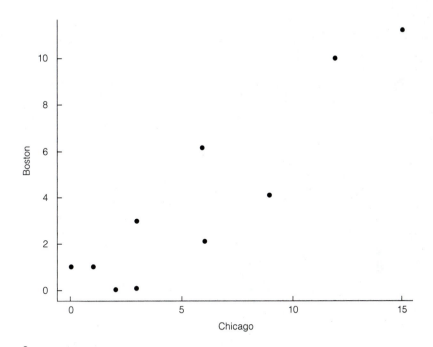

a

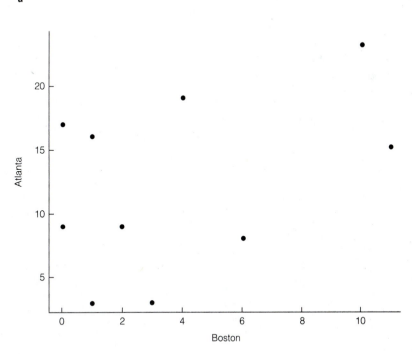

b

F I G U R E **1.13**
PSI Association Between 1990
and 1981

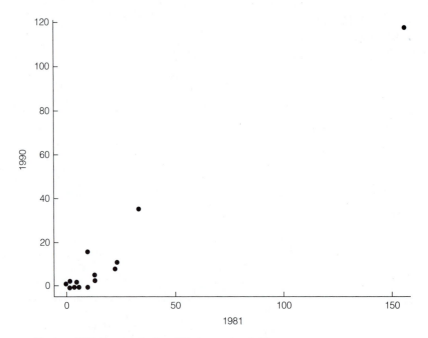

a   Number of PSI days greater than 100 at ozone trend sites

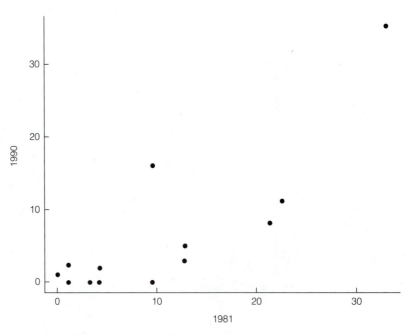

b   Number of PSI days greater than 100 at ozone trend sites with Los Angeles removed

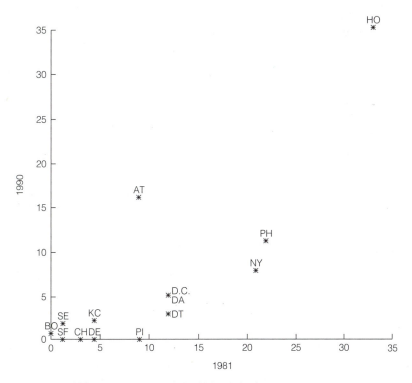

c Number of PSI days greater than 100, with labeled points

Recall that one important aspect of problem solving is to look carefully at the interrelationships among all factors affecting the outcome of interest. Perhaps there are associations among the four pollutants making up the PSI that would allow some economy in their treatment. The scatterplots in Figure 1.14 show some relationships between pollutant concentrations paired on selected sites around the country. The measurements are peak concentrations for 1990, with particulate matter measured in $\mu g/m^3$ (microgram per cubic meter) and the other pollutants measured in ppm (parts per million). The plot of particulate matter against carbon monoxide shows some positive association, with an increase in variability among PM measurements as the level of CO increases. On further checking, we discover that particulate matter comes mostly from factories, construction sites, and motor vehicles, while carbon monoxide comes mostly from motor vehicles. Since both involve motor vehicles and since construction sites are usually in populated areas, perhaps both of these pollutants can, to some extent, be controlled simultaneously.

On checking the relationship between sulfur dioxide and carbon monoxide (Figure 1.14b), we see a *negative association*. Sites with high carbon monoxide levels tend to have low sulfur dioxide levels. A partial explanation, sought out after the negative association was observed, might be that sulfur dioxide comes mainly from coal and oil combustion, refineries, and paper mills, which tend to be in lightly populated areas with fewer motor vehicles. Thus, different policies will have to be considered for various geographic regions.

F I G U R E **1.14**
Associations Between Pollutants

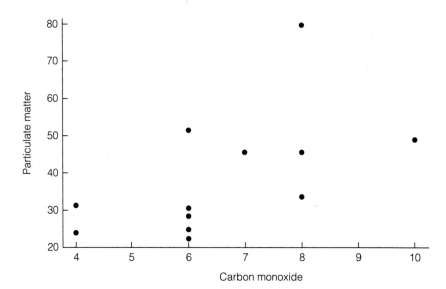

a

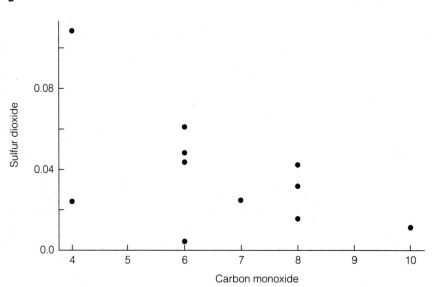

b

FIGURE **1.14**
(Continued)

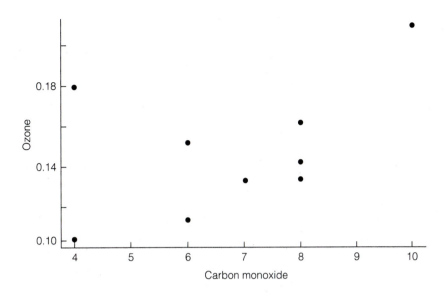

c

What about the relationship between ozone and carbon monoxide? The scatterplot of Figure 1.14c suggests that no strong association exists between these two pollutants in either direction. We will have to treat them separately in our policy analysis.

Time Series Plot   One key feature of the background data on air quality remains to be studied. What are the trends in the emission of pollutants over time? Such trends are most easily spotted on a time series plot, which is similar to a scatterplot with time displayed on the horizontal axis. Figure 1.15a shows the time series plot for the total emission estimates for carbon monoxide and volatile organic compounds across the decade of the study. (The data are from Table 1.1.) Notice that carbon monoxide decreases over time, except for a slight upturn in 1987 and 1988. Can you think of any reason for this? Do any other values appear to be slightly out of line? The contributors to ozone also decrease over the time period, but at a slower rate.

The time series plot of the total PSI days over 100 from Table 1.2 is shown in Figure 1.15b. The two points that deviate greatly from the decreasing trend (1983 and 1988) immediately catch the eye. What went wrong? A little more checking reveals that these two years had unusually hot, dry summers with much stagnant air, ideal conditions for increasing ozone production. Thus, there are some conditions, like the weather, that policies cannot control, but knowledge of this relationship between weather and ozone production is important in setting standards and guiding manufacturers as to when to be on extreme alert.

Let's briefly review some of what we've learned about air quality through our tour of basic statistical tools. The goal is to formulate clear questions and provide as many answers as possible through data collection, analysis, and interpretation. The remaining steps, verifying proposed policy changes and deciding on the next course of action, are beyond the scope of study at this point.

F I G U R E **1.15**
Trends over Time

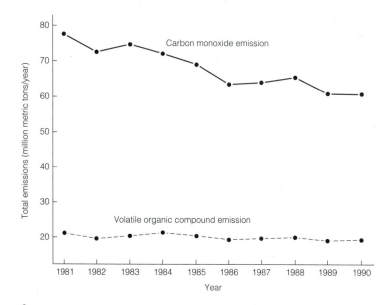

a

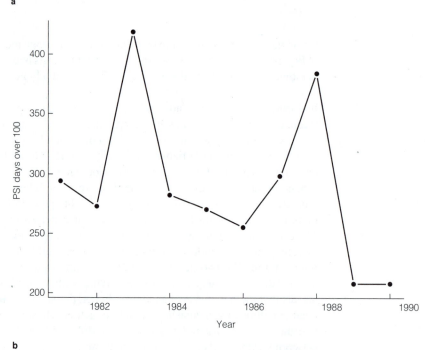

b

Carbon monoxide comes mostly from transportation sources; ozone's chief source has changed from transportation to industrial processes (Pareto analysis). Cities differ considerably with regard to PSI days, but most have realized improvement over the decade; Los Angeles and Houston are outliers deserving special attention (stemplots and boxplots). Some cities have positively associated patterns in their PSI measurements and can be treated similarly; some of the pollutants are likewise positively associated and some are negatively associated, requiring different policies (scatterplots). The main pollutants have emission rates that are decreasing with time, but uncontrollable variables (such as weather) may cause dramatic changes in the pattern (time series plots).

1.5 Refinements of the Data Analysis Tools

The basic tools used above give general impressions about each variable in the study, as well as some rough ideas on possible associations between variables. More refined tools are needed, however, if we are to be more specific in describing a data set or in comparing two data sets. Refined tools make possible answers to refined questions, such as "How does New York compare to the other cities in 1981, and how does this companion change by 1990?" Most of the new tools introduced here will be numerical, rather than graphical. Numerical summaries of data are generally divided into measures of center and measures of variability (or spread).

Measures of Center A basic measure of center is the *median*, introduced earlier. More commonly, perhaps, the center of a set of data points is measured by the arithmetic average, or *mean*. You do not have to look hard to find many uses of the mean; examples include average age, average income, and average grade on an exam. In fact, your grade point average is a mean of sorts that attempts to measure where you "center" as a student. Let us now compare these two measures of center.

We want to describe the PSI data for 1981, from Table 1.2. In the order of the cities on the table, the measurements are

$$9 \quad 0 \quad 3 \quad 12 \quad 4 \quad 12 \quad 33 \quad 4 \quad 152 \quad 21 \quad 22 \quad 9 \quad 1 \quad 1 \quad 12$$

From earlier work, we know that Los Angeles (152) and Houston (33) are both outliers and that the data are highly skewed in the positive direction. The median of these data is 9.0, and the mean is 19.7. (You might want to check the calculation of the mean on your own calculator.) Why are these measures so different? Perhaps the outliers have something to do with this discrepancy, so let's remove them. The reduced data set (reduced to 13 observations by removal of the outliers) has a median of 9.0 and a mean of 8.5. Thus, one important conclusion can be drawn: The mean is very sensitive to one or more outliers, while the median is not. *Line plots* for these two data sets are shown in Figure 1.16. (A *line plot*, or dotplot, simply shows the locations of the data points on a real number line.) Marking the means and medians on these plots, we can see that the mean for the complete data set falls to the right of the main cluster of points, while the median is in the middle of that cluster. Both the mean and the median are in the middle of the cluster for the reduced data set. (Why is the mean slightly smaller than the median for the reduced data set?) In general, the median will provide a better description of the center of a data set if the distribution of the data is highly skewed.

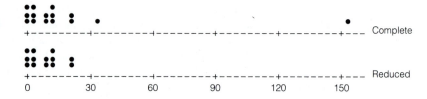

Measures of Variability   Either the median or the mean will do an adequate job of describing center, but how can we succinctly describe variation in the data around the center? If "center" is defined by the median, then the interquartile range (IQR) may provide an adequate summary of the variability. Recall that the IQR is the length of the box in a boxplot; if the box is short, the data set has little variability, and if the box is relatively long, the data set has much variability. In other words, the location of the quartiles relative to the median tells us whether there is a little or a lot of variability. The boxplots for the complete and reduced data sets on 1981 PSI values are provided in Figure 1.17. Note that the complete data set has much more variability than the reduced data set.

F I G U R E **1.17**
1981 PSI Days: Boxplots

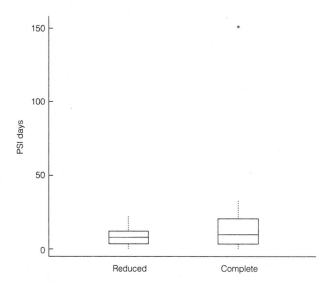

If "center" is defined by the mean, then it would be nice to have a measure that describes the "typical" deviation from the mean. We will now construct such a measure. The contribution that a single data point makes to the variation of the data set can be measured by the difference between that data value and the mean. Thus, n data points will produce n deviations from the mean. How should we combine these n deviations into a single number that measures a typical deviation? Averaging the deviations will not work. (Why not?) Averaging the absolute values of the deviations might work, but absolute value functions are difficult to work with. Instead, it is customary to square the deviations and combine the squared deviations into a single measure of variation called the *standard deviation*.

The basic construction of the standard deviation, using the reduced data set for 1981 PSI values, is shown in Table 1.3. Notationally, x_i is a generic notation for any data point, and $\bar{x}$ denotes the mean of the data points. Concentrate on the column of mean deviations $(x_i - \bar{x})$. Disregarding the sign of these deviations, what would you choose to be a "typical" deviation? The standard deviation is found by "averaging" the squared mean deviations and then taking the square root of the result so as to preserve the measurement scale. For reasons that will become clear later in the text, the divisor of the "average" is generally taken to be $n - 1$ instead of n, where n is the number of data points in the calculation. For the data in Table 1.3, the standard deviation, denoted by s, is given by

$$s = \sqrt{\frac{\sum_{i=1}^{n}(x_i - \bar{x})^2}{n-1}}$$

$$= \sqrt{\frac{631.25}{12}}$$

$$= \sqrt{52.60} = 7.25$$

Is this value, 7.25, close to your chosen "typical" deviation from the mean? The definition and computations for the mean and standard deviation are summarized below.

T A B L E **1.3**
Construction of Standard Deviation

x_i	$\bar{x}$	$x_i - \bar{x}$	$(x_i - \bar{x})^2$
9	8.5	0.5	0.25
0	8.5	−8.5	72.25
3	8.5	−5.5	30.25
12	8.5	3.5	12.25
4	8.5	−4.5	20.25
12	8.5	3.5	12.25
4	8.5	−4.5	20.25
21	8.5	12.5	156.25
22	8.5	13.5	182.25
9	8.5	0.5	0.25
1	8.5	−7.5	56.25
1	8.5	−7.5	56.25
12	8.5	3.5	12.25
Total			631.25

Summary of Mean and Standard Deviation

For n data values labeled $x_1, x_2, \ldots, x_n$, the **mean** is given by

$$\bar{x} = \frac{x_1 + x_2 + \cdots + x_n}{n} = \frac{1}{n}\sum_{i=1}^{n} x_i$$

and the **standard deviation** is given by

$$s = \sqrt{\frac{\sum_{i=1}^{n}(x_i - \bar{x})^2}{n-1}}$$

For the complete data set on 1981 PSI values, the mean is 19.7 and the standard deviation is $s = 37.75$, or nearly 38. (You might want to calculate the standard deviation here, following the steps used for the reduced data.) Why is s so large for the complete data set? It appears that the standard deviation is sensitive to a few outliers, too.

To get a better view of how the two different values of s compare as "typical" deviations, let's study the line plots of the deviations from the mean shown in Figure 1.18. The deviations from the mean for the reduced data set split nearly evenly into

F I G U R E **1.18**
Deviations from the Mean

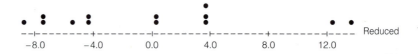

those above zero and those below, with a fairly even spread. The deviations for the complete data set are mostly below zero, with many clustered close to zero on the negative side. If we go one "typical" deviation (7.25) above or below zero for the reduced data set, we are still in the midst of the deviations. So, $s = 7.25$ looks like a good value for describing deviations from the mean. If, on the other hand, we go one typical deviation (38) above zero for the complete data set, we are in a region that contains no data points. Similarly, if we go 38 units below zero, we are below all the negative deviations. So, $s = 38$ is *not* a good value for describing deviations from the mean for the complete data set.

In general, center and variability of data are better described by the median and the interquartile range. If the data distribution is fairly symmetric and mound-shaped, then the center and variability are well described by the mean and standard deviation. We will see that many data sets are nearly symmetric and mound-shaped (or can be transformed to be so) and that the mean and standard deviation are, in fact, widely used measures. Because these measures will be used extensively, it is assumed that anyone working through this book will have ready access to a calculator or computer that will make such calculations electronically.

Standardized Values: z-Scores   Suppose the PSI days from Table 1.2 are a good representation of those values from all sites around the country. Dropping Los Angeles and Houston (known outliers), we can describe the 1981 PSI values as centering at a mean of 8.5 with a standard deviation of 7.25. The 1990 PSI values reduced by the same outliers can be described as centering at a mean of 4.1 with a standard deviation of 4.94. How do New York's PSI days for the two years compare with the other areas of the country? New York went from 21 to 8, which looks like a sizable reduction. Is New York doing better than other cities at improving air quality? One way to answer this question is to put the two values, 21 and 8, on a comparable scale by looking at the data points in terms of standardized distance from the mean. For 1981, New York's *standardized value*, or *z-score*, is calculated as

$$z = \frac{x_i - \bar{x}}{s}$$

$$= \frac{21 - 8.5}{7.25}$$

$$= 1.72$$

A similar calculation shows that New York's z-score for 1990 is 0.79. Thus, New York has done a little better than the typical city, since it went from about 1.7 standard deviations above the mean to only about 0.8 standard deviations above the mean. Atlanta, on the other hand, went from a z-score of 0.07 in 1981 to a z-score of 2.41 in 1990. What does this tell you about Atlanta's air quality?

It is difficult to see the full importance of z-scores, or standardized values, with such a small data set. Imagine, though, that you had similar data for 100 sites around the country. Then the z-score of any one site would be very helpful in determining where it was located relative to the rest of the pack. Much use will be made of z-scores in later chapters.

The Empirical Rule   The mean and standard deviation are a good way to summarize center and spread for a mound-shaped, nearly symmetric distribution of data. For such distributions, these measures actually give us much more information. Armed with knowledge of only the mean and the standard deviation, we can determine something about the relative frequency behavior of the data set. As an example, consider the peak 1990 values of carbon monoxide and ozone from numerous sites around the United States as shown in the line plots in Figure 1.19.

The means and standard deviations are shown in Table 1.4. We will now calculate the proportion of data points in each distribution that lie within one and two standard

FIGURE **1.19**
Peak Carbon Monoxide and
Ozone Values, 1990

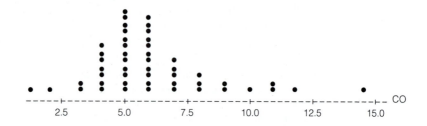

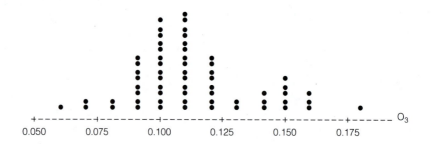

TABLE **1.4**
Summary Statistics

	Mean	Standard Deviation
Carbon monoxide	6.1	2.5
Ozone	0.11	0.024

Note: All measurements are in ppm.

deviations of the mean to see if a pattern emerges. For carbon monoxide, the one-standard-deviation interval is

$$6.1 \pm 2.5 \quad \text{or} \quad (3.6, \ 8.6)$$

and this interval contains 38/49, or about 78%, of the data points. For ozone, the interval is

$$0.11 \pm 0.024 \quad \text{or} \quad (0.086, \ 0.134)$$

and this interval contains 45/62, or about 72%, of the data points. Going to two standard deviations to either side of the mean, the carbon monoxide data give

$$6.1 \pm 5.0 \quad \text{or} \quad (1.1, \ 11.1)$$

which contains 46/49, or about 94%, of the data points. Similarly, the ozone data give an interval of

$$0.11 \pm 0.048 \quad \text{or} \quad (0.062, 0.158)$$

which contains 57/62, or about 92%, of the data points.

Based on these examples, it appears that around 70% of the data should lie within one standard deviation of the mean and over 90% should lie within two standard deviations of the mean. A similar result will hold for *any* mound-shaped, symmetric distribution of data and is stated more precisely below as the *empirical rule*. The empirical rule is a very useful result in that it allows the standard deviation to be given a relative frequency interpretation. If we know the mean and the standard deviation, then we know in what range most of the data will lie. Perhaps more important, we know that few observations will lie more than two standard deviations from the mean in mound-shaped, symmetric data sets.

The Empirical Rule

For any mound-shaped, nearly symmetric distribution of data, the interval $\bar{x} \pm s$ contains approximately 68% of the data points, the interval $\bar{x} \pm 2s$ contains approximately 95% of the data points, and the interval $\bar{x} \pm 3s$ usually contains all the data points.

1.6 Activities for Students: Quality of Life

Scenario

How often have you said something like "I'd really like to improve my living habits so that I would feel better and have more energy. I'd also like to control my use of time a little better." Laudable goals, probably, but the statements are too vague to act upon. Besides, how can we know where to make changes until we clearly understand the current situation? Let's see if we can use some of the basic techniques of quality management to focus on the key issues and then measure the key factors pertaining to these issues. The goal is to form a basis for making rational decisions about some aspects of our life-style, with a view toward making improvements.

Statistical Objective and Pareto Analysis

1   Use brainstorming and Pareto analysis to develop a list of pertinent factors.

2   Develop numerical measures for these factors.

3   Plot data across time and observe patterns and trends.

Materials

Data recording sheet (similar to the one on page 37).

Laboratory Activity

With the class as a whole (or a group of students, if the class is too large), work through the following steps.

1 Defining factors and their measures.

    a Brainstorm until general agreement is reached on a list of important life-style factors, such as study habits, exercise habits, and eating habits.

    b Have each student select his or her top priority from this list. Construct a Pareto chart based on these preferences, and choose the top five or six factors for data collection and analysis.

    c Discuss how each of the factors selected in part (b) should be measured. Also, agree on how often each factor should be measured and on the total time period for data collection.

2 Data collection.

    a Write the agreed upon method of measurement on your data sheet after each factor being measured.

    b Collect the data according to the plan agreed upon above.

3 Data analysis.

    a Construct a time series plot for the measurements from each factor, with the numerical value of the measurement on the vertical axis and time on the horizontal axis. (It is best to add the points to the plot as the data are collected, rather than waiting until all data are collected.)

    b Write a description of any patterns and trends that you see.

    c If you see any undesirable trends, suggest how you might change something in your life-style to improve the pattern for the future.

    d After all data have been collected for each factor, calculate $\bar{x} + 2s$ and $\bar{x} - 2s$, where $\bar{x}$ is the sample mean and s is the sample standard deviation. On each time series plot, draw lines parallel to the time axis that intersect the vertical axis at these two points. Do any of the data points fall above the top line or below the bottom line? If so, what can you say about such points?

Extension

1 At the beginning of each week, each student should report his or her data to the lab assistant for the class. Then the data will be consolidated so that each student has a full class set (without names).

2 For the class data set:

    a Construct stemplots or boxplots for each variable, within each time period, and compare the plots over time.

    b Construct a time series plot of the class averages for each variable, and describe the patterns over time.

3 Discuss the differences and similarities between the patterns seen in the class data and the patterns that were observed in the data for individual students.

DATA SHEET

Sex: M _____   F _____

Class: Fresh. _____   Soph. _____   Junior _____   Senior _____   Other _____

Factor	Description	1	2	3	4	5	6	7	8	9	10	11	12
Study habits													
Sleep habits													
Exercise habits													
Eating habits													
_____													
_____													

1.7 Summary

The modern industrial and business approach to quality improvement (often referred to as total quality management) provides a philosophy for problem solving, as well as a model for problem solving and a set of statistical tools to accomplish the task. The philosophy centers on identification of the *real* problem, the use of simple tools to understand the problem and clearly convey the message to others, and the importance of teamwork in identifying the problem and proposing a solution. The simple tools include *Pareto charts* to identify the most important factors that should be improved first; *cause and effect diagrams* to systematically list factors and show their interrelationships; *stemplots and histograms* to show the shapes of data distributions, observe symmetry and skewness, and spot clusters and gaps in the data; *boxplots* to graphically display medians, quartiles, and interquartile ranges and to conveniently compare essential features of two or more data sets; *scatterplots* to show association (correlation) between two measured variables; and *time series plots* to display trends over time.

Refinements to the above tools include the *mean* as a measure of center to complement the *median* introduced with boxplots, and the *standard deviation* as a measure of variability to complement the *interquartile range*. The mean and standard deviation (most appropriate for mound-shaped, approximately symmetric data distributions) can be used to standardize values so that data measured on different scales can be compared. These two measures also lead to the empirical rule for connecting the mean and standard deviation to a relative frequency interpretation of the underlying data for mound-shaped distributions.

A sad but important example of the failure to look carefully at data will close this chapter. On 28 January 1986, the Space Shuttle *Challenger* exploded shortly after takeoff from the Kennedy Space Center. This disaster was thoroughly investigated by the Rogers Commission, which concluded that the cause was a gas leak through a joint in one of the booster rockets sealed by a device called an O-ring. Why was this potential problem not detected given that data were available on O-ring performance for 23 flights prior to the *Challenger*? At least part of the answer lies in the way the data were analyzed. Engineers looked at the data displayed in Figure 1.20a, which shows the number of incidents of O-ring thermal distress as related to temperature at launch. The engineers involved saw no apparent association between number of incidents and temperature. But these data show only 7 flights. What about the other 16? The answer is that the other 16 flights showed no distressed O-rings and hence were thought to contain no useful information. When added to the plot (Figure 1.20b), these points can be seen to add considerable information. Now, it is clear that a negative association exists between number of incidents and launch temperature. In fact, the Rogers Commission concluded that a mistake was made in the analysis of the thermal distress data; the negative correlation should have been revealed prior to the launch of the *Challenger*, which took place at a temperature of $31°$ F. Careful analysis of all the data pertinent to a problem is essential for sound decision making. The results of improper data collection and analysis can be disastrous.

FIGURE **1.20**
The *Challenger* Data

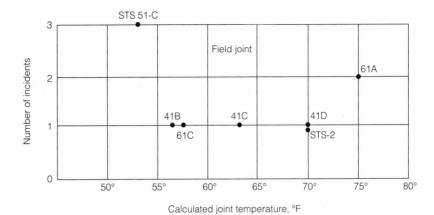

a

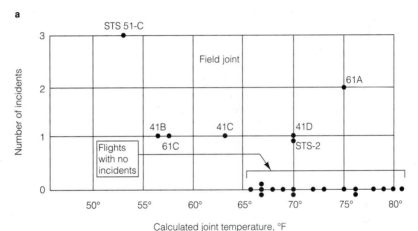

b

Exercises

1.1   Sulfur oxide is one of the major components of the PSI measure of air quality. Table 1.5 shows sulfur oxide emission estimates for a decade, along with major source categories. Obviously, the most important source is fuel combustion, primarily coal and oil combustion in refineries and mills.

  a   Produce a Pareto chart of source categories for 1990. Would the Pareto chart for 1981 look appreciably different?

  b   Construct a time series plot for the total emission estimates across years. Comment on any trends you observe.

1.2   Lead is a major pollutant in both air and water, but it is not part of the PSI. Has the quality of our environment improved in recent years with regard to lead? Table 1.6 shows lead emission estimates and source categories for the period 1981 through 1990.

  a   Sketch a Pareto chart of source categories for 1981 and 1990. How do these compare?

T A B L E **1.5**   National Sulfur Oxide Emission Estimates, 1981–1990 (million metric tons/year)

Source Category	1981	1982	1983	1984	1985	1986	1987	1988	1989	1990
Transportation	0.9	0.8	0.8	0.8	0.9	0.9	0.9	0.9	1.0	0.9
Fuel combustion	17.8	17.3	16.7	17.4	17.0	16.9	16.6	16.6	16.8	17.1
Industrial processes	3.8	3.1	3.1	3.2	3.2	3.2	3.0	3.1	3.0	3.1
Solid waste	0.0	0.0	0.0	0.0	0.0	0.0	0.0	0.0	0.0	0.0
Miscellaneous	0.0	0.0	0.0	0.0	0.0	0.0	0.0	0.0	0.0	0.0
Total[†]	22.5	21.2	20.6	21.5	21.1	20.9	20.5	20.6	20.8	21.2

Source: U.S. Environmental Protection Agency, Office of Air Quality Planning and Standards.

[†]The sums of subcategories may not equal total due to rounding.

T A B L E **1.6**   National Lead Emission Estimates, 1981–1990 (thousand metric tons/year)

Source Category	1981	1982	1983	1984	1985	1986	1987	1988	1989	1990
Transportation	46.5	47.0	40.8	34.7	14.7	3.5	3.0	2.6	2.2	2.2
Fuel combustion	2.8	1.7	0.6	0.5	0.5	0.5	0.5	0.5	0.5	0.5
Industrial processes	3.0	2.7	2.4	2.3	2.3	1.9	1.9	2.0	2.3	2.2
Solid waste	3.7	3.1	2.7	2.7	2.6	2.6	2.6	2.5	2.3	2.2
Miscellaneous	0.0	0.0	0.0	0.0	0.0	0.0	0.0	0.0	0.0	0.0
Total [‡]	56.0	54.5	46.6	40.2	20.1	8.4	8.0	7.6	7.2	7.1

[‡]The sums of subcategories may not equal total due to rounding.

b   Construct a time series plot for total lead emission and another for the transportation category alone. Comment on the trends in these plots. Are the trends similar in the two plots?

c   Can you offer a plausible explanation for the major jumps in the time series plots? (Transportation here is mostly motor vehicles.)

1.3   A simple histogram should not be underrated as a quality improvement tool. Figure 1.21 shows the frequency histogram of quality measurements on the diameters of 500 steel rods. These rods had a lower specification limit (LSL) of 1.0 centimeter, and rods with diameters under the LSL were declared defective. Do you see anything unusual in this histogram? Can you offer a possible explanation of what is happening here?

1.4   The amount of hydrogen chloride in the stratosphere is important because it may catalyze the removal of ozone. Data on the amount of hydrogen chloride in sections of the stratosphere before and after the eruptions of El Chichon volcano in Mexico are given in Table 1.7.

a   Construct back-to-back stem-and-leaf plots for the pre-eruption and posteruption data.

b   Construct parallel boxplots for the pre-eruption and posteruption data.

c   What are the most striking features of the two data sets?

1.5   Are highways getting safer? How can we begin to measure possible improvements in the quality of travel by motor vehicle? One possibility is to study traffic safety data for any discernable patterns or trends. The data in Table 1.8 give the deaths per 100 million miles driven in the 50 states plus the District of Columbia.

a   Construct separate stemplots for the 1985 data and the 1989 data. What differences and similarities do you see?

b   Construct parallel boxplots for the 1985 and 1989 data. How do they compare? Are there any outliers?

F I G U R E **1.21**   Diameters of 500 Steel Rods

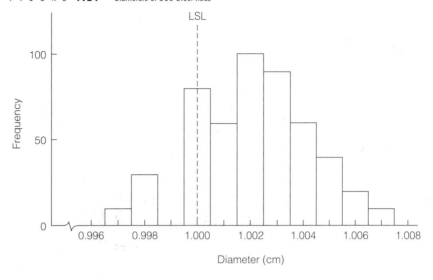

T A B L E **1.7**   Amount of HCl in Columns
Above 12 km at Various Locations

(in 10^{15} molecules per square centimeter)

Pre-eruption	Posteruption
0.78	1.46
1.26	1.40
0.75	1.29
0.69	1.22
0.88	1.63
1.56	1.90
1.18	1.24
	1.45
	1.79

Source: W. G. Mankin and M. T. Coffey (1984). *Science*,
226, October, p. 171.

c   For the 1989 data, construct parallel boxplots for states east and west of the Mississippi River. Which geographic region seems to have the higher motor vehicle death rate?

d   Construct a scatterplot of the 1989 values plotted against the 1985 values. How would you describe the association between the two sets of values? Do you see any states far away from the others? If so, what causes them to stand out? Are there any interesting clustering patterns?

1.6   The back-to-back histogram (often called a population pyramid) shown in Figure 1.22 displays the age distributions of the U.S. population in 1890 and 1990. Is the age distribution of the U.S. population changing? If so, what might that say about our quality of life in future years?

a   Approximately what percentage of the population was under the age of 30 in 1890? in 1990?

T A B L E **1.8** Motor Vehicle Deaths per 100 Million Vehicle Miles

State	1985	1989	State	1985	1989
U.S.	2.6	2.3	MO	2.6	2.3
AL	2.9	2.4	MT	3.1	2.3
AK	3.2	2.3	NE	2.1	2.2
AZ	4.4	2.6	NV	3.9	3.3
AR	3.4	3.3	NH	2.6	2.0
CA	2.6	2.1	NJ	1.9	1.5
CO	2.4	1.9	NM	4.2	3.5
CT	2.0	1.6	NY	2.3	2.1
DE	2.2	1.8	NC	3.1	2.5
DC	3.0	2.2	ND	2.2	1.4
FL	3.4	2.8	OH	2.1	2.1
GA	2.7	2.5	OK	2.5	2.0
HI	2.0	1.9	OR	2.8	2.5
ID	3.5	2.8	PA	2.4	2.3
IL	2.3	2.2	RI	2.1	1.7
IN	2.6	1.9	SC	3.5	3.2
IA	2.4	2.3	SD	2.3	2.3
KS	2.6	2.1	TN	3.4	2.3
KY	2.6	2.4	TX	2.7	2.2
LA	3.0	2.4	UT	2.8	2.4
ME	2.4	1.7	VT	2.5	2.1
MD	2.3	2.0	VA	2.1	1.7
MA	1.9	1.7	WA	2.3	1.8
MI	2.4	2.1	WV	3.6	3.3
MN	2.0	1.6	WI	2.1	1.9
MS	3.6	3.3	WY	2.7	2.3

Source: Statistical Abstract of the United States, 1991.

b How would you describe the major changes in the age distribution over this century? What problems are arising as a result of these changes?

c What age class contained the median age in 1890? How had this changed by 1990?

1.7 Does education increase your chances of employment? The data in Table 1.9 show the employed and unemployed civilian labor force (25 years of age and over, in thousands) according to years of schooling completed.

a Construct a scatterplot of unemployment rate (percent of the labor force unemployed) against years of schooling completed. In order to do this, you must first assign a numerical value to the years-of-schooling categories presented as intervals. What is a reasonable way to do this?

b Comment on the nature of the association you see in the scatterplot.

c Find another way to plot these data. Do you like your technique better than the scatterplot? Why or why not?

1.8 Ease of getting to work is one measure of quality of life in metropolitan areas. Staggered starting times sometimes help alleviate morning traffic congestion, and such a policy is often preferred by workers. The graphs in Figure 1.23 show the results of a study of work start times in Pittsburgh. The plots are cumulative percentages similar to the cumulative plots we've seen on Pareto charts, but without the bar graph.

F I G U R E **1.22**   Percent Distribution of Age, 1890 and 1990

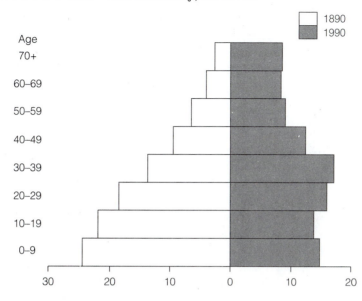

TABLE **1.9**

Employment Status	Pop. 25 Years (and Over)	Elementary		High School		College		
		Less Than 5 Years	5–8 Years	1–3 Years	4 Years	1–3 Years	4 Years	5 Years
Total	154,144	3,645	13,914	17,420	59,208	27,384	18,897	13,676
Civilian labor force	101,736	1,119	4,586	8,849	39,934	20,662	15,149	11,437
Employed	97,621	1,042	4,257	8,145	38,171	19,991	14,786	11,229
Unemployed	4,115	77	329	704	1,763	671	363	208
Not in labor force	52,408	2,526	9,327	8,571	19,274	6,722	3,748	2,239

Source: U.S. Bureau of Labor Statistics.

a Do official start times seem to bunch together more than actual arrival times?

b How would you compare the distribution of desired work start times to the distribution of arrival times? to that of official work start times?

c Give a concise explanation of what this plot shows.

1.9 The following data show the percent change in production of crude petroleum from 1976 to 1977 for selected countries in America, Western Europe, and the Middle East (Source: *The World Almanac and Book of Facts*, 1979):

$$-1.4, +0.3, +8.0, -13.6, -4.1, -2.1, +8.1, -7.5, -2.0,$$
$$+4.5, 0.0, +205.2, -0.2, -6.5, -7.0, -7.6, +7.7, +4.5$$

a Construct a stem-and-leaf display and boxplot.

b Calculate $\bar{x}$ and s, the mean and standard deviation.

c Construct a relative frequency histogram for the data.

d Do $\bar{x}$ and s offer good descriptions of center and variability? Why or why not?

F I G U R E **1.23**   Work Starts, Arrivals, and Desired Starts, Pittsburgh Central Business District

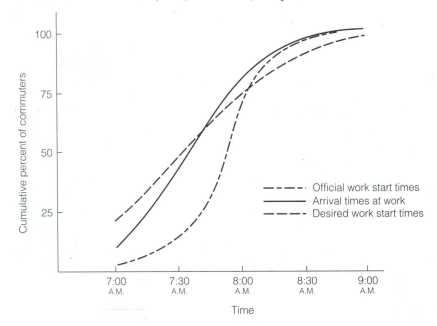

1.10 Refer to Exercise 1.9. The figure +205.2 is from the United Kingdom and is exceptionally high because a major new source of petroleum was located in the North Sea. Eliminate this unusual figure, and answer the questions of Exercise 1.9 with the reduced data set. Note the effect this large value has on the mean and standard deviation.

1.11 The *Los Angeles Times* of 20 September 1981 contains the following statement: "According to the latest enrollment analysis by age-categories, half of the [Los Angeles Community College] district's 128,000 students are over the age of 24. The average student is 29." Can this statement be correct? Explain.

1.12 "In the region we are traveling west of Whitney, precipitation drops off and the average snow depth on April 1 for the southern Sierra is a modest 5 to 6 feet. And two winters out of three, the snow pack is below average." (E. Bowen, *The High Sierra*, Time-Life Books, 1972, p.142.) Explain how this statement can be true.

1.13 Data on the average life of AA batteries under continuous use are given in Table 1.10.

  a Construct back-to-back stem-and-leaf plots of the alkaline versus the nonalkaline batteries. (Group the heavy-duty and regular batteries together.)

  b Construct parallel boxplots for the alkaline and nonalkaline batteries. What patterns do you detect?

  c Calculate $\bar{x}$ and s for each group of batteries. What do these values tell you about center and variability for the two groups?

1.14 Environmental impact studies on the quality of water often measure the level of chemical toxicants by the LC50, the concentration lethal to 50 percent of the organisms of a particular species under a given set of conditions and specified time period. Two common toxicants are methyl parathion and Baytex. LC50 measurements on water samples for each of these toxicants are as follows:

Methyl parathion (MP)   2.8, 3.1, 4.7, 5.2, 5.2, 5.3, 5.7, 5.7, 6.6, 7.1, 8.9, 9.0
Baytex (B)   0.9, 1.2, 1.3, 1.3, 1.4, 1.5, 1.6, 1.6, 1.7, 1.9, 2.4, 3.4

T A B L E **1.10**

Brand and Model	Average Life (in hours)
Alkaline AA	
Duracell MM 1500	4.1
Eveready Energizer E91	4.5
K-Mart Super Cell	4.3
Panasonic AM3	3.8
Radio Shack 23552	4.8
Ray-O-Vac 815	4.4
Sears 3090	4.7
Heavy-duty AA	
Eveready 1215	1.8
Radio Shack 23582	2.0
Ray-O-Vac 5AA	0.6
Sears 344633	0.7
Regular AA	
Eveready 1015	0.8
K-Mart K15S	0.7
Radio Shack 23468	1.0
Ray-O-Vac 7AA	0.5

Source: Consumer Reports, November 1983, p. 592.

a Construct stemplots for these data sets. Comment on the shapes of these distributions.

b Construct parallel boxplots for the two toxicants. How would you describe the differences between the two boxplots?

c Calculate $\bar{x}$ and s for each data set. Do they provide good measures of center and variability?

1.15 Is the system wearing out? Sometimes evidence of decreasing quality can be found by the careful study of measurements taken over time. Time intervals between successive failures of the air conditioning system on a Boeing 720 jet airplane were recorded in order. (Source: F. Proschan, *Technometrics*, 5, no. 3, 1963, p. 376.) The 30 measures are divided into two groups of 15 to see if the later time intervals seem shorter than the earlier ones (possibly indicating wearing out of the system). The data (in hours) are as follows:

Group I   23, 261, 87, 7, 120, 14, 62, 47, 225, 71, 246, 21, 20, 5, 42
Group II   12, 120, 11, 3, 14, 71, 11, 14, 11, 16, 90, 1, 16, 52, 95

How would you analyze these data to build a case for (or against) the possibility of the air conditioning systems wearing out? What is your conclusion?

1.16 Table 1.11 gives the total tax revenues (in billions of dollars) and per capita tax revenues (in dollars) for selected countries in 1988.

a What plots and numerical summaries best describe the total tax revenues?

b What plots and numerical summaries best describe the per capita tax revenues?

c Write brief descriptions of the distributions of each data set.

1.17 Quality of education is often assessed by scores on standardized exams. The data in Table 1.12 show the national average SAT and ACT mathematics scores for the years 1970–1989. The

T A B L E **1.11**

Country	Total Tax Revenue ($ billion)	Per Capita
United States	1,409.2	$5,721
Australia	80.9	4,893
Austria	53.3	7,015
Belgium	69.3	7,014
Canada	169.4	6,529
Denmark	56.0	10,897
Finland	39.9	8,068
France	421.4	7,542
Greece	18.8	1,883
Ireland	13.5	3,810
Italy	307.9	5,360
Japan	903.5	7,368
Luxembourg	3.3	8,739
Netherlands	109.9	7,446
New Zealand	15.8	4,765
Norway	42.8	10,159
Portugal	14.4	1,476
Spain	113.1	2,900
Sweden	100.5	11,914
Switzerland	59.6	8,958
Turkey	16.2	299
United Kingdom	306.7	5,372

Source: Statistical Abstract of the United States, 1991, p. 846.

main problem involved in comparing these scores is that they are measured on different scales, so standardized values might make comparisons more meaningful.

a Plot SAT scores against year and ACT scores against year on the same plot. Do you like the result?

b Calculate the mean and standard deviation for each data set.

T A B L E **1.12**

Year	SAT	ACT	Year	SAT	ACT
1970	488	20.0	1980	466	17.4
1971	488	19.1	1981	466	17.3
1972	484	18.8	1982	467	17.2
1973	481	19.1	1983	468	16.9
1974	480	18.3	1984	471	17.3
1975	472	17.6	1985	475	17.2
1976	472	17.5	1986	475	17.3
1977	470	17.4	1987	476	17.2
1978	468	17.5	1988	476	17.2
1979	467	17.5	1989	476	17.2

Source: U.S. Department of Education.

c Calculate the appropriate z-score for each data point.

d Construct a time series plot of standardized SAT scores and standardized ACT scores on the same graph. Does the result show any interesting patterns?

e Construct a scatterplot of standardized SAT scores against standardized ACT scores. Comment on any patterns you see.

f Suppose this year's SAT average is 482. To what ACT average would this be comparable?

1.18 The health and quality of the economy are sometimes measured by the consumer price index (CPI). Shown in Table 1.13 are the CPIs for selected areas of the United States for the years 1980 and 1990. What can we say about the changes in the CPIs over those years?

TABLE **1.13**

Area	1980	1990
Anchorage, AK MSA	85.5	118.6
Atlanta, GA MSA	80.3	131.7
Baltimore, MD MSA	83.7	130.8
Boston-Lawrence-Salem, MA-NIH CMSA	82.6	138.9
Buffalo-NiagaraFalls, NY CMSA	83.5	127.7
Chicago-Gary-Lake County, IL-IN-WI CMSA	82.2	131.7
Cincinnati-Hamilton, OH-KY-IN CMSA	82.1	126.5
Cleveland-Akron-Lorain, OH CMSA	78.9	129.0
Dallas-Fort Worth, TX CMSA	81.5	125.1
Denver-Boulder, CO CMSA	78.4	120.9
Detroit-Ann Arbor, MI CMSA	85.3	128.6
Honolulu, HI MSA	83.0	138.1
Houston-Galveston-Brazonia, TX CMSA	82.7	120.6
Kansas City, MO-KS CMSA	83.6	126.0
Los Angeles-Anaheim-Riverside, CA CMSA	83.7	135.9
Miami-Fort Lauderdale, FL CMSA	81.1	128.0
Milwaukee, WI PMSA	81.4	126.2
Minneapolis-St. Paul, MN-WI MSA	78.9	127.0
New York-Northern New Jersey-Long Island, NY-NJ-CT CMSA	82.1	138.5
Philadelphia-Wilmington-Trenton, PA-NJ-DE-MD CMSA	83.6	135.8
Pittsburgh-Beaver Valley, PA CMSA	81.0	126.2
Portland-Vancouver, OR-WA CMSA	87.2	127.4
San Diego, CA MSA	79.4	138.4
San Francisco-Oakland-San Jose, CA CMSA	80.4	132.1
Seattle-Tacoma, WA CMSA	82.7	126.8
St. Louis-East St. Louis, MO-IL CMSA	82.5	128.1
Washington, DC-MD-VA MSA	82.9	135.6

Source: Statistical Abstract of the United States, 1991.

a Construct appropriate plots of the 1980 and 1990 data separately. How would you describe these distributions?

b Does the empirical rule hold better for 1980 data or 1990 data? (Check both cases.) Can you give a reason for your answer based on the plots in (a)?

c Calculate the difference between the 1990 CPI and the 1980 CPI for each location. Plot the differences against the 1980 CPIs on a scatterplot. Comment on the result.

d   Construct a scatterplot of the differences against the 1990 CPIs. Comment on any patterns you see.

e   Which cities seem to be doing well against inflation? Which plot helps you spot these cities?

1.19   Quality in baseball is exemplified by the league batting champions. Table 1.14 shows the American League batting champions and their batting averages for the years 1941–1991. Let's see what patterns we can observe.

a   Construct a histogram or stemplot for these data. Describe the distribution.

b   Are there any outliers in the data? (Use the IQR definition of "outlier.")

c   Does the empirical rule work well for these data?

d   Suppose we remove from the data set any outliers found in (b). Does the empirical rule now work better? Why or why not?

1.20   The home run outputs of four contenders for greatest New York Yankee home run hitter are shown in Table 1.15. Who appears to be the best home run hitter? Who appears to be second best? What considerations affect your choice? What statistical tools help you decide?

T A B L E **1.14**   Batting Averages of American League Batting Champions

Year	Player and Club	Ave.	Year	Player and Club	Ave.
1941	Ted Williams, Boston	.406	1967	Carl Yastrzemski, Boston	.326
1942	Ted Williams, Boston	.356	1968	Carl Yastrzemski, Boston	.301
1943	Lucius Appling, Chicago	.328	1969	Rodney Carew, Minnesota	.332
1944	Louis Boudreau, Cleveland	.327	1970	Alexander Johnson, California	.329
1945	George Stirnweiss, New York	.309	1971	Tony Oliva, Minnesota	.337
1946	Mickey Vernon, Washington	.353	1972	Rodney Carew, Minnesota	.318
1947	Ted Williams, Boston	.343	1973	Rodney Carew, Minnesota	.350
1948	Ted Williams, Boston	.369	1974	Rodney Carew, Minnesota	.364
1949	George Kell, Detroit	.343	1975	Rodney Carew, Minnesota	.359
1950	William Godman, Boston	.354	1976	George Brett, Kansas City	.333
1951	Ferris Fain, Philadelphia	.344	1977	Rodney Carew, Minnesota	.388
1952	Ferris Fain, Philadelphia	.327	1978	Rodney Carew, Minnesota	.333
1953	Mickey Vernon, Washington	.337	1979	Fredric Lynn, Boston	.333
1954	Roberto Avila, Cleveland	.341	1980	George Brett, Kansas City	.390
1955	Albert Kaline, Detroit	.340	1981	Carney Lansford, Boston	.336
1956	Mickey Mantle, New York	.353	1982	Willie Wilson, Kansas City	.332
1957	Ted Williams, Boston	.388	1983	Wade Boggs, Boston	.361
1958	Ted Williams, Boston,	.328	1984	Donald Mattingly, New York	.343
1959	Harvey Kuenn, Detroit	.353	1985	Wade Boggs, Boston	.368
1960	Pete Runnel, Boston	.320	1986	Wade Boggs, Boston	.357
1961	Norman Cash, Detroit	.361	1987	Wade Boggs, Boston	.363
1962	Pete Runnels, Boston	.326	1988	Wade Boggs, Boston	.366
1963	Carl Yastrzemski, Boston	.321	1989	Kirby Puckett, Minnesota	.339
1964	Tony Oliva, Minnesota	.323	1990	George Brett, Kansas City	.329
1965	Tony Oliva, Minnesota	.321	1991	Julio Franco, Texas	.341
1966	Frank Robinson, Baltimore	.316			

Source: The Complete Baseball Record Book 1992, Sporting News.

T A B L E **1.15**

Babe Ruth		Lou Gehrig		Mickey Mantle		Roger Maris	
Year	Home Runs	Year	Home Runs	Year	Home Runs	Year	Home Runs
1920	54	1923	1	1951	13	1960	39
1921	59	1924	0	1952	23	1961	61
1922	35	1925	20	1953	21	1962	33
1923	41	1926	16	1954	27	1963	23
1924	46	1927	47	1955	37	1964	26
1925	25	1928	27	1956	52	1965	8
1926	47	1929	35	1957	34	1966	13
1927	60	1930	41	1958	42		
1928	54	1931	46	1959	31		
1929	46	1932	34	1960	40		
1930	49	1933	32	1961	54		
1931	46	1934	49	1962	30		
1932	41	1935	30	1963	15		
1933	34	1936	49	1964	35		
1934	22	1937	37	1965	19		
		1938	29	1966	23		
		1939	0	1967	22		
				1968	18		

Source: The Baseball Encyclopedia, 4th ed., Joseph L. Reicher, ed., 1978.

From Data Tables to Discrete Probability

About This Chapter

Using tools of statistics to make decisions from data in an organized way is the basis of improving the quality of any process. But how do we obtain good data on which to base these decisions? Most good plans for collecting data make use of randomization, and randomization is tied to probability. This chapter, then, looks at ways to understand randomness and to use randomness to obtain data with a predictable pattern. The concept and tools of probability enter the scene as ways of discussing random events. Probability as a relative frequency has a natural connection to data summarized as frequency counts, and this parallelism is explored throughout the chapter. A variety of examples related to the improvement of quality of life are used to illustrate the concepts introduced.

Contents

2.1 Understanding Randomness: An Intuitive Notion of Probability

2.1.1 Randomness with Known Structure

At the start of a football game, a balanced coin is flipped into the air to decide who will receive the ball first. What is the chance that the coin will land heads up? Most of us would say that this chance, or probability, is 0.5, or something very close to that. But what is the meaning of this number, 0.5? If the coin is tossed ten times, will it come up heads exactly five times? Upon deeper reflection, most would agree that ten tosses need not result in exactly five heads, but in repeated flipping the coin should land heads up approximately one-half of the time. From this point, the reasoning begins to get fuzzier. Will 50 tosses result in exactly 25 heads? Will 1,000 tosses result in exactly 500 heads? Not necessarily, but the fraction of heads should be close to one-half after "many" flips of the coin. Thus, 0.5 is regarded as a *long-run* or *limiting relative frequency* as the number of flips gets large.

John Kerrich, a mathematician interned in Denmark during World War II, actually flipped a coin 10,000 times, keeping a careful tally of the number of heads. After 10 tosses, he had 4 heads, a relative frequency of 0.4; after 100 tosses, he had 44 heads (0.44); after 1,000 tosses, he had 502 heads (0.502); and after 10,000 tosses, he had 5,067 heads (0.5067). The relative frequency of heads remained very close to 0.5 after 1,000 tosses, although the actual figure at 10,000 tosses was slightly farther from 0.5 than was the figure at 1,000 tosses.

In the long run, Kerrich obtained a relative frequency of heads close to 0.5. For that reason, the number 0.5 can be called the *probability* of obtaining a head on the toss of a balanced coin. Another way of expressing this result is to say that Kerrich *expected* to see about 5,000 heads among the outcomes of his 10,000 tosses. He actually came close to his expectations, as have others who have repeated the coin-tossing study. The idea of a stabilizing relative frequency after many trials is at the heart of random behavior; another example that illustrates this idea comes from the study of properties of random digits.

A table of random digits (such as Table 1 in the Appendix) or random digits generated by computer are produced according to the following model. Think of ten equal-size chips numbered from 0 through 9, with one number per chip, and thoroughly mixed in a box. Without looking, someone reaches into the box and pulls out a chip, recording the number on the chip. That process is a single draw of a random digit. Putting the chip back into the box, mixing the chips, and then drawing another chip produces a second random digit. A random number table is the result of hundreds of such draws, each from the same group of thoroughly mixed chips.

What is the probability of selecting a digit that is a multiple of 3 from a random number table? We can find this probability, at least approximately, by selecting digits from a random number table and counting the number of multiples of 3 (3, 6, or 9). Figure 2.1 shows the results of two different attempts at doing this, with 100 digits selected in each attempt. The relative frequencies are recorded as a function of the number of digits selected.

Notice two important features of the graph in Figure 2.1. First, both sample paths oscillate greatly for small numbers of observations and then settle down around a value close to 0.30 (actually, about 0.34). Second, the variation between the two

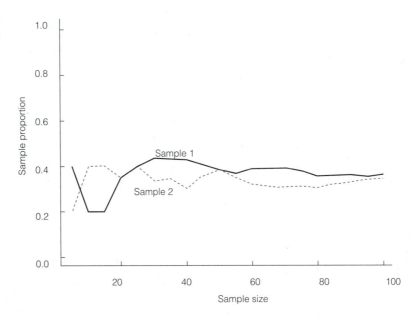

sample fractions is quite great for small sample sizes and quite small for larger sample sizes. From these data, we can approximate the probability of selecting a multiple of 3 from a random number table as 0.34, with the added knowledge that the approximation tends to be more accurate as the sample size increases. It should not be surprising that we would see about 34 multiples of 3 in a selection of 100 random digits.

2.1.2   Randomness with Unknown Structure

For each of the two examples used above, we knew what the resulting long-run relative frequency should be. Now we will use an example for which the result was not obvious (at least to the author) before the collection of data. A standard paper cup, with the open end slightly larger than the closed end, was tossed in the air and allowed to land on the floor. The goal was to approximate the probability that the cup would land on the open end. (You might want to generate your own data here before looking at the results presented below.) After 100 tosses in each of two trials, the sample paths looked like those shown in Figure 2.2.

Notice that the pattern observed in the earlier examples is still apparent; both sample paths seem to stabilize around 0.2, and the variability between the paths is much greater for small sample sizes than for large ones. We can now say that the probability of a tossed cup (like the one used here) landing on its open end is approximately 0.2; in 100 tosses of this cup, we would expect to see it land on its open end about 20 times. Figure 2.2 also includes two sample paths for the outcome "landed on side," which seem to stabilize at slightly under 0.6. The basic notion of probability as a long-run relative frequency works just as well here as it did in the cases for which we had a good theoretical idea about what the relative frequency should be!

F I G U R E **2.2**
Proportions of Cup Landings

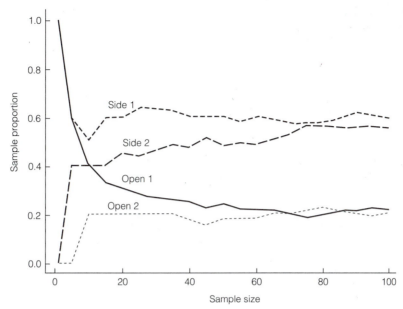

2.1.3   Sampling a Finite Universe

There is no limit to the number of times the coin could have been tossed, the random digits could have been selected, or the paper cup could have been tossed. The set of possible sample values is both infinite and conceptual (it doesn't exist on a list somewhere) in all three cases. (A list of random digits may exist, but it could always be made larger.) Now, however, suppose I have a jar of differently colored beads on my desk from which I plan to take samples of various sizes with the goal of estimating the proportion of yellow beads in the jar. After mixing the beads, I take a sample of five, then five more to make a total of ten, and so on, up to a total of 100 beads. Those sampled are *not* returned to the jar before the next group is selected. Will the relative frequency of yellow beads stabilize here as it did for the infinite, conceptual universes sampled above? Figure 2.3 shows two actual sample paths for this type of sampling from a jar containing over 600 beads. The same properties observed earlier in the infinite cases do, indeed, hold here as well. The probability of randomly sampling a yellow bead is about 0.2; a sample of 100 beads is expected to contain about 20 yellows (or the jar of 600 beads is expected to contain over 120 yellows). Actually, the proportion of yellow beads in the jar is 0.196, so the sampling came very close to "truth."

In general, as long as there is a random mechanism at work that remains the same for all selections, the sample proportion for a certain specified event will eventually stabilize at a specific value that can be called the *probability* for the event in question. Will this limit always converge? If so, can we determine what it will converge to without actually doing the sampling? In order to answer these and similar questions, a more precise mathematical definition of probability must be given, and we will do this later. For now, the intuitive notion of probability will serve us well

F I G U R E **2.3**
Proportions of Yellow Beads

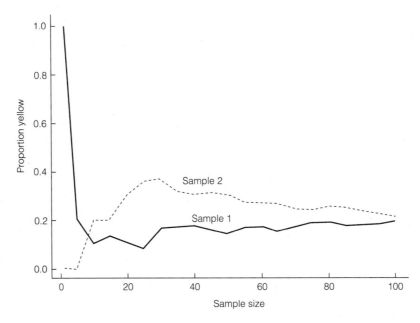

F I G U R E **2.3**
Proportions of Yellow Beads

as we investigate a few real examples of the uses of randomization in collecting and interpreting data.

2.2   Using Randomness: How to Obtain Good Data

The key to decisions is objective data; the key to good decisions is good objective data. It is not enough just to have data; the data must be reliable in that they actually measure what they are supposed to measure, to a reasonable degree of accuracy. This section gives a brief introduction to the four common ways of obtaining data in studies: census, sample survey, experiment, and observational study. These methods will be developed further in later sections.

2.2.1   Census

Everyone has heard of the U.S. Census, which attempts to count everyone living in the United States and to measure various other features of the population. Certainly, the U.S. Census is not completely accurate, but it still does a good job of describing our population. A *census* means simply that information in a study was obtained from every element in the *target population*. If you want to list all the books you own, you would take a census of your books. When a firm takes inventory, it is taking a census of everything in stock. The computerized records of all university students on your campus are, in effect, a census of the student body. So the target population might be your books, the stock of a firm, or the students on campus, but the key that identifies a census is that information is available on each element of that population. Done

	Count	Percent
Total	248,710	
Female	126,842	51
Male	121,868	49
White	199,686	80
Black	29,986	12
Hispanic	22,354	9

T A B L E **2.1**
U.S. Population: 1990 Census
(counts in thousands)

Source: U.S. Bureau of the Census.

correctly, taking a census is a good way to obtain data. Table 2.1 provides a little of the information collected in the 1990 U.S. Census.

No randomization was used in the collection of these data (although the Census Bureau does use some randomization for certain questions). Nevertheless, these data are good reference figures for checking the accuracy of data collected by random methods. A typical Gallup poll, for example, randomly selects around 1,200 respondents from across the United States, a process that has the same key features as selecting beads from a jar. It would not be surprising, then, to see about 51% females in most Gallup polls that sample the U.S. population at large, since the probability of selecting a female on any one random draw is about 0.51. The Gallup organization would expect to see about 960 whites in such polls. The connection between data tables and probability, in this case, is that the table defines the probabilities and the expected outcomes.

2.2.2   Sample Survey

Your school administrators want to know how many students will want parking spaces for automobiles next year. How can we get reliable information on this question? One way is to ask the question of all returning students, but this procedure would be somewhat inaccurate (Why?) and very time consuming. We could take the number of spaces in use this year and assume that next year's needs will be about the same, but this method would have inaccuracies as well. A simple technique that works very well in many cases is to select a *sample* from those students who will be attending the school next year and ask each of them if they will be requesting a parking space. From the proportion of "yes" answers, an estimate of the number of spaces required can be obtained.

The scenario outlined above has all the elements of a typical *sample survey* problem. There is a question of "How many?" or "How much?" to be determined for a specific group of objects called a *target population*, and an approximate answer is to be derived from a *sample* of data extracted from the population of interest. Of key importance is the fact that the approximate answer will be a good approximation only if the sample truly represents the population under study. *Randomization* plays a vital role in the selection of samples that truly represent a population and, hence, produce good approximations. It is clear that we would not want to sample only our classmates or friends on the parking issue. It is less clear, but still true, that virtually any sampling scheme that depends upon subjective judgments as to who should be included will suffer from *sampling bias*.

Once we know who is to be in the sample, we still need to get the pertinent information from them. The *method of measurement*, that is, the questions or measuring devices we use to obtain the data, should be designed to produce the most accurate data possible and should be free of *measurement bias*. We could choose a number of ways to ask students about their parking needs for next year. Here are a few suggested questions:

Do you plan to drive to school next year?

Do you plan to drive to school on more than half of the school days next year?

Do you have regular access to a car for travel to school next year?

Will you drive to school next year if the cost of parking increases?

Which of these questions do you think might bias the results, and in which direction? A few moments of reflection should convince you that measurement bias can be serious in even the simplest of surveys!

Because it is so difficult to get good information in a survey, every survey should be *pretested* on a small group of subjects similar to those that might appear in the final sample. The pretest not only helps improve the questionnaire or measurement procedure, but also helps determine a good plan for *data collection* and *data management*. Can we, for example, list a mutually exclusive and exhaustive set of meaningful options on the parking question so that responses can be easily coded for the *data analysis* phase? The data analysis should lead to clearly stated *conclusions* that relate to the original purpose of the study. The goal of the parking study is to focus on the number of spaces needed, not the types of cars students drive or the fact that auto theft may be a problem.

Key elements of any sample survey, consistent with the problem-solving model of Chapter 1, include the following:

1   State the *objectives* clearly.

2   Define the *target population* carefully.

3   Design the *sample selection* plan using *randomization*, so as to reduce *sampling bias*.

4   Decide on a *method of measurement* that will minimize *measurement bias*.

5   Use a *pretest* to try out the plan.

6   Organize the *data collection* and *data management*.

7   Plan for careful and thorough *data analysis*.

8   Write *conclusions* in light of the original objectives.

Let's see how these points are realized in a real survey.

A serious quality-of-life issue in today's world is the threat of HIV infection and AIDS. Good decisions on how to battle this threat must come from sound data, but data on such a personal issue are difficult to obtain. Improvements are being made in data collection, however, as is seen in a recent study reported in *Science*, 13 November

1992 (J. Catania et al., *"Prevalence of AIDS-Related Risk Factors and Condom Use in the United States"*). In this study, a random sample of 2,673 residents of the United States between the ages of 18 and 75 was selected by random digit dialing (essentially, randomly dialing telephone numbers). A larger sample of 8,263 residents was randomly selected from high-risk cities. Some demographic characteristics of the resulting samples are shown in Table 2.2.

One of the main purposes of using randomization is to ensure that the sample is representative of the target population, of which this study contains two. For the target population of the United States as a whole, some aspects of this representativeness can be checked by comparing the demographic percentages of the national random sample with those in the population, as described in the census data of Table 2.1. The national sample has a few more women and a few less whites than it should, but overall it is quite representative. (Why are the percentages for the target population of high-risk cities much different from the census figures of Table 2.1?) This agreement may allow us to have more confidence in the main results of the survey—the percentages of various HIV-related risk groups, as shown in Table 2.3.

For the nation as a whole, it looks as if about 15% of those between the ages of 18 and 75 are at risk for HIV infection, with the percentage increasing to about 20% for the high-risk cities. (Actually, these percentages go much higher if sexual practices over more than the past year are taken into account.) The percentages can be used as *estimates* of the probabilities of what might happen in certain situations. For example, if a national firm is to hire 1,000 workers across the country, a good

TABLE 2.2
Demographic Characteristics of the National AIDS Behavioral Surveys

	High-Risk Cities Percent	High-Risk Cities Number	National Percent	National Number
Women	57.9	4,785	58.4	1,561
Men	42.1	3,478	41.6	1,112
African-American	33.8	2,795	13.5	360
Hispanic	20.7	1,711	8.3	222
White	42.7	3,525	75.9	2,030
Other	2.8	230	2.3	61

TABLE 2.3
Prevalence of HIV-Related Risk Groups Among Adult Heterosexuals, National and High-Risk Cities Samples

Risk Group	National Percent	National Number	High-Risk Cities Percent	High-Risk Cities Number
Multiple partners[†]	7.0	170	9.5	651
Risky partner	3.2	76	3.7	258
Transfusion recipient	2.3	55	2.1	144
Multiple partner and risky partner	1.7	41	3.0	209
Multiple partner and transfusion recipient	0.0	1	0.3	20
Risky partner and transfusion recipient	0.2	4	0.3	19
All others	0.7	16	0.7	51
No risk	84.9	2,045	80.4	5,539

[†]Past 12 months.

guess as to the number at risk for HIV infection would be 15%, or 150 workers. What if the firm is to hire all 1,000 workers in high-risk cities?

Here, the notion of probability was used in two directions. First, random sampling (a probabilistic idea) was used to obtain a representative sample, and the random digit dialing appears to have done a good job. Second, the main results of the study (percentages of various risk groups) can themselves serve as estimates of probabilities for making future decisions. The data in Table 2.3 are, at once, a description of two different target populations and a set of probabilities by which to weigh future outcomes.

2.2.3 Experiment

How many times has a parent or a teacher admonished a student to turn off the radio while doing homework? Does listening to music while doing homework help or hinder a student? To answer specific questions such as this, we must conduct carefully planned *experiments*. Suppose our class has a physics lesson to study for tomorrow. We could have some students study with the radio on and some study with the radio off. But the time of day that the studying takes place could affect the outcomes as well. Therefore, we have some students study in the afternoon with the radio on and some study in the afternoon with the radio off. Other students study in the evening, some with the radio on and some with the radio off. The measurements on which the issue will be decided (for now) are the scores on tomorrow's quiz.

Males might produce different results from females, so perhaps we should *control* for gender by making sure that both males and females are selected for each of the four *treatment* groups. On thinking about this *design* a bit longer, we conclude that the native ability of the students might also have some effect on the outcome. All the students have similar academic backgrounds, so differentiating on ability is difficult. Therefore, we will *randomly* assign students to the four treatment groups in the hope that any undetected differences in ability will balance out in the long run.

The above outline of a study has most of the key elements of a *designed experiment*. The goal of an experiment is to measure the effect of one or more *treatments* on *experimental units* appropriate to the study to see if one treatment is better than another. Here, there are two main treatments, the radio and the time of day that studying occurs. Another *variable* of interest is the gender of the student (the experimental unit in this case), but this variable is directly *controlled* in the design by making sure we have data from both sexes for all treatments. The variable "ability" cannot be controlled so easily, so we *randomize* the assignment of students to treatments to reduce the possible biasing effect of ability on the response comparisons.

Key elements of any experiment, consistent with the problem-solving outline of Chapter 1, include the following:

1 Clearly define the *question* to be investigated.

2 Identify the key variables to be used as *treatments*.

3 Identify other important variables that can be *controlled*.

4 Identify important background (lurking) variables that cannot be controlled but should be balanced by *randomization*.

5 Randomly assign treatments to the *experimental units*.

6 Decide on a *method of measurement* that will minimize *measurement bias*.

7 Organize the *data collection* and *data management*.

8 Plan for careful and thorough *data analysis*.

9 Write *conclusions* in light of the original question.

10 Plan a *follow-up* study to answer the question more completely or to answer the next logical question related to the issue at hand.

We will now see how these steps are followed in a real experiment of practical significance for improving the quality of life.

An aspirin a day keeps heart attacks away—or does it? This question has been extensively investigated in recent years through planned experiments. Note that the goal here is not to estimate the proportion of people subject to heart attacks but rather to see if a treatment (aspirin) is better than no treatment in preventing heart attacks. Thus, the answer is provided by an experiment rather than by a sample survey. (*Source:* "The Final Report on the Aspirin Component of the Ongoing Physicians' Health Study," *The New England Journal of Medicine*, vol. 231, no. 3, 1989, pp. 129–135.)

During the 1980s, approximately 22,000 physicians over the age of 40 agreed to participate in a long-term health study in which one important *question* was to determine whether or not aspirin helps lower the rate of heart attacks (myocardial infarctions). The *treatment* for this part of the study was aspirin, and the *control* was a placebo. Physicians were *randomly* assigned to one treatment or the other as they entered the study so as to minimize *bias* caused by uncontrolled factors. The method of assignment was equivalent to tossing a coin and sending the physician to the aspirin arm of the study if a head appeared on the coin. After the assignment, neither the participating physicians nor the medical personnel who treated them knew who was taking aspirin and who was taking a placebo. This is called a *double-blind* experiment. (Why is the double-blinding important in a study such as this?) The *method of measurement* was to observe the physicians carefully for an extended period of time and record all heart attacks, as well as other problems, that might occur.

Many variables other than aspirin could have an effect on the rate of heart attacks for the two groups of physicians. For example, the amount of exercise they get and whether or not they smoke are two prime examples of variables that should be *controlled* in the study so that the true effect of aspirin can be measured. Table 2.4 shows how the subjects eventually divided according to exercise and cigarette smoking. Do you think the randomization scheme did a good job in controlling these variables? Would you be concerned about the results for aspirin being unduly influenced by the fact that most of the aspirin takers were also nonsmokers? Would you be concerned about the placebo group possibly having too many who do not exercise?

The *data analysis* for this study reports that 139 heart attacks developed among the aspirin users and 239 heart attacks developed in the placebo group. This result was said to significantly favor aspirin as a possible preventative for heart attacks; the issue of statistical significance will be addressed later in this book.

T A B L E **2.4**
Some Results from the
Physicians' Health Study

	Aspirin	**Placebo**
Exercise Vigorously		
Yes	7,910	7,861
No	2,997	3,060
Cigarette Smoking		
Never	5,431	5,488
Past	4,373	4,301
Current	1,213	1,225

The 139 heart attacks among the over 11,000 physicians in the aspirin arm of the study could be used to form an estimate of the probability of a heart attack among aspirin users. This estimate could be misleading, however, unless great care was taken to account for age, socioeconomic status, and other factors pertinent to the study. Estimating probabilities is *not* the goal of an experiment; the goal is to compare treatments. The role of probability is primarily in the design (randomization) so that data that allow fair comparisons can be obtained.

2.2.4 Observational Study

"The weather will be warm this week; make sure you keep your doors locked and be careful where you walk at night." The implication in this statement is that crime rates are higher in warm weather than in cool weather. Is this really the case? One way to answer the *question* is to collect data on the number of crimes committed for the two *treatment* groups, warm days and cool days, and compare the results. Such an investigation looks somewhat like an experiment, but there is a key difference. In an experiment, treatments are randomly assigned to experimental units, and, in the randomization process, other factors of importance become balanced among the treatment groups. In the investigation suggested above, "warm" and "cool" cannot be randomly assigned to days; therefore, the other factors that might affect crime rates are not necessarily balanced in the study. All the investigator can do is *observe* what is happening in the world around her. Thus, such studies are called *observational studies*.

Other important variables can and should be identified, however, and some of them can be partially *controlled* by stratification. Crime rates vary according to geographic region, so the data can be *stratified* according to the region of the city or county in which the crimes took place. Holidays and major events (sporting events, conventions, and so on) have an effect on crime, and, with enough data, the investigator might be fortunate enough to have some of these events on warm days and some on cool days. No matter how much stratification is accomplished, some potentially important variables that cannot be balanced according to weather will always remain and will *confound* the results, in the sense that one never knows for sure if the difference in crime rates is due to temperature or to some other factor. This is the main reason why *observational studies* are less desirable than experiments, but, if done carefully, they are often better than no study at all.

Key elements of an observational study are very similar to those for an experiment and can be listed as follows:

1 Clearly define the *question* being investigated. (What is a crime?)

2 Identify the key variables to be used as *treatments*. (What is a warm day?)

3 Identify important variables that can be *controlled by stratification*.

4 Identify potential *confounding* variables that cannot be controlled.

5 Select *measurement units* in as *balanced* a way as possible, even though randomization is not used.

6 Decide on a *method of measurement* that will minimize measurement bias.

7 Organize the *data collection* and *data management*.

8 Plan a careful and thorough *data analysis*.

9 Write *conclusions* in light of the original question.

10 Plan a *follow-up* study to answer the question more completely or to answer the next logical question related to the issue at hand.

F I G U R E **2.4**
Comparison of the Cost and Frequency of Patient X-rays Ordered by Doctors Who Have Their Own X-ray Equipment and Doctors Who Send Their Patients to a Radiologist (based on a study of 65,517 doctor visits by privately insured patients treated for colds and low back pain, women treated for pregnancy, and men treated for difficulty in urinating)

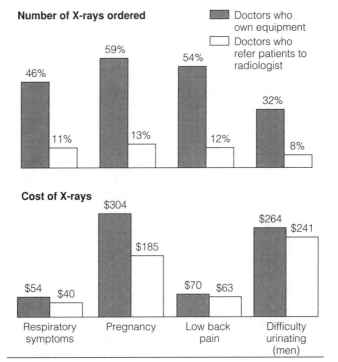

Source: Gainesville Sun, 6 December 1990.

The graphic in Figure 2.4 appeared in a newspaper with little additional comment. Since no randomization is suggested, the data can be thought to have come from an observational study. As you study the graphic display, identify the treatments and the conclusions that the designer of the graphic is expecting you to draw. Are there any potential problems in making good decisions based on these data?

Exercises

2.1 For each of the following situations, discuss whether or not the term probability seems to be used correctly and, if so, how such a *probability* might have been determined.

a A physician prescribes a medication for an infection but says that there is a probability of 0.2 that it will cause an upset stomach.

b An engineer who has worked on the development of a robot says that it should work correctly for at least 1,000 hours with probability 0.99.

c A friend says that the probability of extraterrestrial life is 0.1.

2.2 How good is your intuition about probability? Look at the following situations and answer the questions without making any calculations. Rely on your intuition, and see what happens.

a A six-sided balanced die has four sides colored green and two colored red. Three sequences of tosses resulted in the following:

RGRRR

RGRRRG

GRRRRR

Order the sequences from most likely to least likely.

b If a balanced coin is tossed six times, which of the following sequences is more likely, HTHTTH or HHHTTT? Which of the following is more likely, HTHTTH or HHHHTH?

Activities like these were used by D. Kahneman, P. Slovic, and A. Tversky to study how people reason probabilistically, as reported in their book *Judgment Under Uncertainty: Heuristics and Biases*, Cambridge University Press, 1982.

2.3 *Bias* is the tendency for a whole set of responses to read high or low because of some inherent difficulty with the measurement process. (A chipped die may have a bias toward 6's, in that it comes up 6 much more often than we would expect.) In the study of HIV-related risk groups discussed above, people in the survey were asked intimate questions about their personal lives as part of a telephone interview. Is there a possibility of bias in the responses? If so, in which direction? Will randomization in selecting the respondents help reduce potential bias related to the sensitive questions? Will randomization in selecting the respondents help reduce any potential bias?

2.4 Readers of the magazine *Popular Science* were asked to phone in (on a 900 number) their responses to the following question: "Should the United States build more fossil-fuel generating plants or the new so-called safe nuclear generators to meet the energy crisis of the 90s?" Of the total call-ins, 86% chose the nuclear option. What do you think about the way the poll was conducted? What do you think about the way the question was worded? Do you think the results are a good estimate of the prevailing mood of the country? (See *Popular Science*, August 1990, for details.)

2.5 "Food survey data all wrong?" This was the headline of a newspaper article on a report from the General Accounting Office of the U.S. Government related to the Nationwide Food Consumption Survey. The survey of 6,000 households of all incomes and 3,600 low-income households is intended to be the leading authority on who consumes what foods. Even though the original households were randomly selected, the GAO said the results were questionable because only 34% of the sampled households responded. Do you agree? What is the nature of the biases that could be caused by the low response rate? (The article was printed in the *Gainesville Sun*, 11 September 1991.)

2.6 "Why did they take my favorite show off the air?" The answer lies, no doubt, in low Nielsen ratings. What is this powerful rating system, anyway? Of the 92.1 million households in America, Nielsen Media Research randomly samples 4,000 on which to base their ratings. The sampling design is rather complex, but at the last two stages it involves randomly selecting city blocks (or equivalent units in rural areas) and then randomly selecting one household per block to be the Nielsen household. The *rating* for a program is the percentage of the sampled households that have a TV set on and tuned to the program. The *share* for a program is the percentage of the *viewing households* that have a TV set tuned to the program, where a viewing household is a household that has at least one TV set turned on.

a How many households are equivalent to one rating point?

b Is a share going to be larger or smaller than a rating point?

c For the week of 19 April 1992, "60 Minutes" was the top-rated show, with a rating of 21.7. Explain what this rating means.

d Discuss potential biases in the Nielsen ratings, even with the randomization in the selection of households carefully built in.

2.7 How does Nielsen determine who is watching what show? The determination comes from the data recorded in a journal by members of a Nielsen household. When a person begins to watch a show, he or she is supposed to log on. Computer vision researchers at the University of Florida are developing a peoplemeter that uses computer image recognition to passively, silently, and automatically record who is watching each show. Discuss the potential for this electronic device to reduce bias in the Nielsen ratings. Are there any new problems that might be caused by this device?

2.8 Stress increases your chances of a cold, according to an article in the *New England Journal of Medicine* for 29 August 1991. A total of 394 volunteers was randomly assigned to one of five groups, and each group was exposed to one of five types of virus. Overall, among those under the most stress, 47% caught colds, while among those under little stress, 27% caught colds. The higher rate for those under stress held true for all five viruses.

a Is this study a sample survey or an experiment? Why?

b Can the 47% be used as a good estimate of the rate at which those under stress might catch colds for the U.S. population as a whole? Why?

c The article refers to the fact that personality differences could not explain the findings. Could this be a result of the randomization? How? Is this fact a positive or negative feature of the study?

2.9 Love is not blind! This is the message from a study of how well people recognize their romantic partner by touch. Blindfolded participants were asked to distinguish their lover from two decoys by touching either the back of the hand or the forehead. In the forehead touch, 58% of the 72 blindfolded people in the study correctly chose their partner. In the hand touch, the women correctly identified their man's hand 69% of the time.

a Suppose everyone was guessing. What is the expected percentage of correct decisions? Does it look as if the "love is not blind" message is justified?

b Given the brief sketch presented above, would you call this study an experiment or an observational study? Why? (This study was reported in the *Gainesville Sun*, 22 June 1992.)

2.3 Tables and Sets

The clarity with which patterns in data or interpretations of probability can be seen is often dependent upon the manner in which the quantitative information is displayed.

Figure 2.5 on net new workers is an attractive pie chart, but it is not very useful for enabling us to see the data clearly or see associations in the data. More useful displays are one-way and two-way tables and the Venn diagram.

F I G U R E **2.5** Net New Workers

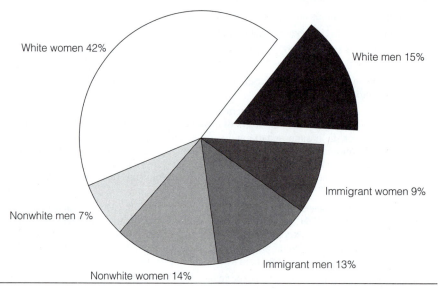

Net New Workers, 1985–2000
25,000,000

White women 42%

White men 15%

Immigrant women 9%

Nonwhite men 7%

Immigrant men 13%

Nonwhite women 14%

Source: U.S. Department of Labor.

2.3.1 One-way Table

A one-way table of the relative frequencies (percentages) in Figure 2.5 is provided in Table 2.5. It is simply a listing of the categories with their relative frequencies, which could be arrayed either in a column (as here) or in a row. One advantage of this array is that it would be easy to add another column to show the frequency counts (out of the 25 million net new workers expected by the year 2000). Another advantage is that it makes it easier to see meaningful ways to combine the categories; whites, nonwhites, and immigrants could be combined easily to reduce the number of rows to three. For the information in a table to be clearly understood, it is important that the row (or column) categories be mutually exclusive (that is, an element in the study cannot be in two rows simultaneously) and exhaustive (that is, every element in the study fits into one of the rows). This caution is actually violated in Table 2.5, which makes this table a little harder to read than it need be.

T A B L E **2.5**
Net New Workers, 1985–2000

Category	Percent
White women	42
White men	15
Nonwhite women	14
Nonwhite men	7
Immigrant women[†]	9
Immigrant men[†]	13

[†]Immigrants contain both white and nonwhite.

2.3.2   Two-way Table

A closer look at the data in Figure 2.5 reveals that two categorical factors are actually defined there, gender and status as regards color and residency. Thus, the data can be conveniently displayed on a two-way table that shows how these two factors are related. Such a display (called a 3×2 table in this case) is shown in Table 2.6.

Looking at the percentages in the cells and the marginal totals (Does adding across rows or columns give numbers that are interpretable?), we see that many more women than men will be entering the workforce and that nearly half of the net new workers will be nonwhites or immigrants. Also, the pattern of the women across the color/immigrant status factor is quite different from the pattern of men. All these features are more easily spotted on the two-way table than on the original figure or the one-way table.

2.3.3   Venn Diagram

Sometimes it is helpful to simply sketch the nature of the relationships among factors under study as set diagrams, usually referred to as Venn diagrams. In Figure 2.6a, the entire rectangle represents the number of net new workers, and the section labeled "women" shows that the category of women is one of the subsets. Figure 2.6b shows schematically that the white and nonwhite subsets do not overlap, whereas Figure 2.6c shows that the women and white subsets do overlap. Venn diagrams are most useful in allowing us to see how probabilities can be combined.

Before going into a formal discussion of probability, we must first outline the *set notation* we will use. Suppose we have a *set S* consisting of points labeled 1, 2, 3, and 4. We denote this by $S = \{1, 2, 3, 4\}$. If $A = \{1, 2\}$ and $B = \{2, 3, 4\}$, then A and B are *subsets* of S, denoted $A \subset S$ and $B \subset S$ (B "is contained in" S). We denote the fact that 2 is an *element* of A by $2 \in A$. The *union* of A and B is the set consisting of

T A B L E **2.6**
Net New Workers, 1985–2000

	Women	Men	
White	42%	15%	57%
Nonwhite	14%	7%	21%
Immigrant	9%	13%	22%
	65%	35%	100%

F I G U R E **2.6**
Venn Diagrams

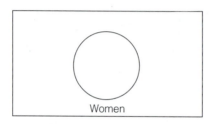

a

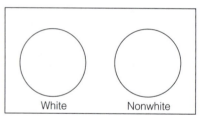

b

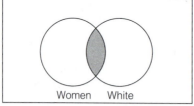

c

all points that are either in A or B or in both. This is denoted by $A \cup B = \{1, 2, 3, 4\}$. If $C = \{4\}$, then $A \cup C = \{1, 2, 4\}$. The *intersection* of two sets A and B is the set consisting of all points that are in both A and B; it is denoted by $A \cap B$, or merely AB. For the above example, $A \cap B = AB = \{2\}$ and $AC = \phi$, where ϕ denotes the null set, or the set consisting of no points.

The *complement* of A, with respect to S, is the set of all points in S that are not in A and is denoted by $\overline{A}$. For the specific sets given above, $\overline{A} = \{3, 4\}$. Two sets are said to be *mutually exclusive*, or *disjoint*, if they have no points in common, as in A and C above.

Venn diagrams can be used to portray effectively the concepts of union, intersection, complement, and disjoint sets, as shown in Figure 2.7. We can easily see from Figure 2.7 that

$$\overline{A} \cup A = S$$

for any set A. Other important relationships among events are the *distributive laws*,

$$A(B \cup C) = AB \cup AC$$
$$A \cup (BC) = (A \cup B)(A \cup C)$$

and *DeMorgan's Laws*,

$$\overline{A \cup B} = \overline{A}\,\overline{B}$$
$$\overline{AB} = \overline{A} \cup \overline{B}$$

It is important to be able to relate descriptions of sets to their symbolic notation, using the symbols given above, and to be able to list correctly or count the elements in sets of interest, as the following example illustrates.

F I G U R E **2.7**
Venn Diagrams of Set Relations

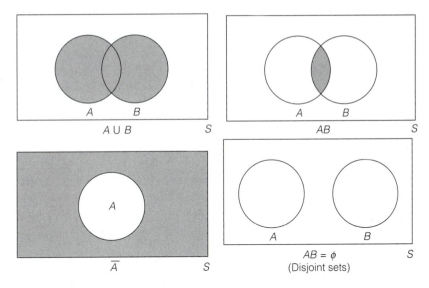

F I G U R E **2.7**
Venn Diagrams of Set Relations

E X A M P L E   **2.1** Twenty electric motors are pulled from an assembly line and inspected for defects. Eleven of the motors are free of defects, eight have defects on the exterior finish, and three have defects in their assembly and will not run. Let A denote the set of motors having assembly defects and F the set having defects on their finish. Using A and F, write a symbolic notation for the following:

a the set of motors having both types of defects

b the set of motors having at least one type of defect

c the set of motors having no defects

d the set of motors having exactly one type of defect

Then give the number of motors in each set.

Solution **a** The motors with both types of defects must be in A and F; thus, this event can be written AF. Since only nine motors have defects, while A contains three and F contains eight motors, two motors must be in AF. (See Figure 2.8.)

b The motors having at least one type of defect must have either an assembly defect or a finish defect. Hence, this set can be written $A \cup F$.

F I G U R E **2.8**
Venn Diagram for Example 2.1
(numbers of motors shown for
each set)

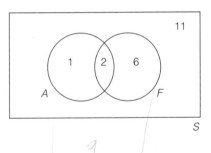

Since 11 motors have no defects, 9 must have at least one defect.

c   The set of motors having no defects is the complement of the set having at least one defect and is written $\overline{A \cup F} = \overline{A}\,\overline{F}$ (by DeMorgan's Law). Clearly, 11 motors fall into this set.

d   The set of motors having exactly one type of defect must be either in A and not in F or in F and not in A. This set can be written $A\overline{F} \cup \overline{A}F$, and seven motors fall into the set.   ∎

Exercises

2.10   Of the 2.7 million engineers in the United States (as of 1988), 95.5% are male. In addition, 90.6% are white, 1.6% are black, 5.6% are Asian, and 0.4% are Native American. Of the 0.7 million computer specialists, 69.1% are male, 88.3% are white, 3.7% are black, 6.6% are Asian, and 0.1% are Native American.

  a   Construct a meaningful table to compare number of engineers and number of computer scientists by sex.

  b   Construct a meaningful table to compare number of engineers and number of computer scientists by racial group.

2.11   Of 25 microcomputers available in a supply room, 10 have circuit boards for a printer, 5 have circuit boards for a modem, and 13 have neither type of board. Using P to denote those that have printer boards and M to denote those that have modem boards, symbolically denote the following sets, and give the number of microcomputers in each set.

  a   Those that have both boards.

  b   Those that have neither board.

  c   Those that have printer boards only.

  d   Those that have exactly one of the boards.

2.12   Construct a table to display the data in Exercise 2.11.

2.13   Five applicants (Jim, Don, Mary, Sue, and Nancy) are available for two identical jobs. A supervisor selects two applicants to fill these jobs.

  a   List all possible ways in which the jobs can be filled. (That is, list all possible selections of two applicants from the five.)

  b   Let A denote the set of selections containing *at least* one male. How many elements are in A?

  c   Let B denote the set of selections containing *exactly* one male. How many elements are in B?

  d   Write the set containing two females in terms of A and B.

  e   List the elements in $\overline{A}$, AB, $A \cup B$, and $\overline{AB}$.

2.14   Use Venn diagrams to verify the distributive laws.

2.15   Use Venn diagrams to verify DeMorgan's Laws.

2.4 Definition of Probability

Probability, as we have seen, requires a target population (conceptual or real) from which observable outcomes are obtained, meaningful categorizations of these outcomes, and a random mechanism for generating outcomes. A conceptually infinite population of coin tosses can be categorized as "heads" or "tails." If the tosses are random, a meaningful statement can be made about the probability of "heads" on the next toss.

To change the setting a bit, suppose a regular six-sided die is tossed onto a table and the number on the upper face is observed. This is a probabilistic situation, since the number occurring on the upper face cannot be determined in advance. We will analyze the components of the situation and arrive at a definition of probability that will allow us to model mathematically what happens in die tosses, as well as in many similar situations.

First, we might toss the die several times in order to collect data on possible outcomes. This data-generation phase allows us to see the nature of outcomes, which we can list in a *sample space*.

DEFINITION   **2.1**

A **sample space** S is a set that includes all possible outcomes for a random selection from a specified population, listed in a mutually exclusive and exhaustive manner.   ∎

In Definition 2.1, "mutually exclusive" means that the elements of the set do not overlap, and "exhaustive" means that the list contains all possible outcomes.

For the die toss, we could write a sample space as

$$S_1 = \{1, 2, 3, 4, 5, 6\}$$

where the integers indicate the possible numbers of dots on the upper face, or as

$$S_2 = \{\text{even, odd}\}$$

Both S_1 and S_2 satisfy Definition 2.1, but S_1 seems the better choice because it gives all the necessary details. S_2 has three possible upper-face outcomes for each listed element, whereas S_1 has only one possible outcome per element.

As another example, suppose an inspector is measuring the length of a machined rod. (This measurement process constitutes the random selection.) A sample space could be listed as

$$S_3 = \{1, 2, 3, \ldots, 50, 51, 52, \ldots, 70, 71, 72, \ldots\}$$

if the length is rounded to the closest integer number of inches. On the other hand, an appropriate sample space could be

$$S_4 = \{x \mid x > 0\}$$

which is read "the set of all real numbers x such that $x > 0$." Whether S_3 or S_4 should be used in a particular problem depends on the nature of the measurement process. If the measurement is to be precise, we need S_4. If only integer measurements are to be used, S_3 will suffice. The point is, then, that sample spaces for a particular process are not unique; they must be selected so as to provide all pertinent information for a given situation.

Let us go back to our first example, the toss of a die. Suppose player A can have the first turn at a board game if he rolls a 6. Therefore, the event "roll a 6" is important to him. Other possible events of interest in the die-tossing experiment are "roll an even number," "roll a number greater than 4," and so on.

DEFINITION   **2.2**   An **event** is any subset of a sample space.   ∎

Definition 2.2 holds as stated for any sample space that has a finite or countable number of elements. Some subsets must be ruled out if a sample space covers a continuum of real numbers, such as S_4 given above, but any subset likely to occur in practice can be called an event.

From this definition, we see that an event is a collection of elements from the sample space. For example, in the die-tossing experiment let us define

> A to be "an even number"
>
> B to be "an odd number"
>
> C to be "a number greater than 4"
>
> E_1 to be "observe a 1"

and, in general,

> E_i to be "observe integer i"

Then, if $S = \{1, 2, 3, 4, 5, 6\}$,

$$A = \{2, 4, 6\}$$
$$B = \{1, 3, 5\}$$
$$C = \{5, 6\}$$
$$E_1 = \{1\}$$

and

$$E_i = \{i\}, \qquad i = 1, 2, 3, 4, 5, 6$$

Sample spaces and events can often be conveniently displayed in Venn diagrams. Some events for the die-tossing experiment are shown in Figure 2.9.

We now know how to establish a sample space and to list appropriate events. The next step is to define a probability for those events. We have seen already that the intuitive idea of probability is related to relative frequency of occurrence. A regular die should show an even number, when tossed, about one-half of the time and a 3 about one-sixth of the time; therefore, probabilities should be fractions between 0

FIGURE **2.9**
Venn Diagram for a Die Toss

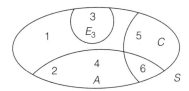

and 1. One of the integers 1, 2, 3, 4, 5, or 6 must occur every time the die is tossed, so the total probability associated with the sample space must be 1. In repeated tosses of the die, if a "1" occurs $1/6$ of the time, then a 1 or 2 must occur $1/6 + 1/6 = 1/3$ of the time. Since relative frequencies for mutually exclusive events add, then so must the probabilities. These considerations lead to the following definition.

DEFINITION **2.3**

Suppose a random selection from a specified population has associated with it a sample space S. A **probability** is a numerically valued function that assigns a number $P(A)$ to every event A so that the following axioms hold:

1 $P(A) \geq 0$

2 $P(S) = 1$

3 If $A_1, A_2, \ldots$ is a sequence of mutually exclusive events (that is, $A_i A_j = \phi$ for any $i \neq j$), then

$$P\left(\bigcup_{i=1}^{\infty} A_1\right) = \sum_{i=1}^{\infty} P(A_i) \quad \blacksquare$$

From axiom 3 it follows that if A and B are mutually exclusive events

$$P(A \cup B) = P(A) + P(B)$$

This is similar to the addition of relative frequencies in the die-tossing example discussed above.

It is now easy to see that if $A \subset B$, then $P(A) \leq P(B)$. To see this, write

$$B = A \cup \overline{A}B$$

then

$$P(B) = P(A \cup \overline{A}B) = P(A) + P(\overline{A}B)$$

Since $P(\overline{A}B) \geq 0$ by axiom 1, it follows that $P(A) \leq P(B)$. In particular, since $A \subset S$ for any even A and $P(S) = 1$, then $P(A) \leq 1$.

In a similar way, we can show that $P(\phi) = 0$. Since S and ϕ are disjoint with $S \cup \phi = S$,

$$1 = P(S) = P(S \cup \phi) = P(S) + P(\phi)$$

The definition of probability tells us only the axioms such a function must obey; it does not tell us what numbers to assign to specific events. The actual assignment of numbers usually comes about from empirical evidence or from careful thought about the selection process. If a die is balanced, we could toss it a few times to see if the upper faces all seem equally likely to occur. Or we could simply assume this would happen and assign a probability of $1/6$ to each of the six elements in S as follows: $P(E_i) = 1/6$, $i = 1, 2, \ldots, 6$. Once we've done this, the model is complete, since by axiom 3 we can now find the probability of any event. For example, for the events defined on page 70 and in Figure 2.9,

$$
\begin{aligned}
P(A) &= P(E_2 \cup E_4 \cup E_6) \\
&= P(E_2) + P(E_4) + P(E_6) \\
&= \frac{1}{6} + \frac{1}{6} + \frac{1}{6} = \frac{1}{2}
\end{aligned}
$$

and

$$
\begin{aligned}
P(C) &= P(E_5 \cup E_6) \\
&= P(E_5) + P(E_6) \\
&= \frac{1}{6} + \frac{1}{6} = \frac{1}{3}
\end{aligned}
$$

It is important to remember that Definition 2.3 and the actual assignment of probabilities to events provide a probabilistic model for a random selection process. If $P(E_i) = 1/6$ is used for the die-toss, the model is good or bad depending on how close the long-run relative frequencies for each outcome actually come to these numbers suggested by theory. If the die is not balanced, then the model is bad, and other probabilities should be substituted for $P(E_i)$. Throughout the remainder of this book, we will develop many specific models based on this underlying definition and will discuss practical situations in which such models work well. No model is perfect, but many are adequate for describing real-world probabilistic phenomena.

The die-toss example assigns equal probabilities to the elements of a sample space, but such is not usually the case. If you have a quarter and a penny in your pocket and pull out the first one you touch, then the quarter may have a higher probability of being chosen because of its larger size. In the discussion of Section 2.2, we saw a number of examples in which empirical evidence suggested unequal probabilities for certain events. In such cases, the data provide only approximations to the true probabilities, but these approximations are often quite good and are usually the only information we have on the events of interest, as in Example 2.2 below.

E X A M P L E   **2.2**   The data in Figure 2.10 on marital status of the unemployed in the United States are provided by the Bureau of Labor Statistics.

Suppose you had met an unemployed worker in 1956. What is the approximate probability that the unemployed person would have been (a) a married woman (b) single (c) married? Also answer parts (a), (b), and (c) for an unemployed worker met in 1982.

FIGURE **2.10**
Marital Status of the
Unemployed

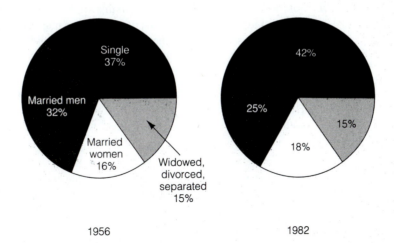

1956   1982

Solution   We assume that nothing else is known about the unemployed person that you met. (The person is, in effect, randomly chosen from the population of unemployed workers.) We see directly from the chart for 1956 that

a P(married woman) $= 0.16$
b P(single) $= 0.37$
c P(married) $= P$(married woman) $+ P$(married man)
                                        $= 0.16 + 0.32 = 0.48$

From the 1982 chart, we have

a P(married woman) $= 0.18$
b P(single) $= 0.42$
c P(married) $= 0.18 + 0.25 = 0.43$

Note that we *cannot* answer some potentially interesting questions from this chart. For example, we cannot find the probability that the unemployed person is a single woman.   ∎

EXAMPLE   **2.3**   Consider the national percentages for HIV-related risk groups given in Table 2.3. A firm hires a new worker after a national advertising campaign.

a   What is the probability that the worker falls in the "risky partner" category?
b   What is the probability that the worker is in at least one of the risk groups?
c   If the firm hires 1,000 workers, how many are expected to be at risk if the 1,000 come from the population at large?
d   If the firm hires 1,000 workers, how many are expected to be at risk if the 1,000 come from high-risk cities?

Solution   **a**   Solutions to practical problems of this type always involve assumptions. To answer parts (a) and (b) with the data given, we must assume that the new worker is randomly selected from the national population, which implies

$$P(\text{risky partner}) = 0.032$$

b   Let's label the seven risk groups, in the order listed in the table, $E_1, E_2, \ldots E_7$. Then the event of being in *at least one* of the groups can be written as the union of these seven, that is,

$$E_1 \cup E_2 \cup \ldots \cup E_7$$

The event "at least one" is the same as the event "E_1, or E_2 or E_3 or $\ldots$ or E_7." Since these seven groups are listed in mutually exclusive fashion,

$$
\begin{aligned}
P(E_1 \cup E_2 \cup \ldots \cup E_7) &= P(E_1) + P(E_2) + \cdots + P(E_7) \\
&= 0.070 + 0.032 + 0.023 + 0.017 + 0.000 \\
&\quad + 0.002 + 0.007 \\
&= 0.151
\end{aligned}
$$

Note that the mutually exclusive property is essential here; otherwise, we could not simply add the probabilities.

c   We assume that all 1,000 new hires are randomly selected from the national population (or a subpopulation of the same makeup). Then 15.1% of the 1,000, or 151 workers, are expected to be at risk.

d   We assume that the 1,000 workers are randomly selected from high-risk cities. Then 19.6%, or 196 workers, are expected to be at risk.

Do you think the assumptions necessary in order to answer the probability questions are reasonable here?   ∎

Exercises

2.16   A vehicle arriving at an intersection can turn left or continue straight ahead. Suppose an experiment consists of observing the movement of one vehicle at this intersection, and do the following.

a   List the elements of a sample space.

b   Attach probabilities to these elements if all possible outcomes are equally likely.

c   Find the probability that the vehicle turns, under the probabilistic model of part (b).

2.17   A manufacturing company has two retail outlets. It is known that 30% of the potential customers buy products from outlet I alone, 50% buy from outlet II alone, 10% buy from both I and II, and 10% buy from neither. Let A denote the event that a potential customer, randomly chosen, buys from I and B denote the event that the customer buys from II. Find the following probabilities.

a $P(A)$ **b** $P(A \cup B)$

c $P(\overline{B})$ **d** $P(AB)$

e $P(A \cup \overline{B})$ **f** $P(\overline{A}\,\overline{B})$

g $P(\overline{A \cup B})$

2.18   For volunteers coming into a blood center, 1 in 3 have O⁺ blood, 1 in 15 have O⁻, 1 in 3 have A⁺, and 1 in 16 have A⁻. What is the probability that the first person who shows up tomorrow to donate blood has

a   type O⁺ blood?   **b**   type O blood?

c   type A blood?   **d**   either type A⁺ or O⁺ blood?

2.19   Information on modes of transportation for coal leaving the Appalachian region is shown in Figure 2.11. If coal arriving at a certain power plant comes from this region, find the probability that it was transported out of the region

a   by truck to rail.

b   by water only.

c   at least partially by truck.

d   at least partially by rail.

e   by modes not involving water. (Assume "other" does not involve water.)

F I G U R E **2.11**

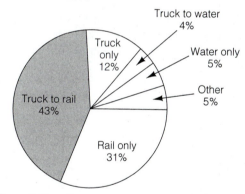

Source: G. Elmes, *Transportation Research*, 18A, no. 1, 1984, p. 19.

2.20   Hydraulic assemblies for landing gear coming from an aircraft rework facility are inspected for defects. History shows that 8% have defects in the shafts alone, 6% have defects in the bushings alone, and 2% have defects in both the shafts and the bushings. If a randomly chosen assembly is to be used on an aircraft, find the probability that it has

a   a bushing defect.

b   a shaft or bushing defect.

c   only one of the two type of defects.

d   no defects in shafts or bushings.

2.5 Counting Rules Useful in Probability

Let's look at the die toss from a slightly different perspective. Since there are six outcomes that should be equally likely for a balanced die, then the probability of A, observe an even number, is

$$P(A) = \frac{3}{6} = \frac{\text{number of outcomes favorable to } A}{\text{total number of equally likely outcomes}}$$

This "definition" of probability will work for any random phenomenon resulting in a finite sample space with *equally likely* outcomes. Thus, it is important to be able to count the number of possible outcomes for a random selection. The number of outcomes can easily become quite large, and counting them is difficult without a few counting rules. Four such rules are presented as theorems in this section.

Suppose a quality control inspector examines two manufactured items selected from a production line. Item 1 can be defective or nondefective, as can item 2. How many possible outcomes are possible for this experiment? In this case, listing the possible outcomes is easy. Using D_i to denote that the ith item is defective and N_i to denote that the ith item is nondefective, the possible outcomes are

$$D_1 D_2 \qquad D_1 N_2 \qquad N_1 D_2 \qquad N_1 N_2$$

These four outcomes could be placed in a two-way table, as in Figure 2.12. This table helps us see that the four outcomes arise from the fact that the first item has two possible outcomes and the second item has two possible outcomes, and hence the experiment of looking at both items has $2 \times 2 = 4$ outcomes. This is an example of the *product rule*, given as Theorem 2.1.

F I G U R E **2.12**

Possible Outcomes for Inspecting Two Items (D_i denotes that the ith item is defective; N_i denotes that the ith item is nondefective.)

First item		Second item	
		D_2	N_2
	D_1	$D_1 D_2$	$D_1 N_2$
	N_1	$N_1 D_2$	$N_1 N_2$

T H E O R E M **2.1**

If the first task of an experiment can result in n_1 possible outcomes and, for each such outcome, the second task can result in n_2 possible outcomes, then there are $n_1 n_2$ possible outcomes for the two tasks together. ∎

The product rule extends to more than two tasks in a sequence. If, for example, three items were inspected and each could be defective or nondefective, then there would be $2 \times 2 \times 2 = 8$ possible outcomes.

Tree diagrams are also helpful in verifying the product rule and in listing possible outcomes. Suppose a firm is deciding where to build two new plants, one in the east and one in the west. Four eastern cities and two western cities are possibilities. Thus, there are $n_1 n_2 = 4 \times 2 = 8$ possibilities for locating the two plants. Figure 2.13 shows the listing of these possibilities on a tree diagram.

F I G U R E **2.13**
Possible Outcomes for Locating Two Plants (A, B, C, and D denote eastern cities; E and F denote western cities.)

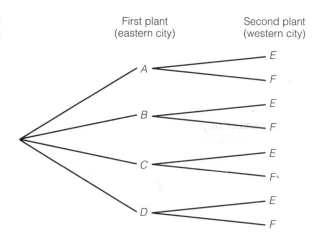

First plant
(eastern city)

Second plant
(western city)

The product rule (Theorem 2.1) helps only in finding the number of elements in a sample space. We must still assign probabilities to these elements to complete our probabilistic model. This is done for the site selection problem in Example 2.4.

E X A M P L E **2.4**   In the case of the firm that plans to build two new plants, the eight possible outcomes are shown in Figure 2.13. If all eight choices are equally likely (that is, one of the pairs of cities is selected at random), find the probability that city E gets selected.

Solution   City E can be selected in four different ways, since there are four possible eastern cities to pair with it. Thus,

$$\{E \text{ gets selected}\} = \{AE\} \cup \{BE\} \cup \{CE\} \cup \{DE\}$$

Each of the eight outcomes has probability $1/8$, since the eight events are assumed to be equally likely. Since these eight events are mutually exclusive,

$$P(E \text{ gets selected}) = P(AE) + P(BE) + P(CE) + P(DE)$$
$$= \frac{1}{8} + \frac{1}{8} + \frac{1}{8} + \frac{1}{8} = \frac{1}{2} \quad \blacksquare$$

E X A M P L E **2.5**   Five motors (numbered 1 through 5) are available for use, and motor 2 is defective. Motors 1 and 2 come from supplier I, and motors 3, 4, and 5 come from supplier II. Suppose two motors are randomly selected for use on a particular day. Let A denote the event that the defective motor is selected and B the event that at least one motor comes from supplier I. Find $P(A)$ and $P(B)$.

Solution   We can see on the tree diagram in Figure 2.14 that there are 20 possible outcomes for this experiment, which agrees with our calculation using the product rule. That is, there are 20 events of the form {1, 2}, {1, 3}, and so forth. Since the motors are randomly selected, each of the 20 outcomes has probability 1/20. Thus,

$$P(A) = P(\{1, 2\} \cup \{2, 1\} \cup \{2, 3\} \cup \{2, 4\} \cup \{2, 5\} \cup \{3, 2\} \cup \{4, 2\} \cup \{5, 2\})$$
$$= \frac{8}{20} = 0.4$$

since the probability of the union is the sum of the probabilities of the events in the union.

The reader can show that B contains 14 of the 20 outcomes and that

$$P(B) = \frac{14}{20} = 0.7 \quad \blacksquare$$

FIGURE **2.14**
Outcomes for the Experiment of
Example 2.5

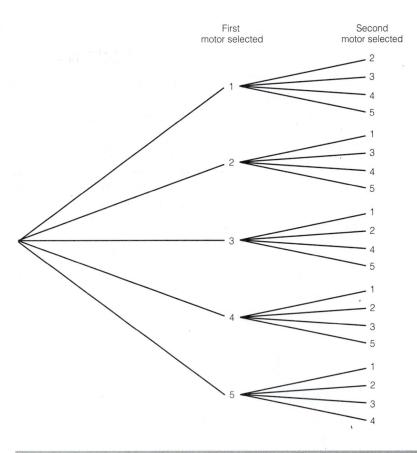

The product rule is often used to develop other counting rules. Suppose that from three pilots a crew of two is to be selected to form a pilot-copilot team. To count the number of ways this can be done, observe that the pilot's seat can be filled in three ways and the copilot's in two ways (after the pilot is selected), so there are $3 \times 2 = 6$ ways of forming the team. This is an example of a *permutation*, for which a general result is given in Theorem 2.2.

THEOREM **2.2**   The number of ordered arrangements, or permutations, of r objects selected from n distinct objects $(r \leq n)$ is given by

$$P_r^n = n(n-1)\cdots(n-r+1) = \frac{n!}{(n-r)!}$$

Proof   The basic idea of a permutation can be thought of as filling r slots in a line with one object in each slot by drawing these objects one at a time from a pool of n distinct objects. The first slot can be filled in n ways, but the second can be filled in only $(n-1)$ ways after the first is filled. Thus, by the product rule, the first two slots can be filled in $n(n-1)$ ways. Extending this reasoning to r slots, we have that the number of ways of filling all r slots is

$$n(n-1)\cdots(n-r+1) = \frac{n!}{(n-r)!} = P_r^n$$

Hence the theorem is proved. We illustrate the use of Theorem 2.2 with the examples that follow.   ▪

EXAMPLE **2.6**   From among ten employees, three are to be selected for travel to three out-of-town plants, A, B, and C, with one employee traveling to each plant. Since the plants are in different cities, the order of assigning the employees to the plants is an important consideration. The first person selected might, for instance, go to plant A and the second to plant B. In how many ways can the assignments be made?

Solution   Since order is important, the number of possible distinct assignments is

$$P_3^{10} = \frac{10!}{7!} = 10(9)(8) = 720$$

In other words there are ten choices for plant A but only nine for plant B and eight for plant C. This gives a total of $10(9)(8)$ ways of assigning employees to the plants.   ▪

EXAMPLE **2.7**   An assembly operation in a manufacturing plant involves four steps, which can be performed in any order. If the manufacturer wishes to experimentally compare the assembly times for each possible ordering of the steps, how many orderings will the experiment involve?

Solution   The number of orderings is the permutation of $n = 4$ things taken $r = 4$ at a time. (All steps must be accomplished each time.) This turns out to be

$$P_4^4 = \frac{4!}{0!} = 4! = 4 \cdot 3 \cdot 2 \cdot 1 = 24$$

since $0! = 1$ by definition. (In fact, $P_r^r = r!$ for any integer r.)   ■

E X A M P L E   **2.8**   A manager is asked to rank four divisions of a firm with respect to their ability to adapt to new technology. If the divisions are labeled D_1, D_2, D_3, and D_4, what is the probability that D_2 gets ranked highest?

Solution   The question has no answer as posited; there must be a random mechanism in operation in order to discuss probability. If the manager has no preferences and is merely ranking the firms randomly, then the question can be answered. There are now

$$P_4^4 = 4! = 24$$

equally likely rankings. If D_2 has the number one ranking, then there remain $3! = 6$ ways to rank the remaining divisions. Thus,

$$P(D_2 \text{ ranks first}) = \frac{3!}{4!} = \frac{6}{24} = \frac{1}{4}$$

Does this look reasonable? The question here could be extended to "What is the probability that D_2 ranks first and D_4 ranks second?" Once the first two ranks are taken, there remain only $2! = 2$ ways to fill out ranks three and four. Thus,

$$P(D_2 \text{ ranks first and } D_3 \text{ ranks second}) = \frac{2!}{4!} = \frac{2}{24} = \frac{1}{12}$$

Note that this answer is *not* $(\frac{1}{4})(\frac{1}{4})$.   ■

Sometimes order is not important, and we are interested only in the number of subsets of a certain size that can be selected from a given set. The general form of such a *combination* is given in Theorem 2.3.

T H E O R E M   **2.3**   The number of distinct subsets, or combinations, of size r that can be selected from n distinct objects ($r \le n$) is given by

$$\binom{n}{r} = \frac{n!}{r!(n - r)!}$$

Proof   The number of ordered subsets of size r, selected from n distinct objects, is given by P_r^n. The number of unordered subsets of size r is denoted by $\binom{n}{r}$.

Since any particular set of r objects can be ordered among themselves in $P_r^r = r!$ ways, it follows that

$$\binom{n}{r} r! = P_r^n$$

or

$$\binom{n}{r} = \frac{1}{r!} P_r^n = \frac{n!}{r!(n-r)!} \quad \blacksquare$$

E X A M P L E **2.9** Suppose that three employees are to be selected from ten to visit a new plant. In how many ways can the selection be made?

Solution Here, order is not important; we merely want to know how many subsets of size $r = 3$ can be selected from $n = 10$ people. The result is

$$\binom{10}{3} = \frac{10!}{3!7!} = \frac{10 \cdot 9 \cdot 8}{1 \cdot 2 \cdot 3} = 120 \quad \blacksquare$$

E X A M P L E **2.10** Refer to Example 2.9. If two of the ten employees are female and eight are male, what is the probability that exactly one female gets selected among the three?

Solution We have seen that there are $\binom{10}{3} = 120$ ways to select three employees from ten. Similarly, there are $\binom{2}{1} = 2$ ways to select one female from the two available and $\binom{8}{2} = 28$ ways to select two males from the eight available. If selections are made at random (that is, if all subsets of three employees are equally likely to be chosen), then the probability of selecting exactly one female is

$$\frac{\binom{2}{1}\binom{8}{2}}{\binom{10}{3}} = \frac{2(28)}{120} = \frac{7}{15} \quad \blacksquare$$

E X A M P L E **2.11** Five applicants for a job are ranked according to ability, with applicant number 1 being the best, number 2 second best, and so on. These rankings are unknown to an employer, who simply hires two applicants at random. What is the probability that this employer hires exactly one of the two best applicants?

Solution The number of possible outcomes for the process of selecting two applicants from five is

$$\binom{5}{2} = \frac{5!}{2!3!} = 10$$

If one of the two best is selected, the selection can be done in

$$\binom{2}{1} = \frac{2!}{1!1!} = 2$$

ways. The other selected applicant must come from among the three lowest-ranking applicants, which can be done in

$$\binom{3}{1} = \frac{3!}{1!2!} = 3$$

ways. Thus, the event of interest (hiring one of the two best applicants) can come about in $2 \times 3 = 6$ ways. The probability of this event is thus $6/10 = 0.6$.   ∎

T H E O R E M   **2.4**

The number of ways of partitioning n distinct objects into k groups containing $n_1, n_2, \ldots, n_k$ objects, respectively, is

$$\frac{n!}{n_1! n_2! \cdots n_k!}$$

where

$$\sum_{i=1}^{k} n_i = n$$

Proof

The partitioning of n objects into k groups can be done by first selecting a subset of size n_1 from the n objects, then selecting a subset of size n_2 from the $n - n_1$ objects that remain, and so on until all groups are filled. The number of ways of doing this is

$$\binom{n}{n_1}\binom{n - n_1}{n_2} \cdots \binom{n - n_1 - \cdots - n_{k-1}}{n_k}$$

$$= \frac{n!}{n_1!(n - n_1)!} \cdot \frac{(n - n_1)!}{n_2!(n - n_1 - n_2)!} \cdots \frac{(n - n_1 - \cdots - n_{k-1})!}{n_k! 0!}$$

$$= \frac{n!}{n_1! n_2! \cdots n_k!} \qquad ∎$$

E X A M P L E   **2.12**

Suppose that ten employees are to be divided among three jobs, with three employees going to job I, four to job II, and three to job III. In how many ways can the job assignment be made?

Solution   This problem involves a partitioning of the $n = 10$ employees into groups of size $n_1 = 3$, $n_2 = 4$, and $n_3 = 3$, and it can be accomplished in

$$\frac{n!}{n_1!n_2!n_3!} = \frac{10!}{3!4!3!} = \frac{10 \cdot 9 \cdot 8 \cdot 7 \cdot 6 \cdot 5}{3 \cdot 2 \cdot 1 \cdot 3 \cdot 2 \cdot 1} = 4,200$$

ways. (Notice the large number of ways this task can be accomplished!)   ▪

EXAMPLE   **2.13**   In the setting of Example 2.12, suppose the only three employees of a certain ethnic group all get assigned to job I. What is the probability of this happening under a random assignment of employees to jobs?

Solution   We have seen in Example 2.12 that there are 4,200 ways of assigning the ten workers to three jobs. The event of interest assigns three specified employees to job I. It remains to be determined how many ways the other seven employees can be assigned to jobs II and III, which is

$$\frac{7!}{4!3!} = \frac{7(6)(5)}{3(2)(1)} = 35$$

Thus, the chance of assigning three specific workers to job I is

$$\frac{35}{4,200} = \frac{1}{120}$$

which is very small, indeed!   ▪

Exercises

2.21   Two vehicles in succession are observed moving through the intersection of two streets.

  a   List the possible outcomes, assuming each vehicle can go straight, turn right, or turn left.

  b   Assuming the outcomes to be equally likely, find the probability that at least one vehicle turns left. (Would this assumption always be reasonable?)

  c   Assuming the outcomes to be equally likely, find the probability that at most one vehicle makes a turn.

2.22   A commercial building is designed with two entrances, door I and door II. Two customers arrive and enter the building.

  a   List the elements of a sample space for this observational experiment.

  b   If all elements in (a) are equally likely, find the probability that both customers use door I; then find the probability that both customers use the same door.

2.23   A corporation has two construction contracts that are to be assigned to one or more of three firms bidding for these contracts. (One firm could receive both contracts.)

  a   List the possible outcomes for the assignment of contracts to the firms.

b   If all outcomes are equally likely, find the probability that both contracts go to the same firm.

c   Under the assumptions of (b), find the probability that one specific firm, say firm I, gets at least one contract.

2.24   Among five portable generators produced by an assembly line in one day, there are two defectives. If two generators are selected for sale, find the probability that both will be nondefective. (Assume the two selected for sale are chosen so that every possible sample of size two has the same probability of being selected.)

2.25   Seven applicants have applied for two jobs. How many ways can the jobs be filled if

a   the first person chosen receives a higher salary than the second?

b   there are no differences between the jobs?

2.26   A package of six light bulbs contains two defective bulbs. If three bulbs are selected for use, find the probability that none is defective.

2.27   How many four-digit serial numbers can be formed if no digit is to be repeated within any one number? (The first digit may be a zero.)

2.28   A fleet of eight taxis is to be randomly assigned to three airports A, B, and C, with two going to A, five to B, and one to C.

a   In how many ways can this be done?

b   What is the probability that the specific cab driven by Jones is assigned to airport C?

2.29   Show that $\binom{n}{r} = \binom{n-1}{r-1} + \binom{n-1}{r}$, $1 \leq r \leq n$.

2.30   Five employees of a firm are ranked from 1 to 5 based on their ability to program a computer. Three of these employees are selected to fill equivalent programming jobs. If all possible choices of three (out of the five) are equally likely, find the probability that

a   the employee ranked number 1 is selected.

b   the highest-ranked employee among those selected has rank 2 or lower.

c   the employees ranked 4 and 5 are selected.

2.31   For a certain style of new automobile, the colors blue, white, black, and green are in equal demand. Three successive orders are placed for automobiles of this style. Find the probability that

a   one blue, one white, and one green are ordered.

b   two blues are ordered.

c   at least one black is ordered.

d   exactly two of the orders are for the same color.

2.32   A firm is placing three orders for supplies among five different distributors. Each order is randomly assigned to one of the distributors, and a distributor can receive multiple orders. Find the probability that

a   all orders go to different distributors.

b   all orders go to the same distributor.

c   exactly two of the three orders go to one particular distributor.

2.33   An assembly operation for a computer circuit board consists of four operations that can be performed in any order.

a   In how many ways can the assembly operation be performed?

b   One of the operations involves soldering wire to a microchip. If all possible assembly orderings are equally likely, what is the probability that the soldering comes first or second?

2.34   Nine impact wrenches are to be divided evenly among three assembly lines.

a   In how many ways can this be done?

b   Two of the wrenches are used, and seven are new. What is the probability that a particular line (line *A*) gets both used wrenches?

2.6   Conditional Probability and Independence: Narrowing the Table

2.6.1   Conditional Probability

Building on the connection between analysis of frequency data in tables and probability, let's consider the employment data in Table 2.7. A common summary of these data is given by the "employment rate," which is the percentage of unemployed workers given by

$$\frac{4,115}{101,735}(100) = 4.0$$

(This figure does not include those no longer actively seeking work.) But the overall unemployment rate does not tell us anything about the association between employment and education. To investigate the association, we must calculate unemployment rates separately for each education category (each row of the table). Narrowing the focus to a single row is often referred to as *conditioning* on the row factor.

The conditional relative frequencies are given in Table 2.8. Now it is apparent that unemployment is associated with educational level; those with less education have higher unemployment rates. The conditional relative frequencies relate directly to *conditional probability*. If a national pool samples 1,000 people from the national labor force, the expected number of unemployed workers is about 4% of 1,000, or 40. If, however, the 1,000 people all have a college education, the expected number of unemployed drops to 2%, or 20 people.

T A B L E **2.7**
Civilian Labor Force in the U.S.;
1989 (over 25 years of age;
figures in thousands)

Education	Employed	Unemployed	
Elementary school	5,299	406	
High school, 1–3 years	8,144	705	
High school, 4 years	38,171	1,763	
College, 1–3 years	19,991	671	
College, 4 or more years	26,015	570	
Total	97,620	4,115	101,735

Source: U.S. Bureau of Labor Statistics.

T A B L E **2.8**
Unemployment Rates by
Education

Education	Employed	Unemployed
Elementary school	93%	7%
High School, 1–3 years	92%	8%
High school, 4 years	96%	4%
College, 1–3 years	97%	3%
College, 4 or more years	98%	2%

E X A M P L E   **2.14**   Table 2.6, which showed percentages of net new workers in the labor force, is reproduced below.

	Women	Men	
White	42%	15%	57%
Nonwhite	14%	7%	21%
Immigrant	9%	13%	22%
	65%	35%	100%

How do the relative frequencies for the three racial/immigrant categories compare between women and men?

Solution   Even though the data are given in terms of percentages rather than frequencies, the relative frequencies can still be computed. Conditioning on women, the total number represents 65% of the population, while the number of whites represents 42% of the population. Therefore, 42/65 represents the proportion of whites among the women. Proceeding similarly across the other categories produces the *two* conditional distributions (one for women and one for men) shown below:

	Women	Men
White	65%	43%
Nonwhite	21%	20%
Immigrant	14%	37%
	100%	100%

Note that there is a fairly strong association between the two factors of sex and racial/immigrant status. Among the new members of the labor force, women will be mostly white, and men will have a high proportion of immigrants.   ∎

Conditioning can also be represented in Venn diagrams. Suppose that of 100 students completing an introductory statistics course, 20 were business majors. Ten students received A's in the course, and three of these students were business majors. These facts are easily displayed on a Venn diagram such as that shown in Figure 2.15, where A represents those students receiving A's and B represents business majors.

For a randomly selected student from this class, $P(A) = 0.1$ and $P(B) = 0.2$. But suppose we know that a randomly selected student is a business major. Then we might want to know the probability that the student received an A, *given* that she is a business major. Among the 20 business majors, 3 received A's. Thus, $P(A$ *given* $B)$, written $P(A|B)$, is 3/20.

FIGURE **2.15**

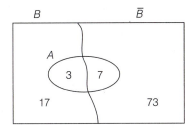

From the Venn diagram, we see that the conditional (or given) information reduces the effective sample space to just the 20 business majors. Among them, 3 received A's. Note that

$$P(A|B) = \frac{3}{20} = \frac{P(AB)}{P(B)} = \frac{3/100}{20/100}$$

This relationship motivates Definition 2.4.

DEFINITION **2.4**

If A and B are any two events, then the **conditional probability** of A given B, denoted by $P(A|B)$, is

$$P(A|B) = \frac{P(AB)}{P(B)}$$

provided $P(B) > 0$. ∎

EXAMPLE **2.15** From five motors, of which one is defective, two motors are to be selected at random for use on a particular day. Find the probability that the second motor selected is nondefective, given that the first was nondefective.

Solution Let N_i denote that the ith motor selected is nondefective. We want $P(N_2|N_1)$. From Definition 2.4 we have

$$P(N_2|N_1) = \frac{P(N_1 N_2)}{P(N_1)}$$

Looking at the 20 possible outcomes given in Figure 2.14, we can see that event N_1 contains 16 of these outcomes and $N_1 N_2$ contains 12. Thus, since the 20 outcomes are equally likely,

$$P(N_2|N_1) = \frac{P(N_1 N_2)}{P(N_1)} = \frac{12/20}{16/20} = \frac{12}{16} = \frac{3}{4}$$

Does this result seem intuitively reasonable? ∎

Conditional probabilities satisfy the three axioms of probability (Definition 2.3), as can easily be shown. First, since $AB \subset B$, then $P(AB) \leq P(B)$. Also, $P(AB) \geq 0$ and $P(B) \geq 0$, so

$$0 \leq P(A|B) = \frac{P(AB)}{P(B)} \leq 1$$

Second,

$$P(S|B) = \frac{P(SB)}{P(B)} = \frac{P(B)}{P(B)} = 1$$

Third, if $A_1, A_2, \ldots$ are mutually exclusive events, then so are $A_1B, A_2B, \ldots$, and

$$P\left(\bigcup_{i=1}^{\infty} A_i|B\right) = \frac{P\left(\left(\bigcup_{i=1}^{\infty} A_i\right)B\right)}{P(B)}$$

$$= \frac{P\left(\bigcup_{i=1}^{\infty}(A_iB)\right)}{P(B)} = \frac{\sum_{i=1}^{\infty} P(A_iB)}{P(B)}$$

$$= \sum_{i=1}^{\infty} \frac{P(A_iB)}{P(B)} = \sum_{i=1}^{\infty} P(A_i|B)$$

Conditional probability plays a key role in many practical applications of probability. In these applications, important conditional probabilities are often drastically affected by seemingly small changes in the basic information from which the probabilities are derived. The following discussion of a medical application of probability illustrates the point.

Screening tests, which indicate the presence or absence of disease, are often used by physicians to detect diseases. Virtually all screening tests, however, have errors associated with their use. On thinking about the possible errors, it is clear that two different kinds of errors are possible: the test could show a person to have the disease when it is, in fact, absent (false positive) or fail to show that a person has the disease when it is present (false negative). Measures of these two types of errors are conditional probabilities called *sensitivity* and *specificity*.

A diagram will help in defining and interpreting these measures, where a plus sign indicates presence of the disease under study and a minus sign indicates absence of the disease. The true diagnosis may never be known, but often it can be determined by more intensive follow-up tests.

		True diagnosis		
		$+$	$-$	
Test result	$+$	a	b	$a+b$
	$-$	c	d	
		$a+c$	$b+d$	$a+b+c+d=n$

In this scenario, n people are tested and $a + b$ are shown by the test to have the disease. Of these, a actually have the disease, and b do not have the disease (false positives). Of the $c + d$ who test negative, c actually have the disease (false negatives). Using these labels,

$$\text{sensitivity} = \frac{a}{a + c}$$

the conditional probability of a positive test given that the person has the disease, and

$$\text{specificity} = \frac{d}{b + d}$$

the conditional probability of a negative test given that the person does not have the disease.

Obviously, a good test should have both sensitivity and specificity close to 1. If sensitivity is close to 1, then c (the number of false negatives) must be small. If specificity is close to 1, then b (the number of false positives) must be small. Even when sensitivity and specificity are both close to 1, a screening test can produce misleading results if not carefully applied. To see this, we look at one other important measure, the *predictive value* of a test, given by

$$\text{predictive value} = \frac{a}{a + b}$$

The predictive value is the conditional probability of the person actually having the disease given that he or she tested positive. Clearly, a good test should have a high predictive value, but this is not always possible even for highly sensitive and specific tests. The reason that all three measures cannot always be close to 1 simultaneously lies in the fact that predictive value is affected by the *prevalence rate* of the disease (that is, the proportion of the population under study that actually has the disease). We illustrate with three numerical situations, given as I, II, and III in the following diagrams.

True diagnosis

		+	−	
I Test result	+	90	10	100
	−	10	90	
		100	100	200

True diagnosis

		+	−	
II Test result	+	90	100	190
	−	10	900	
		100	1,000	1,100

True diagnosis

		+	−	
III Test result	+	90	1,000	1,090
	−	10	9,000	
		100	10,000	10,100

Among the 200 people under study in I, 100 have the disease (a prevalence rate of 50%). The test has sensitivity and specificity each equal to 0.90, and the predictive value is $90/100 = 0.90$. This is a good situation; the test is a good one.

In II, the prevalence rate changes to $100/1{,}100$, or 9%. Even though the sensitivity and specificity are still 0.90, the predictive value has dropped to $90/190 = 0.47$. In III, the prevalence rate is $100/10{,}100$, or about 1%, and the predictive value has dropped further to 0.08. Thus, only 8% of those tested positive actually have the disease, even though the test has high sensitivity and specificity. What does this imply about the use of screening tests on large populations in which the prevalence rate for the disease being studied is low? An assessment of the answer to this question involves a careful look at conditional probabilities.

E X A M P L E   **2.16**   The ELISA test for the presence of HIV antibodies was developed in the mid-1980s to screen blood samples. For test cases involving samples known to be contaminated, ELISA was correct 98% of the time. In testing samples known to be clean, ELISA declared 7% of the samples to be HIV positive.

Suppose a firm has 10,000 employees, all to be screened for HIV by ELISA. What can you say about the predictive value of such a screening procedure?

Solution   The result will depend on what is assumed about the rate of HIV-positive cases among the 10,000 employees. Using the 15% at-risk figure from Table 2.3 and assuming *all* were actually positive (an extremely rare case), the table of expected outcomes would be as follows:

True diagnosis

		+	−	
Test result	+	1,470	595	2,065
	−	30	7,905	7,935
		1,500	8,500	10,000

For this table, the predictive value is

$$\frac{1{,}470}{2{,}065} = 0.71$$

Of those testing positive, 71% are expected to actually be HIV-positive.

A much more conservative view might be to select around 5% of the employees to be potentially HIV-positive. The table of expected outcomes then becomes as follows:

		True diagnosis		
		+	−	
Test result	+	490	665	1,155
	−	10	8,835	8,845
		500	9,500	10,000

The predictive value for this table is only 0.42. What is the danger of actually running a procedure with such a low predictive value?   ■

2.6.2   Independence

Probabilities are usually very sensitive to the conditioning information. Sometimes, however, a probability does *not* change when conditioning information is supplied. If the extra information derived from knowing that an event B has occurred does not change the probability of A—that is, if $P(A|B) = P(A)$—then events A and B are said to be *independent*. Since

$$P(A|B) = \frac{P(AB)}{P(B)}$$

the condition $P(A|B) = P(A)$ is equivalent to

$$\frac{P(AB)}{P(B)} = P(A)$$

or

$$P(AB) = P(A)P(B)$$

DEFINITION **2.5**

Two events A and B are said to be **independent** if

$$P(A|B) = P(A)$$

or

$$P(B|A) = P(B)$$

This is equivalent to stating that

$$P(AB) = P(A)P(B) \quad ■$$

E X A M P L E   **2.17**   Suppose a foreman must select one worker for a special job from a pool of four available workers, numbered 1, 2, 3, and 4. He selects the worker by mixing the four names and randomly selecting one. Let A denote the event that worker 1 or 2 is selected, B the event that worker 1 or 3 is selected, and C the event that worker 1 is selected. Are A and B independent? Are A and C independent?

Solution   Because the name is selected at random, a reasonable assumption for the probabilistic model is to assign a probability of 1/4 to each individual worker. Then $P(A) = 1/2$, $P(B) = 1/2$, and $P(C) = 1/4$. Since the intersection AB contains only worker 1, $P(AB) = 1/4$. Now $P(AB) = 1/4 = P(A)P(B)$, so A and B are independent.

Since AC also contains only worker 1, $P(AC) = 1/4$. But $P(AC) = 1/4 \neq P(A)P(C)$, so A and C *are not* independent. A and C are said to be *dependent*, because the fact that C occurs changes the probability that A occurs.   ∎

Most situations in which independence issues arise are not like that portrayed in Example 2.17, in which events are well defined and one merely calculates probabilities to check the definition. Often, independence is *assumed* for two events in order to calculate their joint probability. Suppose, for example, A denotes that machine A does not break down today, and B denotes that machine B will not break down today. $P(A)$ and $P(B)$ can be approximated from the repair records of the machines. What is $P(AB)$, the probability that neither machine breaks down today? If we assume independence, $P(AB) = P(A)P(B)$, a straightforward calculation. However, if we do not assume independence, we cannot calculate $P(AB)$ unless we form a model for their dependence structure or collect data on their joint performance. Is independence a reasonable assumption? It may be if the operation of one machine is not affected by the other, but it may not be if the machines share the same room, power supply, or job overseer. Independence is often used as a simplifying assumption and may not hold precisely in all cases for which it is assumed. Remember, probabilistic models are simply models, and they do not always precisely mirror reality. This should not be a major concern, since all branches of science make simplifying assumptions when developing their models, whether probabilistic or deterministic.

Now we will show how these definitions aid us in establishing rules for computing probabilities of composite events.

2.7   Rules of Probability

Rules that aid in the calculation of probabilities will now be developed. In each case, there is a similar rule for relative frequency data that will be illustrated using the *net* new worker data last visited in Example 2.14. To review, the unconditional percentages are given in the following table:

	Women	Men	
White	42%	15%	57%
Nonwhite	14%	7%	
Immigrant	9%	13%	
	65%	35%	100%

The conditional percentages for men and women are as follows:

	Women	Men
White	65%	43%
Nonwhite	21%	20%
Immigrant	14%	37%
	100%	100%

Categories like "women" and "men" are *complementary* in the sense that a worker must be in one or the other. Since the percentage of women among new workers is 65%, the percentage of men must be 35%; the two together must add to 100%.

When we add percentages (or relative frequencies), we must take care to preserve the mutually exclusive character of the events. The percentage of whites, men or women, is clearly 42% + 15% = 57%. The percentage of new workers that are either white *or* women is *not* 65% + 57%, however. To find the fraction of those who are white or women, the 42% that are *both* white *and* women must be subtracted from the above total. So the percentage of whites or women can be written as

$$65\% + 57\% - 42\% = 80\%$$

which is the same as accumulating the percentages among the appropriate mutually exclusive categories:

$$9\% + 14\% + 42\% + 15\% = 80\%$$

What about the 15% that are *both* men *and* white? Another way to view this type of question is to use the conditional percentages. If we know that 35% of the workers are men and 43% of the men are white, then 43% of the 35%, or 15%, must represent the percentage of white men among the new workers.

2.7.1 Complementary Events

In probabilistic terms, recall that the complement $\overline{A}$ of an event A is the set of all outcomes in a sample space S that are not in A. Thus, $\overline{A}$ and A are mutually exclusive, and their union is S. That is,

$$\overline{A} \cup A = S$$

It follows that

$$P(\overline{A} \cup A) = P(\overline{A}) + P(A) = P(S) = 1$$

or

$$P(\overline{A}) = 1 - P(A)$$

Thus, we have established the following theorem.

THEOREM **2.5**

If $\overline{A}$ is the complement of an event A in a sample space S, then

$$P(\overline{A}) = 1 - P(A) \quad \blacksquare$$

EXAMPLE **2.18**   A quality-control inspector has ten assembly lines from which to choose products for testing. Each morning of a five-day week, she randomly selects one of the lines to work on for the day. Find the probability that a line is chosen more than once during the week.

Solution   It is easier here to think in terms of complements and first find the probability that no line is chosen more than once. If no line is repeated, five different lines must be chosen on successive days, which can be done in

$$P_5^{10} = \frac{10!}{5!} = 10(9)(8)(7)(6)$$

ways. The total number of possible outcomes for the selection of five lines without restriction is $(10)^5$, by an extension of the multiplication rule. Thus,

$$P(\text{no line is chosen more than once}) = \frac{P_5^{10}}{(10)^5}$$
$$= \frac{10(9)(8)(7)(6)}{(10)^5} = 0.30$$

and

$$P(\text{a line is chosen more than once}) = 1.0 - 0.30 = 0.70 \quad \blacksquare$$

2.7.2 Additive Rule

Axiom 3 applies to $P(A \cup B)$ if A and B are disjoint. But what happens when A and B are not disjoint? Theorem 2.6 gives the answer.

THEOREM **2.6**   If A and B are any two events, then

$$P(A \cup B) = P(A) + P(B) - P(AB)$$

If A and B are mutually exclusive, then

$$P(A \cup B) = P(A) + P(B)$$

Proof   From a Venn diagram for the union of A and B (see Figure 2.7), it is easy to see that

$$A \cup B = A\overline{B} \cup \overline{A}B \cup AB$$

and that the three events on the right-hand side of the equality are mutually exclusive. Hence,

$$P(A \cup B) = P(A\overline{B}) + P(\overline{A}B) + P(AB)$$

Now

$$A = A\overline{B} \cup AB$$

and

$$B = \overline{A}B \cup AB$$

so we have

$$P(A) = P(A\overline{B}) + P(AB)$$

and

$$P(B) = P(\overline{A}B) + P(AB)$$

It follows that

$$P(A\overline{B}) = P(A) - P(AB)$$

and

$$P(\overline{A}B) = P(B) - P(AB)$$

Substituting into the first equation for $P(A \cup B)$, we have

$$P(A \cup B) = P(A) - P(AB) + P(B) - P(AB) + P(AB)$$
$$= P(A) + P(B) - P(AB)$$

and the proof is complete.   ∎

The formula for the probability of the union of k events $A_1, A_2, \ldots, A_k$ is derived in similar fashion and is given by

$$P(A_1 \cup A_2 \cup \cdots \cup A_k) = \sum_{i=1}^{k} P(A_i) - \sum \sum_{i<j} P(A_i A_j) + \sum \sum \sum_{i<j<l} P(A_i A_j A_l)$$
$$- \cdots + \cdots - (-1)^k P(A_1 A_2 \ldots A_k)$$

2.7.3 Multiplicative Rule

The next rule is actually just a rearrangement of the definition of conditional probability for the case in which a conditional probability may be known and we want to find the probability of an intersection.

T H E O R E M   **2.7**

If A and B are any two events, then

$$P(AB) = P(A)P(B|A)$$
$$= P(B)P(A|B)$$

If A and B are independent, then

$$P(AB) = P(A)P(B) \quad \blacksquare$$

We illustrate the use of these three theorems in the following examples.

E X A M P L E   **2.19**   Records indicate that for the parts coming out of a hydraulic repair shop at an airplane rework facility, 20% will have a shaft defect, 10% will have a bushing defect, and 75% will be defect-free. For an item chosen at random from this output, find the probability of the following:

A: The item has at least one type of defect.

B: The item has only a shaft defect.

Solution   The percentages given imply that 5% of the items have both a shaft defect and a bushing defect. Let D_1 denote the event that an item has a shaft defect and D_2 the event that it has a bushing defect. Then

$$A = D_1 \cup D_2$$

and

$$P(A) = P(D_1 \cup D_2)$$
$$= P(D_1) + P(D_2) - P(D_1 D_2)$$
$$= 0.20 + 0.10 - 0.05$$
$$= 0.25$$

Another possible solution is to observe that the complement of A is the event that an item has no defects. Thus,

$$P(A) = 1 - P(\overline{A})$$
$$= 1 - 0.75 = 0.25$$

To find $P(B)$, note that the event D_1 that the item has a shaft defect is the union of the event that it has *only* a shaft defect (B) and the event that it has *both* defects ($D_1 D_2$). That is,

$$D_1 = B \cup D_1 D_2$$

where B and $D_1 D_2$ are mutually exclusive. Therefore,

$$P(D_1) = P(B) + P(D_1 D_2)$$

or

$$P(B) = P(D_1) - P(D_1 D_2)$$
$$= 0.20 - 0.05 = 0.15$$

You should sketch these events on a Venn diagram and verify the results derived above. ∎

EXAMPLE   **2.20**   A section of an electrical circuit has two relays in parallel, as shown in Figure 2.16. The relays operate independently, and when a switch is thrown each will close properly with probability only 0.8. If the relays are both open, find the probability that current will flow from s to t when the switch is thrown.

Solution   Let O denote an open relay and C a closed relay. The four outcomes for this experiment are shown by the following:

$$
\begin{array}{ccc}
 & \text{Relay} & \text{Relay} \\
 & 1 & 2 \\
E_1 = \{(O & , & O)\} \\
E_2 = \{(O & , & C)\} \\
E_3 = \{(C & , & O)\} \\
E_4 = \{(C & , & C)\}
\end{array}
$$

FIGURE 2.16
Two Relays in Parallel

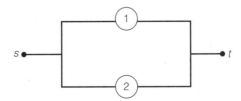

Since the relays operate independently, we can find the probability for these outcomes as follows:

$$P(E_1) = P(O)P(O) = (0.2)(0.2) = 0.04$$
$$P(E_2) = P(O)P(C) = (0.2)(0.8) = 0.16$$
$$P(E_3) = P(C)P(O) = (0.8)(0.2) = 0.16$$
$$P(E_4) = P(C)P(C) = (0.8)(0.8) = 0.64$$

If A denotes the event that current will flow from s to t, then

$$A = E_2 \cup E_3 \cup E_4$$

or

$$\overline{A} = E_1$$

(At least one of the relays must close for current to flow.) Thus,

$$P(A) = 1 - P(\overline{A})$$
$$= 1 - P(E_1)$$
$$= 1 - 0.04 = 0.96$$

which is the same as $P(E_2) + P(E_3) + P(E_4)$. ∎

EXAMPLE **2.21**
Three different orders are to be mailed to three suppliers. However, an absent-minded secretary gets the orders mixed up and just sends them randomly to suppliers. If a match refers to the fact that a supplier gets the correct order, find the probability of (a) no matches (b) exactly one match.

Solution
This problem can be solved by listing outcomes, since only three orders and suppliers are involved, but a more general method of solution will be illustrated. Define the following events:

A_1: match for supplier I
A_2: match for supplier II
A_3: match for supplier III

There are $3! = 6$ equally likely ways of randomly sending the orders to suppliers. There are only $2! = 2$ ways of sending the orders to suppliers if one particular supplier is required to have a match. Hence,

$$P(A_1) = P(A_2) = P(A_3) = 2/6 = 1/3$$

Similarly, it follows that

$$P(A_1 A_2) = P(A_1 A_3) = P(A_2 A_3)$$
$$= P(A_1 A_2 A_3) = 1/6$$

a Note that

$$P(\text{no matches}) = 1 - P(\text{at least one match})$$
$$= 1 - P(A_1 \cup A_2 \cup A_3)$$
$$= 1 - [P(A_1) + P(A_2) + P(A_3)$$
$$\quad - P(A_1 A_2) - P(A_1 A_3)$$
$$\quad - P(A_2 A_3) + P(A_1 A_2 A_3)]$$
$$= 1 - [3(1/3) - 3(1/6) + (1/6)]$$
$$= 1/3$$

b We leave it to the reader to show that

$$P(\text{exactly one match}) = P(A_1) + P(A_2) + P(A_3) - 2[P(A_1 A_2) + P(A_1 A_3)$$
$$\quad + P(A_2 A_3)] + 3P(A_1 A_2 A_3)$$
$$= 3(1/3) - 2(3)(1/6) + 3(1/6) = 1/2 \quad \blacksquare$$

2.7.4 Bayes' Rule

The fourth rule we present in this section is based on the notion of a partition of a sample space. Events $B_1, B_2, \ldots, B_k$ are said to partition a sample space S if two conditions exist:

1 $B_i B_j = \phi$ for any pair i and j (ϕ denotes the null set)
2 $B_1 \cup B_2 \cup \cdots \cup B_k = S$

For example, the set of tires in an auto assembly warehouse might be partitioned according to suppliers, or employees of a firm might be partitioned according to level of education. We illustrate a partition for the case $k = 2$ in Figure 2.17. The key idea involving a partition is the observation that an event A can be written as the union of mutually exclusive events AB_1 and AB_2.

FIGURE **2.17**
A Partition of S into B_1 and B_2

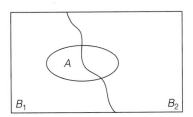

S

That is,

$$A = AB_1 \cup AB_2$$

and thus

$$P(A) = P(AB_1) + P(AB_2)$$

If conditional probabilities $P(A|B_1)$ and $P(A|B_2)$ are known, then $P(A)$ can be found by writing

$$P(A) = P(B_1)P(A|B_1) + P(B_2)P(A|B_2)$$

In problems dealing with partitions, it is frequently of interest to find probabilities of the form $P(B_1|A)$, which can be written

$$P(B_1|A) = \frac{P(B_1 A)}{P(A)}$$

$$= \frac{P(B_1)P(A|B_1)}{P(B_1)P(A|B_1) + P(B_2)P(A|B_2)}$$

This result is a special case of *Bayes' Rule*, of which Theorem 2.8 is a general statement.

THEOREM **2.8**   **Bayes' Rule**   If $B_1, B_2, \ldots, B_k$ form a partition of S, and if A is any event in S, then

$$P(B_j|A) = \frac{P(B_j)P(A|B_j)}{\displaystyle\sum_{i=1}^{k} P(B_i)P(A|B_i)}$$

The proof is an extension of the results derived above for $k = 2$.   ∎

EXAMPLE **2.22**   A company buys tires from two suppliers, 1 and 2. Supplier 1 has a record of delivering tires containing 10% defectives, whereas supplier 2 has a defective rate of only 5%. Suppose 40% of the current supply came from supplier 1. If a tire is taken from this supply and observed to be defective, find the probability that it came from supplier 1.

Solution   Let B_i denote the event that a tire comes from supplier i, $i = 1, 2$, and note that B_1 and B_2 form a partition of the sample space for the experiment of selecting one tire. Let A denote the event that the selected tire is defective. Then

$$P(B_1|A) = \frac{P(B_1)P(A|B_1)}{P(B_1)P(A|B_1) + P(B_2)P(A|B_2)}$$

$$= \frac{0.40(0.10)}{0.40(0.10) + 0.60(0.05)}$$

$$= \frac{0.04}{0.04 + 0.03} = \frac{4}{7}$$

Supplier 1 has a greater probability of being the party supplying the defective tire than does supplier 2.   ▪

Let's return now to tables of real data. Recall that such data may not provide exact probabilities but that the approximations are frequently good enough to provide a basis for clearer insights into the problem at hand.

E X A M P L E   **2.23**   The U.S. Bureau of Labor Statistics provides the data below as a summary of employment in the United States for 1989.

Employment in the United States, 1989
(25 years and over; figures in millions)

Civilian noninstitutional population	154
Civilian labor force	102
Employed	98
Unemployed	4
Not in the labor force	52

If an arbitrarily selected U.S. resident was asked, in 1989, to fill out a questionnaire on employment, find the probability that the resident would have been in the following categories:

a   in the labor force

b   employed

c   employed and in the labor force

d   employed given that he or she was known to be in the labor force

e   either not in the labor force or unemployed

Solution   Let L denote the event that the resident is in the labor force and E that he or she is employed.

a   Since 154 million people comprise the population under study and 102 million are in the labor force,

$$P(L) = \frac{102}{154}$$

b   Similarly, 98 million are employed; thus,

$$P(E) = \frac{98}{154}$$

c   The employed persons are a subset of those in the labor force; in other words, $EL = E$. Hence,

$$P(EL) = P(E) = \frac{98}{154}$$

d   Among the 102 million people known to be in the labor force, 98 million are employed. Therefore,

$$P(E|L) = \frac{98}{102}$$

Note that this value can also be found by using Definition 2.4, as follows:

$$P(E|L) = \frac{P(EL)}{P(L)} = \frac{98/154}{102/154} = \frac{98}{102}$$

e   The event that the resident is not in the labor force, $\overline{L}$, is mutually exclusive from the event that he or she is unemployed. Therefore,

$$P(\overline{L} \cup \overline{E}) = P(\overline{L}) + P(\overline{E})$$
$$= \frac{52}{154} + \frac{4}{154} = \frac{56}{154} \quad \blacksquare$$

The next example uses a data set that is a little more complicated in structure.

E X A M P L E   2.24   The National Fire Incident Reporting Service gave the information in Table 2.9 on fires reported in 1978. All figures in the table are percents, and the main body of the table shows percentages according to cause. For example, the 22 in the upper left corner shows that 22% of fires in family homes were caused by heating.

T A B L E 2.9
Causes of Fires in Residences, 1978 (in percent)

Cause of Fire	Family Homes	Apartments	Mobile Homes	Hotels/ Motels	Other	All Locations
Heating	22	6	22	8	45	19
Cooking	15	24	13	7	0	16
Incendiary substance	10	15	7	16	8	11
Smoking	7	18	6	36	19	10
Electrical	8	5	15	7	28	8
Other	38	32	37	26	0	36
All causes	73	20	3	2	2	

If a residential fire is called into a fire station, find the probability that it is the following:

a a fire caused by heating

b a fire in a family home

c a fire caused by heating given that it was in an apartment

d an apartment fire caused by heating

e a fire in a family home given that it was caused by heating

Solution **a** Over all locations, 19% of fires are caused by heating. (See the right-hand column of Table 2.9.) Thus,

$$P(\text{heating fire}) = 0.19$$

b Over all causes, 73% of the fires are in family homes. (See the bottom row of Table 2.9.) Thus,

$$P(\text{family-home fire}) = 0.73$$

c The second column from the left deals only with apartment fires. Thus, 6% of all apartment fires are caused by heating. Note that this is a conditional probability and can be written

$$P(\text{heating fire}|\text{apartment fire}) = 0.06$$

d In part (c), we know the fire was in an apartment, and the probability in question was conditional on that information. Here we must find the probability of an intersection between "apartment fire" (say, event A) and "heating fire" (say, event H). We know $P(H|A)$ directly from the table, so

$$P(AH) = P(A)P(H|A)$$
$$= (0.20)(0.06) = 0.012$$

In other words, only 1.2% of all reported fires are caused by heating in apartments.

e The figures in the body of Table 2.9 give probabilities of causes, given the location of the fire. We are now asked to find the probability of a location given the cause. This is exactly the type of situation to which Bayes' Rule (Theorem 2.8) applies. The locations form a partition of the set of all fires into five different groups. We have, then, with obvious shortcuts in wording,

$$P(\text{family}|\text{heating}) = \frac{P(\text{family})P(\text{heating}|\text{family})}{P(\text{heating})}$$

Now $P(\text{heating}) = 0.19$ from Table 2.9. However, it might be informative to show how this figure is derived from the other columns of the table. We have

$$P(\text{heating}) = P(\text{family})P(\text{heating}|\text{family})$$
$$+ P(\text{apartment})P(\text{heating}|\text{apartment})$$
$$+ P(\text{mobile})P(\text{heating}|\text{mobile})$$
$$+ P(\text{hotel})P(\text{heating}|\text{hotel}) + P(\text{other})P(\text{heating}|\text{other})$$

$$= (0.73)(0.22) + (0.20)(0.06) + (0.03)(0.22) + (0.02)(0.08)$$
$$+ (0.02)(0.45)$$
$$= 0.19$$

Then

$$P(\text{family}|\text{heating}) = \frac{(0.73)(0.22)}{0.19} = 0.85$$

In other words, 85% of heating fires are found in family homes. ∎

E X A M P L E   **2.25**   Data from past studies are often used to anticipate results of future studies, as we have seen. Suppose three new employees are hired from a national pool of potential employees. Find the probability of each of the following:

a   Exactly one is at risk for HIV.

b   At least one is at risk for HIV.

c   Exactly two are at risk when it is known that at least one is at risk.

Solution   As usual, some assumptions must be made in order to get a random mechanism into the problem. Assuming the three selected employees behave like a random sample from the general population with regard to HIV risk groups, we can assign a probability of 0.15 (see Table 2.3) to the event that any one new employee is at risk. If the three selections of new employees are unrelated to each other, they can be regarded as independent events.

Let R_i denote the event that person i is at risk ($i = 1,2,3$). A tree diagram (as in Figure 2.18) can aid in showing the structure of outcomes for selecting three new employees. The probabilities along connected branches can be multiplied because of the independence assumption.

a   By the rules of probability, P (exactly one at risk)

$$= P(\overline{R}_1 \, \overline{R}_2 R_3 \cup \overline{R}_1 R_2 \overline{R}_3 \cup R_1 \overline{R}_2 \, \overline{R}_3)$$
$$= P(\overline{R}_1 \, \overline{R}_2 R_3) + P(\overline{R}_1 R_2 \overline{R}_3) + P(R_1 \overline{R}_2 \, \overline{R}_3)$$
$$= P(\overline{R}_1)P(\overline{R}_2)P(R_3) + P(\overline{R}_1)P(R_2)P(\overline{R}_3) + P(R_1)P(\overline{R}_2)P(\overline{R}_3)$$
$$= (0.85)(0.85)(0.15) + (0.85)(0.15)(0.85) + (0.15)(0.85)(0.85)$$
$$= 3(0.15)(0.85)^2 = 0.325$$

b   The probability of at least one at risk could be found by repeating part (a) for "exactly two at risk" and "exactly three at risk" and then adding the results. An easier method is to use complements, so that

$$P(\text{at least one at risk}) = 1 - P(\text{none at risk})$$
$$= 1 - P(\overline{R}_1 \, \overline{R}_2 \, \overline{R}_3) = 1 - (0.85)^3 = 0.386$$

F I G U R E **2.18**   A Tree Diagram for Independent Selections

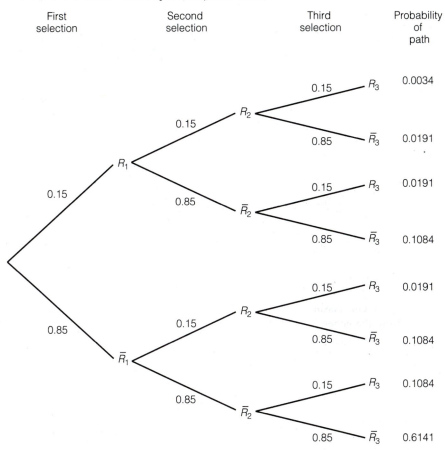

First selection	Second selection	Third selection	Probability of path

c   This problem involves the conditional probability

$$P(\text{exactly two at risk}|\text{at least one at risk}) = \frac{P(\text{exactly two at risk})}{P(\text{at least one at risk})}$$

$$= \frac{3(0.15)^2(0.85)}{0.386} = 0.149$$

Note that the intersection of the two events in question is just the first event, "exactly two at risk." The probability in the numerator is found in a manner analogous to that in part (a).

Suppose the three new hires all come from high-risk cities, in which about 20% of the population is at risk for HIV. How would that fact change these probabilities?   ▪

Exercises

2.35 Vehicles coming into an intersection can turn left or right or go straight ahead. Two vehicles enter an intersection in succession. Find the probability that at least one of the two vehicles turns left given that at least one of the two vehicles turns. What assumptions have you made?

2.36 A purchasing office is to assign a contract for computer paper and a contract for microcomputer disks to any one of three firms bidding for these contracts. (Any one firm could receive both contracts.) Find the probability that

a firm I receives a contract given that both contracts do not go to the same firm.

b firm I receives both contracts.

c firm I receives the contract for paper given that it does not receive the contract for disks.

What assumptions have you made?

2.37 The data in Table 2.10 give the number of accidental deaths overall and for three specific causes for the United States in 1984.

T A B L E **2.10**

	All Types	Motor Vehicle	Falls	Drowning
All ages	92,911	46,263	11,937	5,388
Under 5	3,652	1,132	114	638
5–14	4,198	2,263	68	532
15–24	19,801	14,738	399	1,353
25–44	25,498	15,036	963	1,549
45–64	15,273	6,954	1,624	763
65–74	8,424	3,020	1,702	281
75 and over	16,065	3,114	7,067	272
Male	64,053	32,949	6,210	4,420
Female	28,858	13,314	5,727	968

Source: The World Almanac, 1988.

You are told that a certain person recently died in an accident. Approximate the probability that

a it was a motor vehicle accident.

b it was a motor vehicle accident if you know the person to be male.

c it was a motor vehicle accident if you know the person to be between 15 and 24 years of age.

d it was a fall if you know the person to be over age 75.

e the person was male.

2.38 The data in Table 2.11 show the distribution of arrival times at work by mode of travel for workers in the central business district of a large city. The figures are percentages. (The columns should add to 100, but some do not because of rounding.) A randomly selected worker is asked about his or her travel to work.

a Find the probability that the worker arrives before 7:15 given that the worker drives alone.

b Find the probability that the worker arrives at or after 7:15 given that the worker drives alone.

TABLE **2.11**

Arrival Time	Transit	Drove Alone	Shared Ride with Family Member	Car Pool	All
Before 7:15	17	16	16	19	18
7:15–7:45	35	30	30	42	34
7:45–8:15	32	31	43	35	33
8:15–8:45	10	14	8	2	10
After 8:45	5	11	3	2	6

Source: C. Hendrickson and E. Plank, *Transportation Research*, 18A, no. 1, 1984.

c Find the probability that the worker arrives before 8:15 given that the worker rides in a car pool.

d Can you find the probability that the worker drives alone, using only these data?

2.39 Table 2.9 shows percentages of fires by cause for certain residential locations. If a fire is reported to be residential, find the probability that it was

a caused by smoking.

b in a mobile home.

c caused by smoking given that it was in a mobile home.

d in a mobile home given that it was caused by smoking.

2.40 An incoming lot of silicon wafers is to be inspected for defectives by an engineer in a microchip manufacturing plant. In a tray containing 20 wafers, assume four are defective. Two wafers are to be randomly selected for inspection. Find the probability that

a both are nondefective.

b at least one of the two is nondefective.

c both are nondefective given that at least one is nondefective.

2.41 In the setting of Exercise 2.40, find the same three probabilities if only two among the 20 wafers are assumed to be defective.

2.42 Among five applicants for chemical engineering positions in a firm, two are rated excellent and the other three are rated good. A manager randomly chooses two of these applicants to interview. Find the probability that the manager chooses

a the two rated excellent.

b at least one of those rated excellent.

c the two rated excellent given that one of the chosen two is already known to be excellent.

2.43 Resistors produced by a certain firm are marketed as 10-ohm resistors. However, the actual number of ohms of resistance produced by the resistors may vary. It is observed that 5% of the values are below 9.5 ohms and 10% are above 10.5 ohms. If two of these resistors, randomly selected, are used in a system, find the probability that

a both have actual values between 9.5 and 10.5.

b at least one has an actual value in excess of 10.5.

2.44 Consider the following segment of an electric circuit with three relays. Current will flow from a to b if there is at least one closed path when the relays are switched to "closed." However, the relays may malfunction. Suppose they close properly only with probability 0.9 when the switch is thrown, and suppose they operate independently of one another. Let A denote the event that current will flow from a to b when the relays are switched to "closed."

a Find $P(A)$.

b Find the probability that relay 1 is closed properly, given that current is known to be flowing from *a* to *b*.

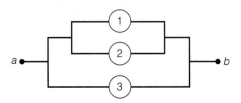

2.45 With relays operating as in Exercise 2.44, compare the probability of current flowing from *a* to *b* in the series system

with the probability of flow in the parallel system

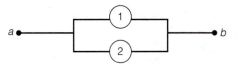

2.46 Electric motors coming off two assembly lines are pooled for storage in a common stockroom, and the room contains an equal number of motors from each line. Motors are periodically sampled from that room and tested. It is known that 10% of the motors from line I are defective and 15% of the motors from line II are defective. If a motor is randomly selected from the stockroom and found to be defective, find the probability that it came from line I.

2.47 Two methods, *A* and *B*, are available for teaching a certain industrial skill. The failure rate is 20% for *A* and 10% for *B*. However, *B* is more expensive and hence is used only 30% of the time. (*A* is used the other 70%.) A worker is taught the skill by one of the methods but fails to learn it correctly. What is the probability that he was taught by method *A*?

2.48 A diagnostic test for a certain disease is said to be 90% accurate in that, if a person has the disease, the test will detect it with probability 0.9. Also, if a person does not have the disease, the test will report that he or she doesn't have it with probability 0.9. Only 1% of the population has the disease in question. If a person is chosen at random from the population and the diagnostic test reports him to have the disease, what is the conditional probability that he does, in fact, have the disease? Are you surprised by the size of the answer? Would you call this diagnostic test reliable?

2.49 Table 2.12 gives estimated percentages of sports footwear purchased by those in various age groups across the United States.

a Explain the meaning of the 4.7 under aerobic shoes. Explain the meaning of the 67.0 under walking shoes.

T A B L E **2.12**   Consumer Purchases of Sports Footwear (in percent)

Characteristic	Total Households	Footwear			
		Aerobic Shoes	Gym Shoes/ Sneakers	Jogging/ Running Shoes	Walking Shoes
Total	100.0	100.0	100.0	100.0	100.0
Age of user:					
Under 14 years old	20.3	4.7	34.6	8.3	2.8
14–17 years old	5.4	5.4	13.8	10.5	1.9
18–24 years old	10.7	11.5	9.7	10.0	3.3
25–34 years old	17.7	30.9	16.9	28.5	12.9
35–44 years old	14.7	22.5	10.3	24.0	19.7
45–64 years old	18.7	17.6	10.7	15.2	35.7
65 years old and over	12.5	7.4	4.0	3.5	23.7
Sex of user:					
Male	48.8	15.7	53.0	64.9	33.0
Female	51.2	84.3	47.0	35.1	67.0

Source: Statistical Abstract of the United States, 1991.

b   Suppose the percentages provided in the table are to be used as probabilities in discussing anticipated purchases for next year. Explain the 28.5 under jogging/running shoes as a probability.

c   Your firm anticipates selling 100,000 pairs of gym shoes/sneakers across the country. How many of these would you expect to be purchased by children under 14 years of age? How many by 18–24-year-olds?

d   You want to target advertising to 25–34-year-olds. If someone in this age group buys sports footwear next year, can you find the probability that he or she will purchase aerobic shoes?

e   Write a brief description of the patterns you see in sales of sports footwear.

2.50   Study the data on employed scientists and engineers provided in Table 2.13. Think of the percentages as probabilities to aid in describing the anticipated employment pattern for scientists and engineers for this year.

T A B L E **2.13**   Estimated Characteristics of Employed Scientists and Engineers, 1988

	Total	Engineers	Life Scientists	Computer Specialists	Physical Scientists	Social Scientists				Environ- mental Scientists	Mathe- matical Scientists
						Total	Economists	Other Social Scientists	Psychol- ogists		
Total (in thousands)	5,286.4	2,718.6	458.6	708.3	312.0	531.0	219.8	311.2	275.9	113.4	168.6
Type of employer:											
Industry	68.0	79.5	37.9	78.7	56.1	52.5	64.0	44.4	40.7	58.4	42.0
Educ. institutions	13.5	4.3	35.2	6.6	25.4	24.5	19.0	28.5	31.0	16.0	44.1
Nonprofit organ.	3.5	1.5	7.3	2.4	3.3	6.0	2.3	8.7	17.6	1.2	2.0
Federal government	7.5	7.5	10.1	6.5	10.2	5.5	7.8	3.9	2.6	15.6	8.2
Military	0.8	1.0	0.4	0.7	0.4	0.4	0.2	0.5	0.2	1.4	0.8
Other government	5.1	4.6	7.5	3.6	2.9	9.6	5.2	12.7	6.1	5.6	1.9
Other and unknown	1.6	1.7	1.6	1.5	1.8	1.3	1.5	1.2	1.7	1.7	0.9

a Explain the conditional nature of the columns of percentages.

b Suppose 3 million engineers are employed this year. How many would you expect to be in industry?

c Combine life scientists, computer specialists, physical scientists, environmental scientists, and mathematical scientists into one category. Find the percentage distribution across employment types for this new category. How does this distribution compare to that for engineers?

d Suppose industry adds 1 million new scientists this year. How many would you expect to be engineers?

e Describe any differences you see between the employment pattern of physical scientists and engineers and that of social scientists.

2.51 The segmented bar graph, as shown in Figure 2.19, is another way to display relative frequency data.

F I G U R E **2.19**

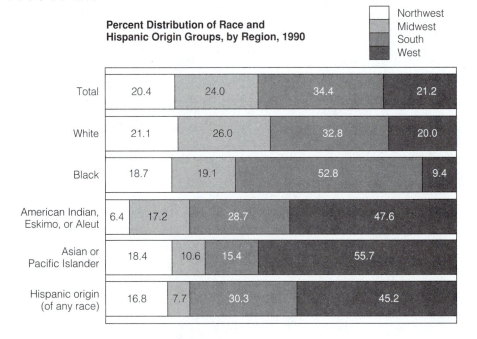

Percent Distribution of Race and Hispanic Origin Groups, by Region, 1990

a Explain the percentages in the bar labeled "Black" as conditional probabilities.

b It appears that most of the people living in the South are black. Is that a correct interpretation of these data?

c Describe in words the pattern of racial and ethnic diversity across the United States, based on these data.

2.8 Odds, Odds Ratios, and Relative Risk

"What are the odds that our team will win today?" This is a common way of talking about events whose unknown outcomes have probabilistic interpretations. The *odds in favor* of an event A are the ratio of the probability of A to the probability of $\overline{A}$. That is,

$$\text{odds in favor of } A = \frac{P(A)}{P(\overline{A})}$$

The odds in favor of a balanced coin coming up heads when flipped are

$$P(H)/P(T) = \frac{1}{2}/\frac{1}{2} = 1$$

often written as 1:1 (one to one). The odds in favor of a randomly selected person from a high-risk city being at risk for HIV are $(0.20/0.80) = 0.25$, or 1:4. The odds *against* that person being at risk are $0.80/0.20 = 4$, or 4:1. The odds against an event B are the same as the odds in favor of $\overline{B}$.

Odds are not just a matter of betting and sports. They are a serious part of the analysis of frequency data, especially when categorical variables in two-way frequency tables are being compared.

Returning to the Physicians' Health Study on the effects of aspirin on the incidence of heart attacks (Table 2.4 and following), we can summarize the data on myocardial infarctions (M.I.'s) as follows:

	M.I.	No M.I.	Total
Aspirin	139	10,898	11,037
Placebo	239	10,795	11,034
			22,071

For the aspirin group, the odds in favor of M.I. are

$$\frac{P(\text{M.I.})}{P(\overline{\text{M.I.}})} = \frac{139/11{,}037}{10{,}898/11{,}037} = \frac{139}{10{,}898} = 0.013$$

For the placebo group, the odds in favor of M.I. are

$$\frac{P(\text{M.I.})}{P(\overline{\text{M.I.}})} = \frac{239/11{,}034}{10{,}795/11{,}034} = \frac{239}{10{,}795} = 0.022$$

In such studies, odds are often interpreted as *risk*. Thus, the above results show the risk of a heart attack with the placebo to be considerably higher than the risk with aspirin. More specifically, the ratio of the two odds (risks) is called the relative risk:

$$\begin{aligned}
\text{relative risk of M.I.} &= \frac{\text{risk of M.I. with aspirin}}{\text{risk of M.I. without aspirin}} \\
&= \frac{\text{odds of M.I. with aspirin}}{\text{odds of M.I. without aspirin}}
\end{aligned}$$

$$= \frac{0.013}{0.022} = 0.59$$

The risk of an M.I. for the aspirin group is 59% of the risk for the placebo group. Odds ratios form a very useful single-number summary of the frequencies in a 2×2 table.

On closer inspection, it is observed that the odds ratio (relative risk) has a simple form for any 2×2 table, which can be written generically as

	I	II
A	a	b
B	c	d

The odds in favor of I for the A group are a/b, and the odds in favor of I for the B group are c/d. Therefore, the odds ratio is simply

$$\frac{a/b}{c/d} = \frac{ad}{bc}$$

which is the ratio of the products of the diagonal elements.

E X A M P L E **2.26**   Data on the civilian labor force (see Table 2.7) can be rewritten into a number of meaningful 2×2 tables. Two such tables show employment status by sex and employment status by education (elementary school only versus at least some high school).

	Employed	Unemployed
Male	54,039	2,207
Female	43,581	1,908

	Employed	Unemployed
High School	46,315	2,468
Elementary School	5,299	406

Source: U.S. Bureau of Labor Statistics.

Interpret these tables by using odds ratios.

Solution   For the first table, the odds ratio is

$$\frac{(54{,}039)(1{,}908)}{(43{,}581)(2{,}207)} = 1.07$$

The odds in favor of being employed is about the same for males as for females. This result might be restated by saying that the risk of being unemployed is close to the same for males and females.

For the second table, the odds ratio is

$$\frac{(46,315)(406)}{(5,299)(2,468)} = 1.44$$

Conditioning on these two education groups, the odds in favor of employment for the high school group are about 1.4 times as great as for the elementary school group. Education does seem to have a significant impact. In terms of risk, it is correct to say that the risk of unemployment for those with some high school education is $1/1.44 = 0.69$ or 69% of the risk of unemployment for those with some elementary education. (Can you show that the latter odds ratio of 0.69 is correct?)

In short, the odds and odds ratio, though easy to compute, serve as a useful summary of the tabled data.   ▪

Exercises

2.52   From the results of the Physicians' Health Study discussed in Section 2.2, cholesterol level seems to be an important factor in myocardial infarctions. The data below give the number of M.I.'s over the number in the cholesterol group for each arm of the study.

Cholesterol Level (mg per 100 ml)	Aspirin Group	Placebo Group
≤ 159	2/382	9/406
160–209	12/1,587	37/1,511
210–259	26/1,435	43/1,444
≥260	14/582	23/570

a   Did the randomization in the study seem to do a good job of balancing the cholesterol levels between the two groups? Explain.

b   Construct a 2×2 table of aspirin or placebo versus M.I. response for each of the four cholesterol levels. Reduce the data in each table to the odds ratio.

c   Compare the four odds ratios found in (b). Comment on the relationship between the effect of aspirin on heart attacks and the cholesterol levels. Do you see why odds ratios are handy tools for summarizing data in a 2×2 table?

2.53   Is the race of a defendant associated with the defendant's chance of receiving the death penalty? This is a controversial issue that has been studied by many. One important data set was collected on 326 cases in which the defendant was convicted of homicide. The death penalty was given in 36 of these cases. The following table shows the defendant's race, the victim's race, and whether or not the death penalty was imposed.

	White Defendant		Black Defendant	
	Yes	No	Yes	No
White Victim	19	132	11	52
Black Victim	0	9	6	97

Source: M. Radelet, "Racial Characteristics and Imposition of the Death Penalty," *American Sociological Review*, 46, 1981, pp. 918–927.

a   Construct a single 2 × 2 table showing penalty versus defendant's race, across all victims. Calculate the odds ratio and interpret it.

b   Decompose the table in (a) into two 2 × 2 tables of penalty versus defendant's race, one for white victims and one for black victims. Calculate the odds ratio for each table and interpret each one.

c   Do you see any inconsistency between the results of (a) and the results of (b)? Can you explain the apparent paradox?

2.54   A proficiency examination for a certain skill was given to 100 employees of a firm. Forty of the employees were male. Sixty of the employees passed the examination, in that they scored above a preset level for satisfactory performance. The breakdown among males and females was as follows:

	Male M	Female F
Pass P	24	36
Fail $\overline{P}$	16	24

Suppose an employee is randomly selected from the 100 who took the examination.

a   Find the probability that the employee passed given that he was a male.

b   Find the probability that the employee was male given that the employee passed.

c   Are events P and M independent?

d   Are events P and F independent?

e   Construct and interpret a meaningful odds ratio for these data.

2.9 Activities for Students: Simulation

The long-run stability of relative frequencies for randomly generated events (Section 2.1) can be used to estimate probabilities through *simulation*. We will illustrate the techniques with a simple example before suggesting activities.

What is the probability that a three-child family has exactly two girls and one boy? Simulation of this event involves four steps:

1   Identify a random mechanism to generate the probability of a key component.

2   Define the number of key components in one trial. Call this number n.

3   Generate a large number of trials, say t.

4   Estimate the probability in question by the ratio of the number of successful trials, s, to the total number of trials, t.

For the question posed above, a key component is the sex of a child, which has $P(G) = P(B) = 0.5$, approximately. Probabilities of 0.5 can be generated by considering whether a random digit is even. So $P(G) = P(\text{even digit}) = 0.5$. (Coins could be used as well.) Since a three-child family was specified, there are $n = 3$ key components (children) in a trial. We must generate groups of three random digits and observe the number of even digits. If this number is 2, the trial is a success. (That's what we are looking for.) For the illustration, we will take $t = 50$ trials from the random number table (Table 1 of the Appendix).

Beginning with the last three columns in the upper right-hand corner of the table (each person should choose his or her own random starting point), the first set of three random digits is 700. Since zero is considered even, this is a successful trial. The next few sets of random digits are

505

629

379

613

of which only 629 is a success (it contains two evens and one odd). After 50 trials (50 rows of the table), the number of successful trials is $s = 15$. Thus, the simulation estimate of the probability in question is $(s/t) = (15/50) = 0.30$. You might want to compare this number with the probability calculated according to the rules of this chapter.

Activity 1: What's the Chance of Driving to School?

A study of the need for parking spaces around the engineering buildings begins by estimating the proportion of students who drive to school. It continues by using these data to anticipate what might happen next year.

1 Conduct a small survey among engineering students to estimate what proportion drives to school. Carefully consider the questions to be asked. Use a randomization device to select the respondents.

2 Using your estimated probability for a student driving to school, design a simulation to approximate the probability that, of five students sampled from next year's engineering student body, at least two will drive to school. (The numbers here are kept small for simplicity. The idea generalizes to any sample size.)

Activity 2: Where Are the Defectives?

An assembly line is thought to produce about 10% defective items. You want to examine a defective to find the cause. If you randomly sample items coming off this line, what is the probability that the first defective you see will occur only after four items have been inspected? What is the probability that seeing two defectives will require inspection of six or more items?

Construct a simulation to approximate these probabilities. Use the four steps given above, and pay particular attention to how a trial is defined.

2.10 Summary

Data are the key to making sound, objective decisions, and *randomness* is the key to obtaining good data in *sample surveys* and *experiments*. Surveys attempt to estimate population characteristics and experiments compare treatments. The importance of randomness derives from the fact that relative frequencies of random events tend to stabilize in the long run; this long-run relative frequency is called *probability*.

A more formal definition of probability allows for its use as a mathematical modeling tool to help explain and anticipate outcomes of events not yet seen. *Conditional probability*, *complements*, and rules for *addition* and *multiplication* of probabilities are essential parts of the modeling process.

The relative frequency notion of probability and the resulting rules of probability calculations have direct parallels in the analysis of frequency data recorded in tables. Even with these rules, it is often more efficient to find an approximate answer through *simulation* than to labor over a theoretical probability calculation.

■ ■ ■ ■ ■ ■

Supplementary Exercises

2.55   A coin is tossed four times, and the outcome is recorded for each toss.

  a   List the outcomes for the experiment.

  b   Let A be the event that the experiment yields three heads. List the outcomes in A.

  c   Make a reasonable assignment of probabilities to the outcomes and find $P(A)$.

2.56   A hydraulic rework shop in a factory turns out seven rebuilt pumps today. Suppose three pumps are still defective. Two of the seven are selected for thorough testing and then classified as defective or nondefective.

  a   List the outcomes for this experiment.

  b   Let A be the event that the selection includes no defectives. List the outcomes in A.

  c   Assign probabilities to the outcomes and find $P(A)$.

2.57   The National Maximum Speed Limit (NMSL) of 55 miles per hour has been in force in the United States since early 1974. The data in Table 2.14 show the percentage of vehicles found to travel at various speeds for three types of highways in 1973 (before the NMSL), 1974 (the year the NMSL was put in force), and 1975.

  a   Find the probability that a randomly observed car on a rural interstate was traveling less than 55 mph in 1973; less than 55 mph in 1974; less than 55 mph in 1975.

  b   Answer the questions in (a) for a randomly observed truck.

  c   Answer the questions in (a) for a randomly observed car on a rural secondary road.

2.58   An experiment consists of tossing a pair of dice.

  a   Use the combinatorial theorems to determine the number of outcomes in the sample space S.

  b   Find the probability that the sum of the numbers appearing on the dice is equal to 7.

2.59   Show that $\binom{3}{0} + \binom{3}{1} + \binom{3}{2} + \binom{3}{3} = 2^3$. Show that, in general, $\sum_{i=0}^{n} \binom{n}{i} = 2^n$.

T A B L E **2.14**

Vehicle Speed (mph)	Rural Interstate						Rural Primary						Rural Secondary					
	Car			Truck			Car			Truck			Car			Truck		
	73	74	75	73	74	75	73	74	75	73	74	75	73	74	75	73	74	75
30–35	0	0	0	0	0	0	0	1	0	1	1	1	4	4	2	6	7	5
35–40	0	0	0	1	1	0	3	2	2	5	5	3	6	9	6	9	11	7
40–45	0	1	1	2	2	2	5	7	5	9	10	7	11	14	10	15	16	13
45–50	2	7	5	5	11	8	13	18	14	20	21	19	19	25	21	22	25	23
50–55	5	24	23	15	29	29	16	29	29	21	30	30	19	23	26	19	21	26
55–60	13	37	41	27	36	40	22	27	32	22	23	27	19	16	22	17	14	19
60–65	21	21	22	25	15	16	19	11	12	14	7	10	11	6	9	7	4	5
65–70	29	7	6	18	5	4	13	3	5	6	2	2	7	2	3	4	2	1
70–75	19	2	2	5	1	1	6	2	1	1	1	1	3	1	1	0	0	1
75–80	7	1	0	2	0	0	1	0	0	0	0	0	1	0	0	1	0	0
80–85	4	0	0	0	0	0	0	0	0	0	0	0	0	0	0	0	0	0

Source: D. B. Kamerud, *Transportation Research*, vol. 17A, no. 1, p. 61.

2.60 Of the persons arriving at a small airport, 60% fly on major airlines, 30% fly on privately owned airplanes, and 10% fly on commercially owned airplanes not belonging to an airline. Of the persons arriving on major airlines, 50% are traveling for business reasons, while this figure is 60% for those arriving on private planes and 90% for those arriving on other commercially owned planes. For a person randomly selected from a group of arrivals, find the probability that

a the person is traveling on business.

b the person is traveling on business and on a private airplane.

c the person is traveling on business given that he arrived on a commercial airliner.

d the person arrived on a private plane given that he is traveling on business.

2.61 In how many ways can a committee of three be selected from ten people?

2.62 How many different seven-digit telephone numbers can be formed if the first digit cannot be zero?

2.63 A personnel director for a corporation has hired ten new engineers. If three (distinctly different) positions are open at a particular plant, in how many ways can he fill the positions?

2.64 An experimenter wants to investigate the effect of three variables—pressure, temperature, and type of catalyst—on the yield in a refining process. If the experimenter intends to use three settings each for temperature and pressure and two types of catalysts, how many experimental runs will have to be conducted if she wants to run all possible combinations of pressure, temperature, and type of catalysts?

2.65 An inspector must perform eight tests on a randomly selected keyboard coming off an assembly line. The sequence in which the tests are conducted is important, because the time lost between tests will vary. If an efficiency expert were to study all possible sequences to find the one that required the minimum length of time, how many sequences would be included in his study?

2.66 a Two cards are drawn from a 52-card deck. What is the probability that the draw will yield an ace and a face card in either order?

b Five cards are drawn from a 52-card deck. What is the probability that all five cards will be spades? that all five cards will be of the same suit?

2.67   The quarterback on a certain football team completes 60% of his passes. If he tries three passes, assumed to be independent, in a given quarter, what is the probability that he will complete

    a   all three?

    b   at least one?

    c   at least two?

2.68   Two men each tossing a balanced coin obtain a "match" if both coins are heads or if both are tails. The process is repeated three times.

    a   What is the probability of three matches?

    b   What is the probability that all six tosses (three for each man) result in "tails"?

    c   Coin tossing provides a model for many practical experiments. Suppose the "coin tosses" represented the answers given by two students to three specific true-false questions on an examination. If the two students gave three matches for answers, would the low probability determined in part (a) suggest collusion?

2.69   Refer to Exercise 2.68. What is the probability that the pair of coins is tossed four times before a match occurs (that is, they match for the first time on the fourth toss)?

2.70   Suppose the probability of exposure to the flu during an epidemic is 0.6. Experience has shown that a serum is 80% successful in preventing an inoculated person from acquiring the flu if exposed. A person not inoculated faces a probability of 0.90 of acquiring the flu if exposed. Two persons, one inoculated and one not, are capable of performing a highly specialized task in a business. Assume that they are not at the same location, are not in contact with the same people, and cannot expose each other. What is the probability that at least one will get the flu?

2.71   Two gamblers bet $1 each on the successive tosses of a coin. Each has a bank of $6.

    a   What is the probability that they break even after six tosses of the coin?

    b   What is the probability that one player, say Jones, wins all the money on the tenth toss of the coin?

2.72   Suppose the streets of a city are laid out in a grid, with streets running north–south and east–west. Consider the following scheme for patrolling an area of 16 blocks by 16 blocks: A patrolman commences walking at the intersection in the center of the area. At the corner of each block, he randomly elects to go north, south, east, or west.

    a   What is the probability that he will reach the boundary of his patrol area by the time he walks the first eight blocks?

    b   What is the probability that he will return to the starting point after walking exactly four blocks?

2.73   Consider two mutually exclusive events, A and B, such that $P(A) > 0$ and $P(B) > 0$. Are A and B independent? Give a proof with your answer.

2.74   An accident victim will die unless she receives in the next 10 minutes an amount of type A Rh$^+$ blood, which can be supplied by a single donor. It requires 2 minutes to "type" a prospective donor's blood and 2 minutes to complete the transfer of blood. A large number of untyped donors are available, and 40% of them have type A Rh$^+$ blood. What is the probability that the accident victim will be saved if only one blood-typing kit is available?

2.75   An assembler of electric fans uses motors from two sources. Company A supplies 90% of the motors, and company B supplies the other 10%. Suppose it is known that 5% of the motors supplied by company A are defective and 3% of the motors supplied by company B are defective. An assembled fan is found to have a defective motor. What is the probability that this motor was supplied by company B?

2.76   Show that for three events A, B, and C

$$P[(A \cup B)|C] = P(A|C) + P(B|C) - P[(A \cap B|C)]$$

2.77   If A and B are independent events, show that A and $\overline{B}$ are also independent.

2.78   Three events A, B, and C are said to be independent if

$$P(AB) = P(A)P(B)$$
$$P(AC) = P(A)P(C)$$
$$P(BC) = P(B)P(C)$$

and

$$P(ABC) = P(A)P(B)P(C)$$

Suppose a balanced coin is independently tossed two times. Define the following events:

   A: Heads appears on the first toss.

   B: Heads appears on the second toss.

   C: Both tosses yield the same outcome.

Are A, B, and C independent?

2.79   A line from a to b has midpoint c. A point is chosen at random on the line and marked x (the point x being chosen at random implies that x is equally likely to fall in any subinterval of fixed length l). Find the probability that the line segments ax, bx, and ac can be joined to form a triangle.

2.80   Eight tires of different brands are ranked from 1 to 8 (best to worst) according to mileage performance. If four of these tires are chosen at random by a customer, find the probability that the best tire among those selected by the customer is actually ranked third among the original eight.

2.81   Suppose that n indistinguishable balls are to be arranged in N distinguishable boxes so that each distinguishable arrangement is equally likely. If $n \geq N$, show that the probability that no box will be empty is given by

$$\frac{\binom{n-1}{N-1}}{\binom{N+n-1}{N-1}}$$

2.82   Relays in a section of an electrical circuit operate independently, and each one closes properly with probability 0.9 when a switch is thrown. The following two designs, each involving four relays, are presented for a section of a new circuit. Which design has the higher probability of current flowing from a to b when the switch is thrown?

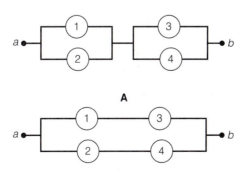

2.83   A blood test for hepatitis has the following accuracy:

	Test Result	
	+	−
Patient with Hepatitis	0.90	0.10
Patient without Hepatitis	0.01	0.99

The disease rate in the general population is 1 in 10,000.

a   What is the probability that a person actually has hepatitis if he or she gets a positive blood test result?

b   A patient is sent for a blood test because he has lost his appetite and has jaundice. The physician knows that this type of patient has probability 0.5 of having hepatitis. If this patient gets a positive result on his blood test, what is the probability that he has hepatitis?

2.84   Show that the following are true for any events A and B.

a $P(AB) \geq P(A) + P(B) - 1$

b   The probability that exactly one of the events occurs is $P(A) + P(B) - 2P(AB)$.

2.85   Using Venn diagrams or similar arguments, show that for events A, B, and C

$$P(A \cup B \cup C) = P(A) + P(B) + P(C) - P(AB) - P(AC) - P(BC) + P(ABC)$$

2.86   Using the definition of conditional probability, show that

$$P(ABC) = P(A)P(B|A)P(C|AB)$$

2.87   There are 23 students in a classroom. What is the probability that at least two of them have the same birthday (day and month)? Assume that the year has 365 days. State your assumptions.

Discrete Probability Distributions

About This Chapter

Quality improvement depends on data; the collection of good data and the proper interpretation of data as a guide to anticipating outcomes yet unseen depend on the rules of probability. In Chapter 2, probability was applied only to categorical outcomes, such as employment status. In this chapter, probability will be applied to *numerical* outcomes that take the form of counts, such as the number of automobiles per family, the number of nonconforming parts manufactured per day, or the number of bacteria per cubic centimeter of water. These discrete counts vary from element to element in a sample, and hence they are called *discrete random variables*. Variables that take the form of a continuous measurement, such as the diameter of a machined rod, are considered in Chapter 4.

Contents

[†]Optional section.

3.1 Random Variables and Their Probability Distributions

Most of the processes we encounter generate outcomes that can be interpreted in terms of real numbers, such as heights of children, number of voters favoring a certain candidate, tensile strengths of wires, and numbers of accidents at specified intersections. These numerical outcomes, with values that can change from trial to trial, are called *random variables*. We will look at an illustrative example of a random variable before we give a more formal definition.

A section of an electrical circuit has two relays, numbered 1 and 2, operating in parallel. The current will flow when a switch is thrown if either one or both of the relays close. The probability of a relay closing properly is 0.8 for each relay. We assume that the relays operate independently. Let E_i denote the event that relay i closes properly when the switch is thrown. Then $P(E_i) = 0.8$.

When the switch is thrown, a numerical event of some interest to the operator of this system is X, the number of relays that close properly. Now X can take on only three possible values, since the number of relays that close must be 0, 1, or 2. We can find the probabilities associated with these values of X by relating them to the underlying events E_i. Thus, we have

$$
\begin{aligned}
P(X = 0) &= P(\overline{E}_1 \overline{E}_2) \\
&= P(\overline{E}_1)P(\overline{E}_2) \\
&= (0.2)(0.2) = 0.04
\end{aligned}
$$

since $X = 0$ means neither relay closes and the relays operate independently. Similarly,

$$
\begin{aligned}
P(X = 1) &= P(E_1 \overline{E}_2 \cup \overline{E}_1 E_2) \\
&= P(E_1 \overline{E}_2) + P(\overline{E}_1 E_2) \\
&= P(E_1)P(\overline{E}_2) + P(\overline{E}_1)P(E_2) \\
&= (0.8)(0.2) + (0.2)(0.8) \\
&= 0.32
\end{aligned}
$$

and

$$
\begin{aligned}
P(X = 2) &= P(E_1 E_2) \\
&= P(E_1)P(E_2) \\
&= (0.8)(0.8) = 0.64
\end{aligned}
$$

The values of the random variable X along with their probabilities are more useful for keeping track of the operation of this system than are the underlying events E_i, for it is the *number* of properly closing relays that is the key to whether or not the system will work. The current will flow if X is at least 1, and this event has probability

$$
\begin{aligned}
P(X \geq 1) &= P(X = 1 \text{ or } X = 2) \\
&= P(X = 1) + P(X = 2) \\
&= 0.32 + 0.64 \\
&= 0.96
\end{aligned}
$$

Note that we have mapped the outcomes of observing a process into a set of three meaningful real numbers and have attached a probability to each. Such situations provide the motivation for Definitions 3.1 and 3.2.

DEFINITION   **3.1**

A random variable is a real-valued function whose domain is a sample space.   ∎

Random variables will be denoted by uppercase letters, such as X, Y, and Z. The actual numerical values that a random variable can assume will be denoted by lowercase letters, such as x, y, and z. We can talk about "the probability that X takes on the value x," $P(X = x)$, denoted by $p(x)$.

In the relay example, the random variable X has only three possible values, and it is a relatively simple matter to assign probabilities to these values. Such a random variable is called *discrete*.

DEFINITION   **3.2**

A random variable X is said to be **discrete** if it can take on only a finite number, or a countable infinity, of possible values x.

In this case,

1 $P(X = x) = p(x) \geq 0$

2 $\sum_x P(X = x) = 1$, where the sum is over all possible values x

The function $p(x)$ is called the **probability function** of X.   ∎

The probability function is sometimes called the *probability mass function* of X to denote the idea that a mass of probability is piled up at discrete points.

It is often convenient to list the probabilities for a discrete random variable in a table. With X defined as the number of closed relays in the problem discussed above, the table is as follows:

x	$p(x)$
0	0.04
1	0.32
2	0.64
Total	1.00

This listing is one way of representing the *probability distribution* of X.

Note that the probability function $p(x)$ for any random variable satisfies two properties:

1 $0 \leq p(x) \leq 1$ for any x

2 $\sum_x p(x) = 1$, where the sum is over all possible values of x

Functional forms for the probability function $p(x)$ will be given in later sections. We now illustrate another method for arriving at a tabular presentation of a discrete probability distribution.

E X A M P L E **3.1**   The output of circuit boards from two assembly lines set up to produce identical boards is mixed into one storage tray. As inspectors examine the boards, it is difficult to determine whether a board comes from line A or line B. A probabilistic assessment of this question is often helpful.

Suppose a storage tray contains ten circuit boards, of which six came from line A and four from line B. An inspector selects two of these identical-appearing boards for inspection. He is interested in X, the number of inspected boards from line A. Find the probability distribution for X.

Solution   The experiment consists of two selections, each of which can result in two outcomes. Let A_i denote the event that the ith board comes from line A, and B_i the event that it comes from line B. Then the probability of selecting two boards from line $A(X = 2)$ is

$$P(X = 2) = P(A_1 A_2) = P(A \text{ on first}) P(A \text{ on second}|A \text{ on first})$$

The multiplicative law of probability is used, and the probability of the second selection depends on what happened on the first selection. There are other possibilities for outcomes that will result in other values of X. These outcomes are conveniently listed on the tree in Figure 3.1. The probabilities for the various selections are given on the branches of the tree.

F I G U R E **3.1**
Outcomes for Example 3.1

First selection	Second selection	Probability	Value of X
	$\frac{5}{9}$ A	$\frac{30}{90}$	2
$\frac{6}{10}$ A	$\frac{4}{9}$ B	$\frac{24}{90}$	1
$\frac{4}{10}$ B	$\frac{6}{9}$ A	$\frac{24}{90}$	1
	$\frac{3}{9}$ B	$\frac{12}{90}$	0

It is easily seen that X has three possible outcomes, with the following probabilities:

x	$p(x)$
0	12/90
1	48/90
2	30/90
Total	1.00

The reader should envision this concept extended to more selections from trays of various structure. ∎

We sometimes study the behavior of random variables by looking at their *cumulative probabilities*. That is, for any random variable X we may look at $P(X \leq b)$ for any real number b. This is the cumulative probability for X evaluated at b. Thus, we can define a function $F(b)$ as

$$F(b) = P(X \leq b)$$

DEFINITION 3.3

The **distribution function** $F(b)$ for a random variable X is defined as

$$F(b) = P(X \leq b)$$

If X is discrete,

$$F(b) = \sum_{x=-\infty}^{b} p(x)$$

where $p(x)$ is the probability function.

The distribution function is sometimes called the **cumulative distribution function** (cdf). ∎

The random variable X, denoting the number of relays closing properly (defined at the beginning of this section), has probability distribution given by

$$P(X = 0) = 0.04$$
$$P(X = 1) = 0.32$$
$$P(X = 2) = 0.64$$

Note that

$$P(X \leq 1.5) = P(X \leq 1.9) = P(X \leq 1) = 0.36$$

The distribution function for this random variable has the form

$$F(b) = \begin{cases} 0 & b < 0 \\ 0.04 & 0 \le b < 1 \\ 0.36 & 1 \le b < 2 \\ 1.00 & 2 \le b \end{cases}$$

This function is graphed in Figure 3.2.

F I G U R E **3.2**
A Distribution Function for a
Discrete Random Variable

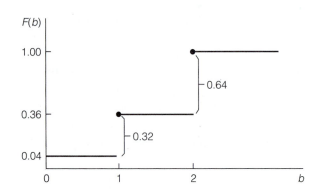

Exercises

3.1 Among ten applicants for an open position, six are females and four are males. Suppose three applicants are randomly selected from the applicant pool for final interviews. Find the probability distribution for X, the number of female applicants among the final three.

3.2 The median annual income for household heads in a certain city is $28,000. Four such household heads are randomly selected for an opinion poll.

    a Find the probability distribution of X, the number (out of the four) that have annual incomes below $28,000.

    b Would you say that it is unusual to see all four below $28,000 in this type of poll? (What is the probability of this event?)

3.3 Wade Boggs of the Boston Red Sox hit .363 in 1987. (He got a hit on 36.3% of his official times at bat.) In a typical game, he was up to bat three official times. Find the probability distribution for X, the number of hits in a typical game. What assumptions are involved in the answer? Are the assumptions reasonable? Is it unusual for a good hitter like Wade Boggs to go zero for three in one game?

3.4 A commercial building has two entrances, numbered I and II. Three people enter the building at 9:00 A.M. Let X denote the number that select entrance I. Assuming the people choose entrances independently, find the probability distribution for X. Were any additional assumptions necessary in arriving at your answer?

3.5 Table 2.9 on page 102 gives information on causes of residential fires. Suppose four independent residential fires are reported in one day, and let X denote the number out of the four that are in family homes.

    a Find the probability distribution for X in tabular form.

    b Find the probability that at least one of the four fires is in a family home.

3.6   It was observed that 40% of the vehicles crossing a certain toll bridge are commercial trucks. Four vehicles will cross the bridge in the next minute. Find the probability distribution for X, the number of commercial trucks among the four, if the vehicle types are independent of one another.

3.7   Of the people entering a blood bank to donate blood, 1 in 3 have type O^+ blood, and 1 in 15 have type O^- blood. For the next three people entering the blood bank, let X denote the number with O^+ blood and Y the number with O^- blood. Assuming independence among the people with respect to blood type, find the probability distributions for X and Y. Also find the probability distribution for $X + Y$, the number of people with type O blood.

3.8   Daily sales records for a computer manufacturing firm show that it will sell 0, 1, or 2 mainframe computer systems with probabilities as listed:

Number of sales	0	1	2
Probability	0.7	0.2	0.1

   a   Find the probability distribution for X, the number of sales in a two-day period, assuming that sales are independent from day to day.

   b   Find the probability that at least one sale is made in the two-day period.

3.9   Four microchips are to be placed in a computer. Two of the four chips are randomly selected for inspection before assembly of the computer. Let X denote the number of defective chips found among the two chips inspected. Find the probability distribution for X if

   a   two of the microchips were defective.

   b   one of the four microchips was defective.

   c   none of the microchips was defective.

3.10   When turned on, each of the three switches in the diagram below works properly with probability 0.9. If a switch is working properly, current can flow through it when it is turned on. Find the probability distribution for Y, the number of closed paths from a to b, when all three switches are turned on.

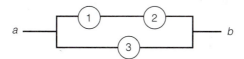

3.2   Expected Values of Random Variables

Since a probability can be thought of as the long-run relative frequency of occurrence for an event, a probability distribution can be interpreted as showing the long-run relative frequency of occurrence for numerical outcomes associated with a random variable. Suppose, for example, that you and a friend are matching balanced coins. Each of you tosses a coin. If the upper faces match, you win $1; if they do not match, you lose $1 (and your friend wins $1). The probability of a match is 0.5 and, in the long run, you will win about half the time. Thus, a relative frequency distribution

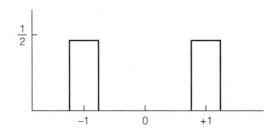

of your winnings should look like Figure 3.3. (The negative sign indicates a loss to you.)

On the average, how much will you win per game over the long run? If Figure 3.3 is a correct display of your winnings, you win -1 half the time and $+1$ half the time, for an average of

$$(-1)(\tfrac{1}{2}) + (1)(\tfrac{1}{2}) = 0$$

This average is sometimes called your expected winnings per game, or the *expected value* of your winnings. (An expected value of 0 indicates that this is a *fair game*.) The general definition of expected value is given in Definition 3.4. Note that the average is another measure of center, similar to the median introduced in Chapter 1.

D E F I N I T I O N   **3.4**

The **expected value** of a discrete random variable X having probability function $p(x)$ is given by[†]

$$E(X) = \sum_x xp(x)$$

(The sum is over all values of x for which $p(x) > 0$.) We sometimes use the notation

$$E(X) = \mu \quad \blacksquare$$

Now payday has arrived, and you and your friend up the stakes to $10 per game of matching coins. You now win -10 or $+10$ with equal probability; your expected winnings per game are

$$(-10)(\tfrac{1}{2}) + (10)(\tfrac{1}{2}) = 0$$

and the game is still fair. The new stakes can be thought of as a function of the old in the sense that if X represents your winnings per game when you were playing for $1, then $10X$ represents your winnings per game when you play for $10. Such functions

---

[†]We assume absolute convergence when the range of X is countable; we talk about an expectation only when it is assumed to exist.

of random variables arise often, and the extension of the definition of expected value to cover these cases is given in Theorem 3.1.

THEOREM **3.1**

If X is a discrete random variable with probability distribution $p(x)$ and if $g(x)$ is any real-valued function of X, then

$$E[g(X)] = \sum_x g(x)p(x)$$

(The proof of this theorem will not be given here.) ∎

You and your friend decide to complicate the payoff picture in the coin-matching game by allowing you to win \$1 if the match is tails and \$2 if the match is heads. You still lose \$1 if the coins do not match. You quickly see that this is not a fair game, since your expected winnings are

$$(-1)\left(\frac{1}{2}\right) + (1)\left(\frac{1}{4}\right) + (2)\left(\frac{1}{4}\right) = 0.25$$

You compensate for this by paying your friend \$1.50 if the coins do not match. Then your expected winnings per game are

$$(-1.5)\left(\frac{1}{2}\right) + (1)\left(\frac{1}{4}\right) + (2)\left(\frac{1}{4}\right) = 0$$

and the game is now fair. What is the difference between this game and the original one in which payoffs were \$1? The difference certainly cannot be explained by the expected value, since both games are fair. You can win more but also lose more with the new payoffs, and the difference between the two games can be partially explained in terms of the variability of your winnings across many games. This increased variability can be seen in Figure 3.4, the relative frequency for your winnings in the new game, which is more spread out than the one shown in Figure 3.3. Formally, variation is often measured by the *variance* and a related quantity called the *standard deviation*, first introduced in Chapter 1 to describe variability in data.

FIGURE **3.4**
Relative Frequency of Winnings

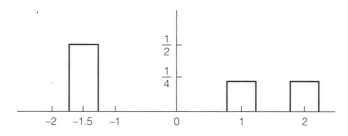

DEFINITION   **3.5**

The **variance** of a random variable X with expected value μ is given by

$$V(X) = E[(X - \mu)^2]$$

We sometimes use the notation

$$E[(X - \mu)^2] = \sigma^2 \quad \blacksquare$$

The smallest value that σ^2 can assume is zero, which would occur if all the probability was at a single point (that is, if X takes on a constant value with probability 1). The variance will become larger as the points with positive probability spread out more.

Observe that the variance squares the units in which we are measuring. A measure of variation that preserves the original units is the *standard deviation*.

DEFINITION   **3.6**

The **standard deviation** of a random variable X is the square root of the variance, given by

$$\sigma = \sqrt{\sigma^2} = \sqrt{E[(X - \mu)^2]} \quad \blacksquare$$

For the game depicted in Figure 3.3, the variance of your winnings (with $\mu = 0$) is

$$\sigma^2 = E[(X - \mu)^2]$$
$$= (-1)^2 \left(\frac{1}{2}\right) + (1)^2 \left(\frac{1}{2}\right) = 1$$

It follows that $\sigma = 1$ as well. For the game depicted in Figure 3.4, the variance of your winnings is

$$\sigma^2 = (-1.5)^2 \left(\frac{1}{2}\right) + (1)^2 \left(\frac{1}{4}\right) + (2)^2 \left(\frac{1}{4}\right)$$
$$= 2.375$$

and

$$\sigma = 1.54$$

Which game would you rather play?

The standard deviation can be thought of as the size of a "typical" deviation between an observed outcome and the expected value. In Figure 3.3, each outcome (-1 or $+1$) does deviate precisely one standard deviation from the expected value. In Figure 3.4, the positive values deviate on average 1.5 units from the expected value of 0 (as does the negative value), which is approximately one standard deviation.

The mean and standard deviation are useful summaries of large data sets, as we saw in Chapter 1. In similar fashion, the mean and standard deviation often provide a useful summary of the probability distribution for a random variable that can assume many values. An illustration is provided by the age distribution of the U.S. population for 1990 and 2050 (projected), as shown in Table 3.1.

Age is actually a continuous measurement; because it is reported in categories, however, we can treat it as a discrete random variable in order to approximate its key functions. To move from continuous age intervals to discrete age classes, we assign each interval the value of its midpoint (rounded). Thus, the data in Table 3.1 are interpreted as reporting that 7.6% of the 1990 population was around three years of age and 22.5% of the 2050 population is anticipated to be around 55 years of age. (The open interval at the upper end was stopped at 100 for convenience.)

With the percentages interpreted as probabilities, the mean age for 1990 is approximated by

$$\mu = \Sigma x p(x)$$
$$= 3(0.076) + 9(0.128) + \ldots + 92(0.012)$$
$$= 35.5$$

(How does this result compare with the median age for 1990, as approximated from Table 3.1?) For 2050, the mean age is approximated by

$$\mu = \Sigma x p(x)$$
$$= 3(0.064) + 9(0.116) + \ldots + 92(0.046)$$
$$= 41.2$$

The mean age is increasing rather markedly (as is the median age).

The variations in the two age distributions can be approximated by the standard deviations. For 1990, this is

$$\sigma = \sqrt{\Sigma(x - \mu)^2 p(x)}$$
$$= \sqrt{(3 - 35.5)^2(0.076) + \ldots + (92 - 35.5)^2(0.012)}$$
$$= 22.5$$

T A B L E **3.1**
Age Distribution of U.S. Population (in percent)

Age Interval	Midpoint	1990	2050
Under 5	3	7.6	6.4
5–13	9	12.8	11.6
14–17	16	5.3	5.2
18–24	21	10.8	9.0
25–34	30	17.3	12.5
35–44	40	15.1	12.2
45–64	55	18.6	22.5
65–84	75	11.3	16.0
85 and over	92	1.2	4.6

Source: U.S. Bureau of the Census.

A similar calculation for the 2050 data yields a standard deviation of 25.4. These results are summarized in Table 3.2.

TABLE **3.2**
Age Distribution of U.S.
Population: Summary

	1990	2050
Mean	35.5	41.2
Standard Deviation	22.5	25.4

The population is not only getting older, on the average, but its variability is increasing. What are some of the implications of these trends? We will return to these data later.

We now provide additional examples and extensions of these basic results.

E X A M P L E   **3.2**   The manager of a stockroom in a factory knows from her study of records that the daily demand (number of times used) for a certain tool has the following probability distribution:

Demand	0	1	2
Probability	0.1	0.5	0.4

(That is, 50% of the daily records show that the tool was used one time.) If X denotes the daily demand, find $E(X)$ and $V(X)$.

Solution   From Definition 3.4, we see that

$$E(X) = \sum_x xp(x)$$
$$= 0(0.1) + 1(0.5) + 2(0.4) = 1.3$$

The tool is used an average of 1.3 times per day.
From Definition 3.5, we see that

$$V(X) = E[(X - \mu)^2]$$
$$= \sum_x (x - \mu)^2 p(x)$$
$$= (0 - 1.3)^2(0.1) + (1 - 1.3)^2(0.5) + (2 - 1.3)^2(0.4)$$
$$= (1.69)(0.1) + (0.09)(0.5) + (0.49)(0.4)$$
$$= 0.41$$

(We will use this value again later in the chapter.)   ▪

Our work in manipulating expected values is greatly facilitated by use of the two results given in Theorem 3.2.

THEOREM 3.2

For any random variable X and constants a and b,

1   $E(aX + b) = aE(X) + b$

2   $V(aX + b) = a^2 V(X)$

Proof

We sketch a proof of this theorem for a discrete random variable X having a probability distribution given by $p(x)$. By Definition 3.4,

$$E(aX + b) = \sum_x (ax + b)p(x)$$
$$= \sum_x [(ax)p(x) + bp(x)]$$
$$= \sum_x axp(x) + \sum_x bp(x)$$
$$= a \sum_x xp(x) + b \sum_x p(x)$$
$$= aE(X) + b$$

(Note that $\sum_x p(x)$ must equal 1.) Also, by Definition 3.5,

$$V(aX + b) = E[(aX + b) - E(aX + b)]^2$$
$$= E[aX + b - (aE(X) + b)]^2$$
$$= E[aX - aE(X)]^2$$
$$= E[a^2(X - E(X))^2]$$
$$= a^2 E[X - E(X)]^2$$
$$= a^2 V(X) \quad \blacksquare$$

We illustrate the use of these results in the following example.

EXAMPLE 3.3

In Example 3.2, suppose it costs the factory $10 each time the tool is used. Find the mean and variance of the daily costs for use of this tool.

Solution

Recall that X in Example 3.2 denotes the daily demand. Then the daily cost of using this tool is $10X$. We have, by Theorem 3.2,

$$E(10X) = 10E(X) = 10(1.3)$$
$$= 13$$

The factory should budget $13 per day to cover the cost of using the tool.
    Also by Theorem 3.2,

$$V(10X) = (10)^2 V(X) = 100(0.41)$$
$$= 41$$

(We will use this value in a later example.) ∎

Theorem 3.2 leads us to a more efficient computational formula for variance, as given in Theorem 3.3.

THEOREM **3.3**

If X is a random variable with mean μ, then

$$V(X) = E(X^2) - \mu^2$$

Proof

Starting with the definition of variance, we have

$$
\begin{aligned}
V(X) = E[(X - \mu)^2] &= E(X^2 - 2X\mu + \mu^2) \\
&= E(X^2) - E(2X\mu) + E(\mu^2) \\
&= E(X^2) - 2\mu E(X) + \mu^2 \\
&= E(X^2) - 2\mu^2 + \mu^2 \\
&= E(X^2) - \mu^2 \quad \blacksquare
\end{aligned}
$$

EXAMPLE **3.4**

Use the result of Theorem 3.3 to compute the variance of X as given in Example 3.2.

Solution

In Example 3.2, X has the probability distribution given by

x	0	1	2
$p(x)$	0.1	0.5	0.4

and we saw that $E(X) = 1.3$. Now

$$
\begin{aligned}
E(X^2) = \sum_x x^2 p(x) \\
= (0)^2(0.1) + (1)^2(0.5) + (2)^2(0.4) \\
= 0 + 0.5 + 1.6 \\
= 2.1
\end{aligned}
$$

By Theorem 3.3,

$$
\begin{aligned}
V(X) = E(X^2) - \mu^2 \\
= 2.1 - (1.3)^2 = 0.41 \quad \blacksquare
\end{aligned}
$$

We have computed means and variances for a number of probability distributions and have argued that these two quantities give us some useful information about the center and spread of the probability mass. Now suppose we know only the mean and variance for a probability distribution. Can we say anything specific about probabilities for certain intervals? The answer is "yes," and two useful results on the relationship between mean, standard deviation and relative frequency will now be discussed.

The empirical rule introduced in Chapter 1 as a data analysis tool has a counterpart in probability. Recall the rule: For a mound-shaped, somewhat symmetric data distribution, about 68% of the data points lie within one standard deviation of the mean, and about 95% lie within two standard deviations of the mean. The rule still holds when the data distribution is interpreted as a probability distribution.

Refer to Tables 3.1 and 3.2 on the age distribution of the U.S. population. For 1990, the interval $\mu \pm \sigma$ becomes (35.5 ± 22.5), or $(13.0, 58.0)$. The percentage of the population inside this interval cannot be determined precisely, but 13 is at the upper end of the 5–13 age interval, and 58 is inside the 45–64 age interval. Adding the percentages above 13 and below 64 yields 67.1%, very close to the 68% of the empirical rule. The interval $\mu \pm 2\sigma$ becomes (35.5 ± 45.0), or $(0, 80.5)$. This interval contains somewhat less than 98.8% of the population.

For the 2050 age distribution, $\mu \pm \sigma$ becomes (41.2 ± 25.4), or $(15.8, 66.6)$. Now 66.6 is toward the low end of the 65–84 age interval, and, depending on how you want to treat this category, $\mu \pm \sigma$ contains somewhere between 61.4% and 77.4% of the population. The interval $\mu \pm 2\sigma$ becomes $(0, 92)$, which includes somewhat less than 100% of the population. So the empirical rule does work for probability distributions as well as for data sets. As is the case for data analysis, it works very well for symmetric, mound-shaped probability distributions.

The second useful result is a theorem relating to μ, σ, and relative frequencies (probabilities) that gives very rough approximations but works in all cases.

THEOREM   **3.4**

Tchebysheff's Theorem   Let X be a random variable with mean μ and variance σ^2. Then for any positive k

$$P(|X - \mu| < k\sigma) \geq 1 - \frac{1}{k^2}$$

Proof

The proof of this theorem begins with the definition of $V(X)$ and then makes substitutions in the sum defining this quantity.

$$V(X) = \sigma^2 = \sum_{-\infty}^{\infty} (x - \mu)^2 p(x)$$

$$= \sum_{-\infty}^{\mu - k\sigma} (x - \mu)^2 p(x) + \sum_{\mu - k\sigma}^{\mu + k\sigma} (x - \mu)^2 p(x) + \sum_{\mu + k\sigma}^{\infty} (x - \mu)^2 p(x)$$

The first sum stops at the largest value of x smaller than $\mu - k\sigma$, and the third sum begins at the smallest value of x larger than $\mu + k\sigma$; the middle sum collects the remaining terms. Observe that the middle sum is always nonnegative, and for both of the outside sums

$$(x - \mu)^2 \geq k^2 \sigma^2$$

Eliminating the middle sum and substituting for $(x - \mu)^2$ in the other two yields

$$\sigma^2 \geq \sum_{-\infty}^{\mu - k\sigma} k^2 \sigma^2 p(x) + \sum_{\mu + k\sigma}^{\infty} k^2 \sigma^2 p(x)$$

or

$$\sigma^2 \geq k^2 \sigma^2 \left[\sum_{-\infty}^{\mu - k\sigma} p(x) + \sum_{\mu + k\sigma}^{\infty} p(x) \right]$$

or

$$\sigma^2 \geq k^2 \sigma^2 P(|X - \mu| \geq k\sigma)$$

It follows that

$$P(|X - \mu| \geq k\sigma) \leq \frac{1}{k^2}$$

or

$$P(|X - \mu| < k\sigma) \geq 1 - \frac{1}{k^2} \quad \blacksquare$$

The inequality in the statement of Theorem 3.4 is equivalent to

$$P(\mu - k\sigma < X < \mu + k\sigma) \geq 1 - \frac{1}{k^2}$$

To interpret this result, let $k = 2$, for example. Then the interval from $\mu - 2\sigma$ to $\mu + 2\sigma$ must contain at least $1 - 1/k^2 = 1 - 1/4 = 3/4$ of the probability mass for the random variable.

We give more specific illustrations in the following two examples.

E X A M P L E   **3.5**   The daily production of electric motors at a certain factory averaged 120 with a standard deviation of 10.

a   What fraction of days will have a production level between 100 and 140?

b   Find the shortest interval certain to contain at least 90% of the daily production levels.

Solution   **a**   The interval 100 to 140 is $\mu - 2\sigma$ to $\mu + 2\sigma$, with $\mu = 120$ and $\sigma = 10$. Thus, $k = 2$, and

$$1 - \frac{1}{k^2} = 1 - \frac{1}{4} = \frac{3}{4}$$

At least 75% of the days will have total production in this interval. (This result could be closer to 95% if the daily production figures showed a mound-shaped, symmetric relative frequency distribution.)

b   To find k, we must set $(1 - 1/k^2)$ equal to 0.9 and solve for k. That is,

$$1 - \frac{1}{k^2} = 0.9$$

$$\frac{1}{k^2} = 0.1$$

$$k^2 = 10$$

or

$$k = \sqrt{10} = 3.16$$

The interval

$$\mu - 3.16\sigma \text{ to } \mu + 3.16\sigma$$

or

$$120 - 3.16(10) \text{ to } 120 + 3.16(10)$$

or

$$88.4 \text{ to } 151.6$$

will contain at least 90% of the daily production levels.   ▪

EXAMPLE   **3.6**   The daily cost for use of a certain tool has a mean of $13 and a variance of 41. (See Example 3.3.) How often will this cost exceed $30?

Solution   First we must find the distance between the mean and 30, in terms of the standard deviation of the distribution of costs. We have

$$\frac{30 - \mu}{\sqrt{\sigma^2}} = \frac{30 - 13}{\sqrt{41}} = \frac{17}{6.4} = 2.66$$

So 30 is 2.66 standard deviations above the mean. Letting $k = 2.66$ in Theorem 3.4, we have that the interval

$$\mu - 2.66\sigma \text{ to } \mu + 2.66\sigma$$

or

$$13 - 2.66(6.4) \text{ to } 13 + 2.66(6.4)$$

or

$$-4 \text{ to } 30$$

must contain at least

$$1 - \frac{1}{k^2} = 1 - \frac{1}{(2.66)^2} = 1 - 0.14 = 0.86$$

of the probability. Since the daily cost cannot be negative, at most 0.14 of the probability mass can exceed 30. Thus, the cost cannot exceed $30 more than 14% of the time.   ▪

Exercises

3.11 You are to pay $1 to play a game consisting of drawing one ticket at random from a box of numbered tickets. You win the amount (in dollars) of the number on the ticket you draw. Two boxes are available with numbered tickets as shown below:

I | 0, 1, 2   II | 0, 0, 0, 1, 4

a Find the expected value and variance of your *net* gain per play with box I.

b Repeat part (a) for box II.

c Given that you have decided to play, which box would you choose and why?

3.12 The size distribution of U.S. families is as follows:

Number of Persons	Percentage
1	24.5
2	32.3
3	17.3
4	15.5
5	6.7
6	2.3
7 or more	1.4

Source: U.S. Bureau of the Census.

a Calculate the mean and standard deviation of family size. Are these exact values or approximations?

b How does the mean family size compare to the median family size?

c Does the empirical rule hold for the family size distribution?

3.13 The graph in Figure 3.5 shows the age distribution for AIDS deaths in the United States through 1989. Approximate the mean and standard deviation of this age distribution. How does the mean age compare to the approximate median age?

FIGURE **3.5**

Distribution of AIDS Deaths, by Age, 1982 Through 1989

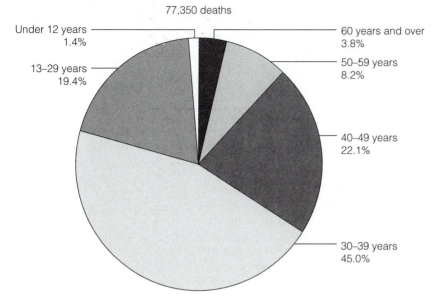

77,350 deaths

Under 12 years
1.4%

13–29 years
19.4%

60 years and over
3.8%

50–59 years
8.2%

40–49 years
22.1%

30–39 years
45.0%

Source: U.S. Bureau of the Census.

3.14 How old are our drivers? The table below gives the age distribution of licensed drivers in the United States.

T A B L E **3.3** Licensed Drivers, 1989

Age	Number (in millions)
19 and under	9.7
20–24	16.9
25–29	20.6
30–34	20.5
35–39	18.6
40–44	16.1
45–49	12.6
50–54	10.2
55–59	9.5
60–64	9.3
65–69	8.3
70 and over	13.3
Total	165.6

Source: U.S. Department of Transportation

a Describe this age distribution in terms of median, mean, and standard deviation.

b Does the empirical rule work well for this distribution?

3.15 Daily sales records for a computer manufacturing firm show that it will sell 0, 1, or 2 mainframe computer systems with probabilities as listed:

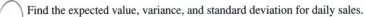

Number of Sales	0	1	2
Probability	0.7	0.2	0.1

Find the expected value, variance, and standard deviation for daily sales.

3.16 Approximately 10% of the glass bottles coming off a production line have serious defects in the glass. If two bottles are randomly selected for inspection, find the expected value and variance of the number of inspected bottles with serious defects.

3.17 The number of breakdowns for a university computer system is closely monitored by the director of the computing center, since it is critical to the efficient operation of the center. The number averages 4 per week, with a standard deviation of 0.8 per week.

a Find an interval that must include at least 90% of the weekly figures on number of breakdowns.

b The center director promises that the number of breakdowns will rarely exceed 8 in a one-week period. Is the director safe in making this claim? Why or why not?

3.18 Keeping an adequate supply of spare parts on hand is an important function of the parts department of a large electronics firm. The monthly demand for microcomputer printer boards was studied for some months and found to average 28, with a standard deviation of 4. How many printer boards should be stocked at the beginning of each month to ensure that the demand will exceed the supply with probability less than 0.10?

3.19 An important feature of golf cart batteries is the number of minutes they will perform before needing to be recharged. A certain manufacturer advertises batteries that will run, under a 75-amp discharge test, for an average of 100 minutes, with a standard deviation of 5 minutes.

a Find an interval that must contain at least 90% of the performance periods for batteries of this type.

b Would you expect many batteries to die out in less than 80 minutes? Why or why not?

3.20 Costs of equipment maintenance are an important part of a firm's budget. Each visit by a field representative to check out a malfunction in a word processing system costs $40. The word processing system is expected to malfunction approximately five times per month, and the standard deviation of the number of malfunctions is 2.

a Find the expected value and standard deviation of the monthly cost of visits by the field representative.

b How much should the firm budget per month to ensure that the cost of these visits will be covered at least 75% of the time?

3.3 The Bernoulli Distribution

At this point it may seem like every problem has its own unique probability distribution and that we must start from the basics to construct such a distribution each time a new problem comes up. Fortunately, that is not the case. Certain basic probability distributions can be developed that will serve as models for a large number of practical problems. In the remainder of this chapter we consider some fundamental discrete distributions, looking at the theoretical assumptions that underlie these distributions as well as the means, variances, and applications of the distributions.

Consider the inspection of a single item taken from an assembly line. Suppose 0 is recorded if the item is nondefective and 1 is recorded if the item is defective. If X is the random variable denoting the condition of the inspected item, then $X = 0$ with probability $(1 - p)$ and $X = 1$ with probability p, where p denotes the probability of observing a defective item. The probability distribution of X is then given by

$$p(x) = p^x (1 - p)^{1-x} \qquad x = 0, 1$$

where $p(x)$ denotes the probability that $X = x$. Such a random variable is said to have a *Bernoulli distribution* or to represent the outcome of a single Bernoulli trial. Any random variable denoting the presence or absence of a certain condition in an observed phenomenon will possess this type of distribution. Frequently, one of the outcomes is termed a "success" and the other a "failure." Note that $p(x) \geq 0$ and $\sum p(x) = 1$.

Suppose we repeatedly observe items of this type and record a value of X for each item observed. What average value of X would we expect to see? By Definition 3.4, we have that the expected value of X is given by

$$E(X) = \sum_x xp(x)$$
$$= 0p(0) + 1p(1)$$
$$= 0(1 - p) + 1(p) = p$$

Thus, if 10% of the items are defective, we expect to observe an *average* of 0.1 defective per item inspected. (In other words, we would expect to see one defective for every ten items inspected.)

For the Bernoulli random variable X, the variance (see Theorem 3.3) is

$$V(X) = E(X^2) - [E(X)]^2$$
$$= \sum_x x^2 p(x) - p^2$$
$$= 0(1 - p) + 1(p) - p^2$$
$$= p - p^2 = p(1 - p)$$

The Bernoulli random variable will be used as a building block to form other probability distributions, such as the binomial distribution (Section 3.4).

The properties of the Bernoulli distribution are summarized here.

The Bernoulli Distribution

$$p(x) = p^x (1 - p)^{1-x} \qquad x = 0, 1 \qquad \text{for } 0 \leq p \leq 1$$

$$E(X) = p \qquad V(X) = p(1 - p)$$

3.4 The Binomial Distribution

3.4.1 Probability Function

Instead of inspecting a single item, as with the Bernoulli random variable, suppose we now independently inspect n items and record values for $X_1, X_2, \ldots, X_n$, where $X_i = 1$ if the ith inspected item is defective and $X_i = 0$ otherwise. We have, in fact, observed a sequence of n independent Bernoulli random variables. One especially interesting function of $X_1, \ldots, X_n$ is the sum

$$Y = \sum_{i=1}^{n} X_i$$

which denotes the number of defectives among the n sampled items.

We can easily find the probability distribution for Y under the assumption that $P(X_i = 1) = p$, where p remains constant over all trials. For simplicity, let's look at the specific case of $n = 3$. The random variable Y can then take on four possible values—0, 1, 2, and 3. For Y to be 0, all three X_i's must be 0. Thus,

$$
\begin{aligned}
P(Y = 0) &= P(X_1 = 0,\, X_2 = 0,\, X_3 = 0) \\
&= P(X_1 = 0)P(X_2 = 0)P(X_3 = 0) \\
&= (1 - p)^3
\end{aligned}
$$

Now if $Y = 1$, then exactly one of the X_i's is 1 and the other two are 0. Thus,

$$
\begin{aligned}
P(Y = 1) &= P[(X_1 = 1,\, X_2 = 0,\, X_3 = 0) \cup (X_1 = 0,\, X_2 = 1,\, X_3 = 0) \\
&\quad \cup (X_1 = 0,\, X_2 = 0,\, X_3 = 1)] \\
&= P(X_1 = 1,\, X_2 = 0,\, X_3 = 0) + P(X_1 = 0,\, X_2 = 1,\, X_3 = 0) \\
&\quad + P(X_1 = 0,\, X_2 = 0,\, X_3 = 1)
\end{aligned}
$$

(since the three possibilities are mutually exclusive)

$$
\begin{aligned}
&= P(X_1 = 1)P(X_2 = 0)P(X_3 = 0) \\
&\quad + P(X_1 = 0)P(X_2 = 1)P(X_3 = 0) \\
&\quad + P(X_1 = 0)P(X_2 = 0)P(X_3 = 1) \\
&= p(1 - p)^2 + p(1 - p)^2 + p(1 - p)^2 \\
&= 3p(1 - p)^2
\end{aligned}
$$

For $Y = 2$, two X_i's must be 1 and one must be 0, which can also occur in three mutually exclusive ways. Hence,

$$
\begin{aligned}
P(Y = 2) &= P(X_1 = 1,\, X_2 = 1,\, X_3 = 0) + P(X_1 = 1,\, X_2 = 0,\, X_3 = 1) \\
&\quad + P(X_1 = 0,\, X_2 = 1,\, X_3 = 1) \\
&= 3p^2(1 - p)
\end{aligned}
$$

The event $Y = 3$ can occur only if all X_i's are 1, so

$$
\begin{aligned}
P(Y = 3) &= P(X_1 = 1, X_2 = 1, X_3 = 1) \\
&= P(X_1 = 1)P(X_2 = 1)P(X_3 = 1) \\
&= p^3
\end{aligned}
$$

Notice that the coefficient in each of the expressions for $P(Y = y)$ is the number of ways of selecting y positions, in the sequence, in which to place 1's. Since there are three positions in the sequence, this number amounts to $\binom{3}{y}$. Thus, we can write

$$
P(Y = y) = \binom{3}{y} p^y (1 - p)^{3-y} \qquad y = 0, 1, 2, 3, \text{ when } n = 3
$$

For general n, the probability that Y takes on a specific value, say y, is given by the term $p^y (1 - p)^{n-y}$ multiplied by the number of possible outcomes resulting in exactly y defectives being observed. This number is the number of ways of selecting y positions for defectives in the n possible positions of the sequence and is given by

$$
\binom{n}{y} = \frac{n!}{y!(n - y)!}
$$

where $n! = n(n - 1) \cdots 1$ and $0! = 1$. Thus, in general,

$$
P(Y = y) = p(y) = \binom{n}{y} p^y (1 - p)^{n-y} \qquad y = 0, 1, \ldots, n
$$

The probability distribution given above is referred to as the *binomial distribution*.

To summarize, a random variable Y possesses a binomial distribution if five conditions exist:

1 The experiment consists of a fixed number n of identical trials.

2 Each trial can result in one of only two possible outcomes, called "success" or "failure."

3 The probability of "success," p, is constant from trial to trial.

4 The trials are independent.

5 Y is defined to be the number of successes among the n trials.

Many experimental and sample survey situations result in a random variable that can be adequately modeled by the binomial distribution. In addition to counts of the number of defectives in a sample of n items, examples include counts of the number of employees favoring a certain retirement policy out of n employees interviewed, the number of pistons in an eight-cylinder engine that are misfiring, and the number of electronic systems sold this week out of the n that are manufactured.

A general formula for $p(x)$ identifies a family of distributions indexed by certain constants called *parameters*. Thus, n and p are the parameters of the binomial distribution.

To demonstrate how choices for the parameters affect the probability function $p(x)$ for the binomial distribution, Figures 3.6 through 3.10 show graphs of $p(x)$ for various choices of n and p. For $n = 20$ and $p = 0.1$ (Figure 3.6), the graph of $p(x)$ has a peak around $x = 1$ and $x = 2$ and then drops off sharply. For $p = 0.3$ and the same n (Figure), the graph becomes more symmetric, with a peak at $x = 6$. With $p = 0.5$ (Figure 3.8), the graph is perfectly symmetric around $x = 10$; with $p = 0.8$ (Figure 3.9), the graph is mound-shaped but not quite symmetric around $x = 16$. To see what happens when n is increased, we observe that for $n = 50$ and $p = 0.3$ (Figure 3.10) the graph is a little more symmetric than it was in the case where $n = 20$ and $p = 0.3$. Also, Figure 3.10 centers at $x = 15$ rather than $x = 6$.

FIGURE **3.6**
Binomial Probability Function;
$n = 20, p = 0.1$

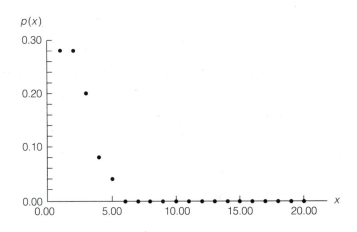

FIGURE **3.7**
Binomial Probability Function;
$n = 20, p = 0.3$

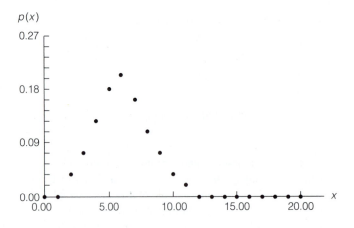

F I G U R E **3.8**
Binomial Probability Function;
$n = 20, p = 0.5$

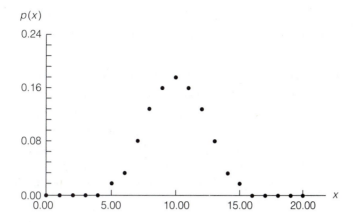

F I G U R E **3.8**
Binomial Probability Function;
$n = 20, p = 0.5$

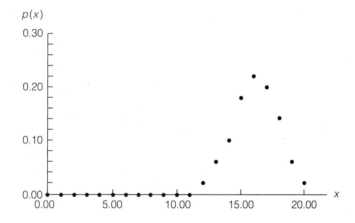

F I G U R E **3.9**
Binomial Probability Function;
$n = 20, p = 0.8$

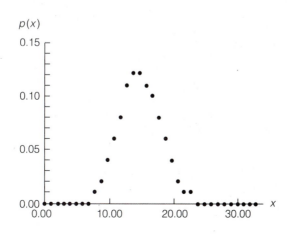

F I G U R E **3.10**
Binomial Probability Function;
$n = 50, p = 0.3$

E X A M P L E   **3.7**   Suppose a large lot of fuses contains 10% defectives. Four fuses are randomly sampled from the lot,

    a   Find the probability that exactly one fuse is defective.

    b   Find the probability that at least one fuse in the sample of four is defective.

Solution   **a**   We assume that the four trials are independent and that the probability of observing a defective is the same (0.1) for each trial. This will be approximately true if the lot is indeed *large*. (If the lot contained only a few fuses, removal of one fuse would substantially change the probability of observing a defective on the second draw.) Thus, the binomial distribution provides a reasonable model for this experiment, and we have, with Y denoting the number of defectives,

$$p(1) = \binom{4}{1}(0.1)^1(0.9)^3 = 0.2916$$

    b   To find $P(Y \geq 1)$, observe that

$$P(Y \geq 1) = 1 - P(Y = 0) = 1 - p(0)$$
$$= 1 - \binom{4}{0}(0.1)^0(0.9)^4$$
$$= 1 - (0.9)^4$$
$$= 0.3439 \quad \blacksquare$$

Discrete distributions, like the binomial distribution, can arise in situations where the underlying problem involves a continuous random variable that is not discrete. The following example provides an illustration.

E X A M P L E   **3.8**   In a study of lifelengths for a certain type of battery, it was found that the probability of a lifelength X exceeding 4 hours is 0.135. If three such batteries are in use in independently operating systems, find the probability that only one of the batteries lasts 4 hours or more.

Solution   Letting Y denote the number of batteries lasting 4 hours or more, we can reasonably assume that Y has a binomial distribution with $p = 0.135$. Hence,

$$P(Y = 1) = p(1) = \binom{3}{1}(0.135)^1(0.865)^2$$
$$= 0.303 \quad \blacksquare$$

3.4.2   Mean and Variance

There are numerous ways of finding $E(Y)$ and $V(Y)$ for a binomial distributed random variable Y. We might use the basic definition and compute

$$E(Y) = \sum_x yp(y)$$

$$= \sum_{y=0}^{n} y \binom{n}{y} p^y(1-p)^{n-y}$$

but direct evaluation of this expression is tricky. Another approach is to make use of the results for linear functions of random variables. We will show in Chapter 5 that, since the binomial Y arose as a sum of independent Bernoulli random variables, $X_1, \ldots, X_n$,

$$E(Y) = E\left[\sum_{i=1}^{n} X_i\right] = \sum_{i=1}^{n} E(X_i)$$

$$= \sum_{i=1}^{n} p = np$$

and

$$V(Y) = \sum_{i=1}^{n} V(x_i) = \sum_{i=1}^{n} p(1-p) = np(1-p)$$

E X A M P L E   **3.9**   Refer to Example 3.7. Suppose the four fuses sampled from the lot were shipped to a customer before being tested, on a guarantee basis. Assume that the cost of making the shipment good is given by $C = 3Y^2$, where Y denotes the number of defectives in the shipment of four. Find the expected repair cost.

Solution   We know that

$$E(C) = E(3Y^2) = 3E(Y^2)$$

and it now remains to find $E(Y^2)$. We have, from Theorem 3.3,

$$V(Y) = E(Y - \mu)^2 = E(Y^2) - \mu^2$$

Since $V(Y) = np(1-p)$ and $\mu = E(Y) = np$, we see that

$$E(Y^2) = V(Y) + \mu^2$$
$$= np(1-p) + (np)^2$$

For Example 3.7, $p = 0.1$ and $n = 4$, and hence

$$E(C) = 3E(Y^2) = 3[np(1-p) + (np)^2]$$
$$= 3[(4)(0.1)(0.9) + (4)^2(0.1)^2]$$
$$= 1.56$$

If the cost was originally in dollars, we could expect to pay an average of $1.56 in repair costs for each shipment of four fuses. ▪

3.4.3 Application

Table 2 of the Appendix gives cumulative binomial probabilities for selected values of n and p. The entries in the table are values of

$$\sum_{y=0}^{a} p(y) = \sum_{y=0}^{a} \binom{n}{y} p^{y}(1-p)^{n-y}$$

The following example illustrates the use of Table 2.

E X A M P L E 3.10 An industrial firm supplies ten manufacturing plants with a certain chemical. The probability that any one firm calls in an order on a given day is 0.2, and this probability is the same for all ten plants. Find the probability that, on a given day, the number of plants calling in orders is (a) at most 3, (b) at least 3, (c) exactly 3.

Solution Let Y denote the number of plants calling in orders on the day in question. If the plants order independently, then Y can be modeled to have a binomial distribution with $p = 0.2$.

a We then have

$$P(Y \leq 3) = \sum_{y=0}^{3} p(y)$$

$$= \sum_{y=0}^{3} \binom{10}{y} (0.2)^{y}(0.8)^{10-y}$$

$$= 0.879$$

from Table 2(b).

b Also,

$$P(Y \geq 3) = 1 - P(Y \leq 2)$$

$$= 1 - \sum_{y=0}^{2} \binom{10}{y} (0.2)^{y}(0.8)^{10-y}$$

$$= 1 - 0.678 = 0.322$$

c Observe that

$$P(Y = 3) = P(Y \leq 3) - P(Y \leq 2)$$

$$= 0.879 - 0.678 = 0.201$$

from results just established. ▪

The examples used to this point have specified n and p as the basis for calculating probabilities or expected values. Sometimes, however, it is necessary to choose n so as to achieve a specific probability. Example 3.11 illustrates this case.

E X A M P L E   **3.11**   The guidance system for a rocket operates correctly with probability p when called upon. Independent but identical backup systems are installed in the rocket so that the probability that at least one system will operate correctly when called upon is no less than 0.99. Let n denote the number of guidance systems in the rocket. How large must n be to achieve the specified probability of at least one guidance system operating if (a) $p = 0.9$; (b) $p = 0.8$?

Solution   Let Y denote the number of correctly operating systems. If the sustems are identical and independent, Y has a binomial distribution. Thus,

$$P(Y \geq 1) = 1 - P(Y = 0)$$
$$= 1 - \binom{n}{0} p^0 (1 - p)^n$$
$$= 1 - (1 - p)^n$$

The conditions specify that n must be such that $P(Y \geq 1) = 0.99$ or more.

a   When $p = 0.9$,

$$P(Y \geq 1) = 1 - (1 - 0.9)^n \geq 0.99$$

results in

$$1 - (0.1)^n \geq 0.99$$

or

$$(0.1)^n \leq 1 - 0.99 = 0.01$$

so $n = 2$. That is, installing two guidance systems will satisfy the specifications.

b   When $p = 0.8$,

$$P(Y \geq 1) = 1 - (1 - 0.8)^n \geq 0.99$$

results in

$$(0.2)^n \leq 0.01$$

Now $(0.2)^2 = 0.04$ and $(0.2)^3 = 0.008$, and we must go to $n = 3$ systems so that

$$P(Y \geq 1) = 1 - (0.2)^3 = 0.992 > 0.99$$

[*Note:* We cannot achieve the 0.99 probability exactly, since Y can assume only integer values.]   ∎

E X A M P L E   **3.12**   Virtually any process can be improved by use of statistics, including the law. A much publicized case that involved a debate about probability was the Collins case, which began in 1964. An incident of purse snatching in the Los Angeles area led to the arrest of Michael and Janet Collins. At their trial, an "expert" presented the following probabilities on characteristics possessed by a couple seen running from the crime:

Man with beard	$\dfrac{1}{10}$
Blond woman	$\dfrac{1}{4}$
Yellow car	$\dfrac{1}{10}$
Woman with ponytail	$\dfrac{1}{10}$
Man with mustache	$\dfrac{1}{3}$
Interracial couple	$\dfrac{1}{1,000}$

The chance that a couple had all these characteristics together is 1 in 12 million. Since the Collinses had the characteristics, the Collinses must be guilty. Do you see any problem with this line of reasoning?

Solution   First, no background data are given to support the probabilities used. Second, the six events are not independent of one another, and therefore the probabilities cannot be multiplied. Third, and most interesting, the wrong question is being addressed. The question of interest is not "What is the probability of finding a couple with these characteristics?" Since one such couple has been found (the Collinses), the question is "What is the probability that *another* such couple exists, given that we have found one?"

Here is where the binomial distribution comes into play. In the binomial model, let

n = number of couples who could have commiteed the crime

p = probability that any one couple possesses the six listed characteristics

x = number of couples that possess the six characteristics

From the binomial distribution,

$$P(X = 0) = (1 - p)^n$$
$$P(X = 1) = np(1 - p)^{n-1}$$
$$P(X \geq 1) = 1 - (1 - p)^n$$

Then the answer to the conditional question posed above is

$$P(X > 1 | X \geq 1) = \frac{1 - (1 - p)^n - np(1 - p)^{n-1}}{1 - (1 - p)^n}$$

Using $p = 1/12{,}000{,}000$ and $n = 12{,}000{,}000$, which are plausible but not well justified guesses,

$$P(X > 1 \mid X \geq 1) = 0.42$$

So the probability of seeing another such couple, given that we've already seen one, is much larger than the probability of seeing such a couple in the first place. This holds true even if the numbers are dramatically changed. For instance, if n is reduced to 1 million, the conditional probability is 0.05, which is still much larger than $1/12{,}000{,}000$.   ▪

The important lessons illustrated here are that the correct probability question is sometimes difficult to determine and that conditional probabilities are *very* sensitive to the conditions.

We now move on to a discussion of other discrete random variables, but the binomial, summarized here, will be used frequently throughout the text.

The Binomial Distribution

$$p(y) = \binom{n}{y} p^y (1 - p)^{n-y} \qquad y = 0, 1, \ldots, n \quad \text{for } 0 \leq p \leq 1$$

$$E(Y) = np \qquad V(Y) = np(1 - p) \quad \blacksquare$$

Exercises

3.21 Let X denote a random variable having a binomial distribution with $p = 0.2$ and $n = 4$. Find the following.

  a $P(X = 2)$ **b** $P(X \geq 2)$

  c $P(X \leq 2)$ **d** $E(X)$

  e $V(X)$

3.22 Let X denote a random variable having a binomial distribution with $p = 0.4$ and $n = 20$. Use Table 2 of the Appendix to evaluate the following.

  a $P(X \leq 6)$ **b** $P(X \geq 12)$

  c $P(X = 8)$

3.23 A machine that fills boxes of cereal underfills a certain proportion p. If 25 boxes are randomly selected from the output of this machine, find the probability that no more than two are underfilled when

  a $p = 0.1$ **b** $p = 0.2$

3.24   In testing the lethal concentration of a chemical found in polluted water, it is found that a certain concentration will kill 20% of the fish that are subjected to it for 24 hours. If 20 fish are placed in a tank containing this concentration of chemical, find the probability that after 24 hours

  a   exactly 14 survive.   **b**   at least 10 survive.

  c   at most 16 survive.

3.25   Refer to Exercise 3.24.

  a   Find the number expected to survive out of 20.

  b   Find the variance of the number of survivors out of 20.

3.26   Among persons donating blood to a clinic, 80% are Rh^+ (that is, have the Rhesus factor present in their blood). Five people donate blood at the clinic on a particular day.

  a   Find the probability that at least one of the five does not have the Rh factor.

  b   Find the probability that at most four of the five have Rh^+ blood.

3.27   Refer to Exercise 3.26. The clinic needs five Rh^+ donors on a certain day. How many people must donate blood to have the probability of at least five RH^+ donors over 0.90?

3.28   The *U. S. Statistical Abstract* reports that the median family income in the United States for 1989 was $34,200. Among four randomly selected families, find the probability that

  a   all four had incomes above $34,200 in 1989.

  b   one of the four had an income below $34,200 in 1989.

3.29   According to an article by B. E. Sullivan in *Transportation Research* (18A, no. 2, 1984, p. 119), 55% of U.S. corporations say that one of the most important factors in locating a corporate headquarters is the "quality of life" for the employees. If five firms are contacted by the governor of Florida concerning possible relocation to that state, find the probability that at least three say the "quality of life" is an important factor in their decision. What assumptions have you made in arriving at this answer?

3.30   A. Goranson and J. Hall (*Aeronautical Journal*, November 1980, pp. 279–280) explain that the probability of detecting a crack in an airplane wing is the product of p_1, the probability of inspecting a plane with a wing crack; p_2, the probability of inspecting the detail in which the crack is located; and p_3, the probability of detecting the damage.

  a   What assumptions justify the multiplication of these probabilities?

  b   Suppose $p_1 = 0.9$, $p_2 = 0.8$, and $p_3 = 0.5$ for a certain fleet of planes. If three planes are inspected from this fleet, find the probability that a wing crack will be detected in at least one of them.

3.31   A missile protection system consists of n radar sets operating independently, each with probability 0.9 of detecting an aircraft entering a specified zone. (All radar sets cover the same zone.) If an airplane enters the zone, find the probability that it will be detected if

  a $n = 2$ **b** $n = 4$

3.32   Refer to Exercise 3.31. How large must n be if it is desired to have probability 0.99 of detecting an aircraft entering the zone?

3.33   A complex electronic system is built with a certain number of backup components in its subsystems. One subsystem has four identical components, each with probability 0.2 of failing in less than 1,000 hours. The subsystem will operate if any two or more of the four components are operating. Assuming the components operate independently, find the probability that

  a   exactly two of the four components last longer than 1,000 hours.

  b   the subsystem operates longer than 1,000 hours.

3.34   An oil exploration firm is to drill ten wells, with each well having probability 0.1 of successfully producing oil. It costs the firm $10,000 to drill each well. A successful well will bring in oil worth $500,000.

   a   Find the firm's expected gain from the ten wells.

   b   Find the standard deviation of the firm's gain.

3.35   A firm sells four items randomly selected from a large lot known to contain 10% defectives. Let Y denote the number of defectives among the four sold. The purchaser of the item will return the defectives for repair, and the repair cost is given by

$$C = 3Y^2 + Y + 2$$

Find the expected repair cost.

3.36   From a large lot of new tires, n are to be sampled by a potential buyer, and the number of defectives X is to be observed. If at least one defective is observed in the sample of n tires, the entire lot is to be rejected by the potential buyer. Find n so that the probability of detecting at least one defective is approximately 0.90 if

   a   10% of the lot is defective.

   b   5% of the lot is defective.

3.37   Ten motors are packaged for sale in a certain warehouse. The motors sell for $100 each, but a "double-your-money-back" guarantee is in effect for any defectives the purchaser might receive. Find the expected net gain for the seller if the probability of any one motor being defective is 0.08. (Assume that the quality of any one motor is independent of the quality of the others.)

3.5 The Geometric Distribution

3.5.1 Probability Function

Suppose a series of test firings of a rocket engine can be represented by a sequence of independent Bernoulli random variables with $X_i = 1$ of the ith trial results in a successful firing and $X_i = 0$ otherwise. Assume that the probability of a successful firing is constant for the trials, and let this probability be denoted by p. For this problem, we might be interested in the number of the trial on which the first successful firing occurs. If Y denotes the number of the trial on which the first success occurs, then

$$P(Y = y) = p(y) = P(X_1 = 0, X_2 = 0, \ldots, X_{y-1} = 0, X_y = 1)$$
$$= P(X_1 = 0)P(X_2 = 0) \cdots P(X_{y-1} = 0)P(X_y = 1)$$
$$= (1 - p)^{y-1}p \qquad y = 1, 2, \ldots$$

because of the independence of the trials. This formula is referred to as the *geometric probability distribution*. Note that this random variable can take on a countably infinite number of possible values.

   In addition to the rocket-firing example just given, other situations result in a random variable whose probability can be modeled by a geometric distribution, including the number of customers contacted before the first sale is made, the number of years a dam is in service before it overflows, and the number of automobiles going through a radar check before the first speeder is detected.

The following example illustrates the use of the geometric distribution.

E X A M P L E   **3.13**   A recruiting firm finds that 30% of the applicants for a certain industrial job have advanced training in computer programming. Applicants are selected at random from the pool and are interviewed sequentially. Find the probability that the first applicant having advanced training in programming is found on the fifth interview.

Solution   The probability of finding a suitably trained applicant will remain relatively constant from trial to trial if the pool of applicants is reasonably large. It then makes sense to define Y as the number of the trial on which the first applicant having advanced training in programming is found and to model Y as having a geometric distribution. Thus,

$$P(Y = 5) = p(5) = (0.7)^4 (0.3)$$
$$= 0.072 \quad \blacksquare$$

3.5.2   Mean and Variance

From the basic definition,

$$E(Y) = \sum_y y p(y) = \sum_{y=1}^{\infty} y p (1-p)^{y-1}$$
$$= p \sum_{y=1}^{\infty} y (1-p)^{y-1}$$
$$= p[1 + 2(1-p) + 3(1-p)^2 + \cdots]$$

The infinite series can be split up into a triangular array of series as follows:

$$E(Y) = p[1 + (1-p) + (1-p)^2 + \cdots$$
$$+ (1-p) + (1-p)^2 + \cdots$$
$$+ (1-p)^2 + \cdots$$
$$+ \cdots]$$

Each line on the right side is an infinite, decreasing geometric progression with common ratio $(1-p)$. Thus the first line inside the bracket sums to $1/p$, the second to $(1-p)/p$, the third to $(1-p)^2/p$, and so on.[†] On accumulating these totals, we have

$$E(Y) = p \left[\frac{1}{p} + \frac{1-p}{p} + \frac{(1-p)^2}{p} + \cdots \right]$$

[†]Recall that $a + ax + ax^2 + ax^3 + \cdots = \dfrac{a}{1-x}$ if $|x| < 1$.

$$= 1 + (1 - p) + (1 - p)^2 + \cdots$$
$$= \frac{1}{1 - (1 - p)} = \frac{1}{p}$$

This answer for $E(Y)$ should seem intuitively realistic. For example, if 10% of a certain lot of items is defective and if an inspector looks at randomly selected items one at a time, then he should expect to wait until the tenth trial to see the first defective. The variance of the geometric distribution will be derived in Section 3.9 by one method and in Chapter 5 by another. The result is

$$V(Y) = \frac{1 - p}{p^2}$$

E X A M P L E   **3.14**   Refer to Example 3.13 and let Y denote the number of the trial on which the first applicant having advanced training in computer programming is found. Suppose the first applicant with advanced training is offered the position, and the applicant accepts. If each interview costs $30, find the expected value and variance of the total cost of interviewing incurred before the job is filled. Within what interval would this cost be expected to fall?

Solution   Since Y is the number of the trial on which the interviewing process ends, the total cost of interviewing is $C = 30Y$. Now

$$E(C) = 30E(Y) = 30\left(\frac{1}{p}\right)$$
$$= 30\left(\frac{1}{0.3}\right) = 100$$

and

$$V(C) = (30)^2 V(Y) = \frac{900(1 - p)}{p^2}$$
$$= \frac{900(0.7)}{(0.3)^2}$$
$$= 7{,}000$$

The standard deviation of C is then $\sqrt{V(C)} = \sqrt{7{,}000} = 83.67$. Tchebysheff's Theorem (see Section 3.2) says that C will lie within two standard deviations of its mean at least 75% of the time. Thus, it is quite likely that C will be between

$$100 - 2(83.67) \text{ and } 100 + 2(83.67)$$

or

$$-67.34 \text{ and } 267.34$$

Since the lower bound is negative, that end of the interval is meaningless. However, we can still say that it is quite likely that the total cost of the interviewing process will be less than $267.34. ∎

The Geometric Distribution

$$p(y) = p(1 - p)^{y-1} \qquad y = 1, 2, \ldots \quad \text{for } 0 < p < 1$$

$$E(Y) = \frac{1}{p} \qquad V(Y) = \frac{1-p}{p^2}$$

3.6 The Negative Binomial Distribution

3.6.1 Probability Function

In Section 3.5, we saw that the geometric distribution models the probabilistic behavior of the number of the trial on which the *first success* occurs in a sequence of independent Bernoulli trials. But what if we were interested in the number of the trial for the second success, or the third success, or, in general, the rth success? The distribution governing the probabilistic behavior in these cases is called the *negative binomial distribution*.

Let Y denote the number of the trial on which the rth success occurs in a sequence of independent Bernoulli trials with p denoting the common probability of "success." We can derive the distribution of Y from known facts. Now

$$P(Y = y) = P[\text{first } (y - 1) \text{ trials contain } (r - 1) \text{ successes and } y\text{th trial is a success}]$$
$$= P[\text{first } (y - 1) \text{ trials contain } (r - 1) \text{ successes}] P[y\text{th trial is a success}]$$

Since the trials are independent, the joint probability can be written as a product of probabilities. The first probability statement is identical to that resulting in a binomial model, and hence

$$P(Y = y) = p(y) = \binom{y - 1}{r - 1} p^{r-1}(1 - p)^{y-r} \times p$$

$$= \binom{y - 1}{r - 1} p^r (1 - p)^{y-r} \qquad y = r, r + 1, \ldots$$

EXAMPLE **3.15** As in Example 3.13, 30% of the applicants for a certain position have advanced training in computer programming. Suppose three jobs requiring advanced programming training are open. Find the probability that the third qualified applicant is found on the fifth interview, if the applicants are interviewed sequentially and at random.

Solution   Again, we assume independent trials, with 0.3 being the probability of finding a qualified candidate on any one trial. Let Y denote the number of the trial on which the third qualified candidate is found. Then Y can reasonably be assumed to have a negative binomial distribution, so

$$
\begin{aligned}
P(Y = 5) = p(5) &= \binom{4}{2}(0.3)^3(0.7)^2 \\
&= 6(0.3)^3(0.7)^2 \\
&= 0.079 \quad \blacksquare
\end{aligned}
$$

3.6.2   Mean and Variance

The expected value, or mean, and the variance for a negative binomially distributed Y are found easily by analogy with the geometric distribution. Recall that Y denotes the number of the trial on which the rth success occurs. Let W_1 denote the number of the trial on which the first success occurs; W_2 the number of trials between the first success and the second success, including the trial of the second success; W_3 the number of trials between the second success and the third success; and so forth. The results of the trials are then as shown next (with F standing for failure and S for success):

$$
\underbrace{FF\cdots FS}_{W_1}\ \underbrace{F\cdots FS}_{W_2}\ \underbrace{F\cdots FS}_{W_3}
$$

It is easily observed that $Y = \sum_{i=1}^{r} W_i$, where the W_i's are independent and each has a geometric distribution. By results to be derived in Chapter 5,

$$
E(Y) = \sum_{i=1}^{r} E(W_i) = \sum_{i=1}^{r} \left(\frac{1}{p}\right) = \frac{r}{p}
$$

and

$$
V(Y) = \sum_{i=1}^{r} V(W_i) = \sum_{i=1}^{r} \frac{1-p}{p^2} = \frac{r(1-p)}{p^2}
$$

EXAMPLE   **3.16**   Of a large stockpile of used pumps, 20% are unusable and need repair. A repairman is sent to the stockpile with three repair kits. He selects pumps at random and tests them one at a time. If a pump works, he goes on to the next one. If a pump doesn't work, he uses one of his repair kits on it. Suppose it takes 10 minutes to test a pump that works and 30 minutes to test and repair a pump that doesn't work. Find the expected value and variance of the total time it takes the repairman to use up his three kits.

Solution   Letting Y denote the number of the trial on which the third defective pump is found, we see that Y has a negative binomial distribution with $p = 0.2$. The total time T taken to use up the three repair kits is

$$T = 10(Y - 3) + 3(30) = 10Y + 3(20)$$

(Each test takes 10 minutes, but the repair takes 20 extra minutes.)
It follows that

$$E(T) = 10E(Y) + 3(20)$$
$$= 10\left(\frac{3}{0.2}\right) + 3(20)$$
$$= 150 + 60 = 210$$

and

$$V(T) = (10)^2 V(Y)$$
$$= (10)^2 \left[\frac{3(0.8)}{(0.2)^2}\right]$$
$$= 100(60)$$
$$= 6{,}000$$

Thus, the total time to use up the kits has an expected value of 210 minutes, with a standard deviation of $\sqrt{6{,}000} = 77.46$ minutes. ∎

The negative binomial distribution is used to model a wide variety of phenomena, from number of defects per square yard in fabrics to number of individuals in an insect population after many generations.

The Negative Binomial Distribution

$$p(y) = \binom{y-1}{r-1} p^r (1-p)^{y-r} \qquad y = r, r+1, \ldots \qquad \text{for } 0 < p < 1$$

$$E(Y) = \frac{r}{p} \qquad V(Y) = \frac{r(1-p)}{p^2}$$

Exercises

3.38 Let Y denote a random variable having a geometric distribution, with probability of success on any trial denoted by p.

a Find $P(Y \geq 2)$ if $p = 0.1$.

b Find $P(Y > 4 | Y > 2)$ for general p. Compare the result with the unconditional probability $P(Y > 2)$.

3.39 Let Y denote a negative binomial random variable with $p = 0.4$. Find $P(Y \geq 4)$ if

a $r = 2$ **b** $r = 4$

3.40 Suppose 10% of the engines manufactured on a certain assembly line are defective. If engines are randomly selected one at a time and tested, find the probability that the first nondefective engine is found on the second trial.

3.41   Refer to Exercise 3.40. Find the probability that the third nondefective engine is found

    a   on the fifth trial.

    b   on or before the fifth trial.

3.42   Refer to Exercise 3.40. Given that the first two engines are defective, find the probability that at least two more engines must be tested before the first nondefective is found.

3.43   Refer to Exercise 3.40. Find the mean and variance of the number of the trial on which

    a   the first nondefective engine is found.

    b   the third nondefective engine is found.

3.44   The employees of a firm that manufactures insulation are being tested for indications of asbestos in their lungs. The firm is requested to send three employees who have positive indications of asbestos to a medical center for further testing. If 40% of the employees have positive indications of asbestos in their lungs, find the probability that ten employees must be tested to find three positives.

3.45   Refer to Exercise 3.44. If each test costs $20, find the expected value and variance of the total cost of conducting the tests to locate three positives. Do you think it is highly likely that the cost of completing these tests would exceed $350?

3.46   If one-third of the persons donating blood at a clinic have O^+ blood, find the probability that

    a   the first O^+ donor is the fourth donor of the day.

    b   the second O^+ donor is the fourth donor of the day.

3.47   A geological study indicates that an exploratory oil well drilled in a certain region should strike oil with probability 0.2. Find the probability that

    a   the first strike of oil comes on the third well drilled.

    b   the third strike of oil comes on the fifth well drilled.

What assumptions are necessary for your answers to be correct?

3.48   In the setting of Exercise 3.47, suppose a company wants to set up three producing wells. Find the expected value and variance of the number of wells that must be drilled to find three successful ones.

3.49   A large lot of tires contains 10% defectives. Four are to be chosen for placement on a car.

    a   Find the probability that six tires must be selected from the lot to get four good ones.

    b   Find the expected value and variance of the number of selections that must be made to get four good tires.

3.50   The telephone lines coming into an airline reservation office are all occupied about 60% of the time.

    a   If you are calling this office, what is the probability that you complete your call on the first try? the second try? the third try?

    b   If you and a friend both must complete separate calls to this reservation office, what is the probability that it takes a total of four tries for the two of you to complete your calls?

3.51   An appliance comes in two colors, white and brown, which are in equal demand. A certain dealer in these appliances has three of each color in stock, although this is not known to the customers. Customers arrive and independently order these appliances. Find the probability that

    a   the third white is ordered by the fifth customer.

    b   the third brown is ordered by the fifth customer.

    c   all of the whites are ordered before any of the browns.

    d   all of the whites are ordered before all of the browns.

3.7 The Poisson Distribution

3.7.1 Probability Function

A number of probability distributions come about through the application of limiting arguments to other distributions. One such very useful distribution is the *Poisson distribution*.

Consider the development of a probabilistic model for the number of accidents occurring at a particular highway intersection over a period of one week. We can think of the time interval as being split up into n subintervals such that

$$P(\text{one accident in a subinterval}) = p$$
$$P(\text{no accidents in a subinterval}) = 1 - p$$

Note that we are assuming that the same value of p holds for all subintervals and that the probability of more than one accident in any one subinterval is zero. If the occurrence of accidents can be regarded as independent from one subinterval to another, then the total number of accidents in the time period (which equals the total number of subintervals containing one accident) will have a binomial distribution.

Although there is no unique way to choose the subintervals and we therefore know neither n nor p, it seems reasonable to assume that as n increases, p should decrease. Thus, we want to look at the limit of the binomial probability distribution as $n \rightarrow \infty$ and $p \rightarrow 0$.

To simplify matters, we take the limit under the restriction that the mean, np in the binomial case, remains constant at a value we will call λ. Now with $np = \lambda$, or $p = \lambda/n$, we have

$$\lim_{n \to \infty} \binom{n}{y} \left(\frac{\lambda}{n}\right)^y \left(1 - \frac{\lambda}{n}\right)^{n-y}$$

$$= \lim_{n \to \infty} \frac{\lambda^y}{y!} \left(1 - \frac{\lambda}{n}\right)^n \frac{n(n-1)\cdots(n-y+1)}{n^y} \left(1 - \frac{\lambda}{n}\right)^{-y}$$

$$= \frac{\lambda^y}{y!} \lim_{n \to \infty} \left(1 - \frac{\lambda}{n}\right)^n \left(1 - \frac{\lambda}{n}\right)^{-y} \left(1 - \frac{1}{n}\right) \left(1 - \frac{2}{n}\right) \cdots \left(1 - \frac{y-1}{n}\right)$$

Noting that

$$\lim_{n \to \infty} \left(1 - \frac{\lambda}{n}\right)^n = e^{-\lambda}$$

and that all other terms involving n tend to unity, we have the limiting distribution

$$p(y) = \frac{\lambda^y}{y!} e^{-\lambda} \qquad y = 0, 1, 2, \ldots$$

Recall that λ denotes the mean number of occurrences in one time period (a week, for the example under consideration), and hence if t nonoverlapping time periods were considered, the mean would be λt.

The above distribution, called the *Poisson with parameter* λ, can be used to model counts involving areas or volumes, as well as those involving time. For example, we may use this distribution to model the number of flaws in a square yard of textile, the number of bacteria colonies in a cubic centimeter of water, or the number of times a machine fails in the course of a workday.

To see how various choices of λ affect the shape of the Poisson probability function, we show graphs of this function for $\lambda = 1$ (Figure 3.11) and $\lambda = 5$ (Figure 3.12). Note that the probability function is very asymmetric for $\lambda = 1$ but is mound-shaped and quite symmetric when λ gets as large as 5. We illustrate the use of the Poisson distribution in the following example.

F I G U R E **3.11**
Poisson Probability Function;
$\lambda \equiv 1$

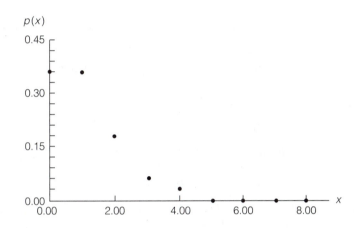

F I G U R E **3.12**
Poisson Probability Function;
$\lambda \equiv 5$

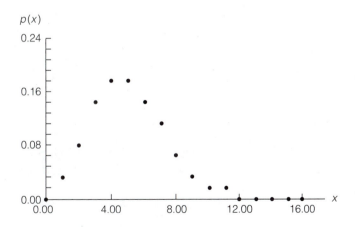

EXAMPLE 3.17 For a certain manufacturing industry, the number of industrial accidents averages three per week. Find the probability that no accidents will occur in a given week.

Solution If accidents tend to occur independently of one another and if they occur at a constant rate over time, the Poisson model provides an adequate representation of the probabilities. Thus,

$$p(0) = \frac{3^0}{0!}e^{-3} = e^{-3} = 0.05 \quad \blacksquare$$

Table 3 of the Appendix gives values for cumulative Poisson probabilities of the form

$$\sum_{y=0}^{a} e^{-\lambda}\frac{\lambda^y}{y!}$$

The following example illustrates the use of Table 3.

EXAMPLE 3.18 Refer to Example 3.17 and let Y denote the number of accidents in the given week. Find $P(Y \leq 4)$, $P(Y \geq 4)$, and $P(Y = 4)$.

Solution From Table 3, we have

$$P(Y \leq 4) = \sum_{y=0}^{4} \frac{(3)^y}{y!}e^{-3} = 0.815$$

Also,

$$P(Y \geq 4) = 1 - P(Y \leq 3)$$
$$= 1 - 0.647 = 0.353$$

and

$$P(Y = 4) = P(Y \leq 4) - P(Y \leq 3)$$
$$= 0.815 - 0.647 = 0.168 \quad \blacksquare$$

3.7.2 Mean and Variance

We can intuitively determine what the mean and variance of a Poisson distribution should be by recalling the mean and variance of a binomial distribution and the relationship between the two distributions. A binomial distribution has mean np and variance $np(1 - p) = np - (np)p$. Now if n gets large and p remains at $np = \lambda$, the variance $np - (np)p = \lambda - \lambda p$ should tend toward λ. In fact, the Poisson distribution does have both mean and variance equal to λ.

The mean of the Poisson distribution is easily derived formally if one remembers the *Taylor series* expansion of e^x, namely,

$$e^x = 1 + x + \frac{x^2}{2!} + \frac{x^3}{3!} + \cdots$$

Then

$$E(Y) = \sum_y y p(y) = \sum_{y=0}^{\infty} y \frac{\lambda^y}{y!} e^{-\lambda} = \sum_{y=1}^{\infty} y \frac{\lambda^y}{y!} e^{-\lambda}$$

$$= \lambda e^{-\lambda} \sum_{y=1}^{\infty} \frac{\lambda^{y-1}}{(y-1)!}$$

$$= \lambda e^{-\lambda} (1 + \lambda + \frac{\lambda^2}{2!} + \frac{\lambda^3}{3!} + \cdots)$$

$$= \lambda e^{-\lambda} e^{\lambda} = \lambda$$

The formal derivation of the fact that

$$V(Y) = \lambda$$

is left as an exercise for the interested reader. [*Hint:* First find $E(Y^2) = E[Y(Y-1)] + E(Y)$.]

EXAMPLE   **3.19**   The manager of an industrial plant is planning to buy a new machine of either type A or type B. For each day's operation, the number of repairs X that machine A requires is a Poisson random variable with mean $0.10t$, where t denotes the time (in hours) of daily operation. The number of daily repairs Y for machine B is a Poisson random variable with mean $0.12t$. The daily cost of operating A is $C_A(t) = 10t + 30X^2$; for B the cost is $C_B(t) = 8t + 30Y^2$. Assume that the repairs take negligible time and that the machines are to be cleaned each night so they operate like new machines at the start of each day. Which machine minimizes the expected daily cost if a day consists of (a) 10 hours? (b) 20 hours?

Solution   The expected cost for A is

$$E[C_A(t)] = 10t + 30E(X^2)$$

$$= 10t + 30[V(X) + (E(X))^2]$$

$$= 10t + 30[0.10t + 0.01t^2]$$

$$= 13t + 0.3t^2$$

Similarly,

$$E[C_B(t)] = 8t + 30E(Y^2)$$

$$= 8t + 30[0.12t + 0.0144t^2]$$

$$= 11.6t + 0.432t^2$$

a
$$E[C_A(10)] = 13(10) + 0.3(10)^2 = 160$$
and
$$E[C_B(10)] = 11.6(10) + 0.432(10)^2 = 159.2$$

which results in the choice of machine B.

b
$$E[C_A(20)] = 380$$
and
$$E[C_B(20)] = 404.8$$

which results in the choice of machine A.

In conclusion, B is more economical for short time periods because of its smaller hourly operating cost. However, for long time periods A is more economical because it needs to be repaired less frequently. ∎

The Poisson Distribution

$$p(y) = \frac{\lambda^y}{y!}e^{-\lambda} \qquad y = 0, 1, 2, \ldots$$

$$E(Y) = \lambda \qquad V(Y) = \lambda$$

Exercises

3.52 Let Y denote a random variable having a Poisson distribution with mean $\lambda = 2$. Find the following.

a $P(Y = 4)$ **b** $P(Y \geq 4)$

c $P(Y < 4)$ **d** $P(Y \geq 4 | Y \geq 2)$

3.53 The number of telephone calls coming into the central switchboard of an office building averages four per minute.

a Find the probability that no calls will arrive in a given 1-minute period.

b Find the probability that at least two calls will arrive in a given 1-minute period.

c Find the probability that at least two calls will arrive in a given 2-minute period.

3.54 The quality of computer disks is measured by sending the disks through a certifier that counts the number of missing pulses. A certain brand of computer disk has averaged 0.1 missing pulse per disk.

a Find the probability that the next inspected disk will have no missing pulse.

b Find the probability that the next inspected disk will have more than one missing pulse.

c   Find the probability that neither of the next two inspected disks will contain any missing pulse.

3.55   The National Maximum Speed Limit (NMSL) of 55 miles per hour has been in force in the United States since early 1974. The benefits of this law have been studied by D. B. Kamerud (*Transportation Research*, 17A, no. 1, 1983, pp. 51–64), who reports that the fatality rate for interstate highways with the NMSL in 1975 is approximately 16 per 10^9 vehicle miles.

a   Find the probability of at most 15 fatalities occurring in 10^9 vehicle miles.

b   Find the probability of at least 20 fatalities occurring in 10^9 vehicle miles.

(Assume that the number of fatalities per vehicle mile follows a Poisson distribution.)

3.56   In the article cited in Exercise 3.55, the projected fatality rate for 1975 if the NMSL had not been in effect was 25 per 10^9 vehicle miles. Under these conditions,

a   find the probability of at most 15 fatalities occurring in 10^9 vehicle miles.

b   find the probability of at least 20 fatalities occurring in 10^9 vehicle miles.

c   compare the answers in parts (a) and (b) to those in Exercise 3.55.

3.57   In a time-sharing computer system, the number of teleport inquiries averages 0.2 per millisecond and follows a Poisson distribution.

a   Find the probability that no inquiries are made during the next millisecond.

b   Find the probability that no inquiries are made during the next 3 milliseconds.

3.58   Rebuilt ignition systems leave an aircraft rework facility at the rate of three per hour on the average. The assembly line needs four ignition systems in the next hour. What is the probability that they will be available?

3.59   Customer arrivals at a checkout counter in a department store have a Poisson distribution with an average of eight per hour. For a given hour, find the probability that

a   exactly eight customers arrive.

b   no more than three customers arrive.

c   at least two customers arrive.

3.60   Refer to Exercise 3.59. If it takes approximately 10 minutes to service each customer, find the mean and variance of the total service time connected to the customer arrivals for 1 hour. (Assume that an unlimited number of servers are available, so no customer has to wait for service.) Is it highly likely that total service time would exceed 200 minutes?

3.61   Refer to Exercise 3.59. Find the probability that exactly two customers arrive in the 2-hour period of time

a   between 2:00 P.M. and 4:00 P.M. (one continuous 2-hour period).

b   between 1:00 P.M. and 2:00 P.M. and between 3:00 P.M. and 4:00 P.M. (two separate 1-hour periods for a total of 2 hours).

3.62   The number of imperfections in the weave of a certain textile has a Poisson distribution with a mean of four per square yard.

a   Find the probability that a 1-square-yard sample will contain at least one imperfection.

b   Find the probability that a 3-square-yard sample will contain at least one imperfection.

3.63   Refer to Exercise 3.62. The cost of repairing the imperfections in the weave is $10 per imperfection. Find the mean and standard deviation of the repair costs for an 8-square-yard bolt of the textile in question.

3.64   The number of bacteria colonies of a certain type in samples of polluted water has a Poisson distribution with a mean of two per cubic centimeter.

a   If four 1-cubic-centimeter samples are independently selected from this water, find the probability that at least one sample will contain one or more bacteria colonies.

b   How many 1-cubic-centimeter samples should be selected to have a probability of approximately 0.95 of seeing at least one bacteria colony?

3.65   Let Y have a Poisson distribution with mean λ. Find $E[Y(Y-1)]$ and use the result to show that $V(Y) = \lambda$.

3.66   A food manufacturer uses an extruder (a machine that produces bite-size foods such as cookies and many snack foods) that produces revenue for the firm at the rate of $200 per hour when in operation. However, the extruder breaks down an average of two times for every 10 hours of operation. If Y denotes the number of breakdowns during the time of operation, the revenue generated by the machine is given by

$$R = 200t - 50Y^2$$

where t denotes hours of operation. The extruder is shut down for routine maintenance on a regular schedule and operates like a new machine after this maintenance. Find the optimal maintenance interval t_0 to maximize the expected revenue between shutdowns.

3.67   The number of cars entering a parking lot is a random variable having a Poisson distribution with a mean of 4 per hour. The lot holds only 12 cars.

a   Find the probability that the lot fills up in the first hour. (Assume all cars stay in the lot longer than 1 hour.)

b   Find the probability that fewer than 12 cars arrive during an 8-hour day.

3.8 The Hypergeometric Distribution

3.8.1   Probability Function

The distributions already discussed in this chapter have as their basic building block a series of *independent* Bernoulli trials. The examples, such as sampling from large lots, depict situations in which the trials of the experiment generate, for all practical purposes, independent outcomes.

Suppose we have a relatively small lot consisting of N items, of which k are defective. If two items are sampled sequentially, then the outcome for the second draw is very much influenced by what happened on the first draw, provided that the first item drawn remains out of the lot. A new distribution must be developed to handle this situation involving *dependent* trials.

In general, suppose a lot consists of N items, of which k are of one type (called successes) and $N - k$ are of another type (called failures). Suppose n items are sampled randomly and sequentially from the lot, with none of the sampled items being replaced. (This is called *sampling without replacement*.) Let $X_i = 1$ if the ith draw results in a success and $X_i = 0$ otherwise, $i = 1, \ldots, n$, and let Y denote the total number of successes among the n sampled items. To develop the probability distribution for Y, let us start by looking at a special case for $Y = y$. One way for y successes to occur is to have

$$X_1 = 1, X_2 = 1, \ldots, X_y = 1, X_{y+1} = 0, \ldots, X_n = 0$$

We know that

$$P(X_1 = 1, X_2 = 2) = P(X_1 = 1)P(X_2 = 1 | X_1 = 1)$$

and this result can be extended to give

$$P(X_1 = 1, X_2 = 1, \ldots, X_y = 1, X_{y+1} = 0, \ldots, X_n = 0)$$

assumed that two will have improperly drilled holes. Five gear boxes must be selected from the 20 available for installation in the next five robots in line.

a Find the probability that all five gear boxes will fit properly.

b Find the expected value, variance, and standard deviation of the time it takes to install these five gear boxes.

Solution **a** In this problem, $N = 20$ and the number of nonconforming boxes is assumed to be $k = 2$, according to the manufacturer's usual standards. Let Y denote the number of nonconforming boxes (the number with improperly drilled holes) in the sample of five. Then

$$P(Y = 0) = \frac{\binom{2}{0}\binom{18}{5}}{\binom{20}{5}}$$

$$= \frac{(1)(8{,}568)}{15{,}504} = 0.55$$

b The total time T taken to install the boxes (in minutes) is

$$T = 10Y + (5 - Y)$$
$$= 9Y + 5$$

since each of Y nonconforming boxes takes 10 minutes to install, and the others take only 1 minute. To find $E(T)$ and $V(T)$, we first need $E(Y)$ and $V(Y)$.

$$E(Y) = n\left(\frac{k}{N}\right) = 5\left(\frac{2}{20}\right) = 0.5$$

and

$$V(Y) = n\left(\frac{k}{N}\right)\left(1 - \frac{k}{N}\right)\left(\frac{N-n}{N-1}\right)$$
$$= 5(0.1)(1 - 0.1)\left(\frac{20-5}{20-1}\right)$$
$$= 0.355$$

It follows that

$$E(T) = 9E(Y) + 5$$
$$= 9(0.5) + 5 = 9.5$$

and

$$V(T) = (9)^2 V(Y)$$
$$= 81(0.355) = 28.755$$

Thus, installation time should average 9.5 minutes, with a standard deviation of $\sqrt{28.755} = 5.4$ minutes.   ∎

The Hypergeometric Distribution

$$p(y) = \frac{\binom{k}{y}\binom{N-k}{n-y}}{\binom{N}{n}} \qquad y = 0, 1, \ldots, k, \ \text{with} \binom{b}{a} = 0 \text{ if } a > b$$

$$E(Y) = n\frac{k}{N} \qquad V(Y) = n\frac{k}{N}\left(1 - \frac{k}{N}\right)\left(\frac{N-n}{N-1}\right)$$

Exercises

3.68 From a box containing four white and three red balls, two balls are selected at random without replacement. Find the probability that

  a exactly one white ball is selected.

  b at least one white ball is selected.

  c two white balls are selected given that at least one white ball is selected.

  d the second ball drawn is white.

3.69 A warehouse contains ten printing machines, four of which are defective. A company randomly selects five of the machines for purchase. What is the probability that all five of the machines are nondefective?

3.70 Refer to Exercise 3.69. The company purchasing the machines returns the defective ones for repair. If it costs $50 to repair each machine, find the mean and variance of the total repair cost. In what interval would you expect the repair costs on these five machines to lie? [*Hint:* Use Tchebysheff's Theorem.]

3.71 A corporation has a pool of six firms, four of which are local, from which it can purchase certain supplies. If three firms are randomly selected without replacement, find the probability that

  a at least one selected firm is not local.

  b all three selected firms are local.

3.72 A foreman has ten employees from whom he must select four to perform a certain undesirable task. Among the ten employees, three belong to a minority ethnic group. The foreman selected all three minority employees (plus one other) to perform the undesirable task. The members of the minority group then protested to the union steward that they had been discriminated against by the foreman.

  The foreman claimed that the selection was completely at random. What do you think?

3.73   Specifications call for a type of thermistor to test out at between 9,000 and 10,000 ohms at 25°C. From ten thermistors available, three are to be selected for use. Let Y denote the number among the three that do not conform to specifications. Find the probability distribution for Y (tabular form) if

a   the ten contain two thermistors not conforming to specifications.

b   the ten contain four thermistors not conforming to specifications.

3.74   Used photocopying machines are returned to the supplier, cleaned, and then sent back out on lease agreements. Major repairs are not made, and, as a result, some customers receive malfunctioning machines. Among eight used photocopiers in supply today, three are malfunctioning. A customer wants to lease four of these machines immediately. Hence four machines are quickly selected and sent out, with no further checking. Find the probability that the customer receives

a   no malfunctioning machines.

b   at least one malfunctioning machine.

c   three malfunctioning machines.

3.75   An eight-cylinder automobile engine has two misfiring spark plugs. If all four plugs are removed from one side of the engine, what is the probability that the two misfiring ones are among them?

3.76   The "worst-case" requirements are defined in the design objectives for a brand of computer terminal. A quick preliminary test indicates that four out of a lot of ten such terminals failed the "worst-case" requirements. Five of the ten are randomly selected for further testing. Let Y denote the number, among the five, that failed the preliminary test. Find the following.

a $P(Y \geq 1)$ **b** $P(Y \geq 3)$

c $P(Y \geq 4)$ **d** $P(Y \geq 5)$

3.77   An auditor checking the accounting practices of a firm samples three accounts from an accounts receivable list of eight accounts. Find the probability that the auditor sees at least one past-due account if there are

a   two such accounts among the eight.

b   four such accounts among the eight.

c   seven such accounts among the eight.

3.78   A group of six software packages available to solve a linear programming problem has been ranked from 1 to 6 (best to worst). An engineering firm selects two of these packages for purchase without looking at the ratings. Let Y denote the number of packages purchased by the firm that are ranked 3, 4, 5, or 6. Show the probability distribution for Y in tabular form.

3.79   Lot acceptance sampling procedures for an electronics manufacturing firm call for sampling n items from a lot of N items and accepting the lot if $Y \leq c$, where Y is the number of nonconforming items in the sample. From an incoming lot of 20 printer covers, 5 are to be sampled. Find the probability of accepting the lot if $c = 1$ and the actual number of nonconforming covers in the lot is

a   0   **b**   1   **c**   2   **d**   3   **e**   4

3.80   Given the setting and terminology of Exercise 3.79, answer parts (a) through (e) if $c = 2$.

3.81   Two assembly lines (I and II) have the same rate of defectives in their production of voltage regulators. Five regulators are sampled from each line and tested. Among the total of ten tested regulators, there were four defectives. Find the probability that exactly two of the defectives came from line I.

3.9 The Moment-Generating Function[†]

We have seen in earlier sections that if $g(Y)$ is a function of a random variable Y with probability distribution given by $p(y)$, then

$$E[g(Y)] = \sum_y g(y)p(y)$$

A special function with many theoretical uses in probability theory is the expected value of e^{tY}, for a random variable Y; this expected value is called the *moment-generating function* (mgf). We denote mgf's by $M(t)$, and thus

$$M(t) = E(e^{tY})$$

The expected values of powers of a random variable are often called *moments*. Thus, $E(Y)$ is the first moment and $E(Y^2)$ the second moment of Y. One use for the moment-generating function is that it does, in fact, generate moments of Y. When $M(t)$ exists, it is differentiable in a neighborhood of the origin $t = 0$, and derivatives may be taken inside the expectation. Thus,

$$M^{(1)}(t) = \frac{dM(t)}{dt} = \frac{d}{dt} E\left[e^{tY}\right]$$
$$= E\left[\frac{d}{dt}e^{tY}\right] = E\left[Ye^{tY}\right]$$

Now if we set $t = 0$, we have

$$M^{(1)}(0) = E(Y)$$

Going on to the second derivative,

$$M^{(2)}(t) = E(Y^2 e^{tY})$$

and

$$M^{(2)}(0) = E(Y^2)$$

In general,

$$M^{(k)}(0) = E(Y^k)$$

It is often easier to evaluate $M(t)$ and its derivatives than to find the moments of the random variable directly. Other theoretical uses of the mgf will be seen in later chapters.

[†]Optional section.

E X A M P L E   **3.22**   Evaluate the moment-generating function for the geometric distribution and use it to find the mean and variance of this distribution.

Solution   We have, for the geometric random variable Y,

$$M(t) = E(e^{tY}) = \sum_{y=1}^{\infty} e^{ty} p(1-p)^{y-1}$$

$$= pe^t \sum_{y=1}^{\infty} (1-p)^{y-1}(e^t)^{y-1}$$

$$= pe^t \sum_{y=1}^{\infty} [(1-p)e^t]^{y-1}$$

$$= pe^t \{1 + [(1-p)e^t] + [(1-p)e^t]^2 + \cdots$$

$$= pe^t \left[\frac{1}{1-(1-p)e^t} \right]$$

since the series is geometric with common ratio e^t.

To evaluate the mean, we have

$$M^{(1)}(t) = \frac{[1-(1-p)e^t]pe^t - pe^t[-(1-p)e^t]}{[1-(1-p)e^t]^2}$$

$$= \frac{pe^t}{[1-(1-p)e^t]^2}$$

and

$$M^{(1)}(0) = \frac{p}{[1-(1-p)]^2} = \frac{1}{p}$$

To evaluate the variance, we first need

$$E(Y^2) = M^{(2)}(0)$$

Now

$$M^{(2)}(t) = \frac{[1-(1-p)e^t]^2 pe^t - pe^t \{2[1-(1-p)e^t](-1)(1-p)e^t\}}{[1-(1-p)e^t]^4}$$

and

$$M^{(2)}(0) = \frac{p^3 + 2p^2(1-p)}{p^4} = \frac{p + 2(1-p)}{p^2}$$

Hence,

$$V(Y) = E(Y^2) - [E(Y)]^2$$

$$= \frac{p + 2(1-p)}{p^2} - \frac{1}{p^2} = \frac{1-p}{p^2} \quad \blacksquare$$

Moment-generating functions have some very important properties that make them extremely useful in finding expected values and determining the probability distributions of random variables. These properties will be discussed in detail in Chapters 4 and 5; one such property is given below in Exercise 3.85.

Exercises

3.82   Find the moment-generating function for the Bernoulli random variable.

3.83   Show that the moment-generating function for the binomial random variable is given by

$$M(t) = [pe^t + (1 - p)]^n$$

Use this result to derive the mean and variance for the binomial distribution.

3.84   Show that the moment-generating function for the Poisson random variable with mean λ is given by

$$M(t) = e^{\lambda(e^t - 1)}$$

Use this result to derive the mean and variance for the Poisson distribution.

3.85   If X is a random variable with moment-generating function $M(t)$, and Y is a function of X given by $Y = aX + b$, show that the moment-generating function for Y is $e^{tb} M(at)$.

3.86   Use the result of Exercise 3.85 to show that

$$E(Y) = aE(X) + b$$

and

$$V(Y) = a^2 V(X)$$

3.10   Activities for Students: Simulating Probability Distributions

Computers lend themselves nicely to use in the area of probability. They can be used not only to calculate probabilities, but also to simulate random variables from specified probability distributions. A *simulation* can be thought of as an experiment performed on the computer. Simulation is used to analyze both theoretical and applied problems. A simulated model attempts to copy the behavior of the situation under consideration. Some practical applications of simulation might be to model inventory control problems, queuing systems, production lines, medical systems, and flight patterns of major jets. Simulation can also be used to determine the behavior of a complicated random variable whose precise probability distribution function is difficult to evaluate mathematically.

The generation of observations from a probability distribution is based upon random numbers on the interval [0, 1]. A random number R_i on [0, 1] is one that is selected on the condition that each number between 0 and 1 has the same probability of being selected. A sequence of numbers that appears to follow a certain pattern or trend would not be considered random. One way to generate random numbers on

[0, 1] is to take the five-digit groupings of random digits in Table 1 of the Appendix and place decimal points in front of them. Most computer languages have built-in random generators that will give a number on [0, 1]. If a built-in generator is not available, algorithms are available for setting up one.

We will now give a brief description of generating discrete random variables for the distributions discussed in this chapter given that a random number R_i on [0, 1] can be generated.

Bernoulli Let p = probability of success. If $R_i \leq p$, then $X_i = 1$. Otherwise, $X_i = 0$.

Binomial A binomial random variable X_i can be expressed as the sum of n independent Bernoulli random variables Y_j; that is, $X_i = \sum Y_j$, $j = 1, \ldots, n$. Thus, to simulate X_i with parameters n and p, simulate n Bernoulli random variables as stated previously. X_i = sum of the n Bernoulli variables.

Geometric Let X_i = the number of trials necessary for the first success with p = probability of success. $X_i = m$, where m = the number of R_i's generated until the condition $R_i \leq p$ is met.

Negative Binomial Let X_i = the number of trials necessary until the rth success with p = probability of success. A negative binomial random variable X_i can be expressed as the sum of r independent geometric random variables Y_j; that is, $X_i = \sum Y_j$, $j = 1, \ldots, r$. Thus, to simulate X_i with parameter p, simulate r geometric random variables as stated previously. X_i = sum of the r geometric variables.

Poisson Generating Poisson random variables will be discussed at the end of Chapter 4. Let's consider some simple examples of possible uses for simulating discrete random variables. Suppose that n_1 items are to be inspected from one production line and n_2 items are to be inspected from another production line. Let p_1 = probability of a defective from line 1 and p_2 = probability of a defective from line 2. Let X be a binomial random variable with parameters n_1 and p_1. Let Y be a binomial random variable with parameters n_2 and p_2. A variable of interest is W, the total number of defective items observed in both production lines. Let $W = X + Y$. Unless $p_1 = p_2$, the distribution of W will not be binomial. To see how the distribution of W will behave, a simulation could be performed. Useful information could be obtained from the simulation by looking at a histogram of the W_i's generated and also by considering the values of the sample mean and sample variance. Let's consider the following random variables X and Y: X is binomial with $n_1 = 7$ and $p_1 = 0.2$, and Y is binomial with $n_1 = 8$ and $p_2 = 0.6$. Defining $W = X + Y$, we performed a simulation that produced the histogram shown in Figure 3.13.

The sample mean was 6.2, with a sample standard deviation of 1.76. In Chapter 5, we will be able to show that these values are very close to the expected value of $\mu_w = 6.2$ and $\sigma_w = 1.74$. (This calculation will be possible after linear functions of random variables are discussed.) From the histogram, the probability that the total number of defective items is at least nine is given by 0.09.

Another example of interest is the coupon-collector problem, which incorporates the geometric distribution. Suppose there are n distinct colors of coupons. Each time

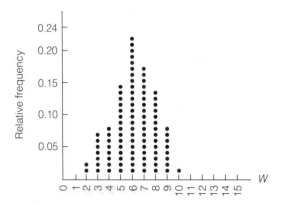

an individual obtains a coupon, we assume that it is equally likely to be any one of the n colors and that the selection of a coupon is independent of a previously obtained coupon. Suppose one can redeem a set of coupons for a prize if each possible color of coupon is represented in a set. We define the random variable X = the number of necessary coupons to be selected in order that one has a complete set of each color coupon. Questions of interest might include the following:

1   What is the expected number of coupons needed in order to obtain this complete set; that is, what is $E(X)$?

2   What is the standard deviation of X?

3   What is the probability that one must select at most x coupons to obtain this complete set?

Instead of answering the above questions by deriving the distribution function of X, we might try simulation. Two simulations are presented in Figures 3.14 and 3.15. The first histogram (Figure 3.14) represents a simulation where n, the number of different-color coupons, is equal to 5. The sample mean was computed to be 11.06, with a sample standard deviation of 4.65. Suppose we are interested in finding $P(X \leq 10)$. Using the results of the simulation, we get a relative frequency probability of 0.555. It might also be noted that, from this simulation, the largest number of coupons needed to obtain the complete set was 31.

The second histogram (Figure 3.15) represents a simulation where $n = 10$. The sample mean was computed to be 29.07, with a sample standard deviation of 10.16. $P(X \leq 20) = 0.22$ from the simulated values. In this simulation, the largest number of coupons necessary to obtain the complete set was 71.

Binomial Proportions

Count data are often more conveniently discussed as a proportion, as in the proportion of inspected cars that have paint defects or the proportion of sampled voters who favor a certain candidate. For a binomial random variable X with n trials and probability p of success on any one trial, the proportion of successes is X/n.

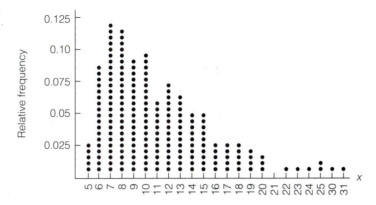

FIGURE **3.14**
Simulating Sums of Geometric
Random Variables

FIGURE **3.15**   Simulating Sums of Geometric Random Variables

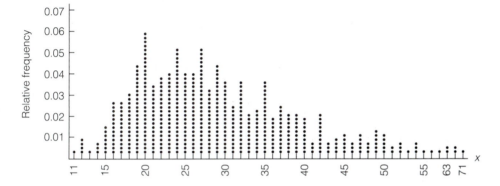

Generate 100 values of X/n and plot them on a line plot or histogram for each of the following cases:

1 $n = 20, p = 0.2$

2 $n = 40, p = 0.2$

3 $n = 20, p = 0.5$

4 $n = 40, p = 0.5$

Compare the four plots and discuss any patterns seen in their centering, variation, and symmetry. Can you propose a formula for the variance of X/n?

Waiting for Blood

A local blood bank knows that about 40% of its donors have A^+ blood. It needs $k = 3$ A^+ donors today. Generate a distribution of values for X, the number of donors sequentially tested in order to find the three A^+ donors. What are the approximate expected value and standard deviation of X from your data? Do these values agree with the theory? What is the estimated probability that the blood bank will need to test ten or more people to find the three A^+ donors? How will the answers to these questions change if k is increased to 4?

3.11 Summary

The outcomes of interest in most investigations involving random events are numerical in nature. The simplest type of numbered outcomes to model are counts, such as the number of nonconforming parts in a shipment, the number of sunny days in a month, or the number of water samples that contain a pollutant. One of the amazing results of probability theory is the fact that a small number of theoretical distributions can cover a wide array of applications. Six of the most useful discrete probability distributions were introduced in this chapter.

The *Bernoulli* is simply an indicator random variable, using a numerical code to indicate presence or absence of a characteristic.

The *binomial* random variable counts the number of "successes" among a fixed number, n, of independent trials, each with the same probability of success.

The *geometric* random variable counts the number of trials one needs to conduct, in sequence, until the first "success" is seen.

The *negative binomial* random variable counts the number of trials one needs to conduct until the rth "success" is seen.

The *Poisson* random variable arises from counts in a restricted domain of time, area, or volume and is most useful for counting fairly rare outcomes.

The *hypergeometric* random variable counts the number of "successes" in sampling from a finite population, which makes the sequential selections dependent upon one another.

These theoretical distributions serve as models for real data that might arise in our quest to improve a process. Each involves assumptions that should be checked carefully before application is made to a particular case.

Supplementary Exercises

3.87 Construct probability histograms for the binomial probability distribution for $n = 5$ and $p = 0.1$, 0.5, and 0.9. (Table 2 of the Appendix will reduce the amount of calculation.) Note the symmetry for $p = 0.5$ and the direction of skewness for $p = 0.1$ and 0.9.

3.88 Use Table 2 of the Appendix to construct a probability histogram for the binomial probability distribution for $n = 20$ and $p = 0.5$. Note that almost all the probability falls in the interval $5 \le y \le 15$.

3.89 The probability that a single radar set will detect an airplane is 0.9. If we have five radar sets, what is the probability that exactly four sets will detect the plane? at least one set? (Assume that the sets operate independently of each other.)

3.90 Suppose that the four engines of a commercial aircraft were arranged to operate independently and that the probability of in-flight failure of a single engine is 0.01. What is the probability that, on a given flight,

a no failures are observed?

b no more than one failure is observed?

3.91 Sampling for defectives from large lots of a manufactured product yields a number of defectives Y that follows a binomial probability distribution. A sampling plan involves specifying the number of items to be included in a sample n and an acceptance number a. The lot is accepted if $Y \le a$ and rejected if $Y > a$. Let p denote the proportion of defectives in the lot. For $n = 5$ and $a = 0$, calculate the probability of lot acceptance if

 a $p = 0$ **b** $p = 0.1$ **c** $p = 0.3$

 d $p = 0.5$ **e** $p = 1.0$

A graph showing the probability of lot acceptance as a function of lot fraction defective is called the *operating characteristic curve* for the sample plan. Construct this curve for the plan $n = 5$, $a = 0$. Note that a sampling plan is an example of statistical inference. Accepting or rejecting a lot based on information contained in the sample is equivalent to concluding that the lot is either "good" or "bad," respectively. "Good" implies that a low fraction is defective and that the lot is therefore suitable for shipment.

3.92 Refer to Exercise 3.91. Use Table 2 of the Appendix to construct the operating characteristic curve for a sampling plan with

 a $n = 10, a = 0$ **b** $n = 10, a = 1$ **c** $n = 10, a = 2$

For each plan, calculate P(lot acceptance) for $p = 0, 0.05, 0.1, 0.3, 0.5$, and 1.0. Our intuition suggests that sampling plan (a) would be much less likely to lead to acceptance of bad lots than plans (b) and (c). A visual comparison of the operating characteristic curves will confirm this supposition.

3.93 A quality-control engineer wants to study the alternative sampling plans $n = 5$, $a = 1$ and $n = 25$, $a = 5$. On a sheet of graph paper, construct the operating characteristic curves for both plans; use acceptance probabilities at $p = 0.05$, $p = 0.10$, $p = 0.20$, $p = 0.30$, and $p = 0.40$ in each case.

 a If you were a seller producing lots with fraction defective ranging from $p = 0$ to $p = 0.10$, which of the two sampling plans would you prefer?

 b If you were a buyer wishing to be protected against accepting lots with fraction defective exceeding $p = 0.30$, which of the two sampling plans would you prefer?

3.94 For a certain section of a pine forest, the number of diseased trees per acre Y has a Poisson distribution with mean $\lambda = 10$. The diseased trees are sprayed with an insecticide at a cost of \$3 per tree, plus a fixed overhead cost for equipment rental of \$50. Letting C denote the total spraying cost for a randomly selected acre, find the expected value and standard deviation for C. Within what interval would you expect C to lie with probability at least 0.75?

3.95 In checking river water samples for bacteria, water is placed in a culture medium so that certain bacteria colonies can grow if those bacteria are present. The number of colonies per dish averages 12 for water samples from a certain river.

 a Find the probability that the next dish observed will have at least ten colonies.

 b Find the mean and standard deviation of the number of colonies per dish.

 c Without calculating exact Poisson probabilities, find an interval in which at least 75% of the colony count measurements should lie.

3.96 The number of vehicles passing a specified point on a highway averages ten per minute.

 a Find the probability that at least 15 vehicles pass this point in the next minute.

 b Find the probability that at least 15 vehicles pass this point in the next 2 minutes.

3.97 A production line produces a variable number N of items each day. Suppose each item produced has the same probability p of not conforming to manufacturing standards. If N has a Poisson distribution with mean λ, then the number of nonconforming items in one day's production Y has a Poisson distribution with mean λp. The average number of resistors produced by a facility in one day has a Poisson distribution with mean 100. Typically, 5% of the resistors produced do not meet specifications.

 a Find the expected number of resistors not meeting specifications on a given day.

b   Find the probability that all resistors will meet the specifications on a given day.

c   Find the probability that more than five resistors fail to meet specifications on a given day.

3.98   A certain type of bacteria cell divides at a constant rate λ over time. (That is, the probability that a cell will divide in a small interval of time t is approximately λt.) Given that a population starts out at time zero with k cells of this type and that cell divisions are independent of one another, the size of the population at time t, $Y(t)$, has the probability distribution

$$P[Y(t) = n] = \binom{N-1}{k-1} e^{-\lambda k t} (1 - e^{-\lambda t})^{n-k}$$

a   Find the expected value of $Y(t)$ in terms of λ and t.

b   If, for a certain type of bacteria cell, $\lambda = 0.1$ per second and the population starts out with two cells at time zero, find the expected population size after 5 seconds.

3.99   The probability that any one vehicle will turn left at a particular intersection is 0.2. The left-turn lane at this intersection has room for three vehicles. If five vehicles arrive at this intersection while the light is red, find the probability that the left-turn lane will hold all the vehicles that want to turn left.

3.100   Refer to Exercise 3.99. Find the probability that six cars must arrive at the intersection while the light is red to fill up the left-turn lane.

3.101   For any probability function $p(y)$, $\sum yp(y) = 1$ if the sum is taken over all possible values y that the random variable in question can assume. Show that this is true for

a   the binomial distribution.

b   the geometric distribution.

c   the Poisson distribution.

3.102   The supply office for a large construction firm has three welding units of Brand A in stock. If a welding unit is requested, the probability is 0.7 that the request will be for this particular brand. On a typical day, five requests for welding units come to the office. Find the probability that all three Brand A units will be in use on that day.

3.103   Refer to Exercise 3.102. If the supply office also stocks three welding units that are not Brand A, find the probability that exactly one of these units will be left immediately after the third Brand A unit is requested.

3.104   The probability of a customer arrival at a grocery service counter in any 1-second interval is equal to 0.1. Assume that customers arrive in a random stream and hence that the arrival at any one second is independent of any other.

a   Find the probability that the first arrival will occur during the third 1-second interval.

b   Find the probability that the first arrival will not occur until at least the third 1-second interval.

3.105   Sixty percent of a population of consumers is reputed to prefer Brand A toothpaste. If a group of consumers is interviewed, what is the probability that exactly five people must be interviewed before a consumer is encountered who prefers Brand A? at least five people?

3.106   The mean number of automobiles entering a mountain tunnel per 2-minute period is one. An excessive number of cars entering the tunnel during a brief period of time produces a hazardous situation.

a   Find the probability that the number of autos n entering the tunnel during a 2-minute period exceeds three.

b   Assume that the tunnel is observed during ten 2-minute intervals, thus giving ten independent observations, $Y_1, Y_2, \ldots, Y_{10}$, on a Poisson random variable. Find the probability that $Y > 3$ during at least one of the ten 2-minute intervals.

3.107   Suppose that 10% of a brand of microcomputers will fail before their guarantee has expired. If 1,000 computers are sold this month, find the expected value and variance of Y, the number that have not failed during the guarantee period. Within what limit would Y be expected to fall? [*Hint:* Use Tchebysheff's Theorem.]

3.108 **a** Consider a binomial experiment for $n = 20$, $p = 0.05$. Use Table 2 of the Appendix to calculate the binomial probabilities for $Y = 0, 1, 2, 3, 4$.

b Calculate the same probabilities as in (a), using the Poisson approximation with $\lambda = np$. Compare your results.

3.109 The manufacturer of a low-calorie dairy drink wishes to compare the taste appeal of a new formula (B) with that of the standard formula (A). Each of four judges is given three glasses in random order, two containing formula A and the other containing formula B. Each judge is asked to state which glass he or she most enjoyed. Suppose the two formulas are equally attractive. Let Y be the number of judges stating a preference for the new formula.

a Find the probability function for Y.

b What is the probability that at least three of the four judges state a preference for the new formula?

c Find the expected value of Y.

d Find the variance of Y.

3.110 Show that the hypergeometric probability function approaches the binomial in the limit as $N \to \infty$ and $p = r/N$ remains constant. That is, show that

$$\lim_{N \to \infty} \frac{\binom{r}{y}\binom{N-r}{n-y}}{\binom{N}{n}} = \binom{n}{y} p^y q^{n-y}$$

for $p = r/N$ constant and $q = 1 - p$.

3.111 A lot of $N = 100$ industrial products contains 40 defectives. Let Y be the number of defectives in a random sample of size 20. Find $p(10)$ by using

a the hypergeometric probability distribution.

b the binomial probability distribution.

Is N large enough so that the binomial probability function is a good approximation to the hypergeometric probability function?

3.112 For simplicity, let us assume that there are two kinds of drivers. The safe drivers, which comprise 70 percent of the population, have a probability of 0.1 of causing an accident in a year. The rest of the population consists of accident makers, who have a probability of 0.5 of causing an accident in a year. The insurance premium is $400 times one's probability of causing an accident in the following year. A new subscriber has an accident during the first year. What should be this driver's insurance premium for the next year?

3.113 A merchant stocks a certain perishable item. She knows that on any given day she will have a demand for either two, three, or four of these items with probabilities 0.1, 0.4, and 0.5, respectively. She buys the items for $1.00 each and sells them for $1.20 each. Any that are left at the end of the day represent a total loss. How many items should the merchant stock to maximize her expected daily profit?

3.114 It is known that 5% of a population have disease A, which can be detected by a blood test. Suppose that N (a large number) people are to be tested. This can be done in two ways: Either (1) each person is tested separately or (2) the blood samples of k people are pooled together and analyzed. (Assume that $N = nk$, with n an integer.) In method 2, if the test is negative, all the people in the pool are healthy (that is, just this one test is needed). If the test is positive, each of the k persons must be tested separately (that is, a total of $k + 1$ tests is needed).

a For fixed k, what is the expected number of tests needed in method 2?

b Find the k that will minimize the expected number of tests in method 2.

c How many tests does part (b) save in comparison with part (a)?

3.115   Four possible winning numbers for a lottery—AB-4536, NH-7812, SQ-7855, and ZY-3221—are given to you. You will win a prize if one of your numbers matches one of the winning numbers. You are told that one first prize of $100,000, two second prizes of $50,000 each, and 10 third prizes of $1,000 each will be awarded. The only thing you need to do is mail the coupon back. No purchase is required. From the structure of the numbers you have received, it is obvious that the entire list consists of all the permutations of two letters followed by four digits. Is the coupon worth mailing back for 29¢ postage?

Continuous Probability Distributions

About This Chapter

Random variables that are not discrete, such as measurements of lifelength or weight, can often be classified as *continuous*. This is the second category of random variables we will discuss. Probability distributions commonly used to model continuous random variables are presented in this chapter, along with their means and variances. Applications of continuous models to reliability and other aspects of quality improvement are presented as well.

Contents

[†]Optional section.

4.1 Continuous Random Variables and Their Probability Distributions

All the random variables discussed in Chapter 3 were discrete; each could assume only a finite number or countable infinity of values. But many of the random variables seen in practice have more than a countable collection of possible values. Weights of patients coming into a clinic may be anywhere from, say, 80 to 300 pounds. Diameters of machined rods from a certain industrial process may be anywhere from 1.2 to 1.5 centimeters. Proportions of impurities in ore samples may run from 0.10 to 0.80. These random variables can take on any value in an interval of real numbers. That is not to say that every value in the interval can be found in sample data if one looks long enough; one may never observe a patient weighing exactly 172.38 pounds. Rather, no one value can be ruled out as a possible observation; one could possibly have a patient weighing 172.38 pounds, so this number must be considered in the set of possible outcomes. Since random variables of this type have a continuum of possible values, they are called *continuous random variables*. We will now develop probability distributions for continuous random variables, presenting the basic ideas in the context of an experiment on lifelengths.

An experimenter is measuring the lifelength X of a transistor. In this case, there is an infinite number of possible values that X can assume. We cannot assign a positive probability to each possible outcome of the experiment because, no matter how small we might make the individual probabilities, they would sum to a value greater than 1 when accumulated over the entire sample space. We can, however, assign positive probabilities to *intervals* of real numbers in a manner consistent with the axioms of probability. To introduce the basic ideas involved here, we will consider a specific example in some detail.

Suppose we have conducted an experiment to measure the lifelengths of 50 batteries of a certain type, selected from a larger population of such batteries. The experiment is completed, and the observed lifelengths are as given in Table 4.1.

The relative frequency histogram for these data (Figure 4.1) shows clearly that most of the lifelengths are near zero and that the frequency drops off rather smoothly as we look at longer lifelengths. Here, 32% of the 50 observations fall into the first subinterval and another 22% fall into the second. There is a decline in frequency as

T A B L E **4.1** Lifelengths of Batteries (in hundreds of hours)				
0.406	0.685	4.778	1.725	8.223
2.343	1.401	1.507	0.294	2.230
0.538	0.234	4.025	3.323	2.920
5.088	1.458	1.064	0.774	0.761
5.587	0.517	3.246	2.330	1.064
2.563	0.511	2.782	6.426	0.836
0.023	0.225	1.514	3.214	3.810
3.334	2.325	0.333	7.514	0.968
3.491	2.921	1.624	0.334	4.490
1.267	1.702	2.634	1.849	0.186

F I G U R E **4.1**
Relative Frequency Histogram of
Data from Table 4.1

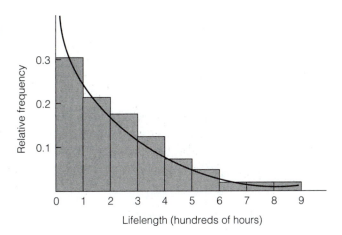

we proceed across the subintervals, until the last subinterval (8 to 9) contains a single observation.

This sample relative frequency histogram not only allows us to picture how the sample behaves but also gives us some insight into a possible probabilistic model for the random variable X. The histogram of Figure 4.1 looks as if it could be approximated quite closely by a negative exponential curve. The particular function

$$f(x) = \frac{1}{2}e^{-x/2} \qquad x > 0$$

is sketched through the histogram in Figure 4.1 and seems to fit reasonably well. Thus, we could take this function as a mathematical model for the behavior of the random variable X. Note that the area under $f(x)$ equals 1. If we want to use a battery of this type in the future, we might want to know the probability that it will last longer than 400 hours. This probability can be approximated by the area under the curve to the right of the value 4, that is, by

$$\int_4^\infty \frac{1}{2}e^{-x/2}\, dx = 0.135$$

Note that this figure is quite close to the observed sample fraction of lifetimes that exceed 4, namely, $8/50 = 0.16$. One might suggest that, since the sample fraction 0.16 is available, we do not really need the model. But there are other questions for which the model would give more satisfactory answers than could otherwise be obtained. For example, suppose we are interested in the probability that X is greater than 9. Then the model suggests the answer

$$\int_9^\infty \frac{1}{2}e^{-x/2}\, dx = 0.011$$

whereas the sample shows no observations in excess of 9. These example are quite simple, but we will see many examples of more involved questions for which a model is essential.

Why did we choose the exponential function as a model here? Wouldn't some other models do just as well? The choice of a model is a fundamental problem, and we will spend considerable time in later sections delving into theoretical and practical reasons for such choices. In this early discussion, we will merely suggest some models that look as if they might do the job.

The function $f(x)$, which models the relative frequency behavior of X, is called the *probability density function*.

DEFINITION   **4.1**

A random variable X is said to be **continuous** if it can take on the infinite number of possible values associated with intervals of real numbers and if there is a function $f(x)$, called the **probability density function**, such that

1   $f(x) \geq 0$ for all x
2   $\int_{-\infty}^{\infty} f(x)\,dx = 1$
3   $P(a \leq X \leq b) = \int_{a}^{b} f(x)\,dx$ ∎

Note that for a continuous random variable X,

$$P(X = a) = \int_{a}^{a} f(x)\,dx = 0$$

for any specific value a. The fact that we must assign zero probability to any specific value should not disturb us, since there is an infinite number of possible values that X can assume. For example, out of all the possible values that the lifelength of a transistor can take on, what is the probability that the transistor we are using will last exactly 497.392 hours? Assigning probability zero to this event does not rule out 497.392 as a possible lifelength, but it does say that the chance of observing this particular lifelength is extremely small.

EXAMPLE   **4.1**

Refer to the random variable X of the lifelength example, which has associated with it a probability density function of the form

$$f(x) = \begin{cases} \dfrac{1}{2}e^{-x/2} & x > 0 \\ 0 & \text{elsewhere} \end{cases}$$

Find the probability that the lifelength of a particular battery of this type is less than 200 or greater than 400 hours.

Solution   Let A denote the event that X is less than 2 and B the event that X is greater than 4. Then since A and B are mutually exclusive,

$$\begin{aligned}
P(A \cup B) &= P(A) + P(B) \\
&= \int_{0}^{2} \frac{1}{2}e^{-x/2}\,dx + \int_{4}^{\infty} \frac{1}{2}e^{-x/2}\,dx \\
&= (1 - e^{-1}) + (e^{-2}) \\
&= 1 - 0.368 + 0.135 \\
&= 0.767 \quad ∎
\end{aligned}$$

E X A M P L E   **4.2**   Refer to Example 4.1. Find the probability that a battery of this type lasts more than 300 hours given that it has already been in use for more than 200 hours.

Solution   We are interested in $P(X > 3 | X > 2)$ and, by the definition of conditional probability, we have

$$P(X > 3 | X > 2) = \frac{P(X > 3)}{P(X > 2)}$$

since the intersection of the events $(X > 3)$ $(X > 2)$ is the event $(X > 3)$. Now

$$\frac{P(X > 3)}{P(X > 2)} = \frac{\int_3^\infty \frac{1}{2}e^{-x/2}\,dx}{\int_2^\infty \frac{1}{2}e^{-x/2}\,dx} = \frac{e^{-3/2}}{e^{-1}} = e^{-1/2} = 0.606 \quad \blacksquare$$

Sometimes it is convenient to look at cumulative probabilities of the form $P(X \leq b)$. To do this, we can make use of the distribution function.

D E F I N I T I O N   **4.2**   The **distribution function** for a random variable X is defined as

$$F(b) = P(X \leq b)$$

If X is continuous with probability density function $f(x)$, then

$$F(b) = \int_{-\infty}^b f(x)\,dx$$

Note that $F'(x) = f(x)$. $\blacksquare$

In the lifelength example, X has a probability density function given by

$$f(x) = \begin{cases} \dfrac{1}{2}e^{-x/2} & x > 0 \\ 0 & \text{elsewhere} \end{cases}$$

Thus,

$$F(b) = P(X \leq b) = \int_0^b \frac{1}{2}e^{-x/2}\,dx$$
$$= -e^{-x/2}\big|_0^b$$
$$= 1 - e^{-b/2} \quad b > 0$$
$$= 0 \quad\quad b \leq 0$$

The function is shown graphically in Figure 4.2.

F I G U R E **4.2**
A Distribution Function for a
Continuous Random Variable

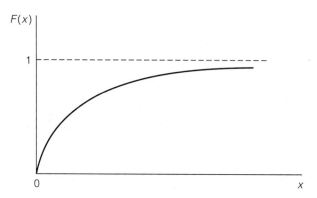

F I G U R E **4.2**
A Distribution Function for a
Continuous Random Variable

F I G U R E **4.3**
$f(x)$ for Example 4.3

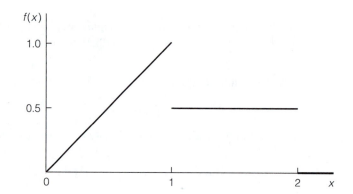

E X A M P L E   **4.3**   A supplier of kerosene has a 200-gallon tank filled at the beginning of each week. His weekly demands show a relative frequency behavior that increases steadily up to 100 gallons and then levels off between 100 and 200 gallons. Letting X denote weekly demand in hundreds of gallons, suppose the relative frequencies for demand are modeled adequately by

$$f(x) = \begin{cases} 0 & x < 0 \\ x & 0 \le x \le 1 \\ 1/2 & 1 < x \le 2 \\ 0 & x > 2 \end{cases}$$

This function has the graphical form shown in Figure 4.3. Find $F(b)$ for this random variable. Use $F(b)$ to find the probability that demand will exceed 150 gallons on a given week.

Solution   From the definition,

$$F(b) = \int_{-\infty}^{b} f(x)\,dx$$

$$= 0 \qquad\qquad b < 0$$

$$= \int_{0}^{b} x\,dx = \frac{b^2}{2} \bigg| \qquad 0 \le b \le 1$$

$$= \frac{1}{2} + \int_{1}^{b} \frac{1}{2}\,dx$$

$$= \frac{1}{2} + \frac{b-1}{2} = \frac{b}{2} \qquad 1 < b \le 2$$

$$= 1 \qquad\qquad b > 2$$

This function is graphed in Figure 4.4. Note that $F(b)$ is continuous over the whole real line, even though $f(b)$ has two discontinuities. The probability that demand will exceed 150 gallons is given by

$$P(X > 1.5) = 1 - P(X \le 1.5) = 1 - F(1.5)$$

$$= 1 - \frac{1.5}{2} = 0.25 \quad \blacksquare$$

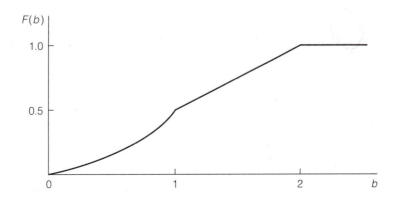

FIGURE **4.4**
$F(b)$ for Example 4.3

Exercises

4.1   For each of the following situations, define an appropriate random variable and state whether it is continuous or discrete.

   a   An environmental engineer is looking at ten field plots to determine whether or not they contain a certain type of insect.

b   A quality-control technician samples a continuously produced fabric in square-yard sections and counts the number of defects she observes for each sampled section.

c   A metallurgist counts the number of grains seen in a cross-sectional sample of aluminum.

d   The metallurgist of part (c) measures the area proportion covered by grains of a certain size rather than simply counting them.

4.2   Suppose a random variable X has a probability density function given by

$$f(x) = \begin{cases} kx(1-x) & 0 \le x \le 1 \\ 0 & \text{elsewhere} \end{cases}$$

a   Find the value of k that makes this a probability density function.

b   Find $P(0.4 \le X \le 1)$.

c   Find $P(X \le 0.4 | X \le 0.8)$.

d   Find $F(b) = P(X \le b)$ and sketch the graph of this function.

4.3   The effectiveness of solar-energy heating units depends on the amount of radiation available from the sun. For a typical October, daily total solar radiation in Tampa, Florida, approximately follows the probability density function given below (units are hundreds of calories):

$$f(x) = \begin{cases} \dfrac{3}{32}(x-2)(6-x) & 2 \le x \le 6 \\ 0 & \text{elsewhere} \end{cases}$$

a   Find the probability that solar radiation will exceed 300 calories on a typical October day.

b   What amount of solar radiation is exceeded on exactly 50% of the October days, according to this model?

4.4   An accounting firm that does not have its own computing facilities rents time from a consulting company. The firm must plan its computing budget carefully and hence has studied the weekly use of CPU time quite thoroughly. The weekly use of CPU time approximately follows the probability density function given below (measurements in hours):

$$f(x) = \begin{cases} \dfrac{3}{64}x^2(4-x) & 0 \le x \le 4 \\ 0 & \text{elsewhere} \end{cases}$$

a   Find the distribution function $F(x)$ for weekly CPU time X.

b   Find the probability that CPU time used by the firm will exceed 2 hours for a selected week.

c   The current budget of the firm covers only 3 hours of CPU time per week. How often will the budgeted figure be exceeded?

d   How much CPU time should be budgeted per week if this figure is to be exceeded with probability only 0.10?

4.5   The pH, a measure of the acidity of water, is important in studies of acid rain. For a certain Florida lake, baseline measurements on acidity are made so any changes caused by acid rain can be noted. The pH of water samples from a lake is a random variable X with probability density function

$$f(x) = \begin{cases} \dfrac{3}{8}(7-x)^2 & 5 \le x \le 7 \\ 0 & \text{elsewhere} \end{cases}$$

a   Sketch the curve of $f(x)$.

b   Find the distribution function $F(x)$ for X.

c   Find the probability that the pH will be less than 6 for a water sample from this lake.

d Find the probability that the pH of a water sample from this lake will be less than 5.5 given that it is known to be less than 6.

4.6 The "on" temperature of a thermostatically controlled switch for an air conditioning system is set at 60°, but the actual temperature X at which the switch turns on is a random variable having probability density function

$$f(x) = \begin{cases} \dfrac{1}{2} & 59 \leq x \leq 61 \\ 0 & \text{elsewhere} \end{cases}$$

a Find the probability that it takes a temperature in excess of 60° to turn the switch on.

b If two such switches are used independently, find the probability that both require a temperature in excess of 60° to turn on.

4.7 The proportion of time, during a 40-hour workweek, that an industrial robot was in operation was measured for a large number of weeks, and the measurements can be modeled by the probability density function

$$f(x) = \begin{cases} 2x & 0 \leq x \leq 1 \\ 0 & \text{elsewhere} \end{cases}$$

If X denotes the proportion of time this robot will be in operation during a coming week, find the following:

a $P(X > 1/2)$ **b** $P(X > 1/2 \,|\, X > 1/4)$

c $P(X > 1/4 \,|\, X > 1/2)$ **d** Find $F(x)$ and graph this function. Is $F(x)$ continuous?

4.8 The proportion of impurities X in certain copper ore samples is a random variable having probability density function

$$f(x) = \begin{cases} 12x^2(1 - x) & 0 \leq x \leq 1 \\ 0 & \text{elsewhere} \end{cases}$$

If four such samples are independently selected, find the probability that

a exactly one has a proportion of impurities exceeding 0.5.

b at least one has a proportion of impurities exceeding 0.5.

4.2 Expected Values of Continuous Random Variables

As in the discrete case, we often want to summarize the information contained in a probability distribution by calculating expected values for the random variable and certain functions of the random variable.

DEFINITION 4.3

The **expected value** of a continuous random variable X having probability density function $f(x)$ is given by [†]

$$E(X) = \int_{-\infty}^{\infty} x f(x) \, dx \quad \blacksquare$$

[†]We assume absolute convergence of the integrals.

THEOREM **4.1**

If X is a continuous random variable with probability distribution $f(x)$ and if $g(x)$ is any real-valued function of X, then

$$E[g(X)] = \int_{-\infty}^{+\infty} g(x)f(x)\,dx$$

The proof of Theorem 4.1 will not be given here. ∎

The definitions of variance and standard deviation and the properties given in Theorems 3.2 and 3.3 hold for the continuous case as well.

For a random variable X with probability density function $f(x)$, the variance of X is given by

$$V(X) = E(X - \mu)^2 = \int_{-\infty}^{+\infty} (x - \mu)^2 f(x)\,dx$$

$$= E(X^2) - \mu^2$$

where $\mu = E(X)$. For constants a and b,

$$E(aX + b) = aE(X) + b$$

and

$$V(aX + b) = a^2 V(X)$$

We illustrate the expectations of continuous random variables in the following examples.

EXAMPLE **4.4**

For a lathe in a machine shop, let X denote the percentage of time out of a 40-hour workweek that the lathe is actually in use. Suppose X has a probability density function given by

$$f(x) = \begin{cases} 3x^2 & 0 \le x \le 1 \\ 0 & \text{elsewhere} \end{cases}$$

Find the mean and variance of X.

Solution From Definition 4.3, we have

$$E(X) = \int_{-\infty}^{\infty} xf(x)\,dx$$

$$= \int_{0}^{1} x(3x^2)\,dx$$

$$= \int_{0}^{1} 3x^3\,dx$$

$$= 3 \left[\frac{x^4}{4} \right]_0^1 = \frac{3}{4} = 0.75$$

Thus, on the average, the lathe is in use 75% of the time.

To compute $V(X)$, we first find $E(X^2)$:

$$E(X^2) = \int_{-\infty}^{\infty} x^2 f(x)\, dx$$
$$= \int_0^1 x^2 (3x^2)\, dx$$
$$= \int_0^1 3x^4\, dx$$
$$= 3 \left[\frac{x^5}{5} \right]_0^1 = \frac{3}{5} = 0.60$$

Then

$$V(X) = E(X^2) - \mu^2$$
$$= 0.60 - (0.75)^2$$
$$= 0.60 - 0.5625 = 0.0375 \quad \blacksquare$$

E X A M P L E   **4.5**   The weekly demand X for kerosene at a certain supply station has a probability density function given by

$$f(x) = \begin{cases} x & 0 \le x \le 1 \\ 1/2 & 1 < x \le 2 \\ 0 & \text{elsewhere} \end{cases}$$

Find the expected weekly demand.

Solution   Using Definition 4.3 to find $E(X)$, we must now carefully observe that $f(x)$ has different nonzero forms over two disjoint regions. Thus,

$$E(X) = \int_{-\infty}^{\infty} x f(x)\, dx$$
$$= \int_0^1 x(x)\, dx + \int_1^2 x(1/2)\, dx$$
$$= \int_0^1 x^2\, dx + \int_1^2 x(1/2)\, dx$$
$$= \left[\frac{x^3}{3} \right]_0^1 + \frac{1}{2} \left[\frac{x^2}{2} \right]_1^2$$
$$= \frac{1}{3} + \frac{1}{2} \left[2 - \frac{1}{2} \right]$$
$$= \frac{1}{3} + \frac{3}{4} = \frac{13}{12} = 1.08$$

That is, the expected weekly demand is 108 gallons.   ∎

Tchebysheff's Theorem (Theorem 3.4) holds for continuous random variables as well as for discrete. Thus, if X is continuous with mean μ and standard deviation σ, then

$$P(|X - \mu| < k\sigma) \geq 1 - \frac{1}{k^2}$$

for any positive number k. We illustrate the use of this result in the next example.

E X A M P L E   **4.6**   The weekly amount Y spent for chemicals in a certain firm has a mean of $445 and a variance of $236. Within what interval would these weekly costs for chemicals be expected to lie at least 75% of the time?

Solution   To find an interval guaranteed to contain at least 75% of the probability mass for Y, we get

$$1 - \frac{1}{k^2} = 0.75$$

which gives

$$\frac{1}{k^2} = 0.25$$

$$k^2 = \frac{1}{0.25} = 4$$

or

$$k = 2$$

Thus, the interval $\mu - 2\sigma$ to $\mu + 2\sigma$ will contain at least 75% of the probability. This interval is given by

$$445 - 2\sqrt{236} \quad \text{to} \quad 445 + 2\sqrt{236}$$
$$445 - 30.72 \quad \text{to} \quad 445 + 30.72$$

or

$$414.28 \quad \text{to} \quad 475.72 \quad \blacksquare$$

Exercises

 4.9 The temperature X at which a thermostatically controlled switch turns on has probability density function

$$f(x) = \begin{cases} \dfrac{1}{2} & 59 \le x \le 61 \\ 0 & \text{elsewhere} \end{cases}$$

Find $E(X)$ and $V(X)$.

 4.10 The proportion of time X that an industrial robot is in operation during a 40-hour work week is a random variable with probability density function

$$f(x) = \begin{cases} 2x & 0 \le x \le 1 \\ 0 & \text{elsewhere} \end{cases}$$

a Find $E(X)$ and $V(X)$.

b For the robot under study, the profit Y for a week is given by

$$Y = 200X - 60$$

Find $E(Y)$ and $V(Y)$.

c Find an interval in which the profit should lie for at least 75% of the weeks that the robot is in use. [*Hint*: Use Tchebysheff's Theorem.]

4.11 Daily total solar radiation for a certain location in Florida in October has probability density function

$$f(x) = \begin{cases} \dfrac{3}{32}(x-2)(6-x) & 2 \le x \le 6 \\ 0 & \text{elsewhere} \end{cases}$$

 with measurements in hundreds of calories. Find the expected daily solar radiation for October.

4.12 Weekly CPU time used by an accounting firm has probability density function (measured in hours)

$$f(x) = \begin{cases} \dfrac{3}{64}x^2(4-x) & 0 \le x \le 4 \\ 0 & \text{elsewhere} \end{cases}$$

a Find the expected value and variance of weekly CPU time.

b The CPU time costs the firm $200 per hour. Find the expected value and variance of the weekly cost for CPU time.

 c Will the weekly cost exceed $600 very often? Why or why not?

4.13 The pH of water samples from a specific lake is a random variable X with probability density function

$$f(x) = \begin{cases} \dfrac{3}{8}(7-x)^2 & 5 \le x \le 7 \\ 0 & \text{elsewhere} \end{cases}$$

a Find $E(X)$ and $V(X)$.

b Find an interval shorter than $(5, 7)$ in which at least 3/4 of the pH measurements must lie.

c Would you expect to see a pH measurement below 5.5 very often? Why?

4.14   A retail grocer has a daily demand X for a certain food sold by the pound, such that X (measured in hundreds of pounds) has probability density function

$$f(x) = \begin{cases} 3x^2 & 0 \leq x \leq 1 \\ 0 & \text{elsewhere} \end{cases}$$

(He cannot stock over 100 pounds.) The grocer wants to order $100k$ pounds of food on a certain day. He buys the food at 6¢ per pound and sells it at 10¢ per pound. What value of k will maximize his expected daily profit?

4.3   The Uniform Distribution

4.3.1   Probability Density Function

We now move from a general discussion of continuous random variables to discussion of specific models found useful in practice. Consider an experiment that consists of observing events occurring in a certain time frame, such as buses arriving at a bus stop or telephone calls coming into a switchboard. Suppose we know that one such event has occurred in the time interval (a, b). (A bus arrived between 8:00 and 8:10.) It may then be of interest to place a probability distribution on the actual time of occurrence of the event under observation, which we will denote by X. A very simple model assumes that X is equally likely to lie in any small subinterval, say of length d, no matter where that subinterval lies within (a, b). This assumption leads to the *uniform* probability distribution, which has probability density function given by

$$f(x) = \begin{cases} \dfrac{1}{b-a} & a \leq x \leq b \\ 0 & \text{elsewhere} \end{cases}$$

This density function is graphed in Figure 4.5.

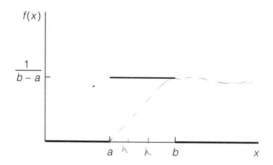

The distribution function for a uniformly distributed X is given by

$$F(x) = \int_a^x \frac{1}{b-a}\, dx = \frac{x-a}{b-a} \qquad a \leq x \leq b$$

If we consider a subinterval $(c, c+d)$ contained entirely within (a, b), we have

$$P(c \leq X \leq c+d) = F(c+d) - F(c)$$

The Uniform Distribution

$$f(x) = \begin{cases} \dfrac{1}{b-a} & a \leq x \leq b \\ 0 & \text{elsewhere} \end{cases}$$

$$E(X) = \frac{a+b}{2} \qquad V(X) = \frac{(b-a)^2}{12}$$

Exercises

4.15   Suppose X has a uniform distribution over the interval (a, b).

a   Find $F(x)$.

b   Find $P(X > c)$ for some point c between a and b.

c   If $a \leq c \leq d \leq b$, find $P(X > d | X > c)$.

4.16   Upon studying low bids for shipping contracts, a microcomputer manufacturing firm finds that intrastate contracts have low bids that are uniformly distributed between 20 and 25, in units of thousands of dollars.

a   Find the probability that the low bid on the next intrastate shipping contract is below $22,000.

b   Find the probability that the low bid is in excess of $24,000.

c   Find the average cost of low bids on contracts of this type.

4.17   If a point is *randomly* located in an interval (a, b) and if X denotes its distance from a, then X will be assumed to have a uniform distribution over $(0, b - a)$.

A plant efficiency expert randomly picks a spot along a 500-foot assembly line from which to observe work habits. Find the probability that she is

a   within 25 feet of the end of the line.

b   within 25 feet of the beginning of the line.

c   closer to the beginning than to the end of the line.

4.18   A bomb is to be dropped along a mile-long line that stretches across a practice target. The target center is at the midpoint of the line. The target will be destroyed if the bomb falls within a tenth of a mile on either side of the center. Find the probability that the target is destroyed if the bomb falls randomly along the line.

4.19   A telephone call arrived at a switchboard at a random time within a 1-minute interval. The switchboard was fully busy for 15 seconds into this 1-minute period. Find the probability that the call arrived when the switchboard was not fully occupied.

4.20   Beginning at 12:00 midnight, a computer center is up for 1 hour and down for 2 hours on a regular cycle. A person who doesn't know the schedule dials the center at a random time between 12:00 midnight and 5:00 A.M. What is the probability that the center will be operating when he dials in?

4.21   The number of defective circuit boards among those coming out of a soldering machine follows a Poisson distribution. On a particular 8-hour workday, one defective board is found.

a   Find the probability that it was produced during the first hour of operation of that day.

    b   Find the probability that it was produced during the last hour of operation for that day.

    c   Given that no defective boards were seen during the first 4 hours of operation, find the probability that the defective board was produced during the fifth hour.

4.22   In determining the range of an acoustic source by triangulation, the time at which the spherical wave front arrives at a receiving sensor must be measured accurately. According to an article by J. Perruzzi and E. Hilliard (*Journal of the Acoustical Society of America*, 75(1), 1984, pp. 197–201), measurement errors in these times can be modeled as having uniform distributions. Suppose measurement errors are uniformly distributed from −0.05 to +0.05 microsecond.

    a   Find the probability that a particular arrival-time measurement will be in error by less than 0.01 microsecond.

    b   Find the mean and variance of these measurement errors.

4.23   In the setting of Exercise 4.22, suppose the measurement errors are uniformly distributed from −0.02 to +0.05 microsecond.

    a   Find the probability that a particular arrival-time measurement will be in error by less than 0.01 microsecond.

    b   Find the mean and variance of these measurement errors.

4.24   According to Y. Zimmels (*AIChE Journal*, 29(4), 1983, pp. 669–676), the sizes of particles used in sedimentation experiments often have uniform distributions. It is important to study both the mean and variance of particle sizes, since in sedimentation with mixtures of various-size particles the larger particles hinder the movements of the smaller ones. Suppose spherical particles have diameters uniformly distributed between 0.01 and 0.05 centimeter. Find the mean and variance of the *volumes* of these particles. (Recall that the volume of a sphere is $\frac{4}{3}\pi r^3$.)

4.25   Arrivals of customers at a certain checkout counter follow a Poisson distribution. It is known that during a given 30-minute period one customer arrived at the counter. Find the probability that she arrived during the last 5 minutes of the 30-minute period.

4.26   A customer's arrival at a counter is uniformly distributed over a 30-minute period. Find the conditional probability that the customer arrived during the last 5 minutes of the 30-minute period given that there were no arrivals during the first 10 minutes of the period.

4.27   In tests of stopping distances for automobiles, those automobiles traveling at 30 miles per hour before the brakes are applied tend to travel distances that appear to be uniformly distributed between two points a and b. Find the probability that one of these automobiles

    a   stops closer to a than to b.

    b   stops so that the distance to a is more than three times the distance to b.

4.28   Suppose three automobiles are used in a test of the type discussed in Exercise 4.27. Find the probability that exactly one of the three travels past the midpoint between a and b.

4.29   The cycle times for trucks hauling concrete to a highway construction site are uniformly distributed over the interval 50 to 70 minutes.

    a   Find the expected value and variance for these cycle times.

    b   How many trucks would you expect to have to assign to this job so that a truckload of concrete can be dumped at the site every 15 minutes?

4.4   The Exponential Distribution

4.4.1   Probability Density Function

The lifelength data of Section 4.1 did not display a probabilistic behavior that was uniform but rather described one in which the probability over intervals of constant length decreased as the intervals moved further and further to the right. We saw that an exponential curve seemed to fit these data rather well; we now discuss the exponential probability distribution in more detail. In general, the exponential density function is given by

$$f(x) = \begin{cases} \dfrac{1}{\theta} e^{-x/\theta} & x \geq 0 \\ 0 & \text{elsewhere} \end{cases}$$

where the parameter θ is a constant that determines the rate at which the curve decreases.

An exponential density function with $\theta = 2$ was sketched in Figure 4.1, and, in general, exponential functions have the form shown in Figure 4.6.

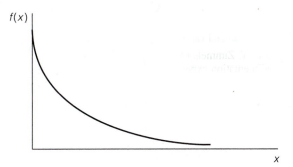

Many random variables occurring in engineering and the sciences can be appropriately modeled as having exponential distributions. Figure 4.7 shows two examples of relative frequency distributions for times between arrivals (interarrival times) of vehicles at a fixed point on a one-directional roadway. Both of these relative frequency histograms can be modeled quite nicely by exponential functions. Note that the higher traffic density causes shorter interarrival times to be more frequent.

4.4.2   Mean and Variance

Finding expected values for the exponential distribution is simplified by an understanding of a certain type of integral called a *gamma* (Γ) *function*. The function $\Gamma(\alpha)$, for $a \geq 1$, is defined by

$$\Gamma(\alpha) = \int_0^\infty x^{\alpha-1} e^{-x} \, dx$$

Integration by parts can be used to show that $\Gamma(\alpha + 1) = \alpha\Gamma(\alpha)$. It follows that $\Gamma(n) = (n-1)!$ for any positive integer n. The integral

$$\int_0^\infty x^{\alpha-1} e^{-x/\beta} \, dx$$

F I G U R E **4.7** Interarrival Times of Vehicles on a One-Directional Roadway

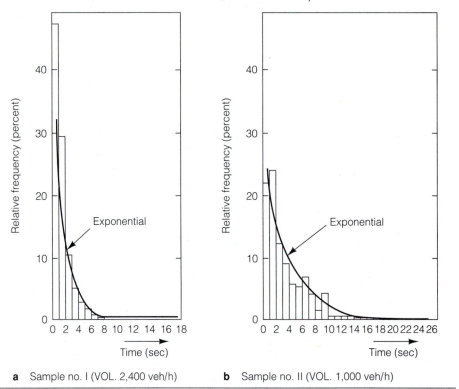

a Sample no. I (VOL. 2,400 veh/h)   **b** Sample no. II (VOL. 1,000 veh/h)

Source: D. Mahalel and A. S. Hakkert, *Transportation Research,* 17A, no. 4, 1983, p. 267.

for positive constants α and β can be evaluated by making the transformation $y = x/\beta$, or $x = \beta y$, $dx = \beta\, dy$. We then have

$$\int_0^\infty (\beta y)^{\alpha-1} e^{-y} \beta\, dy = \beta^\alpha \int_0^\infty y^{\alpha-1} e^{-y}\, dy = \beta^\alpha \Gamma(\alpha)$$

It is useful to note that $\Gamma(1/2) = \sqrt{\pi}$.

Using the above result, we see that, for the exponential distribution,

$$
\begin{aligned}
E(X) &= \int_{-\infty}^\infty x f(x)\, dx = \int_0^\infty x \left(\frac{1}{\theta}\right) e^{-x/\theta}\, dx \\
&= \frac{1}{\theta} \int_0^\infty x e^{-x/\theta}\, dx \\
&= \frac{1}{\theta} \Gamma(2)\theta^2 = \theta
\end{aligned}
$$

Thus, the parameter θ is actually the mean of the distribution.

To evaluate the variance of the exponential distribution, first we can find

$$E(X^2) = \int_0^\infty x^2 \frac{1}{\theta} e^{-x/\theta} \, dx$$

$$= \frac{1}{\theta} \Gamma(3)\theta^3 = 2\theta^2$$

It follows that

$$V(X) = E(X^2) - \mu^2$$

$$= 2\theta^2 - \theta^2 = \theta^2$$

and θ becomes the standard deviation as well as the mean.

The distribution function for the exponential case has a simple form, seen to be

$$F(t) = 0 \qquad \text{for } t < 0$$

$$F(t) = P(X \le t) = \int_0^t \frac{1}{\theta} e^{-x/\theta} \, dx$$

$$= -e^{-x/\theta} |_0^t = 1 - e^{-t/\theta} \qquad \text{for } t \ge 0$$

E X A M P L E   **4.8**   A sugar refinery has three processing plants, all receiving raw sugar in bulk. The amount of sugar that one plant can process in one day can be modeled as having an exponential distribution, with a mean of 4 (measurements in tons), for each of the three plants. If the plants operate independently, find the probability that exactly two of the three plants process more than 4 tons on a given day.

Solution   The probability that any given plant processes more than 4 tons is, with X denoting the amount used,

$$P(X > 4) = \int_4^\infty f(x) \, dx = \int_4^\infty \frac{1}{4} e^{-x/4} \, dx$$

$$= -e^{-x/4} |_4^\infty = e^{-1} = 0.37$$

[*Note:* $P(X > 4) \ne 0.5$ even though the mean of X is 4.]

Knowledge of the distribution function could allow us to evaluate this probability immediately as

$$P(X > 4) = 1 - P(X \le 4) = 1 - (1 - e^{-4/4})$$

$$= e^{-1}$$

Assuming the three plants operate independently, the problem is to find the probability of two successes out of three tries, where 0.37 denotes the probability of success. This is a binomial problem, and the solution is

$$P(\text{exactly two use more than 4 tons}) = \binom{3}{2}(0.37)^2(0.63)$$

$$= 3(0.37)^2(0.63)$$

$$= 0.26 \quad \blacksquare$$

M P L E   **4.9**   Consider a particular plant in Example 4.8. How much raw sugar should be stocked for that plant each day so that the chance of running out of the product is only 0.05?

Solution   Let a denote the amount to be stocked. Since the amount to be used, X, has an exponential distribution, we have

$$P(X > a) = \int_a^\infty \frac{1}{4} e^{-x/4} \, dx = e^{-a/4}$$

We want to choose a so that

$$P(X > a) = e^{-a/4} = 0.05$$

and solving this equation yields

$$a = 11.98 \quad \blacksquare$$

As in the uniform case, there is a relationship between the exponential distribution and the Poisson distribution. Suppose events are occurring in time according to a Poisson distribution with a rate of λ events per hour. Thus, in t hours the number of events, say Y, will have a Poisson distribution with mean value λt. Suppose we start at time zero and ask the question "How long do I have to wait to see the first event occur?" Let X denote the length of time until this first event. Then

$$P(X > t) = P[Y = 0 \text{ on the interval } (0, t)]$$
$$= (\lambda t)^0 e^{-\lambda t}/0! = e^{-\lambda t}$$

and

$$P(X \leq t) = 1 - P(X > t) = 1 - e^{-\lambda t}$$

We see that $P(X \leq t) = F(t)$, the distribution function for X, has the form of an exponential distribution function with $\lambda = (1/\theta)$. Upon differentiating, the probability density function of X is given by

$$f(t) = \frac{dF(t)}{dt} = \frac{d(1 - e^{-\lambda t})}{dt}$$
$$= \lambda e^{-\lambda t}$$
$$= \frac{1}{\theta} e^{-t/\theta} \qquad t > 0$$

and X has an exponential distribution. Actually, we need not start at time zero, for it can be shown that the waiting time from the occurrence of any one event until the occurrence of the next event will have an exponential distribution for events occurring according to a Poisson distribution.

We summarize the properties of the exponential distribution here.

The Exponential Distribution

$$f(x) = \begin{cases} \dfrac{1}{\theta}e^{-x/\theta} & x > 0 \\ 0 & \text{elsewhere} \end{cases}$$

$$E(X) = \theta \qquad V(X) = \theta^2$$

Exercises

4.30   Suppose Y has an exponential density function with mean θ. Show that $P(Y > a + b | Y > a) = P(Y > b)$. This is referred to as the "memoryless" property of the exponential distribution.

4.31   The magnitudes of earthquakes recorded in a region of North America can be modeled by an exponential distribution with mean 2.4 as measured on the Richter scale. Find the probability that the next earthquake to strike this region will

a   exceed 3.0 on the Richter scale.

b   fall between 2.0 and 3.0 on the Richter scale.

4.32   Refer to Exercise 4.31. Of the next ten earthquakes to strike this region, find the probability that at least one will exceed 5.0 on the Richter scale.

4.33   A pumping station operator observes that the demand for water at a certain hour of the day can be modeled as an exponential random variable with a mean of 100 cfs (cubic feet per second).

a   Find the probability that the demand will exceed 200 cfs on a randomly selected day.

b   What is the maximum water-producing capacity that the station should keep on line for this hour so that the demand will exceed this production capacity with a probability of only 0.01?

4.34   Suppose customers arrive at a certain checkout counter at the rate of two every minute.

a   Find the mean and variance of the waiting time between successive customer arrivals.

b   If a clerk takes 3 minutes to serve the first customer arriving at the counter, what is the probability that at least one more customer is waiting when the service of the first customer is completed?

4.35   The length of time X to complete a certain key task in house construction is an exponentially distributed random variable with a mean of 10 hours. The cost C of completing this task is related to the square of the time to completion by the formula

$$C = 100 + 40X + 3X^2$$

a   Find the expected value and variance of C.

b   Would you expect C to exceed 2,000 very often?

4.36   The interaccident times (times between accidents) for all fatal accidents on scheduled American domestic passenger air flights, 1948–1961, were found to follow an exponential distribution with a mean of approximately 44 days (R. Pyke, *Journal of the Royal Statistical Society* (B), 27, 1968, p. 426).

a   If one of those accidents occurred on July 1, find the probability that another one occurred in that same month.

b   Find the variance of the interaccident times.

c   What does the information given above suggest about the clumping of airline accidents?

4.37 The lifelengths of automobile tires of a certain brand, under average driving conditions, are found to follow an exponential distribution with mean 30 (in thousands of miles). Find the probability that one of these tires bought today will last

a   over 30,000 miles.

b   over 30,000 miles given that it already has gone 15,000 miles.

4.38   The dial-up connections from remote terminals come into a computing center at the rate of four per minute. The callers follow a Poisson distribution. If a call arrives at the beginning of a 1-minute period, find the probability that a second call will not arrive in the next 20 seconds.

4.39   The breakdowns of an industrial robot follow a Poisson distribution with an average of 0.5 breakdown per 8-hour workday. The robot is placed in service at the beginning of the day.

a   Find the probability that it will not break down during the day.

b   Find the probability that it will work for at least 4 hours without breaking down.

c   Does what happened the day before have any effect on your answers above? Why?

4.40   One-hour carbon monoxide concentrations in air samples from a large city are found to have an exponential distribution with a mean of 3.6 ppm (J. Zamurs, *Air Pollution Control Association Journal*, 34(6), 1984, p. 637).

a   Find the probability that a concentration will exceed 9 ppm.

b   A traffic control strategy reduced the mean to 2.5 ppm. Now find the probability that a concentration will exceed 9 ppm.

4.41   The weekly rainfall totals for a section of the midwestern United States follow an exponential distribution with a mean of 1.6 inches.

a   Find the probability that a weekly rainfall total in this section will exceed 2 inches.

b   Find the probability that the weekly rainfall totals will not exceed 2 inches in either of the next two weeks.

4.42   The service times at teller windows in a bank were found to follow an exponential distribution with a mean of 3.2 minutes. A customer arrives at a window at 4:00 P.M.

a   Find the probability that he will still be there at 4:02 P.M.

b   Find the probability that he will still be there at 4:04 P.M. given that he was there at 4:02 P.M.

4.43   In deciding how many customer service representatives to hire and in planning their schedules, it is important for a firm marketing electronic typewriters to study repair times for the machines. Such a study revealed that repair times have approximately an exponential distribution with a mean of 22 minutes.

a   Find the probability that a repair time will last less than 10 minutes.

b   The charge for typewriter repairs is $50 for each half hour or part thereof. What is the probability that a repair job will result in a charge of $100?

c   In planning schedules, how much time should be allowed for each repair so that the chance of any one repair time exceeding this allowed time is only 0.10?

4.44   Explosive devices used in a mining operation cause nearly circular craters to form in a rocky surface. The radii of these craters are exponentially distributed with a mean of 10 feet. Find the mean and variance of the area covered by such a crater.

4.5 The Gamma Distribution

4.5.1 Probability Density Function

Many sets of data, of course, will not have relative frequency curves with the smooth decreasing trend found in the exponential model. It is perhaps more common to see distributions that have low probabilities for intervals close to zero, with the probability increasing for a while as the interval moves to the right (in the positive direction) and then decreasing as the interval moves to the extreme positive side. That is, the relative frequency curves appear as in Figure 4.8. In the case of electronic components, for example, few will have very short lifelengths, many will have something close to an average lifelength, and very few will have extraordinarily long lifelengths.

F I G U R E **4.8**
A Common Relative Frequency Curve

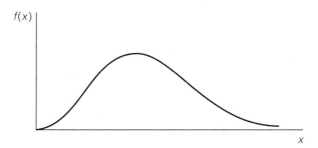

A class of functions that serve as good models for this type of behavior is the *gamma* class. The gamma probability density function is given by

$$f(x) = \begin{cases} \dfrac{1}{\Gamma(\alpha)\beta^{\alpha}} x^{\alpha-1} e^{-x/\beta} & x \geq 0 \\ 0 & \text{elsewhere} \end{cases}$$

where α and β are parameters that determine the specific shape of the curve. Note immediately that the gamma density function reduces to the exponential when $\alpha = 1$. The parameters α and β must be positive but need not be integers. The symbol $\Gamma(\alpha)$ is defined by

$$\Gamma(\alpha) = \int_0^\infty x^{\alpha-1} e^{-x}\, dx$$

Since we have already seen that

$$\int_0^\infty x^{\alpha-1} e^{-x/\beta}\, dx = \beta^\alpha \Gamma(\alpha)$$

it follows that the gamma density function will integrate to 1. Some typical gamma densities are shown in Figure 4.9.

An example of a real data set that closely follows a gamma distribution is shown in Figure 4.10. The data are six-week summer rainfall totals for Ames, Iowa. Notice that many totals are from 2 to 8 inches, but occasionally a rainfall total goes well beyond 8 inches. Of course, no rainfall measurements can be negative.

FIGURE **4.9**
The Gamma Density Function,
$\beta = 1$

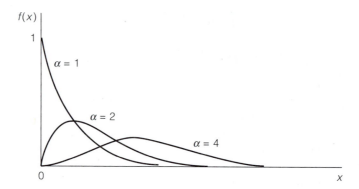

FIGURE **4.10**
Summer Rainfall (six-week
totals) for Ames, Iowa

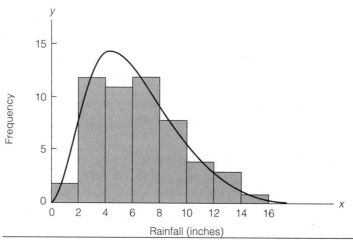

Source: G. L. Barger and H. C. S. Thom, *Agronomy Journal,* 41, 1949, p. 521.

4.5.2 Mean and Variance

The derivation of expectations here is very similar to the exponential case in Section 4.4. We have

$$E(X) = \int_{-\infty}^{\infty} xf(x)\,dx = \int_{0}^{\infty} x \frac{1}{\Gamma(\alpha)\beta^{\alpha}} x^{\alpha-1} e^{-x/\beta}\,dx$$

$$= \frac{1}{\Gamma(\alpha)\beta^{\alpha}} \int_{0}^{\infty} x^{\alpha} e^{-x/\beta}\,dx$$

$$= \frac{1}{\Gamma(\alpha)\beta^{\alpha}} \Gamma(\alpha+1)\beta^{\alpha+1} = \alpha\beta$$

Similar manipulations yield $E(X^2) = \alpha(\alpha+1)\beta^2$, and hence

$$V(X) = E(X^2) - \mu^2$$

$$= \alpha(\alpha+1)\beta^2 - \alpha^2\beta^2 = \alpha\beta^2$$

A simple and often used property of sums of identically distributed, independent gamma random variables will be stated, but not proved, at this point. Suppose $X_1, X_2, \ldots, X_n$ represent independent gamma random variables with parameters α and β as used above. If

$$Y = \sum_{i=1}^{n} X_i$$

then Y also has a gamma distribution with parameters $n\alpha$ and β. Thus, one can immediately see that

$$E(Y) = n\alpha\beta$$

and

$$V(Y) = n\alpha\beta^2$$

E X A M P L E   4.10   A certain electronic system having lifelength X_1 with an exponential distribution and mean 400 hours is supported by an identical backup system with lifelength X_2. The backup system takes over immediately when the primary system fails. If the systems operate independently, find the probability distribution and expected value for the total lifelength of the primary and backup systems.

Solution   Letting Y denote the total lifelength, we have $Y = X_1 + X_2$, where X_1 and X_2 are independent exponential random variables, each with mean $\beta = 400$. By the results stated above, Y will then have a gamma distribution with $\alpha = 2$ and $\beta = 400$; that is,

$$f_Y(y) = \frac{1}{\Gamma(2)(400)^2} y e^{-y/400} \qquad y > 0$$

The mean value is given by

$$E(Y) = \alpha\beta = 2(400) = 800$$

which is intuitively reasonable.   ∎

E X A M P L E   4.11   Suppose that the length of time Y to conduct a periodic maintenance check (from previous experience) on a dictating machine follows a gamma-type distribution with $\alpha = 3$ and $\beta = 2$ (minutes). Suppose that a new repairman requires 20 minutes to check a machine. Does it appear that his time to perform a maintenance check disagrees with prior experience?

Solution   The mean and variance for the length of maintenance times (prior experience) are

$$\mu = \alpha\beta \qquad \text{and} \qquad \sigma^2 = \alpha\beta^2$$

Then, for our example,

$$\mu = \alpha\beta = (3)(2) = 6 \qquad \sigma^2 = \alpha\beta^2 = (3)(2)^2 = 12 \qquad \sigma = \sqrt{12} = 3.46$$

and the observed deviation $(Y - \mu)$ is $20 - 6 = 14$ minutes.

For our example, $y = 20$ minutes exceeds the mean $\mu = 6$ by $k = 14/3.46$ standard deviations. Then, from Tchebysheff's Theorem,

$$P(|Y - \mu| \geq k\sigma) \leq \frac{1}{k^2}$$

or

$$P(|Y - 6| \geq 14) \leq \frac{1}{k^2} = \frac{(3.46)^2}{(14)^2} = 0.06$$

Note that this probability is based on the assumption that the distribution of maintenance times has not changed from prior experience. Then observing that $P(Y \geq 20$ minutes$)$ is small, we must conclude that either our new maintenance man has generated a lengthy maintenance time that occurs with low probability or he is somewhat slower than his predecessors. Noting the low probability for $P(Y \geq 20)$, we would be inclined to favor the latter view. ■

The Gamma Distribution

$$f(x) = \begin{cases} \dfrac{1}{\Gamma(\alpha)\beta^\alpha} x^{\alpha-1} e^{-x/\beta} & x > 0 \\ 0 & \text{elsewhere} \end{cases}$$

$$E(X) = \alpha\beta \qquad V(X) = \alpha\beta^2$$

Exercises

4.45 Four-week summer rainfall totals in a certain section of the midwestern United States have a relative frequency histogram that appears to fit closely to a gamma distribution with $\alpha = 1.6$ and $\beta = 2.0$.

a Find the mean and variance of this distribution of four-week rainfall totals.

b Find an interval that will include the rainfall total for a selected four-week period with probability at least 0.75.

4.46 Annual incomes for engineers in a certain industry have approximately a gamma distribution with $\alpha = 600$ and $\beta = 50$.

a Find the mean and variance of these incomes.

b Would you expect to find many engineers in this industry with an annual income exceeding $35,000?

4.47   The weekly downtime Y (in hours) for a certain industrial machine has approximately a gamma distribution with $\alpha = 3$ and $\beta = 2$. The loss, in dollars, to the industrial operation as a result of this downtime is given by

$$L = 30Y + 2Y^2$$

a   Find the expected value and variance of L.

b   Find an interval that will contain L on approximately 89% of the weeks that the machine is in use.

4.48   Customers arrive at a checkout counter according to a Poisson process with a rate of two per minute. Find the mean, variance, and probability density function of the waiting time between the opening of the counter and

a   the arrival of the second customer.

b   the arrival of the third customer.

4.49   Suppose two houses are to be built and each will involve the completion of a certain key task. The task has an exponentially distributed time to completion with a mean of 10 hours. Assuming the completion times are independent for the two houses, find the expected value and variance of

a   the total time to complete both tasks.

b   the average time to complete the two tasks.

4.50   The total sustained load on the concrete footing of a planned building is the sum of the dead load plus the occupancy load. Suppose the dead load X_1 has a gamma distribution with $\alpha_1 = 50$ and $\beta_1 = 2$, while the occupancy load X_2 has a gamma distribution with $\alpha_2 = 20$ and $\beta_2 = 2$. (Units are in kips, or thousands of pounds.)

a   Find the mean, variance, and probability density function of the total sustained load on the footing.

b   Find a value for the sustained load that should be exceeded only with probability less than 1/16.

4.51   A 40-year history of maximum river flows for a certain small river in the United States shows a relative frequency histogram that can be modeled by a gamma density function with $\alpha = 1.6$ and $\beta = 150$. (Measurements are in cubic feet per second.)

a   Find the mean and standard deviation of the annual maximum river flows.

b   Within what interval will the maximum annual flow be contained with probability at least 8/9?

4.52   The time intervals between dial-up connections to a computer center from remote terminals are exponentially distributed with a mean of 15 seconds. Find the mean, variance, and probability distribution of the waiting time from the opening of the computer center until the fourth dial-up connection from a remote terminal.

4.53   If service times at a teller window of a bank are exponentially distributed with a mean of 3.2 minutes, find the probability distribution, mean, and variance of the time taken to serve three waiting customers.

4.54   The response times at an on-line terminal have approximately a gamma distribution with a mean of 4 seconds and a variance of 8. Write the probability density function for these response times.

4.6 The Normal Distribution

4.6.1 Normal Probability Density Function

Perhaps the most widely used of all the continuous probability distributions is the one referred to as the *normal distribution*. The normal probability density function has the familiar symmetric "bell" shape shown in Figure 4.11. The curve is centered at the mean value μ, and its spread is measured by the standard deviation σ. These two parameters, μ and σ, completely determine the shape and location of the normal density function, whose functional form is given by

$$f(x) = \frac{1}{\sigma\sqrt{2\pi}}e^{-(x-\mu)^2/2\sigma^2} \qquad -\infty < x < \infty$$

F I G U R E **4.11**
Normal Density Functions

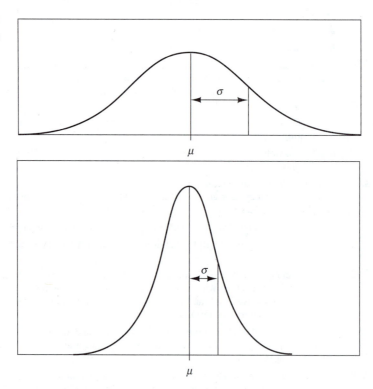

The basic reason why the normal distribution works well as a model for many different types of measurements generated in real experiments will be discussed in some detail in Chapter 6. For now, we simply say that any time that responses tend to be averages of independent qualities, the normal distribution will quite likely provide a reasonably good model for their relative frequency behavior. Many naturally occurring measurements tend to have relative frequency distributions closely resembling the normal curve, probably because nature tends to "average out" the effects of the many variables that relate to a particular response. For example, heights of adult American males tend to have a distribution that shows many measurements clumped closely about a mean height, with relatively few very short or very tall males in the population. In other words, the relative frequency distribution is close to normal.

In contrast, lifelengths of biological organisms or electronic components tend to have relative frequency distributions that are neither normal nor close to normal, due to the fact that lifelength measurements often are a product of "extreme" behavior, not "average" behavior. A component may fail because of one extremely hard shock rather than the average effect of many shocks. Thus, the normal distribution is not often used to model lifelengths.

A naturally occurring example of the normal distribution is seen in A. Michelson's measures of the speed of light. A histogram of these measurements is given in Figure 4.12. The distribution is not perfectly symmetrical but still exhibits an approximately normal shape.

F I G U R E **4.12**
Michelson's 100 Measures of
the Speed of Light in Air
($-299,000$)

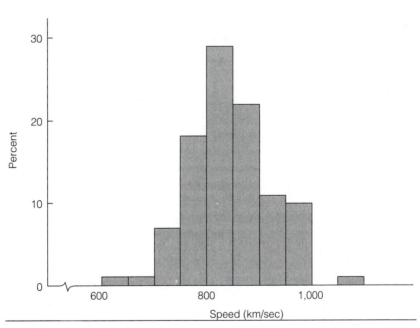

Source: *Astronomical Papers*, 1881, p. 231

4.6.2   Mean and Variance

A very important property of the normal distribution, which will be proved in Section 4.10, is that any linear function of a normally distributed random variable is also normally distributed. That is, if X has a normal distribution with mean μ and variance σ^2 and $Y = aX + b$ for constants a and b, then Y is also normally distributed. It is easily seen that

$$E(Y) = a\mu + b \qquad V(Y) = a^2\sigma^2$$

Suppose Z has a normal distribution with $\mu = 0$ and $\sigma = 1$. This random variable Z is said to have a *standard normal distribution*. Direct integration will show that $E(Z) = 0$ and $V(Z) = 1$. We have

$$E(Z) = \int_{-\infty}^{\infty} z \frac{1}{\sqrt{2\pi}} e^{-z^2/2} \, dz$$

$$= \frac{1}{\sqrt{2\pi}} \int_{-\infty}^{\infty} e^{-z^2/2} z \, dz$$

$$= \frac{1}{\sqrt{2\pi}} [-e^{-z^2/2}]_{-\infty}^{\infty} = 0$$

Similarly,

$$E(Z^2) = \int_{-\infty}^{\infty} \frac{1}{\sqrt{2\pi}} z^2 e^{-z^2/2} \, dz$$

$$= \frac{1}{\sqrt{2\pi}} (2) \int_0^{\infty} z^2 e^{-z^2/2} \, dz$$

On making the transformation $u = z^2$, the integral becomes

$$\frac{1}{\sqrt{2\pi}} \int_0^{\infty} u^{1/2} e^{-u/2} \, du = \frac{1}{\sqrt{2\pi}} \Gamma\left(\frac{3}{2}\right) (2)^{3/2}$$

$$= \frac{1}{\sqrt{2\pi}} (2)^{3/2} \left(\frac{1}{2}\right) \Gamma\left(\frac{1}{2}\right) = 1$$

since $\Gamma(\frac{1}{2}) = \sqrt{\pi}$. Therefore, $E(Z) = 0$ and $V(Z) = E(Z^2) - \mu^2 = E(Z^2) = 1$.

For any normally distributed random variable X, with parameters μ and σ^2,

$$Z = \frac{X - \mu}{\sigma}$$

will have a standard normal distribution. Note that Z is the probability version of the z-score introduced in Chapter 1. Now

$$X = Z\sigma + \mu$$
$$E(X) = \sigma E(Z) + \mu = \mu$$

and

$$V(X) = \sigma^2 V(Z) = \sigma^2$$

The parameters μ and σ^2 do, indeed, measure the mean and variance of the distribution.

4.6.3   Calculating Normal Probabilities

Since any normally distributed random variable can be transformed to the standard normal, probabilities can be evaluated for any normal distribution simply by having a table of standard normal integrals available. Such a table is given in Table 4 of the Appendix, which gives numerical values for

$$P(0 \leq Z \leq z) = \int_0^z \frac{1}{\sqrt{2\pi}} e^{-x^2/2} \, dx$$

Values of the integral are given for z between 0.00 and 3.09.

We will now show how to use Table 4 to find $P(-0.5 \leq Z \leq 1.5)$ for a standard normal variable Z. Figure 4.13 will help you visualize the necessary areas.

F I G U R E **4.13**
A Standard Normal Density
Function

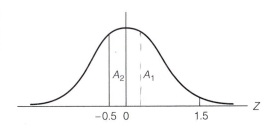

We first must write the probability in terms of intervals to the left and right of zero, the mean of the distribution. This produces

$$P(-0.5 \leq Z \leq 1.5) = P(0 \leq Z \leq 1.5) + P(-0.5 \leq Z \leq 0)$$

Now $(0 \leq Z \leq 1.5) = A_1$ in Figure 4.13 and is found by looking up $z = 1.5$ in Table 4. The result is $A_1 = 0.4332$. Similarly, $P(-0.5 \leq Z \leq 0) = A_2$ in Figure 4.13 and is found by looking up $z = 0.5$ in Table 4. Areas under the standard normal curve for negative z-values are equal to those for corresponding positive z-values, since the curve is symmetric around zero. We find $A_2 = 0.1915$. It follows that

$$P(-0.5 \leq Z \leq 1.5) = A_1 + A_2 = 0.4332 + 0.1915 = 0.6247$$

E X A M P L E   **4.12**   If Z denotes a standard normal variable, find

a $P(Z \leq 1)$

b $P(Z > 1)$

c $P(Z < -1.5)$

d $P(-1.5 \leq Z \leq 0.5)$

Also find a value of z, say z_0, such that $P(0 \leq Z \leq z_0) = 0.49$.

Solution   This example provides practice in using Table 4. We see the following:

a
$$P(Z \leq 1) = P(Z \leq 0) + P(0 \leq Z \leq 1)$$
$$= 0.5 + 0.3413 = 0.8413$$

b
$$P(Z > 1) = 0.5 - P(0 \leq Z \leq 1)$$
$$= 0.5 - 0.3413 = 0.1587$$

c
$$P(Z < -1.5) = P(Z > 1.5)$$
$$= 0.5 - P(0 \le Z \le 1.5)$$
$$= 0.5 - 0.4332 = 0.0668$$

d
$$P(-1.5 \le Z \le 0.5) = P(-1.5 \le Z \le 0) + P(0 \le Z \le 0.5)$$
$$= P(0 \le Z \le 1.5) + P(0 \le Z \le 0.5)$$
$$= 0.4332 + 0.1915$$
$$= 0.6247$$

To find the value of z_0, we must look for the given probability of 0.49 on the area side of Table 4. The closest we can come is 0.4901, which corresponds to a z-value of 2.33. Hence, $z_0 = 2.33$. ∎

Study of the table of normal curve areas for z-scores of 1, 2, and 3 shows how the percentages used in the empirical rule were determined. These percentages actually represent areas under the standard normal curve, as depicted in Figure 4.14.

FIGURE **4.14**
Justification of the Empirical Rule

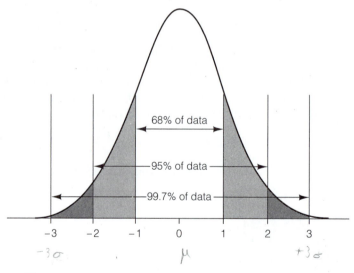

The next example illustrates how the standardization works to allow Table 4 to be used for any normally distributed random variable.

EXAMPLE **4.13**   A firm that manufactures and bottles apple juice has a machine that automatically fills 16-ounce bottles. There is, however, some variation in the number of ounces of liquid dispensed into each bottle by the machine. Over a long period of time, the average amount dispensed into the bottles was 16 ounces, but there is a standard deviation of 1 ounce in these measurements. If the ounces of fill per bottle can be assumed to be normally distributed, find the probability that the machine will dispense more than 17 ounces of liquid in any one bottle.

Solution   Let X denote the ounces of liquid dispensed into one bottle by the filling machine. Then X is assumed to be normally distributed with mean 16 and standard deviation 1. Hence,

$$P(X > 17) = P\left(\frac{X - \mu}{\sigma} > \frac{17 - \mu}{\sigma}\right)$$

$$= P\left(Z > \frac{17 - 16}{1}\right) = P(Z > 1) = 0.1587$$

The answer is found in Table 4, since $Z = (X - \mu)/\sigma$ has a *standard* normal distribution.   ▪

E X A M P L E   4.14   Suppose that another machine, similar to the one of Example 4.13, operates so that ounces of fill have a mean equal to the dial setting for "amount of liquid" but have a standard deviation of 1.2 ounces. Find the proper setting for the dial so that 17-ounce bottles will overflow only 5% of the time. Assume that the amounts dispensed have a normal distribution.

Solution   Letting X denote the amount of liquid dispensed, we are now looking for a value of μ such that

$$P(X > 17) = 0.05$$

as depicted in Figure 4.15. Now

$$P(X > 17) = P\left(\frac{X - \mu}{\sigma} > \frac{17 - \mu}{\sigma}\right)$$

$$= P\left(Z > \frac{17 - \mu}{1.2}\right)$$

From Table 4, we know that if

$$P(Z > z_0) = 0.05$$

then $z_0 = 1.645$. Thus, it must be that

$$\frac{17 - \mu}{1.2} = 1.645$$

and

$$\mu = 17 - 1.2(1.645) = 15.026$$ ▪

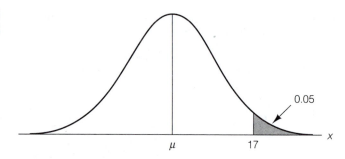

4.6.4 Applications to Real Data

The practical value of probability distributions as related to data analysis is that the probability models help explain the key features of the data succinctly and aid in the constructive use of data to help predict future outcomes. Patterns that appeared regularly in the past are expected to appear again in the future, and if they do not, something of importance may have happened to disturb the process under study. Describing patterns in data through probability distributions is a very useful data analysis tool.

E X A M P L E **4.15** The SAT and ACT college entrance exams are taken by thousands of students each year. The scores on the exam for any one year produce a histogram that looks very much like a normal curve. Thus, we can say that the scores are approximately normally distributed. In recent years, the SAT mathematics scores have averaged around 480 with a standard deviation of 100. The ACT mathematics scores have averaged around 18 with a standard deviation of 6.

a An engineering school sets 550 as the minimum SAT math score for new students. What percent of students would score less than 550 in a typical year? (This percentage is called the *percentile* score equivalent for 550.)

b What would the engineering school set as a comparable standard on the ACT math test?

c What is the probability that a randomly selected student will score over 700 on the SAT math test?

Solution **a** The percentile score corresponding to a raw score of 550 is the area under the normal curve to the left of 550, as shown in Figure 4.16. The area marked *A* can be found from Table 4 (or your computer). The area to the left of the mean is 0.5. To find *A*, observe that the *z*-score for 550 is

$$z = \frac{x - \mu}{\sigma} = \frac{550 - 480}{100} = 0.7$$

F I G U R E **4.16**
Diagram for Example 4.15(a)

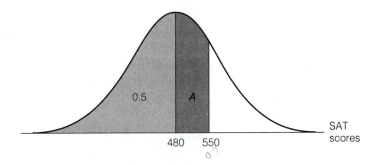

This gives $A = 0.258$, and the percentile score is $0.5 + 0.258 = 0.758$. The score of 550 is almost the 76th percentile. About 75.8% of the students taking the SAT should have scores below this value.

b   The question now is "What is the 75.8th percentile of the ACT math score distribution?" From the calculation above, percentile 75.8 for any normal distribution will be 0.7 standard deviations above the mean. Therefore, using the ACT distribution,

$$x = \mu + z\sigma = 18 + (0.7)6$$
$$= 22.2$$

A score of 22.2 on the ACT should be equivalent to a score of 550 on the SAT.

c   A score of 700 corresponds to a z-score of

$$z = \frac{x - \mu}{\sigma} = \frac{700 - 480}{100} = 2.2$$

which, in turn, gives an area of $A = 0.4861$ from Table 4. (See Figure 4.17.) Thus, the area above 700 is $0.50 - 0.4861 = 0.0139$. The chance of a randomly selected student scoring above 700 is quite small (around 0.014).   ■

F I G U R E **4.17**
Diagram for Example 4.15(c)

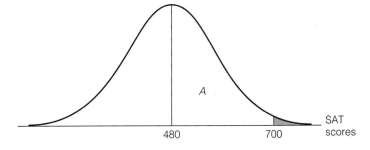

E X A M P L E   **4.16**   The batting averages of the American League batting champions from 1941 through 1991 are graphed on the histogram in Figure 4.18 (see Exercise 1.19 for the data). This graph looks somewhat normal in shape but has a little skewness toward the high values. The mean is .344, and the standard deviation is 0.022 for these data.

F I G U R E **4.18**
Batting Averages, American
League Champions

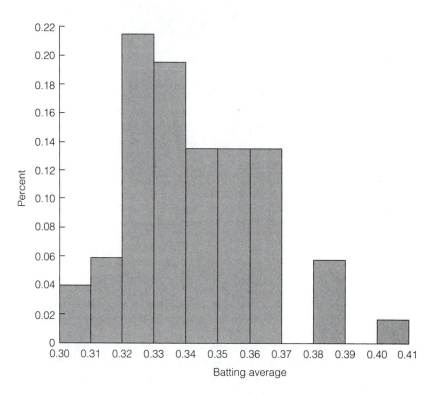

a   Ted Williams batted .406 in 1941, and George Brett batted .390 in 1980. How would you compare these performances?

b   Is there a good chance of anyone in the American League hitting over .400 in any one year?

Solution   a   Obviously, .406 is better than .390, but how much better? One way to describe how these performances compare, and how they compare to the remaining data points in the distribution, is to look at z-scores and percentile scores. Ted Williams has a z-score of

$$z = \frac{0.406 - 0.344}{0.022} = 2.82$$

and a percentile score of

$$0.50 + 0.4976 = 0.9976$$

George Brett has a z-score of

$$z = \frac{0.390 - 0.344}{0.022} = 2.09$$

and a percentile score of

$$0.50 + 0.4817 = 0.9817$$

Both are above the 98th percentile and are far above average. In that sense, both performances are outstanding and, perhaps, not very far apart.

b The chance of the league leader hitting over .400 in a given year can be approximated by looking at a z-score of

$$z = \frac{0.400 - 0.344}{0.022} = 2.54$$

This translates to a probability of hitting above .400 of

$$0.50 - 0.4945 = 0.0055$$

or about six chances out of 1,000. (This is the probability for the league leader. What would be the chances for any other specified player?) ∎

What happens if the normal model is used to describe skewed distributions? To answer this question, let's look at two data sets on comparable variables and see how good the normal approximations are. Figure 4.19 shows dotplots of cancer mortality rates for white males during the entire decade of the 1970s. (These rates are deaths per 100,000.) The top graph shows data for the 67 counties of Florida, and the bottom graph shows data for the 93 counties of Nebraska. Summary statistics are as follows:

	Mean	Standard Deviation
Florida	199.3	21.7
Nebraska	192.5	95.7

On the average, the states perform about the same, but the distributions are quite different. Key features of that difference are demonstrated by looking at empirical versus theoretical relative frequencies, as shown below:

Interval	Observed Proportion		Theoretical Proportion
	Florida	Nebraska	
$\bar{x} \pm s$	$\frac{46}{67} = 0.687$	$\frac{85}{93} = 0.914$	0.68
$\bar{x} \pm 2s$	$\frac{66}{67} = 0.985$	$\frac{87}{93} = 0.935$	0.95

For Florida, observation and theory are close together, although the two-standard-deviation interval does pick up a few too many data points. For Nebraska, observation and theory are not close. In this case, the large outliers inflate the standard deviation so that the one-standard-deviation interval is far too long to agree with normal theory.

F I G U R E **4.19**   Cancer Mortality Rates (per 100,000) for White Males, 1970s

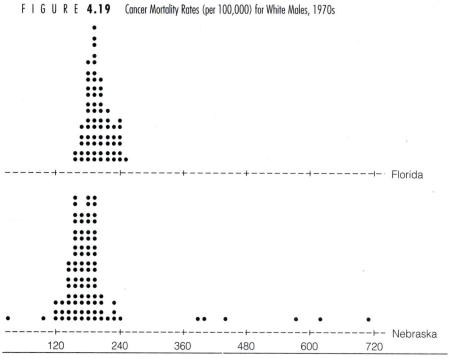

Source: National Cancer Institute.

The two-standard-deviation interval still doesn't reach the outliers, so the observed relative frequency is still smaller than expected. This is typical of the performance of relative frequencies in highly skewed situations; be careful in interpreting standard deviation under skewed conditions.

4.6.5   Quantile-Quantile (Q-Q) Plots

The normal model is popular for describing data distributions, and, as seen above, it often works well. How can we spot when it is *not* going to work well? One way, as also seen above, is to look carefully at histograms, dotplots, and stemplots to visually gauge symmetry and outliers. Another way, presented in this section, is to take advantage of the unique properties of z-scores for the normal case.

If X has a normal (μ, σ) distribution, then

$$X = \mu + \sigma Z$$

There is a perfect linear relationship between X and Z. Now suppose we observe n measurements and order them so that $x_1 \le x_2 \le x_3 \le \cdots \le x_n$. The value x_k has k/n values less than or equal to it, so it is the (k/n)th sample percentile. If the observations come from a normal distribution, x_k should approximate the (k/n)th percentile from the normal distribution and, therefore, be linearly related to the corresponding z-score, z_k. The cancer data used above (Figure 4.19) will serve to illustrate this point. The sample percentiles corresponding to each state were determined for 25%, 50%,

75%, 95%, and 99%, as shown in the table below. The z-scores corresponding to these percentiles for the normal distribution are also shown.

Percentile	Florida Sample	Nebraska Sample	z-scores
25	185	155	−0.680
50	195	174	0.000
75	214	192	0.680
95	237	230	1.645
99	242	622	2.330

In other words, 25% of the Florida data fell on or below 185, and 95% of the Nebraska data fell on or below 230. For the normal distribution, 25% of the area will fall below a point with z-score −0.68.

Plots of the sample percentiles against the z-scores are shown in Figure 4.20. Such plots are called quantile-quantile, or Q-Q, plots. For Florida, the sample percentiles and z-scores fall nearly on a straight line, indicating a good fit of the data to a normal distribution. The slope of the line is about 20, which is close to the standard deviation for these data, and the intercept at $z = 0$ is about 195, close to the sample mean.

For Nebraska, four points fall on a line, or nearly so, but the fifth point is far off the line. The observed 99th percentile is much *larger* than that expected from the normal distribution, indicating a skewness toward the larger data points. Figures 4.20c and 4.20d show the plots for all the sample data, with the same general result. (Integers in the plots indicate the number of points at that position.)

When we are dealing with small samples, it is better to think of x_k as the $k/(n+1)$th sample percentile. The reason for this is that $x_1 \leq x_2 \leq \ldots \leq x_n$ actually divide the population distribution into $(n+1)$ segments, with all segments expected to process roughly equal probability masses.

FIGURE **4.20**
Q-Q Plot: Cancer Mortality Rates

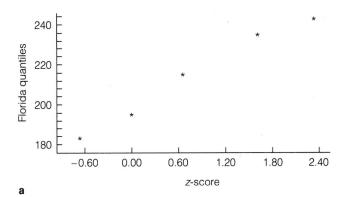

a

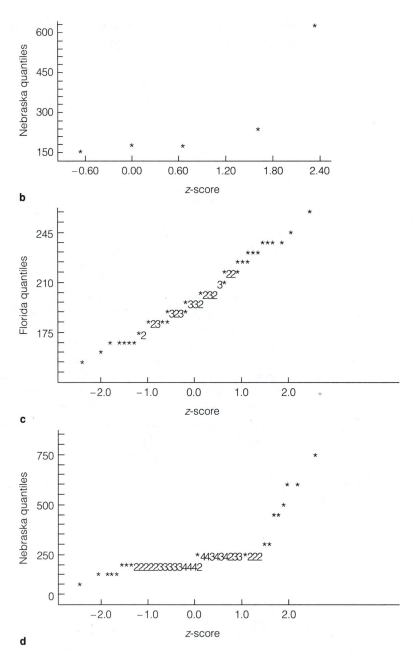

b

c

d

<hr />

E X A M P L E   **4.17**   In the interest of predicting future peak particulate matter values, 12 observations on this variable were obtained from various sites across the United States. (See Chapter 1 for more details.) The ordered measurements, in $\mu g/m^3$, are

22, 24, 24, 28, 30, 31, 33, 45, 45, 48, 51, 79

Should the normal probability model be used to anticipate peak particulate matter values in the future?

Solution   The table below provides the key components of the analysis:

i	X_i	$i/(n+1)$	z-score
1	22	0.077	−1.43
2	24	0.153	−1.02
3	24	0.231	−0.73
4	28	0.308	−0.50
5	30	0.385	−0.29
6	31	0.462	−0.10
7	33	0.538	0.10
8	45	0.615	0.29
9	45	0.692	0.50
10	48	0.769	0.74
11	51	0.846	1.02
12	79	0.923	1.43

Recall that the z-scores are the standard normal values corresponding to the percentiles in the $i/(n+1)$ column. For example, about 69% of the normal curve's area is below a z-score of 0.50. The Q-Q plot of x_i versus z_i in Figure 4.21 shows that the data depart from normality at both ends. The lower values have too short a tail, and the higher values have too long a tail. The data points appear to come from a highly skewed distribution; predicting future values by using a normal distribution would be a poor decision.   ■

FIGURE 4.21
Q-Q Plot: Peak Particulate
Matter Values, 1990

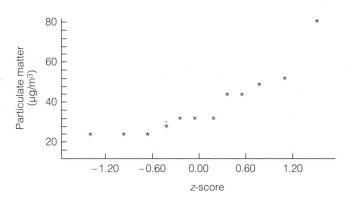

Q-Q plots are cumbersome to construct by hand, especially if the data set is large. Most computer programs for statistical analysis will generate the essential parts quite easily. In Minitab, for example, the NSCORES command (for normal scores) will produce the z-scores corresponding to a set of sample data.

E X A M P L E **4.18** The heights of males between the ages of 18 and 24 have the following *cumulative relative frequency* distribution:

Height (inches)	Cumulative Percentage
61	0.18
62	0.34
63	0.61
64	2.37
65	3.85
66	8.24
67	16.18
68	26.68
69	38.89
70	53.66
71	68.25
72	80.14
73	88.54
74	92.74
75	96.17
76	98.40

Source: Statistical Abstract of the United States, 1991.

Should the normal distribution be used to model the male height distribution? If so, what mean and standard deviation should be used?

Solution Here, the percentiles are easily derived from the cumulative percentages by simply dividing by 100. Using a normal curve area table, or an inverse normal probability function (going from probabilities to z-scores) on a computer, the corresponding z-scores are found to be as follows:

$$-3.29, -2.63, -2.35, -2.01, -1.72, -1.33, -0.92, -0.55,$$
$$-0.21, 0.20, 0.67, 1.05, 1.42, 1.78, 2.13, 2.50$$

The Q-Q plot of heights versus z-scores is provided in Figure 4.22. The plot shows nearly a straight line; the normal distribution can legitimately be used to model these data.

Laying a straightedge through these points to approximate a best-fitting straight line produces a slope around 2.9 and a y-intercept (at $z = 0$) of about 70. Thus, the male heights can be modeled by a normal distribution with mean 70 inches and standard deviation 2.9 inches. (Can you verify the mean and standard deviation approximations by another method?) ▪

As mentioned earlier, we will make much more use of the normal distribution in later chapters. The properties of the normal distribution are summarized here.

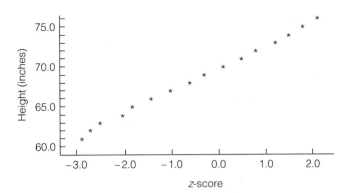

FIGURE **4.22**
Q-Q Plot: Male Heights (18–24 years old)

The Normal Distribution

$$f(x) = \frac{1}{\sigma\sqrt{2\pi}}e^{-(x-\mu)^2/2\sigma^2} \qquad -\infty < x < \infty$$

$$E(X) = \mu \qquad V(X) = \sigma^2$$

Exercises

4.55 Use Table 4 of the Appendix to find the following probabilities for a standard normal random variable Z.

a $P(0 \le Z \le 1.2)$ **b** $P(-0.9 \le Z \le 0)$

c $P(0.3 \le Z \le 1.56)$ **d** $P(-0.2 \le Z \le 0.2)$

e $P(-2.00 \le Z \le -1.56)$

4.56 For a standard normal random variable Z, use Table 4 of the Appendix to find a number z_0 such that

a $P(Z \le z_0) = 0.5$ **b** $P(Z \le z_0) = 0.8749$

c $P(Z \ge z_0) = 0.117$ **d** $P(Z \ge z_0) = 0.617$

e $P(-z_0 \le Z \le z_0) = 0.90$ **f** $P(-z_0 \le Z \le z_0) = 0.95$

4.57 The weekly amount spent for maintenance and repairs in a certain company has approximately a normal distribution with a mean of $400 and a standard deviation of $20. If $450 is budgeted to cover repairs for next week, what is the probability that the actual costs will exceed the budgeted amount?

4.58 In the setting of Exercise 4.57, how much should be budgeted weekly for maintenance and repairs so that the budgeted amount will be exceeded with probability only 0.1?

4.59 A machining operation produces steel shafts having diameters that are normally distributed with a mean of 1.005 inches and a standard deviation of 0.01 inch. Specifications call for diameters to fall within the interval 1.00 ± 0.02 inches. What percentage of the output of this operation will fail to meet specifications?

4.60 Refer to Exercise 4.59. What should be the mean diameter of the shafts produced to minimize the fraction not meeting specifications?

4.61 Wires manufactured for use in a certain computer system are specified to have resistances between 0.12 and 0.14 ohm. The actual measured resistances of the wires produced by Company A have a normal probability distribution with a mean of 0.13 ohm and a standard deviation of 0.005 ohm.

a What is the probability that a randomly selected wire from Company A's production will meet the specifications?

b If four such wires are used in the system and all are selected from Company A, what is the probability that all four will meet the specifications?

4.62 At a temperature of 25°C, the resistances of a type of thermistor are normally distributed with a mean of 10,000 ohms and a standard deviation of 4,000 ohms. The thermistors are to be sorted, with those having resistances between 8,000 and 15,000 ohms being shipped to a vendor. What fraction of these thermistors will be shipped?

4.63 A vehicle driver gauges the relative speed of a preceding vehicle by the speed with which the image of the width of that vehicle varies. This speed is proportional to the speed X of variation of the angle at which the eye subtends this width. According to P. Ferrani and others (*Transportation Research*, 18A, 1984, pp. 50–51), a study of many drivers revealed X to be normally distributed with a mean of zero and a standard deviation of $10(10^{-4})$ radian per second. What fraction of these measurements is more than five units away from zero? What fraction is more than ten units away from zero?

4.64 A type of capacitor has resistances that vary according to a normal distribution with a mean of 800 megohms and a standard deviation of 200 megohms. (See W. Nelson, *Industrial Quality Control*, 1967, pp. 261–268, for a more thorough discussion.) A certain application specifies capacitors with resistances between 900 and 1,000 megohms.

a What proportion of these capacitors will meet this specification?

b If two capacitors are randomly chosen from a lot of capacitors of this type, what is the probability that both will satisfy the specification?

4.65 Sick-leave time used by employees of a firm in one month has approximately a normal distribution with a mean of 200 hours and a variance of 400.

a Find the probability that total sick leave for next month will be less than 150 hours.

b In planning schedules for next month, how much time should be budgeted for sick leave if that amount is to be exceeded with a probability of only 0.10?

4.66 The times of first failure of a unit of a brand of ink jet printers are approximately normally distributed with a mean of 1,500 hours and a standard deviation of 200 hours.

a What fraction of these printers will fail before 1,000 hours?

b What should be the guarantee time for these printers if the manufacturer wants only 5% to fail within the guarantee period?

4.67 A machine for filling cereal boxes has a standard deviation of 1 ounce on ounces of fill per box. What setting of the mean ounces of fill per box will allow 16-ounce boxes to overflow only 1% of the time? Assume that the ounces of fill per box are normally distributed.

4.68 Refer to Exercise 4.67. Suppose the standard deviation σ is not known but can be fixed at certain levels by carefully adjusting the machine. What is the largest value of σ that will allow the actual value dispensed to be within 1 ounce of the mean with probability at least 0.95?

4.69 The histogram in Figure 4.23 depicts the total points scored per game for every NCAA tournament basketball game between 1939 and 1992 ($n = 1,521$). The mean is 143 and the standard deviation is 26.

a Does it appear that total points can be modeled as a normally distributed random variable?

b Show that the empirical rule holds for these data.

c Would you expect a total score to top 200 very often? 250? Why or why not?

d There are 64 games to be played in this year's tournament. How many would you expect to have total scores below 100 points?

F I G U R E **4.23**   Histogram of Total Scores of NCAA Tournament Games, 1939–1992

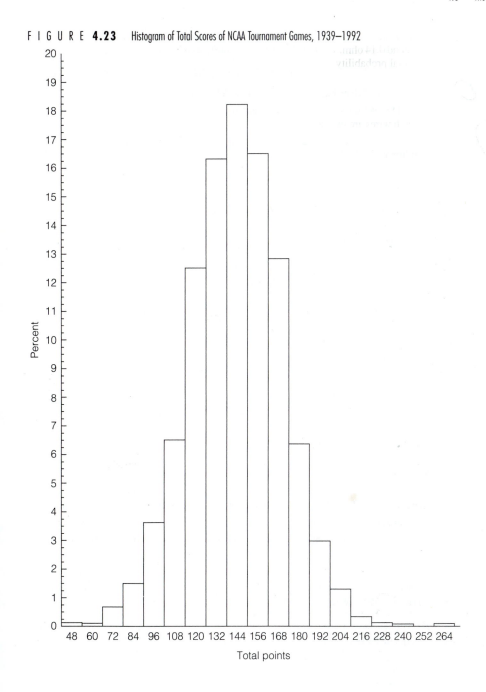

4.70 Another contributor to air pollution is sulfur dioxide. Shown below are twelve peak sulfur dioxide measurements (in ppm) for 1990 from randomly selected U.S. sites:

0.003, 0.010, 0.014, 0.024, 0.024, 0.032, 0.038, 0.042, 0.043, 0.044, 0.047, 0.061

a Should the normal distribution be used as a model for these measurements?

b From a Q-Q plot for these data, approximate the mean and standard deviation. Check the approximation by direct calculation.

4.71 The heights of females aged 18–24 are distributed according to the following cumulative percentages:

Height (inches)	Cumulative Percentage
56	0.05
57	0.43
58	0.94
59	2.22
60	4.22
61	9.13
62	17.75
63	29.06
64	41.81
65	58.09
66	74.76
67	85.37
68	92.30
69	96.23
70	98.34
71	99.38

Source: Statistical Abstract of the United States, 1991.

a Produce a Q-Q plot for these data; discuss their goodness of fit to a normal distribution.

b Approximate the mean and standard deviation of these heights from the Q-Q plot.

4.72 The cumulative proportions of age groups of U.S. residents are shown below for 1900 and 2000 (projected):

Age	Cumulative Proportion, 1900	Cumulative Proportion, 2000
5	0.121	0.066
15	0.344	0.209
25	0.540	0.344
35	0.700	0.480
45	0.822	0.643
55	0.906	0.781
65	0.959	0.870
100	0.990	0.990

a   Construct a Q-Q plot for each year. Use these plots as a basis for discussing key differences between the two age distributions.

b   Each of the Q-Q plots should show some departures from normality. Explain the nature of these departures.

4.7 The Beta Distribution

4.7.1 Probability Density Function

Except for the uniform distribution of Section 4.3, the continuous distributions discussed thus far are defined as nonzero functions over an infinite interval. The beta distribution is very useful for modeling the probabilistic behavior of certain random variables, such as proportions, constrained to fall in the interval (0,1). The beta distribution has the functional form

$$f(x) = \begin{cases} \dfrac{\Gamma(\alpha + \beta)}{\Gamma(\alpha)\Gamma(\beta)} x^{\alpha-1}(1 - x)^{\beta-1} & 0 < x < 1 \\ 0 & \text{elsewhere} \end{cases}$$

where α and β are positive constants. The constant term in $f(x)$ is necessary so that

$$\int_0^1 f(x)dx = 1$$

In other words,

$$\int_0^1 x^{\alpha-1}(1 - x)^{\beta-1}dx = \frac{\Gamma(\alpha)\Gamma(\beta)}{\Gamma(\alpha + \beta)}$$

for positive α and β. This is a handy result to keep in mind. The graphs of some common beta density functions are shown in Figure 4.24.

FIGURE **4.24**
The Beta Density Function

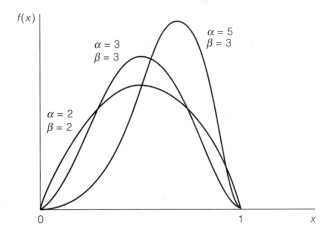

One measurement of interest in the process of sintering copper is the proportion of the volume that is solid, as opposed to the proportion made up of voids. (The proportion due to voids is sometimes called the *porosity* of the solid.) Figure 4.25 shows a relative frequency histogram of proportions of solid copper in samples from a sintering process. This distribution could be modeled with a beta distribution having a large α and small β.

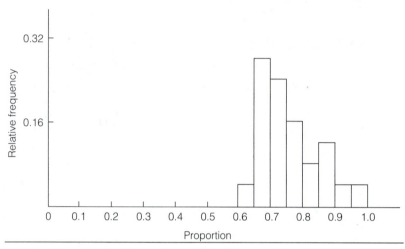

Source: Department of Materials Science, University of Florida.

4.7.2 Mean and Variance

The expected value of a beta random variable is easily found.

$$
\begin{aligned}
E(X) &= \int_0^1 x \frac{\Gamma(\alpha + \beta)}{\Gamma(\alpha)\Gamma(\beta)} x^{\alpha-1}(1-x)^{\beta-1} dx \\
&= \frac{\Gamma(\alpha + \beta)}{\Gamma(\alpha)\Gamma(\beta)} \int_0^1 x^{\alpha}(1-x)^{\beta-1} dx \\
&= \frac{\Gamma(\alpha + \beta)}{\Gamma(\alpha)\Gamma(\beta)} \times \frac{\Gamma(\alpha + 1)\Gamma(\beta)}{\Gamma(\alpha + \beta + 1)} \\
&= \frac{\alpha}{\alpha + \beta}
\end{aligned}
$$

[Recall that $\Gamma(n + 1) = n\Gamma(n)$.] Similar manipulations reveal that

$$
V(X) = \frac{\alpha\beta}{(\alpha + \beta)^2(\alpha + \beta + 1)}
$$

We illustrate the use of this density function in an example.

E X A M P L E   **4.19**   A gasoline wholesale distributor has bulk storage tanks holding a fixed supply. The tanks are filled every Monday. Of interest to the wholesaler is the proportion of this supply that is sold during the week. Over many weeks, this proportion has been observed to be modeled fairly well by a beta distribution with $\alpha = 4$ and $\beta = 2$. Find the expected value of this proportion. Is it highly likely that the wholesaler will sell at least 90% of the stock in a given week?

Solution   By the results given above, with X denoting the proportion of the total supply sold in a given week,

$$E(X) = \frac{\alpha}{\alpha + \beta} = \frac{4}{6} = \frac{2}{3}$$

For the second part, we are interested in

$$P(X > 0.9) = \int_{0.9}^{1} \frac{\Gamma(4+2)}{\Gamma(4)\Gamma(2)} x^3(1-x)dx$$

$$= 20 \int_{0.9}^{1} (x^3 - x^4)dx$$

$$= 20(0.004) = 0.08$$

It is not very likely that 90% of the stock will be sold in a given week.   ∎

The basic properties of the beta distribution are summarized here.

The Beta Distribution

$$f(x) = \begin{cases} \dfrac{\Gamma(\alpha + \beta)}{\Gamma(\alpha)\Gamma(\beta)} x^{\alpha-1}(1-x)^{\beta-1} & 0 < x < 1 \\ 0 & \text{elsewhere} \end{cases}$$

$$E(X) = \frac{\alpha}{\alpha + \beta} \qquad V(X) = \frac{\alpha\beta}{(\alpha + \beta)^2(\alpha + \beta + 1)}$$

Exercises

4.73   Suppose X has a probability density function given by

$$f(x) = \begin{cases} kx^3(1-x)^2 & 0 \le x \le 1 \\ 0 & \text{elsewhere} \end{cases}$$

a   Find the value of k that makes this a probability density function.

b   Find $E(X)$ and $V(X)$.

4.74   If X has a beta distribution with parameters α and β, show that

$$V(X) = \frac{\alpha\beta}{(\alpha + \beta)^2(\alpha + \beta + 1)}$$

4.75   During an 8-hour shift, the proportion of time X that a sheet-metal stamping machine is down for maintenance or repairs has a beta distribution with $\alpha = 1$ and $\beta = 2$. That is,

$$f(x) = \begin{cases} 2(1 - x) & 0 \le x \le 1 \\ 0 & \text{elsewhere} \end{cases}$$

The cost (in hundreds of dollars) of this downtime, due to lost production and cost of maintenance and repair, is given by

$$C = 10 + 20X + 4X^2$$

a   Find the mean and variance of C.

b   Find an interval in which C will lie with probability at least 0.75.

4.76   The percentage of impurities per batch in a certain type of industrial chemical is a random variable X having the probability density function

$$f(x) = \begin{cases} 12x^2(1 - x) & 0 \le x \le 1 \\ 0 & \text{elsewhere} \end{cases}$$

a   Suppose a batch with more than 40% impurities cannot be sold. What is the probability that a randomly selected batch will not be allowed to be sold?

b   Suppose the dollar value of each batch is given by

$$V = 5 - 0.5X$$

Find the expected value and variance of V.

4.77   To study the disposal of pollutants emerging from a power plant, the prevailing wind direction was measured for a large number of days. The direction is measured on a scale of $0°$ to $360°$, but by dividing each daily direction by 360, the measurements can be rescaled to the interval $(0,1)$. These rescaled measurements X are found to follow a beta distribution with $\alpha = 4$ and $\beta = 2$. Find $E(X)$. To what angle does this mean correspond?

4.78   Errors in measuring the arrival time of a wave front from an acoustic source can sometimes be modeled by a beta distribution. (See J. J. Perruzzi and E. J. Hilliard, *Journal of the Acoustical Society of America*, 75(1), 1984, p. 197.)

Suppose these errors have a beta distribution with $\alpha = 1$ and $\beta = 2$, with measurements in microseconds.

a   Find the probability that such a measurement error will be less than 0.5 microsecond.

b   Find the mean and standard deviation of these error measurements.

4.79   In blending fine and coarse powders prior to copper sintering, proper blending is necessary for uniformity in the finished product. One way to check the blending is to select many small samples of the blended powders and measure the weight fractions of the fine particles. These measurements should be relatively constant if good blending has been achieved.

a   Suppose the weight fractions have a beta distribution with $\alpha = \beta = 3$. Find their mean and variance.

b   Repeat (a) for $\alpha = \beta = 2$.

c   Repeat (a) for $\alpha = \beta = 1$.

d   Which of the three cases, (a), (b), or (c), would exemplify the best blending?

4.80　The proportion of pure iron in certain ore samples has a beta distribution with $\alpha = 3$ and $\beta = 1$.

　　a　Find the probability that one of these samples will have more than 50% pure iron.

　　b　Find the probability that two out of three samples will have less than 30% pure iron.

4.8　The Weibull Distribution

4.8.1　Probability Density Function

We have suggested that the gamma distribution can often serve as a probabilistic model for lifetimes of systems components, but other distributions often provide better models for lifelength data. One such distribution is the *Weibull*, which is explored in this section.

A Weibull density function has the form

$$f(x) = \begin{cases} \dfrac{\gamma}{\theta} x^{\gamma-1} e^{-x^{\gamma}/\theta} & x > 0 \\ 0 & \text{elsewhere} \end{cases}$$

for positive parameters θ and γ. For $\gamma = 1$, this becomes an exponential density. For $\gamma > 1$, the functions look something like the gamma functions of Section 4.5 but have somewhat different mathematical properties. We can integrate directly to see that

$$F(x) = \int_0^x \frac{\gamma}{\theta} t^{\gamma-1} e^{-t^{\gamma}/\theta} dt$$

$$= -e^{-t^{\gamma}/\theta} \Big|_0^x = 1 - e^{-x^{\gamma}/\theta} \qquad x > 0$$

A convenient way to look at properties of the Weibull density function is to use the transformation $Y = X^{\gamma}$. Then

$$F_Y(y) = P(Y \le y) = P(X^{\gamma} \le y) = P(X \le y^{1/\gamma})$$

$$= F_X(y^{1/\gamma}) = 1 - e^{-(y^{1/\gamma})^{\gamma}/\theta}$$

$$= 1 - e^{-y/\theta} \qquad y > 0$$

Hence,

$$f_Y(y) = \frac{dF_Y(y)}{dy} = \frac{1}{\theta} e^{-y/\theta} \qquad y > 0$$

and Y has the familiar exponential density.

4.8.2　Mean and Variance

If we want to find $E(X)$ for an X having the Weibull distribution, we have

$$E(X) = E(Y^{1/\gamma}) = \int_0^{\infty} y^{1/\gamma} \frac{1}{\theta} e^{-y/\theta} dy$$

$$= \frac{1}{\theta} \int_0^{\infty} y^{1/\gamma} e^{-y/\theta} dy$$

$$= \frac{1}{\theta} \int_0^\infty y^{1/\gamma} e^{-y/\theta} dy$$

$$= \frac{1}{\theta} \Gamma \left(1 + \frac{1}{\gamma}\right) \theta^{(1+1/\gamma)} = \theta^{1/\gamma} \Gamma \left(1 + \frac{1}{\gamma}\right)$$

The above result follows from recognizing the integral to be of the gamma type.

If we let $\gamma = 2$ in the Weibull density, we see that $Y = X^2$ has an exponential distribution. To reverse the idea outlined above, if we start with an exponentially distributed random variable Y, then the square root of Y will have a Weibull distribution with $\gamma = 2$. We can illustrate empirically by taking the square roots of the data from the exponential distribution given in Table 4.1. These square roots are given in Table 4.2.

A relative frequency histogram for these data is given in Figure 4.26a. Notice that the exponential form has now disappeared and that the curve given by the Weibull density with $\gamma = 2$ and $\theta = 2$ (seen in Figure 4.26b) is a much more plausible model for these observations.

We illustrate the use of the Weibull distribution with the following example.

E X A M P L E   **4.20**   The length of service time during which a certain type of thermistor produces resistances within its specifications has been observed to follow a Weibull distribution with $\theta = 50$ and $\gamma = 2$ (measurements in thousands of hours).

a   Find the probability that one of these thermistors, to be installed in a system today, will function properly for over 10,000 hours.

b   Find the expected lifelength for thermistors of this type.

Solution   a   The Weibull distribution has a closed-form expression for $F(x)$. Thus, if X represents the lifelength of the thermistor in question,

$$P(X > 10) = 1 - F(10)$$
$$= 1 - [1 - e^{-(10)^2/50}]$$
$$= e^{-(10)^2/50} = e^{-2} = 0.14$$

since $\theta = 50$ and $\gamma = 2$.

T A B L E **4.2**
Square Roots of the Lifelengths
of Table 4.1

0.637	0.828	2.186	1.313	2.868
1.531	1.184	1.228	0.542	1.493
0.733	0.484	2.006	1.823	1.709
2.256	1.207	1.032	0.880	0.872
2.364	0.719	1.802	1.526	1.032
1.601	0.715	1.668	2.535	0.914
0.152	0.474	1.230	1.793	1.952
1.826	1.525	0.577	2.741	0.984
1.868	1.709	1.274	0.578	2.119
1.126	1.305	1.623	1.360	0.431

F I G U R E **4.26**
Relative Frequency Histogram
and Weibull Density Function
($\gamma = 2$, $\theta = 2$) for the
Data of Table 4.2

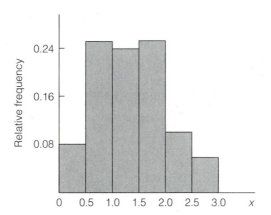

a

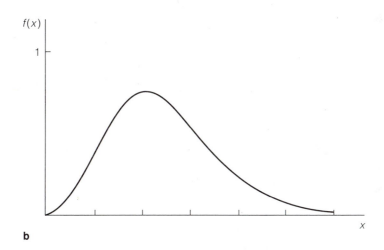

b

b   We know from above that

$$E(X) = \theta^{1/\gamma} \Gamma \left(1 + \frac{1}{\gamma} \right)$$

$$= (50)^{1/2} \Gamma \left(\frac{3}{2} \right)$$

$$= (50)^{1/2} \frac{1}{2} \Gamma \left(\frac{1}{2} \right)$$

$$= (50)^{1/2} \left(\frac{1}{2} \right) \sqrt{\pi} = 6.27$$

Thus, the average service time for these thermistors is 6,270 hours.   ∎

4.8.3 Applications to Real Data

The Weibull distribution is versatile enough to be a good model for many distributions of data that are mound-shaped but skewed. Another advantage of the Weibull distribution over, say, the gamma distribution is that a number of relatively straightforward techniques exist for actually fitting the model to data. We illustrate one such technique here.

For the Weibull distribution,

$$1 - F(x) = e^{-x^{\gamma}/\theta} \qquad x > 0$$

and therefore (with ln denoting natural logarithm)

$$\ln \left[\frac{1}{1 - F(x)} \right] = x^{\gamma}/\theta$$

and

$$\ln \ln \left[\frac{1}{1 - F(x)} \right] = \gamma \ln(x) - \ln(\theta)$$

For simplicity, call the double ln expression on the left LF(x). Plotting LF(x) as a function of ln(x) produces a straight line with slope γ and intercept ln(θ).

Let's see how this works with real data. Peak particulate matter counts from locations across the United States did not appear to fit the normal distribution. Do they fit the Weibull? The counts (in μg/m^3) used above were (in order)

$$22, 24, 24, 28, 30, 31, 33, 45, 45, 48, 51, 79$$

Checking for outliers (using a boxplot, for example) reveals that 79 is an extreme outlier. No smooth distribution will do a good job of picking it up, so we remove it from the data set. Proceeding as in the normal case to produce sample percentiles as approximations to $F(x)$, we have the following data:

i	x	ln(x)	$i/(n+1)$	Approx.LF(x)
1	22	3.09104	0.083333	−2.44172
2	24	3.17805	0.166667	−1.70198
3	24	3.17805	0.250000	−1.24590
4	28	3.33220	0.333333	−0.90272
5	30	3.40120	0.416667	−0.61805
6	31	3.43399	0.500000	−0.36651
7	33	3.49651	0.583333	−0.13300
8	45	3.80666	0.666667	0.09405
9	45	3.80666	0.750000	0.32663
10	48	3.87120	0.833333	0.58320
11	51	3.93183	0.916667	0.91024

FIGURE **4.27**

Plot of LF(x) Versus ln(x)

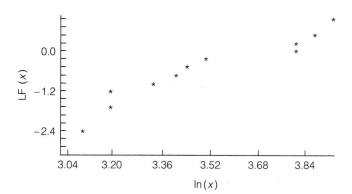

Here $i/(n+1)$ is used to approximate $F(X_i)$ in LF(x). Figure 4.27 shows the plot of LF(x) versus ln(x). The points lie rather close to a straight line that has slope about 3.2 and intercept about -11.6. (The line can be approximated by laying a straightedge through the middle of the scatterplot.) Thus, the particulate counts appear to be adequately modeled by a Weibull distribution with γ approximately 3.2 and ln(θ) approximately 11.6.

Note that fitting an exponential distribution to data could follow a similar procedure with $\gamma = 1$. In this case,

$$\ln\left[\frac{1}{1-F(x)}\right] = x/\theta$$

so that a plot of an estimate of the left side against x should reveal a straight line through the origin with slope $1/\theta$.

The Weibull Distribution

$$f(x) = \begin{cases} \dfrac{\gamma}{\theta} x^{\gamma-1} e^{-x^{\gamma}/\theta} & x > 0 \\ 0 & \text{elsewhere} \end{cases}$$

$$E(X) = \theta^{1/\gamma} \Gamma\left(1 + \frac{1}{\gamma}\right)$$

$$V(X) = \theta^{2/\gamma} \left\{ \Gamma\left(1 + \frac{2}{\gamma}\right) - \left[\Gamma\left(1 + \frac{1}{\gamma}\right)\right]^2 \right\}$$

Exercises

4.81   Fatigue life, in hundreds of hours, for a certain type of bearing has approximately a Weibull distribution with $\gamma = 2$ and $\theta = 4$.

    a   Find the probability that a bearing of this type fails in less than 200 hours.

b   Find the expected value of the fatigue life for these bearings.

4.82   The maximum flood levels, in millions of cubic feet per second, for a certain U.S. river have a Weibull distribution with $\gamma = 1.5$ and $\theta = 0.6$. (See A. C. Cohen, B. Whitten, and Y. Ding, *Journal of Quality Technology*, 16(3), 1984, p. 165, for more details.) Find the probability that the maximum flood level for next year

a   will exceed 0.5.

b   will be less than 0.8.

4.83   The time necessary to achieve proper blending of copper powders before sintering was found to have a Weibull distribution with $\gamma = 1.1$ and $\theta = 2$ (measurements in minutes). Find the probability that a proper blending takes less than 2 minutes.

4.84   The ultimate tensile strength of steel wire used to wrap concrete pipe was found to have a Weibull distribution with $\gamma = 1.2$ and $\theta = 270$ (measurements in thousands of pounds). Pressure in the pipe at a certain point may require an ultimate tensile strength of at least 300,000 pounds. What is the probability that a wire will possess this required strength?

4.85   The yield strengths of certain steel beams have a Weibull distribution with $\gamma = 2$ and $\theta = 3,600$ (measurements in pounds per square inch). Two such beams are used in a construction project that calls for yield strengths in excess of 70,000 psi. Find the probability that both beams meet the specifications for the project.

4.86   The pressure, in thousand psi, exerted on the tank of a steam boiler has a Weibull distribution with $\gamma = 1.8$ and $\theta = 1.5$. The tank is built to withstand pressures of 2,000 psi. Find the probability that this limit will be exceeded.

4.87   Resistors being used in the construction of an aircraft guidance system have lifelengths that follow a Weibull distribution with $\gamma = 2$ and $\theta = 10$ (measurements in thousands of hours).

a   Find the probability that a randomly selected resistor of this type has a lifelength that exceeds 5,000 hours.

b   If three resistors of this type are operating independently, find the probability that exactly one of the three resistors burns out prior to 5,000 hours of use.

c   Find the mean and variance of the lifelength of such a resistor.

4.88   Failures to the bleed systems in jet engines were causing some concern at an air base. It was decided to model the failure-time distribution for these systems so that future failures could be anticipated a little better. Ten observed failure times (in operating hours since installation) are given below (from *Weibull Analysis Handbook, Pratt & Whitney Aircraft*, 1983):

$$1,198, 884, 1,251, 1,249, 708, 1,082, 884, 1,105, 828, 1,013$$

Does a Weibull distribution appear to be a good model for these data? If so, what values should be used for γ and θ?

4.89   Maximum wind-gust velocities in summer thunderstorms were found to follow a Weibull distribution with $\gamma = 2$ and $\theta = 400$ (measurements in feet per second). Engineers designing structures in the areas in which these thunderstorms are found are interested in finding a gust velocity that will be exceeded only with probability 0.01. Find such a value.

4.90   The velocities of gas particles can be modeled by the Maxwell distribution, with probability density function given by

$$f(v) = 4\pi \left(\frac{m}{2\pi KT} \right)^{3/2} v^2 e^{-v^2(m/2KT)} \qquad v > 0$$

where m is the mass of the particle, K is Boltzmann's constant, and T is the absolute temperature.

a   Find the mean velocity of these particles.

b   The kinetic energy of a particle is given by $(1/2)mV^2$. Find the mean kinetic energy for a particle.

4.9 Reliability

One important measure of the quality of products is their *reliability*, or probability of working for a specified period of time. We want products, whether they be cars, TV sets, or shoes, that do not break down or wear out for some time, and we want to know what this time period is expected to be. The study of reliability is a probabilistic exercise, because data or models on component lifetimes are used to predict future behavior—how long a process will operate before it fails.

In reliability studies, the underlying random variable of interest, X, is usually lifetime.

> If a component has lifetime X with distribution function F, then the **reliability** of the component is
>
> $$R(t) = P(X > t) = 1 - F(t)$$

It follows that, for exponentially distributed lifetimes,

$$R(t) = e^{-t/\theta} \quad t \geq 0$$

and for lifetimes following a Weibull distribution

$$R(t) = e^{-t^{\gamma}/\theta} \quad t \geq 0$$

Reliability functions for gamma and normal distributions do not exist in closed form. In fact, the normal model is not often used to model lifelength data, since length of life tends to exhibit positively skewed behavior.

4.9.1 Failure Rate Function

In addition to probability density and distribution functions and the reliability function, another function is of use in working with lifelength data. Suppose X denotes the lifelength of a component with density function $f(x)$ and distribution function $F(x)$. The *failure rate function* $r(t)$ is defined as

$$r(t) = \frac{f(t)}{1 - F(t)} \quad t > 0, \ F(t) < 1$$

For an intuitive look at what $r(t)$ is measuring, suppose dt denotes a very small interval around the point t. Then $f(t)\,dt$ is approximately the probability that X takes on a value in $(t, t + dt)$. Also, $1 - F(t) = P(X > t)$. Thus,

$$r(t)\,dt = \frac{f(t)\,dt}{1 - F(t)}$$
$$\approx P[X \in (t, t + dt) | X > t]$$

In other words, $r(t)\,dt$ represents the probability of failure during the time interval $(t, t + dt)$ given that the component has survived up to time t.

For the exponential case,

$$r(t) = \frac{f(t)}{1 - F(t)} = \frac{\frac{1}{\theta} e^{-t/\theta}}{e^{-t/\theta}} = \frac{1}{\theta}$$

or X has a *constant* failure rate. It is unlikely that many individual components have a constant failure rate over time (most fail more frequently as they age), but it may be true of some systems that undergo regular preventive maintenance. For the Weibull distribution, the failure rate function is

$$r(t) = \frac{\gamma}{\theta} t^{\gamma - 1}$$

which possesses a rich variety of shapes for $\alpha > 1$.

The failure rate function $r(t)$ for the gamma case is not easily displayed, since $F(t)$ does not have a simple closed form. However, for $\alpha > 1$, this function will increase but is always bounded above by $1/\beta$. A typical form is shown in Figure 4.28.

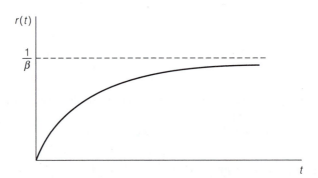

F I G U R E 4.28
The Failure Rate Function for the Gamma Distribution ($\alpha > 1$)

4.9.2   Series and Parallel Systems

A system is made up of a number of components, such as relays in an electrical system or check valves in a water system. The reliability of systems depends critically on how the components are networked into the system. A *series* system (Figure 4.29a) fails as soon as any one component fails. A *parallel* system (Figure 4.29b) fails only when all components have failed.

Suppose the system components in Figure 4.29 each have reliability function $R(t)$, and suppose the components operate independently of one another. What are the system reliabilities, $R_s(t)$? For the series system, lifelength will exceed t only if each component has lifelength exceeding t. Thus,

$$R_s(t) = R(t) \cdot R(t) \cdot R(t) = [R(t)]^3$$

FIGURE **4.29**
Series and Parallel Systems

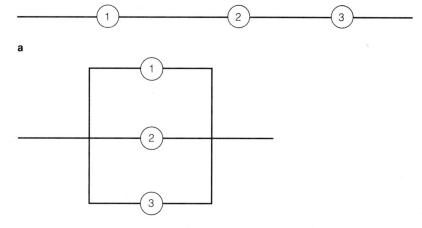

a

b

For the parallel system, lifelength will be less than t only if all components fail before t. Thus,

$$1 - R_s(t) = [1 - R(t)] \cdot [1 - R(t)] \cdot [1 - R(t)]$$

or

$$R_s(t) = 1 - [1 - R(t)]^3$$

Of course, most systems are combinations of components in series and components in parallel. The rules of probability must be used to evaluate the system reliability in each special case, but breaking down the system into series and parallel subsystems often helps simplify this process.

4.9.3 Redundancy

What happens as we add components to a system? For a series of n independent components,

$$R_s(t) = [R(t)]^n$$

and, since $R(t) \leq 1$, adding more components in series will just make things worse! However, for n components operating in parallel,

$$R_s(t) = 1 - [1 - R(t)]^n$$

which will increase with n. Thus, system reliability can be improved by adding backup components in parallel, a practice called *redundancy*.

How many components do we need to achieve a specified system reliability? If we have a fixed value for $R_s(t)$ in mind and have independently operating parallel components, then

$$1 - R_s(t) = [1 - R(t)]^n$$

Letting ln denote the natural logarithm,

$$\ln[1 - R_s(t)] = n \ln[1 - R(t)]$$

and

$$n = \frac{\ln[1 - R_s(t)]}{\ln[1 - R(t)]}$$

This is but a brief introduction to the interesting and important area of reliability. But the key to all reliability problems is a firm understanding of basic probability.

Exercises

4.91 For a component with exponentially distributed lifetime, find the reliability to time $t_1 + t_2$ given that the component has already lived past t_1. Why is the constant failure rate referred to as the "memoryless" property?

4.92 For a series of n components operating independently, each with the same exponential life distribution, find an expression for $R_s(t)$. What is the mean lifetime of the system?

4.93 For independently operating components with identical life distributions, find the system reliability for each of the following:

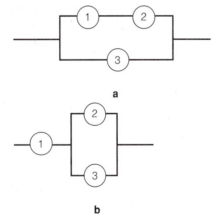

a

b

Which has the higher reliability?

4.94 Suppose each relay in an electrical circuit has reliability 0.9 for a specified operating period of t hours. How could you configure a system of such relays to bring $R_s(t)$ up to 0.999?

4.10 Moment-Generating Functions for Continuous Random Variables[†]

As in the case of discrete distributions, the moment-generating functions of continuous random variables help in finding expected values and in identifying certain properties of probability distributions. We now present a short discussion of moment-generating functions for continuous random variables.

The moment-generating function of a continuous random variable X with probability density function $f(x)$ is given by

$$M(t) = E(e^{tX}) = \int_{-\infty}^{\infty} e^{tx} f(x) dx \quad \blacksquare$$

when the integral exists.

For the exponential distribution, this becomes

$$\begin{aligned}
M(t) &= \int_{0}^{\infty} e^{tx} \frac{1}{\theta} e^{-x/\theta} dx \\
&= \frac{1}{\theta} \int_{0}^{\infty} e^{-x(1/\theta - t)} dx \\
&= \frac{1}{\theta} \int_{0}^{\infty} e^{-x(1-\theta t)/\theta} dx \\
&= \frac{1}{\theta} \Gamma(1) \left(\frac{\theta}{1 - \theta t} \right) = (1 - \theta t)^{-1}
\end{aligned}$$

Since $M(t)$ needs to exist only in a neighborhood of zero, t can be small enough to make $1 - \theta t$ positive.

We can now use $M(t)$ to find $E(X)$, since

$$M'(0) = E(X)$$

and for the exponential distribution we have

$$\begin{aligned}
E(X) = M'(0) &= [-(1 - \theta t)^{-2}(-\theta)]_{t=0} \\
&= \theta
\end{aligned}$$

Similarly, we could find $E(X^2)$ and then $V(X)$ by using the moment-generating function.

An argument analogous to the one used for the exponential distribution will show that, for the gamma distribution,

$$M(t) = (1 - \beta t)^{-\alpha}$$

From this we can see that, if X has a gamma distribution,

$$\begin{aligned}
E(X) = M'(0) &= [-\alpha(1 - \beta t)^{-\alpha-1}(-\beta)]_{t=0} \\
&= \alpha\beta
\end{aligned}$$

[†]Optional section.

$$E(X^2) = M^{(2)}(0) = [\alpha\beta(-\alpha - 1)(1 - \beta t)^{-\alpha-2}(-\beta)]_{t=0}$$
$$= \alpha(\alpha + 1)\beta^2$$
$$E(X^3) = M^{(3)}(0) = [\alpha(\alpha + 1)\beta^2(-\alpha - 2)(1 - \beta t)^{-\alpha-3}(-\beta)]_{t=0}$$
$$= \alpha(\alpha + 1)(\alpha + 2)\beta^3$$

and so on.

E X A M P L E   **4.21**   The force h exerted by a mass m moving at velocity v is

$$h = \frac{mv^2}{2}$$

Consider a device that fires a serrated nail into concrete at a mean velocity of 500 feet per second, where V, the random velocity, possesses a density function

$$f(v) = \frac{v^3 e^{-v/b}}{b^4 \Gamma(4)} \qquad b = 500, \; v \geq 0$$

If each nail possesses mass m, find the expected force exerted by a nail.

Solution

$$E(H) = E\left(\frac{mV^2}{2}\right) = \frac{m}{2} E(V^2)$$

The density function for V is a gamma-type function with $\alpha = 4$ and $\beta = 500$. Therefore,

$$E(V^2) = \alpha(\alpha + 1)\beta^2$$
$$= (4)(5)(500)^2$$
$$= 5{,}000{,}000$$

and

$$E(H) = \frac{m}{2} E(V^2)$$
$$= \frac{m}{2}(5{,}000{,}000) = 2{,}500{,}000m \quad \blacksquare$$

Two important properties of moment-generating functions follow:

1   If a random variable X has moment-generating function $M_X(t)$, then $Y = aX + b$, for constants a and b, has moment-generating function

$$M_Y(t) = e^{tb} M_X(at)$$

2 Moment-generating functions are unique. That is, if two random variables have the same moment-generating function, then they have the same probability distribution. If they have different moment-generating functions, then they have different distributions.

Property 1 is easily shown. The proof of property 2 is beyond the scope of this textbook, but we will use it in identifying distributions, as demonstrated below.

To see something of the usefulness of these properties, suppose X has a gamma distribution with parameters α and β. Then

$$M_X(t) = (1 - \beta t)^{-\alpha}$$

as shown above. Now let

$$Y = aX + b$$

for constants a and b. From property 1 above,

$$M_Y(t) = e^{tb}[1 - \beta(at)]^{-\alpha}$$
$$= e^{tb}[1 - (a\beta)t]^{-\alpha}$$

Since this moment-generating function is *not* of the gamma form (it has an extra multiplier of e^{tb}), Y will *not* have a gamma distribution (from property 2). If, however, $b = 0$ and $Y = aX$, then Y will have a gamma distribution with parameters α and $a\beta$.

The normal random variable is so commonly used that we should investigate its moment-generating function. We do so by first finding the moment-generating function of $X - \mu$, where X is normally distributed with mean μ and variance σ^2. Now

$$E(e^{t(X-\mu)}) = \int_{-\infty}^{\infty} e^{t(x-\mu)} \frac{1}{\sigma\sqrt{2\pi}} e^{-(x-\mu)^2/2\sigma^2} dx$$

Letting $y = x - \mu$, the integral becomes

$$\frac{1}{\sigma\sqrt{2\pi}} \int_{-\infty}^{\infty} e^{ty-y^2/2\sigma^2} = \frac{1}{\sigma\sqrt{2\pi}} \int_{-\infty}^{\infty} \exp\left[-\frac{1}{2\sigma^2}(y^2 - 2\sigma^2 ty)\right] dy$$

Upon the completing of the square in the exponent, we have

$$-\frac{1}{2\sigma^2}(y^2 - 2\sigma^2 ty) = -\frac{1}{2\sigma^2}(y^2 - 2\sigma^2 ty + \sigma^4 t^2) + \frac{1}{2}t^2\sigma^2$$
$$= -\frac{1}{2\sigma^2}(y - \sigma^2 t)^2 + \frac{1}{2}t^2\sigma^2$$

so the integral becomes

$$e^{t^2\sigma^2/2} \frac{1}{\sigma\sqrt{2\pi}} \int_{-\infty}^{\infty} \exp\left[-\frac{1}{2\sigma^2}(y - \sigma^2 t)^2\right] dy = e^{t^2\sigma^2/2}$$

since the remaining integrand forms a normal probability density that integrates to unity.

If $Y = X - \mu$ has a moment-generating function given by

$$M_Y(t) = e^{t^2\sigma^2/2}$$

then

$$X = Y + \mu$$

has a moment-generating function given by

$$
\begin{aligned}
M_X(t) &= e^{t\mu} M_Y(t) \\
&= e^{t\mu} e^{t^2\sigma^2/2} \\
&= e^{t\mu + t^2\sigma^2/2}
\end{aligned}
$$

by property 1 stated above.

For this same normal random variable X, let

$$Z = \frac{X - \mu}{\sigma} = \frac{1}{\sigma}X - \frac{\mu}{\sigma}$$

Then, by property 1,

$$
\begin{aligned}
M_Z(t) &= e^{-\mu t/\sigma} e^{\mu t/\sigma + t^2\sigma^2/2\sigma^2} \\
&= e^{t^2/2}
\end{aligned}
$$

The normal generating function of Z has the form of a moment-generating function for a normal random variable with mean zero and variance 1. Thus, by property 2, Z must have that distribution.

Exercises

4.95 Show that a gamma distribution with parameters α and β has the moment-generating function

$$M(t) = (1 - \beta t)^{-\alpha}$$

4.96 Using the moment-generating function for the exponential distribution with mean θ, find $E(X^2)$. Use this result to show that $V(X) = \theta^2$.

4.97 Let Z denote a standard normal random variable. Find the moment-generating function of Z directly from the definition.

4.98 Let Z denote a standard normal random variable. Find the moment-generating function of Z^2. What does the uniqueness property of the moment-generating function tell you about the distribution of Z^2?

4.11 Activities for the Student: Simulation

When generating random variables from a continuous distribution with a cumulative distribution function (cdf) $F(x)$, the inverse transformation method—or inverse cdf method—may be used. The three-step procedure is as follows:

1 Generate R_i, which is uniform over the interval (0,1). This is denoted as

$$R_i \sim U(0, 1).$$

2 Let $R_i = F(X)$, where $F(x)$ is the cdf for the distribution of x's.

3 Evaluate the equation in (2) for X, which gives $X = F^{-1}(R)$.

Uniform   The probability density function (pdf) for the uniform distribution is defined as $f(x) = 1/(b - a), a \le x \le b$. The cdf is given by $F(x) = (x - a)/(b - a)$. Using the inverse cdf method to generate uniform random numbers x_i, let $r_i = (x_i - a)/(b - a)$. Solving for x_i, we obtain $x_i = a + (b - a)r_i$.

Exponential   The pdf for the exponential distribution is defined as $f(x) = (1/\theta)e^{-x/\theta}$, $x \ge 0$. The cdf is given by $F(x) = 1 - e^{-x/\theta}$. Using the inverse cdf method, we can generate exponential random numbers x_i by letting $r_i = 1 - e^{-x_i/\theta}$. Solving for x_i, $e^{-x_i/\theta} = 1 - r_i$. Taking the natural logarithm on both sides, we obtain $x_i = -\theta \ln(1 - r_i)$. Since $(1 - r_i) \sim U(0, 1)$, we let $x_i = -\theta \ln r_i$.

Gamma   If X_i is a random variable from the gamma distribution with parameters (α, β) and α is an integer n, then X_i is the *convolution* (sum) of n exponential random variables y_j with parameter β. Thus, to simulate a gamma random variable X_i, generate n exponential random variables as stated previously. Then let $X_i = \sum Y_j$, where $j = 1, 2, \ldots, n$.

Normal   In attempting to apply the inverse cdf method to generate normal random variables, a problem arises in that there is no explicit closed-form expression for the cdf of X, $F(x)$, or the inverse function $F^{-1}(R)$. Methods of approximation have been developed that work well in simulating normal random variables, and one of these methods will be discussed in Chapter 6.

Let us now consider a situation involving exponential random variables. In engineering, one is often concerned with a *parallel system*. A system is defined as a parallel system if it functions when at least one of the components of the system is working. Let's consider a parallel system that has two components. We define the random variable $X =$ time until failure for component 1, and the random variable $Y =$ time until failure for component 2. X and Y are independent exponential random variables with the mean for X equal to θ_x and the mean for Y equal to θ_y. A variable of interest is the maximum of X and Y; that is, let $W = \max\{X, Y\}$. By simulation, we can see how the distribution of W will behave. Figure 4.30 gives the results of three simulations for this situation. Note that the distribution of W does *not* look like

F I G U R E **4.30** Simulations of the Distribution of Maximum Exponential Random Variables in a Parallel System

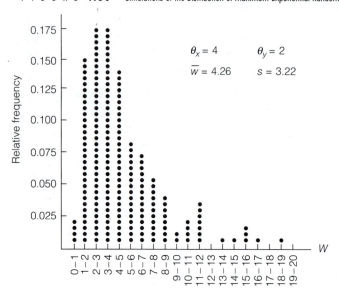

$\theta_x = 4$ $\theta_y = 2$

$\overline{w} = 4.26$ $s = 3.22$

a Simulation 1

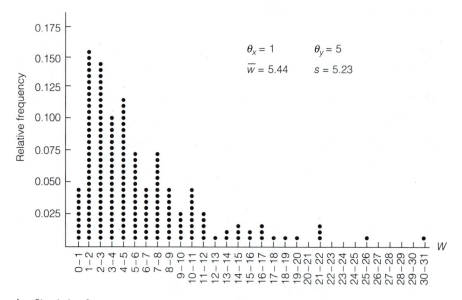

$\theta_x = 1$ $\theta_y = 5$

$\overline{w} = 5.44$ $s = 5.23$

b Simulation 2

FIGURE **4.30**
(Continued)

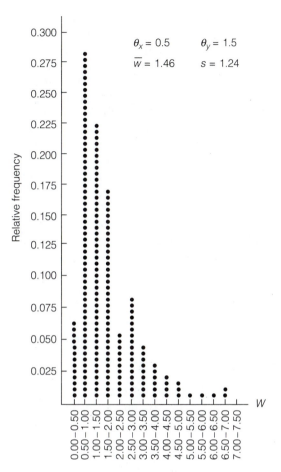

$\theta_x = 0.5$ $\theta_y = 1.5$

$\overline{w} = 1.46$ $s = 1.24$

c   Simulation 3

an exponential distribution. The maximum has relatively low probability of being close to zero but could be quite large compared to most values of X or Y.

It is time for you to try your hand at simulating the behavior of continuous random variables.

1   Suppose a two-component system has components operating independently and in parallel. Each component has the same Weibull distribution of lifelengths. Generate a simulated distribution for the lifelength of the system. (You may choose your own parameter values.)

2   Repeat (1) for the same components operating in series.

4.12 Summary

Improving the quality of processes often requires careful study of the behavior of observations measured on a continuum (time, weight, distance, etc.). Probability distributions for such measurements are characterized by *probability density functions* and cumulative *distribution functions*. The *uniform* distribution models the waiting time for an event that is equally likely to occur anywhere in a finite interval. The *exponential* distribution is a useful model for times between random events. The *gamma* model fits the behavior of sums of exponential variables (and other statistics to be studied later). The most widely used continuous probability model is the *normal* distribution, which exhibits the symmetric, mound-shaped behavior often seen in real data. For data that are mound-shaped but skewed, the *Weibull* distribution provides a useful and versatile model.

All models are approximations to reality, but the ones discussed here are relatively easy to use while providing good approximations in a variety of settings.

Supplementary Exercises

4.99   Let Y possess a density function

$$f(y) = \begin{cases} cy & 0 \le y \le 2 \\ 0 & \text{elsewhere} \end{cases}$$

a   Find c.
b   Find $F(y)$.
c   Graph $f(y)$ and $F(y)$.
d   Use $F(y)$ in part (b) to find $P(1 \le Y \le 2)$.
e   Use the geometric figure for $f(y)$ to calculate $P(1 \le Y \le 2)$.

4.100   Let Y have the density function given by

$$f(y) = \begin{cases} cy^2 + y & 0 \le y \le 2 \\ 0 & \text{elsewhere} \end{cases}$$

a   Find c.
b   Find $F(y)$.
c   Graph $f(y)$ and $F(y)$.
d   Use $F(y)$ in part (b) to find $F(-1)$, $F(0)$, and $F(1)$.
e   Find $P(0 \le Y \le 0.5)$.
f   Find the mean and variance of Y.

4.101   Let Y have the density function given by

$$f(y) = \begin{cases} 0.2 & -1 < y \le 0 \\ 0.2 + cy & 0 < y \le 1 \\ 0 & \text{elsewhere} \end{cases}$$

Answer parts (a) through (f) in Exercise 4.100.

4.102   The grade-point averages of a large population of college students are approximately normally distributed with mean equal to 2.4 and standard deviation equal to 0.5. What fraction of the students will possess a grade-point average in excess of 3.0?

4.103   Refer to Exercise 4.102. If students possessing a grade-point average equal to or less than 1.9 are dropped from college, what percentage of the students will be dropped?

4.104 Refer to Exercise 4.102. Suppose that three students are randomly selected from the student body. What is the probability that all three will possess a grade-point average in excess of 3.0?

4.105 A machine operation produces bearings with diameters that are normally distributed with mean and standard deviation equal to 3.0005 and 0.001, respectively. Customer specifications require the bearing diameters to lie in the interval 3.000 ± 0.0020. Those outside the interval are considered scrap and must be remachined or used as stock for smaller bearings. With the existing machine setting, what fraction of total production will be scrap?

4.106 Refer to Exercise 4.105. Suppose five bearings are drawn from production. What is the probability that at least one will be defective?

4.107 Let Y have density function

$$f(y) = \begin{cases} cye^{-2y} & 0 \le y \le \infty \\ 0 & \text{elsewhere} \end{cases}$$

a Give the mean and variance for Y.

b Give the moment-generating function for Y.

c Find the value of c.

4.108 Find $E(X^k)$ for the beta random variable. Then find the mean and variance for the beta random variable.

4.109 The yield force of a steel reinforcing bar of a certain type is found to be normally distributed with a mean of 8,500 pounds and a standard deviation of 80 pounds. If three such bars are to be used on a certain project, find the probability that all three will have yield forces in excess of 8,700 pounds.

4.110 The lifetime X of a certain electronic component is a random variable with density function

$$f(x) = \begin{cases} (1/100)e^{-x/100} & x > 0 \\ 0 & \text{elsewhere} \end{cases}$$

Three of these components operate independently in a piece of equipment. The equipment fails if at least two of the components fail. Find the probability that the equipment operates for at least 200 hours without failure.

4.111 An engineer has observed that the gap times between vehicles passing a certain point on a highway have an exponential distribution with a mean of 10 seconds.

a Find the probability that the next gap observed will be no longer than 1 minute.

b Find the probability density function for the sum of the next four gap times to be observed. What assumptions are necessary for this answer to be correct?

4.112 The proportion of time, per day, that all checkout counters in a supermarket are busy is a random variable X having probability density function

$$f(x) = \begin{cases} kx^2(1-x)^4 & 0 \le x \le 1 \\ 0 & \text{elsewhere} \end{cases}$$

a Find the value of k that makes this a probability density function.

b Find the mean and variance of X.

4.113 If the lifelength X for a certain type of battery has a Weibull distribution with $\gamma = 2$ and $\theta = 3$ (with measurements in years), find the probability that the battery lasts less than four years given that it is now two years old.

4.114 The time (in hours) it takes a manager to interview an applicant has an exponential distribution with $\theta = 1/2$. Three applicants arrive at 8:00 A.M., and interviews begin. A fourth applicant arrives at 8:45 A.M. What is the probability that he has to wait before seeing the manager?

4.115   The weekly repair cost Y for a certain machine has a probability density function given by

$$f(y) = \begin{cases} 3(1-y)^2 & 0 < y < 1 \\ 0 & \text{elsewhere} \end{cases}$$

with measurements in hundreds of dollars. How much money should be budgeted each week for repair costs so that the actual cost will exceed the budgeted amount only 10% of the time?

4.116   A builder of houses has to order some supplies that have a waiting time for delivery Y uniformly distributed over the interval one to four days. Since she can get by without the supplies for two days, the cost of the delay is fixed at \$100 for any waiting time up to two days. However, after two days the cost of the delay is \$100 plus \$20 per day for any time beyond two days. That is, if the waiting time is 3.5 days, the cost of the delay is \$100 + \$20(1.5) = \$130. Find the expected value of the builder's cost due to waiting for supplies.

4.117   There is a relationship between incomplete gamma integrals and sums of Poisson probabilities, given by

$$\frac{1}{\Gamma(\alpha)} \int_{\lambda}^{\infty} y^{\alpha-1} e^{-y} dy = \sum_{y=0}^{\alpha-1} \frac{\lambda^y e^{-\lambda}}{y!}$$

for integer values of α. If Y has a gamma distribution with $\alpha = 2$ and $\beta = 1$, find $P(Y > 1)$ by using the above equality and Table 3 of the Appendix.

4.118   The weekly downtime in hours for a certain production line has a gamma distribution with $\alpha = 3$ and $\beta = 2$. Find the probability that downtime for a given week will not exceed 10 hours.

4.119   Suppose that plants of a certain species are randomly dispersed over a region, with a mean density of λ plants per unit area. That is, the number of plants in a region of area A has a Poisson distribution with mean λA. For a randomly selected plant in this region, let R denote the distance to the nearest neighboring plant.

a   Find the probability density function for R. [*Hint*: Note that $P(R > r)$ is the same as the probability of seeing no plants in a circle of radius r.]

b   Find $E(R)$.

4.120   A random variable X is said to have a *lognormal* distribution if $Y = \ln(X)$ has a normal distribution. (The symbol "ln" denotes natural logarithm.) In this case, X must be nonnegative. The shape of the lognormal probability density function is similar to that of the gamma distribution, with long tails to the right. The equation of the lognormal density function is

$$f(x) = \begin{cases} \dfrac{1}{\sqrt{2\pi}\sigma x} e^{-(\ln(x)-\mu)^2/2\sigma^2} & x > 0 \\ 0 & \text{elsewhere} \end{cases}$$

Since $\ln(x)$ is a monotonic function of x,

$$P(X \le x) = P[\ln(X) \le \ln(x)] = P[Y \le \ln(x)]$$

where Y has a normal distribution with mean μ and variance σ^2. Thus, probabilities in the lognormal case can be found by transforming them to the normal case. If X has a lognormal distribution with $\mu = 4$ and $\sigma^2 = 1$, find

a $P(X \le 4)$ **b** $P(X > 8)$

4.121   If X has a lognormal distribution with parameters μ and σ^2, then it can be shown that

$$E(X) = e^{\mu + \sigma^2/2}$$

and

$$V(X) = e^{2\mu+\sigma^2}(e^{\sigma^2} - 1)$$

The grains comprising polycrystalline metals tend to have weights that follow a lognormal distribution. For a certain type of aluminum, gram weights have a lognormal distribution with $\mu = 3$ and $\sigma = 4$ (in units of 10^{-2} gram).

a   Find the mean and variance of the grain weights.

b   Find an interval in which at least 75% of the grain weights should lie. [*Hint:* Use Tchebysheff's Theorem.]

c   Find the probability that a randomly chosen grain weighs less than the mean gram weight.

4.122   Let Y denote a random variable with probability density function given by

$$f(y) = (1/2)e^{-|y|} \qquad -\infty < y < \infty$$

Find the moment-generating function of Y and use it to find $E(Y)$.

Multivariate Probability Distributions

About This Chapter

In most investigations, multiple types of observations are made on the same experimental units. Tensile and torsion properties are both measured on the same sample of wire, for example, or amount of cement and porosity are both measured on the same sample of concrete. Physicians studying the effect of aspirin on heart attacks also want to understand its effect on strokes. This chapter provides techniques for modeling the joint behavior of two or more random variables being studied together. The ideas of covariance and correlation are introduced to measure the direction and strength of associations between variables.

Contents

[†]Optional section.

5.1 Bivariate and Marginal Probability Distributions

Chapters 3 and 4 dealt with experiments that produced a single numerical response, or random variable, of interest. We discussed, for example, the lifelength X of a type of battery or the strength Y of a steel casing. Often, however, we want to study the joint behavior of two random variables, such as the joint behavior of lifelength *and* casing strength for these batteries. Perhaps, in such a study, we can identify a region in which there is a combination of lifelength and casing strength that will be optimal in terms of balancing cost of manufacturing with customer satisfaction. To proceed with such a study, we must know how to handle joint probability distributions. When only two random variables are involved, these joint distributions are called *bivariate distributions*. We will discuss the bivariate case in some detail; extensions to more than two variables follow along similar lines.

Other situations in which bivariate probability distributions are important come to mind easily. A physician studies the joint behavior of pulse and exercise. An educator studies the joint behavior of grades and time devoted to study, or the interrelationship of pretest and posttest scores. An economist studies the joint behavior of business volume and profits. In fact, most real problems we come across will have more than one underlying random variable of interest.

The National Highway Traffic Safety Administration, among other groups, is interested in the effect of seat belt use on saving lives. One study reported statistics on children under the age of 5 who were involved in motor vehicle accidents in which at least one fatality occurred. For 7,060 such accidents between 1985 and 1989, the results are as shown in Table 5.1.

T A B L E **5.1**
Children Involved in Fatal Motor Vehicle Accidents

	Survivors	Fatalities	
No belt	1,129	509	1,638
Adult belt	432	73	505
Child seat	733	139	872
	2,294	721	3,015

As seen in earlier chapters, it's more convenient to analyze such data in terms of meaningful random variables than in terms of described categories. For each child, it is important to know whether or not he or she survived and what the seat belt situation was. A variable X_1 defined as

$$X_1 = 0 \text{ if child survived}$$
$$X_1 = 1 \text{ if child did not survive}$$

will keep track of the number of child fatalities. Now child seats usually involve two belts, one the regular adult belt in the vehicle and the other the belt on the seat itself. A variable X_2 defined as

$$X_2 = 0 \text{ if no belt}$$

$$X_2 = 1 \text{ if adult belt used}$$
$$X_2 = 2 \text{ if child seat used}$$

keeps track of the number of belts (properly defined) restraining the child.

The frequencies from Table 5.1 are turned into the relative frequencies of Table 5.2 to produce the *joint probability distribution* of X_1 and X_2. In general, we write

$$P(X_1 = x_1, X_2 = x_2) = p(x_1, x_2)$$

and call $p(x_1, x_2)$ the *joint probability function* of (X_1, X_2). From Table 5.2,

$$P(X_1 = x_1, X_2 = x_2) = p(0, 2) = \frac{733}{3,015} = 0.24$$

represents the approximate probability that a child will both survive *and* be in a child seat when involved in a fatal accident.

T A B L E **5.2**
Joint Probability Distribution

		X_1		
		0	1	
	0	0.38	0.17	0.55
X_2	1	0.14	0.02	0.16
	2	0.24	0.05	0.29
		0.76	0.24	1.00

The probability that a child will be in a child seat is

$$
\begin{aligned}
P(X_2 = 2) &= P(X_1 = 0, X_2 = 2) + P(X_1 = 1, X_2 = 2) \\
&= 0.24 + 0.05 \\
&= 0.29
\end{aligned}
$$

which is one of the *marginal probabilities* for X_2. The univariate distribution along the right margin is the *marginal distribution* for X_2 alone, and the one along the bottom row is the marginal distribution for X_1 alone.

The *conditional probability distribution* for X_1 given X_2 fixes a value of X_2 (a row of Table 5.2) and looks at the relative frequencies for values of X_1 in that row. For example, conditioning on $X_2 = 0$ produces

$$
\begin{aligned}
P(X_1 = 0 | X_2 = 0) &= \frac{P(X_1 = 0, X_2 = 0)}{P(X_2 = 0)} \\
&= \frac{0.38}{0.55} = 0.69
\end{aligned}
$$

and

$$P(X_1 = 1 | X_2 = 0) = 0.31$$

These two values show how survivorship relates to the situation in which no seat belt is in use; 69% of children in such situations survived fatal accidents. That may seem high, but compare it to $P(X_1 = 0 | X_2 = 2)$. Does seat belt use seem to be beneficial?

There is another set of conditional distributions here, namely, those for X_2 conditioning on X_1. In this case,

$$P(X_2 = 0 | X_1 = 0) = \frac{P(X_1 = 0, X_2 = 0)}{P(X_1 = 0)}$$

$$= \frac{0.38}{0.76} = 0.50$$

and

$$P(X_2 = 0 | X_1 = 1) = \frac{0.17}{0.24} = 0.71.$$

What do these two probabilities tell you about the effectiveness of seat belts?
The bivariate discrete case is summarized in Definition 5.1.

DEFINITION   **5.1**

Let X_1 and X_2 be discrete random variables. The **joint probability distribution** of X_1 and X_2 is given by

$$P(x_1, x_2) = P(X_1 = x_1, X_2 = x_2)$$

defined for all real numbers x_1 and x_2. The function $p(x_1, x_2)$ is called the **joint probability function** of X_1 and X_2. The **marginal probability functions** of X_1 and X_2, respectively, are given by

$$p_1(x_1) = \sum_{x_2} p(x_1, x_2)$$

and

$$p_2(x_2) = \sum_{x_1} p(x_1, x_2) \quad \blacksquare$$

EXAMPLE   **5.1**   There are three checkout counters in operation at a local supermarket. Two customers arrive at the counters at different times, when the counters are serving no other customers. Assume that the customers then choose a checkout station at random and independently of one another. Let X_1 denote the number of times counter A is selected and X_2 the number of times counter B is selected by the two customers. Find the joint probability distribution of X_1 and X_2. Find the probability that one of the customers visits counter B given that one of the customers is known to have visited counter A.

Solution   For convenience, let us introduce X_3, defined as the number of customers visiting counter C. Now the event $(X_1 = 0$ and $X_2 = 0)$ is equivalent to the event $(X_1 = 0, X_2 = 0,$ and $X_3 = 2)$. It follows that

$$P(X_1 = 0, X_2 = 0) = P(X_1 = 0, X_2 = 0, X_3 = 2)$$
$$= P(\text{customer I selects counter } C$$
$$\text{and customer II selects counter } C)$$

$$= P(\text{customer I selects counter } C)$$
$$\times P(\text{customer II selects counter } C)$$
$$= \frac{1}{3} \times \frac{1}{3} = \frac{1}{9}$$

since customers' choices are independent and each customer makes a random selection from among the three available counters. The joint probability distribution is shown in Table 5.3.

		X_1			Marginal Probabilities for X_2
		0	1	2	
X_2	0	1/9	2/9	1/9	4/9
	1	2/9	2/9	0	4/9
	2	1/9	0	0	1/9
Marginal Probabilities for X_1		4/9	4/9	1/9	1

It is slightly more complicated to calculate

$$P(X_1 = 1, X_2 = 0) = P(X_1 = 1, X_2 = 0, X_3 = 1)$$

In this event customer I could select counter A and customer II could select counter C, or customer I could select counter C and customer II could select counter A. Thus,

$$P(X_1 = 1, X_2 = 0) = P(\text{I selects } A)P(\text{II selects } C)$$
$$+ P(\text{I selects } C)P(\text{II selects } A)$$
$$= \frac{1}{3} \times \frac{1}{3} + \frac{1}{3} \times \frac{1}{3} = \frac{2}{9}$$

Similar arguments allow us to derive the results in Table 5.3. The second statement in our example asks for

$$P(X_2 = 1 | X_1 = 1) = \frac{P(X_1 = 1, X_2 = 1)}{P(X_1 = 1)}$$
$$= \frac{2/9}{4/9} = \frac{1}{2}$$

Does this answer agree with your intuition? ∎

Before we move to the continuous case, let us first briefly review the situation in one dimension. If $f(x)$ denotes the probability density function of a random variable X, then $f(x)$ represents a relative frequency curve, and probabilities such as $P(a \leq X \leq b)$ are represented as areas under this curve. That is,

$$P(a \leq X \leq b) = \int_a^b f(x)\,dx$$

Now suppose we are interested in the joint behavior of two continuous random variables, say X_1 and X_2, where X_1 and X_2 might, for example, represent the amounts of two different hydrocarbons found in an air sample taken for a pollution study. The relative frequency behavior of these two random variables can be modeled by a bivariate function, $f(x_1, x_2)$, which forms a probability, or relative frequency, surface in three dimensions. Figure 5.1 shows such a surface. The probability that X_1 lies in some interval and X_2 lies in another interval is then represented as a volume under this surface. Thus,

$$P(a_1 \leq X_1 \leq a_2, b_1 \leq X_2 \leq b_2) = \int_{b_1}^{b_2} \int_{a_1}^{a_2} f(x_1, x_2)\,dx_1 dx_2$$

Note that the above integral simply gives the volume under the surface and over the shaded region in Figure 5.1.

FIGURE 5.1
A Bivariate Density Function

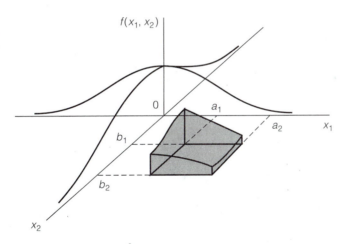

We will illustrate the actual computations involved in such a bivariate problem with a very simple example.

E X A M P L E   **5.2**   A certain process for producing an industrial chemical yields a product containing two predominant types of impurities. For a certain volume of sample from this process, let X_1 denote the proportion of impurities in the sample and let X_2 denote the proportion of type I impurity among all impurities found. Suppose the joint distribution of X_1

and X_2, after investigation of many such samples, can be adequately modeled by the following function:

$$f(x_1, x_2) = \begin{cases} 2(1 - x_1) & 0 \le x_1 \le 1, 0 \le x_2 \le 1 \\ 0 & \text{elsewhere} \end{cases}$$

This function graphs as the surface given in Figure 5.2. Calculate the probability that X_1 is less than 0.5 and that X_2 is between 0.4 and 0.7.

FIGURE **5.2**

Probability Density Function for Example 5.2

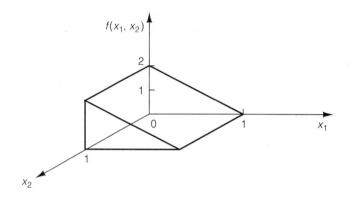

Solution   From the preceding discussion, we see that

$$P(0 \le x_1 \le 0.5, 0.4 \le x_2 \le 0.7) = \int_{0.4}^{0.7} \int_{0}^{0.5} 2(1 - x_1)\, dx_1 dx_2$$

$$= \int_{0.4}^{0.7} [-(1 - x_1)^2]_0^{0.5}\, dx_2$$

$$= \int_{0.4}^{0.7} (0.75)\, dx_2$$

$$= 0.75(0.7 - 0.4) = 0.75(0.3) = 0.225$$

Thus, the fraction of such samples having less than 50% impurities and a relative proportion of type I impurities between 40% and 70% is 0.225.   ∎

Just as the univariate, or marginal, probabilities were computed by summing over rows or columns in the discrete case, the univariate density function for X_1 in the continuous case can be found by integrating ("summing") over values of X_2. Thus, the *marginal density function* of X_1, $f_1(x_1)$, is given by

$$f_1(x_1) = \int_{-\infty}^{\infty} f(x_1, x_2)\, dx_2$$

Similarly, the marginal density function of X_2, $f_2(x_2)$, is given by

$$f_2(x_2) = \int_{-\infty}^{\infty} f(x_1, x_2)\, dx_1$$

E X A M P L E   **5.3**   For the situation in Example 5.2, find the marginal probability density functions for X_1 and X_2.

Solution   Let's first try to visualize what the answers should look like, before going through the integration. To find $f_1(x_1)$, we are to accumulate all probabilities in the x_2 direction. Look at Figure 5.2 and think of collapsing the wedge-shaped figure back onto the $(x_1, f_1(x_1, x_2))$ plane. Much more probability mass will build up toward the zero point of the x_1-axis than toward the unity point. In other words, the function $f_1(x_1)$ should be high at zero and low at 1. Formally,

$$f_1(x_1) = \int_{-\infty}^{\infty} f(x_1, x_2)\, dx_2$$
$$= \int_0^1 2(1 - x_1)\, dx_2$$
$$= 2(1 - x_1) \qquad 0 \le x_1 \le 1$$

The function graphs as in Figure 5.3. Note that our conjecture is correct. Envisioning how $f_2(x_2)$ should look geometrically, suppose the wedge of Figure 5.2 is forced back onto the $(x_2, f(x_1, x_2))$ plane. Then the probability mass should accumulate equally all along the $(0,1)$ interval on the x_2-axis. Mathematically,

$$f_2(x_2) = \int_{-\infty}^{\infty} f(x_1, x_2)\, dx_1$$
$$= \int_0^1 2(1 - x_1)\, dx_1$$
$$= [-(1 - x_1)]_0^1 \qquad 0 < x_2 \le 1$$
$$= 1 \qquad 0 < x_2 \le 1$$

and again our conjecture is verified, as seen in Figure 5.4.   ∎

F I G U R E   **5.3**
Probability Density Function
$f_1(x_1)$ for Example 5.3

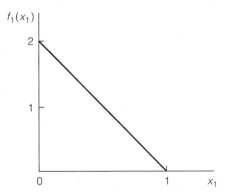

FIGURE **5.4**
Probability Density Function
$f_2(x_2)$ for Example 5.3

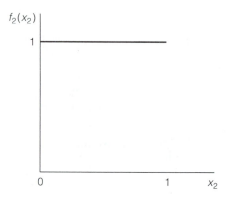

The bivariate continuous case is summarized in Definition 5.2.

DEFINITION   **5.2**

Let X_1 and X_2 be continuous random variables. The **joint probability distribution function** of X_1 and X_2, if it exists, is given by a nonnegative function $f(x_1, x_2)$, which is such that

$$P(a_1 \leq X_1 \leq a_2, b_1 \leq X_2 \leq b_2) = \int_{b_1}^{b_2} \int_{a_1}^{a_2} f(x_1, x_2)\, dx_1 dx_2$$

The **marginal probability density functions** of X_1 and X_2, respectively, are given by

$$f_1(x_1) = \int_{-\infty}^{\infty} f(x_1, x_2)\, dx_2$$

and

$$f_2(x_2) = \int_{-\infty}^{\infty} f(x_1, x_2)\, dx_1 \quad \blacksquare$$

For another (and somewhat more complicated) example, consider the following.

EXAMPLE   **5.4**

Gasoline is to be stocked in a bulk tank once each week and then sold to customers. To keep down the cost of inventory, the supply and demand are carefully monitored. Let X_1 denote the proportion of the tank that is stocked in a particular week and let X_2 denote the proportion of the tank that is sold in that same week. Due to limited supplies, X_1 is not fixed in advance but varies from week to week. Suppose a study over many weeks shows the joint relative frequency behavior of X_1 and X_2 to be such that the following density function provides an adequate model:

$$f(x_1, x_2) = \begin{cases} 3x_1 & 0 \leq x_2 \leq x_1 \leq 1 \\ 0 & \text{elsewhere} \end{cases}$$

Note that X_2 must always be less than or equal to X_1. This density function is graphed in Figure 5.5. Find the probability that X_2 will be between 0.2 and 0.4 for a given week.

Solution   The question refers to the marginal behavior of X_2. Thus, it is necessary to find

$$f_2(x_2) = \int_{-\infty}^{\infty} f(x_1, x_2) \, dx_1$$

$$= \int_{x_2}^{1} 3x_1 \, dx_1 = \frac{3}{2}x_1^2 \bigg|_{x_2}^{1}$$

$$= \frac{3}{2}(1 - x_2^2) \quad 0 < x_2 \le 1$$

It follows directly that

$$P(0.2 \le X_2 \le 0.4) = \int_{0.2}^{0.4} \frac{3}{2}(1 - x_2^2) dx_2$$

$$= \frac{3}{2} \left(x_2 - \frac{x_2^3}{3} \right) \bigg|_{0.2}^{0.4}$$

$$= \frac{3}{2} \left\{ \left[0.4 - \frac{(0.4)^3}{3} \right] - \left[0.2 - \frac{(0.2)^3}{3} \right] \right\}$$

$$= 0.272$$

Note that the marginal density of X_2 graphs as a function that is high at $x_2 = 0$ and then tends to 0 as x_2 tends to 1. Does this agree with your intuition after looking at Figure 5.5?   ∎

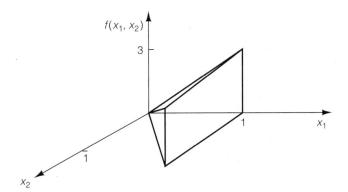

F I G U R E **5.5**
The Joint Density Function for
Example 5.4

5.2 Conditional Probability Distributions

Recall that in the bivariate *discrete* case the conditional probabilities for X_1 for a given X_2 were found by fixing attention on the particular row in which $X_2 = x_2$ and then looking at the relative probabilities within that row. That is, the individual cell probabilities were divided by the marginal total for that row in order to obtain conditional probabilities.

In the bivariate *continuous* case, the form of the probability density function representing the conditional behavior of X_1 for a given value of X_2 is found by slicing through the joint density in the x_1 direction at the particular value of X_2. This function then has to be weighted by the marginal density function for X_2 at that point. We will look at a specific example before giving the general definition of conditional density functions.

E X A M P L E   **5.5**   Refer to the tank inventory problem of Example 5.4. Find the conditional probability that X_2 is less than 0.2 given that X_1 is known to be 0.5.

Solution   Slicing through $f(x_1, x_2)$ in the x_2 direction at $x_1 = 0.5$ yields

$$f(0.5, x_2) = 3(0.5) = 1.5 \quad 0 \le x_2 \le 0.5$$

Thus the conditional behavior of X_2 for a given X_1 of 0.5 is constant over the interval $(0, 0.5)$. The marginal value of $f(x_1)$ at $x_1 = 0.5$ is obtained as follows:

$$
\begin{aligned}
f_1(x_1) &= \int_{-\infty}^{+\infty} f(x_1, x_2)\, dx_2 \\
&= \int_0^{x_1} 3x_1\, dx_2 \\
&= 3x_1^2 \qquad 0 \le x_1 \le 1 \\
f_1(0.5) &= 3(0.5)^2 = 0.75
\end{aligned}
$$

Upon dividing, we see that the conditional behavior of X_2 for a given X_1 of 0.5 is represented by the function

$$
\begin{aligned}
f(x_2 | x_1 = 0.5) &= \frac{f(0.5, x_2)}{f_1(0.5)} \\
&= \frac{1.5}{0.75} = 2 \quad 0 < x_2 < 0.5
\end{aligned}
$$

This function of x_2 has all the properties of a probability density function and is the conditional density function for X_2 at $X_1 = 0.5$. Then

$$
\begin{aligned}
P(X_2 < 0.2 | X_1 = 0.5) &= \int_0^{0.2} f(x_2 | x_1 = 0.5)\, dx_2 \\
&= \int_0^{0.2} 2\, dx_2 = 0.2(2) = 0.4
\end{aligned}
$$

That is, among all weeks in which the tank was half full immediately after stocking, sales amounted to less than 20% of the tank 40% of the time. ▪

We see that the manipulations used above to obtain conditional density functions in the continuous case are analogous to those used to obtain conditional probabilities in the discrete case, except that integrals are used in place of sums.

The formal definition of a conditional probability density function is given below.

DEFINITION **5.3** Let X_1 and X_2 be jointly continuous random variables with joint probability density function $f(x_1, x_2)$ and marginal densities $f(x_1)$ and $f_2(x_2)$, respectively. Then the **conditional probability density function** of X_1 given $X_2 = x_2$ is given by

$$f(x_1|x_2) = \begin{cases} \dfrac{f(x_1, x_2)}{f_2(x_2)} & f_2(x_2) > 0 \\ 0 & \text{elsewhere} \end{cases}$$

and the conditional probability density function of X_2 given $X_1 = x_1$ is given by

$$f(x_2|x_1) = \begin{cases} \dfrac{f(x_1, x_2)}{f_1(x_1)} & f_1(x_1) > 0 \\ 0 & \text{elsewhere} \end{cases} \quad ▪$$

EXAMPLE **5.6** A soft-drink machine has a random amount Y_2 in supply at the beginning of a given day and dispenses a random amount Y_1 during the day (with measurements in gallons). It is not resupplied during the day; hence, $Y_1 \leq Y_2$. It has been observed that Y_1 and Y_2 have joint density

$$f(y_1, y_2) = \begin{cases} 1/2 & 0 \leq y_1 \leq y_2, 0 \leq y_2 \leq 2 \\ 0 & \text{elsewhere} \end{cases}$$

That is, the points (y_1, y_2) are uniformly distributed over the triangle with the given boundaries. Find the conditional probability density of Y_1 given $Y_2 = y_2$. Evaluate the probability that less than 1/2 gallon is sold, given that the machine contains 1 gallon at the start of the day.

Solution The marginal density of Y_2 is given by

$$\begin{aligned} f_2(y_2) &= \int_{-\infty}^{\infty} f(y_1, y_2) \, dy_1 \\ &= \begin{cases} \displaystyle\int_0^{y_2} (1/2) \, dy_1 = (1/2)y_2 & 0 \leq y_2 \leq 2 \\ 0 & \text{elsewhere} \end{cases} \end{aligned}$$

By Definition 5.3,

$$f_1(y_1|y_2) = \frac{f(y_1, y_2)}{f_2(y_2)}$$

$$= \begin{cases} \dfrac{1/2}{(1/2)y_2} = \dfrac{1}{y_2} & 0 \le y_1 \le y_2 \le 2 \\ 0 & \text{elsewhere} \end{cases}$$

The probability of interest is

$$P(Y_1 < 1/2|Y_2 = 1) = \int_{-\infty}^{1/2} f(y_1|y_2 = 1)dy_1$$

$$= \int_0^{1/2} (1)dy_1 = 1/2$$

Note that if the machine had contained 2 gallons at the start of the day, then

$$P(Y_1 \le 1/2|Y_2 = 2) = \int_0^{1/2} \left(\frac{1}{2}\right) dy_1$$

$$= 1/4$$

Thus, the amount sold is highly dependent upon the amount in supply. ▪

5.3 Independent Random Variables

Before defining independent random variables, we will recall once again that two events A and B are independent if $P(AB) = P(A)P(B)$. Somewhat analogously, two discrete random variables are independent if

$$P(X_1 = x_1, X_2 = x_2) = P(X_1 = x_1)P(X_2 = x_2)$$

for all real numbers x_1 and x_2. A similar idea carries over to the continuous case.

DEFINITION 5.4

Discrete random variables X_1 and X_2 are said to be independent if

$$P(X_1 = x_1, X_2 = x_2) = P(X_1 = x_1)P(X_2 = x_2)$$

for all real numbers x_1 and x_2.

Continuous random variables X_1 and X_2 are said to be independent if

$$f(x_1, x_2) = f_1(x_1)f_2(x_2)$$

for all real numbers x_1 and x_2. ▪

The concepts of joint probability density functions and independence extend directly to n random variables, where n is any finite positive integer. The n random variables

$X_1, X_2, \ldots, X_n$ are said to be *independent* if their joint density function $f(x_1, \ldots, x_n)$ is given by

$$f(x_1, \ldots, x_n) = f_1(x_1) f_2(x_2) \cdots f_n(x_n)$$

for all real numbers $x_1, x_2, \ldots, x_n$.

EXAMPLE **5.7**   Show that the random variables having the joint distribution of Table 5.2 are not independent.

Solution   It is necessary to check only one entry in the table. We see that $P(X_1 = 0, X_2 = 0) = 0.38$, whereas $P(X_1 = 0) = 0.76$ and $P(X_2 = 0) = 0.55$. Since

$$P(X_1 = 0, X_2 = 0) \neq P(X_1 = 0) P(X_2 = 0)$$

the random variables cannot be independent.   ∎

EXAMPLE **5.8**   Show that the random variables in Example 5.2, page 261, are independent.

Solution   Here,

$$f(x_1, x_2) = \begin{cases} 2(1 - x_1) & 0 \leq x_1 \leq 1, 0 \leq x_2 \leq 1 \\ 0 & \text{elsewhere} \end{cases}$$

We saw in Example 5.3 that

$$f_1(x_1) = 2(1 - x_1) \qquad 0 \leq x_1 \leq 1$$

and

$$f_2(x_2) = 1 \qquad 0 \leq x_2 \leq 1$$

Thus, $f(x_1, x_2) = f_1(x_1) f_2(x_2)$ for all real numbers x_1 and x_2, and X_1 and X_2 are independent random variables.   ∎

Exercises

5.1   Two construction contracts are to be randomly assigned to one or more of three firms. Numbering the firms I, II, and III, let X_1 be the number of contracts assigned to firm I and X_2 the number assigned to firm II. (A firm may receive more than one contract.)

a   Find the joint probability distribution for X_1 and X_2.

b   Find the marginal probability distribution for X_1.

    c   Find $P(X_1 = 1 | X_2 = 1)$.

5.2   A radioactive particle is randomly located in a square area with sides one unit in length. Let X_1 and X_2 denote the coordinates of the particle. Since the particle is equally likely to fall in any subarea of a fixed size, a reasonable model for (X_1, X_2) is given by

$$f(x_1, x_2) = \begin{cases} 1 & 0 \le x_1 \le 1, 0 \le x_2 \le 1 \\ 0 & \text{elsewhere} \end{cases}$$

    a   Sketch the probability density surface.

    b   Find $P(X_1 \le 0.2, X_2 \le 0.4)$.

    c   Find $P(0.1 \le X_1 \le 0.3, X_2 > 0.4)$.

5.3   The same study that produced the seat belt safety data of Table 5.1 also took into account the age of the child involved in a fatal accident. The results for those children wearing *no* seat belts are shown below. Here, $X_1 = 1$ if the child did not survive and X_2 indicates the age in years. (An age of zero implies that the child was less than 1 year old, an age of 1 implies the child was more than 1 year old but not yet 2, and so on.)

Age	Survivors	Fatalities
0	104	127
1	165	91
2	267	107
3	277	90
4	316	94

    a   Construct an approximate joint probability distribution for X_1 and X_2.

    b   Construct the conditional distribution of X_1 for fixed values of X_2. Discuss the implications of these results.

    c   Construct the conditional distribution of X_2 for fixed values of X_1. Are the implications the same as in part (b)?

5.4   An environmental engineer measures the amount (by weight) of particulate pollution in air samples (of a certain volume) collected over the smokestack of a coal-operated power plant. Let X_1 denote the amount of pollutant per sample when a certain cleaning device on the stack is not operating, and let X_2 denote the amount of pollutant per sample when the cleaning device is operating, under similar environmental conditions. It is observed that X_1 is always greater than $2X_2$, and the relative frequency behavior of (X_1, X_2) can be modeled by

$$f(x_1, x_2) = \begin{cases} k & 0 \le x_1 \le 2, 0 \le x_2 \le 1, 2x_2 \le x_1 \\ 0 & \text{elsewhere} \end{cases}$$

    (That is, X_1 and X_2 are randomly distributed over the region inside the triangle bounded by $x_1 = 2, x_2 = 0$, and $2x_2 = x_1$.)

    a   Find the value of k that makes this a probability density function.

    b   Find $P(X_1 \ge 3X_2)$. (That is, find the probability that the cleaning device will reduce the amount of pollutant by one-third or more.)

5.5   Refer to Exercise 5.2.

    a   Find the marginal density function for X_1.

    b   Find $P(X_1 \le 0.5)$.

    c   Are X_1 and X_2 independent?

5.6   Refer to Exercise 5.4.

    a   Find the marginal density function of X_2.

    b   Find $P(X_2 \le 0.4)$.

    c   Are X_1 and X_2 independent?

    d   Find $P(X_2 \le 1/4 | X_1 = 1)$.

5.7 Let X_1 and X_2 denote the proportion of two different chemicals found in a sample mixture of chemicals used as an insecticide. Suppose X_1 and X_2 have joint probability density given by

$$f(x_1, x_2) = \begin{cases} 2 & 0 \le x_1 \le 1, 0 \le x_2 \le 1, 0 \le x_1 + x_2 \le 1 \\ 0 & \text{elsewhere} \end{cases}$$

(Note that $X_1 + X_2$ must be at most unity since the random variables denote proportions within the same sample.)

  a Find $P(X_1 \le 3/4, X_2 \le 3/4)$.

  b Find $P(X_1 \le 1/2, X_2 \le 1/2)$.

  c Find $P(X_1 \le 1/2 | X_2 \le 1/2)$.

5.8 Refer to Exercise 5.7.

  a Find the marginal density functions for X_1 and X_2.

  b Are X_1 and X_2 independent?

  c Find $P(X_1 > 1/2 | X_2 = 1/4)$.

5.9 Let X_1 and X_2 denote the proportions of time, out of one workweek, that employees I and II, respectively, actually spend performing their assigned tasks. The joint relative frequency behavior of X_1 and X_2 is modeled by the probability density function

$$f(x_1, x_2) = \begin{cases} x_1 + x_2 & 0 \le x_1 \le 1, 0 \le x_2 \le 1 \\ 0 & \text{elsewhere} \end{cases}$$

  a Find $P(X_1 < 1/2, X_2 > 1/4)$.

  b Find $P(X_1 + X_2 \le 1)$.

  c Are X_1 and X_2 independent?

5.10 Refer to Exercise 5.9. Find the probability that employee I spends more than 75% of the week on her assigned task, given that employee II spends exactly 50% of the workweek on his assigned task.

5.11 An electronic surveillance system has one of each of two different types of components in joint operation. Letting X_1 and X_2 denote the random lifelengths of the components of type I and type II, respectively, we have the joint probability density function given by

$$f(x_1, x_2) = \begin{cases} (1/8)x_1 e^{-(x_1 + x_2)/2} & x_1 > 0, x_2 > 0 \\ 0 & \text{elsewhere} \end{cases}$$

(Measurements are in hundreds of hours.)

  a Are X_1 and X_2 independent?

  b Find $P(X_1 > 1, X_2 > 1)$.

5.12 A bus arrives at a bus stop at a randomly selected time within a 1-hour period. A passenger arrives at the bus stop at a randomly selected time within the same hour. The passenger will wait for the bus up to one-quarter of an hour. What is the probability that the passenger will catch the bus? [*Hint*: Let X_1 denote the bus arrival time and X_2 the passenger arrival time. If these arrivals are independent, then

$$f(x_1, x_2) = \begin{cases} 1 & 0 \le x_1 \le 1, 0 \le x_2 \le 1 \\ 0 & \text{elsewhere} \end{cases}$$

Now find $P(X_2 \le X_1 \le X_2 + 1/4)$.]

5.13 Two friends are to meet at a library. Each arrives at independently and randomly selected times within a fixed 1-hour period. Each agrees to wait no more than 10 minutes for the other. Find the probability that they meet. Discuss how your solution generalizes to the study of the distance between two random points in a finite interval.

5.14   Each of two quality-control inspectors interrupts a production line at randomly, but independently, selected times within a given workday (of 8 hours). Find the probability that the two interruptions will be more than 4 hours apart.

5.15   Two telephone calls come into a switchboard at random times in a fixed 1-hour period. If the calls are made independently of each other,

a   find the probability that both are made in the first half hour.

b   find the probability that the two calls are within 5 minutes of each other.

5.16   A bombing target is in the center of a circle with a radius of 1 mile. A bomb falls at a randomly selected point inside that circle. If the bomb destroys everything within 1/2 mile of its landing point, what is the probability that it destroys the target?

5.4   Expected Values of Functions of Random Variables

When we encounter problems that involve more than one random variable, we often combine the variables into a single function. We might be interested in the average lifelength of five different electronic components within the same system or the difference between two different strength-test measurements on the same section of cable. We now discuss how to find expected values of functions of more than one random variable.

Definition 5.5 gives the basic procedure for finding expected values in the bivariate case. The definition can be generalized to higher dimensions.

DEFINITION   **5.5**   Suppose the discrete random variables (X_1, X_2) have a joint probability function given by $p(x_1, x_2)$. If $g(X_1, X_2)$ is any real-valued function of (X_1, X_2), then

$$E[g(X_1, X_2)] = \sum_{x_1} \sum_{x_2} g(x_1, x_2) p(x_1, x_2)$$

The sum is over all values of (x_1, x_2) for which $p(x_1, x_2) > 0$. If (X_1, X_2) are continuous random variables with probability density function $f(x_1, x_2)$, then

$$E[g(X_1, X_2)] = \int_{-\infty}^{\infty} \int_{-\infty}^{\infty} g(x_1, x_2) f(x_1, x_2) dx_1 dx_2 \quad \blacksquare$$

If X_1 and X_2 are independent, then it follows easily from Definition 5.5 that

$$E[g(X_1)h(X_2)] = E[g(X_1)]E[h(X_2)]$$

A function of two variables that is commonly of interest in probabilistic and statistical problems is the *covariance*.

DEFINITION 5.6

The **covariance** between two random variables X_1 and X_2 is given by

$$\text{Cov}(X_1, X_2) = E[(X_1 - \mu_1)(X_2 - \mu_2)]$$

where

$$\mu_1 = E(X_1) \quad \text{and} \quad \mu_2 = E(X_2) \quad \blacksquare$$

The covariance helps us assess the relationship between two variables, in the following sense. If X_2 tends to be large when X_1 is large, and small when X_1 is small, then X_1 and X_2 will have a *positive covariance*. If, on the other hand, X_2 tends to be small when X_1 is large and large when X_1 is small, then the variables will have a *negative covariance*.

While covariance measures the direction of the association between two random variables, *correlation* measures the strength of the association.

DEFINITION 5.7

The **correlation** coefficient between two random variables X_1 and X_2 is given by

$$\rho = \frac{\text{Cov}(X_1, X_2)}{\sqrt{V(X_1) \times V(X_2)}} \quad \blacksquare$$

The correlation coefficient is a unitless quantity taking on values between -1 and $+1$. If $\rho = +1$ or $\rho = -1$, then X_2 must be a linear function of X_1. The next theorem gives a simpler computational form for the covariance.

THEOREM 5.1

If X_1 has mean μ_1 and X_2 has mean μ_2, then

$$\text{Cov}(X_1, X_2) = E(X_1 X_2) - \mu_1 \mu_2$$

[*Note*: The proof of this theorem will be left as an exercise.] $\blacksquare$

If X_1 and X_2 are independent random variables, then $E(X_1 X_2) = E(X_1)E(X_2)$. Using Theorem 5.1, it is then clear that independence between X_1 and X_2 implies that $\text{Cov}(X_1, X_2) = 0$. The converse is not necessarily true. That is, zero covariance does not imply that the variables are independent. To see this, look at the following joint distribution:

$$X_1$$

		−1	0	1	
	−1	1/8	1/8	1/8	3/8
X_2	0	1/8	0	1/8	2/8
	1	1/8	1/8	1/8	3/8
		3/8	2/8	3/8	1

Clearly, $E(X_1) = E(X_2) = 0$, and $E(X_1 X_2) = 0$ as well. Therefore, $\text{Cov}(X_1, X_2) = 0$. On the other hand,

$$P(X_1 = -1, X_2 = -1) = 1/8 \neq P(X_1 = -1)P(X_2 = -1)$$

and so the random variables are dependent.

We will now calculate the covariance in an example.

E X A M P L E   **5.9**   A firm that sells word processing systems keeps track of the number of customers who call on any one day and the number of orders placed on any one day. Let X_1 denote the number of calls, X_2 the number of orders placed, and $p(x_1, x_2)$ the joint probability function for (X_1, X_2); records indicate that

$$
\begin{array}{ll}
p(0, 0) = 0.04 & p(2, 0) = 0.20 \\
p(1, 0) = 0.16 & p(2, 1) = 0.30 \\
p(1, 1) = 0.10 & p(2, 2) = 0.20
\end{array}
$$

That is, for any given day, the probability of, say, two calls and one order is 0.30. Find $\text{Cov}(X_1, X_2)$.

Solution   Theorem 5.1 suggests that we first find (X_1, X_2), which is

$$
\begin{aligned}
E(X_1 X_2) &= \sum_{x_1} \sum_{x_2} x_1 x_2 p(x_1, x_2) \\
&= (0 \times 0)p(0, 0) + (1 \times 0)p(1, 0) + (1 \times 1)p(1, 1) + (2 \times 0)p(2, 0) \\
&\quad + (2 \times 1)p(2, 1) + (2 \times 2)p(2, 2) \\
&= 0(0.4) + 0(0.16) + 1(0.10) + 0(0.20) + 2(0.30) + 4(0.20) \\
&= 1.50
\end{aligned}
$$

Now we must find $E(X_1) = \mu_1$ and $E(X_2) = \mu_2$. The marginal distributions of X_1 and X_2 are given in the following charts:

x_1	$p(x_1)$
0	0.04
1	0.26
2	0.70

x_2	$p(x_2)$
0	0.40
1	0.40
2	0.20

It follows that

$$\mu_1 = 1(0.26) + 2(0.70) = 1.66$$

and

$$\mu_2 = 1(0.40) + 2(0.20) = 0.80$$

Thus,

$$\begin{aligned}
\text{Cov}(X_1, X_2) &= E(X_1 X_2) - \mu_1 \mu_2 \\
&= 1.50 - 1.66(0.80) \\
&= 1.50 - 1.328 = 0.172
\end{aligned}$$

From the marginal distributions of X_1 and X_2, it follows that

$$\begin{aligned}
V(X_1) &= E(X_1^2) - \mu_1^2 \\
&= 3.06 - (1.66)^2 = 0.30
\end{aligned}$$

and

$$\begin{aligned}
V(X_2) &= E(X_2^2) - \mu_2^2 \\
&= 1.2 - (0.8)^2 = 0.56
\end{aligned}$$

Hence,

$$\begin{aligned}
\rho &= \frac{\text{Cov}(X_1, X_2)}{\sqrt{V(X_1) \times V(X_2)}} \\
&= \frac{0.172}{\sqrt{(0.30)(0.56)}} = 0.42
\end{aligned}$$

There is a moderate positive association between the number of calls and the number of orders placed. Do the positive covariance and correlation agree with your intuition? ∎

When a problem involves n random variables $X_1, X_2, \ldots, X_n$, we are often interested in studying linear combinations of those variables. For example, if the random variables measure the quarterly incomes for n plants in a corporation, we might want to look at their sum or average. If X_1 represents the monthly cost of servicing defective plants before a new quality-control system was installed and X_2 denotes that cost after the system was placed in operation, then we might want to study $X_1 - X_2$. Theorem 5.2 gives a general result for the mean and variance of a linear combination of random variables.

THEOREM   **5.2**   Let $Y_1, \ldots, Y_n$ and $X_1, \ldots, X_m$ be random variables with $E(Y_i) = \mu_i$ and $E(X_i) = \xi_i$. Define

$$U_1 = \sum_{i=1}^{n} a_1 Y_1, \qquad U_2 = \sum_{j=1}^{m} b_j X_j$$

for constants $a_1, \ldots, a_n, b_1, \ldots, b_m$. Then

1   $E(U_1) = \sum_{i=1}^{n} a_i \mu_i$

2   $V(U_1) = \sum_{i=1}^{n} a_i^2 V(Y_i) + 2 \sum \sum_{i<j} a_i a_j \mathrm{Cov}(Y_i, Y_j)$

 where the double sum is over all pairs (i, j) with $i < j$

3   $\mathrm{Cov}(U_1, U_2) = \sum_{i=1}^{n} \sum_{j=1}^{m} a_i b_j \mathrm{Cov}(Y_i, X_j)$

Proof   Part (1) follows directly from the definition of expected value and properties of sums or integrals. To prove part (2), we appeal to the definition of variance and write

$$V(U_1) = E[U_1 - E(U_1)]^2$$

$$= E\left[\sum_{i=1}^{n} a_i Y_i - \sum_{i=1}^{n} a_i \mu_i \right]^2$$

$$= E\left[\sum_{i=1}^{n} a_i (Y_i - \mu_i) \right]^2$$

$$= E\left[\sum_{i=1}^{n} a_i^2 (Y_i - \mu_i)^2 + \sum \sum_{i \neq j} a_i a_j (Y_i - \mu_i)(Y_j - \mu_j) \right]$$

$$= \sum_{i=1}^{n} a_i^2 E(y_i - \mu_i)^2 + \sum \sum_{i \neq j} a_i a_j E[(Y_i - \mu_i)(Y_j - \mu_j)]$$

By the definitions of variance and covariance, we then have

$$V(U_1) = \sum_{i=1}^{n} a_i^2 V(Y_i) + \sum \sum_{i \neq j} a_i a_j \mathrm{Cov}(Y_i, Y_j)$$

Note that $\mathrm{Cov}(Y_i, Y_j) = \mathrm{Cov}(Y_j, Y_i)$, and hence we can write

$$V(U_1) = \sum_{i=1}^{n} a_i^2 V(Y_i) + 2 \sum \sum_{i<j} a_i a_j \mathrm{Cov}(Y_i, Y_j)$$

Part (3) is obtained by similar steps. We have

$$\text{Cov}(U_1, U_2) = E\{[U_1 - E(U_1)][U_2 - E(U_2)]\}$$

$$= E\left[\left(\sum_{i=1}^{n} a_i Y_i - \sum_{i=1}^{n} a_i \mu_i\right)\left(\sum_{j=1}^{m} b_j X_j - \sum_{j=1}^{m} b_j \xi_j\right)\right]$$

$$= E\left\{\left[\sum_{i=1}^{n} a_i (Y_i - \mu_i)\right]\left[\sum_{j=1}^{m} b_j (X_j - \xi_j)\right]\right\}$$

$$= E\left[\sum_{i=1}^{n}\sum_{j=1}^{m} a_i b_j (Y_i - \mu_i)(X_j - \xi_j)\right]$$

$$= \sum_{i=1}^{n}\sum_{j=1}^{m} a_i b_j E(Y_i - \mu_i)(X_j - \xi_j)$$

$$= \sum_{i=1}^{n}\sum_{j=1}^{m} a_i b_j \text{Cov}(Y_i, X_j)$$

On observing that $\text{Cov}(Y_i, Y_i) = V(Y_i)$, we can see that part (2) is a special case of part (3). ∎

E X A M P L E   **5.10**   Let $Y_1, Y_2, \ldots, Y_n$ be independent random variables with $E(Y_i) = \mu$ and $V(Y_i) = \sigma^2$. (These variables may denote the outcomes on n independent trials of an experiment.) Defining

$$\overline{Y} = \frac{1}{n}\sum_{i=1}^{n} Y_i$$

show that $E(\overline{Y}) = \mu$ and $V(\overline{Y}) = \sigma^2/n$.

Solution   Note that $\overline{Y}$ is a linear function with all constants a_i equal to $1/n$. That is,

$$\overline{Y} = \left(\frac{1}{n}\right) Y_1 + \cdots + \left(\frac{1}{n}\right) Y_n$$

By Theorem 5.2, part (1),

$$E(\overline{Y}) = \sum_{i=1}^{n} a_i \mu = \mu \sum_{i=1}^{n} a_i$$

$$= \mu \sum_{i=1}^{n} \frac{1}{n} = \frac{n\mu}{n} = \mu$$

By Theorem 5.2, part (2),

$$V(\overline{Y}) = \sum_{i=1}^{n} a_i^2 V(Y_i) + 2\sum\sum_{i<j} a_{ij} \text{Cov}(Y_i, Y_j)$$

but the covariance terms are all zero, since the random variables are independent. Thus,

$$V(\bar{Y}) = \sum_{i=1}^{n} \left(\frac{1}{n}\right)^2 \sigma^2 = \frac{n\sigma^2}{n^2} = \frac{\sigma^2}{n} \qquad \blacksquare$$

E X A M P L E   5.11   With X_1 denoting the amount of gasoline stocked in a bulk tank at the beginning of a week and X_2 the amount sold during the week, $Y = X_1 - X_2$ represents the amount left over at the end of the week. Find the mean and variance of Y if the joint density function of (X_1, X_2) is given by

$$f(x_1, x_2) = \begin{cases} 3x_1 & 0 \le x_2 \le x_1 \le 1 \\ 0 & \text{elsewhere} \end{cases}$$

Solution   We must first find the means and variances of X_1 and X_2. The marginal density function of X_1 is found to be

$$f_1(x_1) = \begin{cases} 3x_1^2 & 0 \le x_1 \le 1 \\ 0 & \text{elsewhere} \end{cases}$$

Thus,

$$E(X_1) = \int_0^1 x_1(3x_1^2)dx_1 = 3\left[\frac{x_1^4}{4}\right]_0^1 = \frac{3}{4}$$

Note that $E(X_1)$ can be found directly from the joint density function by

$$E(X_1) = \int_0^1 \int_0^{x_1} x_1(3x_1)dx_2 dx_1$$

$$= \int_0^1 3x_1^2[x_2]_0^{x_1} dx_1$$

$$= \int_0^1 3x_1^3 dx_1 = \frac{3}{4}[x_1^4]_0^1 = \frac{3}{4}$$

The marginal density of X_2 is found to be

$$f_2(x_2) = \begin{cases} \dfrac{3}{2}(1 - x_2^2) & 0 \le x_2 \le 1 \\ 0 & \text{elsewhere} \end{cases}$$

Thus,

$$E(X_2) = \int_0^1 (x_2)\frac{3}{2}(1 - x_2^2)dx_2$$

$$= \frac{3}{2}\int_0^1 (x_2 - x_2^3)dx_2 = \frac{3}{2}\left\{\left[\frac{x_2^2}{2}\right]_0^1 - \left[\frac{x_2^4}{4}\right]_0^1\right\} = \frac{3}{2}\left\{\frac{1}{2} - \frac{1}{4}\right\} = \frac{3}{8}$$

Using similar arguments, it follows that

$$E(X_1^2) = 3/5$$

$$V(X_1) = 3/5 - (3/4)^2 = 0.0375$$
$$E(X_2^2) = 1/5$$

and

$$V(X_2) = 1/5 - (3/8)^2 = 0.0594$$

The next step is to find $\text{Cov}(X_1, X_2)$. Now

$$E(X_1 X_2) = \int_0^1 \int_0^{x_1} (x_1 x_2) 3x_1 \, dx_2 \, dx_1$$
$$= 3 \int_0^1 \int_0^{x_1} x_1^2 x_2 \, dx_2 \, dx_1$$
$$= 3 \int_0^1 x_1^2 \left[\frac{x_2^2}{2} \right]_0^{x_1} dx_1$$
$$= \frac{3}{2} \int_0^1 x_1^4 \, dx_1$$
$$= \frac{3}{2} \left[\frac{x_1^5}{5} \right]_0^1 = \frac{3}{10}$$

and

$$\text{Cov}(X_1, X_2) = E(X_1 X_2) - \mu_1 \mu_2$$
$$= \frac{3}{10} - \left(\frac{3}{4} \right) \left(\frac{3}{8} \right) = 0.0188$$

From Theorem 5.2,

$$E(Y) = E(X_1) - E(X_2)$$
$$= \frac{3}{4} - \frac{3}{8} = \frac{3}{8} = 0.375$$

and

$$V(Y) = V(X_1) + V(X_2) + 2(1)(-1)\text{Cov}(X_1, X_2)$$
$$= 0.0375 + 0.0594 - 2(0.0188)$$
$$= 0.0593 \quad \blacksquare$$

When the random variables in use are independent, the variance of a linear function simplifies, since the covariances become zero.

E X A M P L E   **5.12**   A firm purchases two types of industrial chemicals. The amount of type I chemical, X_1, purchased per week has $E(X_1) = 40$ gallons with $V(X_1) = 4$. The amount of type II chemical, X_2, purchased has $E(X_2) = 65$ gallons with $V(X_2)$ costs \$3 per gallon, while type II costs \$5 per gallon. Find the mean and variance of the total weekly amount spent for these types of chemicals, assuming that X_1 and X_2 are independent.

Solution   The dollar amount spent per week is given by

$$Y = 3X_1 + 5X_2$$

From Theorem 5.2,

$$E(Y) = 3E(X_1) + 5E(X_2)$$
$$= 3(40) + 5(65)$$
$$= 445$$

and

$$V(Y) = (3)^2 V(X_1) + (5)^2 V(X_2)$$
$$= 9(4) + 25(8)$$
$$= 236$$

The firm can expect to spend \$445 per week on chemicals. The standard deviation of $\sqrt{236} = 15.4$ suggests that this figure will not drop below \$400 or exceed \$500 very often.   ∎

E X A M P L E   **5.13**   Finite population sampling problems can be modeled by selecting balls from urns. Suppose an urn contains r white balls and $(N - r)$ black balls. A random sample of n balls is drawn without replacement, and Y, the number of white balls in the sample, is observed. From Chapter 3, we know that Y has a hypergeometric probability distribution. Find the mean and variance of Y.

Solution   We first observe some characteristics of sampling without replacement. Suppose that the sampling is done sequentially and we observe outcomes for $X_1, X_2, \ldots, X_n$, where

$$X_i = \begin{cases} 1 & \text{if the } i\text{th draw results in a white ball} \\ 0 & \text{otherwise} \end{cases}$$

Unquestionably, $P(X_1 = 1) = r/N$. But it is also true that $P(X_2 = 1) = r/N$, since

$$P(X_2 = 1) = P(X_1 = 1, X_2 = 1) + P(X_1 = 0, X_2 = 1)$$
$$= P(X_1 = 1)P(X_2 = 1 | X_1 = 1) + P(X_1 = 0)P(X_2 = 1 | X_1 = 0)$$
$$= \left(\frac{r}{N}\right)\left(\frac{r-1}{N-1}\right) + \left(\frac{N-r}{N}\right)\left(\frac{r}{N-1}\right)$$
$$= \frac{r(N-1)}{N(N-1)} = \frac{r}{N}$$

The same is true for X_k; that is,

$$P(X_k = 1) = \frac{r}{N} \quad k = 1, \ldots, n$$

Thus, the probability of drawing a white ball on any draw, given no knowledge of the outcomes on previous draws, is r/N. In a similar way, it can be shown that

$$P(X_j = 1, X_k = 1) = \frac{r(r-1)}{N(N-1)} \quad j \neq k$$

Now observe that $Y = \sum_{i=1}^{n} X_i$ and hence

$$E(Y) = \sum_{i=1}^{n} E(X_i) = n\left(\frac{r}{N}\right)$$

To find $V(Y)$, we need $V(X_i)$ and $\text{Cov}(X_i, X_j)$. Since X_i is 1 with probability r/N and 0 with probability $1 - (r/N)$, it follows that

$$V(X_i) = \frac{r}{N}\left(1 - \frac{r}{N}\right)$$

Also,

$$\begin{aligned}
\text{Cov}(X_i, X_j) &= E(X_i X_j) - E(X_i)E(X_j) \\
&= \frac{r(r-1)}{N(N-1)} - \left(\frac{r}{N}\right)^2 \\
&= -\frac{r}{N}\left(1 - \frac{r}{N}\right)\left(\frac{1}{N-1}\right)
\end{aligned}$$

since $X_i X_j = 1$ if and only if $X_i = 1$ and $X_j = 1$. From Theorem 5.2, we have that

$$\begin{aligned}
V(Y) &= \sum_{i=1}^{n} V(X_i) + 2\sum\sum_{i<j}\text{Cov}(X_i, X_j) \\
&= n\left(\frac{r}{N}\right)\left(1 - \frac{r}{N}\right) + 2\sum\sum_{i<j}\left[-\frac{r}{N}\left(1 - \frac{r}{N}\right)\left(\frac{1}{N-1}\right)\right] \\
&= n\left(\frac{r}{N}\right)\left(1 - \frac{r}{N}\right) - n(n-1)\left(\frac{r}{N}\right)\left(1 - \frac{r}{N}\right)\left(\frac{1}{N-1}\right)
\end{aligned}$$

since there are $n(n-1)/2$ terms in the double summation. A little algebra yields

$$V(Y) = n\left(\frac{r}{N}\right)\left(1 - \frac{r}{N}\right)\left(\frac{N-n}{N-1}\right)$$

How strongly associated are X_i and X_j? It is informative to note that their correlation is

$$\begin{aligned}
\rho &= \frac{-\dfrac{r}{N}\left(1 - \dfrac{r}{N}\right)\left(\dfrac{1}{N-1}\right)}{\sqrt{\dfrac{r}{N}\left(1 - \dfrac{r}{N}\right) \times \dfrac{r}{N}\left(1 - \dfrac{r}{N}\right)}} \\
&= -\frac{1}{N-1}
\end{aligned}$$

which tends to zero as N gets large. If N is large relative to the sample size, the hypergeometric model becomes equivalent to the binomial model.   ∎

To appreciate the usefulness of Theorem 5.2, try to find the expected value and variance for the hypergeometric random variable by proceeding directly from the definition of an expectation. The necessary summations are exceedingly difficult to obtain.

Exercises

5.17   Table 5.2 on page 258 shows the joint distribution of fatalities and number of seat belts used by children under age 5. What is the "average" behavior of X_1 and X_2, and how are they associated? Follow the steps below to answer this question.

a   Find $E(X_1)$, $V(X_1)$, $E(X_2)$, and $V(X_2)$.

b   Find $\text{Cov}(X_1, X_2)$. What does this result tell you?

c   Find the correlation coefficient between X_1 and X_2. How would you interpret this result?

5.18   In a study of particulate pollution in air samples over a smokestack, X_1 represents the amount of pollutant per sample when a cleaning device is not operating, and X_2 represents the amount when the cleaning device is operating. Assume that (X_1, X_2) has the joint probability density function

$$f(x_1, x_2) = \begin{cases} 1 & 0 \le x_1 \le 2, 0 \le x_2 \le 1, 2x_2 \le x_1 \\ 0 & \text{elsewhere} \end{cases}$$

The random variable $Y = X_1 - X_2$ represents the amount by which the weight of pollutant can be reduced by using the cleaning device.

a   Find $E(Y)$ and $V(Y)$.

b   Find an interval in which values of Y should lie at least 75% of the time.

5.19   The proportions X_1 and X_2 of two chemicals found in samples of an insecticide have the joint probability density function

$$f(x_1, x_2) = \begin{cases} 2 & 0 \le x_1 \le 1, 0 \le x_2 \le 1, 0 \le x_1 + x_2 \le 1 \\ 0 & \text{elsewhere} \end{cases}$$

The random variable $Y = X_1 + X_2$ denotes the proportion of the insecticide due to both chemicals combined.

a   Find $E(Y)$ and $V(Y)$.

b   Find an interval in which values of Y should lie for at least 50% of the samples of insecticide.

c   Find the correlation between X_1 and X_2 and interpret its meaning.

5.20   For a sheet-metal stamping machine in a certain factory, the time between failures, X_1, has a mean (MTBF) of 56 hours and a variance of 16. The repair time, X_2, has a mean (MTTR) of 5 hours and a variance of 4.

a   If X_1 and X_2 are independent, find the expected value and variance of $Y = X_1 + X_2$, which represents one operation/repair cycle.

b Would you expect an operation/repair cycle to last more than 75 hours? Why or why not?

5.21 A particular fast-food outlet is interested in the joint behavior of the random variables Y_1, the total time between a customer's arrival at the store and his leaving the service window, and Y_2, the time that the customer waits in line before reaching the service window. Since Y_1 contains the time a customer waits in line, we must have $Y_1 \geq Y_2$. The relative frequency distribution of observed values of Y_1 and Y_2 can be modeled by the probability density function

$$f(y_1, y_2) = \begin{cases} e^{-y_1} & 0 \leq y_2 \leq y_1 < \infty \\ 0 & \text{elsewhere} \end{cases}$$

a Find $P(Y_1 < 2, Y_2 > 1)$.

b Find $P(Y_1 \geq 2Y_2)$.

c Find $P(Y_1 - Y_2 \geq 1)$. [*Note:* $Y_1 - Y_2$ denotes the time spent at the service window.]

d Find the marginal density functions for Y_1 and Y_2.

5.22 Refer to Exercise 5.21. If a customer's total waiting time plus service time is known to be more than 2 minutes, find the probability that the customer waited less than 1 minute to be served.

5.23 Refer to Exercise 5.21. The random variable $Y_1 - Y_2$ represents the time spent at the service window.

a Find $E(Y_1 - Y_2)$.

b Find $V(Y_1 - Y_2)$.

c Is it highly likely that a customer would spend more than 2 minutes at the service window?

5.24 Refer to Exercise 5.21. Suppose a customer spends a length of time y_1 at the store. Find the probability that this customer spends less than half of that time at the service window.

5.25 Prove Theorem 5.1 (page 273).

5.5 The Multinomial Distribution

Suppose an experiment consists of n independent trials, much like the binomial case, but each trial can result in any one of k possible outcomes. For example, a customer at a grocery store may choose any one of k checkout counters. Now suppose the probability that a particular trial results in outcome i is denoted by $p_i, i = 1, \ldots, k$, and p_i remains constant from trial to trial. Let $Y_i, i = 1, \ldots, k$, denote the number of the n trials resulting in outcome i. In developing a formula for $P(Y_1 = y_1, \ldots, Y_k = y_k)$, we first call attention to the fact that, because of independence of trials, the probability of having y_1 outcomes of type 1 through y_k outcomes of type k in a particular order will be

$$p_1^{y_1} p_2^{y_2} \cdots p_k^{y_k}$$

It remains only to count the number of such orderings, and this is the number of ways of partitioning the n trials into y_1 type 1 outcomes, y_2 type 2 outcomes, and so on through y_k type k outcomes, or

$$\frac{n!}{y_1! y_2! \cdots y_k!}$$

where

$$\sum_{i=1}^{k} y_i = n$$

Hence,

$$P(Y_1 = y_1, \ldots, Y_k = y_k) = \frac{n!}{y_1! y_2! \cdots y_k!} p_1^{y_1} p_2^{y_2} \cdots p_k^{y_k}$$

This is called the *multinomial probability distribution*. Note that if $k = 2$ we have the binomial case.

The following example illustrates the computations.

E X A M P L E   **5.14**   Items under inspection are subject to two types of defects. About 70% of the items in a large lot are judged to be defect-free, whereas 20% have a type A defect alone and 10% have a type B defect alone. (None have both types of defects.) If six of these items are randomly selected from the lot, find the probability that three have no defects, one has a type A defect, and two have type B defects.

Solution   If we can assume that the outcomes are independent from trial to trial (item to item in our sample), which they nearly would be in a large lot, then the multinomial distribution provides a useful model. Letting Y_1, Y_2, and Y_3 denote the number of trials resulting in zero, type A, and type B defectives, respectively, we have $p_1 = 0.7$, $p_2 = 0.2$, and $p_3 = 0.1$. It follows that

$$P(Y_1 = 3, Y_2 = 1, Y_3 = 2) = \frac{6!}{3!1!2!} (0.7)^3 (0.2)(0.1)^2$$
$$= 0.041 \quad \blacksquare$$

E X A M P L E   **5.15**   Find $E(Y_i)$ and $V(Y_i)$ for the multinomial probability distribution.

Solution   We are concerned with the marginal distribution of Y_i, the number of trials falling in cell i. Imagine all the cells, excluding cell i, combined into a single large cell. Hence, every trial will result in cell i or not with probabilities p_i and $1 - p_i$, respectively, and Y_i possesses a binomial marginal probability distribution. Consequently,

$$E(Y_i) = np_i$$
$$V(Y_i) = np_i q_i \quad \text{where } q_i = 1 - p_i$$

[*Note*: The same results can be obtained by setting up the expectations and evaluating. For example,

$$E(Y_i) = \sum_{y_1} \sum_{y_2} \cdots \sum_{y_k} y_1 \frac{n!}{y_1! y_2! \cdots y_k!} p_1^{y_1} p_2^{y_2} \cdots p_k^{y_k}$$

Since we have already derived the expected value and variance of Y_i, we leave the tedious summation of this expectation to the interested reader.] ∎

E X A M P L E   **5.16**   If $Y_1, \ldots, Y_k$ have the multinomial distribution given in Example 5.15, find $\mathrm{Cov}(Y_s, Y_t)$, $s \neq t$.

Solution   Thinking of the multinomial experiment as a sequence of n independent trials, we define

$$U_i = \begin{cases} 1 & \text{if trial } i \text{ results in class } s \\ 0 & \text{otherwise} \end{cases}$$

and

$$W_i = \begin{cases} 1 & \text{if trial } i \text{ results in class } t \\ 0 & \text{otherwise} \end{cases}$$

Then

$$Y_s = \sum_{i=1}^{n} U_i \quad \text{and} \quad Y_t = \sum_{j=1}^{n} W_j$$

To evaluate $\mathrm{Cov}(Y_s, Y_t)$, we need the following results:

$$E(U_i) = p_s$$
$$E(W_j) = p_t$$
$$\mathrm{Cov}(U_i, W_j) = 0 \quad \text{if } i \neq j \text{ (since the trials are independent)}$$

and

$$\mathrm{Cov}(U_i, W_j) = E(U_i W_i) - E(U_i)E(W_i)$$
$$= 0 - p_s p_t$$

since $U_i W_i$ always equals zero. From Theorem 5.2, we then have

$$\mathrm{Cov}(Y_s, Y_t) = \sum_{i=1}^{n} \sum_{j=1}^{n} \mathrm{Cov}(U_i, W_j)$$
$$= \sum_{i=1}^{n} \mathrm{Cov}(U_i, W_i) + \sum \sum_{i \neq j} \mathrm{Cov}(U_i, W_j)$$
$$= \sum_{i=1}^{n} (-p_s p_t) + 0$$
$$= -n p_s p_t$$

Note that the covariance is negative, which is to be expected since a large number of outcomes in cell s would force the number in cell t to be small. ∎

Multinomial Experiment

1   The experiment consists of n identical trials.

2   The outcome of each trial falls into one of k classes or cells.

3   The probability that the outcome of a single trial will fall in a particular cell, say cell i, is p_i $(i = 1, 2, \ldots, k)$ and remains the same from trial to trial. Note that $p_1 + p_2 + p_3 + \cdots + p_k = 1$.

4   The trials are independent.

5   The random variables of interest are $Y_1, Y_2, \ldots, Y_k$, where Y_i $(i = 1, 2, \ldots, k)$ is equal to the number of trials in which the outcome falls in cell i. Note that $Y_1 + Y_2 + Y_3 + \cdots + Y_k = n$.

The Multinomial Distribution

$$P(Y_1 = y_1, \ldots, Y_k = y_k) = \frac{n!}{y_1! y_2! \cdots y_k!} p_1^{y_1} p_2^{y_2} \cdots p_k^{y_k}$$

$$\text{where} \quad \textstyle\sum_{i=1}^{k} y_i = n \quad \text{and} \quad \textstyle\sum_{i=1}^{k} p_i = 1$$

$$E(Y_i) = np_i \quad V(Y_i) = np_i(1 - p_i) \quad i = 1, \ldots, k$$

$$\text{Cov}(Y_i, Y_j) = -np_i p_j \quad i \neq j$$

Exercises

5.26   The National Fire Incident Reporting Service reports that, among residential fires, approximately 73% are in family homes, 20% are in apartments, and the other 7% are in other types of dwellings. If four fires are independently reported in one day, find the probability that two are in family homes, one is in an apartment, and one is in another type of dwelling.

5.27   The typical cost of damages for a fire in a family home is $20,000, the typical cost for an apartment fire is $10,000, and the typical cost for a fire in other dwellings is only $2,000. Using the information in Exercise 5.26, find the expected total damage cost for four independently reported fires.

5.28   Wing cracks, in the inspection of commercial aircraft, are reported as nonexistent, detectable, or critical. The history of a certain fleet shows that 70% of the planes inspected have no wing cracks, 25% have detectable wing cracks, and 5% have critical wing cracks. For the next five planes inspected, find the probability that

a   one has a critical crack, two have detectable cracks, and two have no cracks.

b   at least one critical crack is observed.

5.29   Given the recent emphasis on solar energy, solar radiation has been carefully monitored at various sites in Florida. For typical July days in Tampa, 30% have total radiation of at most 5 calories, 60% have total radiation of at most 6 calories, and 100% have total radiation of at most 8 calories. A solar collector for a hot water system is to be run for six days. Find the probability that three of the days produce no more than 5 calories, one of the days produces between 5 and 6 calories, and two of the days produce between 6 and 8 calories. What assumptions are you making in order for your answer to be correct?

5.30 The U.S. Bureau of Labor Statistics reports that approximately 21% of the adult population under age 65 is between 18 and 24 years of age, 28% is between 25 and 34, 19% is between 35 and 44, and 32% is between 45 and 64. An automobile manufacturer wants to obtain opinions on a new design from five randomly chosen adults from the group. Of the five so selected, find the approximate probability that two are between 18 and 24, two are between 25 and 44, and one is between 45 and 64.

5.31 Customers leaving a subway station can exit through any one of three gates. Assuming that any one customer is equally likely to select any one of the three gates, find the probability that, among four customers,

   a two select gate A, one selects gate B, and one selects gate C.

   b all four select the same gate.

   c all three gates are used.

5.32 Among a large number of applicants for a certain position, 60% have only a high school education, 30% have some college training, and 10% have completed a college degree. If five applicants are selected to be interviewed, find the probability that at least one will have completed a college degree. What assumptions are necessary for your answer to be valid?

5.33 In a large lot of manufactured items, 10% contain exactly one defect and 5% contain more than one defect. If ten items are randomly selected from this lot for sale, the repair costs total

$$Y_1 + 3Y_2$$

where Y_1 denotes the number among the ten having one defect and Y_2 denotes the number with two or more defects. Find the expected value and variance of the repair costs. Find the variance of the repair costs.

5.34 Refer to Exercise 5.33. If Y denotes the number of items containing at least one defect among the ten sampled items, find the probability that

   a Y is exactly 2.

   b Y is at least 1.

5.35 Vehicles arriving at an intersection can turn right or left or can continue straight ahead. In a study of traffic patterns at this intersection over a long period of time, engineers have noted that 40% of the vehicles turn left, 25% turn right, and the remainder continue straight ahead.

   a For the next five cars entering the intersection, find the probability that one turns left, one turns right, and three continue straight ahead.

   b For the next five cars entering the intersection, find the probability that at least one turns right.

   c If 100 cars enter the intersection in a day, find the expected value and variance of the number turning left. What assumptions are necessary for your answer to be valid?

5.6 More on the Moment-Generating Function[†]

The use of moment-generating functions to identify distributions of random variables works well when we are studying sums of independent random variables. For example, suppose X_1 and X_2 are independent exponential random variables, each with mean θ, and $Y = X_1 + X_2$. The moment-generating function of Y is given by

$$
\begin{aligned}
M_Y(t) = E(e^{tY}) &= E(e^{t(X_1+X_2)}) \\
&= E(e^{tX_1}e^{tX_2}) \\
&= E(e^{tX_1})E(e^{tX_2}) \\
&= M_{X_1}(t)M_{X_2}(t)
\end{aligned}
$$

since X_1 and X_2 are independent. From Chapter 4,

$$
M_{X_1}(t) = M_{X_2}(t) = (1-\theta t)^{-1}
$$

and so

$$
M_Y(t) = (1-\theta t)^{-2}
$$

Upon recognizing the form of this moment-generating function, we can immediately conclude that Y has a gamma distribution with $\alpha = 2$ and $\beta = \theta$. This result can be generalized to the sum of independent gamma random variables with common scale parameter β.

E X A M P L E   **5.17**   Let X_1 denote the number of vehicles passing a particular point on the eastbound lane of a highway in 1 hour. Suppose the Poisson distribution with mean λ_1 is a reasonable model for X_1. Now let X_2 denote the number of vehicles passing a point on the westbound lane of the same highway in 1 hour. Suppose X_2 has a Poisson distribution with mean λ_2. Of interest is $Y = X_1 + X_2$, the total traffic count in both lanes per hour. Find the probability distribution for Y if X_1 and X_2 are assumed to be independent.

Solution   It is known from Chapter 4 that

$$
M_{X_1}(t) = e^{\lambda_1(e^t - 1)}
$$

and

$$
M_{X_2}(t) = e^{\lambda_2(e^t - 1)}
$$

By the property of moment-generating functions seen above,

$$
M_Y(t) = M_{X_1}(t)M_{X_2}(t)
$$

[†]Optional section.

$$= e^{\lambda_1 (e^t - 1)} e^{\lambda_2 (e^t - 1)}$$
$$= e^{(\lambda_1 + \lambda_2)(e^t - 1)}$$

Now the moment-generating function for Y has the form of a Poisson moment-generating function with mean $(\lambda_1 + \lambda_2)$. Thus, by the uniqueness property, Y must have a Poisson distribution with mean $\lambda_1 + \lambda_2$.

The fact that we can add independent Poisson random variables and still retain the Poisson properties is important in many applications. ∎

Exercises

5.36 Find the moment-generating function for the negative binomial random variable. Use it to derive the mean and variance of that distribution.

5.37 There are two entrances to a parking lot. Cars arrive at entrance I according to a Poisson distribution with an average of three per hour and at entrance II according to a Poisson distribution with an average of four per hour. Find the probability that exactly three cars arrive at the parking lot in a given hour.

5.38 Let X_1 and X_2 denote independent normally distributed random variables, not necessarily having the same mean or variance. Show that, for any constants a and b, $Y = aX_1 + bX_2$ is normally distributed.

5.39 Resistors of a certain type have resistances that are normally distributed with a mean of 100 ohms and a standard deviation of 10 ohms. Two such resistors are connected in series, which causes the total resistance in the circuit to be the sum of the individual resistances. Find the probability that the total resistance

a   exceeds 220 ohms.

b   is less than 190 ohms.

5.40 A certain type of elevator has a maximum weight capacity X_1, which is normally distributed with a mean and standard deviation of 5,000 and 300 pounds, respectively. For a certain building equipped with this type of elevator, the elevator loading, X_2, is a normally distributed random variable with a mean and standard deviation of 4,000 and 400 pounds, respectively. For any given time that the elevator is in use, find the probability that it will be overloaded, assuming X_1 and X_2 are independent.

5.7 Conditional Expectations

Section 5.2 contains a discussion of conditional probability functions and conditional density functions; we now relate these functions to *conditional expectations*. Conditional expectations are defined in the same manner as univariate expectations, except that the conditional density function is used in place of the marginal density function.

DEFINITION   **5.8**

If X_1 and X_2 are any two random variables, the **conditional expectation** of X_1 given that $X_2 = x_2$ is defined to be

$$E(X_1|X_2 = x_2) = \int_{-\infty}^{\infty} x_1 f(x_1|x_2)\, dx_1$$

if X_1 and X_2 are jointly continuous and

$$E(X_1|X_2 = x_2) = \sum_{x_i} x_1 p(x_1|x_2)$$

if X_1 and X_2 are jointly discrete.   ∎

EXAMPLE   **5.18**

Let's look again at supply and demand of soft drinks. Refer to the Y_1 and Y_2 of Example 5.6, page 267, which we shall relabel as X_1 and X_2.

$$f(x_1, x_2) = \begin{cases} 1/2 & 0 \le x_1 \le x_2, 0 \le x_2 \le 2 \\ 0 & \text{elsewhere} \end{cases}$$

Find the conditional expectation of amount of sales X_1 given that $X_2 = 1$.

Solution   In Example 5.6, we found that

$$f(x_1|x_2) = \begin{cases} 1/x_2 & 0 \le x_1 \le x_2 \le 2 \\ 0 & \text{elsewhere} \end{cases}$$

Thus, from Definition 5.8,

$$E(X_1|X_2 = 1) = \int_{-\infty}^{\infty} x_1 f(x_1|x_2)\, dx_1$$

$$= \int_0^1 x_1(1)\, dx_1$$

$$= \left. \frac{x_1^2}{2} \right|_0^1 = \frac{1}{2}$$

That is, if the soft-drink machine contains 1 gallon at the start of the day, then the expected sales for that day will be 1/2 gallon.   ∎

The conditional expectation of X_1 given $X_2 = x_2$ is a function of x_2. If we now let X_2 range over all its possible values, we can think of the conditional expectation as a function of the random variable X_2, and hence we can find the expected value of the conditional expectation. The result of this type of iterated expectation is given in Theorem 5.3.

THEOREM   **5.3**

Let X_1 and X_2 denote random variables. Then

$$E(X_1) = E[E(X_1|X_2)]$$

where, on the right-hand side, the inside expectation is with respect to the conditional distribution of X_1 given X_2 and the outside expectation is with respect to the distribution of X_2.

Proof

Let X_1 and X_2 have joint density function $f(x_1, x_2)$ and marginal densities $f_1(x_1)$ and $f_2(x_2)$, respectively. Then

$$
\begin{aligned}
E(X_1) &= \int_{-\infty}^{\infty} x_1 f_1(x_1)\, dx_1 \\
&= \int_{-\infty}^{\infty} \int_{-\infty}^{\infty} x_1 f(x_1, x_2)\, dx_1 dx_2 \\
&= \int_{-\infty}^{\infty} \int_{-\infty}^{\infty} x_1 f_1(x_1|x_2) f_2(x_2)\, dx_1 dx_2 \\
&= \int_{-\infty}^{\infty} \left[\int_{-\infty}^{\infty} x_1 f_1(x_1|x_2)\, dx_1 \right] f_2(x_2) dx_2 \\
&= \int_{-\infty}^{\infty} E(X_1|X_2 = x_2) f_2(x_2) dx_2 \\
&= E[E(X_1|X_2)]
\end{aligned}
$$

The proof is similar for the discrete case.   ∎

EXAMPLE   **5.19**

A quality-control plan for an assembly line involves sampling $n = 10$ finished items per day and counting Y, the number of defectives. If p denotes the probability of observing a defective, then Y has a binomial distribution if the number of items produced by the line is large. However, p varies from day to day and is assumed to have a uniform distribution on the interval 0 to 1/4. Find the expected value of Y for any given day.

Solution

From Theorem 5.3, we know that

$$E(Y) = E[E(Y|p)]$$

For a given p, Y has a binomial distribution, and hence

$$E(Y|p) = np$$

Thus,

$$E(Y) = E(np) = nE(p)$$
$$= n \int_0^{1/4} 4p \, dp$$
$$= n \left(\frac{1}{8} \right)$$

and for $n = 10$

$$E(Y) = \frac{10}{8} = \frac{5}{4}$$

This inspection policy should average 5/4 defectives per day, in the long run. The calculations could be checked by actually finding the unconditional distribution of Y and computing $E(Y)$ directly. ∎

5.8 Compounding and Its Applications

The univariate probability distributions of Chapters 3 and 4 depend on one or more parameters; once the parameters are known, the distributions are completely specified. However, these parameters are frequently unknown and, as in Example 5.19, may sometimes be regarded as random quantities. The process of assigning distributions to these parameters and then finding the marginal distributions of the original random variable is known as *compounding*. This process has theoretical as well as practical uses, as we illustrate next.

EXAMPLE **5.20**   Suppose that Y denotes the number of bacteria per cubic centimeter in a certain liquid and that, for a given location, Y has a Poisson distribution with mean λ. Also assume that λ varies from location to location and, for a location chosen at random, that λ has a gamma distribution with parameters α and β, where α is a positive integer. Find the probability distribution for the bacteria count Y at a randomly selected location.

Solution   Since λ is random, the Poisson assumption applies to the conditional distribution of Y for fixed λ. Thus,

$$p(y|\lambda) = \frac{\lambda^y e^{-\lambda}}{y!} \qquad y = 0, 1, 2, \ldots$$

Also,

$$f(\lambda) = \begin{cases} \dfrac{1}{\Gamma(\alpha)\beta^\alpha} \lambda^{\alpha-1} e^{-\lambda/\beta} & \lambda > 0 \\ 0 & \text{elsewhere} \end{cases}$$

Then the joint distribution of λ and Y is given by

$$g(y, \lambda) = p(y|\lambda)f(\lambda)$$

$$= \frac{1}{y!\Gamma(\alpha)\beta^\alpha} \lambda^{y+\alpha-1} e^{-\lambda[1+(1/\beta)]}$$

The marginal distribution of Y is found by integrating over λ and yields

$$p(y) = \frac{1}{y!\Gamma(\alpha)\beta^\alpha} \int_0^\infty \lambda^{y+\alpha-1} e^{-\lambda[1+(1/\beta)]} \, d\lambda$$

$$= \frac{1}{y!\Gamma(\alpha)\beta^\alpha} \Gamma(y+\alpha) \left(1 + \frac{1}{\beta}\right)^{-(y+\alpha)}$$

Since α is an integer,

$$p(y) = \frac{(y+\alpha-1)!}{(\alpha-1)!y!} \left(\frac{1}{\beta}\right)^\alpha \left(\frac{\beta}{1+\beta}\right)^{y+\alpha}$$

$$= \binom{y+\alpha-1}{\alpha-1} \left(\frac{1}{1+\beta}\right)^\alpha \left(\frac{\beta}{1+\beta}\right)^y$$

If we let $y + \alpha = n$ and $1/(1 + \beta) = p$, then $p(y)$ has the form of a negative binomial distribution. Hence, the negative binomial distribution is a reasonable model for counts in which the mean count may be random. ∎

EXAMPLE **5.21**   Suppose a customer arrives at a checkout counter in a store just as the counter is opening. A random number N of customers will be ahead of him, since some customers may arrive before the counter opens. Suppose this number has the probability distribution

$$p(n) = P(N = n) = pq^n \qquad n = 0, 1, 2, \ldots$$

where $0 < p < 1$ and $q = 1 - p$. (This is a form of the geometric distribution.)

Customer service times are assumed to be independent and identically distributed exponential random variables with mean θ. Find the expected time for this customer to complete his checkout. (This is a general model that could apply to telephone calls and other typical "waiting lines.")

Solution   For a given value of n, the waiting time W is the sum of $n + 1$ independent exponential random variables and thus has a gamma distribution with $\alpha = n + 1$ and $\beta = \theta$. That is,

$$f(w|n) = \frac{1}{\Gamma(n+1)\theta^{n+1}} w^n e^{-w/\theta}$$

Hence,

$$f(w, n) = \frac{p}{\Gamma(n+1)\theta^{n+1}} (qw)^n e^{-w/\theta}$$

and

$$f(w) = \frac{p}{\theta} e^{-w/\theta} \sum_{n=0}^{\infty} \left(\frac{qw}{\theta}\right)^n \frac{1}{n!}$$

$$= \frac{p}{\theta} e^{-w/\theta} e^{qw/\theta}$$

$$= \frac{p}{\theta} e^{-(w/\theta)(1-q)}$$

$$= \frac{p}{\theta} e^{-w(p/\theta)}$$

The waiting time W is still exponential, but the mean is (θ/p).   ∎

5.9 Summary

In Chapter 1, it was noted that process improvement requires careful study of all factors that might affect the process. In such a study, it is important to look not only at the *marginal distributions* of random variables by themselves but also at their *joint distributions* and the *conditional distributions* of one variable for fixed values of another. The *covariance* and *correlation* help assess the direction and strength of the association between two random variables.

Supplementary Exercises

5.41   Let X_1 and X_2 have the joint probability density function given by

$$f(x_1, x_2) = \begin{cases} Kx_1x_2 & 0 \leq x_1 \leq 1, 0 \leq x_2 \leq 1 \\ 0 & \text{elsewhere} \end{cases}$$

a   Find the value of K that makes this a probability density function.
b   Find the marginal densities of X_1 and X_2.
c   Find the joint distribution for X_1 and X_2.
d   Find $P(X_1 < 1/2, X_2 < 3/4)$.
e   Find $P(X_1 \leq 1/2 | X_2 > 3/4)$.

5.42   Let X_1 and X_2 have the joint density function given by

$$f(x_1, x_2) = \begin{cases} 3x_1 & 0 \leq x_2 \leq x_1 \leq 1 \\ 0 & \text{elsewhere} \end{cases}$$

a   Find the marginal density functions of X_1 and X_2.
b   Find $P(X_1 \leq 3/4, X_2 \leq 1/2)$.
c   Find $P(X_1 \leq 1/2 | X_2 \geq 3/4)$.

5.43   From a legislative committee consisting of four Republicans, three Democrats, and two Independents, a subcommittee of three persons is to be randomly selected to discuss budget compromises. Let X_1 denote the number of Republicans and X_2 the number of Democrats on the subcommittee.

    a   Find the joint probability distribution of X_1 and X_2.

    b   Find the marginal distributions of X_1 and X_2.

    c   Find the probability $P(X_1 = 1 | X_2 \geq 1)$.

5.44 For Exercise 5.41, find the conditional density of X_1 given $X_2 = x_2$. Are X_1 and X_2 independent?

5.45 For Exercise 5.42,

    a   Find the conditional density of X_1 given $X_2 = x_2$.

    b   Find the conditional density of X_2 given $X_1 = x_1$.

    c   Show that X_1 and X_2 are dependent.

    d   Find $P(X_1 \leq 3/4 | X_2 = 1/2)$.

5.46 Let X_1 denote the amount of a certain bulk item stocked by a supplier at the beginning of a week, and suppose that X_1 has a uniform distribution over the interval $0 \leq X_1 \leq 1$. Let X_2 denote the amount of this item sold by the supplier during the week, and suppose that X_2 has a uniform distribution over the interval $0 \leq x_2 \leq x_1$, where x_1 is a specific value of X_1.

    a   Find the joint density function for X_1 and X_2.

    b   If the supplier stocks an amount of 1/2, what is the probability that he sells an amount greater than 1/4?

    c   If it is known that the supplier sold an amount equal to 1/4, what is the probability that he had stocked an amount greater than 1/2?

5.47 Let (X_1, X_2) denote the coordinates of a point selected at random inside a unit circle with center at the origin. That is, X_1 and X_2 have the joint density function given by

$$f(x_1, x_2) = \begin{cases} 1/\pi & x_1^2 + x_2^2 \leq 1 \\ 0 & \text{elsewhere} \end{cases}$$

    a   Find the marginal density function of X_1.

    b   Find $P(X_1 \leq X_2)$.

5.48 Let X_1 and X_2 have the joint density function given by

$$f(x_1, x_2) = \begin{cases} x_1 + x_2 & 0 \leq x_1 \leq 1, 0 \leq x_2 \leq 1 \\ 0 & \text{elsewhere} \end{cases}$$

    a   Find the marginal density functions of X_1 and X_2.

    b   Are X_1 and X_2 independent?

    c   Find the conditional density of X_1 given $X_2 = x_2$.

5.49 Let X_1 and X_2 have the joint density function given by

$$f(x_1, x_2) = \begin{cases} K & 0 \leq x_1 \leq 2, 0 \leq x_2 \leq 1; 2x_2 \leq x_1 \\ 0 & \text{elsewhere} \end{cases}$$

    a   Find the value of K that makes the function a probability density function.

    b   Find the marginal densities of X_1 and X_2.

    c   Find the conditional density of X_1 given $X_2 = x_2$.

    d   Find the conditional density of X_2 given $X_1 = x_1$.

    e   Find $P(X_1 \leq 1.5, X_2 \leq 0.5)$.

    f   Find $P(X_2 \leq 0.5 | X_1 \leq 1.5)$.

5.50 Let X_1 and X_2 have a joint distribution that is uniform over the region shaded in the diagram.

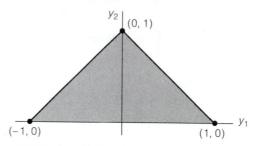

    a Find the marginal density for X_2.

    b Find the marginal density for X_1.

    c Find $P[(X_1 - X_2) \geq 0]$.

5.51 Refer to Exercise 5.41.

    a Find $E(X_1)$.

    b Find $V(X_1)$.

    c Find $\text{Cov}(X_1, X_2)$.

5.52 Refer to Exercise 5.42. Find $\text{Cov}(X_1, X_2)$.

5.53 Refer to Exercise 5.43.

    a Find $\text{Cov}(X_1, X_2)$.

    b Find $E(X_1 + X_2)$ and $V(X_1 + X_2)$ by finding the probability distribution of $X_1 + X_2$.

    c Find $E(X_1 + X_2)$ and $V(X_1 + X_2)$ by using Theorem 5.2.

5.54 Refer to Exercise 5.48.

    a Find $\text{Cov}(X_1, X_2)$.

    b Find $E(3X_1 - 2X_2)$.

    c Find $V(3X_1 - 2X_2)$.

5.55 Refer to Exercise 5.49.

    a Find $E(X_1 + 2X_2)$.

    b Find $V(X_1 + 2X_2)$.

5.56 A quality-control plan calls for randomly selecting three items from the daily production (assumed to be large) of a certain machine and observing the number of defectives. However, the proportion p of defectives produced by the machine varies from day to day and is assumed to have a uniform distribution on the interval $(0,1)$. For a randomly chosen day, find the unconditional probability that exactly two defectives are observed in the sample.

5.57 The number of defects per yard, denoted by X, for a certain fabric is known to have a Poisson distribution with parameter λ. However, λ is not known and is assumed to be random with a probability density function given by

$$f(\lambda) = \begin{cases} e^{-\lambda} & \lambda \geq 0 \\ 0 & \text{elsewhere} \end{cases}$$

Find the unconditional probability function for X.

5.58 The lifelength X of a fuse has a probability density

$$f(x) = \begin{cases} e^{-x/\theta}/\theta & x > 0, \theta > 0 \\ 0 & \text{elsewhere} \end{cases}$$

Three such fuses operate independently. Find the joint density of their lifelengths, X_1, X_2, and X_3.

5.59   A retail grocery merchant figures that her daily gain from sales X is a normally distributed random variable with $\mu = 50$ and $\sigma^2 = 10$ (measurements in dollars). X could be negative if she is forced to dispose of perishable goods. Also, she figures daily overhead costs Y to have a gamma distribution with $\alpha = 4$ and $\beta = 2$. If X and Y are independent, find the expected value and variance of the net daily gain. Would you expect the net gain for tomorrow to go above \$70?

5.60   Refer to Exercises 5.42 and 5.45.

    a   Find $E(X_2|X_1 = x_1)$.

    b   Use Theorem 5.3 to find $E(X_2)$.

    c   Find $E(X_2)$ directly from the marginal density of X_2.

5.61   Refer to Exercise 5.57.

    a   Find $E(X)$ by first finding the conditional expectation of X for given λ and then using Theorem 5.3.

    b   Find $E(X)$ directly from the probability distribution of X.

5.62   Refer to Exercise 5.46. If the supplier stocks an amount equal to 3/4, what is the expected amount sold during the week?

5.63   Let X be a continuous random variable with distribution function $F(x)$ and density function $f(x)$. We can then write, for $x_1 \leq x_2$,

$$P(X \leq x_2 | X \geq x_1) = \frac{F(x_2) - F(x_1)}{1 - F(x_1)}$$

As a function of x_2 for fixed x_1, the right-hand side of this expression is called the conditional distribution function of X given that $X \geq x_1$. On taking the derivative with respect to x_2, we see that the corresponding conditional density function is given by

$$\frac{f(x_2)}{1 - F(x_1)} \qquad x_2 \geq x_1$$

Suppose a certain type of electronic component has lifelength X with the density function (lifelength measured in hours)

$$f(x) = \begin{cases} (1/200)e^{-x/200} & x \geq 0 \\ 0 & \text{elsewhere} \end{cases}$$

Find the expected lifelength for a component of this type that has already been in use for 100 hours.

†5.64   Let X_1, X_2, and X_3 be random variables, either continuous or discrete. The joint moment-generating function of X_1, X_2, and X_3 is defined by

$$M(t_1, t_2, t_3) = E(e^{t_1 X_1 + t_2 X_2 + t_3 X_3})$$

    a   Show that $M(t, t, t)$ gives the moment-generating function of $X_1 + X_2 + X_3$.

    b   Show that $M(t, t, 0)$ gives the moment-generating function of $X_1 + X_2$.

    c   Show that

$$\left. \frac{\partial^{k_1 + k_2 + k_3} M(t_1, t_2, t_3)}{\partial t_1^{k_1} \partial t_2^{k_2} \partial t_3^{k_3}} \right|_{t_1 = t_2 = t_3 = 0} = E(X_1^{k_1} X_2^{k_2} X_3^{k_3})$$

†Optional exercises.

†**5.65**   Let X_1, X_2, and X_3 have a multinomial distribution with the probability function

$$p(x_1, x_2, x_3) = \frac{n!}{x_1!x_2!x_3!}p_1^{x_1}p_2^{x_2}p_3^{x_3} \qquad \sum_{i=1}^{n}x_i = n$$

Use the results of Exercise 5.64 to answer the following:

a   Find the joint moment-generating function of X_1, X_2, and X_3.

b   Use the joint moment-generating function to find $\mathrm{Cov}(X_1, X_2)$.

5.66   The negative binomial variable X was defined as the number of the trial on which the rth success occurs in a sequence of independent trials with constant probability p of success on each trial. Let X_i denote a geometric random variable, defined as the number of the trial on which the first success occurs. Then we can write

$$X = \sum_{i=1}^{n}X_i$$

for independent random variables $X_1, \ldots, X_r$. Use Theorem 5.2 to show that $E(X) = r/p$ and $V(X) = r(1-p)/p^2$.

5.67   A box contains four balls, numbered 1 through 4. One ball is selected at random from this box. Let

$$X_1 = 1 \text{ if ball 1 or ball 2 is drawn}$$
$$X_2 = 1 \text{ if ball 1 or ball 3 is drawn}$$
$$X_3 = 1 \text{ if ball 1 or ball 4 is drawn}$$

and the X_i's are zero otherwise. Show that any two of the random variables X_1, X_2, and X_3 are independent, but the three together are not.

---

†Optional exercises.

Statistics, Sampling Distributions, and Control Charts

About This Chapter

In the process of making an inference from a sample to a population, we usually calculate one or more statistics, such as the mean or variance. Since samples are randomly selected, the values that such statistics assume change from sample to sample. Thus, sample statistics are, themselves, random variables, and their behavior can be modeled by probability distributions. The probability distribution of a sample statistic is called its *sampling distribution*. After studying the properties of sampling distributions, we turn to their use in the construction of a tool of fundamental importance for process improvement—the control chart.

Contents

6.1 Introduction

Chapter 1 contained a discussion of ways to summarize data so that we can derive some useful and important information from them. We now return to this idea of deriving information from data, but from this point on each data set under study will be viewed as a *sample* from a much larger set of data points that could be studied, called the *population*. A sample of 50 household incomes could be selected from all households in your community. A sample of five measurements could be taken of the diameter of a machined rod, these five being a small part of the nearly infinite number of diameter measurements that could have been taken if enough time were available. As you consider these two examples, notice that the first population (households in your community) is real (these households actually exist), while the second population (measurements of rod diameters) is only conceptual (no list of rod-diameter measurements really exists). In either case, however, we can visualize the nature of a population of values of which our sample data constitute only a small part. Our job is to describe populations as accurately as possible given only the data in samples from these populations. In other words, we want to decide how closely our sample mirrors the characteristics of the population from which it was selected. As we have seen, numerical descriptive measures of a *population* are called *parameters*. The *sample* counterparts of these quantities, which form numerical descriptive measures of a sample, are called *statistics*.

D E F I N I T I O N   **6.1**   A **statistic** is a function of sample observations that contains no unknown parameters.   ■

The households in your community actually have an average, or mean, annual income (a parameter), even though it is unknown. A sample of households has a sample mean income (a statistic) that can actually be calculated after the sampling is completed. Will the sample and population means be equal? Probably not, although we'll never be sure. We hope that the sample mean will be close to the population mean, and, as we'll see, if the sampling is done well that hope is generally fulfilled. So we now embark on a study of questions such as "How great is the difference between a sample mean and its corresponding population mean likely to be?" The sample mean is only one of many possible statistics we might study; the median, the quantiles, and the IQR (to name a few) are also important statistics. We begin with a careful study of the mean because of its central role in classical statistical inference procedures.

6.2 The Sample Mean and Variance

To illustrate some essential features of sample means, we will generate samples from a population whose structure is known to us. That population is the set of random digits (0 through 9) in a large random-number table (real) or as generated by a computer (conceptual). If the digits really are randomly generated, an ideal sample of 100 digits (a sample that truly mirrors the population) would look like the sample whose stem-and-leaf plot is given in Figure 6.1. (The leaves here are the tenths digit, all zero in this case.) The mean of this ideal sample is 4.50, right in the middle of [0, 9]. This ideal sample mean can be called the *population mean*, since it reflects the average value of all digits in the population. But a true random sample of 100 digits will actually look more like Figure 6.2. A little irregularity slips into most random samples, and here the sample mean is 4.58, a little off from the ideal 4.50. This mean is based on 100 observations, a pretty large sample! What would happen to the discrepancy between the sample and population means with smaller samples? We investigate that question next.

FIGURE **6.1**
Ideal Sample of Digits 0–9

```
10  0 | 0000000000
20  1 | 0000000000
30  2 | 0000000000
40  3 | 0000000000
50  4 | 0000000000   6|0 denotes 6.0 or 6.
50  5 | 0000000000
40  6 | 0000000000
30  7 | 0000000000
20  8 | 0000000000
10  9 | 0000000000
     Mean = 4.5000
```

FIGURE **6.2**
Real Sample of Digits 0–9

```
11  0 | 00000000000
20  1 | 000000000
27  2 | 0000000
35  3 | 00000000
49  4 | 00000000000000
(8) 5 | 00000000
43  6 | 000000000000
30  7 | 00000000000
19  8 | 00000000000
 8  9 | 00000000
     Mean = 4.5800
```

Figure 6.3 displays the data for 100 *averages* of samples of size 2 from a random digit generator. In other words, two random digits are sampled and their mean (average) is recorded; then the process is repeated for a total of 100 sample averages. Notice first the shape of the distribution. The averages tend to pile up in the middle, with few values close to 0 or 9. But there is still a chance of getting a sample mean more than 4 units from the ideal mean of 4.50.

The same procedure is repeated for samples of size 3, 4, 5, and 10, and the resulting sample means are displayed in Figures 6.4, 6.5, 6.6, and 6.7, respectively.

Notice that the sample means pile up more and more in the middle as the sample size increases. Also, the distribution of the sample means becomes quite symmetric around 4.50, the population mean. By the time the sample size gets to 5, there is virtually no chance of seeing a difference as large as 4 between a sample mean and its population mean. When the sample size reaches 10, there is virtually no chance of seeing a difference larger than 2.

FIGURE **6.3**
Sampling Distribution for Means of 2

```
  5  0 | 55555
  8  1 | 000
 21  2 | 000000555555
 37  3 | 0000000055555555
(19) 4 | 0000000005555555555
 44  5 | 0000000000555555555
 25  6 | 000000555555
 13  7 | 000055
  7  8 | 0005
  3  9 | 000
Mean = 4.4450
```

FIGURE **6.4**
Sampling Distribution for Means of 3

```
  4  1 | 3366
 20  2 | 0000333366666666
 37  3 | 00003333333366666
 49  4 | 000003333666
(26) 5 | 0003333333333333333336666666
 25  6 | 0000003333366666
  9  7 | 03366
  4  8 | 0033
Mean = 4.6367
```

FIGURE **6.5**
Sampling Distribution for Means of 4

```
  4  1 | 2557
 12  2 | 25555777
 36  3 | 002222222255555555577777
(22) 4 | 0222225555557777777777
 42  5 | 000000022255555777777
 20  6 | 00000222257777
  6  7 | 0225
  2  8 | 05
Mean = 4.6075
```

To summarize, the *sampling distributions* of sample means (which is what Figures 6.3 through 6.7 are often called) tend to be symmetric about the true population mean

FIGURE **6.6**
Sampling Distribution for Means of 5

```
  1   1 | 2
 10   2 | 044448888
 27   3 | 00022444444666888
(37)  4 | 0000000222222244444444446666666666688
 36   5 | 0000002224444666688
 17   6 | 000222226888
  5   7 | 00022
Mean = 4.5660
```

FIGURE **6.7**
Sampling Distribution for Means of 10

```
  3   2 | 558
 25   3 | 2222223355667889999999
(41)  4 | 001111111111223334455556666666667777888999
 34   5 | 00000002222222444444444446666666666688
  7   6 | 0123368
Mean = 4.5980
```

and tend to pile up around the population mean, with small tail areas, as the sample size increases. This statement assumes that the samples are taken randomly and independently of one another. The boxplots of Figure 6.8 are another way to show the shrinking variability and the concentration of values around the population mean with increasing sample size. (Notice the outlier for the case $n = 5$.)

We have a principle of centrality for sampling distributions of means: The sample means tend to center around the population mean. Can we find a principle of variability more precise than "the variability among the sample means decreases as the sample size increases?" The study of this problem is more easily presented given some notation.

We have used $\mu = E(X)$ to denote the expected, or mean, value of a random variable X and

$$\sigma^2 = V(X) = E(X - \mu)^2$$

to denote the variance. If the random variable X can take on any of N values with probability $1/N$, then the mean becomes

$$\mu = \frac{1}{N} \sum_{i=1}^{N} x_i$$

and the variance becomes

$$\sigma^2 = \frac{1}{N} \sum_{i=1}^{N} (x_i - \mu)^2$$

where x_i denotes the ith possible value for X. For a sample of size n, the sample mean is

$$\bar{x} = \frac{1}{n} \sum_{i=1}^{n} x_i$$

F I G U R E **6.8**
Boxplots of the Sampling
Distributions of Averages
(sample sizes 2, 3, 4, 5, and
10 from top to bottom)

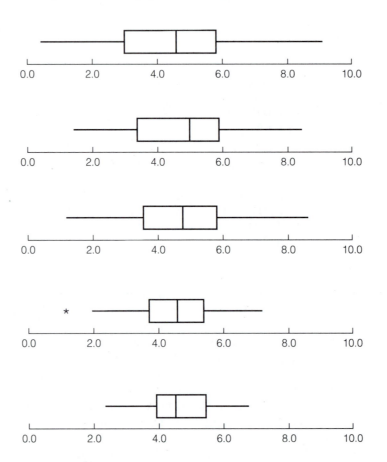

where, in this case, x_i is the ith sample observation. So the sample copy of μ, namely $\bar{x}$, has a distribution in repeated sampling that centers about μ. We might say, then, that $\bar{x}$ is a good estimator of μ. (The average annual household income calculated from a sample of 50 households should be close to the true average for all households in your community.)

To measure variability in a sample, we might try a sample copy of the variance, namely,

$$s^{*2} = \frac{1}{n} \sum_{i=1}^{n} (x_i - \bar{x})^2$$

As it turns out, this is not a very good estimator. Whereas random samples tend to center at their population mean, these same samples tend to have *less* variability than the population that they came from. This should be intuitively reasonable. A relatively small sample is not likely to capture the extremes of the population and, hence, will show less variability than is present in the population. To compensate for this fact, we change the divisor of s^{*2} from n to $n - 1$, thereby making the quantity larger. Thus,

$$s^2 = \frac{1}{n-1} \sum_{i=1}^{n} (x_i - \bar{x})^2$$

is called the *sample variance* and is used to estimate σ^2, the population variance.

To see how s^{*2} and s^2 compare, we calculated each value for 100 samples of size 5 from the random digits 0–9 (the same samples displayed in Figure 6.6). Then we did the same thing for samples of size 10 (the data of Figure 6.7). The results are displayed in Figures 6.9 and 6.10. If any of the digits are equally likely on any one selection, then X, the outcome for that selection, has

$$\mu = E(X) = \frac{1}{10}(0 + 1 + \cdots + 9) = 4.5$$

and

$$\sigma^2 = V(X) = \frac{1}{10}[(0 - 4.5)^2 + \cdots + (9 - 4.5)^2] = 8.25$$

which are the population mean and variance, respectively. Note that s^{*2} seems to be centering around 6 or 7, while s^2 is centering around 8 or more. The mean values of s^2 are quite close to the ideal value of 8.25.

FIGURE **6.9**
Sampling Distributions of s^{*2}

1	0	5				
2	1	4				
11	2	115699999		3	2	278
19	3	23446678		8	3	56688
31	4	001122445689		15	4	0012458
42	5	02234788888		26	5	01244446778
(16)	6	0011112245666668		47	6	000012344456666668889
42	7	000222234477		(15)	7	222223466667888
30	8	245555569		38	8	0000222344566888899
21	9	00348		19	9	0445668
16	10	2222228		12	10	68
9	11	446677		10	11	04469
3	12	49		5	12	044
1	13	4		2	13	24

Mean = 6.6248   Mean = 7.3702
Samples of size 5   Samples of size 10

We now know that s^2 is a good estimator of σ^2; for that reason, s^2 (or s) will be used as the standard measure of variability in a sample. But what about the variability of the sample means? If we let X_i denote the random outcome of the ith sample observation to be selected, then

$$\bar{X} = \frac{1}{n} \sum_{i=1}^{n} X_i$$

FIGURE **6.10**
Sampling Distributions of Sample Variances (s^2) (with boxplots)

```
 1   0 | 7                    1    2 | 4
 2   1 | 8                    4    3 | 029
 4   2 | 77                  13    4 | 012244679
11   3 | 2377777            18    5 | 04677
19   4 | 02335578           32    6 | 00012444677789
29   5 | 0022335578         47    7 | 011223333345667
36   6 | 0235578           (19)   8 | 0000112333457779999
50   7 | 23333355777788     34    9 | 122234566788899
50   8 | 02333335888        19   10 | 0446679
39   9 | 000023377          12   11 | 8
30  10 | 35777778           11   12 | 13779
22  11 | 23378               6   13 | 2478
17  12 | 3888888             2   14 | 79
10  13 | 5
 9  14 | 335577
 3  15 | 5
 2  16 | 28
```

Mean = 8.28   Mean = 8.17
Samples of size 5   Samples of size 10

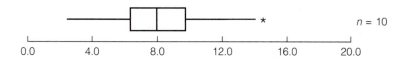

and, from properties of expectations of linear functions,

$$E(\overline{X}) = \frac{1}{n} \sum_{i=1}^{n} E(X_i)$$

$$= \frac{1}{n} \sum_{i=1}^{n} \mu = \mu$$

where $\mu = E(X_i)$, $i = 1, \ldots, n$. If, in addition, the sampled observations are selected independently of one another,

$$V(\overline{X}) = \frac{1}{n^2} V\left(\sum_{i=1}^{n} X_i\right)$$

$$= \frac{1}{n^2} \sum_{i=1}^{n} V(X_i)$$

$$= \frac{1}{n^2} \sum_{i=1}^{n} \sigma^2 = \frac{\sigma^2}{n}$$

Thus, the standard deviation within any one sampling distribution of sample means should be $\sigma/\sqrt{n}$. To check this out for the data generated in Figures 6.3 through 6.7, we calculate $\sigma/\sqrt{n}$ for each case and compared it to the sample standard deviation calculated for the 100 sample means in the corresponding figure. The results are as follows (with $\sigma = \sqrt{8.25} = 2.87$):

Figure	Sample Size	Calculated Standard Deviation	$\sigma/\sqrt{n}$
6.3	2	2.05	2.03
6.4	3	1.74	1.66
6.5	4	1.49	1.44
6.6	5	1.27	1.28
6.7	10	0.92	0.91

We see that in each case the observed variability for the 100 sample means is quite close to the theoretical value of $\sigma/\sqrt{n}$. In summary, then, the variability in the sample means decreases as the sample size increases, with the variance of a sample mean given by σ^2/n. The quantity $\sigma/\sqrt{n}$ is sometimes called the *standard error* of the mean. The variance σ^2 can be estimated by the sample variance s^2.

The properties of the sample mean and sample variance illustrated above hold only for *random samples*. That is, each observation in the sample must be independently selected from the same probability distribution, as was the case in sampling random digits. (Each digit is independently selected from a distribution of digits that places probability 1/10 on each digit 0 through 9.) Suppose n resistors of the same type are placed into service and their lifelengths are measured. If the lifelengths are independent of one another, then the resulting measurements would constitute an observed random sample. The assumption of independence is difficult to justify in some cases, but careful thought should be given to the matter in every statistical problem. If the resistors are all in the same system and fail because of a power surge in the system, then their lifelengths are clearly *not* independent. If, however, they fail only from normal usage, then their lifelengths may be independent even if they are being used in the same system. *Most of the problems and examples in this book will assume that sample measurements come from random samples.*

Random digits and potential measurements of the lifelength of a resistor come from conceptual populations that are infinite. For finite populations, like the employees of a firm or the beads in a bead box, true independence of sample observations is not generally achieved. If a bead is randomly selected from a box and not replaced, then the probability distribution of the remaining beads in the box has changed. The second bead selected comes from a slightly different distribution. If, however, the finite population is large compared to the sample size, the sample observations from a series of random selections are nearly independent. The result, then, will still be called a random sample, and the properties of sample means and variances presented above still hold.

The term *random sample* suggests the manner in which items should be selected to ensure some degree of independence in the resulting measurements; that is, the items should be selected *at random*. One method of random selection involves the use of a random-number table, such as shown in Table 1 of the Appendix. Suppose 250 circuit boards constitute one day's production and ten are to be sampled for quality inspection. The boards are numbered from 1 to 250. Then ten three-digit random numbers between 001 and 250 are selected from a random-number table (or by computer). The sample consists of the ten boards bearing the selected numbers. The ten quality measurements (perhaps the number of properly functioning circuits per board) then make up a random sample of measurements.

E X A M P L E   **6.1**   For random samples of domestic and foreign used car ads in the *Gainesville Sun* (26 February 1992), the year of the car was recorded. The data (with "19" suppressed) are as follows:

Year of Domestic Cars Advertised

83	85	89	88	86	83	84	76	74	83	86	78	91
74	80	87	84	80	89	85	78	87	86	84	89	89
76	86	85	85	68	69	86	80	79	83	84	72	78
81												

Year of Foreign Cars Advertised

88	89	88	88	85	84	91	80	85	89	91	84	88
89	87	87	87	86	85	82	87	87	85	83	68	90
90	87	87	85	86	90	88	83	90	83	76	86	85
79												

a   Calculate the mean and median for each group. What do these values suggest about the data distributions?

b   Calculate the standard deviation for each data set. Do they suggest anything about the data distributions?

c   Will this mean of a sample of size 40 be a good estimate of the mean year of used cars in the Gainesville area? in the United States?

Solution   Straightforward calculations (use your calculator) yield the following:

	Domestic	Foreign
Mean	82.2	85.7
Median	84.0	87.0
Standard Deviation	5.6	4.3

a   The fact that the mean is smaller than the median suggests that the distribution is skewed toward the smaller values (older cars). Does this make sense for this type of data?

b   The standard deviations are relatively large. In fact, adding two standard deviations to the mean takes us beyond 1993 (off the scale for these data collected in 1992). This helps confirm the fact that the distributions must be skewed toward the smaller values. (You might want to construct parallel boxplots or stemplots to show features of the shapes of these data sets.)

c   Even though the individual samples are skewed (as are the populations), sample means for $n = 40$ will tend to have symmetric, mound-shaped sampling distributions centering at the population mean. What is the population in this case? Since the ads came from a small-circulation newspaper, the population of used cars must be quite narrowly defined as those in the area covered by the newspaper. The sample means from these data should be good estimates of the mean for these populations of domestic and foreign used cars. They would not necessarily be good estimates of mean years of used cars across the country. (Why not?)   ∎

Suppose measurements $x_1, x_2, \ldots, x_n$ are transformed to new measurements $y_1, y_2, \ldots, y_n$ by

$$y_i = ax_i + b$$

for constants a and b. If $x_1, \ldots, x_n$ have mean $\bar{x}$ and variance s_x^2, then it is easy to show that $y_1, \ldots, y_n$ have mean $\bar{y}$ and variance s_y^2 given by

$$\bar{y} = a\bar{x} + b$$

and

$$s_y^2 = a^2 s_x^2$$

These properties, along with the calculation of $\bar{x}$ and s^2, are illustrated in Example 6.2.

EXAMPLE   **6.2**   For a simple scaffold structure made of steel, it is important to study the increase in length of tension members under load. For a load of 2,000 kilograms, ten similar tension members showed length increases as follows, with measurements in centimeters:

$$2.5, 2.2, 3.0, 2.1, 2.7, 2.5, 2.8, 1.9, 2.2, 2.6$$

a   Find the mean and standard deviation of these measurements.

b   Suppose the measurements had been taken in meters rather than centimeters. Find the mean and standard deviation of the corresponding measurements in meters.

Solution   a   The mean is given by

$$\bar{x} = \frac{1}{n} \sum_{i=1}^{n} x_i = \frac{1}{10}(2.5 + 2.2 + \cdots + 2.6)$$

$$= \frac{1}{10}(24.5) = 2.45$$

and the variance is given by

$$s_x^2 = \frac{1}{n-1} \sum_{i=1}^{n} (x_i - \bar{x})^2$$

which is equivalent to

$$s_x^2 = \left(\frac{1}{n-1}\right) \left[\sum_{i=1}^{n} x_i^2 - \frac{1}{n} \left(\sum_{i=1}^{n} x_i\right)^2 \right]$$

$$= \frac{1}{9} \left[61.09 - \frac{1}{10}(24.5)^2 \right]$$

$$= \frac{1}{9}(61.09 - 60.025) = 0.1183$$

Then the standard deviation is

$$s_x = \sqrt{s_x^2} = \sqrt{0.1183} = 0.34$$

b   If y_i denotes the corresponding measurement in meters, then

$$y_i = (0.01)x_i$$

It follows that

$$\sum_{i=1}^{10} y_i = \sum_{i=1}^{10} (0.01)x_i = (0.01) \sum_{i=1}^{10} x_i$$

and

$$\bar{y} = (0.01)\bar{x} = (0.01)(2.45) = 0.0245$$

Also,

$$s_y^2 = \frac{1}{9} \sum_{i=1}^{10} (0.01 x_i - 0.01\bar{x})^2$$

$$= \frac{1}{9} \sum_{i=1}^{10} (0.01)^2 (x_i - \bar{x})^2$$

$$= (0.01)^2 \left(\frac{1}{9}\right) \sum_{i=1}^{10} (x_i - \bar{x})^2$$

$$= (0.01)^2 s_x^2$$

Hence,

$$s_y = \sqrt{s_y^2} = (0.01)s_x = (0.01)(0.34) = 0.0034 \quad \blacksquare$$

E X A M P L E   **6.3**   Calculate $\bar{x}$ and s^2 for the 50 lifelength observations in Table 4.1. Also approximate the variance of $\overline{X}$.

Solution   From the data in Table 4.1, we have

$$\bar{x} = \frac{1}{n} \sum_{i=1}^{n} x_i = \frac{1}{50}(113.296) = 2.266$$

Using

$$s^2 = \frac{1}{n-1} \left[\sum_{i=1}^{n} x_i^2 - \frac{1}{n} \left(\sum_{i=1}^{n} x_i \right)^2 \right]$$

we have

$$s^2 = \frac{1}{49} \left[440.2332 - \frac{1}{50}(113.296)^2 \right]$$
$$= 3.745$$

or

$$s = 1.935$$

(If your calculator has a standard deviation key, use it in place of the calculation formula above.)

Note that the exponential model suggested in Section 4.1,

$$f(x) = \frac{1}{2} e^{-x/2} \qquad x > 0$$

gives $E(X) = 2$, which is close to the observed value of $\bar{x}$. Also, from the exponential model $V(X) = \sigma^2 = 4$, which is quite close to the observed s^2 value. $V(\overline{X}) = \sigma^2/n$ can be approximated by s^2/n, or $3.745/50 = 0.075$.   ∎

To further illustrate the behavior of $\bar{x}$ and s^2, we have selected 100 samples each of size $n = 25$, from an exponential distribution with a mean of 10. That is, the probabilistic model for the population is given by

$$f(x) = \begin{cases} \dfrac{1}{10} e^{-x/10} & x > 0 \\ 0 & \text{elsewhere} \end{cases}$$

For this model, $E(X) = \mu = 10$ and $V(X) = \sigma^2 = (10)^2$, or $\sigma = 10$. The values for $\bar{x}$ and s^2 were calculated for each of the 100 samples. The average of the 100 sample means turned out to be 9.88, and the average of the 100 values of s was 9.70. Note that both averages are reasonably close to 10. The standard deviation for the 100 values of $\bar{x}$ was calculated and found to be 2.17. Theoretically, the standard deviation of $\overline{X}$ is

$$\sqrt{V(\overline{X})} = \frac{\sigma}{\sqrt{n}} = \frac{10}{\sqrt{25}} = 2.0$$

which is not far from the observed value of 2.17. A relative frequency histogram for the 100 values of $\bar{x}$ is shown in Figure 6.11. We pursue the notions connected with the shape of this distribution in the following section.

FIGURE **6.11**
Relative Frequency Histogram
for $\bar{x}$ from 100 Samples, Each
of Size 25

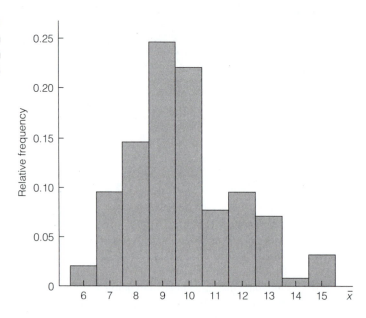

Exercises

6.1 Concentrations of uranium 238 were measured in 12 soil samples from a certain region, with the following results in pCi/g (picoCuries per gram):

0.76, 1.90, 1.84, 2.42, 2.01, 1.77,
1.89, 1.56, 0.98, 2.10, 1.41, 1.32

a Construct a stem-and-leaf display and boxplot.

b Calculate $\bar{x}$ and s^2 for these data.

c If another soil sample is randomly selected from this region, find an interval in which the uranium concentration measurement should lie with probability at least 0.75. [*Hint:* Assume the sample mean and variance are good approximations to the population mean and variance and use Tchebysheff's Theorem.]

6.2 What's the average number of hours that students in your statistics class study in a week? Estimate this average by selecting a random sample of five students and asking each sampled student how many hours he or she studies in a typical week. Use a random-number table in selecting the sample. Collect answers from other students and construct a dotplot of the means from samples of size 5. Describe the shape of the dotplot.

6.3 Obtain a standard six-sided die.

a If the die is balanced, what will the outcomes of an ideal sample of 30 tosses look like?

b Find the mean and variance of the ideal sample outcomes portrayed in (a).

c   Toss your die 30 times and record the upper-face outcomes. How does the distribution of outcomes compare with the ideal sample in (a)?

d   Calculate the sample mean and variance for the data in (c). How do they compare with the results of (b)?

e   Collect the sample means for 30 die tosses from other members of the class. Make a dotplot of the results and comment on its shape.

6.4   The table below gives the average monthly low temperatures for Boston and Paris.

Average Monthly Low Temperature

Month	Boston–°F	Paris–°C
January	22	0
February	22	1
March	30	2
April	39	5
May	49	8
June	58	11
July	64	13
August	64	13
September	56	10
October	47	7
November	37	3
December	26	1

Boston readings are in degrees Fahrenheit, and Paris readings are in degrees Celsius. Degrees C are transformed to degrees F by the equation

$$F = 1.8C + 32$$

a   Find the mean and variance for the Boston average monthly temperatures.

b   Find the mean and variance of the Paris average monthly temperatures.

c   Transform the Paris mean and variance to their corresponding values in degrees F. Compare the results to those of part (a). How would you describe the differences between Boston and Paris low temperatures?

6.5   The dotplot of Figure 6.12 shows the typical 1990 values (in thousands of dollars) of existing one-family homes in 54 metropolitan statistical areas across the United States. These data have a mean of 106.14, a median of 82.95, and a standard deviation of 57.12. Four random samples, each of size $n = 5$, were selected from the 54 values; the sample data are displayed below.

Sample 1	Sample 2	Sample 3	Sample 4
73.0	110.8	86.6	59.7
93.3	86.1	100.9	93.3
183.7	86.1	77.3	84.0
86.6	87.0	100.9	183.7
77.3	58.1	81.5	92.0

Calculate the sample mean and standard deviation for each sample. Compare them to the population mean and standard deviation. Comment on the problems of estimating μ and σ for the type of population shown here.

F I G U R E **6.12** Graph for Exercise 6.5

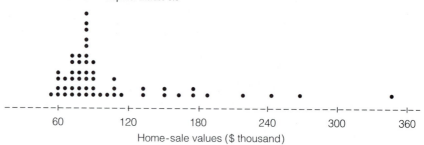

Home-sale values ($ thousand)

6.6 Referring to the home-sale data of Exercise 6.5, calculate the median for each of the four random samples. Compare the sample medians to the population median. Do you think the sample median might be a better estimator of the "center" of the population in this case? Why or why not?

6.7 Show that

$$s^2 = \frac{1}{n-1} \sum_{i=1}^{n} (x_i - \bar{x})^2$$

$$= \frac{1}{n-1} \left[\sum_{i=1}^{n} x_i^2 - \frac{1}{n} \left(\sum_{i=1}^{n} x_i \right)^2 \right]$$

6.8 If $X_1, \ldots, X_n$ is a random sample with $E(X_i) = \mu$ and $V(X_i) = \sigma^2$, show that $E(S^2) = \sigma^2$, where

$$S^2 = \frac{1}{n-1} \sum_{i=1}^{n} (X_i - \overline{X})^2$$

[*Hint:* Write $\sum_{i=1}^{n} (X_i - \overline{X})^2 = \sum_{i=1}^{n} [(X_i - \mu) - (\overline{X} - \mu)]^2$. Square the term in brackets, carry out the summation, and then find the expectations.]

6.3 The Sampling Distribution of $\overline{X}$

We have seen that the approximated sampling distributions of a sample mean (Figures 6.3 through 6.7) tend to have a particular shape, being somewhat symmetric with a single mound in the middle. This result is not coincidental or unique to those particular examples. The fact that sampling distributions for sample means always tend to be approximately normal in shape is a consequence of the *Central Limit Theorem*.

THEOREM **6.1** **The Central Limit Theorem**   If a random sample of size n is drawn from a population with mean μ and variance σ^2, then the sample mean $\overline{X}$ has approximately a normal distribution with mean μ and variance σ^2/n. That is, the distribution function of

$$\frac{\overline{X} - \mu}{\sigma/\sqrt{n}}$$

is approximately a standard normal. The approximation improves as the sample size increases. ∎

Theorem 6.1 will not be proved in this book; the proof belongs in a more advanced course. Note, however, that $\overline{X}$ is a random variable and has a probability distribution.

We sometimes abbreviate the central tenet of Theorem 6.1 to the phrase "$\overline{X}$ is asymptotically normal with mean μ and variance σ^2/n." The practical importance of this result is that, for large n, the sampling distribution of $\overline{X}$ can be closely approximated by a normal distribution. More precisely,

$$P(\overline{X} \le b) = P\left(\frac{\overline{X} - \mu}{\sigma/\sqrt{n}} \le \frac{b - \mu}{\sigma/\sqrt{n}}\right)$$

$$\approx P\left(Z \le \frac{b - \mu}{\sigma/\sqrt{n}}\right)$$

where Z is a standard normal random variable.

We saw some approximate sampling distributions in Figures 6.3 through 6.7; now we will look at more computer simulations of sampling distributions with large sample sizes. Samples of size n were drawn from a population having the probability density function

$$f(x) = \begin{cases} \dfrac{1}{10}e^{-x/10} & x > 0 \\ 0 & \text{elsewhere} \end{cases}$$

The sample mean was computed for each sample. The relative frequency histogram of these mean values for 1,000 samples of size $n = 5$ is shown in Figure 6.13. Figures 6.14 and 6.15 show similar results for 1,000 samples of size $n = 25$ and $n = 100$, respectively. Although all the relative frequency histograms have a sort of bell shape, notice that the tendency toward a symmetric normal curve is better for larger n. A smooth curve drawn through the bar graph of Figure 6.15 would be nearly identical to a normal density function with mean 10 and variance $(10)^2/100 = 1$.

The Central Limit Theorem provides a very useful result for statistical inference, for we now know not only that $\overline{X}$ has mean μ and variance σ^2/n if the population has mean μ and variance σ^2, but also that the probability distribution for $\overline{X}$ is approximately normal. For example, suppose we want to find an interval (a, b) such that

$$P(a \le \overline{X} \le b) = 0.95$$

FIGURE **6.13**
Relative Frequency Histogram
for $\bar{x}$ from 1,000 Samples of
Size $n = 5$

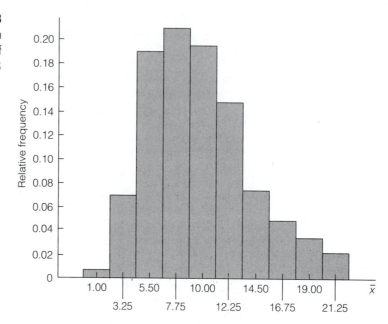

FIGURE **6.14**
Relative Frequency Histogram
for $\bar{x}$ from 1,000 Samples of
Size $n = 25$

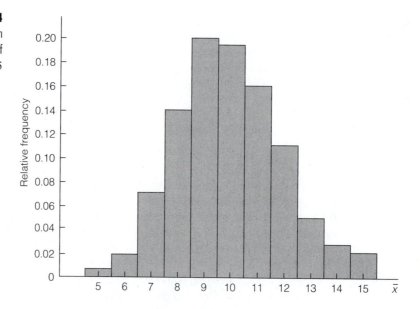

FIGURE **6.15**
Relative Frequency Histogram
for $\bar{x}$ from 1,000 Samples of
Size $n = 100$

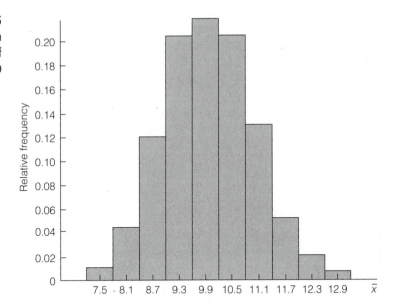

This probability is equivalent to

$$P\left(\frac{a - \mu}{\sigma/\sqrt{n}} \leq \frac{\overline{X} - \mu}{\sigma/\sqrt{n}} \leq \frac{b - \mu}{\sigma/\sqrt{n}}\right) = 0.95$$

for constants μ and σ. Since $(\overline{X} - \mu)/(\sigma/\sqrt{n})$ has approximately a standard normal distribution, the above equality can be approximated by

$$P\left(\frac{a - \mu}{\sigma/\sqrt{n}} \leq Z \leq \frac{b - \mu}{\sigma/\sqrt{n}}\right) = 0.95$$

where Z has a standard normal distribution. From Table 4 in the Appendix, we know that

$$P(-1.96 \leq Z \leq 1.96) = 0.95$$

and hence

$$\frac{a - \mu}{\sigma/\sqrt{n}} = -1.96 \qquad \frac{b - \mu}{\sigma/\sqrt{n}} = 1.96$$

or

$$a = \mu - 1.96\sigma/\sqrt{n} \qquad b = \mu + 1.96\sigma/\sqrt{n}$$

EXAMPLE   **6.4**   The fracture strengths of a certain type of glass average 14 (in thousands of pounds per square inch) and have a standard deviation of 2.

a   What is the probability that the average fracture strength for 100 pieces of this glass exceeds 14.5?

b Find an interval that includes the average fracture strength for 100 pieces of this glass with probability 0.95.

Solution **a** The average strength $\overline{X}$ has approximately a normal distribution with mean $\mu = 14$ and standard deviation

$$\frac{\sigma}{\sqrt{n}} = \frac{2}{\sqrt{100}} = 0.2$$

Thus,

$$P(\overline{X} > 14.5) = P\left(\frac{\overline{X} - \mu}{\sigma/\sqrt{n}} > \frac{14.5 - \mu}{\sigma/\sqrt{n}}\right)$$

is approximately equal to

$$P\left(Z > \frac{14.5 - 14}{0.2}\right) = P\left(Z > \frac{0.5}{0.2}\right)$$
$$= P(Z > 2.5) = 0.5 - 0.4938 = 0.0062$$

from Table 4. The probability of seeing an average value (for $n = 100$) more than 0.5 unit above the population mean is, in this case, very small.

b We have seen that

$$P\left(\mu - 1.96\frac{\sigma}{\sqrt{n}} \leq \overline{X} \leq \mu + 1.96\frac{\sigma}{\sqrt{n}}\right) = 0.95$$

for a normally distributed $\overline{X}$. In this problem,

$$\mu - 1.96\frac{\sigma}{\sqrt{n}} = 14 - 1.96\frac{2}{\sqrt{100}} = 13.6$$

and

$$\mu + 1.96\frac{\sigma}{\sqrt{n}} = 14 + 1.96\frac{2}{\sqrt{100}} = 14.4$$

Approximately 95% of the sample mean fracture strengths, for samples of size 100, should lie between 13.6 and 14.4. ∎

EXAMPLE **6.5** A certain machine that is used to fill bottles with liquid has been observed over a long period of time, and the variance in the amounts of fill is found to be approximately $\sigma^2 = 1$ ounce. However, the mean ounces of fill μ depend on an adjustment that may change from day to day, or from operator to operator. If $n = 25$ observations on ounces of fill dispensed are to be taken on a given day (all with the same machine setting), find the probability that the sample mean will be within 0.3 ounce of the true population mean for that setting.

Solution   We will assume $n = 25$ is large enough for the sample mean $\overline{X}$ to have approximately a normal distribution. Then

$$P(|\overline{X} - \mu| \leq 0.3) = P[-0.3 \leq (\overline{X} - \mu) \leq 0.3]$$

$$= P\left[-\frac{0.3}{\sigma/\sqrt{n}} \leq \frac{\overline{X} - \mu}{\sigma/\sqrt{n}} \leq \frac{0.3}{\sigma/\sqrt{n}}\right]$$

$$= P\left[-0.3\sqrt{25} \leq \frac{\overline{X} - \mu}{\sigma/\sqrt{n}} \leq 0.3\sqrt{25}\right]$$

$$= P\left[-1.5 \leq \frac{\overline{X} - \mu}{\sigma/\sqrt{n}} \leq 1.5\right]$$

Since $(\overline{X} - \mu)/(\sigma/\sqrt{n})$ has approximately a standard normal distribution, the above probability is approximately

$$P[-1.5 \leq Z \leq 1.5] = 0.8664$$

using Table 4 of the Appendix for the standard normal random variable Z. Thus, chances are greater than 86% that the sample mean will fall within 0.3 ounce of the population mean.   ∎

EXAMPLE   **6.6**   In the setting of Example 6.5, how many observations should be taken in the sample so that $\overline{X}$ would be within 0.3 ounce of μ with probability 0.95?

Solution   Now we want

$$P[|\overline{X} - \mu| \leq 0.3] = P[-0.3 \leq (\overline{X} - \mu) \leq 0.3] = 0.95$$

We know that since $\sigma = 1$,

$$P\left[-0.3\sqrt{n} \leq \frac{\overline{X} - \mu}{\sigma/\sqrt{n}} \leq 0.3\sqrt{n}\right]$$

is approximately equal to

$$P\left[-0.3\sqrt{n} \leq Z \leq 0.3\sqrt{n}\right]$$

(on a standard normal random variable Z). But, using Table 4 of the Appendix,

$$P[-1.96 \leq Z \leq 1.96] = 0.95$$

and it must follow that

$$0.3\sqrt{n} = 1.96$$

or

$$n = \left(\frac{1.96}{0.3}\right)^2 = 42.68$$

Thus, 43 observations will be needed for the sample mean to have a 95% chance of being within 0.3 ounce of the population mean.   ∎

Exercises

6.9   Shear-strength measurements for spot welds of a certain type have been found to have a standard deviation of approximately 10 psi. If 100 test welds are to be measured, find the approximate probability that the sample means will be within 1 psi of the true population mean.

6.10   If shear-strength measurements have a standard deviation of 10 psi, how many test welds should be used in the sample if the sample mean is to be within 1 psi of the population mean with probability approximately 0.95?

6.11   The soil acidity is measured by a quantity called the pH, which may range from 0 to 14 for soils ranging from low to high acidity. Many soils have an average pH in the 5 to 8 range. A scientist wants to estimate the average pH for a large field from n randomly selected core samples by measuring the pH in each sample. If the scientist selects $n = 40$ samples, find the approximate probability that the sample mean of the 40 pH measurements will be within 0.2 unit of the true average pH for the field.

6.12   Suppose the scientist of Exercise 6.11 would like the sample mean to be within 0.1 of the true mean with probability 0.90. How many core samples should she take?

6.13   Resistors of a certain type have resistances that average 200 ohms with a standard deviation of 10 ohms. Twenty-five of these resistors are to be used in a circuit.

    a   Find the probability that the average resistance of the 25 resistors is between 199 and 202 ohms.

    b   Find the probability that the *total* resistance of the 25 resistors does not exceed 5,100 ohms. [*Hint*: Note that $P(\sum_{i=1}^{n} X_i > a) = P(n\overline{X} > a) = P(\overline{X} > a/n)$.]

    c   What assumptions are necessary for the answers in (a) and (b) to be good approximations?

6.14   One-hour carbon monoxide concentrations in air samples from a large city average 12 ppm, with a standard deviation of 9 ppm. Find the probability that the average concentration in 100 samples selected randomly will exceed 14 ppm.

6.15   Unaltered bitumens, as commonly found in lead-zinc deposits, have atomic hydrogen/carbon (H/C) ratios that average 1.4 with a standard deviation of 0.05. Find the probability that 25 samples of bitumen have an average H/C ratio below 1.3.

6.16   The downtime per day for a certain computing facility averages 4.0 hours with a standard deviation of 0.8 hour.

    a   Find the probability that the average daily downtime for a period of 30 days is between 1 and 5 hours.

    b   Find the probability that the *total* downtime for the 30 days is less than 115 hours.

    c   What assumptions are necessary for the answers in (a) and (b) to be valid approximations?

6.17   The strength of a thread is a random variable with mean 0.5 pound and standard deviation 0.2 pound. Assume the strength of a rope is the sum of the strengths of the threads in the rope.

    a   Find the probability that a rope consisting of 100 threads will hold 45 pounds.

    b   How many threads are needed for a rope that will hold 50 pounds with 99% assurance?

6.18 Many bulk products, such as iron ore, coal, and raw sugar, are sampled for quality by a method that requires many small samples to be taken periodically as the material is moving along a conveyor belt. The small samples are then aggregated and mixed to form one composite sample. Let Y_i denote the volume of the ith small sample from a particular lot, and suppose $Y_1, \ldots, Y_n$ constitutes a random sample with each Y_i having mean μ and variance σ^2. The average volume of the samples μ can be set by adjusting the size of the sampling device. Suppose the variance of sampling volumes σ^2 is known to be approximately 4 for a particular situation (measurements are in cubic inches). It is required that the total volume of the composite sample exceed 200 cubic inches with probability approximately 0.95 when $n = 50$ small samples are selected. Find a setting for μ that will allow the sampling requirements to be satisfied.

6.19 The service times for customers coming through a checkout counter in a retail store are independent random variables with a mean of 1.5 minutes and a variance of 1.0. Approximate the probability that 100 customers can be serviced in less than 2 hours of total service time by this one checkout counter.

6.20 Refer to Exercise 6.19. Find the number of customers n such that the probability of servicing all n customers in less than 2 hours is approximately 0.1.

6.21 Suppose that $X_1, \ldots, X_{n1}$ and $Y_1, \ldots, Y_{n2}$ constitute independent random samples from populations with means μ_1 and μ_2 and variances σ_1^2 and σ_1^3, respectively. Then the Central Limit Theorem can be extended to show that $\overline{X} - \overline{Y}$ is approximately normally distributed for large n_1 and n_2, with mean $\mu_1 = \mu_2$ and variance $(\sigma_1^2/n_1 + \sigma_2^2/n_2)$.

Water flow through soils depends, among other things, on the porosity (volume proportion due to voids) of the soil. To compare two types of sandy soil, $n_1 = 50$ measurements are to be taken on the porosity of soil A, and $n_2 = 100$ measurements are to be taken on soil B. Assume that $\sigma_1^2 = 0.01$ and $\sigma_2^2 = 0.02$. Find the approximate probability that the difference between the sample means will be within 0.05 unit of the true difference between the population means, $\mu_1 - \mu_2$.

6.22 Refer to Exercise 6.21. Suppose samples are to be selected with $n_1 = n_2 = n$. Find the value of n that will allow the difference between the sample means to be within 0.04 unit of $\mu_1 - \mu_2$ with probability approximately 0.90.

6.23 An experiment is designed to test whether operator A or operator B gets the job of operating a new machine. Each operator is timed on 50 independent trials involving the performance of a certain task on the machine. If the sample means for the 50 trials differ by more than 1 second, the operator with the smaller mean gets the job. Otherwise, the experiment is considered to end in a tie. If the standard deviations of times for both operators are assumed to be 2 seconds, what is the probability that operator A gets the job even though both operators have equal ability?

6.4 The Normal Approximation to the Binomial Distribution

We saw in Chapter 3 that a binomially distributed random variable Y can be written as a sum of independent Bernoulli random variables X_i. That is,

$$Y = \sum_{i=1}^{n} X_i$$

where $X_i = 1$ with probability p and $X_i = 0$ with probability $1 - p$, $i = 1, \ldots, n$. Y can represent the number of successes in a sample of n trials or measurements, such as the number of thermistors conforming to standards in a sample of n thermistors.

Now the *fraction* of successes in the n trials is

$$\frac{Y}{n} = \frac{1}{n} \sum_{i=1}^{n} X_i = \overline{X}$$

so Y/n is a sample mean. In particular, for large n, Y/n has approximately a normal distribution with a mean of

$$E(X_i) = p$$

and a variance of

$$V(Y/n) = \frac{1}{n^2} \sum_{i=1}^{n} V(X_i)$$

$$= \frac{1}{n^2} \sum_{i=1}^{n} p(1-p) = \frac{p(1-p)}{n}$$

The normality follows from the Central Limit Theorem. Since $Y = n\overline{X}$, Y has approximately a normal distribution with mean np and variance $np(1-p)$. Because of the fact that calculation of binomial probabilities is cumbersome for large n, we make extensive use of this normal approximation to the binomial distribution.

Figure 6.16 shows the histogram of a binomial distribution for $n = 20$ and $p = 0.6$. The heights of the bars represent the respective binomial probabilities. For this distribution, the mean is $np = 20(0.6) = 12$, and the variance is $np(1-p) = 20(0.6)(0.4) = 4.8$. Superimposed upon the binomial distribution is a normal distribution with mean $\mu = 12$ and variance $\sigma^2 = 4.8$. Notice how the normal curve closely approximates the binomial histogram.

For the situation displayed in Figure 6.16, suppose we want to find $P(Y \le 10)$. By the exact binomial probabilities found in Table 2 of the Appendix,

$$P(Y \le 10) = 0.245$$

This value is the sum of the heights of the bars from $y = 0$ up to and including $y = 10$.

Looking at the normal curve in Figure 6.16, we can see that the areas in the bars at $y = 10$ and below are best approximated by the area under the curve to the left of 10.5. The extra 0.5 is added so that the total bar at $y = 10$ is included in the area under consideration. Thus, if W represents a normally distributed random variable with $\mu = 12$ and $\sigma^2 = 4.8$ ($\sigma = 2.2$), then

$$P(Y \le 10) \approx P(W \le 10.5)$$

$$= P\left(\frac{W - \mu}{\sigma} \le \frac{10.5 - 12}{2.2}\right) = P(Z \le -0.68)$$

$$= 0.5 - 0.2517 = 0.2483$$

F I G U R E **6.16**   A Binomial Distribution, $n = 20$, $p = 0.6$, and a Normal Distribution, $\mu = 12$, $\sigma^2 = 4.8$

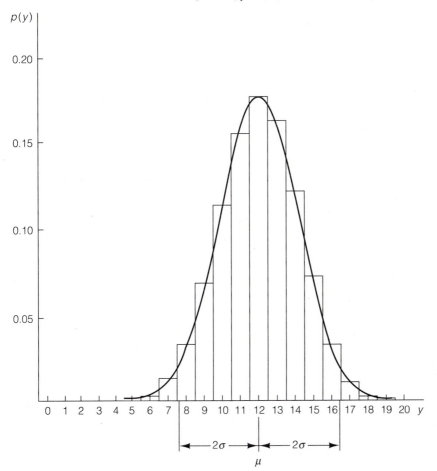

from Table 4 in the Appendix. We see that the normal approximation of 0.248 is close to the exact binomial probability of 0.245. The approximation would be even better if n were larger.

The normal approximation to the binomial distribution works well for even moderately large n, as long as p is not close to zero or 1. A useful rule of thumb is to make sure n is large enough so that $p \pm 2\sqrt{p(1-p)/n}$ lies within the interval $(0, 1)$ before the normal approximation is used. Otherwise, the binomial distribution may be so asymmetric that the symmetric normal distribution cannot provide a good approximation.

E X A M P L E   **6.7**   Silicon wafers coming into a microchip plant are inspected for conformance to specifications. From a large lot of wafers, $n = 100$ are inspected. If the number of nonconformances Y is no more than 12, the lot is accepted. Find the approximate probability of acceptance if the proportion of nonconformances in the lot is $p = 0.2$.

Solution   The number of nonconformances Y has a binomial distribution if the lot is, indeed, large. Before using the normal approximation, we should check to see that

$$p \pm 2\sqrt{\frac{p(1-p)}{n}} = 0.2 \pm 2\sqrt{\frac{(0.2)(0.8)}{100}}$$
$$= 0.2 \pm 0.08$$

is entirely within the interval $(0, 1)$, which it is. Thus, the normal approximation should work well.

Now the probability of accepting the lot is

$$P(Y \leq 12) \approx P(W \leq 12.5)$$

where W is a normally distributed random variable with $\mu = np = 20$ and $\sigma = \sqrt{np(1-p)} = 4$. It follows that

$$P(W \leq 12.5) = P\left(\frac{W - \mu}{\sigma} \leq \frac{12.5 - 20}{4}\right)$$
$$= P(Z \leq -1.88) = 0.5 - 0.4699 = 0.0301$$

There is only a small probability of accepting any lot that has 20% nonconforming wafers.   ▪

Exercises

6.24   The median age of residents of the United States is 33 years. If a survey of 100 randomly selected U.S. residents is taken, find the approximate probability that at least 60 of them will be under 33 years of age.

6.25   A lot acceptance sampling plan for large lots, similar to that of Example 6.7, calls for sampling 50 items and accepting the lot if the number of nonconformances is no more than 5. Find the approximate probability of acceptance if the true proportion of nonconformances in the lot is

a   10%.

b   20%.

c   30%.

6.26   Of the customers entering a showroom for stereo equipment, only 30% make purchases. If 40 customers enter the showroom tomorrow, find the approximate probability that at least 15 make purchases.

6.27   The quality of computer disks is measured by the number of missing pulses. For a certain brand of disk, 80% are generally found to contain no missing pulses. If 100 such disks are inspected, find the approximate probability that 15 or fewer contain missing pulses.

6.28   The capacitances of a certain type of capacitor are normally distributed with a mean of 53 μf (microfarads) and a standard deviation of 2 μf. If 64 such capacitors are to be used in an electronic system, approximate the probability that at least 12 of them will have capacitances below 50 μf.

6.29   The daily water demands for a city pumping station exceed 500,000 gallons with probability only 0.15. Over a 30-day period, find the approximate probability that demand for over 500,000 gallons per day occurs no more than twice.

6.30   At a specific intersection, vehicles entering from the east are equally likely to turn left, turn right, or proceed straight ahead. If 500 vehicles enter this intersection from the east tomorrow, what is the approximate probability that

    a   150 or fewer turn right?

    b   at least 350 turn?

6.31   Waiting times at a service counter in a pharmacy are exponentially distributed with a mean of 10 minutes. If 100 customers come to the service counter in a day, approximate the probability that at least half of them must wait for more than 10 minutes.

6.32   A large construction firm has won 60% of the jobs for which it has bid. Suppose this firm bids on 25 jobs next month.

    a   Approximate the probability that it will win at least 20 of these.

    b   Find the exact binomial probability that it will win at least 20 of these. Compare this probability to your answer in (a).

    c   What assumptions are necessary for your answers in (a) and (b) to be valid?

6.33   An auditor samples 100 of a firm's travel vouchers to check on how many of these vouchers are improperly documented. Find the approximate probability that more than 30% of the sampled vouchers will be found to be improperly documented if, in fact, only 20% of all the firm's vouchers are improperly documented.

6.5   The Sampling Distribution of S^2

The beauty of the Central Limit Theorem lies in the fact that $\overline{X}$ will have approximately a normal sampling distribution no matter what the shape of the probabilistic model for the population, so long as n is large and σ^2 is finite. For many other statistics, additional assumptions are needed before useful sampling distributions can be derived. A common assumption is that the probabilistic model for the population is itself normal. That is, we assume that if the population of measurements of interest could be viewed as a histogram, that histogram would have roughly the shape of a normal curve. This, incidentally, is not a bad assumption for many sets of measurements one is likely to come across in real-world experimentation. Some examples were discussed in Section 4.6.

First, note that if $X_1, \ldots, X_n$ are independent normally distributed random variables with common mean μ and variance σ^2, then $\overline{X}$ will be *precisely* normally distributed with mean μ and variance σ^2/n. No approximating distribution is needed in this case, since linear functions of independent normal random variables are again normal.

Under this normality assumption for the population, a sampling distribution can be derived for S^2, but we do not present the derivation here. It turns out that $(n-1)S^2/\sigma^2$ has a sampling distribution that is a special case of the gamma density function. If we let $(n-1)S^2/\sigma^2 = U$, then U will have the probability density function given by

$$f(u) = \begin{cases} \dfrac{1}{\Gamma\left(\dfrac{n-1}{2}\right) 2^{(n-1)/2}} u^{(n-1)/2-1} e^{-u/2} & u > 0 \\ 0 & \text{elsewhere} \end{cases}$$

The gamma density function with $\alpha = v/2$ and $\beta = 2$ is called a *chi-square* (χ^2) *density function* with parameter v. The parameter v is commonly known as the *degrees of freedom*. Thus when the sampled population is normal, $(n-1)S^2/\sigma^2$ has a chi-square distribution with $n-1$ degrees of freedom.

Specific values that cut off certain right-hand tail areas under the χ^2 density function are given in Table 6 of the Appendix. The value cutting off a tail area of α is denoted by $\chi^2_\alpha(v)$. A typical χ^2 density function is shown in Figure 6.17.

FIGURE **6.17**
A χ^2 Distribution (probability density function)

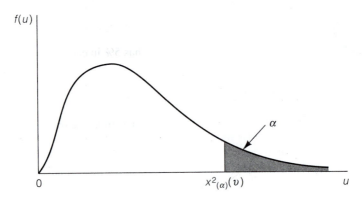

EXAMPLE **6.8** For a machine dispensing liquid into bottles, the variance σ^2 in the ounces of fill was known to be approximately 1. For a sample of size $n = 10$ bottles, find two positive numbers b_1 and b_2 so that the sample variance S^2 among the amounts of fill will satisfy

$$P(b_1 \le S^2 \le b_2) = 0.90$$

Assume that the population of amounts of liquid dispensed per bottle is approximately normally distributed.

Solution Under the normality assumption, $(n-1)S^2/\sigma^2$ has a $\chi^2(n-1)$ distribution. Since

$$P(b_1 \le S^2 \le b_2) = P\left[\frac{(n-1)b_1}{\sigma^2} \le \frac{(n-1)S^2}{\sigma^2} \le \frac{(n-1)b_2}{\sigma^2}\right]$$

the desired values can be found by setting $(n-1)b_2/\sigma^2$ equal to the value that cuts off an area of 0.05 in the upper tail of the χ^2 distribution, and $(n-1)b_1/\sigma^2$ equal to the value that cuts off an area of 0.05 in the lower tail. Using Table 6 of the Appendix yields, with $n-1 = 9$ degrees of freedom,

$$\frac{(n-1)b_2}{\sigma^2} = 16.919 = \chi^2_{0.05}(9)$$

and

$$\frac{(n-1)b_1}{\sigma^2} = 3.32511 = \chi^2_{0.95}(9)$$

Thus,

$$b_2 = 16.919(1)/9 = 1.880$$

and

$$b_1 = 3.32511(1)/9 = 0.369$$

Thus, there is a 90% chance that the sample variance will fall between 0.369 and 1.880. Note that this is not the only interval that would satisfy the desired condition

$$P(b_1 \leq S^2 \leq b_2) = 0.90$$

but it is the convenient interval that has 5% in each tail of the χ^2 distribution.   ∎

We know from Chapter 4 that a gamma random variable having the density function $f(u)$ as given in this section will have a mean of

$$\left(\frac{n-1}{2}\right)2 = n - 1$$

and a variance of

$$\left(\frac{n-1}{2}\right)2^2 = 2(n - 1)$$

Thus, a χ^2 distribution has a mean equal to its degrees of freedom and a variance equal to twice its degrees of freedom. From this information, we can find the mean and variance of S^2.

$$E\left[\frac{(n-1)S^2}{\sigma^2}\right] = (n - 1)$$

$$\frac{(n-1)}{\sigma^2}E(S^2) = (n - 1)$$

or

$$E(S^2) = (n-1)\frac{\sigma^2}{(n-1)} = \sigma^2$$

Actually, S^2 is defined with a denominator of $(n-1)$ so its expected value will equal σ^2, as you saw in Exercise 6.8.

Also,

$$V\left[\frac{(n-1)}{\sigma^2}S^2\right] = 2(n - 1)$$

$$\left(\frac{(n-1)}{\sigma^2}\right)^2 V(S^2) = 2(n - 1)$$

or

$$V(S^2) = 2(n-1)\left(\frac{\sigma^2}{n-1}\right)^2$$

$$= \frac{2\sigma^4}{n-1}$$

The standard deviation of S^2 is $\sqrt{2}\sigma^2/\sqrt{n-1}$. Note that this value is valid for normal populations.

E X A M P L E   **6.9**   In designing mechanisms for hurling projectiles such as rockets at targets, it is very important to study the variance of the distances by which the projectile misses the target center. (Obviously, this variance should be as small as possible.) For a certain launching mechanism, these distances are known to have a normal distribution with variance $\sigma^2 = 100 \, m^2$. An experiment involving $n = 25$ launches is to be conducted. Let S^2 denote the sample variances of the distances between the impact of the projectile and the target center.

a   Approximate $P(S^2 > 50)$.

b   Approximate $P(S^2 > 150)$.

c   Find $E(S^2)$ and $V(S^2)$.

Solution   Let $U = (n-1)S^2/\sigma^2$, which then has a $\chi^2(24)$ distribution for $n = 25$.

a $P(S^2 > 50) = P\left(\frac{n-1}{\sigma^2}S^2 > \frac{24}{100}50\right)$

$= P(U > 12)$

Looking at the row for 24 degrees of freedom in Table 6 of the Appendix, we see

$$P(U > 12.4011) = 0.975$$

and

$$P(U > 10.8564) = 0.990$$

Thus, $P(S^2 > 50)$ is a little larger than 0.975. We cannot find the exact probability, since Table 6 provides only selected tail areas.

b $P(S^2 > 150) = P\left(\frac{n-1}{\sigma^2}S^2 > \frac{24}{100}150\right)$

$= P(U > 36)$

From Table 6, with 24 degrees of freedom,

$$P(U > 36.4151) = 0.05$$

and

$$P(U > 33.1963) = 0.10$$

Thus, $P(S^2 > 150)$ is a little larger than 0.05.

c   We know that

$$E(S^2) = \sigma^2 = 100$$

and

$$V(S^2) = \frac{2\sigma^4}{n-1} = \frac{2(100)^2}{24} = \frac{(100)^2}{12}$$

Following up on this result, the standard deviation of S^2 is $100/\sqrt{12} \approx 29$. Then, by Tchebysheff's Theorem, at least 75% of the values of S^2, in repeated sampling, should lie between

$$100 - 2(29) \text{ and } 100 + 2(29)$$

or between

$$42 \text{ and } 158$$

As a practical consideration, this much range in distance variances might cause the engineers to reassess the design of the launching mechanism.   ∎

Exercises

6.34   The efficiency ratings (in lumens per watt) of light bulbs of a certain type have a population mean of 9.5 and a standard deviation of 0.5, according to production specifications. The specifications for a room in which eight of these bulbs are to be installed call for the average efficiency of the eight bulbs to exceed 10. Find the probability that this specification for the room will be met, assuming efficiency measurements are normally distributed.

6.35   In the setting of Exercise 6.34, what should the mean efficiency per bulb equal if the specification for the room is to be met with probability approximately 0.90? (Assume the standard deviation of efficiency measurements remains at 0.5.)

6.36   The Environmental Protection Agency is concerned with the problem of setting criteria for the amount of certain toxic chemicals to be allowed in freshwater lakes and rivers. A common measure of toxicity for any pollutant is the concentration of the pollutant that will kill half of the test species in a given amount of time (usually 96 hours for fish species). This measure is called the LC50 (lethal concentration killing 50% of the test species).

      Studies of the effects of copper on a certain species of fish (say, species A) show the variance of LC50 measurements to be approximately 1.9, with concentration measured in milligrams per liter. If $n = 10$ studies on LC50 for copper are to be completed, find the probability that the sample mean LC50 will differ from the true population mean by no more than 0.5 unit. Assume that the LC50 measurements are approximately normal in their distribution.

6.37   If, in Exercise 6.36, it is desired that the sample mean differ from the population mean by no more than 0.5 with probability 0.95, how many tests should be run?

6.38   Suppose $n = 20$ observations are to be taken on normally distributed LC50 measurements, with $\sigma^2 = 1.9$. Find two numbers a and b such that $P(a \leq S^2 \leq b) = 0.90$. ($S^2$ is the sample variance of the 20 measurements.)

6.39   Ammeters produced by a certain company are marketed under the specification that the standard deviation of gauge readings be no larger than 0.2 amp. Ten independent readings on a test

$n-10$

circuit of constant current, using one of these ammeters, gave a sample variance of 0.065. Does this suggest that the ammeter used does not meet the company's specification? [*Hint*: Find the approximate probability of a sample variance exceeding 0.065 if the true population variance is 0.04.]

6.40   A certain type of resistor is marketed with the specification that the variance of resistances produced is around 50 ohms. A sample of 15 of these resistors is to be tested for resistances produced.

a   Find the approximate probability that the sample variance S^2 will exceed 80.

b   Find the approximate probability that the sample variance S^2 will be less than 20.

c   Find an interval in which at least 75% of such sample variances should lie.

d   What assumptions are necessary for your answers above to be valid?

6.41   Answer the questions posed in Exercise 6.40 if the sample size is 25 rather than 15.

6.42   In constructing an aptitude test for a job, it is important to plan for a fairly large variance in test scores so the best applicants can be easily identified. For a certain test, scores are assumed to be normally distributed with a mean of 80 and a standard deviation of 10. A dozen applicants are to take the aptitude test. Find the approximate probability that the sample standard deviation of the scores for these applicants will exceed 15.

6.43   For an aptitude test for quality-control technicians in an electronics firm, history shows scores to be normally distributed with a variance of 225. If 20 applicants are to take the test, find an interval in which the sample variance of test scores should lie with probability 0.90.

6.6 Control Charts

The modern approach to improving the quality of a process, due mainly to the work of W. Edwards Deming, J. M. Juran, and their colleagues, emphasizes building quality into every step of the process. A philosophy of "do it right the first time" has replaced the traditional approach of "inspect out the mistakes." Screening out items that do not meet specifications after their completion has proven to be too costly and too ineffective in improving quality. It is far more economical to build quality into a process. Action on the process itself to improve quality is future oriented; it changes the process so that future items will have a better chance of exceeding specifications. Action on the output of a process, as in screening for quality at the factory door, is past oriented; it gives evidence of the quality of items already produced but does not help to assure better quality in the future.

All phases of a process are subject to variation, and, as we have seen since Chapter 1, statistical techniques can help sort and quantify this variation. It is useful to think of variation as being one of two types, that arising from *common causes* and that arising from *special causes*. Common causes are the routine factors that affect a process and cause some variation even though the process itself might be in "statistical control." Special causes, or assignable causes, are not a regular part of the process and may not affect the entire process, but they do cause abnormally high variation when they show up. Consider the time it takes you to get to school each day. Even though you may travel the same route at about the same time each day, there is some variation in travel times due to common causes like traffic, traffic lights, weather conditions, and how you feel. If, however, you arrive late for class on a particular day, it might be due to an extraordinarily long travel time as a result of a special cause such as a traffic accident, a flat tire, or an unusually intense storm.

The special causes are the main focus of attention in this chapter, since statistical techniques are especially good at detecting them. Common-cause variation can be studied by the same techniques, but the process changes required to reduce this variation are not usually obvious or simple. Figure 6.18 diagrams process outputs ("size") for a situation involving only common causes and for another showing special causes. Stable output distributions, or systems "in control," are the goal; statistical techniques can help us achieve that goal by weeding out special causes of variation. The most common tool for assessing process variation is the *control chart*, first introduced by Dr. Walter Shewhart in the 1920s. Commonly used control charts are presented in the subsections that follow.

Imagine an industrial process that is running continuously. A standard quality improvement plan would require sampling one or more items from this process periodically and making the appropriate quality measurements. Usually, more than one item is measured at each time point to increase accuracy and measure variability. The objective in this section is to develop a technique that will help the engineer

FIGURE **6.18**
Variation Due to Common and Special Causes

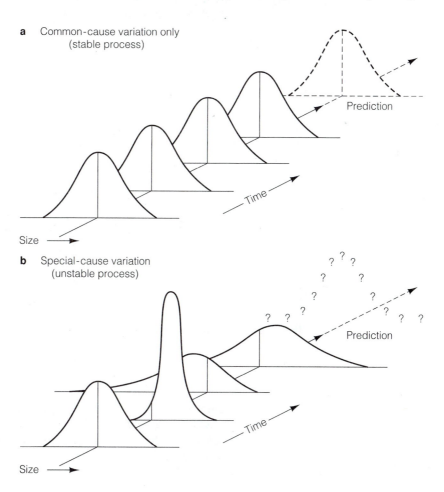

a Common-cause variation only
      (stable process)

Prediction

Size

b Special-cause variation
      (unstable process)

Prediction

Size

decide whether the center (or average location) of the measurements has shifted up or down and whether variation has changed substantially.

Suppose n observations are to be made at each time point in which the process is checked. Let X_i denote the ith observation ($i = 1, \ldots, n$) at the specified time point, and let $\overline{X}_j$ denote the average of the n observations at time point j. If $E(X_i) = \mu$ and $V(X_i) = \sigma^2$, for a process in control, then $\overline{X}_j$ should be approximately normally distributed with $E(\overline{X}_j) = \mu$ and $V(\overline{X}_j) = \sigma^2/n$. As long as the process remains in control, where will most of the $\overline{X}_j$ values lie? Since we know that $\overline{X}_j$ has approximately a normal distribution, we can find an interval that will have a specified probability of containing $\overline{X}_j$. This interval is $\mu \pm z(\sigma/\sqrt{n})$. The usual value for z in most control-chart problems is 3. Thus, the interval $\mu \pm 3(\sigma/\sqrt{n})$ has probability 0.9973 of including $\overline{X}_j$, as long as the process is in control. If μ and σ were known or specified, we could use $\mu - 3(\sigma/\sqrt{n})$ as a lower control limit (LCL) and $\mu + 3(\sigma/\sqrt{n})$ as an upper control limit (UCL). If a value of $\overline{X}_j$ was observed to fall outside of these limits, we would suspect that the process might be out of control (the mean might have shifted to a different value) since this event has a very small probability of occurring (0.0027) when the process is in control. Figure 6.19 shows schematically how the decision process would work.

FIGURE **6.19**
Schematic Representation of
$\overline{X}$-Chart

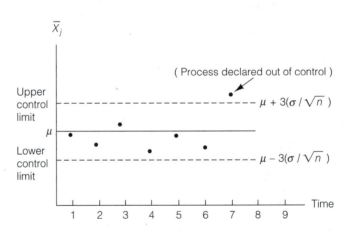

Note that if a process is declared to be out of control because $\overline{X}_j$ fell outside of the control limits, there is a positive probability of making an error. That is, we could declare that the mean of the quality measurements has shifted when, in fact, it has not. However, this probability of error is only 0.0027 for the three-standard deviation limits used above.

6.6.1 The $\overline{X}$- and R-charts

If a control chart is being started for a new process, μ and σ will not be known and hence must be estimated from the data. For establishing control limits, it is generally recommended that at least $k = 20$ time points be sampled before the control limits are calculated. We will now discuss the details of estimating $\mu \pm 3(\sigma/\sqrt{n})$ from k independent samples, each of size n.

For each of k random samples, calculate the sample mean, $\overline{X}_j$, and the sample range, R_j. (Recall that the range is simply the difference between the largest and smallest observations in the sample.) Then calculate the average of the means and ranges,

$$\overline{\overline{X}} = \frac{1}{k} \sum_{j=1}^{k} \overline{X}_j$$

and

$$\overline{R} = \frac{1}{k} \sum_{j=1}^{k} R_j$$

Now $\overline{\overline{X}}$ is a good measure of the center of the process output, since $E(\overline{\overline{X}}) = \mu$. However, $\overline{R}$ by itself is not a good approximation to σ. It can be shown that

$$E(\overline{R}) = d_2 \sigma$$

for a constant, d_2, found in Table 9 of the Appendix. The control limits

$$\mu \pm 3 \frac{\sigma}{\sqrt{n}}$$

can now be approximated by

$$\overline{\overline{X}} \pm \frac{3}{\sqrt{n}} \frac{\overline{R}}{d_2}$$

or

$$\overline{\overline{X}} \pm A_2 \overline{R}$$

where values of A_2 are also found in Table 9 of the Appendix.

The sample ranges serve as a check on sample variation, in addition to their role in estimating σ for control limits on the process mean. It can be shown that

$$V(R_j) = d_3^2 \sigma^2$$

where d_3 can be found in Table 9 of the Appendix. The three-standard deviation interval about the mean of R_j then becomes

$$d_2 \sigma \pm 3 d_3 \sigma$$

or

$$\sigma (d_2 \pm 3 d_3)$$

For any sample, there is a high probability (approximately 0.9973) that its range will fall inside this interval if the process is in control.

As in the case of the $\overline{X}$-chart, if σ is not specified, it must be estimated from the data. The best estimator of σ based on $\overline{R}$ is $\overline{R}/d_2$, and the estimator of $\sigma(d_2 \pm 3d_3)$ then becomes

$$\frac{\overline{R}}{d_2}(d_2 \pm 3d_3)$$

or

$$\overline{R}\left(1 \pm 3\frac{d_3}{d_2}\right)$$

Letting $1 - 3(d_3/d_2) = D_3$ and $1 + 3(d_3/d_2) = D_4$, the control limits take on the form $(\overline{R}D_3, \overline{R}D_4)$. D_3 and D_4 are given in Table 9 of the Appendix.

E X A M P L E   **6.10**   A control chart is to be started for a new machine that fills boxes of cereal by weight. Five observations on amount of fill are taken every hour until 20 such samples are obtained. The data are given in Table 6.1. Construct a control chart for means and another for variation, based on these data. Interpret the results.

Solution   The sample summaries of center and spread are given in Table 6.1. From these data,

$$\overline{\overline{x}} = \frac{1}{20}\sum_{j=1}^{20}\overline{x}_j = 16.32$$

T A B L E **6.1**
Amount of Cereal Dispensed by Filling Machines

Sample	Readings					$\overline{x}_j$	s_j	r_j (range)
1	16.1	16.2	15.9	16.0	16.1	16.06	0.114	0.3
2	16.2	16.4	15.8	16.1	16.2	16.14	0.219	0.6
3	16.0	16.1	15.7	16.3	16.1	16.04	0.219	0.6
4	16.1	16.2	15.9	16.4	16.6	16.24	0.270	0.7
5	16.5	16.1	16.4	16.4	16.2	16.32	0.164	0.4
6	16.8	15.9	16.1	16.3	16.4	16.30	0.339	0.9
7	16.1	16.9	16.2	16.5	16.5	16.44	0.313	0.8
8	15.9	16.2	16.8	16.1	16.4	16.28	0.342	0.9
9	15.7	16.7	16.1	16.4	16.8	16.34	0.451	1.1
10	16.2	16.9	16.1	17.0	16.4	16.52	0.409	0.9
11	16.4	16.9	17.1	16.2	16.1	16.54	0.439	1.0
12	16.5	16.9	17.2	16.1	16.4	16.62	0.432	1.1
13	16.7	16.2	16.4	15.8	16.6	16.34	0.358	0.9
14	17.1	16.2	17.0	16.9	16.1	16.66	0.472	1.0
15	17.0	16.8	16.4	16.5	16.2	16.58	0.319	0.8
16	16.2	16.7	16.6	16.2	17.0	16.54	0.344	0.8
17	17.1	16.9	16.2	16.0	16.1	16.46	0.503	1.1
18	15.8	16.2	17.1	16.9	16.2	16.44	0.541	1.3
19	16.4	16.2	16.7	16.8	16.1	16.44	0.305	0.7
20	15.4	15.1	15.0	15.2	14.9	15.12	0.192	0.5

and

$$\bar{r} = \frac{1}{20} \sum_{j=1}^{20} r_j = 0.82$$

From Table 9 of the Appendix (with $n = 5$), $A_2 = 0.577$, $D_3 = 0$, and $D_4 = 2.114$. Thus, the $\overline{X}$-chart has boundaries

$$\bar{\bar{x}} \pm A_2 \bar{r}$$
$$16.32 \pm (0.577)(0.82)$$
$$16.32 \pm 0.47$$

or

$$(15.85, 16.79)$$

The R-chart has boundaries

$$(D_3 \bar{r}, D_4 \bar{r})$$

$$(0, 2.114(0.82))$$

or

$$(0, 1.73)$$

These boundaries and the sample paths for both $\bar{x}_j$ and r_j are plotted in Figure 6.20. Note that no points are "out of control" among the ranges; the variability of the process appears to be stable. The mean for sample 20 falls below the lower control limit, indicating that the machine seems to be underfilling the boxes by a substantial amount. The operator should look for a special cause, such as changed setting, a change in product flow into the machine, or a blocked passage. ∎

If a sample mean or range is found to be outside the control limits and a special cause is found and corrected, new control limits are calculated on a reduced data set that does not contain the offending sample. Based on the first 19 samples from Example 6.10, the control limits for center become

$$16.38 \pm (0.577)(0.84)$$

or

$$16.38 \pm 0.48$$

Now all sample means are inside the control limits.

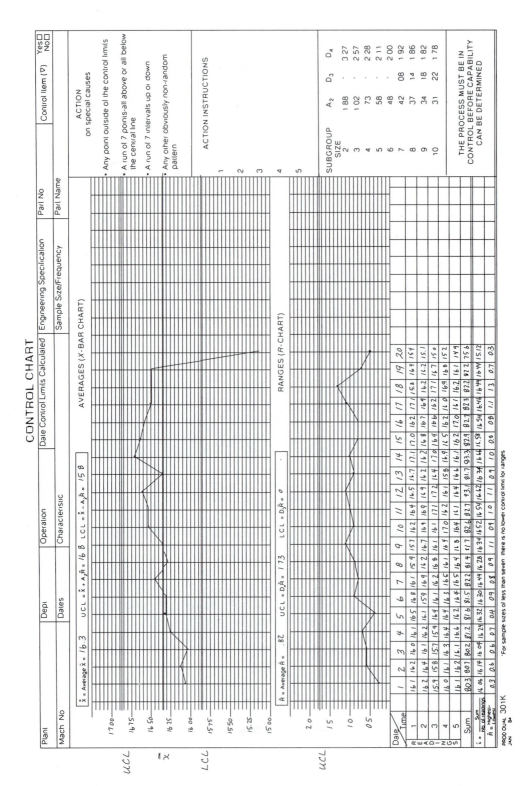

F I G U R E **6.20** $\bar{X}$- and R-Charts for Example 6.10

6.6.2 The $\overline{X}$- and S-Charts

The sample standard deviation, S_1, is a more efficient estimator of σ than is a multiple of R, even though it is more cumbersome to calculate. Now

$$\overline{S} = \frac{1}{k}\sum_{j=1}^{k} S_j$$

can be adjusted by a constant, c_4, to approximate σ, since

$$E(\overline{S}) = E(S) = c_4\sigma$$

where c_4 is found in Table 9. Thus, the control limits

$$\mu \pm 3\frac{\sigma}{\sqrt{n}}$$

can be approximated by

$$\overline{\overline{X}} \pm \frac{3}{\sqrt{n}}\left(\frac{\overline{S}}{c_4}\right)$$

or

$$\overline{\overline{X}} \pm A_3\overline{S}$$

with A_3 values found in Table 9.

What about using $\overline{S}$ as the basis of a chart to track process variability? In the spirit of three-standard-deviation control limits,

$$\overline{S} \pm 3\sqrt{V(S)}$$

would provide control limits for sample standard deviations. Now

$$V(S) = E(S^2) - [E(S)]^2$$
$$= \sigma^2 - (c_4\sigma)^2$$
$$= \sigma^2(1 - c_4^2)$$

Since σ can be estimated by $\overline{S}/c_4$, the control limits become

$$\overline{S} \pm 3\frac{\overline{S}}{c_4}\sqrt{1 - c_4^2}$$

or

$$\overline{S}\left(1 \pm 3\frac{\sqrt{1 - c_4^2}}{c_4}\right)$$

These limits are usually written as

$$(B_3\overline{S}, B_4\overline{S})$$

where B_3 and B_4 can be found in Table 9 of the Appendix.

EXAMPLE **6.11**   Compute an $\overline{X}$-chart and an S-chart for the data of Table 6.1. Interpret the results.

Solution   As before, $\overline{\overline{x}} = 16.32$, so

$$\overline{\overline{x}} \pm A_3\overline{s}$$

becomes

$$16.32 \pm (1.427)(0.337) = 16.32 \pm 0.48$$

or

$$(15.84, 16.80)$$

This is practically identical to the result found above based on ranges, and the twentieth sample mean is still out of control.

For assessing variation,

$$(B_3\overline{s}, B_4\overline{s})$$

becomes,

$$(0, 2.089(0.337))$$

or

$$(0, 0.704)$$

All 20 sample standard deviations are well within this range; the process variation appears to be stable.   ∎

Points outside the control limits are the most obvious indicator of a potential problem with a process. More subtle features of the control chart, however, can serve as warning signs. For the $\overline{X}$-chart, roughly two-thirds of the observed sample means should lie within one standard deviation of the mean and one-third should lie beyond that. That is, on the chart itself, two-thirds of the plotted points should lie within the middle one-third of the region between the control limits. If too many points lie close to the center line, perhaps the limits are incorrectly calculated or the sample data are misrepresenting the process. If too few points lie close to the center line, perhaps the process is going "out of control" but a sample mean has not yet crossed the boundary.

In a similar spirit, long runs of points above or below the center line could indicate a potential problem. Ideally, the sample means should move above and below the center line rather frequently. Many industries take seven to be the magic number of points in a run. That is, if seven sample means in a row are above (or below) the center line, trouble may be brewing.

6.6.3   The p-Chart

The control-charting procedures outlined in the preceding sections of this chapter depend on quality measurements that possess a continuous probability distribution. Sampling such measurements is commonly referred to as "sampling by variables." In many quality-control situations, however, we merely want to assess whether or not a certain item conforms to specifications. We then observe the number of nonconforming items from a particular sample or series of samples. This process is commonly referred to as "sampling by attributes."

As in previous control-charting problems, suppose a series of k independent samples, each of size n, is selected from a production process. Let p denote the proportion of nonconforming (defective) items in the population (total production for a certain time period) for a process in control, and let X_i denote the number of defectives observed in the ith sample. Then X_i has a binomial distribution, assuming random sampling from large lots, with $E(X_i) = np$ and $V(X_i) = np(1 - p)$. We will usually work with sample fractions of defectives (X_i/n) rather than the observed number of defectives. Where will these sample fractions tend to lie for a process in control? As argued previously, most sample fractions should lie within three standard deviations of their mean, or in the interval

$$p \pm 3 \sqrt{\frac{p(1 - p)}{n}}$$

since $E(X_i/n) = p$ and $V(X_i/n) = p(1 - p)/n$.

Since p is unknown, we estimate these control limits, using the data from all k samples, by calculating

$$\overline{P} = \frac{\sum\limits_{i=1}^{k} X_i}{nk} = \frac{\text{total number of defectives observed}}{\text{total sample size}}$$

Since $E(\overline{P}) = p$, approximate control limits are given by

$$\overline{P} \pm 3 \sqrt{\frac{\overline{P}(1 - \overline{P})}{n}}$$

The following example illustrates the calculation of these control limits.

EXAMPLE **6.12**   A process that produces transistors is sampled every 4 hours. At each time point, 50 transistors are randomly sampled, and the number of defectives x_i is observed. The data for 24 samples are given in Table 6.2. Construct a control chart based on these samples.

Solution   From the above data, the observed value of $\overline{P}$ is

$$\bar{p} = \frac{\sum\limits_{i=1}^{k} x_i}{nk} = \frac{52}{50(24)} = 0.04$$

Thus,

$$\bar{p} \pm 3\sqrt{\frac{\bar{p}(1 - \bar{p})}{n}}$$

becomes

$$0.04 \pm 3\sqrt{\frac{(0.04)(0.96)}{50}}$$
$$0.04 \pm 0.08$$

This yields the interval $(-0.04, 0.12)$, but we set the lower control limit at zero since a fraction of defectives cannot be negative, and we obtain $(0, 0.12)$.

All of the observed sample fractions are within these limits, so we would feel comfortable in using them as control limits for future samples. The limits should be recalculated from time to time as new data become available.

The control limits and observed data points are plotted in Figure 6.21.   ■

TABLE **6.2**
Data for Example 6.12

Sample	x_i	x_i/n	Sample	x_i	x_i/n
1	3	0.06	13	1	0.02
2	1	0.02	14	2	0.04
3	4	0.08	15	0	0.00
4	2	0.04	16	3	0.06
5	0	0.00	17	2	0.04
6	2	0.04	18	2	0.04
7	3	0.06	19	4	0.08
8	3	0.06	20	1	0.02
9	5	0.10	21	3	0.06
10	4	0.08	22	0	0.00
11	1	0.02	23	2	0.04
12	1	0.02	24	3	0.06

FIGURE **6.21**
A p-Chart for the Data of Table 6.2

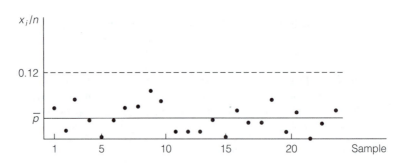

6.6.4 The c-Chart

In many quality-control problems, the particular items being subjected to inspection may have more than one defect, so we may wish to count defects instead of merely classifying an item as defective or nondefective. If C_i denotes the number of defects observed on the ith inspected item, a good working model for many applications is to assume that C_i has a Poisson distribution. We will let the Poisson distribution have a mean of λ for a process in control.

Most of the C_i's should fall within three standard deviations of their mean if the process remains in control. Since $E(C_i) = \lambda$ and $V(C_i) = \lambda$, the control limits become

$$\lambda \pm 3\sqrt{\lambda}$$

If k items are inspected, then λ is estimated by

$$\overline{C} = \frac{1}{k}\sum_{i=1}^{k} C_i$$

The estimated control limits then become

$$\overline{C} \pm 3\sqrt{\overline{C}}$$

The computations are illustrated in Example 6.13.

E X A M P L E **6.13**   Twenty rebuilt pumps were sampled from the hydraulics shop of an aircraft rework facility. The numbers of defects recorded are given in Table 6.3. Use these data to construct control limits for defects per pump.

Solution   Assuming that C_i, the number of defects observed on pump i, has a Poisson distribution, we estimate the mean number of defects per pump by $\overline{C}$ and use

$$\overline{C} \pm 3\sqrt{\overline{C}}$$

T A B L E **6.3**
Data for Example 6.13

Pump	No. of Defects (C_i)	Pump	No. of Defects (C_i)
1	6	11	4
2	3	12	3
3	4	13	2
4	0	14	2
5	2	15	6
6	7	16	5
7	3	17	0
8	1	18	7
9	0	19	2
10	0	20	1

as the control limits. Using the data in Table 6.3, we have

$$\bar{c} = \frac{58}{20} = 2.9$$

and hence

$$\bar{c} \pm 3\sqrt{\bar{c}}$$

becomes

$$2.9 \pm 3\sqrt{2.9}$$
$$2.9 \pm 5.1$$

or

$$(0, \ 8.0)$$

Again the lower limit is raised to zero, since a number of defects cannot be negative.

All of the observed counts are within the control limits, so we could use these limits for future samples. Again, the limits should be recomputed from time to time as new data become available.

The control limits and data points are plotted in Figure 6.22. ∎

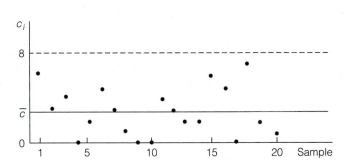

F I G U R E **6.22**
A c-Chart for the Data of Table 6.21

6.6.5 The u-Chart

The c-chart introduced above works well when counts are being made of units of equal size, such as the number of paint defects per automobile or the number of bacteria per cubic centimeter in standard test tube samples of water. If the size of the sampling unit changes, adjustments must be made to the control limits, resulting in a u-chart. Such might be the case, for example, in counting the number of accidents per plant or the number of soldering errors per circuit board. If the plants have varying numbers of employees or the circuit boards have varying numbers of connections to be soldered, this information must be worked into the procedure for calculating control limits.

Suppose the ith sampled unit has size h_i (where h_i could be number of employees, number of electrical connections, volume of water sampled, area of glass examined, and so on) and produces a count of nonconformances C_i. We assume C_i has a Poisson distribution with mean λh_i. That is,

$$E(C_i) = \lambda h_i, \ V(C_i) = \lambda h_i$$

It follows that

$$U_i = C_i / h_i$$

has the properties

$$E(U_i) = \lambda$$

and

$$V(U_i) = (\frac{1}{h_i})^2 (\lambda h_i) = \frac{\lambda}{h_i}$$

If a quality-improvement plan calls for k independently selected samples of this type, then

$$\overline{U} = \frac{\sum_{j=1}^{k} C_j}{\sum_{j=1}^{k} h_j}$$

is a good approximation to λ in the sense that $E(\overline{U}) = \lambda$. For the ith sample, the three-standard-deviation control limits around λ are given by

$$E(U_i) \pm 3\sqrt{V(U_i)}$$

or

$$\lambda \pm 3\sqrt{\frac{\lambda}{h_i}}$$

which can be approximated by

$$\overline{U} \pm 3\sqrt{\frac{\overline{U}}{h_i}}$$

Notice that the control limits might change with each sample if the h_i's change. For simplicity in constructing control limits, h_i may be replaced by

$$\bar{h} = \frac{1}{k} \sum_{j=1}^{k} h_j$$

as long as no individual h_i is more than 25% above or below this average.

EXAMPLE **6.14**   The hydraulics shop of an aircraft rework facility was interested in developing control charts to improve its ongoing quality of repaired parts. The difficulty was that many different kinds of parts, from small pumps to large hydraulic lifts, came through the shop regularly, and each part may have been worked on by more than one person. The decision, ultimately, was to construct a u-chart based on number of defects per hours of labor to complete the repairs. The data for a sequence of 12 sampled parts are given in Table 6.4

Construct a u-chart for these data and interpret the results.

Solution   Following the construction presented above,

$$\bar{u} = \frac{\sum c_i}{\sum h_i} = \frac{28}{2,906.46} = 0.0096$$

and the control limits are given by

$$\bar{u} \pm 3\sqrt{\bar{u}/h_i}$$

(The h_i values are too variable to use $\bar{h}$.) For $i = 1$ and 2, the upper control limits are

$$0.0096 + 3\sqrt{0.0096/58.33} = 0.048$$

and

$$0.0096 + 3\sqrt{0.0096/80.22} = 0.042$$

Now

$$u_1 = 1/58.33 = 0.017$$

and

$$u_2 = 4/80.22 = 0.049$$

TABLE **6.4**
Defects Among Repaired
Hydraulic Parts

Part	Defects	Hours	Part	Defects	Hours
1	1	58.33	7	3	380.20
2	4	80.22	8	5	527.70
3	1	209.24	9	3	319.30
4	2	164.70	10	1	340.20
5	1	253.70	11	2	78.70
6	4	426.90	12	1	67.27

Source: Naval Air Rework Facility, Jacksonville, Florida.

F I G U R E **6.23**
A u-Chart for the Data of
Table 6.4

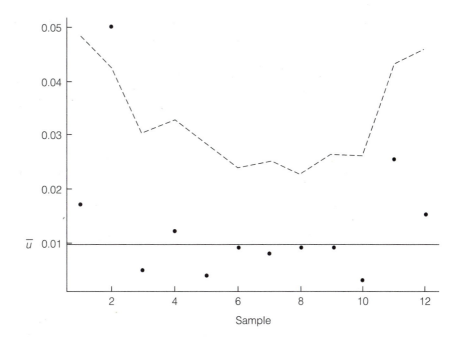

Thus, u_1 is inside its control limit, but u_2 is outside its control limit. All other points are inside the control limits, as shown in Figure 6.23. Note that all lower control limits are rounded up to zero, since they are calculated to be negative.

   The second sampled part should be examined carefully for a special cause, such as the type of part, the training of the worker, the nature of the repairs to be made, the time and date of the repair, and so on.   ▪

6.7 Process Capability

After a process is brought to a state of statistical control, questions may still be asked as to how well the process meets customer demands. Such questions deal with the *capability* of the process to meet specifications. There are many ways to assess process capability, but one of the more common ones is the use of the C_{pk} statistic.

   Suppose a process has an upper specification limit (USL) and a lower specification limit (LSL). In other words, acceptable output values must lie between LSL and USL. In situations for which an $\overline{X}$-chart is appropriate, the center of the process is measured by $\bar{\bar{x}}$. The process capability index C_{pk} measures how far the specification limits are from the process center in terms of process standard deviation. Specifically, we define

$$z_{\text{USL}} = \frac{\text{USL} - \bar{\bar{x}}}{\hat{\sigma}}$$

and

$$z_{LSL} = \frac{\bar{\bar{x}} - LSL}{\hat{\sigma}}$$

where $\hat{\sigma}$ is an estimate of σ, the standard deviation of the outcome measure. Defining z_{min} as the smaller of z_{USL} and z_{LSL}, the capability index C_{pk} is given by

$$C_{pk} = \frac{z_{min}}{3}$$

All else being equal, a large value of C_{pk} ($C_{pk} > 1$) tells us that the process is doing well since the standard deviation is small relative to the distance between specification limits. A small value of C_{pk} may be cause for concern.

In order to calculate C_{pk}, we must estimate σ from the data. This may be done directly from the input to the construction of control limits by either

$$\hat{\sigma}_1 = \bar{r}/d_2$$

or

$$\hat{\sigma}_2 = \bar{s}/c_4$$

E X A M P L E   **6.15**   The cereal weight data of Table 6.1 come from a machine that fills boxes labeled as weighing 16 ounces. Thus, LSL = 16 ounces for this study. To keep a reasonable margin of profit per box, the actual amount of cereal dispensed should not exceed 17 ounces. That is, USL = 17 ounces. Find C_{pk} for this process. Find the total proportion out of specification.

Solution   The estimate of σ for this process should come from the data of Table 6.1 *after* removal of sample 20. (Recall that this sample mean was outside the control limits.) For the remaining 19 samples, $\bar{r} = 0.84$ and $\bar{s} = 0.345$. It follows that

$$\hat{\sigma}_1 = \bar{r}/d_2 = 0.84/2.326 = 0.361$$

and

$$\hat{\sigma}_2 = \bar{s}/c_4 = 0.345/0.940 = 0.367$$

Either estimate will do; we choose to use the estimate based on $\bar{s}$. Then

$$z_{USL} = \frac{USL - \bar{\bar{x}}}{\hat{\sigma}} = \frac{17 - 16.38}{0.367} = 1.69$$

$$z_{LSL} = \frac{\bar{\bar{x}} - LSL}{\hat{\sigma}} = \frac{16.38 - 16}{0.367} = 1.04$$

and

$$z_{min} = \min(z_{LSL}, z_{USL}) = 1.04$$

It follows that

$$C_{pk} = \frac{z_{min}}{3} = \frac{1.04}{3} = 0.35$$

This is quite a small C_{pk}, and it shows that the process is not meeting specifications very well. To make this statement more precise, the actual proportion of product not meeting specifications can be approximated from the normal curve. The area below $z_{LSL} = 1.04$ is $0.5 - 0.3508 = 0.1492$. The area above $z_{USL} = 1.69$ is $0.5 - 0.4545 = 0.0455$. Thus,

$$\text{proportion out of specification} = 0.1492 + 0.0455 = 0.1947$$

Nearly 20% of the boxes are filled outside the specification limits, a situation that is either making customers unhappy or hurting the profit margin. This process should be studied to see how the common-cause variation can be reduced.   ∎

Another commonly used index of process capability, C_p, is defined as

$$C_p = \frac{USL - LSL}{6\hat{\sigma}}$$

The idea behind this index is that most outcome measurements will lie within an interval of six standard deviations. If this six-standard-deviation interval fits inside the specification limits, then the process is meeting specifications most of the time. (C_p is greater than 1.) If the process output is centered at the midpoint between USL and LSL, then $C_p = C_{pk}$. Otherwise, C_p can be a poor index, because a poorly centered process, relative to USL and LSL, could still have a large C_p value.

Exercises

6.44 Production of ammeters is controlled for quality by periodically selecting an ammeter and obtaining four measurements on a test circuit designed to produce 15 amps. The data at the top of page 348 were observed for 15 tested ammeters.

  a   Construct control limits for the mean by using the sample standard deviations.

  b   Do all observed sample means lie within the limits found in (a)? If not, recalculate the control limits, omitting those samples that are "out of control."

  c   Construct control limits for process variation and interpret the results.

6.45 Refer to Exercise 6.44. Repeat parts (a), (b), and (c) using sample ranges instead of sample standard deviations.

6.46 Refer to Exercise 6.44, part (a). How would the control limits change if the probability of a sample mean falling outside the limits, when the process is in control, is to be 0.01?

6.47 Refer to Exercises 6.44 and 6.45. The specifications for these ammeters require individual readings on the test circuit to be within 15.0 ± 0.4. Do the meters seem to be meeting this specification? Calculate C_{pk} and the proportion out of specification.

Ammeter		Reading		
1	15.1	14.9	14.8	15.2
2	15.0	15.0	15.2	14.9
3	15.1	15.1	15.2	15.1
4	15.0	14.7	15.3	15.1
5	14.8	14.9	15.1	15.2
6	14.9	14.9	15.1	15.1
7	14.7	15.0	15.1	15.0
8	14.9	15.0	15.3	14.8
9	14.4	14.5	14.3	14.4
10	15.2	15.3	15.2	15.5
11	15.1	15.0	15.3	15.3
12	15.2	15.6	15.8	15.8
13	14.8	14.8	15.0	15.0
14	14.9	15.1	14.7	14.8
15	15.1	15.2	14.9	15.0

6.48   Bronze castings are controlled for copper content. Ten samples of five specimens each gave the following measurements on percentage of copper.

Sample		Percentage of Copper			
1	82	84	80	86	83
2	90	91	94	90	89
3	86	84	87	83	80
4	92	91	89	91	90
5	84	82	83	81	84
6	82	81	83	84	81
7	80	80	79	83	82
8	79	83	84	82	82
9	81	84	85	79	86
10	81	92	94	79	80

a   Construct control limits for the mean percentage of copper by using the sample ranges.

b   Would you use the limits in (a) as control limits for future samples? Why or why not?

c   Construct control limits for process variability. Does the process appear to be in control with respect to variability?

6.49   Filled bottles of soft drink coming off a production line are checked for underfilling. One hundred bottles are sampled from each half-day's production. The results for 20 such samples are as follows:

Sample	No. of Defectives	Sample	No. of Defectives
1	10	11	15
2	12	12	6
3	8	13	3
4	14	14	2
5	3	15	0
6	7	16	1
7	12	17	1
8	10	18	4
9	9	19	3
10	11	20	6

a   Construct control limits for the proportion of defectives.

b   Do all the observed sample fractions fall within the control limits found in (a)? If not, adjust the limits for future use.

6.50   Refer to Exercise 6.49. Adjust the control limits so that a sample fraction will fall outside the limits with probability 0.05 when the process is in control.

6.51   Bolts being produced in a certain plant are checked for tolerances. Those not meeting tolerance specifications are considered to be defective. Fifty bolts are gauged for tolerance every hour. The results for 30 hours of sampling are as follows:

Sample	No. Defective	Sample	No. Defective	Sample	No. Defective
1	5	11	3	21	6
2	6	12	0	22	4
3	4	13	1	23	9
4	0	14	1	24	3
5	8	15	2	25	0
6	3	16	5	26	6
7	10	17	5	27	4
8	2	18	4	28	2
9	9	19	3	29	1
10	7	20	1	30	2

a   Construct control limits for the proportion of defectives based on these data.

b   Would you use the control limits found in (a) for future samples? Why or why not?

6.52   Woven fabric from a certain loom is controlled for quality by periodically sampling 1-square-meter specimens and counting the number of defects. For 15 such samples, the observed number of defects was as follows:

Sample	No. of Defects
1	3
2	6
3	10
4	4
5	7
6	11
7	3
8	9
9	6
10	14
11	4
12	3
13	1
14	4
15	0

a Construct control limits for the average number of defects per square meter.

b Do all samples observed seem to be "in control"?

6.53 Refer to Exercise 6.52. Adjust the control limits so that a count for a process in control will fall outside the limits with an approximate probability of only 0.01.

6.54 Quality control in glass production involves counting the number of defects observed in samples of a fixed size periodically selected from the production process. For 20 samples, the observed counts of numbers of defects are as follows:

Sample	No. of Defects	Sample	No. of Defects
1	3	11	7
2	0	12	3
3	2	13	0
4	5	14	1
5	3	15	4
6	0	16	2
7	4	17	2
8	1	18	1
9	2	19	0
10	2	20	2

a Construct control limits for defects per sample based on these data.

b Would you adjust the limits in (a) before using them on future samples?

6.55 Suppose variation among defect counts is important in a process like that in Exercise 6.54. Is it necessary to have a separate method to construct control limits for variation? Discuss.

6.56 An aircraft subassembly paint shop has the task of painting a variety of reworked aircraft parts. Some parts are large and require hours to paint, while some are small and can be painted in a few minutes. Rather than looking at paint defects per part as a quality-improvement tool, it was decided to group parts together and look at paint defects over a period of hours worked. Data in this form are given below for a sequence of 20 samples. Construct an appropriate control chart for these data. Does it look as if the painting process is in control?

Sample	Defects	Hours
1	4	51.5
2	1	51.5
3	0	47.1
4	0	47.1
5	2	53.1
6	3	40.4
7	0	57.7
8	0	51.5
9	0	34.8
10	1	51.5
11	1	38.8
12	3	57.8
13	0	39.1
14	2	49.1
15	0	41.6
16	1	51.5
17	1	42.1
18	2	54.3
19	1	36.6
20	0	36.6

6.57 Refer to the capability index discussion in Example 6.15. Suppose the process can be "centered" so that $\bar{\bar{x}}$ is midway between LSL and USL. How will this change C_{pk} and the proportion out of specification?

6.58

  a   Show that $C_p = C_{pk}$ for a centered process.

  b   Construct an example to show that C_p can be a poor index of capability for a process that is not centered.

6.59 For a random sample of size n from a normal distribution, show that $E(S) = c_4\sigma$ where

$$c_4 = \sqrt{\frac{2}{n-1}}\left(\frac{\Gamma\left(\frac{n}{2}\right)}{\Gamma\left(\frac{n-1}{2}\right)}\right)$$

[*Hint:* Make use of proportions of the gamma distribution from Chapter 4.]

6.8 A Useful Method of Approximating Distributions

We saw in the Central Limit Theorem of Section 6.3 that $\sqrt{n}(\overline{X} - \mu)/\sigma$ has approximately a standard normal distribution for large n, under certain general conditions. This notion of asymptotic normality actually extends to a large class of functions of $\overline{X}$ or, for that matter, to functions of *any* asymptotically normal random variable. The result holds because of properties of Taylor series expansions of functions, roughly illustrated as follows. Suppose $\overline{X}$ is a statistic based on a random sample of size n from a population with mean μ and finite variance σ^2. In addition, suppose we are interested in the behavior of a function of $\overline{X}$, say $g(\overline{X})$, where $g(x)$ is a real-valued function such that $g'(x)$ exists and is not zero in a neighborhood of μ. We can then write

$$g(\overline{X}) = g(\mu) + g'(\mu)(\overline{X} - \mu) + R$$

where R denotes the remainder term for the Taylor series. Rearranging terms and multiplying through by $\sqrt{n}$ yields

$$\sqrt{n}[g(\overline{X}) - g(\mu)] = \sqrt{n}g'(\mu)(\overline{X} - \mu) + \sqrt{n}(R)$$

In many cases, $\sqrt{n}(R)$ approaches zero as n increases, and this term can be dropped to give the approximation

$$\sqrt{n}[g(\overline{X}) - g(\mu)] \approx \sqrt{n}(\overline{X} - \mu)g'(\mu)$$
$$= \frac{\sqrt{n}(\overline{X} - \mu)}{\sigma}\sigma g'(\mu)$$

Since $\sqrt{n}(\overline{X} - \mu)/\sigma$ is approximately standard normal, the right-hand side of the equation is approximately normal with mean zero and variance $\sigma^2[g'(\mu)]^2$. Thus, the term on the left side will also tend toward the same normal distribution. In summary,

$$\frac{\sqrt{n}[g(\overline{X}) - g(\mu)]}{|g'(\mu)|\sigma}$$

has a distribution function that will converge to the standard normal distribution function as n approaches infinity. An example of the use of this property is given next.

EXAMPLE   **6.16**   The current I in an electrical circuit is related to the voltage E and the resistance R by Ohm's Law, $I = E/R$. Suppose that for circuits of a certain type E is constant but the resistance R varies slightly from circuit to circuit. The resistance is to be measured independently in n circuits, yielding measurements $X_1, \ldots, X_n$. If $E(X_i) = \mu$ and $V(X_i) = \sigma^2, i = 1, \ldots, n$, approximate the distribution of $E/\overline{X}$, an approximation to the average current.

Solution   In this case, $g(\overline{X}) = E/\overline{X}$ for a constant E. Thus,

$$g'(\mu) = -\frac{E}{\mu^2}$$

and

$$\sqrt{n}\left[\frac{E}{\overline{X}} - \frac{E}{\mu}\right] \Big/ \left(\frac{E\sigma}{\mu^2}\right)$$

has approximately a standard normal distribution for large n. If μ and σ were known, at least approximately, we could make probability statements about the behavior of $E/\overline{X}$ by using the table of normal curve areas.   ∎

Exercises

6.60   Circuits of a certain type have resistances with a mean μ of 6 ohms and a standard deviation σ of 0.5 ohm. Twenty-five such circuits are to be used in a system, each with voltage E of 120. A specification calls for the average current in the 25 circuits to exceed 19 amps. Find the approximate probability that the specification will be met.

6.61   The distance d traveled in t seconds by a particle starting from rest is given by $d = \frac{1}{2}at^2$, where a is the acceleration. An engineer wants to study the acceleration of gravel particles rolling down an incline 10 meters in length. Observations on 100 particles showed that the average time to travel the 10 meters was 20 seconds, and the standard deviation of these measurements was 1.6 seconds. Approximate the mean and variance of the average acceleration for these 100 particles.

6.62   A fuse used in an electric current has a lifelength X that is exponentially distributed with mean θ (measurements in hundreds of hours). The *reliability* of the fuse at time t is given by

$$R(t) = P(X > t) = \int_t^\infty \frac{1}{\theta} e^{-x/\theta}\, dx$$
$$= e^{-t/\theta}$$

If the lifelengths $X_1, \ldots, X_{100}$ of 100 such fuses are to be measured, $e^{-t/\theta}$ can be approximated by $e^{t/\overline{X}}$. Assuming θ is close to 4, approximate the probability that $e^{-t/\overline{X}}$ will be within 0.05 of $e^{-t/\theta}$ for $t = 5$.

6.9   Activities for Students: What Is Random Sampling?

The properties of sample means and proportions developed in this chapter require the samples to be *randomly* selected. Why randomize? This activity is designed to demonstrate that randomness produces predictable sampling distributions but other methods of sampling may not.

1   Judgmental samples

   a   Keep the rectangles in Figure 6.24 covered until the instructor gives the signal to begin. Then look at the page for about 10 seconds and write down your guess as to the average area of the rectangles on the page. (The unit of measure is the underlying square on the graph paper. Thus, rectangle 33 has area $4 \times 3 = 12$.)

   b   Select five rectangles that, in your judgment, are representative of the rectangles on the page. Write down their numbers and their areas. Compute the average of the five areas and compare it to your guess in (a).

c   Now select five more rectangles (all different from the first five) that, in your judgment, represent the rectangles on the page. Write down their numbers and their areas. Compute the average of the second five and compare it to your guess in (a).

d   Compute the average of the areas for the ten rectangles in your two judgmental samples combined. Compare it to your guess in (a). Which estimate of the average area do you like best from among your guess, the two averages of five, and the average of ten? Why?

e   As a class, collect the guesses, the first average of five, the second average of five, and the average of ten from each student. Plot each of the four data sets on a dotplot and discuss any patterns or trends you see in the results.

2   Random samples

a   Using the numbers of the rectangles (100 may be considered as 00) and the random number table, select five rectangles at random (without repetitions). Write down their areas and compute the averages.

b   Now repeat the instructions in 2(a) for another set of five rectangles, all different from those found in 2(a).

c   Compute the average area for all ten randomly selected rectangles. Compare this average, and the averages of five, to the averages obtained in part 1.

d   As in the case of judgmental samples, collect the data from each student in the class and construct dotplots of the two sets of averages of five and the set of averages of ten.

3   Data analysis

a   Compare the dotplots from the guesses, the judgmental samples, and the random samples. Discuss any similarities and differences you see in the patterns. Comment upon bias and variability.

b   To get a better numerical description of variability, calculate the standard deviation for the data in each dotplot. Do you see any patterns emerging?

4   Resolution

a   The true average (mean) value for the areas of rectangles on the page is provided in the answer section. After obtaining this value, discuss the notion of bias in judgmental samples as compared with random samples.

b   If someone told you that he or she did not need to randomly select the respondents to a survey because personal judgment was just as good (or perhaps even better), what would you say?

5   Extension

How would the pattern of the dotplots change if we selected random samples of $n = 15$ and $n = 20$ rectangles and recorded the average area for each sample? Can you think of other ways to sample the rectangles that might be better than simple random samples?

FIGURE **6.24**

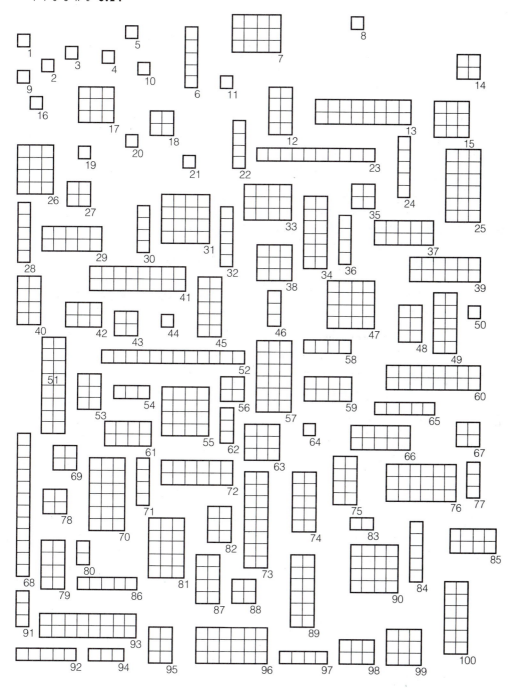

6.10   Summary

For random samples, sample *statistics* (like the mean and variance) assume different values from sample to sample, but these values collectively form predictable distributions, called *sampling distributions*. Most important, sample means and sample proportions have sampling distributions that are approximately normal in shape.

One of the principal applications of sampling distributions to process improvement comes in the construction of *control charts*. Control charts are basic tools that help separate special cause variation from common-cause variation in order to bring process into statistical control. Then *process capability* can be assessed.

It is important to remember that some of the sampling distributions we commonly deal with are exact and some are approximate. If the random sample under consideration comes from a population that can be modeled by a normal distribution, then $\overline{X}$ will have a normal distribution, and $(n-1)S^2/\sigma^2$ will have a χ^2 distribution with *no* approximation involved. However, the Central Limit Theorem states that $\sqrt{n}(\overline{X}-\mu)/\sigma$ will have *approximately* a normal distribution for large n, under almost any probabilistic model for the population itself.

Sampling distributions for statistics will be used throughout the remainder of the text for the purpose of relating sample quantities (statistics) to parameters of the population.

- - - - - - - - - -

Supplementary Exercises

6.63   The U.S. Census Bureau reports median ages of the population and median incomes for households. On the other hand, the College Board reports average scores on the Scholastic Aptitude Test. Do you think each group is using an appropriate measure? Why or why not?

6.64   Twenty-five lamps are connected so that when one lamp fails, another takes over immediately. (Only one lamp is on at any one time.) The lamps operate independently, and each has a mean life of 50 hours and a standard deviation of 4 hours. If the system is not checked for 1,300 hours after the first lamp is turned on, what is the probability that a lamp will be burning at the end of the 1,300-hour period?

6.65   Suppose that $X_1, \ldots, X_{40}$ denotes a random sample of measurements on the proportion of impurities in samples of iron ore. Suppose that each X_t has the probability density function

$$f(x) = \begin{cases} 3x^2 & 0 \le x \le 1 \\ 0 & \text{elsewhere} \end{cases}$$

The ore is to be rejected by a potential buyer if $\overline{X} > 0.7$. Find the approximate probability that the ore will be rejected, based on the 40 measurements.

6.66   Suppose that $X_1, \ldots, X_n$ and $Y_1, \ldots, Y_m$ are independent random samples, with the X_i's having mean μ_1 and variance σ_1^2 and the Y_i's having mean μ_2 and variance σ_2^2. The difference between the sample means, $\overline{X} - \overline{Y}$, will again be approximately normally distributed, since the Central Limit Theorem will apply to this difference.

a   Find $E(\overline{X} - \overline{Y})$.

b   Find $V(\overline{X} - \overline{Y})$.

6.67   A study of the effects of copper on a certain species (say, species A) of fish shows the variance of LC50 measurements (in milligrams per liter) to be 1.9. The effects of copper on a second species (say, species B) of fish produce a variance of LC50 measurements of 0.8. If the population means of LC50s for the two species are equal, find the probability that, with random samples of ten measurements from each species, the sample mean for species A exceeds the sample means for species B by at least one unit.

6.68   If Y has an exponential distribution with mean θ, show that $U = 2Y/\theta$ has a χ^2 distribution with 2 degrees of freedom.

6.69   A plant supervisor is interested in budgeting for weekly repair costs (in dollars) for a certain type of machine. These repair costs over the past ten years tend to have an exponential distribution with a mean of 20 for each machine studied. Let $Y_1, \ldots, Y_5$ denote the repair costs for five of these machines for the next week. Find a number c such that $P(\sum_{i=1}^{5} Y_i > c) = 0.05$, assuming the machines operate independently.

6.70   If Y has a $\chi^2(n)$ distribution, then Y can be represented as

$$Y = \sum_{i=1}^{n} X_i$$

where X_i has a $\chi^2(1)$ distribution and the X_i's are independent. It is not surprising, then, that Y will be approximately normally distributed for large n. Use this fact in the solution of the following problem.

   A machine in a heavy-equipment factory produces steel rods of length Y, where Y is a normally distributed random variable with a mean of 6 inches and a variance of 0.2. The cost C (in dollars) of repairing a rod that is not exactly 6 inches in length is given by

$$C = 4(Y - \mu)^2$$

where $\mu = E(Y)$. If 50 rods with independent lengths are produced in a given day, approximate the probability that the total cost for repairs exceeds $48.

Estimation

About This Chapter

We now introduce one form of statistical inference by looking at the techniques of *estimation*. Sample statistics are employed to estimate population parameters, such as means or proportions. By making use of the sampling distribution of the statistic employed, we construct intervals (confidence intervals) that include the unknown parameter value with high probability.

Contents

[†]Optional section.

7.1 Introduction

We have seen that probabilistic models for populations usually contain some unknown parameters. We have also seen, in Chapter 5, that certain functions of sample observations, called statistics, have sampling distributions that contain some of these same unknown parameters. Thus, it seems that we should be able to use statistics to obtain information on population parameters. For example, suppose a probabilistic model for a population of tensile strength measurements is left unspecified except for the fact that it has mean μ and finite variance σ^2. A random sample $X_1, \ldots, X_n$ from this population gives rise to a sample mean $\overline{X}$ that, according to Section 6.3, will tend to have a normal distribution with mean μ and variance σ^2/n. How can we make use of this information to say something more specific about the unknown parameter μ? In this chapter, we answer this question with a type of statistical inference called *estimation*. Chapter 8 will consider *tests of hypotheses*.

DEFINITION   **7.1**   An **estimator** is a statistic that specifies how to use the sample data to estimate an unknown parameter of the population.   ∎

Note that an estimator is a random variable.

In the above example, it seems obvious that the sample mean $\overline{X}$ could be used as an *estimator* of μ. But $\overline{X}$ will take on different numerical values from sample to sample, so some questions remain. What is the magnitude of the difference between $\overline{X}$ and μ likely to be? How does this difference behave as n gets larger and larger? In short, what properties does $\overline{X}$ have as an estimator of μ? This chapter provides answers to some of these basic questions regarding estimation.

7.2 Properties of Point Estimators

The statistics that will be used as estimators will, of course, have sampling distributions. Sometimes the sampling distribution is known, at least approximately, as in the case of the sample mean from large samples. In other cases, the sampling distribution may not be completely specified, but we can still calculate the mean and variance of the estimator. We might, then, logically ask "What properties would we like this mean and variance to possess?"

Let the symbol θ denote an arbitrary population parameter, such as μ or σ^2, and let $\hat{\theta}$ denote an estimator of θ. We are then concerned with properties of $E(\hat{\theta})$ and $V(\hat{\theta})$. Different samples result in different numerical values for $\hat{\theta}$, but we would hope that some of these values underestimate θ and that others overestimate θ, so that the average value of $\hat{\theta}$ is close to θ. If $E(\hat{\theta}) = \theta$, then $\hat{\theta}$ is said to be an unbiased estimator of θ.

DEFINITION   **7.2**

An estimator $\hat{\theta}$ is **unbiased** for estimating θ if

$$E(\hat{\theta}) = \theta \quad \blacksquare$$

In the sampling distributions simulated in Section 6.2, we saw that the values of $\overline{X}$ tend to center at μ, the true population mean, when random samples are selected from the same population repeatedly. Similarly, values of S^2 centered at σ^2, the true population variance. These are demonstrations that $\overline{X}$ is an unbiased estimator of μ and that S^2 is an unbiased estimator of σ^2.

For an unbiased estimator $\hat{\theta}$, the sampling distribution of the estimator has mean value θ. How do we want the possible values for $\hat{\theta}$ to spread out to either side of θ for this unbiased estimator? Intuitively, it would be desirable for all possible values of $\hat{\theta}$ to be very close to θ. That is, we want the variance of $\hat{\theta}$ to be as small as possible. At the level of this text, it is not possible to prove that some of our estimators do in fact have the smallest variance among all unbiased estimators, but we will use this variance criterion for comparing estimators. That is, if $\hat{\theta}_1$ and $\hat{\theta}_2$ are both unbiased estimators of θ, then we would choose as the better estimator the one possessing the smaller variance (see Figure 7.1).

FIGURE **7.1**
Distributions of Two Unbiased
Estimators, $\hat{\theta}_1$ and $\hat{\theta}_2$, with
$V(\hat{\theta}_2) < V(\hat{\theta}_1)$

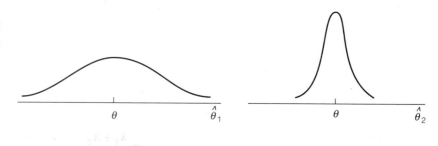

EXAMPLE   **7.1**   Suppose $X_1, \ldots, X_5$ denotes a random sample from some population with $E(X_i) = \mu$ and $V(X_i) = \sigma^2, i = 1, \ldots, 5$. The following are suggested as estimators for μ:

$$\hat{\mu}_1 = X_1 \qquad \hat{\mu}_2 = \frac{1}{2}(X_1 + X_5) \qquad \hat{\mu}_3 = \frac{1}{2}(X_1 + 2X_5)$$

$$\hat{\mu}_4 = \overline{X} = \frac{1}{5}(X_1 + X_2 + \cdots + X_5)$$

Which estimator would you use, and why?

Solution   Looking at the means of these estimators, we have $E(\hat{\mu}_1) = \mu$, $E(\hat{\mu}_2) = \mu$, $E(\hat{\mu}_3) = \frac{3}{2}\mu$, and $E(\hat{\mu}_4) = \mu$. Thus, $\hat{\mu}_1, \hat{\mu}_2$, and $\hat{\mu}_4$ are all unbiased, and we eliminate estimator $\hat{\mu}_3$ because it is biased (it tends to overestimate μ).

Looking at variances of the unbiased estimators, we have

$$V(\hat{\mu}_1) = V(X_1) = \sigma^2$$

$$V(\hat{\mu}_2) = \frac{1}{4}[V(X_1) + V(X_5)] = \frac{\sigma^2}{2}$$

and

$$V(\hat{\mu}_4) = V(\overline{X}) = \frac{\sigma^2}{5}$$

Thus, $\hat{\mu}_4$ would be chosen as the best estimator, since it is unbiased and has the smallest variance among the three unbiased estimators. ∎

We give a method for finding good point estimators in Section 7.7. In the meantime, we discuss further details of estimation for intuitively reasonable point estimators.

Exercises

7.1 Suppose X_1, X_2, X_3 denotes a random sample from the exponential distribution with density function

$$f(x) = \begin{cases} \dfrac{1}{\theta} e^{-x/\theta} & x > 0 \\ 0 & \text{elsewhere} \end{cases}$$

Consider the following four estimators of θ:

$$\hat{\theta}_1 = X_1$$
$$\hat{\theta}_2 = \frac{X_1 + X_2}{2}$$
$$\hat{\theta}_3 = \frac{X_1 + 2X_2}{3}$$
$$\hat{\theta}_4 = \overline{X}$$

a Which of the above estimators are unbiased for θ?

b Among the unbiased estimators of θ, which has the smallest variance?

7.2 The reading on a voltage meter connected to a test circuit is uniformly distributed over the interval $(\theta, \theta + 1)$, where θ is the true but unknown voltage of the circuit. Suppose $X_1, \ldots, X_n$ denotes a random sample of readings from this voltage meter.

a Show that $\overline{X}$ is a biased estimator of θ.

b Find a function of $\overline{X}$ that is an unbiased estimator of θ.

7.3 The number of breakdowns per week for a certain minicomputer is a random variable X having a Poisson distribution with mean λ. A random sample $X_1, \ldots, X_n$ of observations on the number of breakdowns per week is available.

a Find an unbiased estimator of λ.

b The weekly cost of repairing these breakdowns is

$$C = 3Y + Y^2$$

Show that

$$E(C) = 4\lambda + \lambda^2$$

c Find an unbiased estimator of $E(C)$ that makes use of the entire sample, $X_1, \ldots, X_n$.

7.4 The *bias B* of an estimator $\hat{\theta}$ is given by

$$B = |\theta - E(\hat{\theta})|$$

The *mean squared error*, or MSE, of an estimator $\hat{\theta}$ is given by

$$\text{MSE} = E(\hat{\theta} - \theta)^2$$

Show that

$$\text{MSE}(\hat{\theta}) = V(\hat{\theta}) + B^2$$

[*Note:* $\text{MSE}(\hat{\theta}) = V(\hat{\theta})$ if $\hat{\theta}$ is an unbiased estimator of θ. Otherwise, $\text{MSE}(\hat{\theta}) > V(\hat{\theta})$.]

7.5 Refer to Exercise 7.2. Find $\text{MSE}(\overline{X})$ when $\overline{X}$ is used to estimate θ.

7.6 Suppose $X_1, \ldots, X_n$ is a random sample from a normal distribution with mean μ and variance σ^2.

a Show that $S = \sqrt{S^2}$ is a biased estimator of σ.

b Adjust S to form an unbiased estimator of σ.

7.7 For a certain new model of microwave oven, it is desired to set a guarantee period so that only 5% of the ovens sold will have a major failure in this length of time. Assuming the length of time until the first major failure for an oven of this type is normally distributed with mean μ and variance σ^2, the guarantee period should end at $\mu - 1.645\sigma$. Use the results of Exercise 7.6 to find an unbiased estimator of $\mu - 1.645\sigma$ based on a random sample $X_1, \ldots, X_n$ of measurements of time until the first major failure.

7.8 Suppose $\hat{\theta}_1$ and $\hat{\theta}_2$ are each unbiased estimators of θ, with $V(\hat{\theta}_1) = \sigma_1^2$ and $V(\hat{\theta}_2) = \sigma_2^2$. A new unbiased estimator for θ can be formed by

$$\hat{\theta}_3 = a\hat{\theta}_1 + (1 - a)\hat{\theta}_2$$

$(0 \le a \le 1)$. If $\hat{\theta}_1$ and $\hat{\theta}_2$ are independent, how should a be chosen so as to minimize $V(\hat{\theta}_3)$?

7.3 Confidence Intervals: The Single-Sample Case

After knowing something about the mean and variance of $\hat{\theta}$ as an estimator of θ, it would be nice to know something about how small the distance between $\hat{\theta}$ and θ is likely to be. Answers to this question require knowledge of the sampling distribution of $\hat{\theta}$ beyond the behavior of $E(\hat{\theta})$ and $V(\hat{\theta})$ and lead to *confidence intervals*.

The idea behind the construction of confidence intervals is as follows. If $\hat{\theta}$ is an estimator of θ (which has a known sampling distribution) and we can find two quantities that depend on $\hat{\theta}$, say $g_1(\hat{\theta})$ and $g_2(\hat{\theta})$, such that

$$P[g_1(\hat{\theta}) \le \theta \le g_2(\hat{\theta})] = 1 - \alpha$$

for some small positive number α, then we can say that $(g_1(\hat{\theta}), g_2(\hat{\theta}))$ forms an interval that has probability $(1 - \alpha)$ of capturing the true θ. This interval is referred to as a confidence interval with confidence coefficient $(1 - \alpha)$. The quantity $g_1(\hat{\theta})$ is called the *lower confidence limit* (LCL) and $g_2(\hat{\theta})$ the *upper confidence limit* (UCL). We generally want $(1 - \alpha)$ to be near unity and $(g_1(\hat{\theta}), g_2(\hat{\theta}))$ to be as short an interval as possible for a given sample size. We illustrate the construction of confidence intervals for some common parameters here; other cases will be considered in Chapters 9 through 12.

7.3.1   General Distribution: Large-Sample Confidence Interval for μ

Suppose we are interested in estimating a mean μ for a population with variance σ^2 assumed, for the moment, to be known. We select a random sample $X_1, \ldots, X_n$ from this population and compute $\overline{X}$ as a point estimator of μ. If n is large (say, $n \geq 25$ as a rule of thumb), then $\overline{X}$ has approximately a normal distribution with mean μ and variance σ^2/n. From these facts, we can state that the interval $\mu \pm 2\sigma/\sqrt{n}$ contains about 95% of the $\bar{x}$ values that could be generated in repeated random samplings from the population under study. For convenience, let's call this middle 95% the "likely" values of $\bar{x}$. Now suppose we are to observe a single sample producing a single $\bar{x}$. A question of interest is "What possible values for μ would allow this $\bar{x}$ to lie in the likely range of possible sample means?" This set of possible values for μ is the confidence interval with a confidence coefficient of approximately 0.95.

The main idea of confidence interval construction is illustrated in the diagram of Figure 7.2. For any value of μ, the likely sample means are in the interval $\mu \pm 2\sigma/\sqrt{n}$, which shifts to the right as larger values of μ are considered. For a continuum of possible values for μ, the likely sample means fall between the two angled lines (which will have angles of 45° if the same numerical scale is used on both axes). Suppose we observe a single sample mean of $\bar{x}_0$. For what values of μ will $\bar{x}_0$ fall into the likely sample region? The values are precisely those values of μ between the intersections of the dashed line and the angled lines. Considering both axes to have the same numerical scale, we can see that the intersection points are at $\bar{x}_0 - 2\sigma/\sqrt{n}$ and $\bar{x}_0 + 2\sigma/\sqrt{n}$. These two values form an approximate 95% confidence interval for μ (or a confidence interval with confidence coefficient equal to 0.95). Note that we are assuming σ to be a fixed value in this discussion.

More formally, under these conditions,

$$Z = \frac{\overline{X} - \mu}{\sigma/\sqrt{n}}$$

has approximately a standard normal distribution. Now for any prescribed α we can find, from Table 4 in the Appendix, a value $z_{\alpha/2}$ such that

$$P(-z_{\alpha/2} \leq Z \leq +z_{\alpha/2}) = 1 - \alpha$$

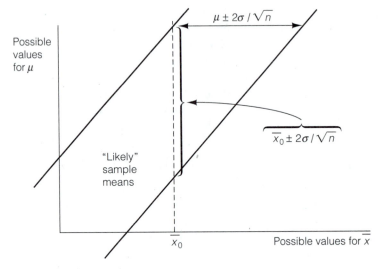

Possible
values
for μ

"Likely"
sample
means

Rewriting this probability statement, we have

$$1 - \alpha = P\left(-z_{\alpha/2} \leq \frac{\overline{X} - \mu}{\sigma/\sqrt{n}} \leq z_{\alpha/2}\right)$$

$$= P\left(-z_{\alpha/2}\frac{\sigma}{\sqrt{n}} \leq \overline{X} - \mu \leq +z_{\alpha/2}\frac{\sigma}{\sqrt{n}}\right)$$

$$= P\left(\overline{X} - z_{\alpha/2}\frac{\sigma}{\sqrt{n}} \leq \mu \leq \overline{X} + z_{\alpha/2}\frac{\sigma}{\sqrt{n}}\right)$$

The interval

$$\left(\bar{x} - z_{\alpha/2}\frac{\sigma}{\sqrt{n}}, \bar{x} + z_{\alpha/2}\frac{\sigma}{\sqrt{n}}\right)$$

forms a realization of a large-sample confidence interval for μ with confidence
coefficient approximately $(1 - \alpha)$.

If σ is unknown, it can be replaced by s, the sample standard deviation, with no
serious loss of accuracy for the large-sample case.

E X A M P L E   **7.2**   Consider once again the 50 lifelength observations of Table 4.1. Using these observations as the sample, find a confidence interval for the mean lifelength of batteries
of this type, with confidence coefficient 0.95.

Solution For these data, $\bar{x} = 2.266$ and $s = 1.935$, with units in 100 hours. Using the confidence interval

$$\bar{x} \pm z_{\alpha/2} \frac{\sigma}{\sqrt{n}}$$

with $(1 - \alpha) = 0.95$, we see that $z_{\alpha/2} = z_{0.025} = 1.96$ (from Table 4 of the Appendix). Substituting s for σ, this interval yields

$$2.266 \pm (1.96) \frac{1.935}{\sqrt{50}}$$

$$2.266 \pm 0.536$$

or

$$(1.730, 2.802)$$

Thus, we are 95% confident that the true mean lies between 1.730 and 2.802 (that is, similarly formed intervals will contain μ about 95% of the time in repeated sampling). ∎

Take a careful look at the interpretation of confidence interval statements. The *interval* is random; the parameter is fixed. Before sampling, there is a probability of $(1 - \alpha)$ that the interval will include the true parameter value. After sampling, the resulting realization of the confidence interval either includes the parameter value or fails to include it, but we are quite confident that it will include the true parameter value if $(1 - \alpha)$ is large. The following computer simulation illustrates the behavior of a confidence interval for a mean in repeated sampling.

We started the simulation by selecting random samples of size $n = 100$ from a population having an exponential distribution with a mean value of $\mu = 10$. That is,

$$f(x) = \begin{cases} \frac{1}{10} e^{-x/10} & x > 0 \\ 0 & \text{elsewhere} \end{cases}$$

For each sample, a 95% confidence interval for μ was constructed by calculating $\bar{x} \pm 1.96 s/\sqrt{n}$. The sample mean, standard deviation, lower confidence limit, and upper confidence limit are given in Table 7.1 for each of 100 samples generated. Seven intervals do *not* include the true mean value. (These intervals are marked with an asterisk.) Thus, we found that 93 out of 100 intervals include the true mean in our simulation, whereas the theory says that 95 out of 100 should include μ. This is fairly good agreement between theory and application. Later in the chapter, we will see other simulations that generate similar results.

We now move to a discussion of sample size selection. Since we know that a confidence interval for a mean μ will be of the form $\overline{X} \pm z_{\alpha/2} \sigma/\sqrt{n}$, we can choose the sample size to ensure a certain accuracy before carrying out an experiment. Suppose the goal is to achieve an interval of length $2B$, so the resulting confidence interval is of the form $\overline{X} \pm B$. We might, for example, want an estimate of the average

T A B L E **7.1** Large-Sample Confidence Intervals ($n = 100$)

Sample	LCL	Mean	UCL	s	Sample	LCL	Mean	UCL	s
1	7.26064	8.6714	10.0822	7.1977	51	7.97829	9.7442	11.5101	9.0098
2	7.59623	9.5291	11.4630	9.8668	52	8.97313	11.2890	13.6049	11.8158
3	7.68938	9.3871	11.0849	8.6619	*53	5.99726	7.7422	9.4871	8.9027
4	8.94686	11.1785	13.4100	11.3856	54	7.51538	9.4089	11.3025	9.6609
5	7.96793	10.1508	12.3336	11.1369	55	8.59019	10.4373	12.2843	9.4238
6	7.61223	10.0101	12.4080	12.2340	56	8.49077	10.3914	12.2920	9.6971
7	7.55475	9.2754	10.9961	8.7791	57	8.12970	10.4056	12.6814	11.6115
8	7.48558	9.1436	10.8016	8.4593	*58	6.85175	8.3456	9.8395	7.6217
9	7.71919	9.5219	11.3245	9.1973	59	8.87405	10.9482	13.0225	10.5827
10	7.95952	9.7012	11.4428	8.8859	60	8.70548	10.2597	11.8140	7.9298
11	7.34959	8.7958	10.2420	7.3785	61	6.37467	8.3475	10.3203	10.0653
12	8.71580	10.8458	12.9759	10.8675	62	8.64835	10.6703	12.6923	10.3161
*13	6.41955	8.1029	9.7863	8.5885	63	8.37157	10.6196	12.8676	11.4696
14	7.27538	8.9598	10.6441	8.5938	64	7.76632	9.6919	11.6174	9.8241
15	8.30285	10.2902	12.2776	10.1398	65	9.19745	11.5565	13.9155	12.0359
16	9.48735	11.5540	13.6206	10.5441	66	7.54964	9.1818	10.8140	8.3275
17	8.35947	10.8521	13.3448	12.7176	67	8.57961	10.8918	13.2039	11.7967
18	7.63421	9.9196	12.2049	11.6599	68	8.23986	10.0088	11.7778	9.0254
19	7.97129	10.1196	12.2679	10.9608	69	8.08091	9.9399	11.7989	9.4846
20	8.86613	10.9223	12.9785	10.4906	70	7.52910	9.2224	10.9156	8.6391
21	7.16061	9.2142	11.2679	10.4778	71	7.52705	9.4949	11.4628	10.0401
22	6.99569	9.2782	11.5607	11.6454	72	8.61466	10.7728	12.9310	11.0110
23	8.13781	10.2701	12.4024	10.8791	73	9.13542	11.2654	13.3953	10.8670
24	7.70300	9.4910	11.2790	9.1226	*74	6.73171	8.1351	9.5386	7.1604
25	9.61738	11.9111	14.2048	11.7026	75	7.78546	10.1938	12.6021	12.2874
26	7.23197	9.2257	11.2194	10.1720	76	8.04925	9.6668	11.2843	8.2526
27	7.85056	9.9053	11.9601	10.4836	77	7.61919	9.6068	11.5943	10.1407
28	8.82373	11.1849	13.5462	12.0470	*78	5.91167	7.7383	9.5649	9.3194
29	8.36303	10.4008	12.4387	10.3970	79	8.36219	10.4003	12.4384	10.3984
30	8.74332	10.4288	12.1144	8.5996	80	8.78276	11.0946	13.4065	11.7953
31	7.80013	9.7064	11.6127	9.7259	81	7.75108	9.4840	11.2170	8.8415
32	7.88003	10.0706	12.2611	11.1763	82	8.41534	10.4168	12.4183	10.2117
33	7.01555	8.6994	10.3833	8.5913	83	7.56210	9.1461	10.7300	8.0815
34	8.31036	10.3074	12.3045	10.1891	*84	6.51658	8.0757	9.6349	7.9549
35	7.91383	9.8344	11.7549	9.7986	85	8.73332	10.7988	12.8643	10.5381
36	8.80795	10.5208	12.2337	8.7390	86	7.10357	8.9350	10.7664	9.3439
37	7.27322	9.1728	11.0725	9.6919	87	7.46386	9.5521	11.6403	10.6541
38	7.47957	9.1333	10.7871	8.4375	88	7.40553	9.0024	10.5993	8.1473
39	7.77926	9.8034	11.8275	10.3272	89	7.70464	9.6524	11.6002	9.9376
40	7.09624	9.0955	11.0947	10.2001	90	8.16395	10.3755	12.5870	11.2834
41	8.32909	10.1639	11.9986	9.3610	91	8.08567	9.8488	11.6120	8.9958
42	8.23895	10.2593	12.2797	10.3080	92	8.45424	10.2642	12.0741	9.2344
43	8.44813	10.4284	12.4087	10.1034	*93	6.78362	8.1950	9.6064	7.2010
44	7.11788	8.8708	10.6237	8.9436	94	8.56442	10.6626	12.7608	10.7050
45	8.14050	10.0992	12.0580	9.9936	95	7.49852	9.2474	10.9962	8.9226
46	7.53466	9.4198	11.3050	9.6183	96	8.01983	10.9016	13.7833	14.7028
47	8.93450	11.0165	13.0986	10.6227	97	7.40014	9.1321	10.8640	8.8363
48	7.43185	8.9012	10.3705	7.4965	98	7.80386	9.6444	11.4849	9.3903
49	7.83028	9.7367	11.6432	9.7268	99	8.47046	10.3403	12.2101	9.5399
50	7.61207	9.3698	11.1275	8.9680	100	8.08183	10.0133	11.9447	9.8542

assembly time for microcomputers to be within 3 minutes of the true average, with high probability. This tells us that B must equal 3 in the final result. If $\overline{X} \pm z_{\alpha/2}\sigma/\sqrt{n}$ must result in $\overline{X} \pm B$, then we must have

$$B = z_{\alpha/2}\frac{\sigma}{\sqrt{n}}$$

or

$$n = \left[\frac{z_{\alpha/2}\sigma}{B}\right]^2$$

Once the confidence coefficient $1 - \alpha$ is established, $z_{\alpha/2}$ is known. The only remaining problem is to find an approximation for σ. This can often be accomplished by prior experience, by taking a preliminary sample and using s to estimate σ, or by using any of a variety of other rough approximations to σ. Even if σ is not known exactly, working through the above equation to determine a sample size that will give desirable results is better than arbitrarily guessing what the sample size should be.

> The sample size for establishing a confidence interval of the form $\overline{X} \pm B$ with confidence coefficient $(1 - \alpha)$ is given by
>
> $$n = \left[\frac{z_{\alpha/2}\sigma}{B}\right]^2$$

EXAMPLE   **7.3**   It is desired to estimate the average distance traveled to work for employees of a large manufacturing firm. Past studies of this type indicate that the standard deviation of these distances should be in the neighborhood of 2 miles. How many employees should be sampled if the estimate is to be within 0.1 mile of the true average, with confidence coefficient 0.95?

Solution   The resulting interval is to be of the form $\overline{X} \pm 0.1$, with $1 - \alpha = 0.95$. Thus, $B = 0.1$ and $z_{0.025} = 1.96$. It follows that

$$n = \left[\frac{z_{\alpha/2}\sigma}{B}\right]^2 = \left[\frac{(1.96)(2)}{0.1}\right]^2$$
$$= 1{,}536.64 \quad \text{or} \quad 1{,}537$$

Thus, 1,537 employees would have to be sampled to achieve the desired result. If a sample of this size is too costly, then either B must be increased or $(1 - \alpha)$ must be decreased.   ∎

The confidence intervals discussed so far have been two-sided intervals. Sometimes we are interested in only a lower or upper limit for a parameter, but not both. A one-sided confidence interval can be formed in such cases.

Suppose in the study of mean lifelength of batteries, as in Example 7.2, we wanted only to establish a lower limit to this mean. That is, we want to find a statistic $g(\overline{X})$ so that

$$P[\mu > g(\overline{X})] = 1 - \alpha$$

Again, we know that

$$P\left(\frac{\overline{X} - \mu}{\sigma/\sqrt{n}} \le z_\alpha\right) = 1 - \alpha$$

and it follows that

$$P\left(\overline{X} - \mu \le z_\alpha \frac{\sigma}{\sqrt{n}}\right) = 1 - \alpha$$

$$P\left(\overline{X} - z_\alpha \frac{\sigma}{\sqrt{n}} \le \mu\right) = 1 - \alpha$$

or

$$\overline{X} - z_\alpha \frac{\sigma}{\sqrt{n}}$$

forms a lower confidence limit for μ with confidence coefficient $(1 - \alpha)$. Using the data from Example 7.2, a lower 95% confidence limit for μ is approximately

$$\overline{X} - z_{0.05} \frac{s}{\sqrt{n}}$$

or

$$2.266 - (1.645)\frac{1.935}{\sqrt{50}} = 2.266 - 0.450 = 1.816$$

Thus, we are 95% confident that the mean battery lifelength exceeds 181.6 hours. Note that this lower confidence limit in the one-sided case is larger than the lower limit in the corresponding two-sided confidence interval.

7.3.2   Binomial Distribution: Large-Sample Confidence Interval for p

Estimating the parameter p for the binomial distribution is analogous to the estimation of a population mean. As we saw in Section 3.3, this is because the random variable Y having a binomial distribution with n trials can be written as

$$Y = \sum_{i=1}^{n} X_i$$

where the X_i's, $i = 1, \ldots, n$, are independent Bernoulli random variables with common mean p. Thus, $E(Y) = np$ or $E(Y/n) = p$, and Y/n is an unbiased estimator of p. Recall that $V(Y) = np(1 - p)$, and hence

$$V(Y/n) = \frac{p(1-p)}{n}$$

Note that $Y/n = \sum_{i=1}^{n} X_i/n = \overline{X}$, with the X_i's having $E(X_i) = p$ and $V(X_i) = p(1 - p)$. The Central Limit Theorem then applies to Y/n, the sample fraction of successes, and we have that Y/n is approximately normally distributed with mean p and variance $p(1 - p)/n$. The large-sample confidence interval for p is then constructed by comparison with the corresponding result for μ. The observed fraction of successes, y/n, will be used as the estimate of p and will be written as $\hat{p}$.

The interval

$$\hat{p} \pm z_{\alpha/2}\sqrt{\frac{\hat{p}(1-\hat{p})}{n}}$$

forms a realization of a large-sample confidence interval for p with confidence coefficient approximately $(1 - \alpha)$, where $\hat{p} = y/n$.

E X A M P L E   **7.4**   In certain water-quality studies, it is important to check for the presence or absence of various types of microorganisms. Suppose 20 out of 100 randomly selected samples of a fixed volume show the presence of a particular microorganism. Estimate the true probability p of finding this microorganism in a sample of the same volume, with a 90% confidence interval.

Solution   We assume that the number of samples showing the presence of the microorganism, out of n randomly selected samples, can be reasonably modeled by the binomial distribution. Using the confidence interval

$$\hat{p} \pm z_{\alpha/2}\sqrt{\frac{\hat{p}(1-\hat{p})}{n}}$$

we have $1 - \alpha = 0.90$, $z_{\alpha/2} = z_{0.05} = 1.645$, $y = 20$, and $n = 100$. Thus, the interval estimate of p is

$$0.20 \pm 1.645\sqrt{\frac{0.2(0.8)}{100}}$$

or

$$0.20 \pm 0.066$$

We are about 90% confident that the true probability p is somewhere between 0.134 and 0.266.   ∎

EXAMPLE **7.5**   A firm wants to estimate the proportion of production-line workers who favor a revised quality-assurance program. The estimate is to be within 0.05 of the true proportion favoring the revised program, with a 90% confidence coefficient. How many workers should be sampled?

Solution   Sample size is determined by

$$n = \left(\frac{z_{\alpha/2}\sigma}{B}\right)^2$$

but, in the binomial case, $\sigma = \sqrt{p(1-p)}$. Thus,

$$n = \left(\frac{z_{\alpha/2}\sqrt{p(1-p)}}{B}\right)^2$$

In this problem, $B = 0.05$ and $1 - \alpha = 0.90$, so $z_{0.05} = 1.645$. It remains to approximate $\sigma = \sqrt{p(1-p)}$. The quantity $p(1-p)$ is maximized at $p = 0.5$. Thus, if no prior knowledge of p is available, we can use $p = 0.5$ in our calculation to determine the largest sample size necessary to ensure the desired results.

We have, under these conditions,

$$n = \left(\frac{1.645\sqrt{(0.5)(0.5)}}{0.05}\right)^2$$
$$= (16.45)^2 = 270.6 \quad \text{or} \quad 271$$

A sample of 271 workers is required in order to estimate the true proportion favoring the revised policy to within 0.05.   ∎

7.3.3   Normal Distribution: Confidence Interval for μ

If the samples we are dealing with are not large enough to ensure an approximately normal sampling distribution for $\overline{X}$, then the above results are not valid. What can we do in that case? One approach is to use techniques that do not depend upon distributional assumptions (so called distribution-free techniques). This approach will be considered in Chapter 12. Another approach, which we will pursue here, is to make an additional assumption concerning the nature of the probabilistic model for the population. In many real problems, it is appropriate to assume that the random variable under study has a normal distribution. We can then develop an exact confidence interval for the mean μ.

The development of the confidence interval for μ in this section parallels the large-sample development in Section 7.3.1. We first must find a function of $\overline{X}$ and μ for which we know the sampling distribution. If $X_1, \ldots, X_n$ are independent normally distributed random variables, each with unknown mean μ and variance σ^2, then we know that

$$Z = \frac{\overline{X} - \mu}{\sigma/\sqrt{n}}$$

has a standard normal distribution. But σ is unknown, and a reasonable estimator is S. We ask, then, what is the sampling distribution of

$$T = \frac{\overline{X} - \mu}{S/\sqrt{n}}$$

Since $E(\overline{X}) = \mu$, the sampling distribution of T should center at zero. Since σ, a constant, was replaced by S, a random variable, the sampling distribution of T should show more variability than the sampling distribution of Z. In fact, the sampling distribution of T is known and is called the *t distribution* (or Student's *t* distribution) with $n - 1$ degrees of freedom. A *t* probability density function is pictured in Figure 7.3. The *t* probability density function is symmetric about zero and has wider tails (is more spread out) than the standard normal distribution. Table 5 of the Appendix contains values t_α that cut off an area of α in the right-hand tail of the distribution, as depicted in Figure 7.3. Only the upper-tail values are given in the table because of the symmetry of the density function. The table values depend on the degrees of freedom, since the distribution of T changes with changes in $(n - 1)$. As n gets large, the *t* distribution approaches the standard normal distribution. Thus, t_α is approximately equal to z_α for large n.

F I G U R E **7.3**
A *t* Distribution (probability density function)

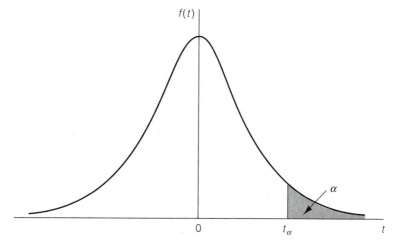

We now wish to construct a confidence interval for μ based on the t distribution. From Table 5 of the Appendix, we can find a value $t_{\alpha/2}$, which cuts off an area of $\alpha/2$ in the right-hand tail of a t distribution with known degrees of freedom. Thus,

$$P\left(-t_{\alpha/2} \leq \frac{\overline{X} - \mu}{S/\sqrt{n}} \leq t_{\alpha/2}\right) = 1 - \alpha$$

for some prescribed α. Reworking this inequality just as in the large-sample case of Section 7.3.1, we have

$$P\left(\overline{X} - t_{\alpha/2}\frac{S}{\sqrt{n}} \leq \mu \leq \overline{X} + t_{\alpha/2}\frac{S}{\sqrt{n}}\right) = 1 - \alpha$$

If $X_1, \ldots, X_n$ denotes a random sample from a normal distribution, then

$$\bar{x} \pm t_{\alpha/2}\frac{s}{\sqrt{n}}$$

provides a realization of an exact confidence interval for μ with confidence coefficient $(1 - \alpha)$. The degrees of freedom are given by $(n - 1)$.

E X A M P L E **7.6** Prestressing wire for wrapping concrete pipe is manufactured in large rolls. A quality-control inspection required five specimens from a roll to be tested for ultimate tensile strength (UTS). The UTS measurements (in 1,000 psi) turned out to be 253, 261, 258, 255, and 256. Use these data to construct a 95% confidence interval estimate of the true mean UTS for the sampled roll.

Solution It is assumed that, if many wire specimens were tested, the relative frequency distribution of UTS measurements would be nearly normal. A confidence interval based on the t distribution can then be employed. With $1 - \alpha = 0.95$ and $n - 1 = 4$ degrees of freedom, $t_{0.025} = 2.776$.
From the observed data,

$$\bar{x} = 256.60 \quad \text{and} \quad s = 3.05$$

Thus,

$$\bar{x} \pm t_{\alpha/2}\frac{s}{\sqrt{n}}$$

becomes

$$256.60 \pm (2.776)\frac{3.05}{\sqrt{5}}$$

or

$$256.60 \pm 3.79$$

We are 95% confident that the interval 252.81 to 260.39 includes the true mean UTS for the role.

If a computer with a statistical package is available to the reader, we would encourage its use to handle computations like those encountered above. The output for solving this problem on Minitab is shown below. The output for other software packages would be similar.

```
N MEAN STDEV SE MEAN 95.0 PERCENT C.I.
5 256.60 3.05 1.36 (252.81, 260.39)   ▪
```

Note that if we are not justified in making the assumption of normality for the population of tensile strengths, then we would be uncertain of the confidence level for this interval. The Central Limit Theorem would not help in this small-sample case.

7.3.4   Normal Distribution: Confidence Interval for σ^2

Still under the assumption that the random sample $X_1, \ldots, X_n$ comes from a normal distribution, we know from Section 5.5 that $(n-1)S^2/\sigma^2$ has a χ^2 distribution, with $v = n - 1$ degrees of freedom. We can employ this fact to establish a confidence interval for σ^2. Using Table 6 of the Appendix, we can find two values, $\chi^2_{\alpha/2}(v)$ and $\chi^2_{1-\alpha/2}(v)$, such that

$$
1 - \alpha = P\left[\chi^2_{1-\alpha/2}(v) \le \frac{(n-1)S^2}{\sigma^2} \le \chi^2_{\alpha/2}(v)\right]
$$

$$
= P\left[\frac{(n-1)S^2}{\chi^2_{\alpha/2}(v)} \le \sigma^2 \le \frac{(n-1)S^2}{\chi^2_{1-\alpha/2}(v)}\right]
$$

If $X_1, \ldots, X_n$ denotes a random sample from a normal distribution, then

$$
\left(\frac{(n-1)s^2}{\chi^2_{\alpha/2}(v)}, \frac{(n-1)s^2}{\chi^2_{1-\alpha/2}(v)}\right)
$$

provides a realization of an exact confidence interval for σ^2 with confidence coefficient $(1 - \alpha)$.

E X A M P L E   **7.7**   In laboratory work, it is desirable to run careful checks on the variability of readings produced on standard samples. In a study of the amount of calcium in drinking water undertaken as part of a water-quality assessment, the same standard was run through the laboratory six times at random intervals. The six readings, in parts per million, were 9.54, 9.61, 9.32, 9.48, 9.70, and 9.26.

Estimate σ^2, the population variance for readings on this standard, using a 90% confidence interval.

Solution   We must first calculate the sample variance s^2 as

$$s^2 = \left(\sum_{i=1}^{n} x_i^2 - n\bar{x}^2 \right) \left(\frac{1}{n-1} \right) = [539.9341 - 6(9.485)^2]\left(\frac{1}{5} \right)$$

$$= (539.9341 - 539.7914)\left(\frac{1}{5} \right) = \frac{0.1427}{5} = 0.0285$$

Using Table 6 of the Appendix, we have that

$$\chi_{0.95}^2(5) = 1.1455 \quad \text{and} \quad \chi_{0.05}^2(5) = 11.0705$$

Thus, the confidence interval for σ^2 becomes

$$\left[\frac{(n-1)s^2}{\chi_{0.05}^2(v)}, \frac{(n-1)s^2}{\chi_{0.95}^2(v)} \right]$$

or

$$\left(\frac{0.1427}{11.0705}, \frac{0.1427}{1.1455} \right)$$

or

$$(0.0129, 0.1246)$$

That is, we are 90% confident that σ^2 is between 0.0129 and 0.1246. Note that this is a fairly wide interval, primarily because n is so small. ▪

Confidence Intervals from Single Samples

General Distribution: Large-sample confidence interval for μ:

$$\bar{x} \pm z_{\alpha/2}\sigma/\sqrt{n} \qquad (\sigma \text{ can be estimated by } s)$$

Binomial Distribution: Large-sample confidence interval for p:

$$\hat{p} \pm z_{\alpha/2}\sqrt{\frac{\hat{p}(1-\hat{p})}{n}} \qquad \text{where } \hat{p} = \frac{y}{n}$$

Normal Distribution: Confidence interval for μ:

$$\bar{x} \pm t_{\alpha/2}s/\sqrt{n}$$

Normal Distribution: Confidence interval for σ^2:

$$\left(\frac{(n-1)s^2}{\chi^2_{\alpha/2}(v)}, \frac{(n-1)s^2}{\chi^2_{1-\alpha/2}(v)} \right) \qquad (v = n - 1)$$

Exercises

7.9 For a random sample of 50 measurements on the breaking strength of cotton threads, the mean breaking strength was found to be 210 grams and the standard deviation 18 grams. Obtain a confidence interval for the true mean breaking strength of cotton threads of this type, with confidence coefficient 0.90.

7.10 A random sample of 40 engineers was selected from among the large number employed by a corporation engaged in seeking new sources of petroleum. The hours worked in a particular week were determined for each engineer selected. These data had a mean of 46 hours and a standard deviation of 3 hours. For that particular week, estimate the mean hours worked for all engineers in the corporation, with a 95% confidence coefficient.

7.11 An important property of plastic clays is the percent of shrinkage on drying. For a certain type of plastic clay, 45 test specimens showed an average shrinkage percentage of 18.4 and a standard deviation of 1.2. Estimate the true average percent of shrinkage for specimens of this type in a 98% confidence interval.

7.12 The breaking strength of threads has a standard deviation of 18 grams. How many measures on breaking strength should be used in the next experiment if the estimate of the mean breaking strength is to be within 4 grams of the true mean breaking strength, with confidence coefficient 0.90?

7.13 In the setting of Exercise 7.10, how many engineers should be sampled if it is desired to estimate the mean number of hours worked to within 0.5 hour with confidence coefficient 0.95?

7.14 Refer to Exercise 7.11. How many specimens should be tested if it is desired to estimate the percent of shrinkage to within 0.2 with confidence coefficient 0.98?

7.15 Upon testing 100 resistors manufactured by Company A, it is found that 12 fail to meet the tolerance specifications. Find a 95% confidence interval for the true fraction of resistors manufactured by Company A that fail to meet the tolerance specification. What assumptions are necessary for your answer to be valid?

7.16 Refer to Exercise 7.15. If it is desired to estimate the true proportion failing to meet tolerance specifications to within 0.05, with confidence coefficient 0.95, how many resistors should be tested?

7.17 Careful inspection of 70 precast concrete supports to be used in a construction project revealed 28 with hairline cracks. Estimate the true proportion of supports of this type with cracks in a 98% confidence interval.

7.18 Refer to Exercise 7.17. Suppose it is desired to estimate the true proportion of cracked supports to within 0.1, with confidence coefficient 0.98. How many supports should be sampled to achieve the desired accuracy?

7.19 In conducting an inventory and audit of parts in a certain stockroom, it was found that, for 60 items sampled, the audit value exceeded the book value on 45 items. Estimate, with confidence coefficient 0.90, the true fraction of items in the stockroom for which the audit value exceeds the book value.

7.20 The Environmental Protection Agency has collected data on the LC50 (concentration killing 50% of the test animals in a specified time interval) measurements for certain chemicals likely to be found in freshwater rivers and lakes. For a certain species of fish, the LC50 measurements (in parts per million) for DDT in 12 experiments yielded the following:

$$16, 5, 21, 19, 10, 5, 8, 2, 7, 2, 4, 9$$

Assuming such LC50 measurements to be approximately normally distributed, estimate the true mean LC50 for DDT with confidence coefficient 0.90.

7.21 The warpwise breaking strength measured on five specimens of a certain cloth gave a sample mean of 180 psi and a standard deviation of 5 psi. Estimate the true mean warpwise breaking strength for cloth of this type in a 95% confidence interval. What assumption is necessary for your answer to be valid?

7.22 Answer Exercise 7.21 if the same sample data resulted from a sample of

a   10 specimens.   **b**   100 specimens.

7.23 Fifteen resistors were randomly selected from the output of a process supposedly producing 10-ohm resistors. The 15 resistors actually showed a sample mean of 9.8 ohms and a sample standard deviation of 0.5 ohm. Find a 95% confidence interval for the true mean resistance of the resistors produced by this process. Assume resistance measurements are approximately normally distributed.

7.24 The variance of LC50 measurements is important, because it may reflect an ability (or inability) to reproduce similar results in identical experiments. Find a 95% confidence interval for σ^2, the true variance of the LC50 measurements for DDT, using the data in Exercise 7.20.

7.25 In producing resistors, variability of the resistances is an important quantity to study, as it reflects the stability of the manufacturing process. Estimate σ^2, the true variance of the resistance measurements, in a 90% confidence interval, if a sample of 15 resistors showed resistances with a standard deviation of 0.5 ohm.

7.26 The increased use of wood for residential heating has caused some concern over the pollutants and irritants released into the air. In one reported study in the northwestern United States, the total suspended particulates (TSP) in nine air samples had a mean of 72 μg/m^3 and a standard deviation of 23. Estimate the true mean TSP in a 95% confidence interval. What assumption is necessary for your answer to be valid? (See J. E. Core, J. A. Cooper, and R. M. Neulicht, *Journal of the Air Pollution Control Association*, 31, no. 2, 1984, pp. 138–143, for more details.)

7.27 The yield stress in steel bars (Grade FeB 400 HWL) reported by P. Booster (*Journal of Quality Technology*, 15, no. 4, 1983, p. 191) gave the following data for one test:

$$n = 150 \qquad \bar{x} = 477 \text{ N/mm}^2 \qquad s = 13$$

Estimate the true yield stress for bars of this grade and size in a 90% confidence interval. Are any assumptions necessary for your answer to be valid?

7.28 In the article quoted in Exercise 7.27, a sample of 12 measurements on the tensile strength of steel bars resulted in a mean of 585 N/mm^2 and a standard deviation of 38. Estimate the true mean tensile strength of bars of this size and type in a 95% confidence interval. List any assumptions you need to make.

7.29 Fatigue behavior of reinforced concrete beams in seawater was studied by T. Hodgkiess, et al. (*Materials Performance*, July 1984, pp. 27–29). The number of cycles to failure in seawater for beams subjected to certain bending and loading stress was as follows (in thousands):

$$774, 633, 477, 268, 407, 576, 659, 963, 193$$

Construct a 90% confidence interval estimate of the average number of cycles to failure for beams of this type.

7.30   Using the data given in Exercise 7.29, construct a 90% confidence interval for the variance of the number of cycles to failure for beams of this type. What assumptions are necessary for your answer to be valid?

7.31   The quantity

$$ z_{\alpha/2} \frac{\sigma}{\sqrt{n}} \left(\text{or } z_{\alpha/2} \sqrt{\frac{p(1-p)}{n}} \right) $$

used in constructing confidence intervals for μ (or p) is sometimes called the *sampling error*. A *Time* (5 April 1993) article on religion in America reported that 54% of those between the ages of 18 and 26 found religion to be a "very important" part of their lives. The article goes on to state that the result comes from a poll of 1,013 people and has a sampling error of 3%. How is the 3% calculated and what is its interpretation? Can we conclude that a majority of people in this age group find religion to be very important?

7.32   The results of a Louis Harris poll state that 36% of Americans list football as their favorite sport. A note then states: "In a sample of this size (1,091 adults) one can say with 95% certainty that the results are within plus or minus three percentage points of what they would be if the entire adult population had been polled." Do you agree? (Source: *Gainesville Sun*, 7 May 1968.)

7.33   A. C. Nielsen Co. has electronic monitors hooked up to about 1,200 of the 80 million American homes. The data from the monitors provide estimates of the proportion of homes tuned to a particular T.V. program. Nielsen offers the following defense of this sample size:

> Mix together 70,000 white beans and 30,000 red beans and then scoop out a sample of 1,000. The mathematical odds are that the number of red beans will be between 270 and 330, or 27 to 33 percent of the sample, which translates to a "rating" of 30, plus or minus three, with a 20 to 1 assurance of statistical reliability. The basic statistical law wouldn't change even if the sampling came from 80 million beans rather than 100,000 (*Sky*, October 1982).

Interpret and justify this statement in terms of the results of this chapter.

7.4   Confidence Intervals: The Multiple-Sample Case

All methods for confidence intervals discussed in the preceding section considered only the case in which a single random sample was selected to estimate a single parameter. In many problems, more than one population, and hence more than one sample, is involved. For example, we might want to estimate the difference between mean daily yields for two industrial processes for producing a certain liquid fertilizer. Or we might want to compare the rates of defectives produced on two or more assembly lines within a factory.

7.4.1   General Distribution: Large-Sample Confidence Interval for Linear Functions of Means

It is commonly of interest to compare two population means, and this is conveniently accomplished by estimating the *difference* between the two means. Suppose the two populations of interest have means and variances denoted by μ_1 and σ_1^2 and μ_2 and σ_2^2, respectively, with $\mu_1 - \mu_2$ the parameter to be estimated. A large sample of size n_1 from the first population has a mean of $\overline{X}_1$ and a variance of S_1^2. An *independent* large sample from the second population has a mean of $\overline{X}_2$ and a variance of S_2^2. If $\overline{X}_1 - \overline{X}_2$ is used to estimate $\mu_1 - \mu_2$, then

$$E(\overline{X}_1 - \overline{X}_2) = \mu_1 - \mu_2$$

(the estimator is unbiased) and

$$V(\overline{X}_1 - \overline{X}_2) = V(\overline{X}_1) + V(\overline{X}_2)$$
$$= \frac{\sigma_1^2}{n_1} + \frac{\sigma_2^2}{n_2}$$

Also, $\overline{X}_1 - \overline{X}_2$ will have approximately a normal sampling distribution. Thus, a confidence interval for $\mu_1 - \mu_2$ can be formed in the same manner as the one for μ in Section 7.3.1.

The interval

$$(\bar{x}_1 - \bar{x}_2) \pm z_{\alpha/2}\sqrt{\frac{\sigma_1^2}{n_1} + \frac{\sigma_2^2}{n_2}}$$

forms a realization of a large-sample confidence interval for $\mu_1 - \mu_2$ with confidence coefficient approximately $(1 - \alpha)$.

EXAMPLE   **7.8**   A farm-equipment manufacturer wants to compare the average daily downtime for two sheet-metal stamping machines located in two different factories. Investigation of company records for 100 randomly selected days on each of the two machines gave the following results:

$$n_1 = 100 \qquad\qquad n_2 = 100$$
$$\bar{x}_1 = 12 \text{ minutes} \qquad \bar{x}_2 = 9 \text{ minutes}$$
$$s_1^2 = 6 \qquad\qquad\quad s_2^2 = 4$$

Construct a 90% confidence interval estimate for $\mu_1 - \mu_2$.

Solution   Using the form of the interval given above, with s_i^2 estimating σ_i^2, we have

$$(\bar{x}_1 - \bar{x}_2) \pm z_{0.05}\sqrt{\frac{s_1^2}{n_1} + \frac{s_2^2}{n_2}}$$

yielding

$$(12 - 9) \pm (1.645)\sqrt{\frac{6}{100} + \frac{4}{100}}$$

or

$$3 \pm 0.52$$

That is, we are about 90% confident that $\mu_1 - \mu_2$ is between $3 - 0.52 = 2.48$ and $3 + 0.52 = 3.52$. This evidence suggests that μ_1 must be larger than μ_2.   ∎

Suppose we are interested in three populations, numbered 1, 2, and 3, with unknown means μ_1, μ_2, and μ_3, respectively, and variances σ_1^2, σ_2^2, and σ_3^2. If a random sample of size n_i is selected from the population with mean μ_i, $i = 1, 2, 3$, then any linear function of the form

$$\theta = a_1 \mu_1 + a_2 \mu_2 + a_3 \mu_3$$

can be estimated unbiasedly by

$$\hat{\theta} = a_1 \overline{X}_1 + a_2 \overline{X}_2 + a_3 \overline{X}_3$$

where $\overline{X}_i$ is the mean of the sample from population i. Also, if the samples are independent of one another,

$$V(\hat{\theta}) = a_1^2 V(\overline{X}_1) + a_2^2 V(\overline{X}_2) + a_3^2 V(\overline{X}_3)$$
$$= a_1^2 \left(\frac{\sigma_1^2}{n_1} \right) + a_2^2 \left(\frac{\sigma_2^2}{n_2} \right) + a_3^2 \left(\frac{\sigma_3^2}{n_3} \right)$$

and as long as all of the n_i's are reasonably large, $\hat{\theta}$ will have approximately a normal distribution.

The interval

$$\hat{\theta} \pm z_{\alpha/2} \sqrt{V(\hat{\theta})}$$

will provide a large-sample confidence interval for θ with confidence coefficient approximately $(1 - \alpha)$.

E X A M P L E   **7.9**   A company has three machines for stamping sheet metal located in three different factories around the country. If μ_i denotes the average downtime (in minutes) per day for the ith machine, $i = 1, 2$, and 3, then the expected daily cost for downtime on these three machines is

$$C = 3\mu_1 + 5\mu_2 + 2\mu_3$$

Investigation of company records for 100 randomly selected days on each of these machines showed the following:

$$
\begin{array}{lll}
n_1 = 100 & n_2 = 100 & n_2 = 100 \\
\bar{x}_1 = 12 & \bar{x}_2 = 9 & \bar{x}_3 = 14 \\
s_1^2 = 6 & s_2^2 = 4 & s_3^2 = 5
\end{array}
$$

Estimate C in a 95% confidence interval.

Solution   We have seen above that an unbiased estimator of C is

$$\hat{C} = 3\overline{X}_1 + 5\overline{X}_2 + 2\overline{X}_3$$

with variance

$$V(\hat{C}) = 9\left(\frac{\sigma_1^2}{100}\right) + 25\left(\frac{\sigma_2^2}{100}\right) + 4\left(\frac{\sigma_3^2}{100}\right)$$

We obtained an observed value of $\hat{C}$ equal to

$$3(12) + 5(9) + 2(14) = 109$$

Since the σ_i^2's are unknown, we can approximate them with s_i^2's and find an approximate variance of $\hat{C}$ to be

$$9\left(\frac{6}{100}\right) + 25\left(\frac{4}{100}\right) + 4\left(\frac{5}{100}\right) = 1.74$$

(This approximation works well only for large samples.) Thus, a realization of the 95% confidence interval for C is

$$109 \pm 1.96\sqrt{1.74}$$

or

$$109 \pm 2.58$$

We are about 95% confident that the mean daily cost is between \$106.42 and \$111.58.   ■

7.4.2   Binomial Distribution: Large-Sample Confidence Interval for Linear Functions of Proportions

Since we have seen that sample proportions behave like sample means and that for large samples the sample proportions will tend to have normal distributions, we can adapt the above methodology to include estimation of linear functions of sample proportions. Again, the most common practical problem concerns the estimation of the difference between two proportions. If Y_i has a binomial distribution with sample size n_i and probability of success p_i, then $p_1 - p_2$ can be estimated by

$Y_1/n_1 - Y_2/n_2$. After the samples are taken, we estimate p_1 by $\hat{p}_1 = y_1/n_1$ and p_2 by $\hat{p}_2 = y_2/n_2$.

The interval

$$(\hat{p}_1 - \hat{p}_2) \pm z_{\alpha/2}\sqrt{\frac{\hat{p}_1(1 - \hat{p}_1)}{n_1} + \frac{\hat{p}_2(1 - \hat{p}_2)}{n_2}}$$

forms a realization of a large-sample confidence interval for $p_1 - p_2$ with confidence coefficient approximately $(1 - \alpha)$, where $\hat{p}_1 = y_1/n_1$ and $\hat{p}_2 = y_2/n_2$.

EXAMPLE **7.10**   We want to compare the proportion of defective electric motors turned out by two shifts of workers. From the large number of motors produced in a given week, $n_1 = 50$ motors were selected from the output of shift I, and $n_2 = 40$ motors were selected from the output of shift II. The sample from shift I revealed four to be defective, and the sample from shift II showed six faulty motors. Estimate the true difference between proportions of defective motors produced in a 95% confidence interval.

Solution   Starting with

$$(\hat{p}_1 - \hat{p}_2) \pm 1.96\sqrt{\frac{\hat{p}_1(1 - \hat{p}_1)}{n_1} + \frac{\hat{p}_2(1 - \hat{p}_2)}{n_2}}$$

we have as an approximate 95% confidence interval

$$\left(\frac{4}{50} - \frac{6}{40}\right) \pm 1.96\sqrt{\frac{0.08(0.92)}{50} + \frac{0.15(0.85)}{40}}$$

or

$$-0.07 \pm 0.13$$

Since the interval contains zero, there does not appear to be any significant difference between the rates of defectives for the two shifts. That is, zero cannot be ruled out as a plausible value of the true difference between proportions of defective motors, which implies that these proportions may be equal.   ∎

More general linear functions of sample proportions are sometimes of interest, as is the case in the following example.

E X A M P L E   **7.11**   The personnel director of a large company wants to compare two different aptitude tests that are supposed to be equivalent in terms of which aptitudes they measure. He has reason to suspect that the degree of difference in test scores is not the same for women and men. Thus, he sets up an experiment in which 100 men and 100 women, of nearly equal aptitude, are selected to take the tests. Fifty men take test I, and 50 independently selected men take test II. Likewise, 50 women take test I, and 50 independently selected women take test II. The proportions, out of 50, receiving passing scores are recorded for each of the four groups, and the data are as follows:

	Test	
	I	II
Male	0.7	0.9
Female	0.8	0.9

If p_{ij} denotes the true probability of passing in row i and column j, the director wants to estimate $(p_{12} - p_{11}) - (p_{22} - p_{21})$ to see if the change in probability for males is different from the corresponding change for females. Estimate this quantity in a 95% confidence interval.

Solution   If $\hat{p}_{ij}$ denotes the observed proportion passing the test, out of 50, in row i and column j, then

$$(p_{12} - p_{11}) - (p_{22} - p_{21})$$

is estimated by

$$(\hat{p}_{12} - \hat{p}_{11}) - (\hat{p}_{22} - \hat{p}_{21})$$

which is observed to be

$$(0.9 - 0.7) - (0.9 - 0.8) = 0.1$$

Because of the independence of the samples, the estimated variance of $(\hat{p}_{12} - \hat{p}_{11}) - (\hat{p}_{22} - \hat{p}_{21})$ is

$$\frac{\hat{p}_{12}(1 - \hat{p}_{12})}{50} + \frac{\hat{p}_{11}(1 - \hat{p}_{11})}{50} + \frac{\hat{p}_{22}(1 - \hat{p}_{22})}{50} + \frac{\hat{p}_{21}(1 - \hat{p}_{21})}{50}$$

which is observed to be

$$\frac{1}{50}[(0.9)(0.1) + (0.7)(0.3) + (0.9)(0.1) + (0.8)(0.2)] = 0.011$$

The interval estimate is then

$$0.1 \pm 1.96\sqrt{0.011}$$

or

$$0.1 \pm 0.20$$

Since this interval contains zero, there is really no reason to suspect that the change in probability for males is different from the change in probability for females.   ■

7.4.3   Normal Distribution: Confidence Interval for Linear Functions of Means

We must make some additional assumptions for the estimation of linear functions of means in small samples (where the Central Limit Theorem does not ensure approximate normality). If we have k populations and k independent random samples, we assume that all populations have approximately *normal distributions with a common unknown variance* σ^2. This assumption should be considered carefully, and if it does not seem reasonable, the methods outlined below should not be employed. The arguments for producing confidence intervals parallel those given for the interval based on the t distribution in Section 7.3.3.

As in the large-sample case of Section 7.4.1, the most commonly occurring function of means that we might want to estimate is a simple difference of the form $\mu_1 - \mu_2$. Suppose two populations of interest are normally distributed with means μ_1 and μ_2 and common variance σ^2. Independent random samples from these populations yield sample statistics of $\overline{X}_1$ and S_1^2 from the first population and $\overline{X}_2$ and S_2^2 from the second.

The obvious unbiased estimator of $\mu_1 - \mu_2$ is still $\overline{X}_1 - \overline{X}_2$. Now

$$V(\overline{X}_1 - \overline{X}_2) = V(\overline{X}_1) + V(\overline{X}_2)$$
$$= \frac{\sigma^2}{n_1} + \frac{\sigma^2}{n_2} = \sigma^2\left(\frac{1}{n_1} + \frac{1}{n_2}\right)$$

because of the common variance assumption. Thus, we must find a good estimator of the *common* variance σ^2. This estimator will be a function of S_1^2 and S_2^2, but the unbiased estimator with the smallest variance is produced by weighting each S_i^2 according to its degrees of freedom. That is, the pooled sample variance

$$\frac{(n_1 - 1)S_1^2 + (n_2 - 1)S_2^2}{n_1 + n_2 - 2} = S_p^2$$

is a good estimator of σ^2. The quantity

$$T = \frac{(\overline{X}_1 - \overline{X}_2) - (\mu_1 - \mu_2)}{S_p\sqrt{\dfrac{1}{n_1} + \dfrac{1}{n_2}}}$$

has a t distribution with $n_1 + n_2 - 2$ degrees of freedom. This quantity can then be turned into a confidence interval for $(\mu_1 - \mu_2)$ by finding $t_{\alpha/2}$ such that

$$P(-t_{\alpha/2} \le T \le t_{\alpha/2}) = 1 - \alpha$$

and reworking the inequality. We have

$$
P\left(-t_{\alpha/2} \le \frac{(\overline{X}_1 - \overline{X}_2) - (\mu_1 - \mu_2)}{S_p\sqrt{\dfrac{1}{n_1} + \dfrac{1}{n_2}}} \le t_{\alpha/2} \right) = 1 - \alpha
$$

$$
P\left(-t_{\alpha/2}S_p\sqrt{\dfrac{1}{n_1} + \dfrac{1}{n_2}} \le (\overline{X}_1 - \overline{X}_2) - (\mu_1 - \mu_2) \right.
$$

$$
\left. \le t_{\alpha/2}S_p\sqrt{\dfrac{1}{n_1} + \dfrac{1}{n_2}} \right) = 1 - \alpha
$$

or

$$
P\left((\overline{X}_1 - \overline{X}_2) - t_{\alpha/2}S_p\sqrt{\dfrac{1}{n_1} + \dfrac{1}{n_2}} \le \mu_1 - \mu_2 \le (\overline{X}_1 - \overline{X}_2) \right.
$$

$$
\left. + t_{\alpha/2}S_p\sqrt{\dfrac{1}{n_1} + \dfrac{1}{n_2}} \right) = 1 - \alpha
$$

which yields the confidence interval for $\mu_1 - \mu_2$.

> If both random samples come from independent normal distributions with common variance, then the interval
>
> $$
> (\bar{x}_1 - \bar{x}_2) \pm t_{\alpha/2}s_p\sqrt{\dfrac{1}{n_1} + \dfrac{1}{n_2}}
> $$
>
> provides a realization of an exact confidence interval for $\mu_1 - \mu_2$, with confidence coefficient $(1 - \alpha)$.

EXAMPLE 7.12 Copper produced by sintering (heating without melting) a powder under certain conditions is then measured for porosity (the volume fraction due to voids) in a certain laboratory. A sample of $n_1 = 4$ independent porosity measurements shows a mean of $\bar{x}_1 = 0.22$ and a variance of $s_1^2 = 0.0010$. A second laboratory repeats the same process on an identical powder and gets $n_2 = 5$ independent porosity measurements with $\bar{x}_2 = 0.17$ and $s_2^2 = 0.0020$. Estimate the true difference between the population means $(\mu_1 - \mu_2)$ for these two laboratories, with confidence coefficient 0.95.

Solution First, we assume that the population of porosity measurements in either laboratory could be modeled by a normal distribution and that the population variances are approximately equal. Then a confidence interval based on the t distribution can be used, with

$$s_p^2 = \frac{(n_1 - 1)s_1^2 + (n_2 - 1)s_2^2}{n_1 + n_2 - 2}$$

$$= \frac{3(0.001) + 4(0.002)}{7} = 0.0016$$

or

$$s_p = 0.04$$

Using

$$(\bar{x}_1 - \bar{x}_2) \pm t_{\alpha/2} s_p \sqrt{\frac{1}{n_1} + \frac{1}{n_2}}$$

we have, with $\alpha/2 = 0.025$ and 7 degrees of freedom,

$$(0.22 - 0.17) \pm (2.365)(0.04)\sqrt{\frac{1}{4} + \frac{1}{5}}$$

or

$$0.05 \pm 0.06$$

Since the interval contains zero, we would say that there is not much evidence of any difference between the two population means. That is, zero is a plausible value for $\mu_1 - \mu_2$, and if $\mu_1 - \mu_2 = 0$, then $\mu_1 = \mu_2$. ∎

To see how a statistical computer package (Minitab in this case) presents the output for a two-sample confidence interval, consider the following classical example of the importance of statistical analysis.

EXAMPLE   **7.13**   Lord Rayleigh was one of the earliest scientists to study the density of nitrogen. In his studies, he noticed something peculiar. The nitrogen densities produced from chemical compounds tended to be smaller than the densities of nitrogen produced from the air. He was working with fairly small samples, but was he correct in his conjecture? (Lord Rayleigh's measurements are given in *Proceedings, Royal Society* (London), 55, 1894, pp. 340–344.)

Solution   Lord Rayleigh's data for chemical and atmospheric measurements are given below, along with the Minitab printout of a two-sample confidence interval with pooled variance estimate. The units here are the mass of nitrogen filling a certain flask under specified pressure and temperature.

Chemical	Atmospheric
2.30143	2.31017
2.29890	2.30986
2.29816	2.31010
2.30182	2.31001
2.29869	2.31024
2.29940	2.31010
2.29849	2.31028
2.29889	2.31163
2.30074	2.30956
2.30054	

```
TWOSAMPLE   T   FOR ATMOSPHE VS CHEMICAL

               N                         MEAN    STDEV   SE MEAN
ATMOSPHE      9                        2.310217  0.000574   0.00019
CHEMICAL     10                        2.29971   0.00131    0.00042
95 PCT CI FOR MU ATMOSPHE  -  MU CHEMICAL: (0.00951, 0.01151)
POOLED STDEV = 0.00103
```

The observed confidence interval is on the positive side of zero, indicating that the mean for atmospheric measurements seems to be larger than the mean for chemical measurements. Rather than suppressing this difference, Lord Rayleigh emphasized it, and that emphasis led to the discovery of the inert gases in the atmosphere. ∎

Suppose we have three populations of interest, numbered 1, 2, and 3, all assumed to be approximately normal in distribution. The populations have means μ_1, μ_2, and μ_3, respectively, and common variance σ^2. We want to estimate a linear function of the form

$$\theta = a_1\mu_1 + a_2\mu_2 + a_3\mu_3$$

A good estimator is

$$\hat{\theta} = a_1\overline{X}_1 + a_2\overline{X}_2 + a_3\overline{X}_3$$

where $\overline{X}_i$ is the mean of the random sample from population i. Now

$$V(\hat{\theta}) = a_1^2 V(\overline{X}_1) + a_2^2 V(\overline{X}_2) + a_3^2 V(\overline{X}_3)$$
$$= \sigma^2 \left(\frac{a_1^2}{n_1} + \frac{a_2^2}{n_2} + \frac{a_3^2}{n_3} \right)$$

The problem now is to choose the best estimator of the common variance σ^2 from the three sample variances S_i^2, based on respective sample sizes n_1, n_2, and n_3. It can be shown that an unbiased and minimum variance estimator of σ^2 is

$$\frac{\sum_{i=1}^{3}(n_i - 1)S_i^2}{\sum_{i=1}^{3}(n_i - 1)} = S_p^2$$

(The subscript p denotes "pooled.") Since $\hat{\theta}$ will have a normal distribution and $\sum_{i=1}^{3}(n_i - 1)S_p^2/\sigma^2$ will have a χ^2 distribution with $\sum_{i=1}^{3}(n_i - 1)$ degrees of freedom,

$$T = \frac{\hat{\theta} - \theta}{S_p\sqrt{\dfrac{a_1^2}{n_1} + \dfrac{a_2^2}{n_2} + \dfrac{a_3^2}{n_3}}}$$

will have a t distribution with $\sum_{i=1}^{3}(n_i - 1)$ degrees of freedom. We can then use this statistic to derive a confidence interval for θ.

If all random samples come from normal distributions, then the interval

$$\hat{\theta} \pm t_{\alpha/2} S_p\sqrt{\frac{a_1^2}{n_1} + \frac{a_2^2}{n_2} + \frac{a_3^2}{n_3}}$$

provides an exact confidence interval for θ, with confidence coefficient $(1 - \alpha)$. Here $\theta = a_1\mu_1 + a_2\mu_2 + a_3\mu_3$ and $\hat{\theta} = a_1\overline{X}_1 + a_2\overline{X}_2 + a_3\overline{X}_3$.

E X A M P L E **7.14** Refer to Example 7.12. A third laboratory sinters a slightly different copper powder and then takes $n_3 = 10$ independent observations on the porosity. These yield a mean of $\bar{x}_3 = 0.12$ and a variance of $s_3^2 = 0.0018$. Compare the population mean for this powder with the mean for the type of powder used in Example 7.12. (Estimate the difference in a 95% confidence interval.)

Solution Again, we assume normality for the population of measurements that could be obtained for this third laboratory, and we assume that the variances for these porosity measurements would be approximately equal for all three laboratories. Since the first two samples are conducted with the same type of powder, we assume that they have a common population mean μ. We estimate this parameter by using a weighted average of the first two sample means, namely,

$$\frac{n_1\bar{x}_1 + n_2\bar{x}_2}{n_1 + n_2}$$

This weighted average is unbiased and will have a smaller variance than the simple average $(\bar{x}_1 + \bar{x}_2)/2$. To estimate $\mu - \mu_3$, we take

$$\frac{n_1}{n_1 + n_2}\bar{x}_1 + \frac{n_2}{n_1 + n_2}\bar{x}_2 - \bar{x}_3 = \frac{4}{9}(0.22) + \frac{5}{9}(0.17) - 0.12 = 0.0722$$

The pooled estimate of σ^2 is

$$s_p^2 = \frac{\sum_{i=1}^{3}(n_i - 1)s_i^2}{\sum_{i=1}^{3}(n_i - 1)} = \frac{3(0.0010) + 4(0.0020) + 9(0.0018)}{16}$$

$$= 0.0017$$

and $s_p = 0.0412$.

Now the confidence interval estimate, with $\alpha/2 = 0.025$ and 16 degrees of freedom, is

$$\left(\frac{4}{9}\bar{x}_1 + \frac{5}{9}\bar{x}_2 - \bar{x}_3\right) \pm t_{\alpha/2}s_p\sqrt{\frac{(4/9)^2}{n_1} + \frac{(5/9)^2}{n_2} + \frac{(-1)^2}{n_3}}$$

yielding

$$0.0722 \pm (2.120)(0.0412)\sqrt{\frac{(4/9)^2}{4} + \frac{(5/9)^2}{5} + \frac{1}{10}}$$

or

$$0.0722 \pm 0.0401$$

This interval would suggest that the second type of powder seems to give a lower mean porosity than the first, since the interval does *not* contain zero. ∎

7.4.4   Normal Distribution: Confidence Interval for σ_2^2/σ_1^2

When two samples from normally distributed populations are available in an experimental situation, it is sometimes of interest to compare the population variances. We generally do this by estimating the *ratio* of the population variances, σ_1^2/σ_2^2.

Suppose two independent random samples from normal distributions with respective sizes n_1 and n_2 yield sample variances of S_1^2 and S_2^2. If the populations in question have variances σ_1^2 and σ_2^2, respectively, then

$$F = \frac{S_1^2/\sigma_1^2}{S_2^2/\sigma_2^2}$$

has a known sampling distribution, called an F *distribution*, with $\nu_1 = n_1 - 1$ and $\nu_2 = n_2 - 1$ degrees of freedom.

A probability density function of a random variable having an F distribution is shown in Figure 7.4. The shape of the distribution depends on both sets of degrees of freedom, ν_1 and ν_2. Values of an F variate that cut off right-hand tail areas of size α

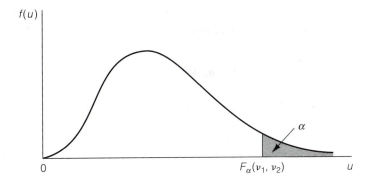

are given in Tables 7 and 8 of the Appendix for $\alpha = 0.05$ and $\alpha = 0.01$, respectively. Left-hand tail values of F can be found by the equation

$$F_{1-\alpha}(v_1, v_2) = \frac{1}{F_{\alpha}(v_2, v_1)}$$

where the subscripts still denote right-hand areas.

Knowledge of the F distribution allows us to construct a confidence interval for the ratio of two variances. We can find two values $F_{1-\alpha/2}(v_1, v_2)$ and $F_{\alpha/2}(v_1, v_2)$ so that

$$P(F_{1-\alpha/2}(v_1, v_2) \leq F \leq F_{\alpha/2}(v_1, v_2)) = 1 - \alpha$$

which yields

$$P\left(F_{1-\alpha/2}(v_1, v_2) \leq \frac{S_1^2}{S_2^2} \cdot \frac{\sigma_2^2}{\sigma_1^2} \leq F_{\alpha/2}(v_1, v_2)\right) = 1 - \alpha$$

$$P\left(\frac{S_2^2}{S_1^2}F_{1-\alpha/2}(v_1, v_2) \leq \frac{\sigma_2^2}{\sigma_1^2} \leq \frac{S_2^2}{S_1^2}F_{\alpha/2}(v_1, v_2)\right) = 1 - \alpha$$

Upon using the identity given above for lower-tailed (left-hand) F values, the confidence interval for σ_2^2/σ_1^2 is established.

If both random samples come from normal distributions, then the interval

$$\left[\frac{s_2^2}{s_1^2}\frac{1}{F_{\alpha/2}(v_2, v_1)}, \; \frac{s_2^2}{s_1^2}F_{\alpha/2}(v_1, v_2)\right]$$

provides a realization of an exact confidence interval for σ_2^2/σ_1^2, with confidence coefficient $(1 - \alpha)$.

EXAMPLE   **7.15**   A random sample of $n_1 = 10$ observations on breaking strengths of a type of glass gave $s_1^2 = 2.31$ (measurements were made in pounds per square inch). An independent random sample of $n_2 = 16$ measurements on a second machine, but with the same kind of glass, gave $s_2^2 = 3.68$. Estimate the true variance ratio, σ_2^2/σ_1^2, in a 90% confidence interval.

Solution   It is, of course, necessary to assume that the measurements in both populations can be modeled by normal distributions. Then the above confidence interval will yield

$$\left[\frac{s_2^2}{s_1^2} \frac{1}{F_{0.05}(15, 9)}, \frac{s_2^2}{s_1^2} F_{0.05}(9, 15) \right]$$

or, using Table 7 of the Appendix,

$$\left[\frac{3.68}{2.31} \left(\frac{1}{3.01} \right), \frac{3.68}{2.31}(2.59) \right]$$

or

$$(0.53, 4.13)$$

Thus, we are about 90% confident that the ratio of σ_2^2 to σ_1^2 is between 0.53 and 4.13. Equality of σ_2^2 and σ_1^2 would imply a ratio of 1.00. Since our confidence interval includes 1.00, there is no real evidence that σ_2^2 differs from σ_1^2.   ∎

Confidence Intervals from Two Independent Samples

General Distribution: Large-sample confidence interval for $\mu_1 - \mu_2$:

$$(\bar{x}_1 - \bar{x}_2) \pm z_{\alpha/2} \sqrt{\frac{\sigma_1^2}{n_1} + \frac{\sigma_2^2}{n_2}}$$

(σ_1 and σ_2 can be estimated by s_1 and s_2, respectively.)

Binomial Distribution: Large-sample confidence interval for $p_1 - p_2$:

$$(\hat{p}_1 - \hat{p}_2) \pm z_{\alpha/2} \sqrt{\frac{\hat{p}_1(1 - \hat{p}_1)}{n_1} + \frac{\hat{p}_2(1 - \hat{p}_2)}{n_2}}$$

where

$$\hat{p}_1 = y_1/n_1 \quad \text{and} \quad \hat{p}_2 = y_2/n_2$$

Normal Distribution with Common Variance: Confidence interval for $\mu_1 - \mu_2$:

$$(\bar{x}_1 - \bar{x}_2) \pm t_{\alpha/2} s_p \sqrt{\frac{1}{n_1} + \frac{1}{n_2}}$$

where

$$s_p^2 = \frac{(n_1 - 1)s_1^2 + (n_2 - 1)s_2^2}{n_1 + n_2 - 2}$$

and $t_{\alpha/2}$ depends on $n_1 + n_2 - 2$ degrees of freedom.

Normal Distribution: Confidence interval for σ_2^2 / σ_1^2:

$$\left(\frac{s_2^2}{s_1^2} \frac{1}{F_{\alpha/2}(v_2, v_1)}, \frac{s_2^2}{s_1^2} F_{\alpha/2}(v_1, v_2) \right)$$

where

$$v_1 = n_1 - 1 \quad \text{and} \quad v_2 = n_2 - 1$$

Exercises

7.34 The abrasive resistance of rubber is increased by adding a silica filler and a coupling agent to chemically bond the filler to the rubber polymer chains. Fifty specimens of rubber made with a type I coupling agent gave a mean resistance measure of 92, the variance of the measurements being 20. Forty specimens of rubber made with a type II coupling agent gave a mean of 98 and a variance of 30 on resistance measurements. Estimate the true difference between mean resistances to abrasion in a 95% confidence interval.

7.35 Refer to Exercise 7.34. Suppose a similar experiment is to be run again with an equal number of specimens from each type of coupling agent. How many specimens should be used if we want to estimate the true difference between mean resistances to within 1 unit, with a confidence coefficient of 0.95?

7.36 Two different types of coating for pipes are to be compared with respect to their ability to aid in resistance to corrosion. The amount of corrosion on a pipe specimen is quantified by measuring the maximum pit depth. For coating A, 35 specimens showed an average maximum pit depth of 0.18 cm. The standard deviation of these maximum pit depths was 0.02 cm. For coating B, the maximum pit depths in 30 specimens had a mean of 0.21 cm and a standard deviation of 0.03 cm. Estimate the true difference between mean depths in a 90% confidence interval. Do you think coating B does a better job of inhibiting corrosion?

7.37 Bacteria in water samples are sometimes difficult to count, but their presence can easily be detected by culturing. In 50 independently selected water samples from a certain lake, 43 contained certain harmful bacteria. After adding a chemical to the lake water, another 50 water samples showed only 22 with the harmful bacteria. Estimate the true difference between the proportions of samples containing the harmful bacteria with a 95% confidence coefficient. Does the chemical appear to be effective in reducing the amount of bacteria?

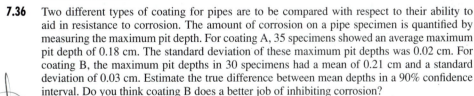

7.38 In studying the proportion of water samples containing harmful bacteria, how many samples should be selected before and after a chemical is added if we want to estimate the true difference between proportions to within 0.1 with a 95% confidence coefficient? (Assume the sample sizes are to be equal.)

7.39 A large firm made up of several companies has instituted a new quality-control inspection policy. Among 30 artisans sampled in Company A, only 5 objected to the new policy. Among 35 artisans sampled in Company B, 10 objected to the policy. Estimate the true difference between the proportions voicing *no* objection to the new policy for the two companies, with confidence coefficient 0.98.

7.40 A measurement of physiological activity important to runners is the rate of oxygen consumption. *Research Quarterly*, May 1979, reports on the differences between oxygen consumption rates for college males trained by two different methods, one involving continuous training for a period of time each day and the other involving intermittent training of about the same overall duration. The sample sizes, means, and standard deviations are as follows (measurements in ml/kg · min.):

Continuous Training	Intermittent Training
$n_1 = 9$	$n_2 = 7$
$\bar{y}_1 = 43.71$	$\bar{y}_2 = 39.63$
$s_1 = 5.88$	$s_2 = 7.68$

If the measurements are assumed to come from normally distributed populations, estimate the difference between the population means with confidence coefficient 0.95. Also find a 90% confidence interval for the ratio of the true variances for the two training methods. Does it appear that intermittent training gives more variable results?

7.41 For a certain species of fish, the LC50 measurements (in parts per million) for DDT in 12 experiments were as follows, according to the EPA:

$$16, 5, 21, 19, 10, 5, 8, 2, 7, 2, 4, 9$$

Another common insecticide, Diazinon, gave LC50 measurements of 7.8, 1.6, and 1.3 in three independent experiments. Estimate the difference between the mean LC50 for DDT and the mean LC50 for Diazinon in a 90% confidence interval. What assumptions are necessary for your answer to be valid? Also estimate the true variance ratio in a 90% confidence interval.

7.42 *Research Quarterly*, May 1979, reports on a study of impulses applied to the ball by tennis rackets of various construction. Three measurements on ball impulses were taken on each type of racket. For a Classic (wood) racket, the mean was 2.41 and the standard deviation was 0.02. For a Yamaha (graphite) racket, the mean was 2.22 and the standard deviation was 0.07. Estimate the difference between true mean impulses for the two rackets with confidence coefficient 0.95. What assumptions are necessary for the method used to be valid?

7.43 Seasonal ranges (in hectares) for alligators were monitored on a lake outside Gainesville, Florida, by biologists from the Florida Game and Fish Commission. Six alligators monitored in the spring showed ranges of 8.0, 12.1, 8.1, 18.1, 18.2, and 31.7. Four different alligators monitored in the summer had ranges of 102.0, 81.7, 54.7, and 50.7. Estimate the difference between mean spring and summer ranges on a 95% confidence interval, assuming the data to come from normally distributed populations with common variance.

7.44 Unaltered, altered, and partly altered bitumens are found in carbonate-hosted lead-zinc deposits and may aid in the production of sulfide necessary to precipitate ore bodies in carbonate rocks. (See T. G. Powell, *Science*, 6 April 1984, p. 63.) The atomic hydrogen/carbon (H/C) ratios for 15 samples of altered bitumen had a mean of 1.02 and a standard deviation of 0.04. The ratios for 7 samples of partly altered bitumen had a mean of 1.16 and a standard deviation of 0.05. Estimate the difference between true mean H/C ratios for altered and partly altered bitumen, in a 98% confidence interval. Assume equal population standard deviations.

7.45   One-hour carbon monoxide concentrations in 45 air samples from a section of a city showed an average of 11.6 ppm and a variance of 82.4. After a traffic control strategy was put into place, 19 air samples showed an average carbon monoxide concentration of 6.2 ppm and a variance of 38.1. Estimate the true difference in average carbon monoxide concentrations in a 95% confidence interval. What assumptions are necessary for your answer to be valid? (See T. Zamurs, *Air Pollution Control Association Journal*, 34, no. 6, 1984, p. 637, for details of the measurements.)

7.46   The total amount of hydrogen chloride (HCl) in columns above an altitude of 12 kilometers was measured at selected sites over the El Chichon volcano both before and after an eruption. (See W. G. Mankin and M. T. Coffey, *Science*, 12 October 1984, p. 170.) Before-and-after data for one location were as follows (units are 10^{15} molecules per square centimeter):

Preeruption	Posteruption
$n_1 = 40$	$n_2 = 20$
$\bar{x}_1 = 1.26$	$\bar{x}_2 = 1.40$
$s_1 = 0.14$	$s_2 = 0.04$

Estimate the difference between mean HCl amounts with a confidence coefficient of 0.98.

7.47   Acid gases must be removed from other refinery gases in chemical production facilities in order to minimize corrosion of the plants. Two methods for removing acid gases produced the corrosion rates (in mm/yr) listed below in experimental tests:

Method A: 0.3, 0.7, 0.5, 0.8, 0.9, 0.7, 0.8

Method B: 0.7, 0.8, 0.7, 0.6, 2.1, 0.6, 1.4, 2.3

Estimate the difference in mean corrosion rates for the two methods, using a confidence coefficient of 0.90. What assumptions must you make for your answer to be valid? (The data come from A. J. Kosseim, et al., *Chemical Engineering Progress*, 80, 1984, p. 64.)

7.48   Using the data presented in Exercise 7.47, estimate the ratio of variances for the two methods of removing acid gases in a 90% confidence interval.

7.49   The number of cycles to failure for reinforced concrete beams was measured in seawater and in air. The data (in thousands) are as follows:

Seawater: 774, 633, 477, 268, 407, 576, 659, 963, 193

Air: 734, 571, 520, 792, 773, 276, 411, 500, 672

Estimate the difference between mean cycles to failure for seawater and air, using a confidence coefficient of 0.95. Does seawater seem to lessen the number of cycles to failure? (See T. Hodgkiess, et al., *Materials Performance*, July 1984, pp. 27–29.)

7.50   The yield stresses were studied for two sizes of steel rods, with results as follows:

10-mm-diameter Rods	14-mm-diameter Rods
$n_1 = 51$	$n_2 = 12$
$\bar{x}_1 = 485 \text{ N/mm}^2$	$\bar{x}_2 = 499 \text{ N/mm}^2$
$s_1 = 17.2$	$s_2 = 26.7$

Estimate the average increase in yield stress of the 14-mm rod over the 10-mm rod, with a confidence coefficient of 0.90.

7.51 Silicon wafers are scored and then broken into the many small microchips that will be mounted into circuits. Two breaking methods are being compared. Out of 400 microchips broken by method A, 32 are unusable because of faulty breaks. Out of 400 microchips broken by method B, only 28 are unusable. Estimate the difference between proportions of improperly broken microchips for the two breaking methods. Use a confidence coefficient of 0.95. Which method of breaking would you recommend?

7.52 Time-Yankelovich surveys, regularly seen in the news magazine *Time*, report on telephone surveys of approximately 1,000 respondents. In December 1983, 60% of the respondents said that they worry about nuclear war. In a similar survey in June 1983, only 50% said that they worry about nuclear war. (See *Time*, 2 January 1984, p. 51.) The article reporting these figures says that when they are compared "the potential sampling error is plus or minus 4.5%." Explain how the 4.5% is obtained and what it means. Then estimate the true difference in these proportions in a 95% confidence interval.

7.53 An electric circuit contains three resistors, each of a different type. Tests on 10 type I resistors showed a sample mean resistance of 9.1 ohms with a sample standard deviation of 0.2 ohm, tests on 8 type II resistors yielded a sample mean of 14.3 ohms and a sample standard deviation of 0.4 ohm, while tests on 12 type III resistors yielded a sample mean of 5.6 ohms and a sample standard deviation of 0.1 ohm. Find a 95% confidence interval for $\mu_I + \mu_{II} + \mu_{III}$, the expected resistance for the circuit. What assumptions are necessary for your answer to be valid?

7.5 Prediction Intervals

Previous sections of this chapter have considered the problem of *estimating* parameters, or population constants. A similar problem involves *predicting* a value for a future observation of a random variable. Given a set of n lifelength measurements on components of a certain type, it may be of interest to form an interval (a prediction interval) in which we think the next observation on the lifelength of a similar component is quite likely to lie.

For independent random variables having a common normal distribution, a prediction interval is easily derived using the t distribution. Suppose we are to observe n independent, identically distributed normal random variables and then use the information to predict where X_{n+1} might lie. We know from previous discussions that $\overline{X}_n - X_{n+1}$ is normally distributed with mean zero and variance $(\sigma^2/n) + \sigma^2 = \sigma^2[1 + (1/n)]$. If we use the variables $X_1, \ldots, X_n$ to calculate S^2 as an estimator of σ^2, then

$$T = \frac{\overline{X}_n - X_{n+1}}{S\sqrt{1 + \frac{1}{n}}}$$

will have a t distribution with $(n - 1)$ degrees of freedom. Thus,

$$1 - \alpha = P\left(-t_{\alpha/2} \le \frac{\overline{X}_n - X_{n+1}}{S\sqrt{1 + \frac{1}{n}}} \le t_{\alpha/2}\right)$$

$$= P\left(\overline{X}_n - t_{\alpha/2}S\sqrt{1 + \frac{1}{n}} \le X_{n+1} \le \overline{X}_n + t_{\alpha/2}S\sqrt{1 + \frac{1}{n}}\right)$$

The interval

$$\bar{x}_n \pm t_{\alpha/2} s \sqrt{1 + \frac{1}{n}}$$

forms a realization of a prediction interval for X_{n+1}, with coverage probability $(1 - \alpha)$.

E X A M P L E   **7.16**   Ten independent observations are taken on bottles coming off a machine designed to fill them to 16 ounces. The $n = 10$ observations show a mean of $\bar{X} - 16.1$ ounces and a standard deviation of $s = 0.01$ ounce. Find a 95% prediction interval for the ounces of fill in the next bottle to be observed.

Solution   Assuming a normal probability model for ounces of fill, the interval

$$\bar{x} \pm t_{0.025} s \sqrt{1 + \frac{1}{n}}$$

will provide the answer. We obtain

$$16.1 \pm (2.262)(0.01)\sqrt{1 + \frac{1}{10}}$$

with $t_{0.025}$ based on 9 degrees of freedom, which yields

$$16.1 \pm 0.024$$

or

$$(16.076, 16.124)$$

We are about 95% confident that the next observation will lie between 16.076 and 16.124.   ∎

Note that the prediction interval will *not* become arbitrarily small with increasing n. In fact, if n is very large, the term $1/n$ might be ignored, resulting in an interval of the form $\bar{x} \pm t_{\alpha/2} s$. There will be a certain amount of error in predicting the value of a random variable no matter how much information is available to estimate μ and σ^2.

This type of prediction interval is sometimes used as the basis for control charts when "sampling by variables." That is, if n observations on some important variable are made while a process is in control, then the next observation should lie in the interval $\bar{x} \pm t_{\alpha/2} s \sqrt{1 + 1/n}$. If it doesn't, then there may be some reason to suspect that the process has gone out of control. (Usually, more than one observation would have to be observed outside the appropriate interval before an investigation into causes of the abnormality would be launched.)

Exercises

7.54 In studying the properties of a certain type of resistor, the actual resistances produced were measured on a sample of 15 such resistors. These resistances had a mean of 9.8 ohms and a standard deviation of 0.5 ohm. One resistor of this type is to be used in a circuit. Find a 95% prediction interval for the resistance it will produce. What assumption is necessary for your answer to be valid?

7.55 For three tests with a certain species of freshwater fish, the LC50 measurements for Diazinon, a common insecticide, were 7.8, 1.6, and 1.3 (in parts per million). Assuming normality of the LC50 measurements, construct a 90% prediction interval for the LC50 measurement of the next test with Diazinon.

7.56 It is extremely important for a business firm to be able to predict the amount of downtime for its computer system over the next month. A study of the past five months has shown the downtime to have a mean of 42 hours and a standard deviation of 3 hours. Predict the downtime for next month in a 95% prediction interval. What assumptions are necessary for your answer to be valid?

7.57 In designing concrete structures, the most important property of the concrete is its compressive strength. Six tested concrete beams showed compressive strengths (in thousand psi) of 3.9, 3.8, 4.4, 4.2, 3.8, and 5.4. Another beam of this type is to be used in a construction project. Predict its compressive strength in a 90% prediction interval. Assume compressive strengths to be approximately normally distributed.

7.58 A particular model of subcompact automobile has been tested for gas mileage 50 times. These mileage figures have a mean of 39.4 mpg and a standard deviation of 2.6 mpg. Predict the gas mileage to be obtained on the next test, with $1 - \alpha = 0.90$.

7.6 Tolerance Intervals

As we saw in Chapter 6, control charts are built around the idea of controlling the center (mean) and variation (variance) of a process variable such as tensile strength of wire or volume of soft drink put into cans. In addition to information on these selected parameters, engineers and technicians often want to determine an interval that contains a certain (high) proportion of the measurements, with a known probability. For example, in the production of resistors it is helpful to have information of the form "we are 95% confident that 90% of the resistances produced are between 490 and 510 ohms." Such an interval is called a *tolerance interval* for 90% of the population with confidence coefficient $1 - \alpha = 0.95$. The lower end of the interval (490, in this case) is called the *lower tolerance limit*, and the upper end (510, in this case) is called the *upper tolerance limit*. We still use the term "confidence coefficient" to reflect the confidence we have in a particular method producing an interval that includes the prescribed portion of the population. That is, a method for constructing 95% tolerance intervals for 90% of a population should produce intervals that contain approximately 90% of the population measurements in 95 out of every 100 times it is used.

Tolerance intervals are sometimes specified by a design engineer or production supervisor, in which case measurements must be taken from time to time to see if the specification is being met. More often, measurements on a process are taken first, and then tolerance limits are calculated to describe how the process is operating. These "natural" tolerance limits are often made available to customers.

Suppose the key measurements (resistance, for example) on a process have been collected and studied over a long period of time and are known to have a normal distribution with mean μ and standard deviation σ. We can then easily construct a tolerance interval for 90% of the population, since the interval

$$(\mu - 1.645\sigma, \ \mu + 1.645\sigma)$$

must contain precisely 90% of the population measurements. The confidence coefficient here is 1.0, since we have no doubt that exactly 90% of the measurements fall in the resulting interval. In most cases, however, μ and σ are not known and must be estimated by $\bar{x}$ and s, which are calculated from sample data. The interval

$$(\bar{x} - 1.645s, \ \bar{x} + 1.645s)$$

will no longer cover exactly 90% of the population because of the errors introduced by the estimators. However, it is possible to find a number K such that

$$(\bar{x} - Ks, \ \bar{x} + Ks)$$

has the property that it covers a proportion δ of the population with confidence coefficient $(1 - \alpha)$. Table 19 in the Appendix gives values of K for three values of δ and two values of $(1 - \alpha)$.

E X A M P L E   **7.17**   A random sample of 45 resistors was tested with a resulting average resistance produced of 498 ohms. The standard deviation of these resistances was 4 ohms. Find a 95% tolerance interval for 90% of the resistance measurements in the population from which this sample was selected. Assume a normal distribution for this population.

Solution   For these data, $n = 45$, $\bar{x} = 498$, and $s = 4$. From Table 19 with $\delta = 0.90$ and $1 - \alpha = 0.95$, we have $K = 2.021$. Thus,

$$(\bar{x} - 2.021s, \ \bar{x} + 2.021s)$$

or

$$[498 - 2.021(4), \ 498 + 2.021(4)]$$

or

$$(489.9, \ 506.1)$$

forms the realization of the 95% tolerance interval. We are 95% confident that 90% of the population resistances lie between 489.9 and 506.1 ohms. (In constructing such intervals, the lower tolerance limit should be rounded down and the upper tolerance limit rounded up.)

Notice that, as n increases, the values of K tend toward the corresponding $z_{\alpha/2}$ values. Thus, the tolerance intervals $\bar{x} \pm Ks$ tend toward the intervals $\mu \pm z_{\alpha/2}\sigma$ as n gets large.   ∎

The tolerance interval formulation discussed in previous paragraphs requires an assumption of normality for the data on which the interval is based. We now present a simple nonparametric technique for constructing tolerance intervals for any continuous distribution. Suppose a random sample of measurements $X_1, X_2, \ldots, X_n$ is ordered from smallest to largest to produce an ordered set $Y_1, Y_2, \ldots, Y_n$. That is, $Y_1 = \min(X_1, \ldots, X_n)$ and $Y_n = \max(X_1, \ldots, X_n)$. It can then be shown that

$$P[(Y_1, Y_n) \text{ covers at least } \delta \text{ of the population}] = 1 - n\delta^{n-1} + (n-1)\delta^n$$

Thus, the interval (Y_1, Y_n) is a tolerance interval for a proportion δ of the population with confidence coefficient $1 - n\delta^{n-1} + (n-1)\delta^n$. For a specified δ, the confidence coefficient is determined by the sample size. We illustrate the use of nonparametric tolerance intervals in the following examples.

E X A M P L E   **7.18**   A random sample of $n = 50$ measurements on the lifelength of a certain type of L.E.D. display showed the minimum to be $y_1 = 2{,}150$ hours and the maximum to be $y_{50} = 2{,}610$ hours. If (y_1, y_{50}) is used as a tolerance interval for 90% of the population measurements on lifelengths, find the confidence coefficient.

Solution   It is specified that $\delta = 0.90$ and $n = 50$. Thus, the confidence coefficient for an interval of the form (Y_1, Y_{50}) is given by

$$1 - n\delta^{n-1} + (n-1)\delta^n = 1 - 50(0.9)^{49} + 49(0.9)^{50} = 0.97$$

We are 97% confident that the interval $(2{,}150, 2{,}610)$ contains at least 90% of the lifelength measurements for the population under study.   ∎

E X A M P L E   **7.19**   In the setting of Example 7.18, suppose the experimenter wants to choose the sample size n so that (Y_1, Y_n) is a 90% tolerance interval for 90% of the population measurements. What value should the experimenter choose for n?

Solution   Here the confidence coefficient is 0.90 and $\delta = 0.90$. So we must solve the equation

$$1 - n(0.9)^{n-1} + (n-1)(0.9)^n = 0.90$$

This can be done quickly by trial and error, with a calculator. The value $n = 37$ gives a confidence coefficient of 0.896, and the value $n = 38$ gives a confidence coefficient of 0.905. Therefore, $n = 38$ should be chosen to guarantee a confidence coefficient of at least 0.90. With $n = 38$ observations, (Y_1, Y_n) is a 90.5% tolerance interval for 90% of the population measurements.   ∎

Exercises

7.59 A bottle-filling machine is set to dispense 10.0 cc of liquid into each bottle. A random sample of 100 filled bottles gives a mean fill of 10.1 cc and a standard deviation of 0.02 cc. Assuming the amounts dispensed to be normally distributed,

    a construct a 95% tolerance interval for 90% of the population measurements.

    b construct a 99% tolerance interval for 90% of the population measurements.

    c construct a 95% tolerance interval for 99% of the population measurements.

7.60 The average thickness of the plastic coating on electrical wire is an important variable in determining the wearing characteristics of the wire. Ten thickness measurements from randomly selected points along a wire gave the following (in thousandths of an inch):

$$8, \ 7, \ 10, \ 5, \ 12, \ 9, \ 8, \ 7, \ 9, \ 10$$

Construct a 99% tolerance interval for 95% of the population measurements on coating thickness. What assumption is necessary for your answer to be valid?

7.61 A cutoff operation is supposed to cut dowel pins to 1.2 inches in length. A random sample of 60 such pins gave an average length of 1.1 inches and a standard deviation of 0.03 inch. Assuming normality of the pin lengths, find a tolerance interval for 90% of the pin lengths in the population from which this sample was selected. Use a confidence coefficient of 0.95.

7.62 In the setting of Exercise 7.61, construct the tolerance interval if the sample standard deviation is

    a 0.015 inch.

    b 0.06 inch.

Compare these answers to that of Exercise 7.61. Does the size of s seem to have a large effect on the length of the tolerance interval?

7.63 In the setting of Exercise 7.61, construct the tolerance interval if the sample size is as follows:

    a $n = 100$

    b $n = 200$

Does the sample size seem to have a great effect on the length of the tolerance interval, as long as it is "large"?

7.64 For the data of Exercise 7.60 and without the assumption of normality,

    a find the confidence coefficient if (y_1, y_n) is to be used as a tolerance interval for 95% of the population.

    b find the confidence coefficient if (y_1, y_n) is used as a tolerance interval for 80% of the population.

7.65 The interval (Y_1, Y_n) is to be used as a method of constructing tolerance intervals for a proportion δ of a population with confidence coefficient $(1 - \alpha)$. Find the minimum sample sizes needed to ensure that

    a $1 - \alpha = 0.90$ for $\delta = 0.95$.

    b $1 - \alpha = 0.90$ for $\delta = 0.80$.

c $1 - \alpha = 0.95$ for $\delta = 0.95$.
d $1 - \alpha = 0.95$ for $\delta = 0.80$.

7.7 The Method of Maximum Likelihood[†]

The estimators presented in previous sections of this chapter were justified merely by the fact that they seemed reasonable. For instance, our intuition says that a sample mean should be a good estimator of a population mean, or that a sample proportion should somehow resemble the corresponding population proportion. There are, however, general methods for deriving estimators for unknown parameters of population models. One widely used method is called the *method of maximum likelihood*.

Suppose, for example, that a box contains four balls, of which an unknown number θ are white ($4 - \theta$ are nonwhite). We are to sample two balls at random and count X, the number of white balls in the sample. We know the probability distribution of X to be given by

$$P(X = x) = p(x) = \frac{\binom{\theta}{x}\binom{4 - \theta}{2 - x}}{\binom{4}{2}}$$

Now suppose we observe that $X = 1$. What value of θ will maximize the probability of this event? From the above distribution, we have

$$p(1|\theta = 0) = 0$$

$$p(1|\theta = 1) = \frac{\binom{1}{1}\binom{3}{1}}{\binom{4}{2}} = \frac{3}{6} = \frac{1}{2}$$

$$p(1|\theta = 2) = \frac{2}{3}$$

$$p(1|\theta = 3) = \frac{1}{2}$$

$$p(1|\theta = 4) = 0$$

Hence, $\theta = 2$ maximizes the probability of the observed sample, so we would choose this value, 2, as the maximum likelihood estimate for θ, given that we observed that $X = 1$. You can show for yourself that if $X = 2$, then 4 will be the maximum likelihood estimate of θ.

In general, suppose the random sample $X_1, X_2, \ldots, X_n$ results in observations $x_1, x_2, \ldots, x_n$. In the discrete case, the *likelihood* L of the sample is the product of the marginal probabilities; that is,

$$L(\theta) = p(x_1; \theta)p(x_2; \theta) \cdots p(x_n; \theta)$$

[†]Optional section.

where

$$P(X = x_i) = p(x_i; \theta)$$

with θ being the parameter of interest. In the continuous case, the likelihood of the sample is the product of the marginal density functions; that is,

$$L(\theta) = f(x_1; \theta) f(x_2; \theta) \cdots f(x_n; \theta)$$

In either case, we maximize L as a function of θ to find the maximum likelihood estimator of θ.

If the form of the probability distribution is not too complicated, we can generally use methods of calculus to find the functional form of the maximum likelihood estimator instead of working out specific numerical cases. We illustrate with the following examples.

E X A M P L E   **7.20**   Suppose that in a sequence of n independent, identical Bernoulli trials, Y successes are observed. Find the maximum likelihood estimator of p, the probability of successes on any single trial.

Solution Y is really the sum of n independent Bernoulli random variables $X_1, X_2, \ldots, X_n$. Thus, the likelihood is given by

$$\begin{aligned} L(p) &= p^{x_1} p^{x_2} (1 - p)^{1-x_1} (1 - p)^{1-x_2} \cdots p^{x_n} (1 - p)^{1-x_n} \\ &= p^{\sum_{i=1}^{n} x_i} (1 - p)^{n - \sum_{i=1}^{n} x_i} \\ &= p^y (1 - p)^{n-y} \end{aligned}$$

Since $\ln L(p)$ is a monotone function of $L(p)$, both $\ln L(p)$ and $L(p)$ will have their maxima at the same value of p. In many cases, it is easier to maximize $\ln L(p)$, so we will follow that approach here.

Now

$$\ln L(p) = y \ln p + (n - y) \ln (1 - p)$$

is a continuous function of p $(0 < p < 1)$, and thus the maximum value can be found by setting the first derivative equal to zero. We have

$$\frac{\partial \ln L(p)}{\partial p} = \frac{y}{p} - \frac{n - y}{1 - p}$$

Setting

$$\frac{y}{\hat{p}} - \frac{n - y}{1 - \hat{p}} = 0$$

we find

$$\hat{p} = \frac{y}{n}$$

and hence we take Y/n as the maximum likelihood estimator of p. In this case, as well as in many others, the maximum likelihood estimator agrees with our intuitive estimator given earlier in this chapter.   ∎

We now look at an example that is a bit more involved.

EXAMPLE   **7.21**   Suppose we are to observe n independent lifelength measurements $X_1, \ldots, X_n$ from components known to have lifelengths exhibiting a Weibull model, given by

$$f(x) = \begin{cases} \dfrac{\gamma x^{\gamma-1}}{\theta} e^{-x^{\gamma}/\theta} & x > 0 \\ 0 & \text{elsewhere} \end{cases}$$

Assuming γ is known, find the maximum likelihood estimator of θ.

Solution   In analogy with the discrete case given above, we write the likelihood $L(\theta)$ as the joint density function of $X_1, \ldots, X_n$, or

$$\begin{aligned} L(\theta) &= f(x_1, x_2, \ldots, x_n) \\ &= f(x_1)f(x_2) \cdots f(x_n) \\ &= \frac{\gamma x_1^{\gamma-1}}{\theta} e^{-x_1^{\gamma}/\theta} \cdots \frac{\gamma x_n^{\gamma-1}}{\theta} e^{-x_n^{\gamma}/\theta} \\ &= \left(\frac{\gamma}{\theta}\right)^n (x_1, x_2, \ldots, x_n)^{\gamma-1} e^{-\sum_{i=1}^n x_i^{\gamma}/\theta} \end{aligned}$$

Now

$$\ln L(\theta) = n \ln \gamma - n \ln \theta + (\gamma - 1) \ln (x_1 \cdots x_n) - \frac{1}{\theta} \sum_{i=1}^n x_i^{\gamma}$$

To find the value of θ that maximizes $L(\theta)$, we take

$$\frac{\partial \ln L(\theta)}{\partial \theta} = -\frac{n}{\theta} + \frac{1}{\theta^2} \sum_{i=1}^n x_i^{\gamma}$$

Setting this derivative equal to zero, we get

$$-\frac{n}{\hat{\theta}} + \frac{1}{\hat{\theta}^2} \sum_{i=1}^n x_i^{\gamma} = 0$$

or

$$\hat{\theta} = \frac{1}{n} \sum_{i=1}^n x_i^{\gamma}$$

Thus, we take $(1/n) \sum_{i=1}^{n} X_i^{\gamma}$ as the maximum likelihood estimator of θ. The interested reader can check to see that this estimator is unbiased for θ.   ∎

The maximum likelihood estimators shown in Examples 7.20 and 7.21 are both unbiased, but this is not always the case. For example, if X has a geometric distribution, as given in Chapter 3, then the maximum likelihood estimator of p, the probability of a success on any one trial, is $1/X$. In this case, $E(1/X) \neq p$, so this estimator is not unbiased for p.

Many times the maximum likelihood estimator is simple enough that we can use its probability distribution to find a confidence interval, either exact or approximate, for the parameter in question. We illustrate the construction of an exact confidence interval for a nonnormal case with the exponential distribution. Suppose $X_1, \ldots, X_n$ represents a random sample from a population modeled by the density function

$$f(x) = \begin{cases} \dfrac{1}{\theta} e^{-x/\theta} & x > 0 \\ 0 & \text{elsewhere} \end{cases}$$

Since this exponential density is a special case of the Weibull with $\gamma = 1$, we know from Example 7.21 that the maximum likelihood estimator of θ is $(1/n) \sum_{i=1}^{n} X_i$. We also know that $\sum_{i=1}^{n} X_i$ will have a gamma distribution, but we do not have percentage points of most gamma distributions in standard sets of tables, except for the special case of the χ^2 distribution. Thus, we will try to transform $\sum_{i=1}^{n} X_i$ into something that has a χ^2 distribution.

Let $U_i = 2X_i/\theta$. Then

$$F_U(u) = P(U_i \leq u) = P\left(\frac{2X_i}{\theta} \leq u\right)$$

$$= P\left(X_i \leq \frac{u\theta}{2}\right) = 1 - e^{-(1/\theta)(u\theta/2)}$$

$$= 1 - e^{-u/2}$$

Hence,

$$f_U(u) = \frac{1}{2} e^{-u/2}$$

and U_i has an exponential distribution with mean 2, which is equivalent to the $\chi^2(2)$ distribution. It then follows that

$$\sum_{i=1}^{n} U_i = \frac{2}{\theta} \sum_{i=1}^{n} X_i$$

has a χ^2 distribution with $2n$ degrees of freedom. Thus, a confidence interval for θ with confidence coefficient $(1 - \alpha)$ can be formed by finding a $\chi^2_{\alpha/2}(2n)$ and a $\chi^2_{1-\alpha/2}(2n)$ value and writing

$$(1 - \alpha) = P\left[\chi^2_{1-\alpha/2}(2n) \leq \frac{2}{\theta}\sum_{i=1}^{n} X_i \leq \chi^2_{\alpha/2}(2n)\right]$$

$$= P\left[\frac{1}{\chi^2_{\alpha/2}(2n)} \leq \frac{\theta}{2\sum_{i=1}^{n} X_i} \leq \frac{1}{\chi^2_{1-\alpha/2}(2n)}\right]$$

$$= P\left[\frac{2\sum_{i=1}^{n} X_i}{\chi^2_{\alpha/2}(2n)} \leq \theta \leq \frac{2\sum_{i=1}^{n} X_i}{\chi^2_{1-\alpha/2}(2n)}\right]$$

E X A M P L E   **7.22**   Consider the first ten observations (in hundreds of hours) from Table 4.1 on lifelengths of batteries: 0.406, 2.343, 0.538, 5.088, 5.587, 2.563, 0.023, 3.334, 3.491, and 1.267. Using these observations as a realization of a random sample of lifelength measurements from an exponential distribution with mean θ, find a 95% confidence interval for θ.

Solution   Using

$$\left[\frac{2\sum_{i=1}^{n} x_i}{\chi^2_{\alpha/2}(2n)}, \frac{2\sum_{i=1}^{n} x_i}{\chi^2_{1-\alpha/2}(2n)}\right]$$

as the confidence interval for θ, we have

$$\sum_{i=1}^{10} x_i = 24.640$$

$$\chi^2_{0.975}(20) = 9.591$$

and

$$\chi^2_{0.025}(20) = 34.170$$

Hence, the realization of the confidence interval becomes

$$\left[\frac{2(24.640)}{34.170}, \frac{2(24.640)}{9.591}\right]$$

or $(1.442, 5.138)$. We are about 95% confident that the true mean lifelength is between 144 and 514 hours. The rather wide interval is a reflection of both our relative lack of information in the sample of only $n = 10$ batteries and the large variability in the data.   ∎

It is informative to compare the exact intervals for the mean θ from the exponential distribution with what would have been generated if we had falsely assumed the random variables to have a normal distribution and used $\bar{x} \pm t_{\alpha/2} s/\sqrt{n}$ as a confidence interval. Table 7.2 shows one hundred 95% confidence intervals generated by exact methods (using the χ^2 table) for samples of size $n = 5$ from an exponential distribution with $\theta = 10$. Note that five of the intervals (marked with an asterisk) do not include the true θ, as expected. Four miss on the low side, and one misses on the high side.

Table 7.3 repeats the process of generating confidence intervals for θ, with the same samples of size $n = 5$, but now using $\bar{x} \pm t_{\alpha/2} s/\sqrt{n}$ with $\alpha/2 = 0.025$ and 4 degrees of freedom. Note that ten of the intervals do not include the true θ, twice the number expected. Also, the intervals that miss the true $\theta = 10$ all miss on the low side. That fact, coupled with the fact that the lower bounds are frequently negative even though $\theta > 0$, seems to indicate that the intervals based on normality are too far left compared to the true intervals based on the χ^2 distribution.

The lesson to be learned here is that *one should be careful of inflicting the normality assumption on data*, especially if a more reasonable and mathematically tractable model is available.

Before we leave the section on maximum likelihood estimation, we now show how we can carry out the estimation of certain functions of parameters. In general, if we have a maximum likelihood estimator, say Y, for a parameter θ, then any continuous function of θ, say $g(\theta)$, will have a maximum likelihood estimator $g(Y)$. Rather than going deeper into the theory here, we illustrate with a particular case.

Suppose measurements $X_1, \ldots, X_n$ again denote independent lifelength measurements from a population modeled by the exponential distribution with mean θ. The *reliability* of each component is defined by

$$R(t) = P(X_i > t)$$

In the exponential case,

$$R(t) = \int_t^\infty \frac{1}{\theta} e^{-x/\theta} dx = e^{-t/\theta}$$

Suppose we want to estimate $R(t)$ from $X_1, \ldots, X_n$. We know that the maximum likelihood estimator of θ is $\overline{X}$, and hence the maximum likelihood estimator of $R(t) = e^{-t/\theta}$ is $e^{-t/\overline{X}}$.

Now we can use the results from Section 5.6 to find an approximate confidence interval for $R(t)$ if n is large. Let

$$g(\theta) = e^{-t/\theta}$$

Then $g'(\theta) = (t/\theta^2)e^{-t/\theta}$ and, since $\overline{X}$ is approximately normally distributed with mean θ and standard deviation $\theta/\sqrt{n}$ (since $V(X_i) = \theta^2$), it follows that

$$\sqrt{n} \frac{[g(\overline{X}) - g(\theta)]}{|g'(\theta)|\theta} = \frac{\sqrt{n}[e^{-t/\overline{X}} - e^{-t/\theta}]}{\left(\dfrac{t}{\theta}\right)e^{-t/\theta}}$$

Sample	LCL	UCL	Sample	LCL	UCL
1	2.03600	12.8439	51	9.10805	57.4570
*2	1.49454	9.4281	52	4.20731	26.5413
3	4.15706	26.2243	53	5.08342	32.0681
4	3.07770	19.4153	54	5.04982	31.8561
5	5.89943	37.2158	55	3.60932	22.7689
6	6.49541	40.9755	56	5.84841	36.8940
7	2.19705	13.8598	57	5.19353	32.7627
8	2.37724	14.9965	58	4.18032	26.3710
9	5.90726	37.2652	59	4.66717	29.4422
10	7.67249	48.4009	60	5.39428	34.0291
11	6.64161	41.8977	61	6.06969	38.2899
12	4.78097	30.1601	62	3.45198	21.7764
13	3.17174	20.0085	63	5.19596	32.7781
*14	1.51473	9.5555	64	4.79768	30.2655
15	3.70163	23.3513	65	5.76487	36.3669
16	3.69998	23.3408	66	4.15674	26.2223
17	6.90343	43.5494	67	8.96348	56.5450
18	2.36005	14.8881	68	6.74431	42.5456
19	6.53294	41.2122	69	5.36919	33.8709
20	4.04756	25.5335	70	3.77090	23.7883
21	5.61737	35.4365	71	5.42996	34.2542
22	4.07102	25.6815	72	2.7016	17.0429
23	4.65284	29.3518	73	4.2962	27.1020
24	4.58284	28.9103	74	2.3106	14.5761
25	3.80908	24.0291	75	7.6502	48.2602
26	4.22984	26.6834	76	7.4693	47.1190
27	5.17128	32.6223	77	8.5458	53.9103
28	8.77035	55.3266	78	2.9840	18.8243
29	2.54689	16.0667	79	9.2678	58.4647
30	5.27866	33.2998	80	4.2078	26.5442
31	7.26216	45.8124	81	8.0267	50.6353
32	2.27762	14.3680	82	9.9686	62.8854
33	4.63549	29.2424	83	2.2581	14.2447
34	3.80281	23.9895	84	9.9971	63.0651
35	1.98777	12.5396	*85	10.2955	64.9478
36	4.63779	29.2569	86	6.0830	38.3740
37	7.36773	46.4784	87	3.0037	18.9483
38	3.55928	22.4533	88	2.3759	14.9881
39	2.99137	18.8707	89	4.0007	25.2376
40	5.79160	36.5356	*90	1.1737	7.4044
41	5.62684	35.4962	91	3.8375	24.2081
42	3.95994	24.9808	92	2.9560	18.6473
43	2.30394	14.5341	93	4.9714	31.3616
44	7.40844	46.7352	94	3.0063	18.9648
45	3.06875	19.3588	95	4.1684	26.2958
46	2.99999	18.9251	96	4.6269	29.1879
47	3.95912	24.9756	97	3.8026	23.9883
48	3.01418	19.0146	98	6.0056	37.8858
49	2.69137	16.9782	*99	1.5153	9.5592
50	4.28299	27.0187	100	7.0408	44.4161

T A B L E **7.3**
Confidence Intervals Based on
the Normal Distribution (n = 5)

Sample	LCL	Mean	UCL	S
*1	1.975	4.1704	6.3656	1.7682
*2	−1.069	3.0613	7.1916	3.3269
3	−1.550	8.5150	18.5800	8.1074
4	−1.019	6.3041	13.6277	5.8991
5	4.183	12.0839	19.9850	6.3644
6	3.838	13.3046	22.7712	7.6253
7	−3.036	4.5002	12.0368	6.0707
*8	0.010	4.8693	9.7289	3.9144
9	1.004	12.0999	23.1954	8.9374
10	4.917	15.7156	26.5146	8.6986
11	8.340	13.6041	18.8683	4.2404
12	−0.397	9.7929	19.9832	8.2083
13	1.042	6.4967	11.9513	4.3937
*14	−0.274	3.1026	6.4790	2.7197
15	2.542	7.5821	12.6217	4.0594
16	3.398	7.5787	11.7595	3.3678
17	6.994	14.1404	21.2868	5.7564
*18	1.244	4.8341	8.4245	2.8920
19	−6.383	13.3815	33.1458	15.9202
20	1.122	8.2907	15.4591	5.7742
21	−3.593	11.5061	26.6049	12.1621
22	−0.433	8.3387	17.1102	7.0654
23	−5.925	9.5305	24.9856	12.4491
24	−11.200	9.3871	29.9740	16.5828
25	−4.166	7.8022	19.7701	9.6402
26	1.564	8.6640	15.7640	5.7191
27	3.818	10.5924	17.3667	5.4567
28	−3.492	17.9644	39.4208	17.2832
29	−0.656	5.2168	11.0900	4.7308
30	1.837	10.8123	19.7876	7.2296
31	2.581	14.8751	27.1690	9.9027
32	−1.941	4.6653	11.2716	5.3214
33	−10.734	9.4949	29.7238	16.2943
34	−2.725	7.7893	18.3035	8.4692
*35	−0.546	4.0716	8.6890	3.7193
36	0.841	9.4996	18.1584	6.9747
37	−1.467	15.0914	31.6496	13.3376
38	2.144	7.2905	12.4370	4.1455
39	−2.991	6.1272	15.2452	7.3446
40	−4.714	11.8630	28.4403	13.3530
41	4.356	11.5255	18.6946	5.7747
42	−4.436	8.1112	20.6585	10.1069
*43	1.221	4.7192	8.2178	2.8181
44	−0.877	15.1748	31.2261	12.9294
45	0.865	6.2858	11.7065	4.3665
46	0.751	6.1449	11.5392	4.3451
47	−1.325	8.1095	17.5442	7.5997
48	−1.116	6.1740	13.4641	5.8722
49	−2.837	5.5128	13.8623	6.7256
50	2.529	8.7729	15.0168	5.0294
51	−3.559	18.6561	40.8713	17.8944

Sample	LCL	Mean	UCL	S
52	−2.949	8.6179	20.1847	9.3171
53	1.648	10.4124	19.1769	7.0598
54	−2.852	10.3436	23.5391	10.6290
55	1.645	7.3930	13.1407	4.6298
56	2.344	11.9794	21.6150	7.7615
57	−8.525	10.6380	29.8011	15.4359
58	−0.781	8.5626	17.9059	7.5260
59	2.364	9.5598	16.7556	5.7962
60	−1.314	11.0491	23.4128	9.9589
61	1.465	12.4326	23.3998	8.8341
62	−2.978	7.0707	17.1193	8.0941
63	−6.065	10.6429	27.3514	13.4586
64	−4.378	9.8271	24.0327	11.4426
65	3.155	11.8082	20.4612	6.9700
66	0.298	8.5143	16.7307	6.6184
67	−4.655	18.3600	41.3751	18.5387
68	3.14	13.8144	24.4798	8.5910
69	−2.948	10.9978	24.9433	11.2331
70	0.332	7.7240	15.1161	5.9544
71	−3.383	11.1222	25.6275	11.8840
72	1.027	5.5338	10.0406	3.6302
73	−1.720	8.7999	19.3194	8.4735
74	−0.755	4.7328	10.2205	4.4203
75	−4.753	15.6699	36.0924	16.4503
76	−12.223	15.2994	42.8221	22.1695
77	−5.102	17.5045	40.1115	18.2099
78	0.504	6.1122	11.7204	4.5175
79	0.042	18.9833	37.9249	15.2575
80	−0.873	8.6188	18.1102	7.6453
81	7.232	16.4411	25.6501	7.4179
82	−0.289	20.4187	41.1267	16.6803
*83	−0.684	4.6252	9.9343	4.2765
84	1.162	20.4771	39.7917	15.5579
85	−13.471	21.0884	55.6473	27.8373
86	−0.086	12.4599	25.0060	10.1059
87	−3.659	6.1524	15.9641	7.9033
88	−4.660	4.8666	14.3930	7.6735
89	0.914	8.1946	15.4755	5.8648
*90	0.826	2.4042	3.9824	1.2712
91	−0.416	7.8603	16.1361	6.6662
92	−0.093	6.0547	12.2023	4.9519
93	2.451	10.1830	17.9150	6.2281
94	−6.613	6.1578	18.9282	10.2866
95	−0.470	8.5382	17.5461	7.2559
96	−0.036	9.4772	18.9902	7.6627
97	−2.201	7.7889	17.7785	8.0466
98	2.279	12.3014	22.3234	8.0727
*99	1.383	3.1038	4.8248	1.3862
100	2.804	14.4218	31.6476	13.8754

is approximately a standard normal random variable if n is large. Thus, for large n, $e^{-t/\overline{X}}$ is approximately normally distributed with mean $e^{-t/\theta}$ and standard deviation $(1/\sqrt{n})(t/\theta)e^{-t/\theta}$. Note that this is only a large-sample approximation, as $E(e^{-t/\overline{X}}) \neq e^{-t/\theta}$. This result can be used to form an approximate confidence interval for $R(t) = e^{-t/\theta}$ of the form

$$e^{-t/\bar{x}} \pm z_{\alpha/2} \frac{1}{\sqrt{n}} \left(\frac{t}{\bar{x}}\right) e^{-t/\bar{x}}$$

using $\bar{x}$ as an approximation to θ in the standard deviation.

E X A M P L E   **7.23**   Using the 50 observations given in Table 4.1 as a realization of a random sample from an exponential distribution, estimate $R(5) = P(X_i > 5) = e^{5/\theta}$ in an approximate 95% confidence interval.

Solution   For $(1 - \alpha) = 0.95$, $z_{\alpha/2} = 1.96$. Using the method indicated above, the interval becomes

$$e^{-t/\bar{x}} \pm (1.96) \frac{1}{\sqrt{n}} \left(\frac{t}{\bar{x}}\right) e^{-t/\bar{x}}$$

or

$$e^{-5/2.267} \pm (1.96) \frac{1}{\sqrt{50}} \left(\frac{5}{2.267}\right) e^{-5/2.267}$$

$$0.110 \pm 0.067$$

$$(0.043, 0.177)$$

That is, we are about 95% confident that the probability is between 0.043 and 0.177. For these data, the true value of θ was 2, and $R(5) = e^{-5/2} = 0.082$ is well within this interval. The interval could be narrowed by selecting a larger sample. ▪

Exercises

7.66   If $X_1, \ldots, X_n$ denotes a random sample from a Poisson distribution with mean λ, find the maximum likelihood estimator of λ.

7.67   Since $V(X_i) = \lambda$ in the Poisson case, it follows from the Central Limit Theorem that $\overline{X}$ will be approximately normally distributed with mean λ and variance λ/n, for large n.

a   Use the above facts to construct a large-sample confidence interval for λ.

b   Suppose that 100 reinforced concrete trusses were examined for cracks. The average number of cracks per truss was observed to be four. Construct an approximate 95%

confidence interval for the true mean number of cracks per truss for trusses of this type. What assumptions are necessary for your answer to be valid?

7.68   Suppose $X_1, \ldots, X_n$ denotes a random sample from the normal distribution with mean μ and variance σ^2. Find the maximum likelihood estimators of μ and σ^2.

7.69   Suppose $X_1, \ldots, X_n$ denotes a random sample from the gamma distribution with a known α but unknown β. Find the maximum likelihood estimator of β.

7.70   The stress resistances for specimens of a certain type of plastic tend to have a gamma distribution with $\alpha = 2$, but β may change with certain changes in the manufacturing process. For eight specimens independently selected from a certain process, the resistances (in psi) were

$$29.2, 28.1, 30.4, 31.7, 28.0, 32.1, 30.1, 29.7$$

Find a 95% confidence interval for β. [*Hint*: Use the methodology outlined for Example 7.21.]

7.71   If X denotes the number of the trial on which the first defective is found in a series of independent quality-control tests, find the maximum likelihood estimator of p, the true probability of observing a defective.

7.72   The absolute errors in the measuring of the diameters of steel rods are uniformly distributed between zero and θ. Three such measurements on a standard rod produced errors of 0.02, 0.06, and 0.05 centimeter. What is the maximum likelihood estimate of θ? [*Hint*: This problem cannot be solved by differentiation. Write down the likelihood, $L(\theta)$, for three independent measurements and think carefully about what value of θ produces a maximum. Remember, each measurement must be between zero and θ.]

7.73   The number of improperly soldered connections per microchip in an electronics manufacturing operation follows a binomial distribution with $n = 20$ and p unknown. The cost of correcting these malfunctions, per microchip, is

$$C = 3X + X^2$$

Find the maximum likelihood estimate of $E(C)$ if $\hat{p}$ is available as an estimate of p.

7.8   Bayes Estimators[†]

In all previous sections on estimation, population parameters were treated as unknown constants. However, it is sometimes convenient and even necessary to treat an unknown parameter as a random variable in its own right. For example, the proportion of defectives p produced by an assembly line may change from day to day, or even hour to hour. It may be possible to model this changing behavior by allowing p to be a random variable with a probability density function, say $g(p)$. Even though we may not be able to choose the correct value of p for a given hour, we can find

$$P(a \leq p \leq b) = \int_a^b g(p)\,dp$$

Similarly, the mean daily cost of production for the assembly line in question may vary from day to day and could be assigned an appropriate probability density function to model the day-to-day variation.

Suppose we are to observe a random sample $Y_1, \ldots, Y_n$ from a probability density function with a single unknown parameter θ. If θ is a constant, then the density function is completely specified as $f(y|\theta)$. Note that $f(y|\theta)$ is now a *conditional* density function for y, *given* a fixed value of θ. The joint conditional density function for the random sample is given by

$$f(y_1, \ldots, y_n|\theta) = f(y_1|\theta) \cdots f(y_n|\theta)$$

Next, suppose θ varies according to a probability density function $g(\theta)$. This density function $g(\theta)$ is referred to as the *prior* density for θ because it is assigned prior to the actual collection of data in a sample.

Knowledge of the above density functions allows us to compute

$$
\begin{aligned}
f(\theta|y_1, \ldots, y_n) &= \frac{f(y_1, \ldots, y_n, \theta)}{f(y_1, \ldots, y_n)} \\
&= \frac{f(y_1, \ldots, y_n|\theta)g(\theta)}{\int_{-\infty}^{\infty} f(y_1, \ldots, y_n|\theta)g(\theta)d\theta}
\end{aligned}
$$

The density function $f(\theta|y_1, \ldots, y_n)$ is the conditional density of θ given the sample data and is called the *posterior* density function for θ. We call the *mean* of the posterior density function the *Bayes estimator* of θ. The computations are illustrated in the following example.

E X A M P L E   **7.24**   Let Y denote the number of defectives observed in a random sample of n items produced by a given machine in one day. The proportion p of defectives produced by this machine varies from day to day according to the probability density function

$$g(p) = \begin{cases} 1 & 0 \le p \le 1 \\ 0 & \text{elsewhere} \end{cases}$$

[In other words, p is uniformly distributed over the interval $(0,1)$.] Find the Bayes estimator of p.

Solution   For a given p, Y will have a binomial distribution given by

$$f(y|p) = \binom{n}{y} p^y (1-p)^{n-y} \qquad y = 0, 1, \ldots, n$$

Then

$$
\begin{aligned}
f(y, p) &= f(y|p)g(p) \\
&= \binom{n}{y} p^y (1-p)^{n-y}(1) \qquad y = 0, 1, \ldots, n \qquad 0 \le p \le 1
\end{aligned}
$$

Now

$$
\begin{aligned}
f(y) &= \int_{-\infty}^{\infty} f(y, p)dp \\
&= \int_0^1 \binom{n}{y} p^y (1-p)^{n-y} dp \\
&= \binom{n}{y} \int_0^1 p^y (1-p)^{n-y} dp
\end{aligned}
$$

$$= \binom{n}{y} \frac{y!(n-y)!}{(n+1)!}$$

$$= \frac{1}{n+1} \quad y = 0, 1, \ldots, n$$

We evaluate the above integral by recognizing it as a beta function described in Chapter 4. Note that $f(y)$ says that each possible value of y has the same unconditional probability of occurring, since nothing is known about p other than the fact that it is in the interval (0,1).

It follows that

$$f(p|y) = \frac{f(y, p)}{f(y)}$$

$$= (n+1)\binom{n}{y} p^y (1-p)^{n-y}$$

$$= \frac{(n+1)!}{y!(n-y)!} p^y (1-p)^{n-y} \quad 0 \le p \le 1$$

which is the beta density function.

The Bayes estimate of p is taken to be the mean of this posterior density function, namely,

$$\int_0^1 pf(p|y)dp = \int_0^1 p \frac{(n+1)!}{y!(n-y)!} p^y (1-p)^{n-y} dp$$

$$= \frac{y+1}{n+2}$$

(Check Section 4.7 to find the mean of a beta density function.) Recall that the simple maximum likelihood estimator of p, when p is fixed but unknown, is Y/n. ∎

Exercises

7.74   Refer to Example 7.24. Suppose that

$$g(p) = \begin{cases} 2 & 0 \le p \le \frac{1}{2} \\ 0 & \text{elsewhere} \end{cases}$$

If two items are produced on a given day and one is defective, find the Bayes estimate of p.

7.75   Let Y denote the number of defects per yard for a certain type of fabric. For a given mean λ, Y has a Poisson distribution. But λ varies from yard to yard according to the density function

$$g(\lambda) = \begin{cases} e^{-\lambda} & \lambda > 0 \\ 0 & \text{elsewhere} \end{cases}$$

Find the Bayes estimator of λ.

7.76   Suppose the lifelength Y of a certain component has probability density function

$$f(y|\theta) = \begin{cases} \dfrac{\theta^{\alpha}}{\Gamma(\alpha)} y^{\alpha-1} e^{-y\theta} & y > 0 \\ 0 & \text{elsewhere} \end{cases}$$

Also, suppose θ has a prior density function given by

$$g(\theta) = \begin{cases} e^{-\theta} & \theta > 0 \\ 0 & \text{elsewhere} \end{cases}$$

For single observation Y, show that the Bayes estimator of θ is given by $(\alpha + 2)/(Y + 1)$.

7.9 Summary

We have developed *confidence intervals* for individual means and proportions, linear functions of means and proportions, individual variances, and ratios of variances. In a few cases, we have shown how to calculate confidence intervals for other functions of parameters. We also touched on the notion of a *prediction interval* for a random variable and a *tolerance interval* for a distribution. Most often, these intervals were two-sided, in the sense that they provided both an upper and a lower bound. Occasionally, a one-sided interval is in order.

Most of the estimators used in the construction of confidence intervals were selected on the basis of intuitive appeal. However, the *method of maximum likelihood* provides a technique for finding estimators that are usually close to being unbiased and have small variance. (Most of the intuitive estimators turned out to be maximum likelihood estimators.)

If it is appropriate to treat a population parameter as a random quantity, then *Bayesian* techniques can be used to find estimates.

Many of the estimators used in this chapter will be used in a slightly different setting in Chapter 8.

Supplementary Exercises

7.77   The diameter measurements of an armored electric cable, taken at ten points along the cable, yield a sample mean of 2.1 centimeters and a sample standard deviation of 0.3 centimeter. Estimate the average diameter of the cable in a confidence interval with a confidence coefficient of 0.90. What assumptions are necessary for your answer to be valid?

7.78   Suppose the sample mean and standard deviation of Exercise 7.77 had come from a sample of 100 measurements. Construct a 90% confidence interval for the average diameter of the cable. What assumption must necessarily be made?

7.79   The Rockwell hardness measure of steel ingots is produced by pressing a diamond point into the steel and measuring the depth of penetration. A sample of 15 Rockwell hardness measurements on specimens of steel gave a sample mean of 65 and a sample variance of 90. Estimate the true mean hardness in a 95% confidence interval, assuming the measurements to come from a normal distribution.

7.80   Twenty specimens of a slightly different steel from that used in Exercise 7.79 were observed and yielded Rockwell hardness measurements with a mean of 72 and a variance of 94. Estimate the difference between the mean hardness for the two varieties of steel in a 95% confidence interval.

7.81   In the fabrication of integrated circuit chips, it is of great importance to form contact windows of precise width. (These contact windows facilitate the interconnections that make up the circuits.) Complementary metaloxide semiconductor (CMOS) circuits are fabricated by using a photolithography process to form windows. The key steps involve applying a photoresist to the silicon wafer, exposing the photoresist to ultraviolet radiation through a mask, and then dissolving away the photoresist from the exposed areas, revealing the oxide surface. The printed windows are then etched through the oxide layers down to the silicon. Key measurements that help determine how well the integrated circuits may function are the window widths before and after the etching process. (See M. S. Phadke, et al., *The Bell System Technical Journal*, 62, no. 5, 1983, pp. 1273–1309, for more details.)

Preetch window widths for a sample of ten test locations are as follows (in μm):

$$2.52, 2.50, 2.66, 2.73, 2.71, 2.67, 2.06, 1.66, 1.78, 2.56$$

Postetch window widths for ten independently selected test locations are

$$3.21, 2.49, 2.94, 4.38, 4.02, 3.82, 3.30, 2.85, 3.34, 3.91$$

One important problem is simply to estimate the true average window width for this process.

a   Estimate the average preetch window width, for the population from which this sample was drawn, in a 95% confidence interval.

b   Estimate the average postetch window width in a 95% confidence interval.

7.82   In the setting of Exercise 7.81, a second important problem is to compare the average window widths before and after etching. Using the data given there, estimate the true difference in average window widths in a 95% confidence interval.

7.83   Window sizes in integrated circuit chips must be of fairly uniform size in order for the circuits to function properly. Thus, the variance of the window width measurements is an important quantity. Refer to the data of Exercise 7.81.

a   Estimate the true variance of the postetch window widths in a 95% confidence interval.

b   Estimate the ratio of the variance of postetch window widths to that of preetch window widths in a 95% confidence interval.

c   What assumptions are necessary for your answers in (a) and (b) to be valid?

7.84   The weights of aluminum grains follow a lognormal distribution, which means that the natural logarithms of the grain weights follow a normal distribution. A sample of 177 aluminum grains has logweights averaging 3.04 with a standard deviation of 0.25. (The original weight measurements were in 10^{-4} gram.) Estimate the true mean logweight of the grains from which this sample was selected, using a confidence coefficient of 0.90. (Source: Department of Materials Science, University of Florida.)

7.85   Two types of aluminum powders are blended before a sintering process is begun to form solid aluminum. The adequacy of the blending is gauged by taking numerous small samples from the blend and measuring the weight proportion of one type of powder. For adequate blending, the weight proportions from sample to sample should have small variability. Thus, the variance becomes a key measure of blending adequacy.

Ten samples from an aluminum blend gave the following weight proportions for one type of powder:

$$0.52, 0.54, 0.49, 0.53, 0.52, 0.56, 0.48, 0.50, 0.52, 0.51$$

a   Estimate the population variance in a 95% confidence interval.

b   What assumptions are necessary for your answer in (a) to be valid? Are the assumptions reasonable for this case?

7.86   It is desired to estimate the proportion of defective items produced by a certain assembly line to within 0.1 with confidence coefficient 0.95. What is the smallest sample size that will guarantee this accuracy no matter where the true proportion of defectives might lie? [*Hint*: Find the value of p that maximizes the variance $p(1-p)/n$, and choose the value of n that corresponds to it.]

7.87   Suppose that, in a large-sample estimate of $\mu_1 - \mu_2$ for two populations with respective variances σ_1^2 and σ_2^2, a *total* of n observations is to be selected. How should these n observations be allocated to the two populations so that the length of the resulting confidence interval will be minimized?

7.88   Suppose the sample variances given in Exercise 7.34 are good estimates of the population variances. Using the allocation scheme of Exercise 7.87, find the number of measurements to be taken on each coupling agent to estimate the true difference in means to within 1 unit with confidence coefficient 0.95. Compare your answer to that of Exercise 7.35.

7.89   A factory operates with two machines of type A and one machine of type B. The weekly repair costs Y for type A machines are normally distributed with mean μ_1 and variance σ^2. The weekly repair costs X for machines of type B are also normally distributed, but with mean μ_2 and variance $3\sigma^2$. The expected repair cost per week for the factory is then $2\mu_1 + \mu_2$. Suppose $Y_1, \ldots, Y_n$ denotes a random sample on costs for type A machines and $X_1, \ldots, X_m$ denotes an independent random sample on costs for type B machines. Use these data to construct a 95% confidence interval for $2\mu_1 + \mu_2$.

7.90   In polycrystalline aluminum, the number of grain nucleation sites per unit volume is modeled as having a Poisson distribution with mean λ. Fifty unit-volume test specimens subjected to annealing regime A showed an average of 20 sites per unit volume. Fifty independently selected unit-volume test specimens subjected to annealing regime B showed an average of 23 sites per volume. Find an approximate 95% confidence interval for the difference between the mean site frequencies for the two annealing regimes. Would you say that regime B tends to increase the number of nucleation sites?

7.91   A random sample of n items is selected from the large number of items produced by a certain production line in one day. The number of defectives X is observed. Find the maximum likelihood estimator of the ratio R of the number of defective to nondefective items.

Hypothesis Testing

About This Chapter

We have learned how to estimate population parameters using sample information. The second formal manner of making inferences from sample to population is through *tests of hypotheses*. The methodology allows the weight of sample evidence for or against the experimenter's hypotheses to be assessed probabilistically and is basic to the scientific method. Again, the sampling distributions of the statistics involved play a key role in the development of the methodology.

Contents

8.1   Introduction

One method of using sample data to formulate inferences about a population parameter, as seen in Chapter 7, is to produce a confidence interval estimate of the parameter in question. But it often happens that an experimenter is interested only in checking a claim, or hypothesis, concerning the value of a parameter and is not really interested in the location or length of the confidence interval itself. For example, if an electronic component is guaranteed to possess a mean lifelength of at least 200 hours, then an investigator might be interested only in checking the hypothesis that the mean really is 200 hours or more against the alternative that the mean is less than 200 hours. A confidence interval on the mean is not, in itself, of great interest, although it can provide a mechanism for checking, or *testing*, the hypothesis of interest.

Hypothesis testing arises as a natural consequence of the scientific method. The scientist observes nature, formulates a theory, and then tests theory against observation. In the context of the problems in this text, the experimenter theorizes that a population parameter takes on a certain value or set of values. A sample is then selected from the population in question, and observation is compared with theory. If the observations disagree seriously with the theory, then the experimenter may reject the theory (hypothesis). If the observations are compatible with the theory, then the experimenter may proceed as if the hypothesis were true. Generally, the process does not stop in either case. New theories are posed, new experiments are completed, and new comparisons between theory and reality are made in a continuing cycle.

Note that hypothesis testing requires a decision when the observed sample is compared with theory. How do we decide whether the sample disagrees with the hypothesis? When should we reject the hypothesis, and when should we not reject it? What is the probability that we will make a wrong decision? What function of the sample observations should we employ in our decision-making process? The answers to these questions lie in the study of statistical hypothesis testing.

8.2   Hypothesis Testing: The Single-Sample Case

In this section, we present tests of hypotheses concerning the mean of an unspecified distribution, the probability of success in a binomial distribution, and the mean and variance of a normal distribution. Our first two tests require large samples and are approximate in nature, while the second two are exact tests for any sample size.

8.2.1   General Distribution: Testing the Mean

One of the most common and, at the same time, easiest hypothesis-testing situations to consider is that of testing a population mean when a large random sample from the population in question is available for observation. Denoting the population mean by μ and the mean of the sample size n by $\overline{X}$, we know from Chapter 7 that $\overline{X}$ is a good estimator of μ. Therefore, it seems intuitively reasonable that our decision about μ should be based on $\overline{X}$, with extra information provided by the population variance σ^2 or its estimator S^2.

Suppose it is claimed that μ takes on the specific value μ_0, and we want to test the validity of this claim. How can we tie this testing notion into the notion of a confidence interval established in Chapter 7? Upon reflection, it may seem reasonable that if the resulting confidence interval for μ does *not* contain μ_0, then we should *reject* the hypothesis that $\mu = \mu_0$. On the other hand, if μ_0 is within the confidence limits, then we cannot reject it as a plausible value for μ. Now if μ_0 is not within the interval $\overline{X} \pm z_{\alpha/2}\sigma/\sqrt{n}$, then either

$$\mu_0 < \overline{X} - z_{\alpha/2}\frac{\sigma}{\sqrt{n}}$$

or

$$\mu_0 > \overline{X} + z_{\alpha/2}\frac{\sigma}{\sqrt{n}}$$

The former equation can be rewritten as

$$\frac{\overline{X} - \mu_0}{\sigma/\sqrt{n}} > z_{\alpha/2}$$

and the latter as

$$\frac{\overline{X} - \mu_0}{\sigma/\sqrt{n}} < -z_{\alpha/2}$$

Assuming a large sample, so that $\overline{X}$ has approximately a normal distribution, we see that the hypothesized value μ_0 is rejected if

$$|Z| = \left|\frac{\overline{X} - \mu_0}{\sigma/\sqrt{n}}\right| > z_{\alpha/2}$$

for some prescribed α, where Z has approximately a standard normal distribution. In other words, we reject μ_0 as a plausible value for μ if $\overline{X}$ is too many standard deviations (more than $z_{\alpha/2}$) from μ_0. Whereas $(1 - \alpha)$ was called the confidence coefficient in estimation problems, α is referred to as the *significance level* in hypothesis-testing problems.

E X A M P L E   **8.1**   The depth setting on a certain drill press is 2 inches. One could then hypothesize that the average depth of all holes drilled by this machine is $\mu = 2$ inches. To check this hypothesis (and the accuracy of the depth gauge), a random sample of $n = 100$ holes drilled by this machine was measured and found to have a sample mean of $\bar{x} = 2.005$ inches with a standard deviation of $s = 0.03$ inch. With $\alpha = 0.05$, can the hypothesis be rejected based on these sample data?

Solution   With $n = 100$, it is assumed that $\overline{X}$ will be approximately normal in distribution. Thus, we reject the hypothesis $\mu = 2$ if

$$|Z| = \left|\frac{\overline{X} - \mu_0}{\sigma/\sqrt{n}}\right| > z_{\alpha/2} = 1.96$$

For large n, S can be substituted for σ, and the approximate normality of Z still holds. Thus, the observed test statistic becomes

$$\frac{\overline{x} - \mu_0}{s/\sqrt{n}} = \frac{2.005 - 2.000}{0.03/\sqrt{100}} = 1.67$$

(This is actually a t statistic, but it can be approximated by z for large n.) Since the observed value of $(\overline{x} - \mu_0)/(s/\sqrt{n})$ is less than 1.96, we cannot reject the hypothesis that 2 is a plausible value of μ. (The 95% confidence interval for μ is 1.991 to 2.011, which includes the value 2.) Note that *not rejecting* the hypothesis that $\mu = 2$ is not the same as *accepting* the hypothesis that $\mu = 2$. When we do not reject the hypothesis $\mu = 2$, we are saying that 2 is a possible value of μ, but there are other equally plausible values for μ. We cannot conclude that μ is equal to 2, and 2 alone. ▪

We now introduce the standard terminology of hypothesis-testing problems. The hypothesis that specifies a particular value for the parameter being studied is called the *null hypothesis* and is denoted by H_0. This hypothesis usually represents the standard operating procedure of a system or known specifications. In Example 8.1, $\mu = 2$ inches is the null hypothesis; it gives a specific value for μ and represents what should happen when the machine is operating according to the specification.

The hypothesis that specifies those values of the parameter that represent an important change from standard operating procedure or known specifications is called the *alternative hypothesis* and is denoted by H_a. In Example 8.1, values of μ less than 2 and values of μ greater than 2 both represent important departures from the specification $\mu = 2$. Thus, the alternative hypothesis is $H_a : \mu \neq 2$.

Sample observations are collected and analyzed to determine whether the evidence supports H_0 or H_a. The sample quantity on which the decision to support H_0 or H_a is based is called the test statistic. The set of values of the test statistic that leads to rejection of the null hypothesis in favor of the alternative hypothesis is called the *rejection region* (or *critical region*). In Example 8.1, the sample mean serves as an appropriate statistic to estimate μ, but we really want to know if μ is larger or smaller than 2. Thus, the test statistic becomes Z, as shown above, since Z actually measures the standardized difference between $\overline{X}$ and 2. We reject H_0 if the absolute value of this difference is large, and the size of this rejection region is determined by α.

An experiment, you may recall, has the objective of deciding if one treatment is better than another. In such settings, one treatment is often new (or "experimental") and the other is a standard. A new process for drawing wire, for example, might lead researchers to believe that the tensile strength can be improved by 10 psi, on the average, over the tensile strength using the standard process. In this context, samples of wire from the new process are tested, and their mean tensile strength is compared to the known mean of the standard process, μ_0. Here, H_0 is $\mu = \mu_0$ and H_a is $\mu = \mu_0 + 10$, where μ is the unknown mean tensile strength for wire from the new process. H_a is the alternative proposed by the researchers and, for that reason, is often called the *research hypothesis*.

Table 8.1 summarizes the terminology of hypothesis testing and provides an example for a large-sample test of a population mean.

Terminology	Example
Null hypothesis, H_0	$H_0 : \mu = \mu_0$
Alternative hypothesis, H_a	$H_a : \mu \neq \mu_0$
Test statistic	$Z = \dfrac{\overline{X} - \mu_0}{\sigma/\sqrt{n}}$
Rejection region	$|Z| > z_{\alpha/2}$

Sometimes, a null hypothesis is enlarged to include an interval of possible values. For example, a can for 12 ounces of cola is designed to hold only slightly more than 12 ounces, so it is important that overfilling not be serious. However, underfilling of the cans could have serious implications. Suppose μ represents the mean ounces of cola per can dispensed by a filling machine. Under a standard operating procedure that meets specifications, μ must be 12 or more, so we set H_0 as $\mu \geq 12$. An important change occurs if μ drops below 12, so H_a becomes $\mu < 12$. Under H_0, the vendor is in compliance with the law, and customers are satisfied. Under H_a, the vendor is violating a law, and customers are unhappy. The vendor wants to sample the process regularly to be reasonably sure that a shift from H_0 to H_a has not taken place.

When H_0 is an interval such as $\mu \geq 12$, the equality at $\mu = 12$ still plays a key role. Sample data will be collected, and a value for $\overline{x}$ will be calculated. If $\overline{x}$ is much less than 12, H_0 will be rejected. If the absolute difference between $\overline{x}$ and 12 is large enough to reject H_0, then the differences between $\overline{x}$ and 12.5 or 12.9 will be even larger and will again lead to rejection of H_0. Thus, we need to consider only the boundary point between H_0 and H_a (12 in this case) when calculating an observed value of a test statistic.

For testing $H_0 : \mu \geq \mu_0$ versus $H_a : \mu < \mu_0$, we consider only the analogous one-sided confidence interval for μ. Here it is important to establish an upper confidence limit for μ, since we want to know how large μ is likely to be. If μ_0 is larger than the upper confidence limit for μ, then we will reject $H_0 : \mu \geq \mu_0$. The one-sided upper limit for μ is $\overline{X} + z_\alpha \sigma/\sqrt{n}$. (See Section 7.3.) We reject H_0 in favor of H_a when $\overline{X} + z_\alpha \sigma/\sqrt{n} < \mu_0$ or, equivalently, when

$$\frac{\overline{X} - \mu_0}{\sigma/\sqrt{n}} < -z_\alpha$$

Notice that we will have the same test statistic but the rejection region has changed somewhat.

The corresponding rejection region for testing $H_0 : \mu \leq \mu_0$ versus $H_a : \mu > \mu_0$ is given by

$$\frac{\overline{X} - \mu_0}{\sigma/\sqrt{n}} > z_\alpha$$

E X A M P L E   **8.2**   A vice-president for a large corporation claims that the number of service calls on equipment sold by that corporation is no more than 15 per week, on the average. To investigate her claim, service records were checked for $n = 36$ randomly selected weeks, with the result $\bar{x} = 17$ and $s^2 = 9$ for the sample data. Does the sample evidence contradict the vice-president's claim at the 5% significance level?

Solution   The vice-president states that, according to her specifications, the mean must be 15 or less. Thus, we are testing

$$H_0 : \mu \leq 15 \quad \text{versus} \quad H_a : \mu > 15$$

Using as the test statistic

$$Z = \frac{\bar{X} - \mu_0}{\sigma/\sqrt{n}}$$

we reject H_0 for $z > z_{0.05} = 1.645$. Substituting s for σ, we calculate the observed statistic to be

$$\frac{\bar{x} - \mu_0}{s/\sqrt{n}} = \frac{17 - 15}{3/\sqrt{36}} = 4$$

That is, the observed sample mean is four standard deviations greater than the hypothesized value of $\mu_0 = 15$. With $\alpha = 0.05$, $z_\alpha = 1.645$. Since the test statistic exceeds z_α, we have sufficient evidence to reject the null hypothesis. It does appear that the mean number of service calls exceeds 15.   ∎

In the preceding hypothesis-testing problems, we see that there are two ways that errors can be made in the decision process. We can reject the null hypothesis when it is true, called a *Type I error*, or we can fail to reject the null hypothesis when it is false, called a *Type II error*. The possible decisions and errors are displayed in Table 8.2.

T A B L E **8.2**
Decisions and Errors in
Hypothesis Testing

	Decision	
	Reject H_0	Do Not Reject H_0
H_0 True	Type I error	Correct decision
H_0 False	Correct decision	Type II error

For the one-sided test $H_0 : \mu \geq \mu_0$ versus $H_a : \mu < \mu_0$, we reject H_0 if

$$\frac{\bar{X} - \mu_0}{\sigma/\sqrt{n}} < -z_\alpha$$

The probability that we reject H_0 when $\mu = \mu_0$ is given by

$$P(\text{reject } H_0 \text{ when } \mu = \mu_0) = P\left(\frac{\overline{X} - \mu_0}{\sigma/\sqrt{n}} < -z_\alpha \text{ when } \mu = \mu_0\right) = \alpha$$

So we see that

$$\alpha = P(\text{Type I error when } \mu = \mu_0)$$

Thus, in the case of a null hypothesis like $\mu \geq \mu_0$, the significance level α is the same as the probability of a Type I error at the point $\mu = \mu_0$.

We will denote the probability of a Type II error by β. That is,

$$\beta = P(\text{Type II error when } \mu = \mu_a)$$
$$= P(\text{do not reject } H_0 \text{ when } \mu = \mu_a)$$

FIGURE **8.1**
α and β for a Statistical Test

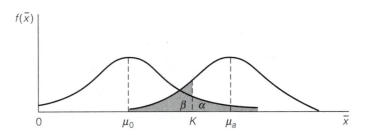

For the large-sample test of a mean, we can show α and β as in Figure 8.1. In this figure, if μ_0 is the true value of μ under H_0 and μ_a is a specific alternative value of interest, then the area to the right of K under the normal curve centered at μ_0 is α. Note that α represents the chance that the null hypothesized value μ_0 will be rejected when, in fact, μ_0 is the true value of μ. The area to the left of K under the curve centered at μ_a is β for that particular alternative. Note that β represents the chance that the null hypothesis will be accepted when, in fact, μ_a is the true value of μ. How do we find the point denoted by K? Recall that for testing $H_0 : \mu = \mu_0$ we reject H_0 if

$$\frac{\overline{X} - \mu_0}{\sigma/\sqrt{n}} > z_\alpha$$

Thus,

$$\alpha = P\left(\frac{\overline{X} - \mu_0}{\sigma/\sqrt{n}} > z_\alpha\right)$$

$$= P\left(\overline{X} > \mu_0 + z_\alpha \frac{\sigma}{\sqrt{n}}\right)$$

and therefore $K = \mu_0 + z_\alpha(\sigma/\sqrt{n})$.

We show how to calculate β in the following example.

E X A M P L E   **8.3**   In a power-generating plant, pressure in a certain line is supposed to maintain an average of 100 psi over any 4-hour period. If the average pressure exceeds 103 psi for a 4-hour period, serious complications can evolve. During a given 4-hour period, $n = 30$ measurements (assumed random) are to be taken. For testing $H_0 : \mu \leq 100$ versus $H_a : \mu = 103$, α is to be 0.01. If $\sigma = 4$ psi for these measurements, calculate the probability of a Type II error.

Solution   For testing $H_0 : \mu \leq 100$ versus $H_a : \mu = 103$, we reject H_0 if

$$\frac{\overline{X} - \mu_0}{\sigma/\sqrt{n}} > z_{0.01} = 2.33$$

or if

$$\overline{X} > \mu_0 + z_{0.01}\frac{\sigma}{\sqrt{n}}$$

$$= 100 + 2.33\left(\frac{4}{\sqrt{30}}\right) = 101.7$$

Now if the true mean is really 103, then

$$\beta = P(\overline{X} < 101.7) = P\left(\frac{\overline{X} - \mu_a}{\sigma/\sqrt{n}} < \frac{101.7 - 103}{4/\sqrt{30}}\right)$$

$$= P(Z < -1.78) = 0.0375$$

Under these conditions, the chance of observing a sample mean that is not in the rejection region when the true average pressure is 103 psi is quite small (less than 4%). Thus, the operation can confidently continue if the sample mean of the tests supports H_0, that is, if $\overline{x}$ is less than 101.7.   ■

Notice that, for a fixed sample size, increasing the size of the rejection region will increase α and decrease β. In many practical problems, a specific value for an alternative will not be known, and consequently β cannot be calculated. In those cases, we want to choose an appropriate significance level α and a test statistic that will make β as small as possible. Fortunately, most of the test statistics we use in this text have the property that, for fixed α and fixed sample size, β is nearly as small as possible for any alternative parameter value. In that sense, we are using the best possible test statistics. Furthermore, we set up our hypotheses so that if the test statistic falls into the rejection region, we reject H_0, knowing that the risk of a Type I error is fixed at α. However, if the test statistic is not in the rejection region for H_0, we hedge by stating that the evidence is insufficient to reject H_0. We *do not* affirmatively accept H_0.

For specified μ_0 and μ_a, we can see from Figure 8.1 that if α is decreased, K will move to the right and β will increase. On the other hand, if α is increased, K will move to the left and β will decrease. However, if the sample size is increased, the sampling distributions will display less variability (they will become narrower and more sharply peaked), and *both* α and β can decrease. We will now show that it is possible to select a sample size to guarantee specified α and β, as long as μ_0 and μ_a are also specified.

For testing $H_0 : \mu = \mu_0$ against $H_a : \mu = \mu_a$, where $\mu_a > \mu_0$, we have seen that H_0 is rejected when $\overline{X} > K$, where $K = \mu_0 + z_\alpha(\sigma/\sqrt{n})$. But from Figure 8.1 it is clear that we also have $K = \mu_a - z_\beta(\sigma/\sqrt{n})$. Thus,

$$\mu_0 + z_\alpha(\sigma/\sqrt{n}) = \mu_a - z_\beta(\sigma/\sqrt{n})$$
$$(z_\alpha + z_\beta)(\sigma/\sqrt{n}) = \mu_a - \mu_0$$

or

$$\sqrt{n} = \frac{(z_\alpha + z_\beta)\sigma}{\mu_a - \mu_0}$$

This allows us to find a value of n that will produce a specified α and β for known μ_0 and μ_a.

The sample size for specified α and β when testing $H_0 : \mu = \mu_0$ versus $H_a : \mu = \mu_a$ is given by

$$n = \frac{(z_\alpha + z_\beta)^2 \sigma^2}{(\mu_a - \mu_0)^2}$$

E X A M P L E   **8.4**   For the power plant of Example 8.3, it is desired to test $H_0 : \mu = 100$ versus $H_a : \mu = 103$ with $\alpha = 0.01$ and $\beta = 0.01$. How many observations are needed to ensure this result?

Solution   Since $\alpha = \beta = 0.01$, then $z_\alpha = z_\beta = 2.33$. Also, $\sigma = 4$ for these measurements. We then have

$$n = \frac{(z_\alpha + z_\beta)^2 \sigma^2}{(\mu_a - \mu_0)^2}$$

$$= \frac{(2.33 + 2.33)^2 (4)^2}{(103 - 100)^2} = 38.6 \text{ or } 39$$

By taking 39 measurements, we can reduce β to 0.01 while also holding α at 0.01.   ∎

The hypothesis-testing procedure we have outlined above involves specifying a significance level α, finding $z_{\alpha/2}$ (or z_α for a one-sided test), calculating a value for the test statistic Z, and rejecting H_0 if $|z| > z_{\alpha/2}$ (or the appropriate one-sided adjustment). In this process, α is somewhat arbitrarily determined. An alternative to specifying α is to find the *smallest* significance level at which the observed result would lead to rejection of H_0. This value is called the *P value* of the test, or the *attained significance level*. We would then reject H_0 for small P values.

In Example 8.1, the data produced $z = 1.67$ as the observed value of the test statistic for testing $H_0 : \mu = 2$ versus $H_a : \mu \neq 2$. The *smallest* significance level α that would lead to rejection of H_0 in this case must be such that $z_{\alpha/2} = 1.67$, since this is a two-sided test. Thus, $\alpha/2$ must be the area under the standard normal curve to the right of 1.67, or $\alpha/2 = 0.5000 - 0.4525 = 0.0475$, which implies that $\alpha = 2(0.0475) = 0.095$. The P value for this test is 0.095, and we would reject H_0 for a significance level greater than or equal to this P value. Note that we did *not* reject H_0 at the 0.05 significance level in Example 8.1 and that 0.05 is less than 0.095.

The P value can be thought of as the weight of evidence in the sample regarding H_0. If the P value is small (close to zero), there is not much evidence in favor of H_0, and it should be rejected. If the P value is large (say, close to 0.5), there is strong evidence in favor of H_0, and it should not be rejected.

8.2.2   Binomial Distribution: Testing the Probability of Success

Just as in the case of interval estimation, the large-sample test for a mean can be easily transformed into a test for a binomial proportion. Recall that if Y has a binomial distribution with mean np, then Y/n is approximately normal for large n, with mean p and variance $p(1 - p)/n$. We illustrate with the following example.

EXAMPLE 8.5   A machine in a certain factory must be repaired if it produces more than 10% defectives among the large lot of items it produces in a day. A random sample of 100 items from the day's production contains 15 defectives, and the foreman says that the machine must be repaired. Does the sample evidence support his decision at the 0.01 significance level? Find the P value for this test.

Solution We want to test $H_0 : p \leq 0.10$ versus the alternative $H_a : p > 0.10$, where p denotes the proportion of defectives in the population. The test statistic will be based on Y/n, where Y denotes the number of defectives observed. We will reject H_0 in favor of H_a if Y/n is suitably large. Following the procedure for testing a mean, we reject H_0 if

$$Y/n > K$$

or

$$Z = \frac{Y/n - p_0}{\sqrt{\dfrac{p_0(1 - p_0)}{n}}} > z_\alpha$$

Note that p_0 can be used in the variance of Y/n, since we are assuming H_0 is true when we perform the test. Our observed value of the test statistic, with $\hat{p} = y/n = 0.15$, is

$$z = \frac{\hat{p} - p_0}{\sqrt{\dfrac{p_0(1 - p_0)}{n}}} = \frac{0.15 - 0.10}{\sqrt{\dfrac{(0.1)(0.9)}{100}}} = \frac{5}{3} = 1.67$$

Now $z_{0.01} = 2.33$, and since $1.67 < 2.33$ we will not reject H_0. It is quite likely that a sample fraction of 15% could occur even if p is in the neighborhood of 0.10. The evidence does not support the foreman's decision at the 0.01 significance level. The α value should be chosen to be quite small here, since a Type I error (rejecting H_0 when it is true) has serious consequences; namely, the machine would be shut down unnecessarily for repairs.

Note that, even though the test statistic did not fall in the rejection region, we did not suggest that the null hypothesis be accepted (i.e., that the true proportion of defectives is 0.10). We simply stated that the sample does not refute the null hypothesis. To be more specific would be to risk a Type II error, and we have not computed a probability β of making such an error for any specified p_a.

The smallest significance level leading to rejection of H_0 in this case is the one producing $z_\alpha = 1.67$, since this is a one-sided test. From Table 4 of the Appendix, the area to the right of 1.67 is 0.0475, which is the P value for this test. We would reject H_0 at any significance level at or above this value. ∎

8.2.3 Normal Distribution: Testing the Mean

When sample sizes are too small for the Central Limit Theorem to provide a good approximation to the distribution of $\overline{X}$, additional assumptions must be made on the nature of the probabilistic model for the population. Just as in the case of estimation, one common procedure is to assume, whenever it can be justified, that the population measurements fit the normal distribution reasonably well. Then $\overline{X}$, the sample mean for a random sample of size n, will have a normal distribution, and

$$\frac{\overline{X} - \mu}{S/\sqrt{n}}$$

will have a t distribution with $(n-1)$ degrees of freedom. Thus, the hypothesis-testing procedures for the population mean will have the same pattern as above, but the test statistic

$$T = \frac{\overline{X} - \mu_0}{S/\sqrt{n}}$$

will have a t distribution, rather than a normal distribution, under $H_0: \mu = \mu_0$. For testing $H_0: \mu = \mu_0$ versus $H_a: \mu \neq \mu_0$, the test statistic $T = (\overline{X} - \mu_0)/(S/\sqrt{n})$ is calculated for the observed sample, and H_0 is rejected if $|T| > t_{\alpha/2}$, where $t_{\alpha/2}$ is the point cutting off an area of $\alpha/2$ in the upper tail of the t distribution with $(n-1)$ degrees of freedom. (See Table 5 of the Appendix.) Corresponding one-sided tests can be conducted in a manner analogous to the large-sample case. The following examples show the methodology.

E X A M P L E **8.6** A corporation sets its annual budget for a new plant on the assumption that the average weekly cost for repairs is to be $\mu = \$1,200$. To see if this claim is realistic, $n = 10$ weekly repair cost figures are obtained from similar plants. The sample is assumed to be random and yields $\overline{x} = 1,290$ and $s = 110$. Since the detection of a departure from the assumed average in either direction would be important for budgeting purposes, it is desired to test $H_0: \mu = 1,200$ versus $H_a: \mu \neq 1,200$. Use $\alpha = 0.05$ and carry out the test.

Solution Assuming normality of weekly repair costs, the test statistic is

$$T = \frac{\overline{X} - \mu_0}{S/\sqrt{n}}$$

and the rejection region starts at $t_{\alpha/2} = t_{0.025} = 2.262$ for $n-1 = 9$ degrees of freedom. The observed value of the test statistic is

$$t = \frac{\overline{x} - \mu_0}{s/\sqrt{n}} = \frac{1,290 - 1,200}{110/\sqrt{10}} = 2.587$$

which is greater than 2.262. Therefore, we reject the null hypothesis. There is reason to suspect that 1,200 is not a good assumed value of μ, and perhaps more investigation into these repair costs should be made before the budget is set. ∎

The following example illustrates a one-sided alternative-testing situation.

E X A M P L E **8.7** Muzzle velocities of eight shells tested with a new gunpowder yield a sample mean of $\overline{x} = 2,959$ feet per second and a standard deviation of $s = 39.4$. The manufacturer claims that the new gunpowder produces an average velocity of no less than 3,000 feet per second. Does the sample provide enough evidence to contradict the manufacturer's claim? Use $\alpha = 0.05$.

Solution Here we are interested in testing $H_0 : \mu \geq 3,000$ versus $H_a : \mu < 3,000$, since we want to see if there is evidence to refute the manufacturer's claim. Assuming that muzzle velocities can be reasonably modeled by a normal probability distribution, we have the test statistic

$$T = \frac{\overline{X} - \mu_0}{S/\sqrt{n}}$$

which is observed to be

$$t = \frac{\overline{x} - \mu_0}{s/\sqrt{n}} = \frac{2,959 - 3,000}{39.4/\sqrt{8}} = -2.943$$

The rejection region starts at $-t_{0.05} = -1.895$ for 7 degrees of freedom. Since $t < -t_\alpha$, we reject the null hypothesis and say that there appears to be good reason to doubt the manufacturer's claim. ∎

8.2.4 Normal Distribution: Testing the Variance

The variance of the underlying normal distributions in the above examples was assumed to be unknown, which is generally the case. If it is desired to test a hypothesis on the variance σ^2, it can be done by making use of the statistics used in constructing the confidence interval for σ^2. Recall that, for a random sample of size n from a normal distribution,

$$\frac{(n-1)S^2}{\sigma^2}$$

has a $\chi^2(n-1)$ distribution. Following the same principles as above for testing $H_0 : \sigma = \sigma_0^2$ versus $H_a : \sigma^2 \neq \sigma_0^2$, we calculate a value for the statistic $(n-1)S^2/\sigma_0^2$ and reject H_0, at level α, if this statistic is larger than $\chi_{\alpha/2}^2(n-1)$ or smaller than $\chi_{1-\alpha/2}^2(n-1)$. Suitable one-sided tests can also be employed.

E X A M P L E **8.8** A machined engine part produced by a certain company is claimed to have a diameter variance no larger than 0.0002 inch. A random sample of ten such parts gave a sample variance of $s^2 = 0.0003$. Assuming normality of diameter measurement, is there evidence here to refute the company's claim? Use $\alpha = 0.05$.

Solution The test statistic is

$$\frac{(n-1)S^2}{\sigma_0^2}$$

which is observed to be

$$\frac{(n-1)s^2}{\sigma_0^2} = \frac{9(0.0003)}{0.0002} = 13.5$$

Here we are testing $H_0 : \sigma^2 \leq 0.0002$ versus $H_a : \sigma^2 > 0.0002$. Thus, we reject H_0 if $(n-1)S^2/\sigma_0^2$ is larger than $\chi_{0.05}^2(9) = 16.919$. Since this is not the case, we do not have sufficient evidence to refute the company's claim.   ∎

Many other exact small-sample tests could be illustrated for other distributions, but these examples give the general idea for some commonly used cases. The single-sample tests of this chapter are summarized in Table 8.3.

T A B L E **8.3**   Summary of Single-Sample Tests

	Null Hypothesis (H_0)	Alternative Hypothesis (H_a)	Test Statistic	Rejection Region
General	$\mu = \mu_0$	$\mu \neq \mu_0$		$|z| > z_{\alpha/2}$
Distribution	$\mu \leq \mu_0$	$\mu > \mu_0$	$Z = \dfrac{\overline{X} - \mu_0}{\sigma\sqrt{n}}$	$z > z_\alpha$
(large sample)	$\mu \geq \mu_0$	$\mu < \mu_0$		$z < -z_\alpha$
Binomial	$p = p_0$	$p \neq p_0$		$|z| > z_{\alpha/2}$
Distribution	$p \leq p_0$	$p > p_0$	$Z = \dfrac{Y/n - p_0}{\sqrt{\dfrac{p_0(1 - p_0)}{n}}}$	$z > z_\alpha$
(large sample)	$p \geq p_0$	$p < p_0$		$z < -z_\alpha$
Normal	$\mu = \mu_0$	$\mu \neq \mu_0$		$|t| > t_{\alpha/2}$
Distribution	$\mu \leq \mu_0$	$\mu > \mu_0$	$T = \dfrac{\overline{X} - \mu_0}{S/\sqrt{n}}$	$t > t_\alpha$
	$\mu \geq \mu_0$	$\mu < \mu_0$	$(n-1)$ df (degrees of freedom)	$t < -t_\alpha$
Normal	$\sigma^2 = \sigma_0^2$	$\sigma^2 \neq \sigma_0^2$		$\dfrac{(n-1)s^2}{\sigma_0^2} < \chi_{1-\alpha/2}^2$ $> \chi_{\alpha/2}^2$
Distribution	$\sigma^2 \leq \sigma_0^2$	$\sigma^2 > \sigma_0^2$	$\dfrac{(n-1)S^2}{\sigma_0^2}$	$\dfrac{(n-1)s^2}{\sigma_0^2} > \chi_\alpha^2$
	$\sigma^2 \geq \sigma_0^2$	$\sigma^2 < \sigma_0^2$	$(n-1)$df	$\dfrac{(n-1)s^2}{\sigma_0^2} < \chi_{1-\alpha}^2$

Exercises

8.1 The output voltage for a certain electric circuit is specified to be 130. A sample of 40 independent readings on the voltage for this circuit gave a sample mean of 128.6 and a standard deviation of 2.1. Test the hypothesis that the average output voltage is 130 against the alternative that it is less than 130. Use a 5% significance level.

8.2 Refer to Exercise 8.1. If the voltage falls as low as 129, serious consequences may result. For testing $H_0 : \mu \geq 130$ versus $H_a : \mu = 129$, find β, the probability of a Type II error, for the rejection region used in Exercise 8.1.

8.3 For testing $H_0 : \mu = 130$ versus $H_a : \mu = 129$ with $\sigma = 2.1$, as in Exercise 8.2, find the sample size that will yield $\alpha = 0.05$ and $\beta = 0.01$.

8.4 The Rockwell hardness index for steel is determined by pressing a diamond point into the steel and measuring the depth of penetration. For 50 specimens of a certain type of steel, the Rockwell hardness index averaged 62 with a standard deviation of 8. The manufacturer claims that this steel has an average hardness index of at least 64. Test this claim at the 1% significance level. Find the P value for this test.

8.5 Steel is sufficiently hard for a certain use as long as the mean Rockwell hardness measure does not drop below 60. Using the rejection region found in Exercise 8.4, find β for the specific alternative $\mu = 60$.

8.6 For testing $H_0 : \mu = 64$ versus $H_a : \mu = 60$ with $\sigma = 8$, as in Exercises 8.4 and 8.5, find the sample size required for $\alpha = 0.01$ and $\beta = 0.05$.

8.7 The pH of water coming out of a certain filtration plant is specified to be 7.0. Thirty water samples independently selected from this plant show a mean pH of 6.8 and a standard deviation of 0.9. Is there any reason to doubt that the plant's specification is being maintained? Use $\alpha = 0.05$. Find the P value for this test.

8.8 A manufacturer of resistors claims that 10% fail to meet the established tolerance limits. A random sample of resistance measurements for 60 such resistors reveals eight to lie outside the tolerance limits. Is there sufficient evidence to refute the manufacturer's claim, at the 5% significance level? Find the P value for this test.

8.9 For a certain type of electronic surveillance system, the specifications state that the system will function for more than 1,000 hours with probability at least 0.90. Checks on 40 such systems show that five failed prior to 1,000 hours of operation. Does this sample provide sufficient information to conclude that the specification is not being met? Use $\alpha = 0.01$.

8.10 The hardness of a certain rubber (in degrees Shore) is claimed to be 65. Fourteen specimens are tested, resulting in an average hardness measure of 63.1 and a standard deviation of 1.4. Is there sufficient evidence to reject the claim, at the 5% level of significance? What assumption is necessary for your answer to be valid?

8.11 Certain rockets are manufactured with a range of 2,500 meters. It is theorized that the range will be reduced after the rockets are in storage for some time. Six of these rockets are stored for a certain period of time and then tested. The ranges found in the tests are as follows: 2,490, 2,510, 2,360, 2,410, 2,300, and 2,440. Does the range appear to be shorter after storage? Test at the 1% significance level.

8.12 For screened coke, the porosity factor is measured by the difference in weight between dry and soaked coke. A certain supply of coke is claimed to have a porosity factor of 1.5 kilograms. Ten samples are tested, resulting in a mean porosity factor of 1.9 kilograms and a variance of 0.04. Is there sufficient evidence to indicate that the coke is more porous than is claimed? Use $\alpha = 0.05$ and assume the porosity measurements are approximately normally distributed.

8.13 The stress resistance of a certain plastic is specified to be 30 psi. The results from ten specimens of this plastic show a mean of 27.4 psi and a standard deviation of 1.1 psi. Is there sufficient evidence to doubt the specification at the 5% significance level? What assumption are you making?

8.14 Yield stress measurements on 51 steel rods with 10-mm diameters gave a mean of 485 N/mm^2 and a standard deviation of 17.2. (See P. Booster, *Journal of Quality Technology*, 1, no. 4, 1983, p. 191, for details.) Suppose the manufacturer claims that the yield stress for these bars is 490. Does the sample information suggest rejecting the manufacturer's claim, at the 5% significance level?

8.15 The widths of contact windows in certain CMOS circuit chips have a design specification of 3.5 μm. (See M. S. Phadke, et al., *The Bell System Technical Journal*, 62, no. 5, 1983, pp. 1273–1309 for details.) Postetch window widths of test specimens were as follows:

$$3.21, 2.49, 2.94, 4.38, 4.02, 3.82, 3.30, 2.85, 3.34, 3.91$$

(These data were also used in Exercise 7.81.) Can we reject the hypothesis that the design specification is being met, at the 5% significance level? What assumptions are necessary for this test to be valid?

 8.16 The variation in window widths for CMOS circuit chips must be controlled at a low level if the circuits are to function properly. Suppose specifications state that $\sigma = 0.30$ for window widths. Can we reject the claim that this specification is being met, using the data of Exercise 8.15? Use $\alpha = 0.05$.

8.17 The Florida Poll of February–March 1984 interviewed 871 adults from around the state. On one question, 53% of the respondents favored strong support of Israel. Would you conclude that a majority of adults in Florida favor strong support of Israel? (Source: *Gainesville Sun*, 1 April 1984.)

 8.18 A Yankelovich, Skelly, and White poll reported upon on November 30, 1984, showed that 54% of 2,207 people surveyed thought the U.S. income tax system was too complicated. Can we conclude safely, at the 5% significance level, that the majority of Americans think the income tax system is too complicated?

 8.19 Wire used for wrapping concrete pipe should have an ultimate tensile strength of 300,000 pounds, according to a design engineer. Forty tests of such wire used on a certain pipe showed a mean ultimate tensile strength of 295,000 pounds and a standard deviation of 10,000. Is there sufficient evidence to suggest that the wire used on the tested pipe does not meet the design specification at the 10% significance level?

 8.20 The break angle in a torsion test of wire used in wrapping concrete pipe is specified to be 40°. Fifty torsion tests of wire wrapping a certain malfunctioning pipe resulted in a mean break angle of 37° and a standard deviation of 6°. Can we say that the wire used on the malfunctioning pipe does not meet the specification at the 5% significance level?

 8.21 Soil pH is an important variable in the design of structures that will contact the soil. The pH at a potential construction site was said to average 6.5. Nine test samples of soil from the site gave readings of

$$7.3, 6.5, 6.4, 6.1, 6.0, 6.5, 6.2, 5.8, 6.7$$

Do these readings cast doubt upon the claimed average? (Test at the 5% significance level.) What assumptions are necessary for this test to be valid?

 8.22 The resistances of a certain type of thermistor are specified to be normally distributed with a mean of 10,000 ohms and a standard deviation of 500 ohms, at a temperature of 25°C. Thus, only about 2.3% of these thermistors should produce resistances in excess of 11,000 ohms. The exact measurement of resistances produced is time-consuming, but it is easy to determine if the resistance is larger than a specified value (say 11,000). For 100 thermistors tested, 5% showed resistances in excess of 11,000 ohms. Do you think the thermistors from which this sample was selected fail to meet the specifications? Why or why not?

8.23 The dispersion, or variance, of haul times on a construction project are of great importance to the project supervisor, since highly variable haul times cause problems in scheduling jobs. The supervisor of the truck crews states that the range of haul times should not exceed 40 minutes. (The range is the difference between the longest and shortest times.) Assuming these haul times to be approximately normally distributed, the project supervisor takes the statement on the range to mean that the standard deviation σ should be approximately 10 minutes. Fifteen haul times are actually measured and show a mean of 142 minutes and a standard deviation of 12 minutes. Can the claim of $\sigma = 10$ be refuted at the 5% significance level?

8.24 Aptitude tests should produce scores with a large amount of variation so that an administrator can distinguish between persons with low aptitude and persons with high aptitude. The standard test used by a certain industry has been producing scores with a standard deviation of 5 points. A new test is tried on 20 prospective employees and produces a sample standard deviation of 8 points. Are scores from the new test significantly more variable than scores from the standard? Use $\alpha = 0.05$.

8.25 For the rockets of Exercise 8.11, the variation in ranges is also of importance. New rockets have a standard deviation of range measurements equal to 20 kilometers. Does it appear that storage increases the variability of these ranges? Use $\alpha = 0.05$.

8.3 Hypothesis Testing: The Multiple-Sample Case

Just as we developed confidence interval estimates for linear functions of means when k samples from k different populations are available, we could develop the corresponding hypothesis tests. However, most tests of population means involve tests of simple differences of the form $\mu_1 - \mu_2$. Thus, we will consider only this case. (The more general case of testing equality of k means will be covered in a discussion of analysis of variance in Chapter 11.)

8.3.1 General Distributions: Testing the Difference Between Two Means

We know from Chapter 7 that, in the large-sample case, the estimator $\overline{X}_1 - \overline{X}_2$ has an approximate normal distribution. Hence, for testing $H_0 : \mu_1 - \mu_2 = D_0$ versus $H_a : \mu_1 - \mu_2 \neq D_0$, we can use as a test statistic

$$Z = \frac{(\overline{X}_1 - \overline{X}_2) - D_0}{\sqrt{\dfrac{\sigma_1^2}{n_1} + \dfrac{\sigma_2^2}{n_2}}}$$

We reject H_0 for $|Z| > z_{\alpha/2}$ for a specified α. Similar one-sided tests can be constructed in the obvious way. If σ_1^2 and σ_2^2 are unknown, they can be estimated by the sample variances S_1^2 and S_2^2, respectively.

EXAMPLE **8.9** A study was conducted to compare the length of time it took men and women to perform a certain assembly-line task. Independent samples of 50 men and 50 women were employed in an experiment in which each person was timed on identical tasks. The results were as follows:

Men	Women
$n_1 = 50$	$n_2 = 50$
$\bar{x}_1 = 42$ sec.	$\bar{x}_2 = 38$ sec.
$s_1^2 = 18$	$s_2^2 = 14$

Do the data present sufficient evidence to suggest a difference between the true mean completion times for men and women at the 5% significance level?

Solution Since we are interested in detecting a difference in either direction, we want to test $H_0 : \mu_1 - \mu_2 = 0$ (no difference) versus $H_a : \mu_1 - \mu_2 \neq 0$. The test statistic is calculated to be (with s_i estimating σ_i)

$$z = \frac{(\bar{x}_1 - \bar{x}_2) - D_0}{\sqrt{\dfrac{s_1^2}{n_1} + \dfrac{s_2^2}{n_2}}} = \frac{42 - 38}{\sqrt{\dfrac{18}{50} + \dfrac{14}{50}}} = 5$$

That is, the difference between the sample mean is five standard deviations from the hypothesized zero difference. Now $z_{0.025} = 1.96$, and since $|z| > 1.96$, we reject H_0. It does appear that the difference in times between men and women is real. ∎

8.3.2 Normal Distributions with Equal Variances: Testing the Difference Between Two Means

As before, the small-sample case necessitates an assumption as to the nature of the probabilistic model for the random variables in question. If both populations seem to have normal distributions and if the two variances are equal ($\sigma_1^2 = \sigma_2^2 = \sigma^2$), then t tests on hypotheses concerning $\mu_1 - \mu_2$ can be constructed. For testing $H_0 : \mu_1 - \mu_2 = D_0$ versus $H_a : \mu_1 - \mu_2 \neq D_0$, the test statistic to be employed is

$$T = \frac{(\overline{X}_1 - \overline{X}_2) - D_0}{S_p \sqrt{\dfrac{1}{n_1} + \dfrac{1}{n_2}}}$$

where

$$S_p^2 = \frac{(n_1 - 1)S_1^2 + (n_2 - 1)S_2^2}{n_1 + n_2 - 2}$$

T has a t distribution with $n_1 + n_2 - 2$ degrees of freedom when H_0 is true. As in the large-sample case, we reject H_0 whenever $|T| > t_{\alpha/2}$. Corresponding one-sided tests are obtained easily.

E X A M P L E 8.10 The designer of a new sheet-metal stamping machine claims that her new machine can turn out a certain product faster than the machine now in use. Nine independent trials of stamping the same item on each machine gave the following results on times to completion:

Standard Machine	New Machine
$n_1 = 9$	$n_2 = 9$
$\bar{x}_1 = 35.22$ sec.	$\bar{x}_2 = 31.56$ sec.
$(n_1 - 1)s_1^2 = 195.50$	$(n_2 - 1)s_2^2 = 160.22$

At the 5% significance level, can the designer's claim be substantiated?

Solution   We are interested here in testing $H_0 : \mu_1 - \mu_2 \leq 0$ versus $H_a : \mu_1 - \mu_2 > 0$ (or $\mu_2 < \mu_1$). The test statistic T is calculated as

$$t = \frac{(\bar{x}_1 - \bar{x}_2) - D_0}{s_p\sqrt{\dfrac{1}{n_1} + \dfrac{1}{n_2}}} = \frac{35.22 - 31.56}{4.71\sqrt{\dfrac{1}{9} + \dfrac{1}{9}}} = 1.65$$

since

$$s_p^2 = \frac{(n-1)s_1^2 + (n_2 - 1)s_2^2}{n_1 + n_2 - 2} = \frac{195.50 + 160.22}{16} = 22.24$$

Since $t_{0.05} = 1.746$ for 16 degrees of freedom, we cannot reject H_0. There is insufficient evidence here to substantiate the designer's claim.   ∎

To see how a typical computer printout displays the computations from a standard two-sample test of equality of means, study the following example.

E X A M P L E   **8.11**   A quality-control inspector compares the ultimate tensile strength (UTS) measurements for class II and class III prestressing wire by taking a sample of five specimens from a roll of each class for laboratory testing. The sample data (in 1,000 psi) are as follows:

Class II: 253, 261, 258, 255, 256

Class III: 274, 275, 271, 277, 276

Do the true mean UTS measurements for the two classes of wire appear to differ?

Solution   The computer output (Minitab) is as follows:

```
TWOSAMPLE T FOR classII VS classIII
           N      MEAN STDEV        SE MEAN
classII    5     256.60   3.05         1.4
classIII   5     274.60   2.30         1.0
TTEST MU classII = MU classIII (VS NE): T=-10.53  P=0.0000  DF=8
POOLED STDEV =    2.70
```

For testing equality of means, the t statistic is -10.53 with a two-sided P value of essentially zero. Thus, there is strong evidence to indicate that the mean UTS measurements differ for class II and class III wire.   ∎

8.3.3   Normal Distributions with Unequal Variances: Testing the Difference Between Two Means

If one suspects that the variances in the two normal populations under study are *not* equal, then the *t* test based on pooling the sample variances should not be used. What should be used in its place? Unfortunately, the answer to that question is not easy, since there is no exact way to solve this problem. There are, however, many approximate solutions; one fairly simple one will be presented here.

For this approximate procedure, the first step is to calculate the T statistic in a manner similar to that for the Z statistic of Section 8.3.1, namely,

$$T = \frac{(\overline{X}_1 - \overline{X}_2) - D_0}{\sqrt{\dfrac{S_1^2}{n_1} + \dfrac{S_2^2}{n_2}}}$$

This statistic has approximately a t distribution under $H_0 : \mu_1 - \mu_2 = D_0$, with degrees of freedom given by the integer part of

$$\mathrm{df} = \frac{[(s_1^2/n_1) + (s_2^2/n_2)]^2}{[(s_1^2/n_1)^2/(n_1 - 1)] + [(s_2^2/n_2)^2/(n_2 - 1)]}$$

where s_1^2 and s_2^2 are the sample variances. It is recommended that these computations be done by computer, so we will show a Minitab calculation for an example of this type.

EXAMPLE **8.12**   The prestressing wire on each of two concrete pipes manufactured at different times was compared for torsion properties. Ten specimens randomly selected from each pipe were twisted in a laboratory apparatus until they broke, with the number of revolutions until complete failure being the measurement of interest. The results are as follows, with C1 and C2 denoting the two concrete pipes:

C1:   6.38, 6.88, 7.00, 4.75, 1.50, 3.37, 7.63, 3.63, 4.00, 4.63
C2:   3.38, 2.81, 3.00, 5.88, 5.25, 4.08, 4.25, 4.50, 4.13, 4.88

Is there evidence to suggest that the true mean revolutions to failure differ for the wire on the two pipes?

Solution   Careful study of the data sets shows that C1 results appear to be more variable than the C2 results. This can be shown more dramatically on a stem-and-leaf plot of the two sets (here truncated to one decimal place):

C1		C2
5	1	
	2	8
63	3	03
760	4	01258
	5	28
83	6	
60	7	

It doesn't seem reasonable here to assume that these two sets of measurements come from populations with the same variance. (We will see how to conduct a test for equal variances in Section 8.3.5.) Thus, we will use the approximate procedure outlined above. The computer printout is shown below:

```
TWOSAMPLE T FOR C1 VS C2
       N       MEAN      STDEV      SE MEAN
C1   10       4.98       1.95         0.62
C2   10       4.216      0.974        0.31
TTEST MU C1=MU C2 (VS NE): T=1.10   P=0.29   DF=13
```

The observed t statistic is 1.10 with 13 degrees of freedom. The P value of 0.29 suggests that we not reject H_0 and state that there is no evidence to suggest the mean torsion values differ for the two pipes.   ■

8.3.4   Normal Distributions: Testing the Difference Between Means for Paired Samples

The two-sample t test given above works only in the case of *independent* samples. On many occasions, however, two samples will arise in a dependent fashion. A commonly occurring situation is that in which repeated observations are taken on the *same* sampling unit, such as counting the number of accidents in various plants both before and after a safety awareness program is effected. The counts in one plant may be independent of the counts in another, but the two counts (before and after) within any one plant will be dependent. Thus, we must develop a mechanism for analyzing measurements that occur in pairs.

Let $(X_1, Y_1), \ldots, (X_n, Y_n)$ denote a random sample of paired observations. That is, (X_i, Y_i) denotes two measurements taken in the same sampling unit, such as counts of accidents within a plant before and after a safety awareness program is put into effect, or lifelengths of two components within the same machine. Suppose it is of interest to test a hypothesis concerning the difference between $E(X_i)$ and $E(Y_i)$. The two-sample tests developed earlier in this chapter cannot be used for this purpose because of the dependence between X_i and Y_i. Observe, however, that $E(X_i) - E(Y_i) = E(X_i - Y_i)$. Thus, the comparison of $E(X_i)$ with $E(Y_i)$ can be made by looking at the mean of the differences, $(X_i - Y_i)$. Letting $X_i - Y_i = D_i$, the hypothesis $E(X_i) - E(Y_i) = 0$ is equivalent to the hypothesis $E(D_i) = 0$.

To construct a test statistic with a known distribution, we assume that $D_1, \ldots, D_n$ is a random sample of *differences*, each possessing a normal distribution with mean μ_D and variance σ_D^2. To test $H_0 : \mu_D = 0$ versus the alternative $H_a : \mu_D \neq 0$, we employ the test statistic

$$T = \frac{\overline{D} - 0}{S_D/\sqrt{n}}$$

where

$$\overline{D} = \frac{1}{n} \sum_{i=1}^{n} D_i$$

and

$$S_D^2 = \frac{1}{n-1} \sum_{i=1}^{n} (D_i - \overline{D})^2$$

Since differences $D_1, \ldots, D_n$ have the same normal distribution, T will have a t distribution with $n-1$ degrees of freedom when H_0 is true. Thus, we have reduced the problem to that of a one-sample t test. Example 8.13 illustrates the use of this procedure.

E X A M P L E   **8.13**   Two methods of determining the percentage of iron in ore samples are to be compared by subjecting 12 ore samples to each method. The results of the experiment are as follows:

Ore Sample	Method A	Method B	d_i
1	38.25	38.27	−0.02
2	31.68	31.71	−0.03
3	26.24	26.22	+0.02
4	41.29	41.33	−0.04
5	44.81	44.80	+0.01
6	46.37	46.39	−0.02
7	35.42	35.46	−0.04
8	38.41	38.39	+0.02
9	42.68	42.72	−0.04
10	46.71	46.76	−0.05
11	29.20	29.18	+0.02
12	30.76	30.79	−0.03

Do the data provide evidence that method B has a higher average percentage than method A? Use $\alpha = 0.05$.

Solution   We want to test $H_0 : \mu_D \geq 0$ versus $H_a : \mu_D < 0$, since μ_D will be negative if method B has the larger mean. For the given data,

$$\overline{d} = \frac{1}{12}(-0.20) = -0.0167$$

and

$$s_D^2 = \frac{\sum_{i=1}^{n} d_i^2 - \frac{1}{n}\left(\sum_{i=1}^{n} d_i\right)^2}{n-1} = \frac{0.0112 - \frac{1}{12}(-0.20)^2}{11} = 0.0007$$

It follows that

$$t = \frac{\overline{d} - 0}{s_d / \sqrt{n}} = \frac{-0.0167}{\sqrt{0.0007}/\sqrt{12}} = -2.1865$$

Since $\alpha = 0.05$ and $n - 1 = 11$ degrees of freedom, the rejection region consists of those values of t smaller than $-t_{0.05} = -1.796$. Hence, we reject the null hypothesis that $\mu_D \geq 0$ and conclude that the data support the alternative, $\mu_D < 0$. Note that the validity of this conclusion, especially the level of α, depends on the assumption that the probabilistic model underlying the differences is normal.   ∎

8.3.5 Normal Distributions: Testing the Ratio of Variances

We can compare the variances of two normal populations by looking at the ratio of sample variances. For testing $H_0 : \sigma_1^2 = \sigma_2^2$ (or $\sigma_1^2/\sigma_2^2 = 1$) versus $H_a : \sigma_1^2 \neq \sigma_2^2$, we calculate a value for S_1^2/S_2^2 and reject H_0 if this statistic is either very large or very small. The precise rejection region can be found by observing that

$$ F = \frac{S_1^2}{S_2^2} $$

has an F distribution when $H_0 : \sigma_1^2 = \sigma_2^2$ is true. Thus, we reject H_0 for $F > F_{\alpha/2}(v_1, v_2)$ or $F < F_{1-\alpha/2}(v_1, v_2)$. A one-sided test can always be constructed with the larger S_i^2 in the numerator, so only the upper-tail critical F value needs to be found.

E X A M P L E   **8.14**   Suppose the machined engine part of Example 8.8 is to be compared, with respect to diameter variance, with a similar part manufactured by a competitor. The former showed $s_1^2 = 0.0003$ for a sample of $n_1 = 10$ diameter measurements. A sample of $n_2 = 20$ of the competitor's parts showed $s_2^2 = 0.0001$. Is there evidence to conclude that $\sigma_2^2 < \sigma_1^2$ at the 5% significance level? Assume normality for both populations.

Solution   We are interested in testing $H_0 : \sigma_1^2 \leq \sigma_2^2$ versus $H_a : \sigma_1^2 > \sigma_2^2$. The outcome of the statistic $F = S_1^2/S_2^2$, which has an F distribution if $\sigma_1^2 = \sigma_2^2$, is calculated to be

$$ \frac{s_1^2}{s_2^2} = \frac{0.0003}{0.0001} = 3 $$

We reject H_0 if $F > F_\alpha(v_1, v_2) = F_{0.05}(9, 19) = 2.42$. Since the observed ratio is 3, we reject H_0. It looks as if the second manufacturer has less variability in the diameter measurements.   ∎

We can employ the F-test prior to the use of a two-sample t test to compare means in order to tell which t test to use (the test for equal population variances or the test for unequal population variances). In Example 8.12, the F ratio for the sample variances is 4.0, which is in the rejection region for the 0.05 significance level. This is further evidence that we were correct in applying the approximate t test that does not require equality of population variances.

A word on the *robustness* of statistical tests of hypotheses is in order. Tests based on T, χ^2, and F assume that the sample measurements come from normally distributed populations. What if the populations are not normal? Will the tests still work? The ability of a test to perform satisfactorily when distributional assumptions are not met is called robustness. In that sense, the t tests are quite robust; t tests perform well even when the normality assumptions are violated by a considerable amount.

On the other hand, tests for variances based on χ^2 and F are not very robust. The worst offender is the F-test for equality of variances. If the population distributions are slightly skewed, this test will reject the null hypothesis of equal variances too often. What is thought, for example, to be a 0.05-level test may, in fact, be a 0.10-level test. The test is actually testing the null hypothesis of equal variances from normal distributions against the alternative of unequal variances from normal distributions *or* distributions that are skewed. So be careful when you interpret the outcome of an F-test on variances; rejection of the null hypothesis may come about for a variety of reasons.

The multiple-sample tests of this section are summarized in Table 8.4.

T A B L E **8.4**   Summary of Two-Sample Tests

	Null Hypothesis (H_0)	Alternative Hypothesis (H_a)	Test Statistic	Rejection Region
General	$\mu_1 - \mu_2 = D_0$	$\mu_1 - \mu_2 \neq D_0$		$|z| > z_{\alpha/2}$
Distribution	$\mu_1 - \mu_2 \leq D_0$	$\mu_1 - \mu_2 > D_0$	$Z = \dfrac{(\overline{X}_1 - \overline{X}_2) - D_0}{\sqrt{\dfrac{\sigma_1^2}{n_1} + \dfrac{\sigma_2^2}{n_2}}}$	$z > z_\alpha$
(large sample)	$\mu_1 - \mu_2 \geq D_0$	$\mu_1 - \mu_2 < D_0$		$z < -z_\alpha$
Normal	$\mu_1 - \mu_2 = D_0$	$\mu_1 - \mu_2 \neq D_0$	$T = \dfrac{(\overline{X}_1 - \overline{X}_2) - D_0}{S_p \sqrt{\dfrac{1}{n_1} + \dfrac{1}{n_2}}}$	$|t| > t_{\alpha/2}$
Distribution (equal variances)	$\mu_1 - \mu_2 \leq D_0$ $\mu_1 - \mu_2 \geq D_0$	$\mu_1 - \mu_2 > D_0$ $\mu_1 - \mu_2 < D_0$	$(n_1 + n_2 - 2)$ df	$t > t_\alpha$ $t < -t_\alpha$
Normal	$\mu_D = \mu_{D_0}$	$\mu_D \neq \mu_{D_0}$	$T = \dfrac{\overline{D} - \mu_{D_0}}{S_D / \sqrt{n}}$	$|t| > t_{\alpha/2}$
Distribution (paired samples)	$\mu_D \leq \mu_{D_0}$ $\mu_D \geq \mu_{D_0}$	$\mu_D > \mu_{D_0}$ $\mu_D < \mu_{D_0}$	$(n - 1)$ df	$t > t_\alpha$ $t < -t_\alpha$
Normal Distribution	$\sigma_1^2 = \sigma_2^2$	$\sigma_1^2 \neq \sigma_2^2$	$F = \dfrac{S_1^2}{S_2^2}$	$\dfrac{S_1^2}{S_2^2} > F_{\alpha/2}(v_1, v_2)$ or $< F_{1-\alpha/2}(v_1, v_2)$
	$\sigma_1^2 \leq \sigma_2^2$	$\sigma_1^2 > \sigma_2^2$		$\dfrac{S_1^2}{S_2^2} > F_\alpha(v_1, v_2)$
	$\sigma_1^2 \geq \sigma_2^2$	$\sigma_1^2 < \sigma_2^2$	$v_1 = (n_1 - 1)$ df $v_2 = (n_2 - 1)$ df	$\dfrac{S_2^2}{S_1^2} > F_\alpha(v_2, v_1)$

Exercises

8.26   Two designs for a laboratory are to be compared with respect to the average amount of light produced on table surfaces. Forty independent measurements (in footcandles) are taken in each laboratory, with the following results:

Design I	Design II
$n_1 = 40$	$n_2 = 40$
$\bar{x}_1 = 28.9$	$\bar{x}_2 = 32.6$
$s_1^2 = 15.1$	$s_2^2 = 15.8$

Is there sufficient evidence to suggest that the designs differ with respect to the average amount of light produced? Use $\alpha = 0.05$.

8.27   Shear-strength measurements derived from unconfined compression tests for two types of soils gave the following results (measurements in tons per square foot):

Soil Type I	Soil Type II
$n_1 = 30$	$n_2 = 35$
$\bar{x}_1 = 1.65$	$\bar{x}_2 = 1.43$
$s_1 = 0.26$	$s_2 = 0.22$

Do the soils appear to differ with respect to average shear strength, at the 1% significance level?

8.28   A study was conducted by the Florida Game and Fish Commission to assess the amounts of chemical residues found in the brain tissue of brown pelicans. For DDT, random samples of $n_1 = 10$ juveniles and $n_2 = 13$ nestlings gave the following results (measurements in parts per million):

Juveniles	Nestlings
$n_1 = 10$	$n_2 = 13$
$\bar{y}_1 = 0.041$	$\bar{y}_2 = 0.026$
$s_1 = 0.017$	$s_2 = 0.016$

Test the hypothesis that there is no difference between mean amounts of DDT found in juveniles and nestlings versus the alternative that the juveniles have a larger mean amount. Use $\alpha = 0.05$. (This test has important implications regarding the build-up of DDT over time.)

8.29   The strength of concrete depends, to some extent, on the method used for drying. Two different drying methods showed the following results for independently tested specimens (measurements in psi):

Method I	Method II
$n_1 = 7$	$n_2 = 10$
$\bar{x}_1 = 3,250$	$\bar{x}_2 = 3,240$
$s_1 = 210$	$s_2 = 190$

Do the methods appear to produce concrete with different mean strengths? Use $\alpha = 0.05$. What assumptions are necessary in order for your answer to be valid?

8.30   The retention of nitrogen in the soil is an important consideration in cultivation practices, including the cultivation of forests. Two methods for preparing plots for planting pine trees after clear-cutting were compared on the basis of the percentage of labeled nitrogen recovered. Method A leaves much of the forest floor intact, while method B removes most of the organic material. It is clear that method B will produce a much lower recovery of nitrogen from the forest floor. The question of interest is whether or not method B will cause more nitrogen to be retained in the microbial biomass, as a compensation for having less organic material available. Percentage of nitrogen recovered in the microbial biomass was measured on six test plots for each method. Method A plots showed a mean of 12 and a standard deviation of 1. Method B plots showed a mean of 15 and a standard deviation of 2. At the 10% significance level, should we say that the mean percentage of recovered nitrogen is larger for method B? (See P. M. Vitousek and P. A. Matson, *Science*, 225, 6 July 1984, p. 51, for more details.)

8.31   In the presence of reverberation, native as well as nonnative speakers of English have some trouble recognizing consonants. Random samples of ten natives and ten nonnatives were given the Modified Rhyme Test, and the percentages of correct responses were recorded. (See A. K. Nabelek and A. M. Donahue, *Journal of the Acoustical Society of America*, 75, 2, 1984, p. 633.) The data are as follows:

Natives: 93, 85, 89, 81, 88, 88, 89, 85, 85, 87
Nonnatives: 76, 84, 78, 73, 78, 76, 70, 82, 79, 77

Is there sufficient evidence to say that nonnative speakers of English have a smaller mean percentage of correct responses at the 5% significance level?

8.32   Biofeedback monitoring devices and techniques in the control of physiologic functions help astronauts control stress. One experiment related to this topic is discussed by G. Rotondo, et al., *Acta Astronautica*, 10, no. 8, 1983, pp. 591–598. Six subjects were placed in a stressful situation (using video games), followed by a period of biofeedback and adaptation. Another group of six subjects was placed under the same stress and then simply told to relax. The first group had an average heart rate of 70.4 and a standard deviation of 15.3. The second group had an average heart rate of 74.9 and a standard deviation of 16.0. At the 10% significance level, can we say that the average heart rate with biofeedback is lower than that without biofeedback? What assumptions are necessary for your answer to be valid?

8.33   The t test for comparing means assumes that the population variances are equal. Is that a valid assumption based on the data of Exercise 8.32? (Test the equality of population variances at the 10% significance level.)

8.34   Data from the U.S. Department of Interior–Geological Survey give flow rates for a small river in North Florida. It is of interest to compare flow rates for March and April, two relatively dry months. Thirty-one measurements for March showed a mean flow rate of 6.85 cubic feet per second and a standard deviation of 1.2. Thirty measurements for April showed a mean of 7.47 and a standard deviation of 2.3. Is there sufficient evidence to say that these two months have different average flow rates, at the 5% significance level?

8.35   An important measure of the performance of a machine is the mean time between failures (MTBF). A certain printer attached to a word processor was observed for a period of time during which ten failures were observed. The times between failures averaged 98 working hours, with a standard deviation of 6 hours. A modified version of this printer was observed for eight failures, with the times between failures averaging 94 working hours, with a standard deviation of 8 hours. Can we say, at the 1% significance level, that the modified version of the printer has a smaller MTBF? What assumptions are made in this test? Do you think the assumptions are reasonable for this situation?

8.36   It is claimed that tensile-strength measurements for a 12-mm-diameter steel rod should, on the average, be at least 8 units (N/mm^2) higher than the tensile-strength measurements for a 10-mm-diameter steel rod. Independent samples of 50 measurements each for the two sizes of rods gave the following results:

10-mm Rod	12-mm Rod
$\bar{x}_1 = 545$	$\bar{x}_2 = 555$
$s_1 = 24$	$s_2 = 18$

Is the claim justified at the 5% significance level?

8.37   Alloying is said to reduce the resistance in a standard type of electrical wire. Ten measurements on a standard wire yielded a mean resistance of 0.19 ohm and a standard deviation of 0.03. Ten independent measurements on an alloyed wire yielded a mean resistance of 0.11 ohm and a standard deviation of 0.02. Does alloying seem to reduce the mean resistance in the wire? Test at the 10% level of significance.

8.38   Two different machines, A and B, used for torsion tests of steel wire were tested on 12 pairs of different types of wire, with one member of each pair tested on each machine. The results (break angle measurements) were as follows:

						Wire Type						
	1	2	3	4	5	6	7	8	9	10	11	12
Machine A	32	35	38	28	40	42	36	29	33	37	22	42
Machine B	30	34	39	26	37	42	35	30	30	32	20	41

a   Is there evidence at the 5% significance level to suggest that machines A and B give different average readings?

b   Is there evidence at the 5% significance level to suggest that machine B gives a lower average reading than machine A?

8.39   Six rockets, nominally with a range of 2,500 meters, were stored for some time and then tested. The ranges found in the tests were 2,490, 2,510, 2,360, 2,410, 2,300, and 2,440. Another group of six rockets, of the same type, was stored for the same length of time but in a different manner. The ranges for these six were 2,410, 2,500, 2,360, 2,290, 2,310, and 2,340. Do the storage methods produce significantly different mean ranges? Use $\alpha = 0.05$ and assume range measurements to be approximately normally distributed with the same variance for each manner of storage.

8.40   The average depth of bedrock at two possible construction sites is to be compared by driving five piles at random locations within each site. The results, with depths in feet, are as follows:

Site A	Site B
$n_1 = 5$	$n_2 = 5$
$\bar{x}_1 = 142$	$\bar{x}_2 = 134$
$s_1 = 14$	$s_2 = 12$

Do the average depths of bedrock differ for the two sites at the 10% significance level? What assumptions are you making?

8.41   Gasoline mileage is to be compared for two automobiles, A and B, by testing each automobile on five brands of gasoline. Each car used one tank of each brand, with the following results (in miles per gallon):

Brand	Auto A	Auto B
1	28.3	29.2
2	27.4	28.4
3	29.1	28.2
4	28.7	28.0
5	29.4	29.6

Is there evidence to suggest a difference between true average mileage figures for the two automobiles? Use a 5% significance level.

8.42 The two drying methods for concrete introduced in Exercise 8.29 were used on seven different mixes, with each mix of concrete subjected to each drying method. The resulting strength-test measurements (in psi) are given below:

Mix	Method I	Method II
A	3,160	3,170
B	3,240	3,220
C	3,190	3,160
D	3,520	3,530
E	3,480	3,440
F	3,220	3,210
G	3,120	3,120

Is there evidence of a difference between average strengths for the two drying methods at the 10% significance level?

8.43 Two procedures for sintering copper are to be compared by testing each procedure on six different types of powder. The measurement of interest is the porosity (volume percentage due to voids) of each test specimen. The results of the tests are as follows:

Powder	Procedure I	Procedure II
1	21	23
2	27	26
3	18	21
4	22	24
5	26	25
6	19	16

Is there evidence of a difference between true average porosity measurements for the two procedures? Use $\alpha = 0.05$.

8.44 Refer to Exercise 8.29. Do the variances of the strength measurements differ for the two drying methods? Test at the 10% significance level.

8.45 Refer to Exercise 8.39. Is there sufficient evidence to say that the variances among range measurements differ for the two storage methods? Use $\alpha = 0.10$.

8.46 Refer to Exercise 8.40. Does site A have significantly more variation among depth measurements than site B? Use $\alpha = 0.05$.

8.4 χ^2 Tests on Frequency Data

In Section 8.2, we saw how to construct tests of hypotheses concerning a binomial parameter p in the large-sample case. In that case, the test statistic was based on Y, the *number* (or *frequency*) of successes in n trials of an experiment. The observation of interest was a frequency count, rather than a continuous measurement such as a lifelength, reaction time, velocity, or weight. Now we take a detailed look at three types of situations in which hypothesis-testing problems arise with frequency, or count, data. All the results in this section are approximations that work well only when samples are reasonably large. What we mean by "large" will be stated later.

8.4.1   Testing Parameters of the Multinomial Distribution

The first of the three situations involves testing a hypothesis concerning the parameters of a multinomial distribution. (See Section 3.8 for a description of this distribution and some of its properties.) Suppose n observations (trials) come from a multinomial distribution with k possible outcomes per trial. Let $X_i, i = 1, \ldots, k$, denote the number of trials resulting in outcome i, with p_i denoting the probability that any one trial will result in outcome i. Recall that $E(X_i) = np_i$. Suppose we want to test the hypothesis that the p_i's have specified values, that is, $H_0 : p_1 = p_{10}, p_2 = p_{20}, \ldots, p_k = p_{k0}$. The alternative will be the most general one, which simply states "at least one equality fails to hold." To test the validity of this hypothesis, we can compare the observed count in cell i, X_i, with what we would expect that count to be if H_0 were true, namely, $E(X_i) = np_{i0}$. So the test statistic is based on $X_i - E(X_i), i = 1, \ldots, k$. We don't know if these differences are large or small unless we can standardize them in some way. It turns out that a good statistic to use squares these differences and divides them by $E(X_i)$, resulting in the test statistic

$$X^2 = \sum_{i=1}^{k} \frac{[X_i - E(X_i)]^2}{E(X_i)}$$

In repeated sampling from a multinomial distribution for which H_0 is true, X^2 will have approximately a $\chi^2(k-1)$ distribution. The null hypothesis is rejected for large values of X^2 ($X^2 > \chi^2_\alpha(k-1)$), since X^2 will be large if there are large discrepancies between X_i and $E(X_i)$.

EXAMPLE   **8.15**   The ratio of the number of items produced in a factory by three shifts, first, second, and third, is 4:2:1, due primarily to the decreased number of employees on the later shifts. This means that 4/7 of the items produced come from the first shift, 2/7 from the second, and 1/7 from the third. It is hypothesized that the number of defectives produced should follow this same ratio. A sample of 50 defective items was tracked back to the shift that produced them, with the following results:

	Shift		
	1	2	3
Number of defectives	20	16	14

Test the hypothesis indicated above with $\alpha = 0.05$.

Solution   Using the X^2 statistic with X_i replaced by its observed value, we have, for $H_0 : p_1 = 4/7, p_2 = 2/7, p_3 = 1/7$,

$$\sum_{i=1}^{3} \frac{[X_i - E(X_i)]^2}{E(X_i)} = \frac{[20 - 50(4/7)]^2}{50(4/7)} + \frac{[16 - 50(2/7)]^2}{50(2/7)}$$

$$+ \frac{[14 - 50(1/7)]^2}{50(1/7)} = 9.367$$

Now $\chi^2_{0.05}(2) = 5.991$, and since our observed X^2 is larger than this critical χ^2 value, we reject H_0.

This chi-square test is always two-tailed, and although the observed 14 is approximately twice the expected $50(1/7)$, the observed 20 is less than the expected $50(4/7)$, and both observed cells are out of line. This test does not indicate which discrepancy is more serious. ■

8.4.2 Testing Equality Among Binomial Parameters

The second situation in which a χ^2 statistic can be used on frequency data is in the testing of equality among binomial parameters for k separate populations. In this case, k independent random samples are selected that result in k binomially distributed random variables $Y_1, \ldots, Y_k$, where Y_i is based on n_i trials with success probability p_i on each trial. The problem is to test the null hypothesis $H_0 : p_1 = p_2 = \cdots = p_k$ against the alternative of at least one inequality. There are now $2k$ cells to consider, as outlined in Figure 8.2.

FIGURE **8.2**
Cells for k Binomial
Observations

				Observation			
		1	2	3	$\cdots$	k	Total
Successes		y_1	y_2	y_3		y_k	y
Failures		$n_1 - y_1$	$n_2 - y_2$	$n_3 - y_3$		$n_k - y_k$	$n - y$
Total		n_1	n_2	n_3		n_k	n

$$n = \sum_{i=1}^{k} n_i \qquad y = \sum_{i=1}^{k} y_i$$

As in the case of testing multinomial parameters, we should construct the test statistic by comparing the observed cell frequencies with the expected cell frequencies. However, the null hypothesis does not specify values for $p_1, p_2, \ldots, p_k$, and hence the expected cell frequencies must be estimated. We now discuss the expected cell frequencies and their estimators.

Since each Y_i has a binomial distribution, we know from Chapter 3 that

$$E(Y_i) = n_i p_i$$

and

$$E(n_i - Y_i) = n_i - n_i p_i = n_i(1 - p_i)$$

Under the null hypothesis that $p_1 = p_2 = \ldots = p_k = p$, we estimate the common value of p by pooling the data from all k samples. The minimum-variance unbiased estimator of p is

$$\frac{1}{n} \sum_{i=1}^{k} Y_i = \frac{Y}{n} = \frac{\text{total number of successes}}{\text{total sample size}}$$

The estimators of the expected cell frequencies are then taken to be

$$\hat{E}(Y_i) = n_i \left(\frac{Y}{n} \right)$$

and

$$\hat{E}(n_i - Y_i) = n_i \left(1 - \frac{Y}{n} \right) = n_i \left(\frac{n - Y}{n} \right)$$

In each case, note that the estimate of an expected cell frequency is found by the following rule:

$$\text{estimated expected cell frequency} = \frac{(\text{column total})(\text{row total})}{\text{overall total}}$$

To test $H_0 : p_1 = \cdots = p_k = p$, we make use of the statistic

$$X^2 = \sum_{i=1}^{k} \left\{ \frac{[Y_i - \hat{E}(Y_i)]^2}{\hat{E}(Y_i)} + \frac{[(n_i - Y_i) - \hat{E}(n_i - y_i)]^2}{\hat{E}(n_i - Y_i)} \right\}$$

This statistic has approximately a $\chi^2(k - 1)$ distribution, as long as the sample sizes are reasonably large.

We illustrate with an example.

E X A M P L E   **8.16**   A chemical company is experimenting with four different mixtures of a chemical designed to kill a certain species of insect. Independent random samples of 200 insects are subjected to one of the four chemicals, and the number of insects dead after 1 hour of exposure is counted. The results are as follows (with estimated expected cell frequencies shown in parentheses):

	1	2	3	4	Total
Dead	124 (141)	147 (141)	141 (141)	152 (141)	564
Not Dead	76 (59)	53 (59)	59 (59)	48 (59)	236
Total	200	200	200	200	800

We want to test the hypothesis that the rate of kill is the same for all four mixtures. Test this claim at the 5% level of significance.

Solution   We are testing $H_0 : p_1 = p_2 = p_3 = p_4 = p$, where p_i denotes the probability that an insect subjected to chemical i dies in the indicated length of time. Now the estimated cell frequencies under H_0 are found by

$$\frac{(n_1)(y)}{n} = \frac{1}{n}(\text{column 1 total})(\text{row 1 total})$$

$$= \frac{1}{800}(200)(564) = \frac{1}{4}(564) = 141$$

$$\frac{(n_1)(n-y)}{n} = \frac{1}{800}(200)(236) = \frac{1}{4}(236) = 59$$

and so on. Note that in this case all estimated frequencies in any one row are equal. Calculating an observed value for X^2, we get

$$\frac{(124 - 141)^2}{141} + \cdots + \frac{(48 - 59)^2}{59} = 10.72$$

Since $\chi^2_{0.05}(3) = 7.815$, we reject the hypothesis of equal kill rates for the four chemicals.   ▪

8.4.3   Contingency Tables

The third type of experimental situation that makes use of a χ^2 test again makes use of the multinomial distribution. However, this time the cells arise from a double classification scheme. For example, employees could be classified according to sex and marital status, which results in four cells, as follows:

	Male	Female
Married		
Unmarried		

A random sample of n employees would then give values of random frequencies to the four cells. In these two-way classifications, a common question to ask is "Does the row criterion depend on the column criterion?" In other words, "Is the row criterion *contingent* upon the column criterion?" In this context, the two-way tables like the one just illustrated are referred to as *contingency tables*. We might want to know whether or not marital status depends on the sex of the employee, or if these two classification criteria seem to be independent of each other.

The null hypothesis in these two-way tables is that the row criterion and the column criterion are independent of each other. The alternative is simply that they are *not* independent. In looking at a 2×2 table, we find that an array such as

5	15
10	30

would support the hypothesis of independence, since within each column the chance of being in row 1 is about 1 out of 3. That is, the chance of being in row 1 does *not* depend on which column you happen to be in. On the other hand, an array like

5	30
10	15

would support the alternative hypothesis of dependence.

In general, the cell frequencies and cell probabilities can be tabled as follows:

		Column				
		1	$\cdots$	j	c	
	1	$X_{11}; p_{11}$			$X_{1c}; p_{1c}$	$X_{1\cdot}$
	$\vdots$	$\vdots$				
Row	i		$\cdots$	$X_{ij}; p_{ij}$		$X_{i\cdot}$
	r	$X_{r1}; p_{r1}$			$X_{rc}; p_{rc}$	
		$X_{\cdot 1}$		$X_{\cdot j}$		n

We will let $X_{i.}$ denote the frequency total for row i and $X_{.j}$ the total for column j. Also, $p_{i.}$ will denote the probability of being in row i, and $p_{.j}$ the probability of being in column j.

Under the null hypothesis of independence between rows and columns, $p_{ij} = p_{i.}p_{.j}$. Also, $E(X_{ij}) = np_{ij} = np_{i.}p_{.j}$. Now $p_{i.}$ and $p_{.j}$ must be estimated, and the best estimators are

$$\hat{p}_{i\cdot} = \frac{X_{i\cdot}}{n}, \qquad \hat{p}_{\cdot j} = \frac{X_{\cdot j}}{n}$$

Thus, the best estimator of an expected cell frequency is

$$\hat{E}(X_{ij}) = \left(n\frac{X_{i\cdot}}{n}\right)\left(\frac{X_{\cdot j}}{n}\right) = \frac{X_{i}.X_{\cdot j}}{n}$$

Once again we see that the estimate of an expected cell frequency is given by

$$\frac{x_{i}.x_{\cdot j}}{n} = \frac{1}{n}(\text{row } i \text{ total})(\text{column } j \text{ total})$$

The test statistic for testing the independence hypothesis is

$$X^2 = \sum_{ij} \frac{[X_{ij} - \hat{E}(X_{ij})]^2}{\hat{E}(X_{ij})}$$

which has approximately a $\chi^2(r-1)(c-1)$ distribution, where r is the number of rows and c is the number of columns in the table.

E X A M P L E   **8.17**   A sample of 200 machined parts is selected from the one-week output of a machine shop that employs three machinists. The parts are inspected to determine whether or not they are defective, and they are categorized according to which machinist did the work. The results are as follows:

	Machinist			
	A	B	C	
Defective	10 (9.92)	8 (10.88)	14 (11.2)	32
Nondefective	52 (52.08)	60 (57.12)	56 (58.8)	168
	62	68	70	200

Is the defective/nondefective classification independent of machinist classification? Conduct a test at the 1% significance level.

Solution   First we must find the estimated expected cell frequencies, as follows:

$$\frac{x_1.x._1}{n} = \frac{(32)(62)}{200} = 9.92$$

$$\frac{x_1.x._2}{n} = \frac{(32)(68)}{200} = 10.88$$

$$\vdots$$

$$\frac{x_2.x._3}{n} = \frac{(168)(70)}{200} = 58.8$$

(These numbers are shown in parentheses in the table.) Now the test statistic is observed to be

$$\frac{(10-9.22)^2}{9.22} + \cdots + \frac{(56-58.8)^2}{58.8} = 1.74$$

In this case, the degrees of freedom are given by

$$(r-1)(c-1) = (2-1)(3-1) = 2$$

and

$$\chi^2_{0.01}(2) = 9.21$$

Thus, we cannot reject the null hypothesis of independence between the machinists and the defective/nondefective classifications. There is not sufficient evidence to say that the rate of defectives produced differs among machinists.   ∎

Recall that all of the above test statistics have only an approximate χ^2 distribution for large samples. A rough guideline for "large" is that each cell should contain an expected frequency count of at least two.

Exercises

8.47 Two types of defects, A and B, are frequently seen in the output of a certain manufacturing process. Each item can be classified into one of the four classes AB, $A\overline{B}$, $\overline{A}B$, $\overline{A}\,\overline{B}$, where $\overline{A}$ denotes the absence of the type-A defect. For 100 inspected items, the following frequencies were observed:

$$
\begin{array}{ll}
AB & 48 \\
A\overline{B} & 18 \\
\overline{A}B & 21 \\
\overline{A}\,\overline{B} & 13
\end{array}
$$

Test the hypothesis that the four categories, in the order listed, occur in the ratio 5:2:2:1. (Use $\alpha = 0.05$.)

8.48 Vehicles can turn right, turn left, or continue straight ahead at a certain intersection. It is hypothesized that half the vehicles entering this intersection will continue straight ahead. Of the other half, equal proportions will turn right and left. Fifty vehicles were observed to have the following behavior:

	Straight	Left Turn	Right Turn
Frequency	28	12	10

Test the stated hypothesis at the 10% level of significance.

 8.49 A manufacturer of stereo amplifiers has three assembly lines. We want to test the hypothesis that the three lines do not differ with respect to the number of defectives produced. Independent samples of 30 amplifiers each are selected from the output of the lines, and the number of defectives is observed. The data are as follows:

	Line		
	I	II	III
Number of Defectives	6	5	9
Sample Size	30	30	30

Conduct a test of the hypothesis given above at the 5% significance level.

8.50 Two inspectors are asked to rate independent samples of textiles from the same loom. Inspector A reports that 18 out of 25 samples fall in the top category, while inspector B reports that 20 out of 25 samples merit the top category. Do the inspectors appear to differ in their assessments? Use $\alpha = 0.05$.

 8.51 Two chemicals, A and B, are designed to protect pine trees from a certain disease. One hundred trees are sprayed with chemical A, and 100 are sprayed with chemical B. All trees are subjected to the disease, with the following results:

	A	B
Infected	20	16
Sample Size	100	100

Do the chemicals appear to differ in their ability to protect the trees? Test at the 1% significance level.

8.52   Refer to Exercise 8.47. Test the hypothesis that the type A defects occur independently of the type B defects. Use $\alpha = 0.05$.

 8.53   The *Sociological Quarterly* for Spring 1978 reports on a study of the relationship between athletic involvement and academic achievement for college students. The 852 students sampled were categorized according to amount of athletic involvement and grade-point averages at graduation, with results as follows:

		Athletic Involvement			
		None	1–3 Semesters	4 or More Semesters	
GPA	Below Mean	290	94	42	
	Above Mean	238	125	63	
		528	219	105	852

a   Do final grade-point averages appear to be independent of athletic involvement? (Use $\alpha = 0.05$.)

b   For students with four or more semesters of athletic involvement, is the proportion with GPAs above the mean significantly different, at the 0.05 level, from the proportion with GPAs below the mean?

 8.54   A new sick-leave policy is being introduced into a firm. A sample of employee opinions showed the following breakdowns by sex and opinion:

	Favor	Oppose	Undecided
Male	31	44	6
Female	42	36	8

Does the reaction to the new policy appear to be related to sex? Test at the 5% significance level.

8.55   A sample of 150 people is observed using one of four entrances to a commercial building. The data are as follows:

Entrance	1	2	3	4
No. of People	42	36	31	41

a   Test the hypothesis that all four entrances are used equally often. Use $\alpha = 0.05$.

b   Entrances 1 and 2 are on a subway level, while entrances 3 and 4 are on ground level. Test the hypothesis that subway and ground-level entrances are used equally often, again with $\alpha = 0.05$.

8.56   The National Maximum Speed Limit (NMSL) of 55 miles per hour was put in force in early 1974. (See D. B. Kamerud, *Transportation Research*, 17A, no. 1, p. 61, for details.) In experimental observation of trucks on rural secondary roads, 71% traveled under 55 mph in 1973, and 80% traveled under 55 mph in 1974. If 100 trucks were observed in each year, can we say that the proportion of trucks traveling under 55 mph was significantly different for the two years? Use $\alpha = 0.01$.

8.57   The use of aerosol dispensers has been somewhat controversial in recent years because of their possible effects on the ozone. According to T. Walsh (*Manufacturing Chemist*, 55, no. 12, 1984, pp. 43–45), a sample of 300 women in 1983 found that 35% feared the aerosol effects on ozone, whereas only 29% of a sample of 300 stated this fear in 1984. Is there a significant difference between the true proportions for the two years, at the 1% level of significance?

8.58 The same article referenced in Exercise 8.57 shows results of a four-year study, with 300 respondents per year. The percents of respondents stating that the aerosol dispensers did not spray properly were as follows:

1981	1982	1983	1984
73%	62%	65%	69

Are there significant differences among these percentages at the 1% level of significance?

8.59 Industrial workers who are required to wear respirators were checked periodically to see that the equipment fit properly. From past records, it was expected that 1.2% of those checked would fail (i.e., their equipment would not fit properly). The data from one series of checks were as follows:

Time After First Measurement (months)	Number of Workers Measured	Observed Failures	Expected Failures (1.2% of col. 2)
0–6	28	1	0.34
6–12	92	3	1.10
12–18	270	1	3.24
18–24	702	9	8.42
24–30	649	10	7.79
30–36	134	0	1.61
>36	93	0	1.12
Total	1,968	24	23.6

Can the assumption of the constant 1.2% failure rate be rejected at the 5% significance level?

8.60 In a study of 1,000 major U.S. firms in 1976, 54% stated that quality of life was a major factor in locating corporate headquarters. A similar study of 1,000 firms in 1981 showed 55% making this statement. (See B. E. Sullivan, *Transportation Research*, 18A, no. 2, 1984, p. 119.) Is this a significant change, at the 5% significance level?

8.61 A Yankelovich, Skelly, and White poll carried out in December 1983 showed 60% of 1,000 respondents saying that they "worry a lot" about the possibility of nuclear war. A similar poll conducted six months earlier showed only 50% responding this way. (See *Time*, 2 January 1984.) Is this a significant difference, at the 1% level of significance?

8.62 A precedent-setting case in the use of statistics in courts of law was the U.S. Supreme Court case of Castaneda v. Partida [430 U.S. 482, 51 L. Ed. 2d 498 (1977)]. A county in Texas was accused of discrimination against Mexican-Americans in the selection of grand juries. Although 79% of the population of the county was Mexican-American, only 339 Mexican-Americans were selected among the 870 summoned grand jurors over an 11-year period. Assuming a binomial model for the number of Mexican-Americans selected, one would expect to see $870(0.79) = 687.3$, or approximately 687, of them on the grand juries over the time in question. Do you think the Supreme Court was correct in stating that the small number of Mexican-Americans selected was probably not due to chance? (Conduct a statistical test at the 5% level of significance.)

8.5 Goodness-of-Fit Tests

Thus far, our main emphasis has been on choosing a probabilistic model for a measurement, or a set of measurements, taken on some natural phenomenon. We have talked about the basic probabilistic manipulations one can make and have discussed estimation of certain unknown parameters, as well as testing hypotheses concerning these unknown parameters. Up to this point, we have not discussed any rigorous way of checking to see whether the data we observe do, in fact, agree with the underlying probabilistic model assumed for these data. That subject, called testing goodness of fit, is discussed in this section. Goodness-of-fit testing is a broad subject, with entire textbooks devoted to it. We merely skim the surface by presenting two techniques that have wide applicability and are fairly easy to calculate: the χ^2 goodness-of-fit test for discrete distributions and the Kolmogorov-Smirnov (K-S) test for continuous distributions.

8.5.1 χ^2 Test

Let Y denote a discrete random variable that can take on values $y_1, y_2, \ldots$. Under the null hypothesis H_0, we assume a probabilistic model for Y. That is, we assume $P(Y = y_i) = p_i$ where p_i may be completely specified or may be a function of other unknown parameters. For example, we could assume a Poisson model for Y, and thus

$$P(Y = y_i) = p_i = \lambda^{y_i} e^{-\lambda}/y_i!$$

In addition, we could assume that λ is specified at λ_0, or we could assume that nothing is known about λ, in which case λ would have to be estimated from the sample data. The alternative hypothesis is the general one that the model in H_0 does not fit.

The sample consists of n independently observed values of the random variable Y. For this sample, let F_i denote the number of times $Y = y_i$ is observed. Thus, F_i is the frequency count for the value y_i. Under the null hypothesis that $P(Y = y_i) = p_i$, the expected frequency can be calculated as $E(F_i) = np_i$.

After the sample is taken, we will have an observed value of F_i in hand. At that point, we can compare the observed value of F_i with what is expected under the null hypothesis. The test statistic will be based on the differences $[F_i - E(F_i)]$, as in previously discussed χ^2 tests. In many cases, one or more parameters will have to be estimated from sample data. In those cases, $E(F_i)$ is also estimated and is referred to as $\hat{E}(F_i)$, the estimated expected frequency. For certain samples, many of the frequency counts F_i may be quite small. (In fact, many may be zero.) When this occurs, consecutive values of y_i are grouped together and the corresponding probabilities added, since

$$P(Y = y_i \text{ or } Y = y_j) = p_i + p_j$$

It follows that the frequencies F_i and F_j can be added, since $E(F_i + F_j) = n(p_i + p_j)$. As a handy rule of thumb, values of Y should be grouped so that the expected frequency count is at least 2 in every cell. After this grouping is completed, let k denote the number of cells obtained.

The test statistic for the null hypothesis that $P(Y = y_i) = p_i, i = 1, 2, \ldots,$ is given by

$$X^2 = \sum_{i=1}^{k} \frac{[F_i - \hat{E}(F_i)]^2}{\hat{E}(F_i)}$$

For large n, this statistic has approximately a χ^2 distribution with degrees of freedom given by

$$(k - 1) - \text{(the number of parameters estimated)}$$

Thus, if one parameter is estimated, the statistic would have $(k - 2)$ degrees of freedom.

We illustrate this procedure with a Poisson example.

E X A M P L E   **8.18**   The number of accidents per week Y in a certain factory was checked for a random sample of $n = 50$ weeks, with the following results:

y	Frequency
0	32
1	12
2	6
3 or more	0

Test the hypothesis that Y follows a Poisson distribution, with $\alpha = 0.05$.

Solution H_0 states that Y has a probability distribution given by

$$P(y_i) = \frac{\lambda^{y_i} e^{-\lambda}}{y_i!} \qquad y_i = 0, 1, 2, \ldots$$

Since λ is unknown, it must be estimated, and the best estimate is $\hat{\lambda} = \overline{Y}$. For the given data, $\overline{y} = [0(32) + 1(12) + 2(6)]/50 = 24/50 = 0.48$.

Using the guideline that each cell in a χ^2 test should have two or more observations, we must group the last two categories into one cell, giving us a total of three cells, indexed by the values $Y = 0$, $Y = 1$, and $Y \geq 2$. The corresponding probabilities for observing a value in these cells are

$$p_0 = P(Y = 0) = e^{-\lambda}$$
$$p_1 = P(Y = 1) = \lambda e^{-\lambda}$$

and

$$p_2 = P(Y \geq 2) = 1 - e^{-\lambda} - \lambda e^{-\lambda}$$

If F_0, F_1, and F_2 denote the observed frequencies in the respective cells, then

$$E(F_i) = np_i$$

and

$$\hat{E}(F_i) = n\hat{p}_i$$

where $\hat{p}_i$ denotes p_i with λ replaced by $\hat{\lambda}$. The observed values of the estimated expected frequencies then become

$$50e^{-0.48} = 30.94$$
$$50(0.48)e^{-0.48} = 14.85$$

and

$$50(1 - e^{-0.48} - 0.48e^{-0.48}) = 4.21$$

The observed value of X^2 is then

$$\frac{(32 - 30.94)^2}{30.94} + \frac{(12 - 14.85)^2}{14.85} + \frac{(6 - 4.21)^2}{4.21} = 1.34$$

The test statistic has 1 degree of freedom, since $k = 3$ and one parameter is estimated. Now $\chi^2_{0.05}(1) = 3.841$, and hence we do not have sufficient evidence to reject H_0. The data do appear to fit the Poisson model reasonably well. ∎

8.5.2 Kolmogonov-Smirnov Test

It is possible to construct χ^2 tests for goodness of fit to continuous distributions, but the procedure is a little more subjective, since the continuous random variable does not provide natural cells into which the data can be grouped. For the continuous case, we choose to use the Kolmogorov-Smirnov (K-S) statistic, which compares the empirical distribution function of a random sample with a hypothesized theoretical distribution function.

Before we define the K-S statistic, let us define the *empirical distribution function*. Suppose Y is a continuous random variable having distribution function $F(y)$. A random sample of n realizations of Y yields the observations $y_1, \ldots, y_n$. It is convenient to reorder these observed values from smallest to largest, and we denote the ordered y_i's by $y_{(1)} \leq y_{(2)} \leq \cdots \leq y_{(n)}$. That is, if $y_1 = 7$, $y_2 = 9$, and $y_3 = 3$, then $y_{(1)} = 3$, $y_{(2)} = 7$, and $y_{(3)} = 9$. Now the empirical distribution function is given by

$$F_n(y) = \text{fraction of the sample less than or equal to } y$$
$$= \begin{cases} \dfrac{(i-1)}{n} & \text{if } y_{(i-1)} \leq y < y_{(i)} \qquad i = 1, \ldots, n \\ 1 & \text{if } y \geq y_{(n)} \end{cases}$$

where we let $y_0 = -\infty$.

Suppose a continuous random variable Y is assumed, under the null hypothesis, to have a distribution function given by $F(y)$. The alternative hypothesis is that $F(y)$ is *not* the true distribution function for Y. After a random sample of n values of Y is observed, $F(y)$ should be "close" to $F_n(y)$ provided the null hypothesis is true. Our

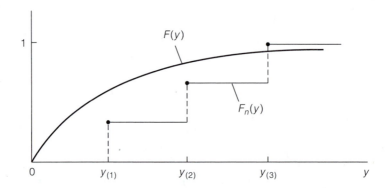

statistic, then, must measure the closeness of $F(y)$ to $F_n(y)$ over the whole range of y-values. [See Figure 8.3 for a typical plot of $F(y)$ and $F_n(y)$.]

The K-S statistic D is based on the *maximum* distance between $F(y)$ and $F_n(y)$. That is,

$$D = \max_y |F(y) - F_n(y)|$$

The null hypothesis is rejected if D is "too large."

Because $F(y)$ and $F_n(y)$ are nondecreasing and $F_n(y)$ is constant between sample observations, the maximum deviation between $F_n(y)$ and $F(y)$ will occur either at one of the observation points $y_1, \ldots, y_n$ or immediately to the left of one of these points. To find the observed value of D, then, it is necessary to check only

$$D^+ = \max_{1 \le i \le n} \left[\frac{i}{n} - F(y_i) \right]$$

and

$$D^- = \max_{1 \le i \le n} \left[F(y_i) - \frac{i-1}{n} \right]$$

since

$$D = \max(D^+, D^-)$$

If H_0 hypothesizes the *form* of $F(y)$ but leaves some parameters unspecified, then these unknown parameters must be estimated from the sample data before the test can be carried out.

Values cutting off upper-tail areas of 0.15, 0.10, 0.05, 0.025, and 0.01 for a modified form of D are given by M. A. Stephens ("EDF Statistics for Goodness of Fit and Some Comparisons," *Journal of the Am. Stat. Assoc.*, vol. 69, no. 347, 1974, pp. 730–737) and are reproduced in Table 8.5 for three cases. These cases are the null hypotheses of a fully specified $F(y)$, a normal $F(y)$ with unknown mean and variance, and an exponential $F(y)$ with unknown mean.

We illustrate the use of all three cases.

T A B L E **8.5** Upper-Tail Percentage Points of Modified D

	Modified Form of D	0.15	0.10	0.05	0.025	0.01
				Tail Area		
Specified $F(y)$	$(D)(\sqrt{n}+0.12+0.11/\sqrt{n})$	1.138	1.224	1.358	1.480	1.626
Normal $F(y)$ with Unknown μ, σ^2	$(D)(\sqrt{n}-0.01+0.85/\sqrt{n})$	0.775	0.819	0.895	0.955	1.035
Exponential $F(y)$ with Unknown θ	$(D-0.2/n)(\sqrt{n}+0.26+0.5/\sqrt{n})$	0.926	0.990	1.094	1.190	1.308

E X A M P L E **8.19** Consider the first ten observations of Table 4.1 to be a random sample from a continuous distribution. Test the hypothesis that these data are from an exponential distribution with mean 2, at the 0.05 significance level.

Solution We must order the ten observations and then find, for each $y_{(i)}$, the value of $F(y_i)$, where H_0 states that $F(y)$ is exponential with $\theta = 2$. Thus,

$$F(y_i) = 1 - e^{-y_i/2}$$

The data and pertinent calculations are given in Table 8.6.

T A B L E **8.6**
Data and Calculations for
Example 8.19

i	$y_{(i)}$	$F(y_i)$	i/n	$(i-1)/n$	$i/n - F(y_i)$	$F(y_i) - (i-1)/n$
1	0.023	0.0114	0.1	0	0.0886	0.0114
2	0.406	0.1837	0.2	0.1	0.0163	0.0837
3	0.538	0.2359	0.3	0.2	0.0641	0.0359
4	1.267	0.4693	0.4	0.3	−0.0693	0.1693
5	2.343	0.6901	0.5	0.4	−0.1901	0.2901
6	2.563	0.7224	0.6	0.5	−0.1224	0.2224
7	3.334	0.8112	0.7	0.6	−0.1112	0.2112
8	3.491	0.8254	0.8	0.7	−0.0254	0.1254
9	5.088	0.9214	0.9	0.8	−0.0214	0.1214
10	5.587	0.9388	1.0	0.9	0.0612	0.0388

D^+ is the maximum value in column 6, and D^- is the maximum in column 7. Thus, $D^+ = 0.0886$ and $D^- = 0.2901$, giving $D = 0.2901$. To find the critical value from Table 8.5, we need to calculate

$$(D)\left(\sqrt{n}+0.12+\frac{0.11}{\sqrt{n}}\right) = (0.2901)(3.317) = 0.9623$$

At the $\alpha = 0.05$ significance level, the rejection region starts at 1.358. Thus, we do not reject the null hypothesis. Notice we are saying that we cannot reject the exponential model ($\theta = 2$) as a plausible model for these data. This does not imply that the exponential model is the *best* model for these data, since numerous other possible null hypotheses would not be rejected either. ∎

EXAMPLE **8.20**   The following data are an ordered sample of observations on the amount of pressure (in pounds per square inch) needed to fracture a certain type of glass. Test the hypothesis that these data fit a Weibull distribution with $\gamma = 2$ but unknown θ.

4.90	8.60	11.42	15.46	19.19	20.69
40.29	41.19	43.55	44.62	53.56	77.61

Solution   We see the form of the Weibull probability density function in Section 4.8. We also see there that when $\gamma = 2$, Y^2 will have an exponential distribution under the null hypothesis that Y has a Weibull distribution. Hence, we want to test the hypothesis that Y^2 has an exponential distribution with unknown θ. The best estimator of θ is the average of the Y^2 values, observed to be 1,446.93. $F(y)$ is then estimated to be

$$\hat{F}(y) = 1 - e^{-y/1,446.93}$$

The information needed to complete the test is given in Table 8.7. From the last two columns, we see that $D = 0.2439$. Thus, the modified D is

$$\left(D - \frac{0.2}{n}\right)\left(\sqrt{n} + 0.26 + \frac{0.5}{\sqrt{n}}\right) = (0.2439 - 0.0167)(3.8684)$$
$$= 0.8789$$

TABLE **8.7**
Data and Calculations for Example 8.20

i	$u_i = y_{(i)}^2$	$\hat{F}(u_i)$	i/n	$i/n - \hat{F}(u_i)$	$\hat{F}(u_i) - (i-1)/n$
1	24.01	0.0165	0.0833	0.0668	0.0165
2	73.96	0.0498	0.1667	0.1169	-0.0335
3	130.42	0.0862	0.2500	0.1638	-0.0805
4	239.01	0.1523	0.3333	0.1810	-0.0977
5	368.26	0.2247	0.4167	0.1920	-0.1086
6	428.08	0.2561	0.5000	0.2439	-0.1606
7	1623.28	0.6743	0.5833	-0.0910	0.1743
8	1696.62	0.6904	0.6667	-0.0237	0.1071
9	1896.60	0.7304	0.7500	0.0196	0.0637
10	1990.94	0.7474	0.8333	0.0859	-0.0026
11	2868.67	0.8623	0.9167	0.0544	0.0290
12	6023.31	0.9844	1.0000	0.0156	0.0677

If we select $\alpha = 0.05$, we have from Table 8.5 that the critical value is 1.094. Thus, we cannot reject the null hypothesis that the data fit a Weibull distribution with $\gamma = 2$.   ∎

EXAMPLE **8.21**   Soil-water flux measurements (in centimeters/day) were taken at 20 experimental plots in a field. The soil-water flux was measured in a draining soil profile in which steady-state soil-water flow conditions had been established. Theory and empirical evidence have suggested that the measurements should fit a lognormal distribution. The data recorded as y_i in Table 8.8 are the logarithms of the actual measurements. Test the hypothesis that the y_i's come from a normal distribution. Use $\alpha = 0.05$.

TABLE **8.8**
Data and Calculations for
Example 8.21

i	$y_{(i)}$	$u_i = \dfrac{y_{(i)} - \bar{y}}{s}$	i/n	$F(u_i)$	$i/n - F(u_i)$	$F(u_i) - (i-1)/n$
1	0.3780	−2.0224	0.0500	0.0216	0.0284	0.0216
2	0.5090	−1.5215	0.1000	0.0641	0.0359	0.0141
3	0.6230	−1.0856	0.1500	0.1388	0.0112	0.0388
4	0.6860	−0.8448	0.2000	0.1991	0.0009	0.0491
5	0.7350	−0.6574	0.2500	0.2555	−0.0055	0.0555
6	0.7520	−0.5924	0.3000	0.2768	0.0232	0.0268
7	0.7580	−0.5695	0.3500	0.2845	0.0655	−0.0155
8	0.8690	−0.1451	0.4000	0.4423	−0.0423	0.0923
9	0.8890	−0.0686	0.4500	0.4726	−0.0226	0.0726
10	0.8890	−0.0686	0.5000	0.4726	0.0274	0.0226
11	0.8990	−0.0304	0.5500	0.4879	0.0621	−0.0121
12	0.9370	0.1149	0.6000	0.5457	0.0543	−0.0043
13	0.9820	0.2869	0.6500	0.6129	0.0371	0.0129
14	1.0220	0.4399	0.7000	0.6700	0.0300	0.0200
15	1.0370	0.4972	0.7500	0.6905	0.0595	−0.0095
16	1.0880	0.6922	0.8000	0.7556	0.0444	0.0056
17	1.1230	0.8260	0.8500	0.7956	0.0544	−0.0044
18	1.2060	1.1434	0.9000	0.8736	0.0264	0.0236
19	1.3340	1.6328	0.9500	0.9487	0.0013	0.0487
20	1.4230	1.9730	1.0000	0.9758	0.0242	0.0258

Solution   The basic idea of the K-S test in this case is to transform the observed data into new observations that should look like standard normal variates, if the null hypothesis is true. If random variable Y has a normal distribution with mean μ and variance σ^2, then $(Y - \mu)/\sigma$ will have a standard normal distribution. Since μ and σ are unknown, we will estimate them by $\bar{y}$ and s, respectively. Then we transform each y_i value to a u_i value, where

$$u_i = \frac{y_i - \bar{y}}{s}$$

and test to see if the u_i's fit a standard normal distribution. The $F(u_i)$ values in Table 8.8 come from cumulative probabilities under a standard normal curve and can be obtained from a table such as Table 4 of the Appendix or from a computer subroutine.

From Table 8.8, we see that $D = 0.0923$. Using Table 8.5, the modified D is

$$(D)\left(\sqrt{n} - 0.01 + \frac{0.85}{\sqrt{n}}\right) = (0.0923)(4.6522) = 0.4294$$

From Table 8.5, the critical value at the 5% significance level is 0.895, and thus we do not reject the null hypothesis. The data appear to fit the normal distribution. Figure 8.4 shows graphically how $F_n(u)$ and $F(u)$ compare. The figure is a plot of the empirical distribution function for $n = 20$ soil-water flux measurements and a normal distribution function. The statistics D^- and D^+ are illustrated in the inset. ∎

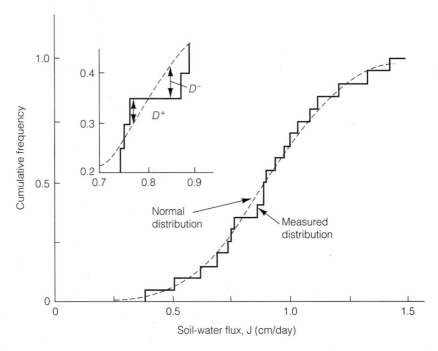

Example 8.21 has a more appropriate sample size for a goodness-of-fit test than our previous examples. Samples of size less than 20 do not allow for such discrimination among distributions; that is, many different distributions may all appear to fit equally well.

There are numerous goodness-of-fit statistics that could be considered in addition to the Kolmogorov-Smirnov, including the Cramer–von Mises, the Anderson-Darling, and the Watson statistics. The interested reader can find information on these procedures in the references given in the bibliography, particularly the 1974 paper by Stephens.

A graphical check for normality, introduced in Section 4.6, is derived from the fact that if X is normally distributed with mean μ and standard deviation σ, then

$$\frac{X - \mu}{\sigma} = Z$$

where Z has a standard normal distribution, or

$$X = \mu + \sigma Z$$

If $x_{(1)} \le x_{(2)} \le \ldots \le x_{(n)}$ represents ordered sample observations, then $x_{(i)}$ can be thought of as the sample percentile $i/(n + 1)$. The corresponding standard normal percentile can be found as a z-score, $z_{(i)}$, called the normal score. Plotting $x_{(i)}$ against $z_{(i)}, i = 1, \ldots, n$, produces something close to a straight line if the data come from a normal distribution. The plot will deviate substantially from a straight line if the data do not come from a normal distribution.

Many computer packages calculate normal scores corresponding to sample data, so the plot of $x_{(i)}$ versus $z_{(i)}$ is easy to produce. One note of caution must be mentioned, however. These packages might use more precise sample percentile calculations than simply $i/(n+1)$. Minitab, for example, calls $x_{(i)}$ the sample percentile

$$\frac{i-\frac{3}{8}}{n+\frac{1}{4}}$$

Figure 8.5 shows the plot (from Minitab) of the sample data versus normal scores for the soil-water flux measurements of Table 8.8. To see how normal scores are calculated, let's look at the second ordered pair ($i = 2$). Here the normal score is the

$$\frac{i-\frac{3}{8}}{n+\frac{1}{4}} = \frac{2-\frac{3}{8}}{20+\frac{1}{4}} = 0.08$$

percentage point of the standard normal distribution, or the z-value that cuts off an area of 0.08 in the lower tail. From Table 4 of the Appendix (or your computer), we find this value to be around -1.40. Thus, the second smallest point on Figure 8.5 is $(-1.40, 0.509)$. The other 19 points are determined in a similar way.

F I G U R E **8.5**
Normal Scores Plot of Data from
Table 8.8

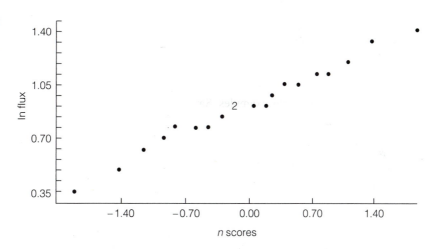

The points in Figure 8.5 do fall rather close to a straight line, which substantiates the claim that the data come from a normal distribution. The intercept (ln flux measurement at a normal score of zero) is approximately 0.90, which is close to the sample mean for the 20 data points of 0.907. We can approximate the slope of the line by looking at two points, such as $(1.40, 1.33)$ and $(-1.40, 0.50)$, which gives a slope of

$$\frac{1.33 - 0.50}{1.40 - (-1.40)} = 0.296$$

This is close to the calculated sample standard deviation of 0.262.

To see that normality is required in order to get the straight-line fit, look at Figure 8.6 which shows the data from Table 8.6 plotted against normal scores. The plot shows large deviations from a straight line, especially at the lower end, and these data do not appear to come from a normal distribution. (We have already shown that these data fit an exponential model.) Such departures from a straight line often appear in plots of data versus normal scores, readily signaling that the data do *not* come from a normal distribution.

F I G U R E **8.6**
Normal Scores Plot of Data from
Table 8.6

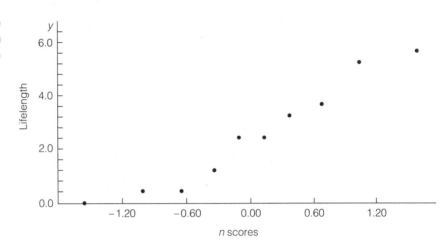

Exercises

8.63   For fabric coming off a certain loom, the number of defects per square yard is counted on 50 sample specimens, each 1 square yard in size. The results are as follows:

Number of Defects	Frequency of Observation
0	0
1	3
2	5
3	10
4	14
5	8
6 or more	10

Test the hypothesis that the data come from a Poisson distribution. Use $\alpha = 0.05$.

8.64   The following data show the frequency counts for 400 observations on the number of bacterial colonies within the field of a microscope, using samples of milk film. (Source: C. Bliss and D. Owens, *Biometrics*, 9, 1953.)

Number of Colonies per Field	Frequency of Observations
0	56
1	104
2	80
3	62
4	42
5	27
6	9
7	9
8	5
9	3
10	2
11	0
19	1
	400

Test the hypothesis that the data fit the Poisson distribution. Use $\alpha = 0.05$.

8.65   The number of accidents experienced by machinists in a certain industry was observed for a certain period of time with the following results. (Source: C. Bliss and R. A. Fisher, *Biometrics*, 9, 1953.)

Accidents per Machinist	0	1	2	3	4	5	6	7	8
Frequency of Observation (number of machinists)	296	74	26	8	4	4	1	0	1

At the 5% level of significance, test the hypothesis that the data came from a Poisson distribution.

8.66   Counts on the number of items per cluster (or colony or group) must necessarily be greater than or equal to 1. Thus, the Poisson distribution does not generally fit these kinds of counts. For modeling counts on phenomena such as number of bacteria per colony, number of people per household, and number of animals per litter, the *logarithmic series* distribution often proves useful. This discrete distribution has the probability function given by

$$p(y) = -\frac{1}{\ln(1 - \alpha)} \left(\frac{\alpha^y}{y} \right) \qquad y = 1, 2, 3, \ldots \qquad 0 < \alpha < 1$$

where α is an unknown parameter.

a   Show that the maximum likelihood estimator, $\hat{\alpha}$, of α satisfies the equation

$$\bar{y} = \frac{\hat{\alpha}}{-(1 - \hat{\alpha}) \ln (1 - \hat{\alpha})}$$

where $\bar{y}$ is the mean of the sampled observations $y_1, \ldots, y_n$.

b   The following data give frequencies of observation for counts on the number of bacteria per colony for a certain type of soil bacteria. (Source: C. Bliss and R. A. Fisher, *Biometrics*, 9, 1953.)

Bacteria per Colony	1	2	3	4	5	6	7 or more
Number of Colonies Observed	359	146	57	41	26	17	29

Test the hypothesis that these data fit a logarithmic series distribution. Use $\alpha = 0.05$. (Note that $\bar{y}$ must be approximated, because we do not have exact information on counts greater than 6. Treat the "7 or more" as "exactly 7.")

8.67   The following data are observed LC50 values on copper in a certain species of fish (measurements in parts per million):

0.075,   0.10,   0.23,   0.46,   0.10,   0.15,   1.30,   0.29,
0.31,   0.32,   0.33,   0.54,   0.85,   1.90,   9.00

It has been hypothesized that the natural logarithms of these data fit the normal distribution. Check this claim at the 5% significance level. Also, make a normal scores plot if a computer is available.

8.68   The time (in seconds) between vehicle arrivals at a certain intersection was measured for a certain time period, with the following results:

9.0,   10.1,   10.2,   9.3,   9.5,   9.8,   14.2,   16.1,
8.9,   10.5,   10.0,   18.1,   10.6,   16.8,   13.6,   11.1

a   Test the hypothesis that these data come from an exponential distribution. Use $\alpha = 0.05$.

b   Test the hypothesis that these data come from an exponential distribution with a mean of 12 seconds. Use $\alpha = 0.05$.

8.69   The weights of copper grains are hypothesized to follow a lognormal distribution, which implies that the logarithms of the weight measurements should follow a normal distribution. Twenty weight measurements, in 10^{-4} gram, are as follows:

2.0,   3.0,   3.1,   4.3,   4.4,   4.8,   4.9,   5.1,   5.4,   5.7,
6.1,   6.6,   7.3,   7.6,   8.3,   9.1,   11.2,   14.4,   16.7,   19.8

Test the hypothesis stated above at the 5% significance level.

8.70   Fatigue life, in hours, for ten bearings of a certain type was as follows:

152.7,   172.0,   172.5,   173.3,   193.0,
204.7,   216.5,   234.9,   262.6,   422.6

Source: A .C. Cohen, et al., *Journal of Quality Technology*, 16, no. 3, 1984, p. 165.

Test the hypothesis, at the 5% significance level, that these data follow a Weibull distribution with $\gamma = 2$.

8.71   The times to failure of eight turbine blades in jet engines, in 10^3 hours, were as follows:

3.2,   4.7,   1.8,   2.4,   3.9,   2.8,   4.4,   3.6

Source: Pratt & Whitney Aircraft, PWA 3001, 1967.)

Do the data appear to fit a Weibull distribution with $\gamma = 3$? Use $\alpha = 0.025$.

8.72   Records were kept on the time of successive failures of the air conditioning system of a Boeing 720 jet airplane. If the air conditioning system has a constant failure rate, then the intervals between successive failures must have an exponential distribution. The observed intervals, in hours, between successive failures are as follows:

23, 261, 87, 7, 120, 14, 62, 47, 225, 71,
246, 21, 42, 20, 5, 12, 120, 11, 3, 14,
71, 11, 14, 11, 16, 90, 1, 16, 52, 95

Source: F. Proschan, *Technometrics,* 5, no. 3, 1963, p. 376.

Do these data seem to follow an exponential distribution at the 5% significance level? If possible, construct a normal scores plot for these data. Does the plot follow a straight line?

8.6   Acceptance Sampling

From Chapter 1 to this point, emphasis has continually been focused on the use of statistical techniques for improving the quality of processes (or total quality management). The old-style quality-control plans consisted of items being inspected as they left the factory, with the good ones being sold and the bad ones being discarded or reworked. That philosophy has given way to a more cost-efficient philosophy of making things right in the first place, and statistical techniques like the tools introduced in Chapter 1 and the control charts of Chapter 6 reflect the modern view.

As a result of this shift in emphasis, acceptance sampling plans (an old-style technique for inspecting lots) have fallen into disfavor. However, many constructive uses of acceptance sampling plans still exist in applications that are "process-centered" rather than "product-centered." Suppose worker A solders wires in circuit boards that are then passed on to worker B for installation into a system. It might be worthwhile to sample soldered connections occasionally, pulling them apart to assess the strength of the soldering, so that the quality of this operation is maintained. (Obviously, every connection cannot be tested due to both the destructive nature of the test and time constraints.) A sampling plan must be designed to efficiently measure the defect rate for the soldering process. Here, worker A is the "producer." Notice that the use of sampling within the process is far different from the old scheme of manufacturing the systems first and then testing a sample of the finished products.

In short, sampling plans still have legitimate uses, and we now turn to discussion of two types of acceptance sampling.

8.6.1   Acceptance Sampling by Attributes

This section discusses the problem of making a decision with regard to the proportion of defective items in a finite lot based on the number of defectives observed in a sample from that lot. Inspection in which a unit of product is classified simply as defective or nondefective is called *inspection by attributes.*

Lots of items, either raw materials or finished products, are sold by a producer to a consumer with some guarantee as to quality. In this section, quality will be determined by the proportion p of defective items in the lot. To check on the quality characteristics of a lot, the consumer will sample some of the items, test them, and observe the number of defectives. Generally, some defectives are allowable, because defective-free lots may be too expensive for the consumer to purchase. But if the number of defectives is too large, the consumer will reject the lot and return it to the producer. In the decision to accept or reject the lot, sampling is generally used, because the cost of inspecting the entire lot may be too high or the inspection process may be destructive.

Before sampling inspection takes place for a lot, the consumer must have in mind a proportion p_0 of defectives that will be acceptable. Thus, if the true proportion of defectives p is no greater than p_0, the consumer will want to accept the lot, and if p is greater than p_0, the consumer will want to reject the lot. This maximum proportion of defectives satisfactory to the consumer is called the *acceptable quality level* (AQL).

Note that we now have a hypothesis-testing problem in which we are testing $H_0 : p \le p_0$ versus $H_a : p > p_0$. A random sample of n items will be selected from the lot and the number of defectives Y observed. The decision concerning rejection or nonrejection of H_0 (rejecting or accepting the lot) will be based on the observed value of Y.

The probability of a Type I error α in this problem is the probability that the lot will be rejected by the consumer when, in fact, the proportion of defectives is satisfactory to the consumer. This is referred to as the *producer's risk*. The value of α calculated for $p = p_0$ is the upper limit of the proportion of good lots rejected by the sampling plan being considered. The probability of a Type II error β calculated for some proportion of defectives p_1, where $p_1 > p_0$, represents the probability that an unsatisfactory lot will be accepted by the consumer. This is referred to as the *consumer's risk*. The value of β calculated for $p = p_1$ is the upper limit of the proportion of bad lots accepted by the sampling plan, for all $p \ge p_1$.

The null hypothesis $H_0 : p \le p_0$ will be rejected if the observed value of Y is larger than some constant a. Since the lot is accepted if $Y \le a$, the constant a is called the *acceptance number*. In this context, the significance level α is given by

$$\alpha = P(\text{reject } H_0 \text{ when } p = p_0)$$
$$= P(Y > a \text{ when } p = p_0)$$

and

$$1 - \alpha = P(\text{not rejecting } H_0 \text{ when } p = p_0)$$
$$= P(\text{accepting the lot when } p = p_0)$$
$$= P(A)$$

Since p_0 may not be known precisely, it is of interest to see how $P(A)$ behaves as a function of the true p, for a given n and a. A plot of these probabilities is called an *operating characteristic curve*, abbreviated OC curve. We illustrate its calculation with the following numerical example.

E X A M P L E   **8.22**   One sampling inspection plan calls for $n = 10$, and $a = 1$, while another calls for $n = 25$ and $a = 3$. Plot the operating characteristic curves for both plans. If the plant using these plans can operate efficiently on 30% defective raw materials, considering the price, but cannot operate efficiently if the proportion gets close to 40%, which plan should be chosen?

Solution   We must calculate $P(A) = P(Y \leq a)$ for various values of p in order to plot the OC curves. For $(n = 10, a = 1)$, we have (using Table 2 of the Appendix)

$p = 0,$	$P(A) = 1$
$p = 0.1,$	$P(A) = 0.736$
$p = 0.2,$	$P(A) = 0.376$
$p = 0.3,$	$P(A) = 0.149$
$p = 0.4,$	$P(A) = 0.046$
$p = 0.6,$	$P(A) = 0.002$

For $(n = 25, a = 3)$, we have

$p = 0,$	$P(A) = 1$
$p = 0.1,$	$P(A) = 0.764$
$p = 0.2,$	$P(A) = 0.234$
$p = 0.3,$	$P(A) = 0.033$
$p = 0.4,$	$P(A) = 0.002$

These values are plotted in Figure 8.7. Note that the OC curve for the $(n = 10, a = 1)$ plan drops slowly and does not get appreciably small until p is in the neighborhood of 0.4. For p around 0.3, this plan still has a fairly high probability of accepting the lot. The other curve $(n = 25, a = 3)$ drops much more rapidly and falls to a small $P(A)$ at $p = 0.3$. The latter plan would be better for the plant, if it can afford to sample $n = 25$ items out of each lot before making a decision.   ∎

Numerous handbooks give operating characteristic curves and related properties for a variety of sampling plans for inspecting by attributes; one of the most widely used sets of plans is Military Standard 105D (MIL-STD-105D).

F I G U R E **8.7**
Operating Characteristic Curves
for Example 8.22

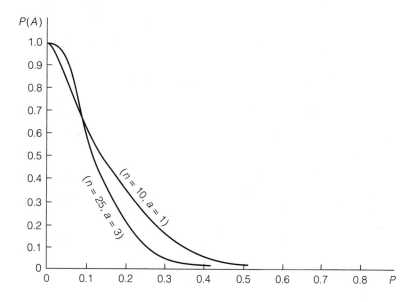

Even though the probability of acceptance, as calculated in Example 8.22, generally does not depend on the lot size as long as the lot is large compared to the sample size, MIL-STD-105D is constructed so that sample size increases with lot size. The practical reasons for the increasing sample sizes are that small random samples are sometimes difficult to obtain from large lots and that discrimination between good and bad lots may be more important for large lots. The adjustments in sample sizes for large lots are based on empirical evidence rather than on strict probabilistic considerations.

MIL-STD-105D contains plans for three sampling levels, labeled I, II, and III. Generally, sampling or inspection level II is used. Level I gives slightly smaller samples and less discrimination, whereas level III gives slightly larger samples and more discrimination. Table 10 in the Appendix gives the code letters for various lot or batch sizes and inspection levels I, II, and III. This code letter is used in entering Table 11, 12, or 13 to find the sample size and acceptance number to be used for specified values of the AQL.

In addition to having three inspection levels, MIL-STD-105D also has sampling plans for normal, tightened, and reduced inspection. These plans are given in Tables 11, 12, and 13, respectively. Normal inspection is designed to protect the producer against the rejection of lots with percent defectives less than the AQL. Tightened inspection is designed to protect the consumer from accepting lots with percent defectives greater than the AQL. Reduced inspection is introduced as an economical plan to be used if the quality history of the lots is good. When sampling a series of lots, rules for switching among normal, tightened, and reduced sampling are given in MIL-STD-105D. When MIL-STD-105D is used on a single lot, the OC curves for various plans, given in the handbook, should be studied carefully before a decision is made on the optimal plan for a particular problem.

EXAMPLE   **8.23**   A lot of size 1,000 is to be sampled with normal inspection at level II. If the AQL of interest is 10%, find the appropriate sample size and acceptance number from MIL-STD-105D.

Solution   Entering Table 10 in the Appendix for a lot of size 1,000, we see that the code letter for level II is J. Since normal inspection is to be used, we enter Table 11 at row J. The sample size is 80, and, moving over to the column headed 10% AQL, we see that the acceptance number is 14. Thus, we sample 80 items and accept the lot if the number of defectives is less than or equal to 14. This plan will have a high probability of accepting any lot with a true percentage of defectives below 10%.   ∎

8.6.2   Acceptance Sampling by Variables

In acceptance sampling by attributes, we merely record, for each sampled item, whether or not the item is defective. The decision to accept or reject the lot then depends on the number of defectives observed in the sample. However, if the characteristic under study involves a measurement on each sampled item, we could base the decision to accept or reject on the measurements themselves, rather than simply on the number of items having measurements that do not meet a certain standard.

For example, suppose a lot of manufactured steel rods is checked for quality, with the quality measurement being the diameter. Rods are not allowed to exceed 10 centimeters in diameter, and a rod with a diameter in excess of 10 centimeters is declared defective for this special case. In attribute sampling, the decision to accept or reject the lot is based on the number of defective rods observed in a sample. But the decision could be made instead on the basis of some function of the actual diameter observations themselves, say, the mean of these observations. If the sample mean is close to or in excess of 10 centimeters, we would expect the lot to contain a high percentage of defectives, whereas if the sample mean is much smaller than 10 centimeters, we would expect the lot to contain relatively few defectives.

Using actual measurements of the quality characteristic as a basis for acceptance or rejection of a lot is called *acceptance sampling by variables*. Sampling by variables often has advantages over sampling by attributes, since the variables method may contain more information for a fixed sample size than the attribute method. Also, when looking at the actual measurements obtained by the variables method, the experimenter may gain some insights into the degree of nonconformance and may quickly be able to suggest methods for improving the product.

For acceptance sampling by variables, let

$$U = \text{upper specification limit}$$

and

$$L = \text{lower specification limit}$$

If X denotes the quality measurement being considered, then U is the maximum allowable value for X on an acceptable item. (U is equal to 10 centimeters in the rod-diameter example.) Similarly, L is the minimum allowable value for X on an acceptable item. Some problems may specify U, others L, and still others both U and L.

As in acceptance sampling by attributes, an AQL is specified for a lot, and a sampling plan is then determined that will give high probability of acceptance to lots with percent defectives lower than the AQL and low probability of acceptance to lots with percent defectives higher than the AQL. In practice, an *acceptability constant k* is determined such that a lot is accepted if

$$\frac{U - \overline{X}}{S} \geq k$$

or

$$\frac{\overline{X} - L}{S} \geq k$$

where $\overline{X}$ is the sample mean and S the sample standard deviation of the quality measurements.

As in the case of sampling by attributes, handbooks of sampling plans are available for acceptance sampling by variables. One of the most widely used of these handbooks is Military Standard 414 (MIL-STD-414). Tables 14 and 15 of the Appendix will help illustrate how MIL-STD-414 is used. Table 14A shows which tabled AQL value to use if the problem calls for an AQL not specifically tabled. Table 14B gives the sample size code letter to use with any possible lot size. Sample sizes are tied to lot sizes, and different levels of sampling are used as in MIL-STD-105D. Inspection level IV is recommended for general use.

Table 15 gives the sample size and value of k for a fixed code letter and AQL. Both normal and tightened inspection procedures can be found in this table. The use of these tables is illustrated in Example 8.24.

E X A M P L E   8.24   The tensile strengths of wires in a certain lot of size 400 are specified to exceed 5 kilograms. We want to set up an acceptance sample plan with AQL = 1.4%. With inspection level IV and normal inspection, find the appropriate plan from MIL-STD-414.

Solution   Looking first at Table 14A, we see that a desired AQL of 1.4% would correspond to a tabled value of 1.5%. Finding the lot size of 400 for inspection level IV in Table 14B yields a code letter of I.

Looking at Table 15 in the row labeled I, we find a sample size of 25. For that same row, the column headed with an AQL of 1.5% gives $k = 1.72$. Thus, we are to sample 25 wires from the lot and accept the lot if

$$\frac{\overline{X} - 5}{S} \geq 1.72$$

where $\overline{X}$ and S are the mean and standard deviation for the 25 tensile strength measurements.   ∎

MIL-STD-414 also gives a table for reduced inspection, much like that in MIL-STD-105D. The handbook provides OC curves for the sampling plans tabled there and, in addition, gives sampling plans based on sample ranges rather than standard deviations.

The theory related to the development of the sampling plans for inspection by variables is more difficult than that for inspection by attributes and cannot be developed fully here. However, we outline the basic approach.

Suppose again that X is the quality measurement under study and U is the upper specification limit. The probability that an item is defective is then given by $P(X > U)$. (This probability also represents the proportion of defective items in the lot.) Now we want to accept only lots for which $P(X > U)$ is small, say, less than or equal to a constant M. But

$$P(X > U) \leq M$$

is equivalent to

$$P\left(\frac{X - \mu}{\sigma} > \frac{U - \mu}{\sigma}\right) \le M$$

where $\mu = E(X)$ and $\sigma^2 = V(X)$. If, in addition, X has a normal distribution, then the latter probability statement is equivalent to

$$P\left(Z > \frac{U - \mu}{\sigma}\right) \le M$$

where Z has a standard normal distribution.

It follows that $P(X > U)$ will be "small" if and only if $(U - \mu)/\sigma$ is "large." Thus,

$$P(X > U) \le M$$

is equivalent to

$$\frac{U - \mu}{\sigma} \ge k^*$$

for some constant k^*. Since μ and σ are unknown, the actual acceptance criterion used in practice becomes

$$\frac{U - \overline{X}}{S} \ge k$$

where k is chosen so that $P(X > U) \le M$.

For normally distributed X, $P(X > U)$ can be estimated by making use of the sample information provided by $\overline{X}$ and S, and these estimates are tabled in MIL-STD-414. These estimates are used in the construction of OC curves. Also, tables are given that show the values of M, in addition to the value of k, for sampling plans designed for specified AQLs.

Exercises

8.73 Construct operating characteristic curves for the following sampling plans, with sampling by attributes.

   a $n = 10, a = 2$

   b $n = 10, a = 4$

   c $n = 20, a = 2$

   d $n = 20, a = 4$

8.74 Suppose a lot of size 3,000 is to be sampled by attributes. An AQL of 4% is specified for acceptable lots. Find the appropriate level II sampling plan under

   a   normal inspection.

   b   tightened inspection.

   c   reduced inspection.

8.75 Fuses of a certain type must not allow the current passing through them to exceed 20 amps. A lot of 1,200 of these fuses is to be checked for quality by a lot-acceptance sampling plan with inspection by variables. The desired AQL is 2%. Find the appropriate sampling plan for level IV sampling and

a normal inspection.

b tightened inspection.

8.76 Refer to Exercise 8.75. If, under normal inspection, the sample mean turned out to be 19.2 amps and the sample standard deviation 0.3 amp, what would you conclude?

8.77 For a lot of 250 one-gallon cans filled with a certain industrial chemical, each gallon is specified to have a percentage of alcohol in excess of L. If the AQL is specified to be 10%, find an appropriate lot-acceptance sampling plan (level IV inspection) under

a normal inspection.

b tightened inspection.

8.7 Summary

Hypothesis testing parallels the components of the scientific method. A new treatment comes onto the scene and is claimed to be better than the standard treatment with respect to some parameter, such as proportion of successful outcomes. An experiment is run to collect data on the new treatment. A decision is then made as to whether the data support the claim. The parameter value for the standard is the *null hypothesis*, and the nature of the change in parameter claimed for the new treatment is the *alternative* or *research hypothesis*. Guides to which decision is most appropriate are provided by the probabilities of *Type I* and *Type II errors*. For the simple cases that we have considered, hypothesis-testing problems parallel estimation problems in that similar statistics are used. Some exceptions were noted, such as the χ^2 tests on frequency data and goodness-of-fit tests. These latter tests are not motivated by confidence intervals.

How does one choose appropriate test statistics? We did not discuss a general method of choosing a test statistic as we did for choosing an estimator (the maximum likelihood principle). General methods of choosing test statistics with good properties do, however, exist. The reader interested in the theory of hypothesis testing should consult a text on mathematical statistics. The tests presented in this chapter have some intuitive appeal, as well as good theoretical properties.

Supplementary Exercises

8.78 An important quality characteristic of automobile batteries is their weight, since that characteristic is sensitive to the amount of lead in the battery plates. A certain brand of heavy-duty battery has a weight specification of 69 pounds. Forty batteries selected from recent production have an average weight of 64.3 pounds and a standard deviation of 1.2 pounds. Would you suspect that something has gone wrong with the production process? Why or why not?

8.79 Quality of automobile batteries can also be measured by cutting them apart and measuring plate thicknesses. A certain brand of batteries is specified to have positive plate thicknesses of 120 thousandths of an inch and negative plate thicknesses of 100 thousandths of an inch. A sample of batteries from recent production was cut apart, and plate thicknesses were measured. Sixteen positive plate thickness measurements averaged 111.6, with a standard deviation of

2.5. Nine negative plate thickness measurements averaged 99.44, with a standard deviation of 3.7. Do the positive plates appear to meet the thickness specification? Do the negative plates appear to meet the thickness specification?

8.80   The mean breaking strength of cotton threads must be at least 215 grams for the thread to be used in a certain garment. A random sample of 50 measurements on a certain thread gave a mean breaking strength of 210 grams and a standard deviation of 18 grams. Should this thread be used on the garments?

8.81   Prospective employees of an engineering firm are told that engineers in the firm work at least 45 hours per week, on the average. A random sample of 40 engineers in the firm showed that, for a particular week, they averaged 44 hours of work with a standard deviation of 3 hours. Are the prospective employees being told the truth?

8.82   Company A claims that no more than 8% of the resistors it produces fail to meet the tolerance specifications. Tests on 100 randomly selected resistors yielded 12 that failed to meet the specifications. Is the company's claim valid? Test at the 10% significance level.

8.83   An approved standard states that the average LC50 for DDT should be ten parts per million for a certain species of fish. Twelve experiments produced LC50 measurements of

$$16, 5, 21, 19, 10, 5, 8, 2, 7, 2, 4, 9$$

Do the data cast doubt on the standard at the 5% significance level?

8.84   A production process is supposed to be producing 10-ohm resistors. Fifteen randomly selected resistors showed a sample mean of 9.8 ohms and a sample standard deviation of 0.5 ohm. Are the specifications of the process being met? Should a two-tailed test be used here?

8.85   For the resistors of Exercise 8.84, the claim is made that the standard deviation of the resistors produced will not exceed 0.4 ohm. Can the claim be refuted at the 10% significance level?

8.86   The abrasive resistance of rubber is increased by adding a silica filler and a coupling agent to chemically bond the filler to the rubber polymer chains. Fifty specimens of rubber made with a type I coupling agent gave a mean resistance measurement of 92, with the variance of the measurements being 20. Forty specimens of rubber made with a type II coupling agent gave a mean of 98 and a variance of 30 on resistance measurements. Is there sufficient evidence to say that the mean resistance to abrasion differs for the two coupling agents at the 5% level of significance?

8.87   Two different types of coating for pipes are to be compared with respect to their ability to aid in resistance to corrosion. The amount of corrosion on a pipe specimen is quantified by measuring the maximum pit depth. For coating A, 35 specimens showed an average maximum pit depth of 0.18 cm. The standard deviation of these maximum pit depths was 0.02 cm. For coating B, the maximum pit depths for 30 specimens had a mean of 0.21 cm and a standard deviation of 0.03 cm. Do the mean pit depths appear to differ for the two types of coating? Use $\alpha = 0.05$.

8.88   Bacteria in water samples are sometimes difficult to count, but their presence can easily be detected by culturing. In 50 independently selected water samples from a lake, 43 contained certain harmful bacteria. After adding a chemical to the lake water, another 50 water samples showed only 22 with the harmful bacteria. Does the addition of the chemical significantly reduce the proportion of samples containing the harmful bacteria? Use $\alpha = 0.025$.

8.89   A large firm made up of several companies has instituted a new quality-control inspection policy. Among 30 artisans sampled from Company A, only 5 objected to the new policy. Among 35 artisans sampled from Company B, 10 objected to the policy. Is there a significant difference, at the 5% significance level, between the proportions voicing *no* objection to the new policy?

8.90   *Research Quarterly*, May 1979, reports on a study of impulses applied to the ball by tennis rackets of various construction. Three measurements on ball impulses were taken on each type of racket. For a Classic (wood) racket, the mean was 2.41 and the standard deviation was 0.02. For a Yamaha (graphite) racket, the mean was 2.22 and the standard deviation was 0.07. Is there evidence of a significant difference between the mean impulses for the two rackets at the 5% significance level?

8.91 For the impulse measurements of Exercise 8.90, is there sufficient evidence to say that the graphite racket gives more variable results than the wood racket? Use $\alpha = 0.05$.

8.92 An interesting and practical use of the χ^2 test comes about in the testing for segregation of species of plants or animals. Suppose that two species of plants, A and B, are growing on a test plot. To assess whether or not the species tend to segregate, n plants are randomly sampled from the plot, and the species of each sampled plant *and* the species of its *nearest* neighbor are recorded. The data are then arranged in a table as follows:

		Nearest Neighbor	
		A	B
Sample Plant	A	a	b
	B	c	d
			n

If a and d are large relative to b and c, we would be inclined to say that the species tend to segregate. (Most of A's neighbors are of type A, and most of B's neighbors are of type B.) If b and c are large compared to a and d, we would say that the species tend to be overly mixed. In either of these cases (segregation or overmixing), a χ^2 test should yield a large value, and the hypothesis of random mixing would be rejected. For each of the following cases, test the hypothesis of random mixing (or, equivalently, the hypothesis that the species of a sampled plant is independent of the species of its nearest neighbor). Use $\alpha = 0.05$ in each case.

a $a = 20, b = 4, c = 8, d = 18$

b $a = 4, b = 20, c = 18, d = 8$

c $a = 20, b = 4, c = 18, d = 8$

8.93 "Love is not blind." An article on this topic in the *Gainesville Sun*, 22 June 1992, states that 72 blindfolded people tried to distinguish their partner from two decoys of similar age by feeling their foreheads. And 58% of them were correct! Is this as amazing a result as the article suggests, or could the result have been achieved by guessing? What is an appropriate null hypothesis here? What is an appropriate alternative hypothesis?

8.94 Is knowledge of right-of-way laws at four-legged intersections associated with similar knowledge for T intersections? D. Montgomery and L. Carstens (*Journal of Transportation Engineering*, vol. 113, no. 3, May 1987, pp. 299–314) show the results of a questionnaire designed to investigate this question (among others). The data are as follows:

Response to T-Intersection Question

		Correct	Incorrect	
Response to Four-Legged-	Correct	141	145	286
Intersection Question	Incorrect	13	188	201
		154	333	487

How would you answer the question posed above?

Simple Regression

About This Chapter

In previous chapters, we estimated and tested hypotheses concerning a population mean by making use of a set of measurements on a single variable from that population. Frequently, however, the mean of one variable is dependent upon one or more related variables. For example, the average amount of energy required to heat houses of a certain size depends on the air temperature during the days of the study. In this chapter, we begin to build models in which the mean of one variable can be written as a linear function of another variable. This procedure is sometimes referred to as the *regression* of one variable upon another. A regression model is one way to characterize the association between two variables. The strength of such a linear association is measured by the correlation coefficient.

Contents

9.1 Introduction

Many engineering applications involve *modeling* the relationships among sets of variables. For example, a chemical engineer might want to model the yield of a chemical process as a function of the temperature and pressure at which the reactions take place. An electrical engineer might be interested in modeling the daily peak load of a power plant as a function of the time of year and number of customers. An environmental engineer might want to model the dissolved oxygen of samples from a large lake as a function of the algal and nitrogen content of the sample.

One method of modeling the relationship between variables is called *regression analysis*. This chapter presents regression models for the association between two variables, such as heights and weights of people. Chapter 10 extends these ideas to more than two variables.

9.2 Probabilistic Models

A problem facing every power plant is the estimation of the daily peak power load. Suppose we wanted to model the peak power load as a function of the maximum temperature for the day. The first question to be answered is: "Does an exact relationship exist between these variables?" In other words, is it possible to predict the exact peak load if the maximum temperature is known? We might see that this is not possible, for several reasons. Peak load depends on other factors besides the maximum temperature. For example, the number of customers the plant services, the geographical location of the plant, the capacity of the plant, and the average temperature over the day all probably affect the daily peak power load. However, even if all these factors were included in a model, it is still unlikely that we would be able to predict the peak load *exactly*. There will almost certainly be some variation in peak load due strictly to *random phenomena* that cannot be modeled or explained.

If we were to construct a model that hypothesized an exact relationship between variables, it would be called a *deterministic model*. For example, if we believe that y, the peak load (in megawatts), will be exactly five times x, the maximum temperature (in degrees Fahrenheit), we write

$$y = 5x$$

This equation represents a deterministic relationship between the variables y and x. It implies that y can always be determined exactly when the value of x is known. There is no allowance for error in this prediction.

The primary usefulness of deterministic models is in the description of physical laws. For example, Ohm's Law describes the relationship between current and resistance in a deterministic manner. Newton's laws of motion are other examples of deterministic models. However, it should be noted that, in all these examples, the laws hold precisely only under ideal conditions. Laboratory experiments rarely reproduce these laws exactly. Usually, random error will be introduced by the experiment, so the laws will provide only approximations to reality.

Returning to the peak power load example, if we believe there will be unexplained variation in peak power load, we will discard the deterministic model and use a model that accounts for this *random error*. This *probabilistic model* includes both a deterministic component and a random error component. For example, if we hypothesize that the peak load Y, now a random variable, is related to the maximum temperature x by

$$Y = 5x + \text{ random error}$$

we are hypothesizing a *probabilistic relationship* between Y and x. Note that the deterministic component of this probabilistic model is $5x$.

Figure 9.1a shows possible peak loads for five different values of x when the model is deterministic. All the peak loads must fall exactly on the line, because the deterministic model leaves no room for error.

F I G U R E **9.1**   Deterministic and Probabilistic Models

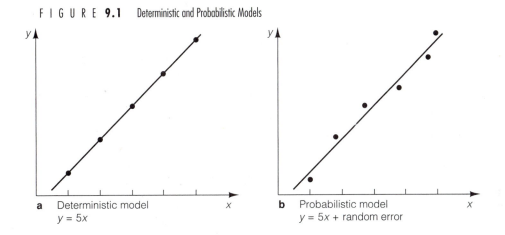

a   Deterministic model
$y = 5x$

b   Probabilistic model
$y = 5x + \text{random error}$

Figure 9.1b shows a possible set of responses for the same values of x when we are using a probabilistic model. Note that the deterministic part of the model (the straight line itself) is the same. Now, however, the inclusion of a random error component allows the peak loads to vary from this line. Since we believe that the peak load will vary randomly for a given value of x, the probabilistic model provides a more realistic model for Y than does the deterministic model.

General Form of Probabilistic Models

$$Y = \text{deterministic component } + \text{ random error}$$

where Y is the random variable to be predicted. We will always assume that the mean value of the random error equals zero. This is equivalent to assuming that the mean value of Y, $E(Y)$, equals the deterministic component of the model:

$$E(Y) = \text{deterministic component}$$

In the previous two chapters, we discussed the simplest form of a probabilistic model. We showed how to make inferences about the mean of Y, $E(Y)$, when

$$E(Y) = \mu$$

where μ is a constant. However, we realized that this did not imply that Y would equal μ exactly, but instead that Y would be equal to μ plus or minus a random error. In particular, if we assume that Y is normally distributed with mean μ and variance σ^2, then we may write the probabilistic model

$$Y = \mu + \varepsilon$$

where the random component ε (epsilon) is normally distributed with mean zero and variance σ^2. Some possible observations generated by this model are shown in Figure 9.2.

FIGURE **9.2**
The Probabilistic Model
$Y = \mu + \varepsilon$

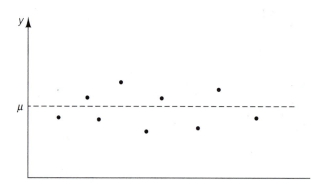

The purpose of this chapter is to generalize the model to allow $E(Y)$ to be a function of other variables. For example, if we want to model the daily peak power load Y as a function of the maximum temperature x for the day, we might hypothesize that the mean of Y is a straight-line function of x, as described in the following box.

The Straight-Line Probabilistic Model

$$Y = \beta_0 + \beta_1 x + \varepsilon$$

where Y = dependent variable (variable to be modeled)

x = independent[†] variable (variable used as a predictor of Y)

ε = random error component

β_0 = y-intercept of the line, that is, the point at which the line intercepts or cuts through the y-axis (see Figure 9.3)

β_1 = slope of the line, that is, amount of increase (or decrease) in the mean of Y for every 1-unit increase in x (see Figure 9.3)

FIGURE 9.3
Straight-Line Probabilistic Model

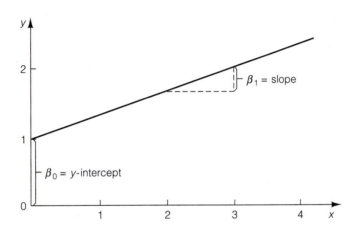

Note that we use the Greek symbols β_0 and β_1 to represent the y-intercept and slope of the model, just as we used the Greek symbol μ to represent the constant mean in the model $Y = \mu + \varepsilon$. In each case, these symbols represent population parameters with numerical values that will need to be estimated using sample data.

It is helpful to think of regression analysis as a five-step procedure:

Step 1.   Hypothesize the form for the mean $E(Y)$ (deterministic component of the model).

Step 2.   Collect sample data and use them to estimate unknown parameters in the model.

Step 3.   Specify the probability distribution of ε, the random error component, and estimate any unknown parameters of this distribution.

Step 4.   Statistically check the adequacy of the model.

Step 5.   When satisfied with the model's adequacy, use it for prediction, estimation, and so on.

[†]The phrase *independent variable* is used in regression analysis to refer to a predictor (or explanatory) variable for the response Y.

In this chapter, we simplify step 1 and introduce the concepts of regression analysis via the straight-line model. In Chapter 10, we discuss how to build more complex models.

9.3 Fitting the Model: The Least-Squares Approach

Suppose we want to estimate the mean daily peak load for a power plant given the sample of peak loads for ten days in Table 9.1. We hypothesize the model

$$Y = \mu + \varepsilon$$

and want to use the sample data to estimate μ. One method, the *least-squares approach*, chooses the estimator that minimizes the sum of squared errors (SSE). That is, we choose the estimator $\hat{\mu}$ so that

$$\text{SSE} = \sum_{i=1}^{n}(y_i - \hat{\mu})^2$$

is minimized. The form of this estimator can be obtained by differentiating SSE with respect to $\hat{\mu}$, setting it equal to zero, and solving for $\hat{\mu}$. Thus,

$$\frac{d(\text{SSE})}{d\hat{\mu}} = -2\sum_{i=1}^{n}(y_i - \hat{\mu}) = 0$$

Simplifying, we have

$$-2\sum_{i=1}^{n} y_i + 2n\hat{\mu} = 0$$

Solving for $\hat{\mu}$,

$$\hat{\mu} = \frac{\sum_{i=1}^{n} y_i}{n} = \bar{y}$$

TABLE **9.1**
Sample of Ten Days' Peak
Power Load (in megawatts)

Day	Peak Load (y_i)
1	214
2	152
3	156
4	129
5	254
6	266
7	210
8	204
9	213
10	150

1948

Thus, the sample mean $\overline{Y}$ is the estimator that minimizes the sum of squared errors and is called the *least-squares estimator* of μ.

For the peak power load data in Table 9.1, we calculate

$$\overline{y} = \frac{1{,}948}{10} = 194.8$$

and

$$\text{SSE} = \sum_{i=1}^{n}(y_i - \overline{y})^2 = 19{,}263.6$$

We know that no other estimate of μ will yield as small an SSE as this, because $\overline{Y}$ is the *least-squares* estimator.

Now suppose we decide that the mean peak power load can be modeled as a function of the maximum temperature x for the day. Specifically, we model the mean peak power load $E(Y)$ as a straight-line function of x. Thus, we record the maximum temperature for the days in Table 9.1 and obtain the data in Table 9.2. A plot of this data, called a *scatterplot*, is shown in Figure 9.4.

We hypothesize the straight-line probabilistic model

$$Y = \beta_0 + \beta_1 x + \varepsilon$$

and want to use the sample data to estimate the y-intercept β_0 and the slope β_1. We use the same principle to estimate β_0 and β_1 in the straight-line model that we used to estimate μ in the constant mean model: the least-squares approach. Thus, we choose the estimate

$$\hat{y} = \hat{\beta}_0 + \hat{\beta}_1 x$$

so that

$$\text{SSE} = \sum_{i=1}^{n}(y_1 - \hat{y})^2 = \sum_{i=1}^{n}(y_i - \hat{\beta}_0 - \hat{\beta}_1 x_i)^2$$

	Maximum	
TABLE 9.2	**Temperature**	**Peak Power Load**
Data for Peak Power Load y and Maximum Temperature x for Ten Days		
Day	**x**	**y**
1	95	214
2	82	152
3	90	156
4	81	129
5	99	254
6	100	266
7	93	210
8	95	204
9	93	213
10	87	150

FIGURE **9.4**
Scatterplot for Data in Table 9.2

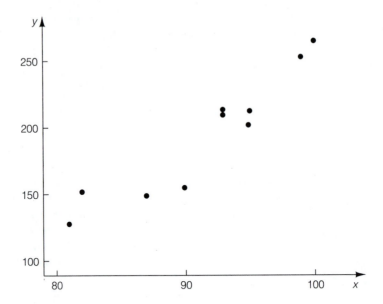

is minimized. We differentiate SSE with respect to $\hat{\beta}_0$ and $\hat{\beta}_1$, set the results equal to zero, and then solve for $\hat{\beta}_0$ and $\hat{\beta}_1$.

$$\frac{\partial(\text{SSE})}{\partial\hat{\beta}_0} = -2\sum_{i=1}^{n}(y_i - \hat{\beta}_0 - \hat{\beta}_1 x_i) = 0$$

$$\frac{\partial(\text{SSE})}{\partial\hat{\beta}_1} = -2\sum_{i=1}^{n}x_i(y_i - \hat{\beta}_0 - \hat{\beta}_1 x_i) = 0$$

and the solution is

$$\hat{\beta}_1 = \frac{\sum_{i=1}^{n}(x_i - \bar{x})(y_i - \bar{y})}{\sum_{i=1}^{n}(x_i - \bar{x})^2} = \frac{\text{SS}_{xy}}{\text{SS}_{xx}}$$

$$\hat{\beta}_0 = \bar{y} - \hat{\beta}_1 \bar{x}$$

For the data in Table 9.2, we find

$$\text{SS}_{xy} = \sum_{i=1}^{n}(x_i - \bar{x})(y_i - \bar{y}) = \sum_{i=1}^{n}x_i y_i - \frac{\left(\sum_{i=1}^{n}x_i\right)\left(\sum_{i=1}^{n}y_i\right)}{n}$$

$$= 180{,}798 - \frac{(915)(1{,}948)}{10} = 2{,}556$$

$$\text{SS}_{xx} = \sum_{i=1}^{n}(x_i - \bar{x})^2 = \sum_{i=1}^{n}x_i^2 - \frac{\left(\sum_{i=1}^{n}x_i\right)^2}{n}$$

$$= 84,103 - \frac{(915)^2}{10} = 380.5$$

and

$$\bar{x} = \frac{915}{10} = 91.5$$
$$\bar{y} = 194.8$$

Then the least-squares estimates are

$$\hat{\beta}_1 = \frac{SS_{xy}}{SS_{xx}} = \frac{2,556}{380.5} = 6.7175$$

$$\hat{\beta}_0 = \bar{y} - \hat{\beta}_1 \bar{x} = 194.8 - (6.7175)(91.5)$$
$$= -419.85$$

The least-squares line is therefore

$$\hat{y} = -419.85 + 6.7175x$$

as shown in Figure 9.5. Note that the errors are the vertical distances between the observed points and the prediction line, $(y_i - \hat{y}_i)$. The predicted values $\hat{y}_i$, the error $(y_i - \hat{y}_i)$, and the squared error $(y_i - \hat{y}_i)^2$ are shown in Table 9.3. You can see that the sum of squared errors, SSE, is 2,093.43. We know that no other straight line will yield as small an SSE as this one.

	x	y	$\hat{y} = \hat{\beta}_0 + \hat{\beta}_1 x$	$(y - \hat{y})$	$(y - \hat{y})^2$
T A B L E 9.3	95	214	218.31	−4.31	18.58
Data, Predicted Values, and	82	152	130.98	21.02	441.84
Errors for the Peak Power Load	90	156	184.72	−28.72	824.84
Data	81	129	124.27	4.73	22.37
	99	254	245.18	8.82	77.79
	100	266	251.90	14.10	198.81
	93	210	204.88	5.12	26.21
	95	204	218.31	−14.31	204.78
	93	213	204.88	8.12	65.93
	87	150	164.57	−14.57	212.28
				SSE =	2,093.43

To summarize, we have defined the best-fitting straight line to be the one that satisfies the least-squares criterion, that is, the line for which the sum of squared errors will be smaller than for any other straight-line model. This line is called the *least-squares line*, and its equation is called the *least-squares prediction equation*. The prediction equation can be determined by substituting the sample values of x and y into the formulas given in the box below. We will discuss the adequacy of the model later.

FIGURE **9.5**
Scatterplot and Least-Squares
Line for the Data in Table 9.2

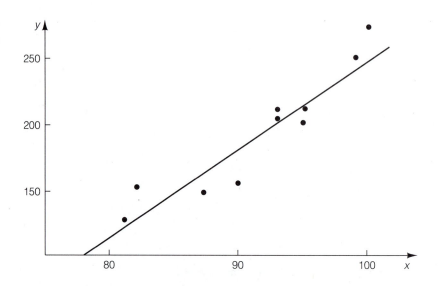

Formulas for the Least-Squares Line

Slope: $\hat{\beta}_1 = \dfrac{SS_{xy}}{SS_{xx}}$

y-intercept: $\hat{\beta}_0 = \bar{y} - \hat{\beta}_1 \bar{x}$

where

$$SS_{xy} = \sum_{i=1}^{n} x_i y_i - \frac{\left(\sum\limits_{i=1}^{n} x_i\right)\left(\sum\limits_{i=1}^{n} y_i\right)}{n} = \sum_{i=1}^{n}(x_i - \bar{x})(y_i - \bar{y})$$

$$SS_{xx} = \sum_{i=1}^{n} x_i^2 - \frac{\left(\sum\limits_{i=1}^{n} x_i\right)^2}{n} = \sum_{i=1}^{n}(x_i - \bar{x})^2$$

n = number of pairs of observations (sample size)

Exercises

9.1   Use the method of least squares to fit a straight line to the following six data points:

x	1	2	3	4	5	6
y	1	2	2	3	5	5

a   What are the least-squares estimates of β_0 and β_1?

b Plot the data points and graph the least-squares line. Does the line pass through the data points?

9.2 Use the method of least squares to fit a straight line to the following data points:

x	−2	−1	0	1	2
y	4	3	3	1	−1

a What are the least-squares estimates of β_0 and β_1?

b Plot the data points and graph the least-squares line. Does the line pass through the data points?

9.3 The elongation of a steel cable is assumed to be linearly related to the amount of force applied. Five identical specimens of cable gave the following results when varying forces were applied:

Force (x)	1.0	1.5	2.0	2.5	3.0
Elongation (y)	3.0	3.8	5.4	6.9	8.4

Use the method of least squares to fit the line

$$Y = \beta_0 + \beta_1 x + \varepsilon$$

9.4 A company wants to model the relationship between its sales and the sales for the industry as a whole. For the following data, fit a straight line by the method of least squares.

Year	Company Sales y ($ million)	Industry Sales x ($ million)
1982	0.5	10
1983	1.0	12
1984	1.0	13
1985	1.4	15
1986	1.3	14
1987	1.6	15

9.5 It is thought that abrasion loss in certain steel specimens should be a linear function of the Rockwell hardness measure. A sample of eight specimens gave the following results:

Rockwell Hardness (x)	60	62	63	67	70	74	79	81
Abrasion Loss (y)	251	245	246	233	221	202	188	17

Fit a straight line to these measurements.

9.6 Due primarily to the price controls of the Organization of Petroleum Exporting Countries (OPEC), a cartel of crude-oil suppliers, the price of crude oil has risen dramatically since the early 1970s. As a result, motorists have been confronted with a similar upward spiral of gasoline prices. The data in the table are typical prices for a gallon of regular leaded gasoline and a barrel of crude oil (refiner acquisition cost) for the indicated years.

a Use these data to calculate the least-squares line that describes the relationship between the price of a gallon of gas and the price of a barrel of crude oil.

b Plot your least-squares line on a scatterplot of the data. Does your least-squares line appear to be an appropriate characterization of the relationship between y and x? Explain.

c If the price of crude oil fell to $20 per barrel, to what level (approximately) would the price of regular gasoline fall? Justify your response.

Year	Gasoline (cents/gal.) y	Crude Oil (dollars/bbl) x
1973	38.8	4.15
1975	56.7	10.38
1976	59.0	10.89
1977	62.2	11.96
1978	62.6	12.46
1979	85.7	17.72
1980	119.1	28.07
1981	133.3	36.11

Source: Statistical Abstract of the United States, 1981, p. 476.

9.7 A study is made of the number of parts assembled as a function of the time spent on the job. Twelve employees are divided into three groups, assigned to three time intervals, with the following results:

Time x	Number of Parts Assembled y
10 minutes	27, 32, 26, 34
15 minutes	35, 30, 42, 47
20 minutes	45, 50, 52, 49

Fit the model

$$Y = \beta_0 + \beta_1 x + \varepsilon$$

by the method of least squares.

9.8 Laboratory experiments designed to measure LC50 values for the effect of certain toxicants on fish are basically run by two different methods. One method has water continuously flowing through laboratory tanks, and the other has static water conditions. For purposes of establishing criteria for toxicants, the Environmental Protection Agency (EPA) wants to adjust all results to the flow-through condition. Thus, a model is needed to relate the two methods. Observations on certain toxicants examined under both static and flow-through conditions yielded the following measurements (in parts per million):

Toxicant	LC50 Flow-Through y	LC50 Static x
1	23.00	39.00
2	22.30	37.50
3	9.40	22.20
4	9.70	17.50
5	0.15	0.64
6	0.28	0.45
7	0.75	2.62
8	0.51	2.36
9	28.00	32.00
10	0.39	0.77

Fit the model $Y = \beta_0 + \beta_1 x + \varepsilon$ by the method of least squares.

9.4 The Probability Distribution of the Random Error Component

We have now completed the first two steps of regression modeling: We have hypothesized the form of $E(Y)$, and we have used the sample data to estimate unknown parameters in the model. The hypothesized model relating peak power load Y to daily maximum temperature x is

$$Y = \beta_0 + \beta_1 x + \varepsilon$$

and the least-squares estimate of $E(Y) = \beta_0 + \beta_1 x$ is

$$\hat{y} = -419.85 + 6.7175x$$

Step 3. Specify the probability distribution of ε, the random error component, and estimate any unknown parameters of this distribution.

Recall that when we wanted to make inferences about a population mean and had only a small sample with which to work, we used a t statistic and assumed that the data were normally distributed. Similarly, when we want to make inferences about the parameters of a regression model, we assume that the error component ε is normally distributed, and the assumption is most important when the sample size is small. The mean of ε is zero, since the deterministic component of the model describes $E(Y)$. Finally, we assume that the variance of ε is σ^2, a constant for all values of x, and that the errors associated with different observations are independent.

The assumptions about the probability distribution of ε are summarized in the box and shown in Figure 9.6.

F I G U R E **9.6**   The Probability Distribution of ε

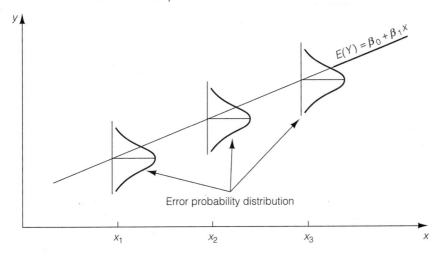

Error probability distribution

Probability Distribution of the Random Error Component ε

The error component is normally distributed with mean zero and constant variance σ^2. The errors associated with different observations are independent.

Various techniques exist for testing the validity of the assumptions, and there are some alternative techniques to be used when the assumptions appear to be invalid. These techniques often require us to return to step 2 and to use a method other than least squares to estimate the model parameters. Much current statistical research is devoted to alternative methodologies. Most are beyond the scope of this text, but we will mention a few in later chapters. In actual practice, the assumptions need not hold exactly for the least-squares techniques to be useful.

Note that the probability distribution of ε would be completely specified if the variance σ^2 were known. (We already know that ε is normally distributed with a mean of zero.) To estimate σ^2, we make use of the SSE for the least-squares model. The estimate s^2 of σ^2 is calculated by dividing SSE by the number of degrees of freedom associated with the error component. We use 2 degrees of freedom to estimate the y-intercept and slope in the straight-line model, leaving $(n - 2)$ degrees of freedom for the error variance estimation. Thus, an unbiased estimator of σ^2 is

$$s^2 = \frac{\text{SSE}}{n - 2}$$

where

$$\text{SSE} = \sum_{i=1}^{n}(y_i - \hat{y}_i)^2 = \text{SS}_{yy} - \hat{\beta}_1 \text{SS}_{xy}$$

and

$$\text{SS}_{yy} = \sum_{i=1}^{n}(y_i - \bar{y})^2 = \sum_{i=1}^{n} y_i^2 - \frac{\left(\sum_{i=1}^{n} y_i\right)^2}{n}$$

In our peak power load example, we calculated SSE $= 2,093.43$ for the least-squares line. Recalling that there were $n = 10$ data points, we have $n - 2 = 8$ degrees of freedom for estimating σ^2. Thus,

$$s^2 = \frac{\text{SSE}}{n - 2} = \frac{2,093.43}{8} = 261.68$$

is the estimated variance, and

$$s = \sqrt{261.68} = 16.18$$

is the estimated standard deviation of ε.

You may be able to obtain an intuitive feeling for s by recalling that about 95% of the observations from a normal distribution lie within two standard deviations of the mean. Thus, we expect approximately 95% of the observations to fall within $2s$ of the estimated mean $\hat{y} = \hat{\beta}_0 + \hat{\beta}_1 x$. For our peak power load example, note that all ten observations fall within $2s = 2(16.18) = 32.36$ of the least-squares line. In Section 9.8, we show how to use the estimated standard deviation to evaluate the prediction error when $\hat{y}$ is used to predict a value of y to be observed for a given value of x.

Exercises

9.9 Calculate SSE and s^2 for the data of Exercises 9.1–9.8.

9.10 J. Matis and T. Wehrly (*Biometrics*, 35, no. 1, March 1979) report the following data on the proportion of green sunfish that survive a fixed level of thermal pollution for varying lengths of time:

Proportion of Survivors (y)	1.00	0.95	0.95	0.90	0.85	0.70	0.65	0.60	0.55	0.40
Scaled Time (x)	0.10	0.15	0.20	0.25	0.30	0.35	0.40	0.45	0.50	0.55

a Fit the linear model $Y = \beta_0 + \beta_1 x + \varepsilon$ by the method of least squares.

b Plot the data points and graph the line found in (a). Does the line fit through the points?

c Compute SSE and s^2.

9.5 Assessing the Adequacy of the Model: Making Inferences About Slope β_1

Step 4. Statistically check the adequacy of the model.

Now that we have specified the probability distribution of ε and estimated the variance σ^2, we are prepared to make statistical inferences about the model's adequacy for describing the mean $E(Y)$ and for predicting the values of Y for given values of x.

Refer again to the peak power load example and suppose that the daily peak load is *completely unrelated* to the maximum temperature on a given day. This implies that the mean $E(Y) = \beta_0 + \beta_1 x$ does not change as x changes. In the straight-line model, this means that the true slope β_1 is equal to zero (see Figure 9.7). Note that if β_1 is equal to zero, the model is simply $Y = \beta_0 + \varepsilon$, the constant mean model used in previous chapters. (Note that the symbol β_0 is the same as the symbol for the mean μ when $\beta_1 = 0$.) Therefore, to test the null hypothesis that x contributes no information regarding the prediction of Y against the alternative hypothesis that these variables are linearly related with a slope differing from zero, we test

$$H_0 : \beta_1 = 0 \quad \text{versus} \quad H_a : \beta_1 \neq 0$$

F I G U R E **9.7**
Graph of the Model
$Y = \beta_0 + \varepsilon$

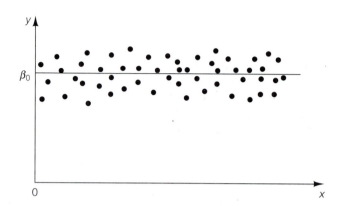

If the data support the alternative hypothesis, we conclude that x does contribute information for the prediction of Y using the straight-line model (although the true relationship between $E(Y)$ and x could be more complex than a straight line). Thus, to some extent this is a test of the adequacy of the hypothesized model.

The appropriate test statistic is found by considering the sampling distribution of $\hat{\beta}_1$, the least-squares estimator of the slope β_1.

Sampling Distribution of $\hat{\beta}_1$

If we assume that the error components are independent normal random variables with mean zero and constant variance σ^2, the sampling distribution of the least-squares estimator $\hat{\beta}_1$ of the slope will be normal, with mean β_1 (the true slope) and standard deviation

$$\sigma_{\hat{\beta}_1} = \frac{\sigma}{\sqrt{\text{SS}_{xx}}} \qquad \text{(see Figure 9.8)}$$

[*Note*: Proof of the unbiasedness of $\hat{\beta}_1$ and a derivation of its standard deviation are given next.]

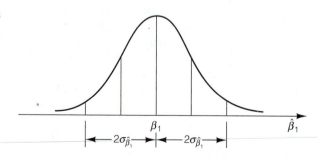

Proof of Unbiasedness of $\hat{\beta}_1$

Recall that

$$\hat{\beta}_1 = \frac{\text{SS}_{xy}}{\text{SS}_{xx}} = \frac{\sum_{i=1}^{n}(x_i - \bar{x})(Y_i - \bar{Y})}{\sum_{i=1}^{n}(x_i - \bar{x})^2}$$

Using the following facts:

1   $Y_i = \beta_0 + \beta_1 x + \varepsilon_1$
2   $\bar{Y} = \beta_0 + \beta_1 \bar{x} + \bar{\varepsilon}$
3   x_i's are fixed[†]

we find

$$E(\hat{\beta}_1) = E\left\{\frac{\sum_{i=1}^{n}(x_i - \bar{x})(Y_i - \bar{Y})}{\sum_{i=1}^{n}(x_i - \bar{x})^2}\right\}$$

$$= \frac{1}{\sum_{i=1}^{n}(x_i - \bar{x})^2}\sum_{i=1}^{n}(x_i - \bar{x})E(Y_i - \bar{Y})$$

$$= \frac{1}{\text{SS}_{xx}}\sum_{i=1}^{n}(x_i - \bar{x})E[(\beta_0 + \beta_1 x_i + \varepsilon_i) - (\beta_0 + \beta_1 \bar{x} + \bar{\varepsilon})]$$

$$= \frac{1}{\text{SS}_{xx}}\sum_{i=1}^{n}(x_i - \bar{x})[\beta_1(x_i - \bar{x})]$$

$$= \frac{\text{SS}_{xx}}{\text{SS}_{xx}}\beta_1$$

$$= \beta_1$$

showing that $\hat{\beta}_1$ is an unbiased estimator of β_1.

[†]If the dependent variable X is random, we use expectations *conditional* on the observed values of X.

Derivation of the Variance of $\hat{\beta}_1$

Using the formula derived for the variance of a linear function,

$$\sigma_{\hat{\beta}_1}^2 = V(\hat{\beta}_1) = V\left\{\frac{\sum\limits_{i=1}^{n}(x_i - \bar{x})(Y_i - \overline{Y})}{\sum\limits_{i=1}^{n}(x_i - \bar{x})^2}\right\}$$

$$= \frac{1}{\left[\sum\limits_{i=1}^{n}(x_i - \bar{x})^2\right]^2} V\left\{\sum\limits_{i=1}^{n}(x_i - \bar{x})(Y_i - \overline{Y})\right\}$$

$$= \frac{1}{SS_{xx}^2}\left[\sum\limits_{i=1}^{n}(x_i - \bar{x})^2 V(Y_i - \overline{Y})\right.$$

$$\left. + \sum\limits_{i=1}^{n}\sum\limits_{\substack{j=1 \\ i \neq j}}^{n}(x_i - \bar{x})(x_j - \bar{x})\,\text{Cov}\left\{(Y_i - \overline{Y}), (Y_j - \overline{Y})\right\}\right]$$

Now

$$V(Y_i - \overline{Y}) = V\{(\beta_0 + \beta_1 x_i + \varepsilon_i) - (\beta_0 + \beta_1 \bar{x} + \bar{\varepsilon})\}$$

$$= V(\varepsilon_i - \bar{\varepsilon}) = \sigma^2 + \frac{\sigma^2}{n} - 2\frac{\sigma^2}{n}$$

$$= \sigma^2 - \frac{\sigma^2}{n}$$

and

$$\text{Cov}\left\{(Y_i - \overline{Y}), (Y_j - \overline{Y})\right\} = \text{Cov}\left\{(\varepsilon_i - \bar{\varepsilon}), (\varepsilon_j - \bar{\varepsilon})\right\}$$

$$= -\text{Cov}(\varepsilon_i, \bar{\varepsilon}) - \text{Cov}(\varepsilon_j, \bar{\varepsilon}) + V(\bar{\varepsilon})$$

$$= -\frac{\sigma^2}{n} - \frac{\sigma^2}{n} + \frac{\sigma^2}{n}$$

$$= -\frac{\sigma^2}{n}$$

where we have used the following facts:

1   $V(\varepsilon_i) = \sigma^2$

2   $\text{Cov}(\varepsilon_i, \varepsilon_j) = 0, i \neq j$

3 $$V(\bar{\varepsilon}) = V\left(\frac{\sum\limits_{i=1}^{n} \varepsilon_i}{n}\right) = \frac{1}{n^2}\sum_{i=1}^{n} V(\varepsilon_i)$$

$$= \frac{n\sigma^2}{n^2} = \frac{\sigma^2}{n}$$

4 $$\mathrm{Cov}(\varepsilon_i, \bar{\varepsilon}) = \mathrm{Cov}\left(\varepsilon_i, \frac{\sum\limits_{j=1}^{n} \varepsilon_j}{n}\right) = \frac{V(\varepsilon_j)}{n} = \frac{\sigma^2}{n}$$

Thus,

$$\sigma_{\hat{\beta}_1}^2 = \frac{1}{SS_{xx}^2}\left[\sum_{i=1}^{n}(x_i - \bar{x})^2\left(\sigma^2 - \frac{\sigma^2}{n}\right) - \sum_{i=1}^{n}\sum_{j=1}^{n}{}_{i\neq j}\,(x_i - \bar{x})(x_j - \bar{x})\right.$$

$$\left. \times \left(-\frac{\sigma^2}{n}\right)\right]$$

$$= \frac{1}{SS_{xx}^2}\left\{SS_{xx}\sigma^2 - \frac{\sigma^2}{n}\left[\sum_{i=1}^{n}(x_i - \bar{x})\right]\right\}$$

$$= \frac{\sigma^2}{SS_{xx}} \qquad \text{using } \sum_{i=1}^{n}(x_i - \bar{x}) = 0$$

The standard error of $\hat{\beta}_1$ is therefore $\sigma_{\hat{\beta}_1} = \sigma/\sqrt{SS_{xx}}$.

Since σ, the standard deviation of ε, is usually unknown, the appropriate test statistic is a Student's t statistic formed as follows:

$$t = \frac{\hat{\beta}_1 - (\text{hypothesized value of } \beta_1)}{s_{\hat{\beta}_1}}$$

where $s_{\hat{\beta}_1} = s/\sqrt{SS_{xx}}$, the estimated standard deviation of the sampling distribution of $\hat{\beta}_1$. Note that the statistic has the usual form of a t-statistic, a normally distributed estimator divided by an estimate of its standard deviation. We will usually be testing the null hypothesis $H_0 : \beta_1 = 0$, so the t statistic becomes

$$t = \frac{\hat{\beta}_1}{s/\sqrt{SS_{XX}}}$$

For the peak power load example with $H_a : \beta_1 \neq 0$, we choose $\alpha = 0.05$, and since $n = 10$, our rejection region is

$$t < -t_{0.025}(10 - 2) = -2.306$$
$$t > t_{0.025}(8) = 2.306$$

We previously calculated $\beta_1 = 6.7175$, $s = 16.18$, and $SS_{xx} = 380.5$. Thus,

$$t = \frac{\hat{\beta}_1}{s/\sqrt{SS_{xx}}} = \frac{6.7175}{16.18/\sqrt{380.5}} = 8.10$$

Since this calculated t value falls in the upper-tail rejection region, we reject the null hypothesis and conclude that the slope β_1 is not zero. The sample evidence indicates that the peak power load tends to increase as a day's maximum temperature increases.

A Test of Model Utility

One-Tailed Test	*Two-Tailed Test*

$H_0 : \beta_1 = 0$

$H_a : \beta_1 < 0$ (or $H_a : \beta_1 > 0$)

$H_0 : \beta_1 = 0$

$H_a : \beta_1 \neq 0$

Test statistic:

$$t = \frac{\hat{\beta}_1}{s_{\hat{\beta}_1}} = \frac{\hat{\beta}_1}{s/\sqrt{SS_{xx}}}$$

Test statistic:

$$t = \frac{\hat{\beta}_1}{s_{\hat{\beta}_1}} = \frac{\hat{\beta}_1}{s/\sqrt{SS_{xx}}}$$

Rejection Region:

$$t < -t_\alpha(n-2)$$

Rejection Region:

$$t < -t_{\alpha/2}(n-2) \quad \text{or}$$

(or $t > t_\alpha(n-2)$ when $H_a : \beta_1 > 0$)

$$t > t_{\alpha/2}(n-2)$$

Assumptions are the four assumptions about ε stated in Section 9.4.

What conclusion can be drawn if the calculated t value does not fall in the rejection region? We know from previous discussions of the philosophy of hypothesis testing that such a t value does not lead us to accept the null hypothesis. That is, we do not conclude that $\beta_1 = 0$. Additional data might indicate that β_1 differs from zero, or a more complex relationship may exist between x and Y, requiring the fitting of a model other than the simple linear model. We discuss several such models in Chapter 10.

Another way to make inferences about the slope β_1 is to estimate it by using a confidence interval. This interval is formed as shown next.

A 100$(1 - \alpha)$ Percent Confidence Interval for the Slope $\hat{\beta}_1$

$$\hat{\beta}_1 \pm t_{\alpha/2}(n-2)s_{\hat{\beta}_1}$$

where

$$s_{\hat{\beta}_1} = \frac{s}{\sqrt{SS_{xx}}}$$

Assumptions are the four assumptions about ε stated in Section 9.4.

For the peak power load example, a 95% confidence interval for the slope β_1 is

$$\hat{\beta}_1 \pm t_{0.025}(8)s_{\hat{\beta}_1} = 6.7175 \pm 2.306 \left(\frac{16.18}{\sqrt{380.5}} \right)$$
$$= 6.7175 \pm 1.913 = (4.80, 8.63)$$

This confidence interval confirms the conclusion arrived at in the test of the null hypothesis $H_0 : \beta_1 = 0$, since it implies that the true slope is between 4.80 and 8.63.

Exercises

9.11   Refer to Exercise 9.1. Is there sufficient evidence to say that the slope of the line is significantly different from zero? Use $\alpha = 0.05$.

9.12   Refer to Exercise 9.2. Estimate β_1 in a confidence interval with confidence coefficient 0.90.

9.13   Refer to Exercise 9.3. Estimate the amount of elongation per unit increase in force. Use a confidence coefficient of 0.95.

9.14   Refer to Exercise 9.4. Do industry sales appear to contribute any information to the prediction of company sales? Test at the 10% level of significance.

9.15   Refer to Exercise 9.5. Does Rockwell hardness appear to be linearly related to abrasion loss? Use $\alpha = 0.05$.

9.16   In Exercise 9.6, the price of a gallon of gasoline y was modeled as a function of the price of a barrel of crude oil x using the following equation: $Y = \beta_0 + \beta_1 x + \varepsilon$. Estimates of β_0 and β_1 were found to be 25.883 and 3.115, respectively.

   a   Estimate the variance and standard deviation of ε.

   b   Suppose the price of a barrel of crude oil is $15. Estimate the mean and standard deviation of the price of a gallon of gas under these circumstances.

   c   Repeat part (b) for a crude oil price of $30.

   d   What assumptions about ε did you make in answering parts (b) and (c)?

9.17   Refer to Exercise 9.7. Estimate the increase in expected number of parts assembled per 5-minute increase in time spent on the job. Use a 95% confidence coefficient.

9.18   Refer to Exercise 9.8. Does it appear that flow-through LC50 values are linearly related to static LC50 values? Use $\alpha = 0.05$.

9.19   It is well known that large bodies of water have a mitigating effect on the temperature of the surrounding land masses. On a cold night in central Florida, temperatures (in degrees Fahrenheit) were recorded at equal distances along a transect running in the downwind direction from a large lake. The data are as follows:

Site (x)	1	2	3	4	5	6	7	8	9	10
Temperature (y)	37.00	36.25	35.41	34.92	34.52	34.45	34.40	34.00	33.62	33.90

a   Fit the linear model $Y = \beta_0 + \beta_1 x + \varepsilon$ by the method of least squares.

b   Find SSE and s^2.

c   Does it appear that temperatures decrease significantly as distance from the lake increases? Use $\alpha = 0.05$.

9.6   The Coefficient of Correlation

In looking for an association between two measurement variables, fitting a simple linear regression model and testing the slope provide evidence as to whether or not a significant association exists. The next step is to measure the strength of the association, which is accomplished through the *correlation coefficient*. Correlation between random variables was introduced in Chapter 5 as

$$\rho = \text{Corr}(X, Y) = \frac{\text{Cov}(X, Y)}{\sqrt{V(X)V(Y)}}$$

Just as $V(X)$ is estimated by $\text{SS}_{xx}/(n-1)$ and $V(Y)$ is estimated by $\text{SS}_{yy}/(n-1)$, $\text{Cov}(X, Y)$ is estimated by $\text{SS}_{xy}/(n-1)$. Thus, a sample estimate of ρ is

$$r = \frac{\text{SS}_{xy}}{\sqrt{\text{SS}_{xx}\text{SS}_{yy}}}$$

(Note that the $(n-1)$ terms cancel out.) Now

$$\hat{\beta}_1 = \frac{\text{SS}_{xy}}{\text{SS}_{xx}}$$

so we can write

$$r = \hat{\beta}_1 \sqrt{\frac{\text{SS}_{xx}}{\text{SS}_{yy}}} = \hat{\beta}_1 \frac{s_x}{s_y}$$

Note that r is computed by using the same quantities used in fitting the least-squares line. Since both r and $\hat{\beta}_1$ provide information about the utility of the model, it is not surprising that there is a similarity in their computational formulas. Particularly note that SS_{xy} appears in the numerators of both expressions and, since both denominators are always positive, r and $\hat{\beta}_1$ will always be of the same sign (either both positive or both negative). A value of r near or equal to zero implies little or no linear relationship between y and x. In contrast, the closer r is to 1 or -1, the stronger the linear relationship between y and x. And if $r = 1$ or $r = -1$, all the points fall exactly on the least-squares line. Positive values of r imply that y increases as x increases; negative values of r imply that y decreases as x increases. Each of these situations is portrayed in Figure 9.9.

FIGURE **9.9**
Values of r and Their
Implications

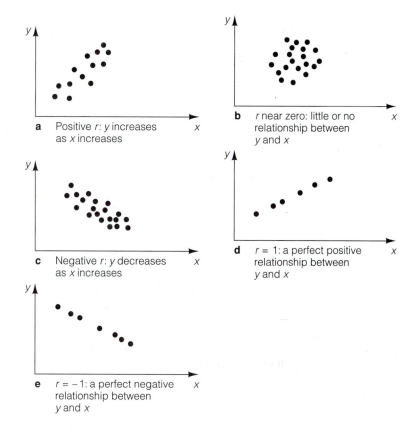

a   Positive r: y increases
as x increases

b r near zero: little or no
relationship between
y and x

c   Negative r: y decreases
as x increases

d $r = 1$: a perfect positive
relationship between
y and x

e $r = -1$: a perfect negative
relationship between
y and x

In the peak power load example, we related peak daily power usage y to the day's maximum temperature x. For these data (Table 9.2), we found

$$SS_{xy} = 2{,}556$$

and

$$SS_{xx} = 380.5$$

where

$$SS_{yy} = \sum_{i=1}^{n}(y_i - \bar{y})^2 = 19{,}263.6$$

Therefore,

$$r = \frac{2{,}556}{\sqrt{(380.5)(19{,}263.6)}} = 0.94$$

The fact that the value of r is positive and near 1 indicates that the peak power load tends to increase as the daily maximum temperature increases *for this sample of ten days*. This is the same conclusion we reached upon finding that the least-squares slope $\hat{\beta}_1$ was positive.

If we want to use the sample information to test the null hypothesis that the normal random variables X and Y are uncorrelated, that is, $H_0 : \rho = 0$, we can use the results of the test that the slope of the straight-line model is zero, that is, $H_0 : \beta_1 = 0$. The test given in Section 9.5 for the latter hypothesis is identical to the test used for the former hypothesis. When we tested the null hypothesis $H_0 : \beta_1 = 0$ in connection with the peak power load example, the data led to a rejection of the null hypothesis at the $\alpha = 0.05$ level. This implies that the null hypothesis of a zero linear correlation between the two variables (maximum temperature and peak power load) can also be rejected at the $\alpha = 0.05$ level. The only real difference between the least-squares slope $\hat{\beta}_1$ and the coefficient of correlation r is the measurement scale. Therefore, the information they provide about the utility of the least-squares model is to some extent redundant. However, the test for correlation requires the additional assumption that the variables X and Y have a bivariate normal distribution, while the test of slope does not require that x be a random variable. For example, we could use the straight-line model to describe the relationship between the yield of a chemical process Y and the temperature at which the reaction takes place x, even if the temperature is controlled and therefore nonrandom. However, the correlation coefficient would have no meaning in this case, since the bivariate normal distribution requires that both variables be marginally normal.

We close this section with a caution: Do not infer a causal relationship on the basis of high sample correlation. Although it is probably true that a high maximum daily temperature *causes* an increase in peak power demand, the same would not necessarily be true of the relationship between availability of oil and peak power load, even though they are probably positively correlated. That is, we would not be willing to state that low availability of oil *causes* a lower peak power load. Rather, the scarcity of oil tends to cause an increase in conservation awareness, which in turn tends to cause a decrease in peak power load. Thus, a large sample correlation indicates only that the two variables tend to change together; causality must be determined after considering other factors in addition to the size of the correlation.

9.7 The Coefficient of Determination

Another way to measure the contribution of x in predicting Y is to consider how much the errors of prediction of Y were reduced by using the information provided by x. If you do not use x, the best prediction for any value of Y would be $\bar{y}$, and the sum of squares of the deviations of the observed y values about $\bar{y}$ is the familiar

$$SS_{yy} = \sum (y_i - \bar{y})^2$$

On the other hand, if you use x to predict Y, the sum of squares of the deviations of the y values about the least-squares line is

$$SSE = \sum (y_i - \hat{y}_i)^2$$

To illustrate, suppose a sample of data has the scatterplot shown in Figure 9.10a. If we assume that x contributes no information for the prediction of y, the best

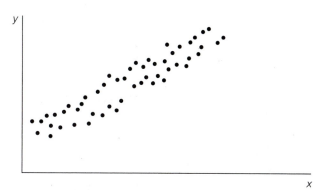

a   Scatterplot of data

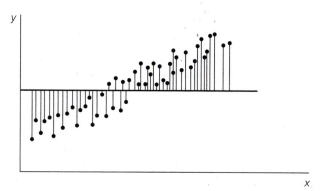

b   Assumption: x contributes no information for predicting y,
$\hat{y} = \bar{y}$

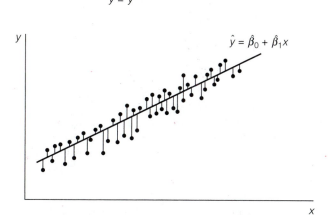

c   Assumption: x contributes information for predicting y,
$\hat{y} = \hat{\beta}_0 + \hat{\beta}_1 x$

prediction for a value of y is the sample mean $\bar{y}$, which is shown as the horizontal line in Figure 9.10b. The vertical line segments in Figure 9.10b are the deviations of the points about the mean $\bar{y}$. Note that the sum of squares of deviations for the model $\hat{y} = \bar{y}$ is

$$SS_{yy} = \sum (y_i - \bar{y})^2$$

Now suppose you fit a least-squares line to the same set of data and locate the deviations of the points about the line, as shown in Figure 9.10c. Compare the deviations about the prediction lines in parts (b) and (c) of Figure 9.10. You can see the following:

1   If x contributes little or no information for the prediction of y, the sums of squares of deviations for the two lines,

$$SS_{yy} = \sum (y_i - \bar{y})^2 \qquad \text{and} \qquad SSE = \sum (y_i - \hat{y}_i)^2$$

will be nearly equal.

2   If x does contribute information for the prediction of y, SSE will be smaller than SS_{yy}, as shown in Figure 9.10. In fact, if all the points fall on the least-squares line, then SSE $= 0$.

It can be shown that

$$\sum (y_i - \bar{y})^2 = \sum (y_i - \hat{y}_i)^2 + \sum (\hat{y}_i - \bar{y})^2$$

We label the third sum, $\sum (\hat{y}_i - \bar{y})^2$, the Sum of Squares for Regression and denote it by SSR. Thus,

$$SS_{yy} = SSE + SSR$$

and it follows that

$$SS_{yy} - SSE \geq 0$$

The SSR assesses differences between $\hat{y}_i$ and $\bar{y}$. If the regression line has slope near zero, then $\hat{y}_i$ and $\bar{y}$ will be close together for any i, and SSR will be small. If, on the other hand, the slope is far from zero, SSR will be large.

Then the reduction in the sum of squares of deviations that can be attributed to x, expressed as a proportion of SS_{yy}, is

$$\frac{SS_{yy} - SSE}{SS_{yy}} = \frac{SSR}{SS_{yy}}$$

It can be shown that this quantity is equal to the square of the simple linear coefficient of correlation.

DEFINITION   9.1

The square of the coefficient of correlation is called the **coefficient of determination**. It represents the proportion of the sum of squares of deviations of the y values about their mean that can be attributed to a linear relation between y and x.

$$r^2 = \frac{\text{SS}_{yy} - \text{SSE}}{\text{SS}_{yy}} = 1 - \frac{\text{SSE}}{\text{SS}_{yy}} = \frac{\text{SSR}}{\text{SS}_{yy}} \quad .$$

Note that r^2 is always between 0 and 1. Thus, $r^2 = 0.60$ means that 60% of the sum of squares of deviations of the observed y values about their mean is attributable to the linear relation between y and x.

EXAMPLE **9.1**   Calculate and interpret the coefficient of determination for the peak power load example.

Solution   We have previously calculated

$$\text{SS}_{yy} = 19{,}263.6$$
$$\text{SSE} = 2{,}093.4$$

so we have

$$r^2 = \frac{\text{SS}_{yy} - \text{SSE}}{\text{SS}_{yy}} = \frac{19{,}263.6 - 2{,}093.4}{19{,}263.6}$$
$$= 0.89$$

Note that we could have obtained $r^2 = (0.94)^2 = 0.89$ more simply by using the correlation coefficient calculated in the previous section. However, use of the above formula is worthwhile because it reminds us that r^2 represents the fraction reduction in variability from SS_{yy} to SSE. In the peak power load example, this fraction is 0.89, which means that the sample variability of the peak loads about their mean is reduced by 89% when the mean peak load is modeled as a linear function of daily high temperature. Remember, though, that this statement holds true only for the sample at hand. We have no way of knowing how much of the population variability can be explained by the linear model.   ∎

Exercises

9.20   Compute r and r^2 for the data of Exercises 9.1–9.8.

9.21   In the summer of 1981, the Minnesota Department of Transportation installed a state-of-the-art weigh-in-motion scale in the concrete surface of the eastbound lanes of Interstate 494 in Bloomington, Minnesota. The system is computerized and monitors traffic continuously. It is capable of distinguishing among 13 different types of vehicles (car, five-axle semi, five-axle twin trailer, etc.). The primary purpose of the system is to provide traffic counts and weights for use in the planning and design of future roadways. Following installation, a study was undertaken to determine whether the scale's readings corresponded with the static weights of the vehicles being monitored. Studies of this type are known as *calibration studies*. After some preliminary comparisons using a two-axle, six-tire truck carrying different loads (see table below), calibration adjustments were made in the software of the weigh-in-motion system, and the weigh-in-motion scales were reevaluated (T. Wright, D. Owen, and A. Pena, "Status of MN/DOT's Weigh-in-Motion Program," Minnesota DOT, 1983).

Trial No.	Static Weight of Truck x	Weigh-in-Motion Reading y_1 (prior to calibration adjustment)	Weigh-in-Motion y_2 (after calibration adjustment)
1	27.9	26.0	27.8
2	29.1	29.9	29.1
3	38.0	39.5	37.8
4	27.0	25.1	27.1
5	30.3	31.6	30.6
6	34.5	36.2	34.3
7	27.8	25.1	26.9
8	29.6	31.0	29.6
9	33.1	35.6	33.0
10	35.5	40.2	35.0

Note: All weights are in thousands of pounds.

a   Construct two scatterplots, one of y_1 versus x, and the other of y_2 versus x.

b   Use the scatterplots of part (a) to evaluate the performance of the weigh-in-motion scale both before and after the calibration adjustment.

c   Calculate the correlation coefficient for both sets of data and interpret their values. Explain how these correlation coefficients can be used to evaluate the weigh-in-motion scale.

d   Suppose the sample correlation coefficient for y_2 and x were 1. Could this happen if the static weights and the weigh-in-motion readings disagreed? Explain.

9.22   A problem of both economic and social concern in the United States is the importation and sale of illicit drugs. The data shown below are a part of a larger body of data collected by the Florida Attorney General's Office in an attempt to relate the incidence of drug seizures and drug arrests to the characteristics of the Florida counties. They show the number y of drug arrests per county in 1982, the density x_1 of the county (population per square mile), and the number x_2 of law enforcement employees. In order to simplify the calculations, we show data for only ten counties.

	County									
	1	2	3	4	5	6	7	8	9	10
Population Density x_1	169	68	278	842	18	42	112	529	276	613
Number of Law Enforcement Employees x_2	498	35	772	5,788	18	57	300	1,762	416	520
Number of Arrests in 1982 y	370	44	716	7,416	25	50	189	1,097	256	432

a   Fit a least-squares line to relate the number y of drug arrests per county in 1982 to the county population density x_1.

b   We might expect the mean number of arrests to increase as the population density increases. Do the data support this theory? Test by using $\alpha = 0.05$.

    c   Calculate the coefficient of determination for this regression analysis and interpret its value.

9.23   Repeat the instructions of Exercise 9.22, except let the independent variable be the number x_2 of county law enforcement employees.

9.24   Refer to Exercise 9.22.

    a   Calculate the correlation coefficient r between the county population density x_1 and the number of law enforcement employees x_2.

    b   Does the correlation between x_1 and x_2 differ significantly from zero? Test by using $\alpha = 0.05$.

    c   Note that the regression lines of Exercises 9.22 and 9.23 both have positive slopes. How might the value for the correlation coefficient, calculated in part (a), be related to this phenomenon?

9.8 Using the Model for Estimation and Prediction

If we are satisfied that a useful model has been found to describe the relationship between peak power load and maximum daily temperature, we are ready to accomplish the original objective in constructing the model.

    Step 5. When satisfied with the model's adequacy, use it for prediction, estimation, and so on.

    The most common uses of a probabilistic model for making inferences can be divided into two categories. The first is the use of the model for estimating the mean value of Y, $E(Y)$, for a specific value of x. For our peak power load example, we might want to estimate the mean peak power load for days during which the maximum temperature is 90°F. The second category involves use of the model to predict a particular value of Y for a given x. That is, we might want to predict the peak power load for a particular day during which the maximum temperature will be 90°F.

    In the first case, we are attempting to estimate the mean value of Y over a very large number of days. In the second case, we are trying to predict the Y value for a single day. Which of these uses of the model, estimating the mean value of Y or predicting an individual value of Y (for the same value of x), can be accomplished with the greater accuracy?

    Before answering this question, we first consider the problem of choosing an estimator (or predictor) of the mean (or individual) Y value. We use the least-squares model

$$\hat{y} = \hat{\beta}_0 + \hat{\beta}_1 x$$

both to estimate the mean value of Y and to predict a particular value of Y for a given value of x. For our example, we found

$$\hat{y} = -419.85 + 6.7175x$$

so the estimated mean peak power load for all days when $x = 90$ (maximum temperature is $90°F$) is

$$\hat{y} = -419.85 + 6.7175(90) = 184.72$$

The identical value is used to predict the Y value when $x = 90$. That is, both the estimated mean and the predicted value of Y are $\hat{y} = 184.72$ when $x = 90$, as shown in Figure 9.11.

F I G U R E **9.11**
Estimated Mean Value and
Predicted Individual Value of
Peak Power Load Y for
$x = 90$

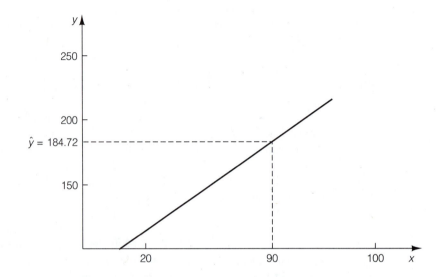

The difference between these two uses of the probabilistic model lies in the relative precision of the estimate and the prediction. The precisions are measured by the expected squared distances between the estimator/predictor $\hat{y}$ and the quantity being estimated or predicted. In the case of estimation, we are trying to estimate

$$E(Y) = \beta_0 + \beta_1 x_p$$

where x_p is the particular value of x at which the estimate is being made. The estimator is

$$\hat{Y} = \hat{\beta}_0 + \hat{\beta}_1 x_p$$

so the sources of error in estimating $E(Y)$ are the estimators $\hat{\beta}_0$ and $\hat{\beta}_1$. The variance can be shown to be

$$\sigma_{\hat{Y}}^2 = E\{[\hat{Y} - E(\hat{Y})]^2\} = \sigma^2 \left[\frac{1}{n} + \frac{(x_p - \bar{x})^2}{SS_{xx}} \right]$$

In the case of predicting a particular y value, we are trying to predict

$$Y_p = \beta_0 + \beta_1 x_p + \varepsilon_p$$

where ε_p is the error associated with the particular Y value Y_p. Thus, the sources of prediction error are the estimators $\hat{\beta}_0$ and $\hat{\beta}_1$ *and* the error ε_p. The variance associated with ε_p is σ^2, and it can be shown that this is the additional term in the expected squared prediction error:

$$\sigma^2_{(Y_p - \hat{Y})} = E[(\hat{Y} - Y_p)^2] = \sigma^2 + \sigma^2_{\hat{Y}}$$

$$= \sigma^2 \left[1 + \frac{1}{n} + \frac{(x_p - \bar{x})^2}{SS_{xx}} \right]$$

The sampling errors associated with the estimator and predictor are summarized next. Figures 9.12 and 9.13 graphically depict the difference between the error of estimating a mean value of Y and the error of predicting a future value of Y for $x = x_p$. Note that the error of estimation is just the vertical difference between the least-squares line and the true line of means, $E(Y) = \beta_0 + \beta_1 x$. However, the error in predicting some future value of Y is the sum of two errors: the error of estimating the mean of Y, $E(Y)$, shown in Figure 9.12, plus the random error ε_p that is a component of the value Y_p to be predicted. Note that both errors will be smallest when $x_p = \bar{x}$. The farther x_p lies from the mean, the larger the errors of estimation and prediction.

Sampling Errors for the Estimator of the Mean of Y and the Predictor of an Individual Y

1 The standard deviation of the sampling distribution of the estimator $\hat{Y}$ of the mean value of Y at a fixed x is

$$\sigma_{\hat{Y}} = \sigma \sqrt{\frac{1}{n} + \frac{(x - \bar{x})^2}{SS_{xx}}}$$

where σ is the standard deviation of the random error ε.

2 The standard deviation of the prediction error for the predictor $\hat{Y}$ of an individual Y-value at a fixed x is

$$\sigma_{(Y - \hat{Y})} = \sigma \sqrt{1 + \frac{1}{n} + \frac{(x - \bar{x})^2}{SS_{xx}}}$$

where σ is the standard deviation of the random error ε.

The true value of σ will rarely be known. Thus, we estimate σ by s and calculate the estimation and prediction intervals as shown next.

FIGURE **9.12**
Error of Estimating the Mean
Value of Y for a Given
Value of x

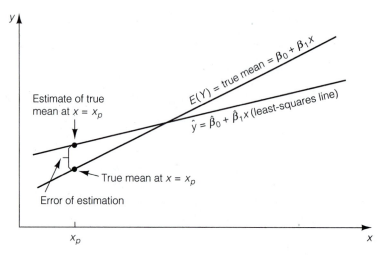

FIGURE **9.13**
Error of Predicting a Future Value
of Y for a Given Value of x

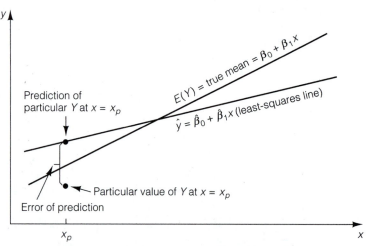

A 100$(1 - \alpha)$ Percent Confidence Interval for the Mean Value of Y at a Fixed x

$$\hat{y} \pm t_{\alpha/2}(n - 2)(\text{estimated standard deviation of } \hat{Y})$$

or

$$\hat{y} \pm t_{\alpha/2}(n - 2)s\sqrt{\frac{1}{n} + \frac{(x - \bar{x})^2}{\mathrm{SS}_{xx}}}$$

E X A M P L E   **9.2**   Find a 95% confidence interval for the mean peak power load when the maximum
daily temperature is 90°F.

Solution A 95% confidence interval for estimating the mean peak power load at a high temperature of 90° is given by

$$\hat{y} \pm t_{\alpha/2}(n-2)s\sqrt{\frac{1}{n} + \frac{(x-\bar{x})^2}{SS_{xx}}}$$

We have previously calculated $\hat{y} = 184.72$, $s = 16.18$, $\bar{x} = 91.5$, and $SS_{xx} = 380.5$. With $n = 10$ and $\alpha = 0.05$, $t_{0.025}(8) = 2.306$, and thus the 95% confidence interval is

$$\hat{y} \pm t_{\alpha/2}(n-2)s\sqrt{\frac{1}{n} + \frac{(x-\bar{x})^2}{SS_{xx}}}$$

$$184.72 \pm (2.306)(16.18)\sqrt{\frac{1}{10} + \frac{(90-91.5)^2}{380.5}}$$

or

$$184.72 \pm 12.14$$

Thus, we can be 95% confident that the *mean* peak power load is between 172.58 and 196.86 megawatts for days with a maximum temperature of 90°F. ∎

A 100(1 − α) Percent Prediction Interval† for an Individual Y at a Fixed x

$$\hat{y} \pm t_{\alpha/2}(n-2)[\text{estimated standard deviation of } (Y - \hat{Y})]$$

or

$$\hat{y} \pm t_{\alpha/2}(n-2)s\sqrt{1 + \frac{1}{n} + \frac{(x-\bar{x})^2}{SS_{xx}}}$$

EXAMPLE **9.3** Predict the peak power load for a day during which the maximum temperature is 90°F.

†The term *prediction interval* is used when the interval is intended to enclose the value of a random variable. The term *confidence interval* is reserved for estimation of population parameters (such as the mean).

Solution Using the same values as in the previous example, we find

$$\hat{y} \pm t_{\alpha/2}(n-2)s\sqrt{1 + \frac{1}{n} + \frac{(x-\bar{x})^2}{\text{SS}_{xx}}}$$

$$184.72 \pm (2.306)(16.18)\sqrt{1 + \frac{1}{10} + \frac{(90-91.5)^2}{380.5}}$$

or

$$184.72 \pm 39.24$$

Thus, we are 95% confident that the peak power load will be between 145.48 and 223.96 megawatts on a particular day when the maximum temperature is 90°F. ∎

A comparison of the confidence interval for the mean peak power load and the prediction interval for a single day's peak power load for a maximum temperature of 90°F ($x = 90$) is illustrated in Figure 9.14. It is important to note that the prediction interval for an individual value of Y will always be wider than the corresponding confidence interval for the mean value of Y.

Be careful not to use the least-squares prediction equation to estimate the mean value of Y or to predict a particular value of Y for values of x that fall outside the range of the values of x contained in your sample data. The model might provide a good fit to the data over the range of x values contained in the sample but a very poor fit for values of x outside this region (see Figure 9.15). Failure to heed this warning may lead to errors of estimation and prediction that are much larger than expected.

F I G U R E **9.14**
A 95% Confidence Interval for Mean Peak Power Load and a Prediction Interval for Peak Power Load When $x = 90$

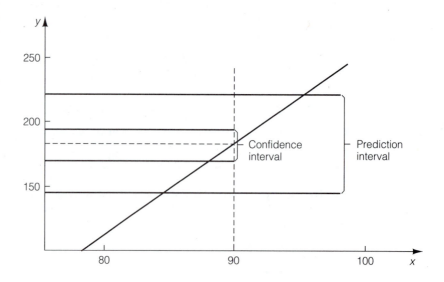

F I G U R E **9.15**
The Danger of Using a Model to
Predict Outside the Range of the
Sample Values of x

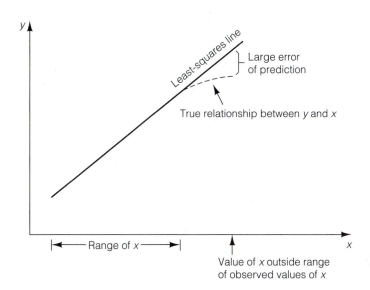

F I G U R E **9.15**
The Danger of Using a Model to
Predict Outside the Range of the
Sample Values of x

Exercises

9.25 Refer to Exercise 9.1.

  a Estimate the mean value of Y for $x = 2$, using a 95% confidence interval.

  b Find a 95% prediction interval for a response Y at $x = 2$.

9.26 Refer to Exercise 9.3. Estimate the expected elongation for a force of 1.8, using a 90% confidence interval.

9.27 Refer to Exercise 9.4. If in 1988 the industry sales figure is 16, find a 95% prediction interval for the company sales figure. Do you see any possible difficulties with the solution to this problem?

9.28 Refer to Exercise 9.5. Estimate, in a 95% confidence interval, the mean abrasion loss for steel specimens with a Rockwell hardness measure of 75. What assumptions are necessary for your answer to be valid?

9.29 Refer again to Exercise 9.5. Predict, in a 95% prediction interval, the abrasion loss for a particular steel specimen with a Rockwell hardness measure of 75. What assumptions are necessary for your answer to be valid?

9.30 Refer to Exercise 9.7. Estimate, in a 90% confidence interval, the expected number of parts assembled by employees working for 12 minutes.

9.31 Refer to Exercise 9.8. Find a 95% prediction interval for a flow-through measurement corresponding to a static measurement of 21.0.

9.32   Refer to Exercise 9.3. If no force is applied, the expected elongation should be zero (that is, $\beta_0 = 0$). Construct a test of $H_0 : \beta_0 = 0$ versus $H_a : \beta_0 \neq 0$, using the data of Exercise 9.3. Use $\alpha = 0.05$.

9.33   In planning for an initial orientation meeting with new electrical engineering majors, the chairman of the Electrical Engineering Department wants to emphasize the importance of doing well in the major courses in getting better-paying jobs after graduation. To support this point, the chairman plans to show that there is a strong positive correlation between starting salaries for recent electrical engineering graduates and their grade-point averages in the major courses. Records for seven of last year's electrical engineering graduates are selected at random and given in the following table:

Grade-Point Average in Major Courses x	Starting Salary y (\$ thousand)
2.58	16.5
3.27	18.8
3.85	19.5
3.50	19.2
3.33	18.5
2.89	16.6
2.23	15.6

a   Find the least-squares prediction equation.

b   Plot the data and graph the line as a check on your calculations. Do the data provide sufficient evidence that grade-point average provides information for predicting starting salary?

c   Find a 95% prediction interval for the starting salary of a graduate whose grade-point average is 3.2.

d   What is the mean starting salary for graduates with grade-point averages equal to 3.0? Use a 95% confidence interval.

9.34   Managers are an important part of any organization's resource base. Accordingly, the organization should be just as concerned with forecasting its future managerial needs as it is with forecasting its needs for, say, the natural resources used in its production processes. According to William F. Glueck (*Management*, Dryden Press, 1977), one commonly used procedure for forecasting the demand for managers is to model the relationship between sales and the number of managers needed, the theory being that "... the demand for managers is the result of the increases and decreases in the demand for products and services that an enterprise offers its customers and clients" (p. 274). To develop this relationship, data such as those shown in the table can be assembled from a firm's records.

Date	Monthly Sales (units) x	Number of Managers on 15th of the Month y
3/80	5	10
6/80	4	11
9/80	8	10
12/80	7	10
3/81	9	9
6/81	15	10
9/81	20	11
12/81	21	17
3/82	25	19
6/82	24	21
9/82	30	22
12/82	31	25
3/83	36	30
6/83	38	30
9/83	40	31
12/83	41	31
3/84	51	32
6/84	40	30
9/84	48	32
12/84	47	32

a   Use simple linear regression to model the relationship between the number of managers and the number of units sold for the data in the table.

b   Plot your least-squares line on a scatterplot of the data. Does it appear that the relationship between x and y is linear? If not, does it appear that your least-squares model will provide a useful approximation to the relationship? Explain.

c   Test the usefulness of your model. Use $\alpha = 0.05$. State your conclusion in the context of the problem.

d   The company projects that next May it will sell 39 units. Use your least-squares model to construct (1) a 90% confidence interval for the mean number of managers needed when sales are at a level of 39 and (2) a 90% prediction interval for the number of managers needed next May.

e   Compare the sizes and the interpretations of the two intervals you constructed in part (d).

9.9 Computation of More Detailed Analyses

As you can see, the computations for a simple regression analysis are quite complicated and cumbersome. This load is lightened by any of a number of readily available statistical computer packages. We now illustrate how computer printouts can produce the analyses explained in the earlier sections of this chapter, and extend them a bit, by the use of Minitab.

The standard computer printout for a regression analysis of the peak power load data looks like the following sample:

```
The regression equation is
y = -420 + 6.72 x

Predictor         Coef     Stdev      t-ratio        p
Constant       -419.85     76.06       -5.52     0.000
x                6.7175    0.8294        8.10     0.000


s = 16.18    R-sq = 89.1%    R-sq(adj) = 87.8%

Analysis of Variance

SOURCE        DF       SS        MS          F        P
Regression     1    17170     17170      65.60    0.000
Error          8     2094       262
Total          9    19264
```

We identify the components of the printout with the corresponding terms used in this chapter in the following list (SS can be read "sum of squares" and MS can be read "mean square"):

$$\hat{\beta}_0 = \text{Predictor constant coef.} = -419.85$$
$$\hat{\beta}_1 = \text{Predictor } x \text{ coef.} = 6.7175$$
$$SS_{yy} = \text{Total SS (under "Analysis of Variance")} = 19{,}264$$
$$SSE = \text{Error SS} = 2{,}094$$
$$SS_{yy} - SSE = SSR = 17{,}170$$
$$s^2 = \text{Error MS} = 262$$
$$s = \sqrt{s^2} = 16.18$$
$$s_{\hat{\beta}_1} = \text{Predictor } x \text{ st. dev.} = 0.8294$$
$$s_{\hat{\beta}_0} = \text{Predictor constant st. dev.} = 76.06$$
$$t = \frac{\hat{\beta}_1}{s_{\hat{\beta}_1}} = \text{Predictor } x \text{ } t\text{-ratio} = 8.10$$

(and the p column gives the observed significance level, 0.000 in this case)

$$r^2 = R\text{-}sq = 0.891 = \frac{SSR}{\text{total SS}}$$

The R-sq(adj.) component will not be used in this book; the F column will be explained in Chapter 10. The regression equation is printed at the top, and all pertinent computations are displayed concisely.

If we ask for an estimate of $E(Y)$ at a particular x, say $x = 90$, and a prediction of a specific Y at $x = 90$, the printout will be as follows:

```
Fit          Stdev. Fit          95% C.I.            95% P.I.
184.72             5.26      (172.58, 196.87)    (145.48,223.97)
```

The "fit" value is $\hat{y}$ at $x = 90$ and

$$\text{St. dev. fit } = s\sqrt{\frac{1}{n} + \frac{(x - \bar{x})^2}{SS_{xx}}}$$

The confidence interval and prediction interval are based on the t statistic, as developed earlier in this chapter.

9.9.1   Residual Analysis

A computer allows for quick and detailed analyses of the "residuals," the deviations $y - \hat{y}$ for each value of x (where y denotes the observed value of the dependent variable). Regression analysis as developed earlier makes certain assumptions about these residuals:

1   They should be "random" quantities.

2   They should have variation that does not depend on x.

3   For estimation and hypothesis testing, they should be approximately normally distributed.

The first two assumptions can sometimes be checked by looking at a plot of the residuals graphed against the values of x. Such a plot for the peak power load data is given in Figure 9.16. These residuals look fairly random, but there is a hint of a pattern in that residuals for the middle x values are mostly negative, while those at either end are positive. We will see a possible reason for this in Chapter 10. The variation in the residuals seems to be about the same for small x values as for large ones, but this is difficult to check here because of the small number of repeated x values.

FIGURE **9.16**
Residuals Versus Values of x

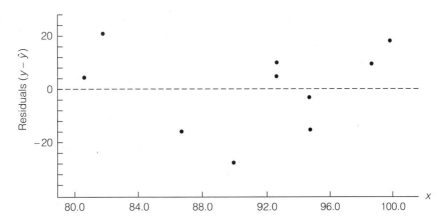

For the normality assumption, we could look at a stem-and-leaf plot of the residuals (Figure 9.17), which does exhibit some evidence of being mound-shaped. A more discriminating method of checking for normality is to use the normal scores plot introduced in Chapter 8. For these residuals, the normal scores plot (Figure 9.18)

F I G U R E **9.17**
Stem-and-Leaf Plot of Residuals

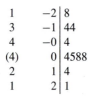

```
 1       −2 | 8
 3       −1 | 44
 4       −0 | 4
(4)       0 | 4588
 2        1 | 4
 1        2 | 1
```

F I G U R E **9.18**
Plot of Residuals Versus Normal
Scores (peak power load data)

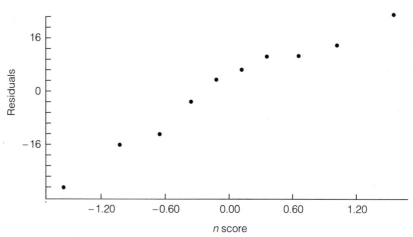

shows quite a good fit to a straight line, which supports the assumption of normality of residuals.

9.9.2   Transformations

What if the assumptions are not met or the simple regression model does not fit? Then we must work harder to find an acceptable probabilistic model. To illustrate one way to improve the model, we will look at a data set for which a transformation will solve the problem.

Table 9.4 shows the average population per square mile in the United States for each census year since 1790. (The table also gives the natural logarithms of these figures, which will be used later.) Since these average population figures increase with the year, we might start building a model of the relationship between population per square mile and year by fitting a simple linear regression model. The Minitab output for such a model is shown below. (In this analysis, the row value—1, 2, ..., 20—was used as a coded year.)

```
The regression equation is
apop = -8.42 + 3.01 x

Predictor       Coef       Stdev      t-ratio         p
Constant      -8.424       2.968      -2.84       0.011
x             3.0071      0.2478      12.14       0.000

s = 6.390    R-sq = 89.1%    R-sq(adj) = 88.5%
```

Analysis of Variance

SOURCE	DF	SS	MS	F	p
Regression	1	6013.2	6013.2	147.28	0.000
Error	18	734.9	40.8		
Total	19	6748.2			

T A B L E **9.4**
Average Population/per Square Mile

Row	Year	Avg. Pop.	In Avg. Pop.
1	1790	4.5	1.50408
2	1800	6.1	1.80829
3	1810	4.3	1.45862
4	1820	5.5	1.70475
5	1830	7.4	2.00148
6	1840	9.8	2.28238
7	1850	7.9	2.06686
8	1860	10.6	2.36085
9	1870	10.9	2.38876
10	1880	14.2	2.65324
11	1890	17.8	2.87920
12	1900	21.5	3.06805
13	1910	26.0	3.25810
14	1920	29.9	3.39786
15	1930	34.7	3.54674
16	1940	37.2	3.61631
17	1950	42.6	3.75185
18	1960	50.6	3.92395
19	1970	57.5	4.05178
20	1980	64.0	4.15888

Source: World Almanac and Book of Facts, 1988.

To the casual observer, the linear model might seem to fit well, with a highly significant positive slope and a large r^2. But a graph of the data and a plot of the residuals will show that the "good" fit is not nearly good enough. The actual data (Figure 9.19) show a definite curvature; the rate of increase seems larger in later years. This curvature is magnified in the plot of residuals from the straight regression line, as seen in Figure 9.20. Here the residuals show a definite pattern; they cannot be regarded as random fluctuations.

Since the rate of increase in the population seems to be growing larger with time, an exponential relationship between population size and year might be appropriate. Exponential growth can be modeled by

$$y = ae^{bx}$$

Then

$$\ln y = \ln a + bx$$

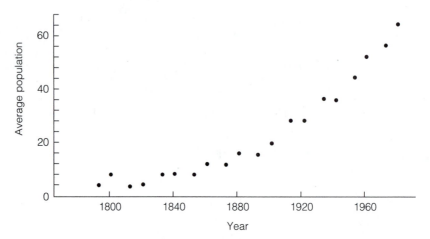

FIGURE **9.19**
Average Population per Square Mile Versus Year

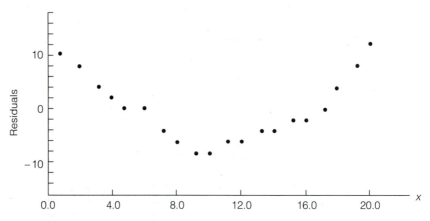

FIGURE **9.20**
Residuals Versus Values of x for the Average Population Data

and there will be a linear relationship between ln y and x. The graph of ln y versus year for the population data is shown in Figure 9.21. This transformation does appear to "linearize" the data.

A simple linear model fit to the transformed data produced the following results:

```
The regression equation is
lnapop = 1.24 + 0.148x

Predictor        Coef        Stdev      t-ratio         p
Constant      1.23582      0.05930       20.84      0.000
x            0.148408      0.004950      29.98      0.000

s = 0.1277     R-sq = 98.0%     R-sq(adj) = 97.9%
```

```
Analysis of Variance

SOURCE          DF          SS          MS          F          p
Regression       1      14.647      14.647      898.84      0.000
Error           18       0.293       0.016
Total           19      14.940
```

F I G U R E **9.21**
Logarithm of Average Population per Square Mile Versus Year

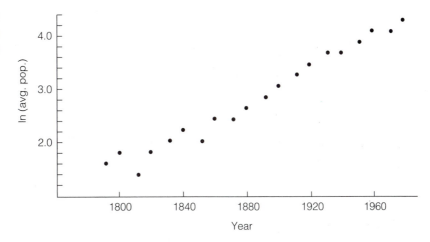

This is a much better fitting straight line, as is shown by the higher value of r^2 and the fact that the pattern shown by the residuals is removed (see Figure 9.22).

The stem-and-leaf plot of the residuals (Figure 9.23) does show some mound-shapedness and symmetry, so the normality assumption might be justified.

The exponential model fits well, and the assumptions for regression analysis seem to be satisfied. However, the fit seems to be better for larger values of x than for smaller ones (looking at Figures 9.21 and 9.22). Can you suggest a historical reason for this?

We were fortunate in that a rather simple transformation solved our "linearization" problem in the population growth example. Sometimes, however, such a transformation on y or x, or both, is much more difficult to find (and may not produce such dramatic improvement) or may not even exist. We will give one more example of a case in which a reasonable transformation makes a small improvement. One technique for handling problems in which a simple linearizing transformation is *not* sufficient will be discussed in Chapter 10.

The Florida Game and Freshwater Fish Commission is interested in developing a model that will allow the accurate prediction of the weight of an alligator from more easily observed data on length. Length and weight data for a sample of 25 alligators are shown in Table 9.5.

A plot of weight versus length (Figure 9.24) shows that a linear model will not fit these data well. If, however, we proceed to fit a simple linear model anyway, r^2 is about 0.83 and the slope is significantly positive.

FIGURE **9.22**
Residuals Versus Values of x for
ln Average Population Data

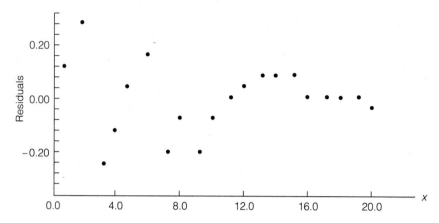

FIGURE **9.23**
Stem-and-Leaf Plot of Residuals
for ln Average Population Data

```
2      -2 | 20
3      -1 | 8
4      -1 | 2
6      -0 | 66
9      -0 | 400
(4)     0 | 0112
7       0 | 5889
3       1 | 1
2       1 | 5
1       2 |
1       2 | 7
```

TABLE **9.5**
Alligator Weights (pounds) and
Lengths (inches)

Weight	Length	Weight	Length
130	94	83	86
51	74	70	88
640	147	61	72
28	58	54	74
80	86	44	61
110	94	106	90
33	63	84	89
90	86	39	68
36	69	42	76
38	72	197	114
366	128	102	90
84	85	57	78
80	82		

FIGURE **9.24**
Weight Versus Length of
Alligators

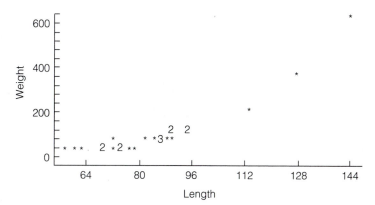

The regression equation is
WEIGHT = -393 + 5.90 LENGTH

```
Predictor        Coef          Stdev      t-ratio        p
Constant      -393.26          47.53        -8.27    0.000
LENGTH         5.9024         0.5448        10.83    0.000
s = 54.01                 R-sq = 83.6%
```

So as not to be completely misled, look carefully at the residual plot (Figure 9.25). This plot shows a definite curved pattern in the residuals, ample evidence of the fact that the straight line is not a good model.

FIGURE **9.25**
Residual Plot for the Simple
Linear Regression Model

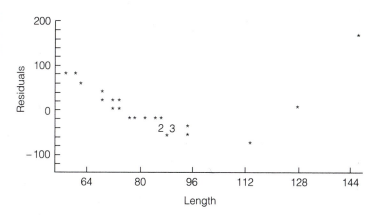

How should we proceed from here? The exponential growth model got us out of trouble before and could work again here. Plotting ln (weight) versus length (Figure 9.26) does seem to produce a scatterplot that could be fit by a straight line. The fit of ln (weight) as a function of length is much better than the fit in the untransformed case.

FIGURE **9.26**
ln (weight) Versus Length of
Alligators

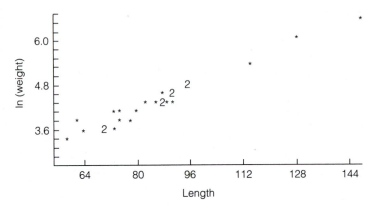

```
The regression equation is
LNWEIGHT = 1.34 + 0.0354 LENGTH

Predictor         Coef         Stdev      t-ratio        p
Constant        1.3353       0.1314        10.16    0.000
LENGTH        0.035416     0.001506        23.52    0.000
s = 0.1493                R-sq = 96.0%
```

However, the residual plot (Figure 9.27) shows that the variation in residuals depends on the explanatory variable, length. The variation in residuals is much greater at the shorter lengths than at the longer lengths, and this condition violates one of the three properties of residuals that a good model should produce.

FIGURE **9.27**
Residual Plot for ln (weight)
Versus Length

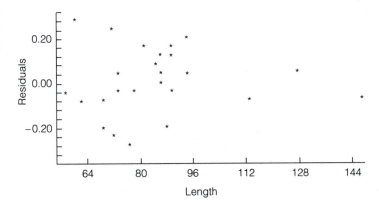

A little reflection on the nature of the physical problem might help at this point. Weight is related to volume, a three-dimensional feature. So it's not surprising that weight does not turn out to be a linear function of length. A model of the form

$$w = al^b$$

allows weight to be modeled as a function of length raised to an appropriate power.

Taking logarithms,

$$\ln(w) = \ln(a) + b\ln(l)$$

and fitting $\ln(w)$ versus $\ln(l)$ will provide an estimate of b. The plot of $\ln(w)$ versus $\ln(l)$ is shown in Figure 9.28, and the scatterplot looks quite linear.

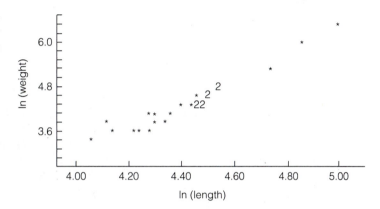

FIGURE **9.28**
ln(weight) Versus ln(length) for
the Alligator Data

```
The regression equation is
LNWEIGHT = 10.2 + 3.29 LENGTH

    Predictor      Coef           Stdev      t-ratio          p
    Constant    -10.1746          0.7316      -13.91      0.000
    LENGTH        3.2860          0.1654       19.87      0.000
    s = 0.1753              R-sq = 94.5%
```

The fitted model is almost as good (in terms of r^2) as the previous models, and the residuals (Figure 9.29) look quite random (although a little curvature might still remain). Finally, the stemplot (Figure 9.30) shows that the residuals from this model have a nicely symmetric, mound-shaped distribution.

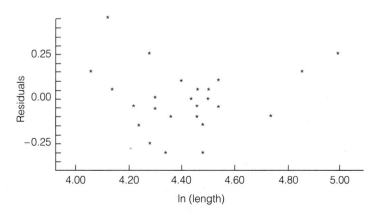

FIGURE **9.29**
Residual Plot for ln (weight)
Versus ln (length)

FIGURE **9.30**
Stemplot of Residuals

```
Stem-and-leaf of Residuals      n = 25
Leaf unit = 0.010
       1      −3 | 1
       3      −2 | 84
       6      −1 | 540
      12      −0 | 985432
      (7)      0 | 0123557
       6       1 | 136
       3       2 | 33
       1       3 |
       1       4 | 5
```

To review, transformations may be found by trial and error, but it is better to think first of transformations that relate to the phenomenon under study. It should not be too surprising that populations grow exponentially (by a multiplicative factor) and that weight is related to length to a power (around 3). A model generated by the physical process is often superior to one found by guesswork.

9.9.3 Influence

Interpreting the significance of regression slopes and the strength of linear associations through r^2 requires great care—and plots of the data. Some points of interpretation are illustrated by budget data for civilian employees of the Executive Department of the U.S. government (Table 9.6).

The goal is to find a model that shows the association between budget outlay y and the number of employees x. A plot of these data (Figure 9.31) shows three points that appear to be far away from the others. If a straight line is fit through these data, the two high points will produce large residuals. The point at the far right (Defense) will pull the regression line to it and, hence, will produce a small residual. The regression line is given on page 523, and the residuals are shown in Figure 9.32.

TABLE **9.6**
Finances and Employment—Executive Departments of the U.S. Government, 1989

Department	Employees (thousands)	Budget Outlay ($ billion)
Agriculture	122.1	48.3
Commerce	45.1	2.6
Defense	1,075.4	318.3
Education	4.7	21.6
Energy	17.1	11.4
Health and Human Services	122.2	399.8
Housing and Urban Development	13.5	19.7
Interior	77.5	5.2
Justice	79.7	6.2
Labor	18.1	22.7
State	25.3	3.7
Transportation	65.6	26.6
Treasury	152.5	230.6
Veterans' Affairs	246.0	30.0

Source: Statistical Abstract of the United States, 1991.

```
The regression equation is
Budget = 40.9 + 0.278 Employee

Predictor        Coef .         Stdev      t-ratio          p
Constant        40.87           43.14         1.24      0.254
Employee        0.2783         0.1125         2.47      0.029
s = 111.7                  R-sq = 33.8%
```

Note that r^2 is small here, even though the slope is significant.

F I G U R E 9.31
Budget Outlay Versus Employees

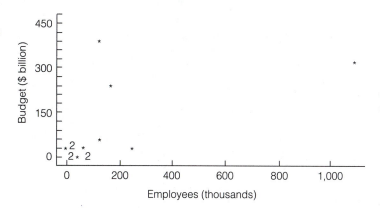

F I G U R E 9.32
Residual Plot for Simple Linear
Model

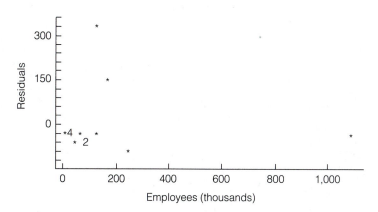

Let's see what happens if we drop the two data points producing the large residuals (Health and Human Services, Treasury). Then the fitted line is given by the following:

```
The regression equation is
Budget-2 = 0.26 + 0.287 Employ-2

Predictor          Coef         Stdev      t-ratio        p
Constant          0.260         6.239         0.04    0.968
Employ-2        0.28668       0.01931        14.84    0.000
s = 19.17                 R-sq = 95.7%
```

Notice how r^2 jumped all the way to 0.957! The residual plot for this model is given in Figure 9.33. Now notice the pattern in the residuals. It appears (since the residuals have a linear trend from positive to negative) that the linear model would have a smaller slope were it not for the point way off to the right. What happens if this point (Defense) is dropped from the analysis? The fitted model follows:

```
The regression equation is
Budget-3 =12.4 + 0.0863 Employ-3

Predictor          Coef         Stdev      t-ratio        p
Constant         12.396         5.601         2.21    0.054
Employ-3        0.08625       0.06008         1.44    0.185
s = 13.32                 R-sq = 18.6%
```

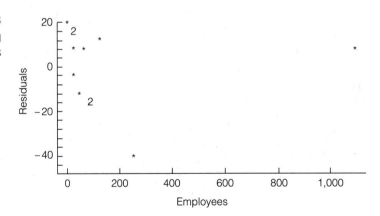

FIGURE **9.33**
Residual Plot After Dropping
Large Residuals

This model shows no significant linear trend at all, which is entirely consistent with the plot of these points in Figure 9.34.

The three points deleted in the course of this analysis were highly *influential*. The two large residuals kept r^2 low (0.338) in the first analysis. The third point (Defense) caused r^2 to jump to 0.957 after the first two were deleted. Upon deleting this value, no significant linear trend shows up. In short, the linear model could have appeared to be highly significant, with a high r^2, simply because of the influence of a single point. To guard against having one or two points unduly affect results, always plot the data and the residuals, and study both plots carefully!

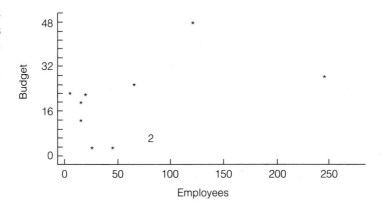

F I G U R E **9.34**
Budget Outlay Versus Employees
(without three influential points)

9.10   Activities for Students: Finding Models

The Top 25

According to *U.S. News and World Report*, the 25 top-rated universities in the country possess the statistics shown in Table 9.7 on six important factors:

average SAT or ACT score

percentage of freshmen in top 10% of high-school class

acceptance rate

student/faculty ratio

spending per student

graduation rate

Enter these data into a computer in order to study associations among these variables. In particular, answer the following questions:

1   What is the best single variable for predicting graduation rate? What is the prediction equation? Find and interpret r^2 for this model?

2   What variable is most closely associated with acceptance rate? Explain the nature of this association.

3   Find a linear model that predicts spending per student from student/faculty ratio. Are there any influential observations here? What happens to the association if the influential observations are removed?

4   Comment on any other associations that look interesting.

Building a Distance Gauge

Most people have some difficulty in accurately judging distances. Since judging distance can be an important skill (think of the distance between you and the approaching stop sign when driving), one might study the relationship between guessed distances to certain objects and the actual distances to these same objects. A model of this

T A B L E **9.7**   U.S. News Top 25 National Universities

Rank/Name	Avg. or Midpoint SAT/ACT Score	Freshmen in Top 10% of HS Class	Acceptance Rate	Student/ Faculty Ratio	Total Spending per Student	Graduation Rate
1. Harvard University	1370	90%	18%	8/1	$50,677	93%
2. Yale University	1365	95	20	6/1	57,879	92
3. Stanford University	1365	90	22	9/1	72,551	89
4. Princeton University	1340	89	17	9/1	50,786	93
5. California Inst. of Tech.	1400	98	30	5/1	106,611	81
6. Mass. Inst. of Tech.	1375	94	32	8/1	63,605	87
7. Duke University	1305	90	33	6/1	40,229	92
8. Dartmouth College	1310	82	26	5/1	42,444	95
9. Columbia University	1275	79	28	6/1	51,262	90
10. University of Chicago	1290	70	46	6/1	49,954	79
11. Johns Hopkins University	1315	67	53	4/1	74,750	87
12. Cornell University	1280	82	30	11/1	43,941	85
13. University of Pennsylvania	1285	83	42	7/1	36,617	87
14. Northwestern University	1245	82	46	8/1	35,394	84
15. Rice University	1323	86	25	9/1	29.533	88
16. Univ. of California–Berkeley	1210	95	38	17/1	24,789	70
17. Brown University	1295	80	23	13/1	27,689	90
18. Washington University	1205	62	59	4/1	57,667	76
19. Vanderbilt University	1190	55	59	5/1	37,564	76
20. Georgetown University	1228	68	29	7/1	28,037	84
21. University of Virginia	1215	72	32	10/1	22,080	88
22. University of Michigan	1180	69	60	12/1	25,486	79
23. Univ. of California–Los Angeles	1130	90	43	23/1	23,306	63
24. Carnegie Mellon University	1225	53	65	9/1	40,844	69
25. Univ. of N. Carolina–Chapel Hill	1110	76	37	10/1	24,853	75

Source: U.S. News and World Report, 30 September 1991, p. 73.

relationship could serve as a personal distance gauge by allowing the "owner" to adjust his or her guessed distances to more accurately reflect the true distances.

1   From a fixed point at which a person is standing, have the person guess the distances to at least ten objects that range from a few yards, to, perhaps, 100 or so yards away. For example, an open area of a college campus might have a street sign close by, some trees at various distances, a flagpole at another distance, and a building at an extreme distance. Record these guesses on a data collection sheet.

2   Now have a team of students measure the actual distances from the point at which the person was standing to the objects in question. Record the actual measurements next to the guessed measurements on the data sheet.

3   Plot the guesses versus the actual measurements on a scatterplot. Describe any patterns or trends.

4   If a linear trend is apparent, fit a simple linear regression model to the data. Remember, a goal of this study is to build a model that will allow prediction of

actual distances from the person's guessed distances. With that in mind, which variable should be explanatory and which should be the response?

5   Interpret the slope of the line obtained above in terms of whether the person appears to guess on the short side or the long side.

6   Verify that the model holds by having the person "owning" the model try it again on new objects not part of the original study. Some of these objects should be relatively close and some relatively far away. Have the person guess the distances and then predict the actual distances from the model. Then measure the actual distances. Does the model work?

7   If the verification in step 6 above shows that the model is not working well, think about why this might be the case and plan a new study to build a better model (that might work well only in selected circumstances).

8   After many students in the class have constructed their personal "distance gauge," provide the class with all the models. Study the models to see if the slopes show any patterns. For example, do most people tend to guess short or long? Do people wearing glasses do better or worse than people without glasses? Do those with athletic or outdoor experience do better than those without such experience?

9.11   Summary

Many engineering problems involve modeling the relationship between a dependent variable Y and one or more independent variables. In this chapter, we have discussed the case of a single independent variable x. The regression analysis presented here considers only the situation in which $E(Y)$ is a linear function of x, but the general steps to follow are these:

Step 1.   Hypothesize the form of $E(Y)$.

Step 2.   Use the sample data to estimate unknown parameters in the model.

Step 3.   Specify the probability distribution of the random error component of the model.

Step 4.   Statistically check the adequacy of the model.

Step 5.   When satisfied with the model's adequacy, use it for prediction and estimation.

After a significant linear association has been found, the strength of the association is measured by the correlation r (or by r^2). Since many subtleties affect such linear associations, data plots and residual plots should be studied carefully, and influential observations should be noted. Sometimes a transformation of one or both variables will help improve the fit of a linear model.

Supplementary Exercises

9.35   Two different elements, nickel and iron, can be used in bonding zircaloy components. An experiment was conducted to determine whether there is a correlation between the strengths of the bonds for the two elements. Two pairs of components were randomly selected from each of seven different batches of the zircaloy components. One pair in each batch was bonded by nickel and the other pair by iron, and the strength of the bond was determined for each pair by measuring the amount of pressure (in thousands of pounds per square inch) required to separate the bonded pair. The data from the experiment are shown below:

Batch	Nickel	Iron
1	71.3	76.2
2	71.8	73.1
3	80.3	86.9
4	77.0	82.4
5	77.4	78.5
6	79.9	88.2
7	70.1	75.1

a   Find the correlation coefficient for the data.

b   Find the coefficient of determination and interpret it.

c   Is there sufficient evidence to indicate that the correlation between the strengths of the bonds of the two elements differs from zero? Use $\alpha = 0.05$.

9.36   Labor and material costs are two basic components in analyzing the cost of construction. Changes in the component costs, of course, will lead to changes in total construction costs.

Month	Index of Construction Cost[†] y	Index of All Construction Materials[‡] x
January	193.2	180.0
February	193.1	181.7
March	193.6	184.1
April	195.1	185.3
May	195.6	185.7
June	198.1	185.9
July	200.9	187.7
August	202.7	189.6

[†]U.S. Department of Commerce, Bureau of the Census.

[‡]U.S. Department of Labor, Bureau of Labor Statistics. Tables E-1 (p. 43) and E-2 (p. 44) in *Construction Review*, U.S. Department of Commerce, October 1976, 22(8).

a   Use the data in the table to find a measure of the importance of the materials component. Do this by determining the fraction of reduction in the variability of the construction cost index that can be explained by a linear relationship between the construction cost index and the material cost index.

b   Do the data provide sufficient evidence to indicate a nonzero correlation between Y and X?

9.37 Use the method of least squares and the sample data in the table below to model the relationship between the number of items produced by a particular manufacturing process and the total variable cost involved in production. Find the coefficient of determination and explain its significance in the context of this problem.

Total Output y	Total Variable Cost x (dollars)
10	10
15	12
20	20
20	21
25	22
30	20
30	19

9.38 Does a linear relationship exist between the Consumer Price Index (CPI) and the Dow Jones Industrial Average (DJA)? A random sample of ten months selected from several years in the 1970s produced the following corresponding DJA and CPI data:

DJA y	CPI x
660	13.0
638	14.2
639	13.7
597	15.1
702	12.6
650	13.8
579	15.7
570	16.0
725	11.3
738	10.4

a Find the least-squares line relating the DJA, y, to the CPI, x.

b Do the data provide sufficient evidence to indicate that x contributes information for the prediction of Y? Test by using $\alpha = 0.05$.

c Find a 95% confidence interval for β_1 and interpret your result.

d Suppose you want to estimate the value of the DJA when the CPI is at 15.0. Should you calculate a 95% prediction interval for a particular value of the DJA or a 95% confidence interval for the mean value of the DJA? Explain the difference.

e Calculate both intervals considered in part (d) when the CPI is 15.0.

9.39 Spiraling energy costs have generated interest in energy conservation in businesses of all sizes. Consequently, firms planning to build new plants or make additions to existing facilities have become very conscious of the energy efficiency of proposed new structures. Such firms are interested in knowing the relationship between a building's yearly energy consumption and the factors that influence heat loss. Some of these factors are the number of building stories above and below ground, the materials used in the construction of the building shell, the number of square feet of building shell, and climatic conditions (American Society of Heating, Refrigerating, and Air Conditioning Engineers [ASHRAE], *Guide and Data Book*, 1968). The table below lists the energy consumption in British thermal units (a BTU is the amount of heat required to raise 1 pound of water 1°F) for 1980 for 22 buildings that were all subjected to the same climatic conditions.

BTU/Year (thousands)	Shell Area (square feet)	BTU/Year (thousands)	Shell Area (square feet)
1,371,000	13,530	337,500	5,650
2,422,000	26,060	567,500	8,001
672,200	6,355	555,300	6,147
233,100	4,576	239,400	2,660
218,900	24,680	2,629,000	19,240
354,000	2,621	1,102,000	10,700
12,220,000	59,660	2,680,000	23,680
3,135,000	23,350	423,500	9,125
1,470,000	18,770	423,500	6,510
1,408,000	12,220	1,691,000	13,530
2,201,000	25,490	1,870,000	18,860

a   Find the least-squares line for BTUs as a function of shell area. (You may round the data to the thousands digit.)

b   Find a 95% confidence interval for the slope of the line estimated in (a) and interpret your result.

9.40   Does the national 55-mile-per-hour highway speed limit provide a substantial savings in fuel? To investigate the relationship between automobile gasoline consumption and driving speed, a small economy car was driven twice over the same stretch of an interstate freeway at each of six different speeds. The numbers of miles per gallon measured for each of the 12 trips are shown below:

Miles per Hour	50	55	60	65	70	75
Miles per Gallon	34.8, 33.6	34.6, 34.1	32.8, 31.9	32.6, 30.0	31.6, 31.8	30.9, 31.7

a   Fit a least-squares line to the data.

b   Is there sufficient evidence to conclude that a linear relationship exists between speed and gasoline consumption?

c   Construct a 90% confidence interval for β_1.

d   Construct a 95% confidence interval for miles per gallon when the speed is 72 miles per hour.

e   Construct a 95% prediction interval for miles per gallon when the speed is 58 miles per hour.

9.41   At temperatures approaching absolute zero (273 degrees below zero Celsius), helium exhibits traits that defy many laws of conventional physics. An experiment has been conducted with helium in solid form at various temperatures near absolute zero. The solid helium is placed in a dilution refrigerator along with a solid impure substance, and the fraction (in weight) of the impurity passing through the solid helium is recorded. (The phenomenon of solids passing directly through solids is known as *quantum tunneling*.) The data are given in the table on page 531.

Temperature $x(°C)$	Proportion of Impurity Passing Through Helium y
−262.0	0.315
−265.0	0.202
−256.0	0.204
−267.0	0.620
−270.0	0.715
−272.0	0.935
−272.4	0.957
−272.7	0.906
−272.8	0.985
−272.9	0.987

a Fit a least-squares line to the data.

b Test the null hypothesis $H_0 : \beta_1 = 0$ against the alternative hypothesis $H_a : \beta_1 < 0$ at the $\alpha = 0.01$ level of significance.

c Compute r^2 and interpret your results.

d Find a 95% prediction interval for the percentage of the solid impurity passing through solid helium at −273°C. (Note that this value of x is outside the experimental region, where use of the model for prediction may be dangerous.)

9.42 The data in the table were collected to calibrate a new instrument for measuring interocular pressure. The interocular pressure for each of ten glaucoma patients was measured by the new instrument and by a standard, reliable, but more time-consuming method.

Patient	Reliable Method x	New Instrument y
1	20.2	20.0
2	16.7	17.1
3	17.1	17.2
4	26.3	25.1
5	22.2	22.0
6	21.8	22.1
7	19.1	18.9
8	22.9	22.2
9	23.5	24.0
10	17.0	18.1

a Fit a least-squares line to the data.

b Calculate r and r^2. Interpret each of these qualities.

c Predict the pressure measured by the new instrument when the reliable method gives a reading of 20.0. Use a 90% prediction interval.

9.43 During June, July, and early August of 1981, a total of ten bids were made by DuPont, Seagram, and Mobil to take over Conoco. Finally, on August 5, DuPont announced that it had succeeded. The total value of the offer accepted by Conoco was $7.54 billion, making it the largest takeover in the history of American business at that time. As part of an analysis of the Conoco takeover, Richard S. Ruback ("The Conoco Takeover and Stockholder Returns," *Sloan Management Review*, 23, 1982) used regression analysis to examine whether movements in the rate of return of each of the above-mentioned companies' common stock could be explained by movements in the rate of return of the stock market as a whole. He used the following model: $Y = \beta_0 + \beta_1 X + \varepsilon$, where Y is the daily rate of return of a stock, X is the daily rate of return of the stock market as a whole as measured by the daily rate of return of Standard & Poor's 500 Composite Index, and ε is assumed to possess the characteristics described in the assumptions of Section 9.4. This model is known in financial literature as the *market model*. Note that the parameter β_1 reflects the sensitivity of the stock's rate of return to movements in the stock market as a whole. Using daily data from the beginning of 1979 through the end of 1980 ($n = 504$), Ruback obtained the following least-squares lines for the four firms in question:

Firm	Estimated Market Model	
Conoco	$\hat{Y} = 0.0010 + 1.40X$	($t = 21.93$)
DuPont	$\hat{Y} = -0.0005 + 1.21X$	($t = 18.76$)
Mobil	$\hat{Y} = 0.0010 + 1.62X$	($t = 16.21$)
Seagram	$\hat{Y} = 0.0013 + 0.76X$	($t = 6.05$)

The t statistics associated with values of $\hat{\beta}_1$ are shown to the right of each least-squares prediction equation.

a For each of the above models, test $H_0 : \beta_1 = 0$ versus $H_a : \beta_1 \neq 0$. Use $\alpha = 0.01$. Draw the appropriate conclusions regarding the usefulness of the market model in each case.

b If the rate of return of Standard & Poor's 500 Composite Index increased by 0.10, how much change would occur in the mean rate of return of Seagram's common stock?

c Which of the two stocks, Conoco or Seagram, appears to be more responsive to changes in the market as a whole? Explain.

9.44 A study was conducted to determine whether there is a linear relationship between the breaking strength y of wooden beams and the specific gravity x of the wood. Ten randomly selected beams of the same cross-sectional dimensions were stressed until they broke. The breaking strengths and the density of the wood are shown below for each of the ten beams:

Beam	Specific Gravity x	Strength y
1	0.499	11.14
2	0.558	12.74
3	0.604	13.13
4	0.441	11.51
5	0.550	12.38
6	0.528	12.60
7	0.418	11.13
8	0.480	11.70
9	0.406	11.02
10	0.467	11.41

a Fit the model $Y = \beta_0 + \beta_1 x + \varepsilon$.

b Test $H_0 : \beta_1 = 0$ against the alternative hypothesis $H_a : \beta_1 \neq 0$.

c Estimate the mean strength for beams with specific gravity 0.590, using a 90% confidence interval.

9.45 A firm's *demand curve* describes the quantity of its product that the firm can sell at different possible prices, other things being equal (R. Leftwich, *The Price System and Resource Allocation*, Dryden Press, 1973). Over a period of a year, a tire company varied the price of one of its radial tires to estimate the firm's demand curve for the tire. The company observed that when the price was set very low or very high, it sold few tires. The latter result was easily understood; the former was determined to be due to consumers' misperception that the tire's low price must be linked to poor quality. The data in the table describe the tire's sales over the experimental period.

Tire Price x (dollars)	Number Sold y (hundreds)
20	13
35	57
45	85
60	43
70	17

a Calculate a least-squares line to approximate the firm's demand function.

b Construct a scatterplot and plot your least-squares line as a check on your calculations.

c Test $H_0 : \beta_1 = 0$, using a two-tailed test and $\alpha = 0.05$. Draw the appropriate conclusions in the context of the problem.

d Does your nonrejection of H_0 in part (c) imply that no relationship exists between tire price and sales volume? Explain.

e Calculate the coefficient of determination for the least-squares line of part (a) and interpret its value in the context of the problem.

9.46 The octane number y of refined petroleum is related to the temperature x of the refining process, but it is also related to the particle size of the catalyst. An experiment with a small-particle catalyst gave a fitted least-squares line of

$$\hat{y} = 9.360 + 0.115x$$

with $n = 31$ and $s_{\hat{\beta}_1} = 0.0225$. An independent experiment with a large-particle catalyst gave

$$\hat{y} = 4.265 + 0.190x$$

with $n = 11$ and $s_{\hat{\beta}_1} = 0.0202$. (Source: D. Gweyson and R. Cheasley, *Petroleum Refiner*, August 1959, p. 135.)

a Test the hypotheses that the slopes are significantly different from zero, testing each at the 5% significance level.

b Test, at the 5% level, that the two types of catalyst produce the same slope in the relationship between octane number and temperature. Assume that the true variances of the slope estimates are equal.

9.47 The data in the accompanying table give the mileages per gallon obtained by a test automobile when using gasolines of varying levels of octane.

a Calculate r and r^2.

b Do the data provide sufficient evidence to indicate a correlation between octane level and miles per gallon for the test automobile?

Mileage y (miles per gallon)	Octane x
13.0	89
13.2	93
13.0	87
13.6	90
13.3	89
13.8	95
14.1	100
14.0	98

For the following data sets, find a good simple linear regression model to answer the problems under study. You should always test the slope of a line for significance and use the r^2 values to judge whether or not one model fits the data better than another. Then study the residuals to assess goodness of fit and to check the underlying assumptions. Some data sets will require transforming the dependent variable, the independent variable, or both. The answers in the back of this book will provide some suggested solutions, but remember that there often is no single "best" answer to any one problem.

9.48 The table below gives average height and weight measurements for children. Find regression models for each of the following:

a predicting a boy's height from his age.

b predicting a boy's age from his height.

c predicting a girl's height from her age.

d predicting a boy's weight from his height.

e predicting a girl's weight from her age.

f predicting a girl's weight from the weight of a boy.

Study the residual plots for each model and comment on any patterns seen there.

Average Height and Weight of Children

	Boys		Girls	
Age (yrs)	Height (cm)	Weight (kg)	Height (cm)	Weight (kg)
1	73.6	9.5	73.6	9.1
2	83.8	11.8	83.8	11.3
3	91.4	14.0	91.4	13.6
4	99.0	15.4	99.0	15.0
5	106.6	17.7	104.1	17.2
6	114.2	20.9	111.7	20.4
7	119.3	23.1	119.3	22.2
8	127.0	25.9	127.0	25.4
9	132.0	28.6	132.0	28.1
10	137.1	31.3	137.1	31.3
11	142.2	34.9	142.2	34.9
12	147.3	37.7	147.3	39.0
13	152.4	41.7	152.4	45.5
14	157.5	48.5	157.5	48.5

Source: World Almanac and Book of Facts, 1988, p. 825.

9.49   The data below show the cumulative frequency of stamps issued by the United States from 1848 to 1988. Fit a model to these data for the purpose of explaining the essential features of stamp growth. Examine the residuals from this model and comment on any patterns seen there.

Cumulative Number of U.S. Postage Stamps Issued, by Ten-Year Intervals

Year	Number of U.S. Stamps Issued, Cumulative
1848	2
1858	30
1868	88
1878	181
1888	218
1898	293
1908	341
1918	529
1928	647
1938	838
1948	980
1958	1,123
1968	1,364
1978	1,769
1988	2,400

Source: Scott Standard Postage Stamp Catalog, 1989.

9.50   The number of years required for a planet to complete a revolution around the sun (called its *sidereal year*) is related to the planet's distance from the sun. The following table shows average distance from the sun and sidereal year measurements for the planets.

Planet	Average Distance from Sun (millions of miles)	Sidereal Year
Mercury	36.0	0.241
Venus	67.0	0.615
Earth	93.0	1.000
Mars	141.5	1.880
Jupiter	483.0	11.900
Saturn	886.0	29.500
Uranus	1782.0	84.000
Neptune	2793.0	165.000
Pluto	3670.0	248.000

Use the data to find a linear regression model relating these two measurements, with sidereal year as the dependent variable. [*Hint*: Knowledge of Kepler's Third Law may help you suggest an appropriate transformation.]

9.51   The data below show the amount of suspended particulate matter (in millions of metric tons) emitted into the air for various years (in the United States).

1940	22.8	1976	9.7
1950	24.5	1977	9.1
1960	21.1	1978	9.2
1970	18.1	1979	9.0
1971	16.7	1980	8.5
1972	15.2	1981	7.9
1973	14.1	1982	7.0
1974	12.4	1983	6.7
1975	10.4	1984	7.0
		1985	7.3

Source: U.S.A. by Numbers, Zero Population Growth, Inc., 1988.

Find a regression model for predicting the amount of particulate matter as a function of the year. Use your model to predict the amount of suspended particulate matter for 1986. Do you see any difficulty with this approach to prediction?

9.52   The Materials Science Department of the University of Florida carried out a research project on the properties of self-lubricating bronze bearings manufactured by sintering copper and tin powders. Four variables of importance are the sintering time, change in weight of the bearings, Rockwell hardness of the bearings, and porosity as measured by the weight of liquid wax taken up by the bearing. Data on these four variables are given below. (Note that not all measurements were taken on each sample.)

Sample Bearing	Sintering Time (min.)	Change in Weight (g)	Rockwell Hardness	Weight of Wax (g)
1	1.0	−0.224		
2	1.0	−0.197		
3	2.0	−0.259		
4	2.0	−0.257		
5	2.0	−0.258		
6	2.0	−0.260		
7	4.0	−0.262		
8	4.0	−0.264		
9	4.0	−0.266		
10	4.0	−0.267		
11	5.5	−0.263		
12	5.5	−0.252		
13	7.0	−0.276	48.18	
14	7.0	−0.276	48.18	0.615
15	7.0	−0.280	48.18	0.606
16	7.0	−0.276	48.18	0.611
17	9.0	−0.281	45.01	
18	9.0	−0.279	45.01	0.586
19	11.0	−0.282	45.45	
20	11.0	−0.282	45.45	0.511
21	11.0	−0.284	45.45	0.454
22	11.0	−0.285	45.45	0.440
23	13.0	−0.289	47.80	
24	13.0	−0.282	47.80	0.393
25	15.0	−0.288	46.06	
26	15.0	−0.292	46.06	0.322
27	15.0	−0.289	46.06	0.343
28	15.0	−0.286	46.06	0.341

Using linear regression techniques, model the following:

a change in weight as a function of sintering time.

b Rockwell hardness as a function of sintering time.

c weight of wax absorbed as a function of sintering time.

Do you see any sample values that are questionable in this study?

9.53 Are women catching up? The data below show median salaries for full-time workers in the United States in both current dollars and constant (1989) dollars.

	Men	Women		Men	Women
Current:			**Constant (1989):**		
1980	19,173	11,591	1980	28,853	17,443
1981	20,692	12,457	1981	28,227	16,993
1982	21,655	13,663	1982	27,826	17,557
1983	22,508	14,479	1983	28,022	18,026
1984	24,004	15,422	1984	28,648	18,405
1985	24,999	16,252	1985	28,809	18,729
1986	25,894	16,843	1986	29,296	19,056
1987	26,722	17,504	1987	29,168	19,106
1988	27,342	18,545	1988	28,659	19,439
1989	28,419	19,638	1989	28,419	19,638

How would you employ regression analysis to answer the question posed above? What is your answer to the question?

9.54 Sulfate concentrations in a shallow unconfined aquifer were measured quarterly for a period of seven years. Similarly, chloride concentrations were measured quarterly for nine years at a different site. The data are shown below. Do these data sets show trends over time? Does either data set appear to fit the normal distribution? Do you think a transformed data set (in either case) would fit the normal distribution more closely?

Sample No.	Concentration(mg/l)	Sample No.	Concentration (mg/l)
Sulfate:			
1	111	14	102
2	107	15	145
3	108.7	16	87
4	108.7	17	112
5	109.3	18	111
6	104.7	19	104
7	104.7	20	151
8	108	21	103
9	108	22	113
10	109.3	23	113
11	114.5	24	125
12	113	25	101
13	51.1	26	108

Sample No.	Concentration (mg/l)	Sample No.	Concentration (mg/l)
Chloride:			
1	38	18	—
2	40	19	41
3	35	20	—
4	37	21	35
5	32	22	49
6	37	23	64
7	37	24	73
8	—	25	67
9	32	26	67
10	45	27	—
11	38	28	59
12	33.8	29	73
13	14	30	—
14	—	31	92.5
15	39	32	45.5
16	46	33	40.4
17	48	34	33.9
		35	28.1

Source: B. Harris, M. Loftis, and D. Montgomery, *Ground Water*, vol. 25, no. 2, 1987, pp. 185–193.

9.55 Hydrogen peroxide may have important consequences in air pollution, but its ambient air concentration is difficult to measure. A new absorption device for H_2O_2 was tested in aqueous solutions with known concentrations of hydrogen peroxide. The results were as follows:

Calibration Data for H_2O_2 in Aqueous Solutions

H_2O_2 Content (μg/ml)	Number of Measurements	Absorbance		
		Mean	S.D.	Rel. S.D.(%)
0.000	25	0.0226	0.0015	6.64
0.060	12	0.0368	0.0033	8.97
0.300	12	0.0514	0.0021	4.09
0.600	12	0.0758	0.0021	2.77
1.200	12	0.1248	0.0033	2.64
3.000	12	0.2683	0.0024	0.89
6.000	12	0.5133	0.0033	0.64
12.000	4	1.0018	0.0020	0.20

Source: H. Hartkamp and R. Bachhausen, *Atmospheric Environment*, vol. 21, no. 10, 1987, pp. 2207–2213.

These data can be used to "calibrate" the new device so that absorbances from unknown samples can be used to predict H_2O_2 content for those samples. Find a linear calibration model relating H_2O_2 content to amount absorbed, using the sample means given above. Would the correlation based on these means increase or decrease if all the original data were used in finding the linear model?

Multiple Regression Analysis

About This Chapter

Following up on the work of Chapter 9, we now proceed to build models in which the mean of one variable can be written as a function of two or more independent variables. For example, the average amount of energy required to heat a house depends not only on the air temperature but also on the size of the house, the amount of insulation, and the type of heating unit, among other things. We will use *multiple regression* models for estimating means and for predicting future values of key variables under study.

Contents

10.1 Introduction

Most practical applications of regression analysis utilize models that are more complex than the simple straight-line model. For example, a realistic model for a power plant's peak power load would include more than just the daily high temperature. Factors such as humidity, day of the week, and season are a few of the many variables that might be related to peak load. Thus, we would want to incorporate these and other potentially important independent variables into the model to make more accurate predictions.

Probabilistic models that include terms involving x^2, x^3 (or higher-order terms), or more than one independent variable are called *multiple regression models*. The general form of these models is

$$Y = \beta_0 + \beta_1 x_1 + \beta_2 x_2 + \cdots + \beta_k x_k + \varepsilon$$

The dependent variable Y is now written as a function of k independent variables $x_1, x_2, \ldots, x_k$. The random error term is added to allow for deviation between the deterministic part of the model, $\beta_0 + \beta_1 x_1 + \cdots + \beta_k x_k$, and the value of the dependent variable Y. The random error component makes the model probabilistic rather than deterministic. The value of the coefficient β_i determines the contribution of the independent variable x_i, and β_0 is the Y-intercept. The coefficients $\beta_0, \beta_1, \ldots, \beta_k$ will usually be unknown, since they represent population parameters.

At first glance, it might appear that the regression model shown above would not allow for anything other than straight-line relationships between Y and the independent variables, but this is not the case. Actually, $x_1, x_2, \ldots, x_k$ can be functions of variables as long as the functions do not contain unknown parameters. For example, the yield Y of a chemical process could be a function of the independent variables

$$x_1 = \text{temperature at which the reaction takes place}$$
$$x_2 = (\text{temperature})^2 = x_1^2$$
$$x_3 = \text{pressure at which the reaction takes place}$$

We might hypothesize the model

$$E(Y) = \beta_0 + \beta_1 x_1 + \beta_2 x_1^2 + \beta_3 x_3$$

Although this model contains a second-order (or "quadratic") term x_1^2 and therefore is curvilinear, the model is still *linear in the unknown parameters* $\beta_0, \beta_1, \beta_2$, and β_3. Multiple regression models that are linear in the unknown parameters (the β's) are called *linear models*, even though they may contain nonlinear independent variables.

The same steps we followed in developing a straight-line model are applicable to the multiple regression model.

Step 1. Hypothesize the form of the model. This involves the choice of the independent variables to be included in the model.

Step 2. Estimate the unknown parameters $\beta_0, \beta_1, \ldots, \beta_k$.

Step 3. Specify the probability distribution of the random error component ε and estimate its variance σ^2.

Step 4. Check the adequacy of the model.

Step 5. Use the fitted model to estimate the mean value of Y or to predict a particular value of Y for given values of the independent variables.

First we consider steps 2–5, leaving the more difficult problem of model construction until last.

10.2 Fitting the Model: The Least-Squares Approach

The method used for fitting multiple regression models is identical to that used for fitting the simple straight-line model—the method of least squares. That is, we choose the estimated model

$$\hat{y} = \hat{\beta}_0 + \hat{\beta}_1 x_1 + \cdots + \hat{\beta}_k x_k$$

that minimizes

$$\text{SSE} = \sum_{i=1}^{n} (y_i - \hat{y}_i)^2$$

As in the case of the simple linear model, the sample estimates $\hat{\beta}_0, \hat{\beta}_1, \ldots, \hat{\beta}_k$ are obtained as a solution of a set of simultaneous linear equations.

The primary difference between fitting the simple and multiple regression models is computational difficulty. The $(k+1)$ simultaneous linear equations that must be solved to find the $(k+1)$ estimated coefficients $\hat{\beta}_0, \hat{\beta}_1, \ldots, \hat{\beta}_k$ are difficult (sometimes nearly impossible) to solve with a pocket or desk calculator. Consequently, we resort to the use of computers. Many computer packages have been developed to fit a multiple regression model using the method of least squares. We present output from some of the more popular computer packages instead of presenting the tedious hand calculations required to fit the models. Most package regression programs have similar output, so you should have little trouble interpreting regression output from other packages. As in previous chapters, we will make most of our computations using Minitab.

Recall the peak power load data from Chapter 9, repeated in Table 10.1. We previously used a straight-line model to describe the peak power load/daily high temperature relationship. Now suppose we want to hypothesize a curvilinear relationship

$$Y = \beta_0 + \beta_1 x + \beta_2 x^2 + \varepsilon$$

Note that the scatterplot in Figure 10.1 provides some support for the inclusion of the term $\beta_2 x^2$ in the model, since there appears to be some curvature present in the relationship.

Part of the output from the Minitab multiple regression routine for the peak power load data is reproduced in Figure 10.2. The least-squares estimates of the β parameters appear in the column labeled Coef. You can see that $\hat{\beta}_0 = 1{,}784.2$, $\hat{\beta}_1 = -42.39$, and $\hat{\beta}_2 = 0.27$. Therefore, the equation that minimizes the SSE for the data is

$$\hat{y} = 1{,}784.2 - 42.39x + 0.27x^2$$

Daily High Temperature x(°F)	Peak Power Load y(megawatts)
95	214
82	152
90	156
81	129
99	254
100	266
93	210
95	204
93	213
87	150

F I G U R E **10.1**
Scatterplot of Power Load Data
from Table 10.1

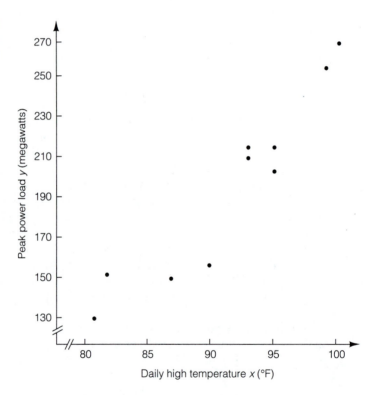

The value of the SSE, 1,175.1, also appears in the printout. We will discuss the rest of the printout as the chapter progresses.

Note that the graph of the multiple regression model (Figure 10.3) provides a good fit to the data of Table 10.1. However, before we can more formally measure the utility of the model, we need to estimate the variance of the error component ε.

FIGURE **10.2**

Computer Output for the Peak
Power Load Example

```
The regression equation is
y = 1784-42.4x+0.272x-sq

Predictor     Coef     Stdev    t-ratio        p
Constant     1784.2     944.1       1.89    0.101
x            -42.39     21.00      -2.02    0.083
x-sq         0.2722    0.1163       2.34    0.052

s=12.96    R-sq=93.9%    R-sq(adj)=92.2%

Analysis of Variance

SOURCE        DF        SS       MS       F       p
Regression     2   18088.5   9044.3   53.88   0.000
Error          7    1175.1    167.9
Total          9   19263.6
```

FIGURE **10.3**
Plot of Curvilinear Model

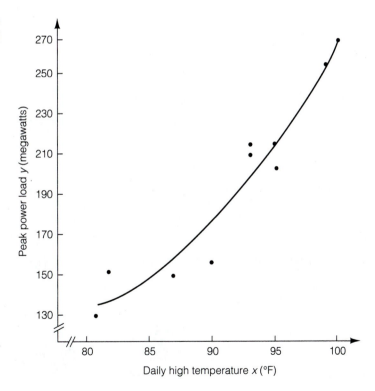

10.3 Estimation of σ^2, the Variance of ε

The specification of the probability distribution of the random error component ε of the multiple regression model follows the same general outline as for the straight-line model. We assume that ε is normally distributed with mean zero and constant variance σ^2 for any set of values for the independent variables $x_1, x_2, \ldots, x_k$. Furthermore, the errors are assumed to be independent. Given these assumptions, the remaining task in specifying the probability distribution of ε is to estimate σ^2.

For example, in the quadratic model describing peak power load as a function of daily high temperature, we found a minimum SSE = 1,175.1. Now we want to use this quantity to estimate the variance of ε. Recall that the estimator for the straight-line model was $s^2 = \text{SSE}/(n - 2)$, and note that the denominator is n minus the number of estimated β parameters, which is $(n - 2)$ in the straight-line model. Since we must estimate one more parameter β_2 for the quadratic model $Y = \beta_0 + \beta_1 x + \beta_2 x^2 + \varepsilon$, the estimate of σ^2 is

$$s^2 = \frac{\text{SSE}}{n - 3}$$

That is, the denominator becomes $(n - 3)$ because there are now three β parameters in the model.

The numerical estimate for our example is

$$s^2 = \frac{\text{SSE}}{10 - 3} = \frac{1,175.1}{7} = 167.87$$

where s^2 is called the mean square for error, or MSE. This estimate of σ^2 is shown in Figure 10.2 in the column labeled MS and the row labeled Error.

For the general multiple regression model

$$Y = \beta_0 + \beta_1 x_1 + \beta_2 x_2 + \cdots + \beta_k x_k + \varepsilon$$

we must estimate the $(k + 1)$ parameters $\beta_0, \beta_1, \beta_2, \ldots, \beta_k$. Thus, the estimator of σ^2 is the SSE divided by the quantity $[n - (\text{number of estimated } \beta \text{ parameters})]$.

Estimator of σ^2 for Multiple Regression Model with k Independent Variables

$$\text{MSE} = \frac{\text{SSE}}{n - (\text{number of estimated } \beta \text{ parameters})}$$
$$= \frac{\text{SSE}}{n - (k + 1)}$$

We use the estimator of σ^2 both to check the adequacy of the model (Sections 10.4 and 10.5) and to provide a measure of the reliability of predictors and estimates when the model is used for those purposes (Section 10.8). Thus, you can see that the estimation of σ^2 plays an important part in the development of a regression model.

10.4 A Test of Model Adequacy: The Coefficient of Determination

To find a statistic that measures how well a multiple regression model fits a set of data, we use the multiple regression equivalent of r^2, the coefficient of determination for the straight-line model (Chapter 9). Thus, we define the *multiple coefficient of determination R^2* as

$$R^2 = 1 - \frac{\sum_{i=1}^{n} (y_i - \hat{y}_i)^2}{\sum_{i=1}^{n} (y_i - \bar{y})^2} = 1 - \frac{SSE}{SS_{yy}}$$

where $\hat{y}_i$ is the predicted value of Y_i for the model. Just as for the simple linear model, R^2 represents the fraction of the sample variation of the y values (measured by SS_{yy}) that is explained by the least-squares prediction equation. Thus, $R^2 = 0$ implies a complete lack of fit of the model to the data, and $R^2 = 1$ implies a perfect fit, with the model passing through every data point. In general, the larger the value of R^2, the better the model fits the data.

To illustrate, the value of $R^2 = 0.939$ for the peak power load data is indicated in Figure 10.2. This value of R^2 implies that using the independent variable daily high temperature in a quadratic model results in a 93.9% reduction in the total *sample variation* (measured by SS_{yy}) of peak power load Y. Thus, R^2 is a sample statistic that tells us how well the model fits the data and thereby represents a measure of the adequacy of the model.

The fact that R^2 is a sample statistic implies that it can be used to make inferences about the utility of the entire model for predicting the population of y values at each setting of the independent variables. In particular, for the peak power load data, the test

$H_0 : \beta_1 = \beta_2 = 0$

H_a : At least one of the coefficients is nonzero

would formally test the global utility of the model. The test statistic used to test this null hypothesis is

$$\text{Test statistic: } F = \frac{R^2/k}{(1 - R^2)/[n - (k + 1)]}$$

where n is the number of data points and k is the number of parameters in the model, not including β_0. The test statistic F will have the F probability distribution with k degrees of freedom in the numerator and $[n - (k + 1)]$ degrees of freedom in the denominator. The tail values of the F distribution are given in Tables 7 and 8 in the Appendix.

The F-test statistic becomes large as the coefficient of determination R^2 becomes large. To determine how large F must be before we can conclude at a given significance level that the model is useful for predicting Y, we set up the rejection region as follows:

$$\text{Rejection region: } F > F_\alpha(k, n - (k + 1))$$

For the electrical usage example ($n = 10$, $k = 2$, $n - (k + 1) = 7$, and $\alpha = 0.05$), we reject $H_0 : \beta_1 = \beta_2 = 0$ if

$$F > F_{0.05}(2, 7)$$

or

$$F > 4.74$$

From the computer printout (Figure 10.2), we find that the computed F is 53.88. Since this value greatly exceeds the tabulated value of 4.74, we conclude that at least one of the model coefficients β_1 and β_2 is nonzero. Therefore, this global F test indicates that the quadratic model $Y = \beta_0 + \beta_1 x + \beta_2 x^2 + \varepsilon$ is useful for predicting peak load.

Before placing full confidence in the value of R^2 and the F-test, however, we must examine residuals to see if there is anything unusual in the data. In the case of fitting y as a function of a single x, the residuals were plotted against x. In multiple regression problems, there may be more than one x against which to plot residuals, and any of these plots could be informative. To standardize procedures a bit, it is customary in such cases to plot the residuals against the predicted values, $\hat{y}$. This plot should look similar to the plot of residuals against x when there is only a single x to consider. Two residual plots for the peak power load example are shown in Figure 10.4, one with predicted values on the horizontal axis and one with x on the horizontal axis. Note that both plots show fairly random patterns, which helps confirm the adequacy of the model. Also, the patterns in the two plots are similar to each other.

Testing the Utility of a Multiple Regression Model: The Global F-Test

$H_0 : \beta_1 = \beta_2 = \cdots = \beta_k = 0$

$H_a :$ At least one of the β parameters does not equal zero

Test statistic: $F = \dfrac{R^2/k}{(1 - R^2)/[n - (k + 1)]}$

Assumptions: See Section 9.4 for the four assumptions about the random component ε.

Rejection region: $F > F_\alpha(k, n - (k + 1))$

where

$n = $ number of data points

$k = $ number of β parameters in the model, excluding β_0

FIGURE **10.4**
Residual Plots for Peak Power
Load Data

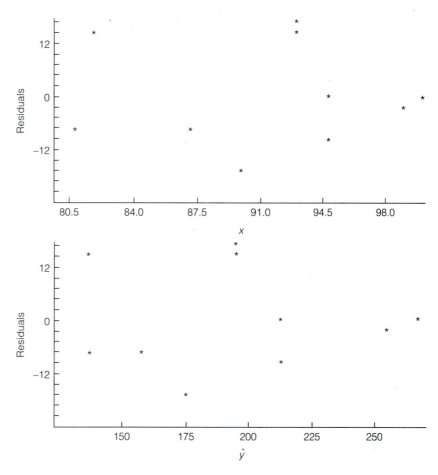

FIGURE **10.4**
Residual Plots for Peak Power
Load Data

EXAMPLE **10.1**   A study is conducted to determine the effects of company size and presence or absence of a safety program on the number of work hours lost due to work-related accidents. A total of 40 companies are selected for the study, 20 randomly chosen from companies with no active safety program and the other 20 from companies that have enacted active safety programs. Each company is monitored for a one-year period, and the following model is proposed:

$$E(Y) = \beta_0 + \beta_1 x_1 + \beta_2 x_2$$

where

$$Y = \text{lost work hours over the one-year study period}$$

$$x_1 = \text{number of employees}$$

$$x_2 = \begin{cases} 1 & \text{if an active safety program is used} \\ 0 & \text{if no active safety program is used} \end{cases}$$

The variable x_2 is called a *dummy* or *indicator variable*. Dummy variables are used to represent categorical or qualitative independent variables, such as "presence or absence of a safety program." Note that the coefficient β_2 of the dummy variable x_2 represents the expected difference in lost work hours between companies with safety programs and those without safety programs, assuming the companies have the same number of employees. For example, the expected number of lost work hours for a company with 500 employees and no safety program is, according to the model,

$$E(Y) = \beta_0 + \beta_1(500) + \beta_2(0)$$
$$= \beta_0 + 500\beta_1$$

while the expected number of lost work hours for a company with 500 employees and an active safety program is

$$E(Y) = \beta_0 + \beta_1(500) + \beta_2(1)$$
$$= \beta_0 + 500\beta_1 + \beta_2$$

You can see that the difference in the means is β_2. Thus, the use of the dummy variable allows us to assess the effects of a qualitative variable on the mean response. The data are shown in Table 10.2.

T A B L E **10.2**   Data for Example 10.1

Number of Employees x_1	Safety Program x_2 ($x_2 = 0$, no program; $x_2 = 1$, active program)	Lost Hours Due to Accidents y (annually, in thousands of hours)	Number of Employees x_1	Safety Program x_2 ($x_2 = 0$, no program; $x_2 = 1$, active program)	Lost Hours to Accidents y (annually, in thousands of hours)
6,490	0	121	3,077	1	44
7,244	0	169	6,600	1	73
7,943	0	172	2,732	1	8
6,478	0	116	7,014	1	90
3,138	0	53	8,321	1	71
8,747	0	177	2,422	1	37
2,020	0	31	9,581	1	111
4,090	0	94	9,326	1	89
3,230	0	72	6,818	1	72
8,786	0	171	4,831	1	35
1,986	0	23	9,630	1	86
9,653	0	177	2,905	1	40
9,429	0	178	6,308	1	44
2,782	0	65	1,908	1	36
8,444	0	146	8,542	1	78
6,316	0	129	4,750	1	47
2,363	0	40	6,056	1	56
7,915	0	167	7,052	1	75
6,928	0	115	7,794	1	46
5,526	0	123	1,701	1	6

a Test the utility of the model by testing $H_0 : \beta_1 = \beta_2 = 0$.

b Give and interpret the least-squares coefficients $\hat{\beta}_1$ and $\hat{\beta}_2$.

Solution **a** The computer printout corresponding to the least-squares fit of the model

$$Y = \beta_0 + \beta_1 x_1 + \beta_2 x_2 + \varepsilon$$

is shown in Figure 10.5. We want to use this information to test

$H_0 : \beta_1 = \beta_2 = 0$

H_a : At least one β is nonzero, that is, the model is useful for predicting Y

Test statistic: $F = \dfrac{R^2/k}{(1 - R^2)/[n - (k + 1)]} = \dfrac{R^2/2}{(1 - R^2)/37}$

Rejection region: for $\alpha = 0.05$, $F > F_{0.05}(2, 37)$

The value of R^2 is 0.86. Then

$$F = \frac{0.86/2}{(1 - 0.86)/37} = 113.6$$

Thus, we can be very confident that this model contributes information for the prediction of lost work hours, since $F = 113.6$ greatly exceeds the tabled value, 3.25. The F value is also given on the printout, so we do not have to perform the calculation. Using the printout value also helps us avoid rounding errors; note the difference between our calculated F value and the printout value of 112.94. We should be careful not to be too excited about this very large F value, since we have concluded only that the model contributes *some* information about Y. The widths of the prediction intervals and confidence intervals are better measures of the amount of information the model provides about Y. Note on the printout that the estimated standard deviation of this model is 19.97. Since most (usually 90% or more) of the values of Y will fall within two standard deviations of the

F I G U R E 10.5
Computer Printout of
Least-Squares Fit for Example
10.1

```
The regression equation is
y = 31.7 + 0.0143 x1 - 58.2 x2

Predictor      Coef      Stdev    t-ratio        p
Constant     31.670      8.560       3.70    0.001
x1         0.014272   0.001222      11.68    0.000
x2          -58.223      6.316      -9.22    0.000

s = 19.97    R-sq = 85.9%     R-sq(adj) = 85.2%

Analysis of Variance

SOURCE         DF        SS       MS        F        p
Regression      2     90058    45029   112.94    0.000
Error          37     14752      399
Total          39    104811
```

predicted value, we have the notion that the model predictions will usually be accurate to within about 40,000 hours. We will make this notion more precise in Section 10.5, but the standard deviation helps us obtain a preliminary idea of the amount of information the model contains about Y.

b   The estimated values of β_1 and β_2 are shown on the printout. The least-squares model is

$$\hat{y} = 31.67 + 0.0143x_1 - 58.22x_2$$

The coefficient $\hat{\beta}_1 = 0.0143$ represents the estimated slope of the number of lost hours (in thousands) versus the number of employees. In other words, we estimate that, on average, each employee loses 14.3 work hours (0.0143 thousand hours) annually due to accidents. Note that this slope applies both to plants with safety programs and to plants without safety programs. We will show how to construct models that allow the slopes to differ later in this chapter.

The coefficient $\hat{\beta}_2 = -58.2$ is *not* a slope, because it is the coefficient of the dummy variable x_2. Instead, $\hat{\beta}_2$ represents the estimated mean change in lost hours between plants with no safety programs ($x_2 = 0$) and plants with safety programs ($x_2 = 1$), assuming both plants have the same number of employees. Our estimate indicates that an average of 58,200 fewer work hours are lost by plants with safety programs than by plants without them.   ∎

10.5   Estimating and Testing Hypotheses About Individual β Parameters

After determining that the model is useful by testing and rejecting $H_0: \beta_1 = \beta_2 = \cdots = \beta_k = 0$, we might be interested in making inferences about particular β parameters that have practical significance. In the peak power load example, we fit the model

$$E(Y) = \beta_0 + \beta_1 x + \beta_2 x^2$$

where Y is the peak load and x is the daily maximum temperature. A test of particular interest would be

$H_0: \beta_2 = 0$ (no quadratic relationship exists)

$H_a: \beta_2 > 0$ (the peak power load increases at an increasing rate as the daily maximum temperature increases)

A test of this hypothesis can be performed using a Student's T-test. The T-test utilizes a test statistic analogous to that used to make inferences about the slope of the simple straight-line model (Section 9.5). The t statistic is formed by dividing the sample estimate $\hat{\beta}_2$ of the population coefficient β_2 by the estimated standard deviation of the repeated sampling distribution of $\hat{\beta}_2$:

$$\text{Test statistic: } t = \frac{\hat{\beta}_2}{s_{\hat{\beta}_2}}$$

We use the symbol $s_{\hat{\beta}_2}$ to represent the estimated standard deviation of $\hat{\beta}_2$. Most computer packages list the estimated standard deviation $s_{\hat{\beta}_i}$ for each of the estimated model coefficients β_i. In addition, they usually give the calculated t values for each coefficient in the model.

The rejection region for the test is found in exactly same way as the rejection regions for the T-tests in previous chapters. That is, we consult the t table (Table 5) in the Appendix to obtain an upper-tail value of t. This is a value t_α such that $P(T > t_\alpha) = \alpha$. Then we can use this value to construct rejection regions for either one- or two-tailed tests. To illustrate with the power load example, the error degrees of freedom is $(n - 3) = 7$, the denominator of the estimate of σ^2. Then the rejection region (shown in Figure 10.6) for a one-tailed test with $\alpha = 0.05$ is

$$\text{Rejection region: } t > t_\alpha(n - 3)$$
$$t > 1.895$$

In Figure 10.7, we again show a portion of the computer printout for the peak power load example. The estimated standard deviations for the model coefficients appear under the column labeled Stdev. The t statistic for testing the null hypothesis that the true coefficients are equal to zero appears under the column headed t-ratio. The t value corresponding to the test of the null hypothesis $H_0 : \beta_2 = 0$ is the last one in the column, that is, $t = 2.34$. Since this value is greater than 1.895, we conclude that the quadratic term $\beta_2 x^2$ makes a contribution to the predicted value of peak power load and that there is evidence that the peak power load increases at an increasing rate as the daily maximum temperature increases.

The Minitab printout shown in Figure 10.7 also lists the observed two-tailed significance levels[†] for each t value. These values appear under the column headed

F I G U R E **10.6**
Rejection Region for Test of
$H_0 : \beta_2 = 0$

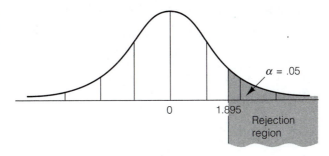

$\alpha = .05$

0   1.895

Rejection region

[†]See Chapter 8 for a discussion of significance levels in tests of hypotheses.

```
Predictor       Coef     Stdev    t-ratio          p
Constant      1784.2     944.1        189      0.101
x             -42.39     21.00      -2.02      0.083
x-sq          0.2722    0.1163       2.34      0.052

s = 12.96    R-sq = 93.9%    R-sq(adj) = 92.2%

Analysis of Variance

SOURCE          DF         SS        MS       F        p
Regression       2    18088.5    9044.3   53.88    0.000
Error            7     1175.1     167.9
Total            9    19263.6
```

p. The significance level 0.052 corresponds to the quadratic term; it implies that we would reject $H_0 : \beta_2 = 0$ in favor of $H_a : \beta_2 \neq 0$ at any α level larger than 0.052. Since our alternative was one-sided, $H_a : \beta_2 > 0$, the significance level is half that given in the printout, that is, $(1/2)(0.052) = 0.026$. Thus, there is evidence at the $\alpha = 0.05$ level that the peak power load increases more quickly per unit increase in daily maximum temperature for high temperatures than for low ones.

We can also form a confidence interval for the parameter β_2 as follows:

$$\hat{\beta}_2 \pm t_{\alpha/2}(n-3)s_{\hat{\beta}_2} = 0.272 \pm (1.895)(0.116)$$

or (0.052, 0.492). Note that the t value 1.895 corresponds to $\alpha/2 = 0.05$ and $(n-3) = 7$ degrees of freedom. This interval constitutes a 90% confidence interval for β_2 and represents an estimate of the rate of curvature in mean peak power load as the daily maximum temperature increases. Note that all values in the interval are positive, reconfirming the conclusion of our test.

Testing a hypothesis about a single β parameter that appears in any multiple regression model is accomplished in exactly the same manner as described for the quadratic electrical usage model. The form of such a test is shown here.

Test of an Individual Parameter Coefficient in the Multiple Regression Model

One-Tailed Test	Two-Tailed Test
$H_0 : \beta_i = 0$	$H_0 : \beta_i = 0$
$H_a : \beta_i < 0$ (or $H_a : \beta_i > 0$)	$H_a : \beta_i \neq 0$
Test Statistic	Test Statistic
$t = \dfrac{\hat{\beta}_i}{s_{\hat{\beta}_i}}$	$t = \dfrac{\hat{\beta}_i}{s_{\hat{\beta}_i}}$
Rejection Region	Rejection Region
$t < -t_\alpha[n - (k+1)]$	$t < -t_{\alpha/2}[n - (k+1)]$
(or $t > t_\alpha[n - (k+1)]$	or
when $H_a : \beta_i > 0$)	$t > t_{\alpha/2}[n - (k+1)]$

> where
>
> $n =$ number of observations
> $k =$ number of β parameters in the model, excluding β_0
>
> Assumptions: See Section 9.4 for the assumptions about the probability distribution for the random error component ε.

If all the tests of model utility indicate that the model is useful for predicting Y, can we conclude that the best prediction model has been found? Unfortunately, we cannot. There is no way of knowing (without further analysis) whether the addition of other independent variables will improve the utility of the model, as Example 10.2 shows.

EXAMPLE   **10.2**   Refer to Example 10.1, in which we modeled a plant's lost work hours as a function of number of employees and the presence or absence of a safety program. Suppose a safety engineer believes that safety programs tend to help larger companies even more than smaller companies. Thus, instead of a relationship like that shown in Figure 10.8a, in which the rate of increase in mean lost work hours per employee is the same for companies with and without safety programs, the engineer believes that the relationship is like that shown in Figure 10.8b. Note that although the mean number of lost work hours is smaller for companies with safety programs no matter how many employees the company has, the magnitude of the difference becomes greater as the number of employees increases. When the slope of the relationship between $E(Y)$ and one independent variable (x_1) depends on the value of a second

FIGURE **10.8**
Examples of No-Interaction and Interaction Models

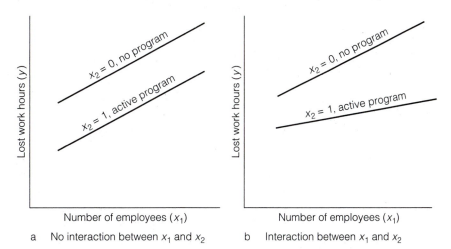

a   No interaction between x_1 and x_2

b   Interaction between x_1 and x_2

independent variable (x_2), as is the case here, we say that x_1 and x_2 *interact*. The model for mean lost work hours that includes interaction is written

$$E(Y) = \beta_0 + \beta_1 x_1 + \beta_2 x_2 + \beta_3 x_1 x_2$$

Note that the increase in mean lost work hours $E(Y)$ for each one-person increase in number of employees x_1 is no longer given by the constant β_1 but is now given by $\beta_1 + \beta_3 x_2$. That is, the amount that $E(Y)$ increases for each 1-unit increase in x_1 is dependent on whether the company has a safety program. Thus, the two variables x_1 and x_2 interact to affect Y.

The 40 data points listed in Table 10.2 were used to fit the model with interaction. A portion of the computer printout is shown in Figure 10.9. Test the hypothesis that the effect of safety programs on mean lost work hours is greater for larger companies.

FIGURE **10.9**

Computer Printout for the Safety
Program Example

```
The regression equation is
y = -0.74 + 0.0197 x1 + 5.4 x2 - 0.0107 x1*x2

Predictor         Coef      Stdev      t-ratio        p
Constant        -0.739      7.927       -0.09    0.926
x1             0.019696   0.001219       16.16    0.000
x2                5.37      11.08         0.48    0.631
x1*x2         -0.010737   0.001715       -6.26    0.000


s = 14.01    R-sq = 93.3%    R-sq(adj) = 92.7%


Analysis of Variance

SOURCE        DF        SS        MS         F       p
Regression     3     97749     32583    166.11   0.000
Error         36      7062       196
Total         39    104811
```

Solution   The interaction model is

$$E(Y) = \beta_0 + \beta_1 x_1 + \beta_2 x_2 + \beta_3 x_1 x_2$$

so the slope of $E(Y)$ versus x_1 is β_1 for companies with no safety program and $\beta_1 + \beta_3$ for those with safety programs. Our research hypothesis is that the slope is smaller for companies with safety programs, so we will test

$$H_0 : \beta_3 = 0 \qquad H_a : \beta_3 < 0$$

$$\text{Test statistic: } t = \frac{\hat{\beta}_3}{s_{\hat{\beta}_3}}$$

Rejection region: for $\alpha = 0.05$,

$$t < -t_{0.05}[n - (k+1)] \quad \text{or} \quad t < -t_{0.05}(36) = -1.645$$

The t value corresponding to the test for β_3 is indicated in Figure 10.9. The value $t = -6.26$ is less than -1.645 and therefore falls in the rejection region. Thus, the safety engineer can conclude that the larger the company, the greater the benefit of the safety program in terms of reducing the mean number of lost work hours. ∎

In comparing the results for Examples 10.1 and 10.2, we see that the model with interaction fits the data better than the model without an interaction term. The R^2 value is greater for the model with interaction, and the interaction coefficient is significantly different from zero. How else might we compare how well these two models fit the data? Thinking back for a moment to Chapter 9, we might want to look at residuals, as we did for the simple straight-line model.

Figure 10.10 shows the residuals plotted against x_1 for the model *without* interaction and the stem-and-leaf plot for these residuals. Figure 10.11 shows the residuals plotted against x_1 and the stem-and-leaf plot of residuals for the model *with* interaction. Note that the plot in Figure 10.11 seems to show more "randomness" in the points and less spread along the vertical axis. The stem-and-leaf plot in Figure 10.11 is more compact and more symmetrically mound-shaped than the stem-and-leaf plot in Figure 10.10. All of these features indicate that the interaction model fits the data better than the model without interaction.

FIGURE **10.10**
Residuals and Stem-and-Leaf Plot for the Model Without Interaction (Example 10.1)

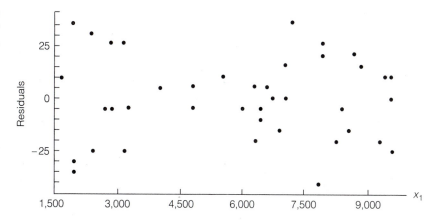

FIGURE **10.10**
(Continued)

Stem-and-leaf of residuals, $N = 40$
Leaf unit = 1.0

```
 2   -3  87
 2   -3
 4   -2  95
 7   -2  431
11   -1  9775
11   -1
16   -0  87665
19   -0  433
(4)   0  0013
17    0  55778
12    1  123
 9    1  6
 8    2  02
 6    2  5668
 2    3  3
 1    3  5
```

FIGURE **10.11**
Residuals and Stem-and-Leaf
Plot for the Model with
Interaction (Example 10.2)

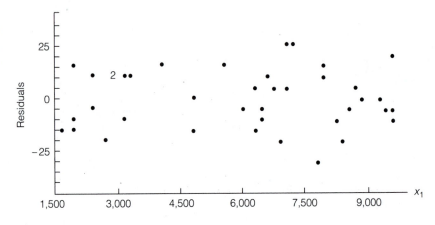

FIGURE **10.11**
(Continued)

Stem-and-leaf of residuals, $N = 40$
Lead unit $= 1.0$

```
  1  −2  8
  3  −2  10
  6  −1  975
 10  −1  3220
 16  −0  888665
 (5) −0  43210
 19   0  0
 18   0  55567999
 11   1  0011444
  4   1  6
  3   2  02
  1   2  7
```

Exercises

10.1   Suppose that you fit the model

$$Y = \beta_0 + \beta_1 x_1 + \beta_2 x_2 + \beta_3 x_1 x_2 + \beta_4 x_1^2 + \beta_5 x_2^2 + \varepsilon$$

to $n = 30$ data points and that

$$\text{SSE} = 0.37 \quad \text{and} \quad R^2 = 0.89$$

a   Do the values of SSE and R^2 suggest that the model provides a good fit to the data? Explain.

b   Is the model of any use in predicting y? Test the null hypothesis that $E(Y) = \beta_0$, that is,

$$H_0 : \beta_1 = \beta_2 = \cdots = \beta_5 = 0$$

against the alternative hypothesis

$$H_a : \text{At least one of the parameters } \beta_1, \beta_2, \ldots, \beta_5 \text{ is nonzero.}$$

Use $\alpha = 0.05$.

10.2   In hopes of increasing the company's share of the fine-food market, researchers for a meat-processing firm that prepares meats for exclusive restaurants are working to improve the quality of its hickory-smoked hams. One of their studies concerns the effect of time spent in the smokehouse on the flavor of the ham. Hams that were in the smokehouse for varying amounts of time were each subjected to a taste test by a panel of ten food experts. The following model was thought by the researchers to be appropriate:

$$Y = \beta_0 + \beta_1 t + \beta_2 t^2 + \varepsilon$$

where

$$Y = \text{mean of the taste scores for the ten experts}$$
$$t = \text{time in the smokehouse (hours)}$$

Assume that the least-squares model estimated using a sample of 20 hams is

$$\hat{y} = 20.3 + 5.2t - 0.0025t^2$$

and that $s_{\hat{\beta}_2} = 0.0011$. The coefficient of determination is $R^2 = 0.79$.

a   Is there evidence to indicate that the overall model is useful? Test at $\alpha = 0.05$.

b   Is there evidence to indicate that the quadratic term is important in this model? Test at $\alpha = 0.05$.

10.3   Because the coefficient of determination R^2 never decreases when a new independent variable is added to the model, it is tempting to include many variables in a model to force R^2 to be near 1. However, doing so reduces the degrees of freedom available for estimating σ^2, which adversely affects our ability to make reliable inferences. As an example, suppose you want to use 18 economic indicators to predict next year's GNP. You fit the model

$$Y = \beta_0 + \beta_1 x_1 + \beta_2 x_2 + \cdots + \beta_{17} x_{17} + \beta_{18} x_{18} + \varepsilon$$

where $Y = \text{GNP}$ and $x_1, x_2, \ldots, x_{18}$ are indicators. Only 20 years of data ($n = 20$) are used to fit the model, and you obtain $R^2 = 0.95$. Test to see whether this impressive-looking R^2 value is large enough for you to infer that this model is useful, that is, that at least one term in the model is important for predicting GNP. Use $\alpha = 0.05$.

10.4   A utility company of a major city reported the average utility bills listed in the table for a standard-size home during the last year.

Month	Average Monthly Temperature $x(^\circ F)$	Average Utility Bill y(dollars)
January	38	99
February	45	91
March	49	78
April	57	61
May	69	55
June	78	63
July	84	80
August	89	95
September	79	65
October	64	56
November	54	74
December	41	93

a   Plot the points in a scatterplot.

b   Use the methods of Chapter 9 to fit the model

$$Y = \beta_0 + \beta_1 x + \varepsilon$$

What do you conclude about the utility of this model?

c   Hypothesize another model that might better describe the relationship between the average utility bill and average temperature. If you have access to a computer package, fit the model and test its utility.

10.5   To run a manufacturing operation efficiently, it is necessary to know how long it takes employees to manufacture the product. Without such information, the cost of making the product cannot be determined. Furthermore, management would not be able to establish an effective incentive plan for its employees, because it would not know how to set work standards (S. Chase and B. Aquilano, *Production and Operations Management*, Irwin, 1979). Estimates of production time are frequently obtained by using time studies. The data in the table were obtained from a recent time study of a sample of 15 employees on an automobile assembly line.

Time to Complete Task y (minutes)	Months of Experience x
10	24
20	1
15	10
11	15
11	17
19	3
11	20
13	9
17	3
18	1
16	7
16	9
17	7
18	5
10	20

a The computer printout for fitting the model $y = \beta_0 + \beta_1 x + \beta_2 x^2 + \varepsilon$ is shown here. Find the least-squares prediction equation.

```
Predictor     Coef      Stdev    t-ratio        p
Constant   20.0911     0.7247      27.72    0.000
x          -0.6705     0.1547      -4.33    0.001
x*x       0.009535  0.006326       1.51    0.158

s=1.091      R-sq=91.6%       R-sq(adj)=90.2%

Analysis of Variance

SOURCE        DF          SS        MS       F        p
Regression     2     156.119    78.060   65.59    0.000
Error         12      14.281     1.190
Total         14     170.400
```

b Plot the fitted equation on a scatterplot of the data. Is there sufficient evidence to support the inclusion of the quadratic term in the model? Explain.

c Test the null hypothesis that $\beta_2 = 0$ against the alternative hypothesis that $\beta_2 \neq 0$. Use $\alpha = 0.01$. Does the quadratic term make an important contribution to the model?

d Your conclusion in part (c) should have been to drop the quadratic term from the model. Do so, and fit the "reduced model" $y = \beta_0 + \beta_1 x + \varepsilon$ to the data.

e Define β_1 in the context of this exercise. Find a 90% confidence interval for β_1 in the reduced model of part (d).

10.6 A researcher wanted to investigate the effects of several factors on production-line supervisors' attitudes toward handicapped workers. A study was conducted involving 40 randomly selected supervisors. The response Y, a supervisor's attitude toward handicapped workers, was measured with a standardized attitude scale. Independent variables used in the study were

$$x_1 = \begin{cases} 1 & \text{if the supervisor is female} \\ 0 & \text{if the supervisor is male} \end{cases}$$

x_2 = number of years of experience in a supervisory job

The researcher fit the model

$$Y = \beta_0 + \beta_1 x_1 + \beta_2 x_2 + \beta_3 x_2^2 + \varepsilon$$

to the data with the following results:

$$\hat{y} = 50 + 5x_1 + 5x_2 - 0.1x_2^2$$
$$s_{\hat{\beta}_3} = 0.03$$

a   Do these data provide sufficient evidence to indicate that the quadratic term, years of experience x_2^2, is useful for predicting attitude score? Use $\alpha = 0.05$.

b   Sketch the predicted attitude score $\hat{y}$ as a function of the number of years of experience x_2 for male supervisors $(x_1 = 0)$. Next, substitute $x_1 = 1$ into the least-squares equation and obtain a plot of the prediction equation for female supervisors. [*Note:* For both males and females, plotting $\hat{y}$ for $x_2 = 0, 2, 4, 6, 8,$ and 10 will produce a good picture of the prediction equations. The vertical distance between the males' and females' prediction curves is the same for all values of x_2.]

10.7   To project personnel needs for the Christmas shopping season, a department store wants to project sales for the season. The sales for the previous Christmas season are an indication of what to expect for the current season. However, the projection should also reflect the current economic environment by taking into consideration sales for a more recent period. The following model might be appropriate:

$$Y = \beta_0 + \beta_1 x_1 + \beta_2 x_2 + \varepsilon$$

where

$$x_1 = \text{previous Christmas sales}$$
$$x_2 = \text{sales for August of current year}$$
$$y = \text{sales for upcoming Christmas}$$

(All units are in thousands of dollars.) Data for ten previous years were used to fit the prediction equation, and the following values were calculated:

$$\hat{\beta}_1 = 0.62 \qquad s_{\hat{\beta}_1} = 0.273$$
$$\hat{\beta}_2 = 0.55 \qquad s_{\hat{\beta}_2} = 0.181$$

Use these results to determine whether there is evidence to indicate that the mean sales this Christmas are related to this year's August sales in the proposed model.

10.8   Suppose you fit the second-order model

$$Y = \beta_0 + \beta_1 x + \beta_2 x^2 + \varepsilon$$

to $n = 30$ data points. Your estimate of β_2 is $\hat{\beta}_2 = 0.35$, and the standard error of the estimate is $s_{\hat{\beta}_2} = 0.13$.

a   Test the null hypothesis that the mean value of Y is related to x by the (first-order) linear model

$$E(Y) = \beta_0 + \beta_1 x$$

Test $H_0 : \beta_2 = 0$ against the alternative hypothesis $(H_a : \beta_2 \neq 0)$ that the true relationship is given by the quadratic model (a second-order linear model)

$$E(Y) = \beta_0 + \beta_1 x + \beta_2 x^2$$

Use $\alpha = 0.05$.

b Suppose you wanted only to determine whether the quadratic curve opens upward, that is, whether the slope of the curve increases as x increases. Give the test statistic and the rejection region for the test for $\alpha = 0.05$. Do the data support the theory that the slope of the curve increases as x increases? Explain.

c What is the value of the F statistic for testing the null hypothesis that $\beta_2 = 0$?

d Could the F statistic in part (c) be used to conduct the tests in parts (a) and (b)? Explain.

10.9 How is the number of degrees of freedom available for estimating σ^2 (the variance of ε) related to the number of independent variables in a regression model?

10.10 An employer has found that factory workers who are with the company longer tend to invest more in a company investment program per year than workers who have less time with the company. The following model is believed to be adequate in modeling the relationship of annual amount invested Y to years working for the company x:

$$Y = \beta_0 + \beta_1 x + \beta_2 x^2 + \varepsilon$$

The employer checks the records for a sample of 50 factory employees for a previous year and fits the above model to get $\hat{\beta}_2 = 0.0015$ and $s_{\hat{\beta}_2} = 0.00712$. The basic shape of a quadratic model depends upon whether $\beta_2 < 0$ or $\beta_2 > 0$. Test to see whether the employer can conclude that $\beta_2 > 0$. Use $\alpha = 0.05$.

10.11 Automobile accidents result in a tragic loss of life and, in addition, represent a serious dollar loss to the nation's economy. Shown in the table are the number of highway deaths (to the nearest hundred) and the number of licensed vehicles (in hundreds of thousands) for the years 1950–1979. (The years are coded 1–30 for convenience.) During the years 1974–1979 (years 25–30 in the table), the nationwide 55-mile-per-hour speed limit was in effect.

Year	Deaths y	Number of Vehicles x_1	Year	Deaths y	Number of Vehicles x_1
1	34.8	49.2	16	49.1	91.8
2	37.0	51.9	17	53.0	95.9
3	37.8	53.3	18	52.9	98.9
4	38.0	56.3	19	54.9	103.1
5	35.6	58.6	20	55.8	107.4
6	38.4	62.8	21	54.6	111.2
7	39.6	65.2	22	54.3	116.3
8	38.7	67.6	23	56.3	122.3
9	37.0	68.8	24	55.5	129.8
10	37.9	72.1	25	46.4	134.9
11	38.1	74.5	26	45.9	137.9
12	38.1	76.4	27	47.0	143.5
13	40.8	79.7	28	49.5	148.8
14	43.6	83.5	29	51.5	153.6
15	47.7	87.3	30	51.9	159.4

Source: U.S. Department of Transportation.

a Write a second-order model relating the number y of highway deaths for a year to the number x_1 of licensed vehicles.

b The computer printout for fitting the model to the data is shown here. Is there sufficient evidence to indicate that the model provides information for the prediction of the number of annual highway deaths? Test by using $\alpha = 0.05$.

```
Predictor          Coef        Stdev    t-ratio       p
Constant         -1.408        6.895      -0.20   0.840
x1                0.8455       0.1462      5.78   0.000
x1*x1          -0.0033220   0.0007093     -4.68   0.000


s=3.706     R-sq=76.7%        R-sq(adj)=75.0%


Analysis of Variance


SOURCE         DF          SS        MS       F       p
Regression      2     1222.16    611.08   44.50   0.000
Error          27      370.79     13.73
Total          29     1592.95
```

c   Give the P-value for the test of part (b) and interpret it.

d   Does the second-order term contribute information for the prediction of y? Test by using $\alpha = 0.05$.

e   Give the P value for the test of part (d) and interpret it.

f   Plot the residuals from the fitted model against x_1. Do you think this is the best model to use here?

10.12   If producers (providers) of goods (services) are able to reduce the unit cost of their goods (services) by increasing the scale of their operation, they are the beneficiaries of an economic force known as *economies of scale*. Economies of scale cause a firm's long-run average costs to decline (W. Ferguson and G. Maurice, *Economics Analysis*, Irwin, 1971). The question of whether economies of scale, diseconomies of scale, or neither (i.e., constant economies of scale) exist in the U.S. motor-freight common carrier industry has been debated for years. In an effort to settle the debate within a specific subsection of the trucking industry, T. Sugrue, M. Ledford, and W. Glaskowsky (*Transportation Journal*, 1982, pp. 27–41) used regression analysis to model the relationship between each of a number of profitability/cost measures and the size of the operation. In one case, they modeled expense per vehicle mile Y_1 as a function of the firm's total revenue X. In another case, they modeled expense per ton mile Y_2 as a function of X. Data were collected from 264 firms, and the following least-squares results were obtained:

Dependent Variable	$\hat{\beta}_0$	$\hat{\beta}_1$	r	F
Expense per vehicle mile	2.279	−0.00000069	−0.00783	1.616
Expense per ton mile	0.1680	−0.000000066	−0.0902	2.148

a   Investigate the usefulness of the two models estimated by Sugrue, Ledford, and Glaskowsky. Use $\alpha = 0.05$. Draw appropriate conclusions in the context of the problem.

b   Are the observed significance levels of the hypothesis tests you conducted in part (a) greater than 0.10 or less than 0.10? Explain.

c   What do your hypothesis tests of part (a) suggest about economies of scale in the subsection of the trucking industry investigated by Sugrue, Ledford, and Glaskowsky—the long-haul, heavy-load, intercity general-freight common carrier section of the industry? Explain.

10.6 Regression Software

A number of different statistical program packages feature regression software. Some of the most popular are BMD, Minitab, SAS, and SPSS. More recently, many regression packages have been offered for microcomputers. You probably have access to one or more of these packages.

The multiple regression computer programs for these packages may differ in what they are programmed to do, how they do it, and the appearance of their computer printouts, but all of them print the basic statistics needed for a regression analysis. For example, some will compute confidence intervals for $E(Y)$ and prediction intervals for Y; others will not. Some test the null hypotheses that the individual β parameters equal zero, using Student's T-tests, while others use F-tests.[†] But all give the least-squares estimates, the values of SSE, s^2, and so forth.

To illustrate, the Minitab, SAS, and SPSS regression analysis computer printouts for Example 10.2 are shown in Figure 10.12. For that example, we fit the model

$$y = \beta_0 + \beta_1 x_1 + \beta_2 x_2 + \beta_3 x_1 x_2 + \varepsilon$$

to $n = 40$ data points. The variables in the model were

$$y = \text{lost hours due to accidents}$$
$$x_1 = \text{number of employees}$$
$$x_2 = \text{safety program (1 if active, 0 if no program)}$$

Notice that the Minitab printout in Figure 10.12a gives the prediction equation at the top of the printout. The independent variables, shown in the prediction equation and listed at the left side of the printout, are x_1, x_2, and $x_1 x_2$. Thus, Minitab treats the product $x_1 x_2$ as a third independent variable x_3, which must be computed before the fitting commences. For this reason, the Minitab prediction equation will always appear on the printout as first-order even though some of the independent variables shown in the prediction equation may actually be the squares or products of other independent variables. The inclusion of the squares or products of independent variables is treated in the same manner in the SPSS program shown in Figure 10.12c. The SAS program is the only one of these three that can be instructed to include such terms automatically; they appear in the printout with an asterisk (∗) that indicates multiplication.

The estimates of the regression coefficients appear opposite the identifying variable in the Minitab column titled Coef, in the SAS column titled ESTIMATE, and in the SPSS column titled B. Compare the estimates given in these three columns. Note that the Minitab printout gives the estimates with a much lesser degree of precision (fewer decimal places) than the SAS and SPSS printouts. (Ignore the column titled BETA in the SPSS printout. These are standardized estimates and will not be discussed in this text.)

[†] A two-tailed Student's T-test based on ν df is equivalent to an F-test where the F statistic possesses 1 df in the numerator and ν df in the denominator.

F I G U R E **10.12**   Computer Printouts for Example 10.2

The regression equation is
y = -0.74 + 0.0197 x1 + 5.4 x2 - 0.0107 x1*x2

Predictor	Coef	Stdev	t-ratio	p
Constant	-0.739	7.927	-0.09	0.926
x1	0.019696	0.001219	16.16	0.000
x2	5.37	11.08	0.48	0.631
x1*x2	-0.010737	0.001715	-6.26	0.000

s = 14.01   R-sq = 93.3%   R-sq(adj) = 92.7%

Analysis of Variance

SOURCE	DF	SS	MS	F	p
Regression	3	97749	32583	166.11	0.000
Error	36	7062	196		
Total	39	104811			

a   Minitab regression printout

SAS
GENERAL LINEAR MODELS PROCEDURE

DEPENDENT VARIABLE, Y

SOURCE	DF	SUM OF SQUARES	MEAN SQUARE	F VALUE
MODEL	3	97749.09184154	32583.03061385	166.11
ERROR	36	7061.68315846	196.15786551	PR>F
CORRECTED TOTAL	39	104810.77500000		0.0001

R-SQUARE	C.V.	ROOT MSE	Y MEAN
0.932624	16.0846	14.00563692	87.07500000

PARAMETER	ESTIMATE	T FOR H0, PARAMETER = 0	PR>|T|	STD ERROR OF ESTIMATE
INTERCEPT	-0.73908475	-0.09	0.9262	7.92730641
X1	0.01969560	16.16	0.0001	0.00121874
X2	5.36689032	0.48	0.6310	11.07978453
X1*X2	-0.01073708	-6.26	0.0001	0.00171477

b   SAS regression printout

The estimated standard errors of the estimates are given in the Minitab column titled Stdev, in the SAS column titled STD ERROR OF ESTIMATE, and in the SPSS column titled SE B.

The values of the test statistics for testing $H_0 : \beta_i = 0$, where $i = 1, 2$, and 3, are shown in the Minitab column titled t-ratio and in the SAS column titled T FOR H0:

F I G U R E **10.12**

(Continued)   SSPS

```
                   **** MULTIPLE REGRESSION ****
LISTWISE DELETION OF MISSING DATA
EQUATION NUMBER 1    DEPENDENT VARIABLE.. Y

BEGINNING BLOCK NUMBER 1.    METHOD,   ENTER
VARIABLE(S) ENTERED ON STEP NUMBER 1.. X1X2
                                     2.. X1
                                     3.. X2

ANALYSIS OF VARIANCE
               DF  SUM OF SQUARES   MEAN SQUARE
REGRESSION      3     97749.09184   32583.03061
RESIDUAL       36      7061.68316     196.15787

F= 166.10616   SIGNIF F= .0000

MULTIPLE R         .96572
R SQUARE           .93262
ADJUSTED R SQUARE .92701
STANDARD ERROR   14.00564

     **** VARIABLES IN THE EQUATION ****
VARIABLE         B        SE B       BETA        T       SIG T
X1X2        -.010737    .001715   -.725998   -6.262     .0000
X1           .019696    .001219    .994050   16.161     .0000
X2          5.366890  11.079785    .052423     .484     .6310
(CONSTANT)  -.739085   7.927306               -.093     .9262

END BLOCK NUMBER  1   ALL REQUESTED VARIABLES ENTERED.
```

c   SPSS regression printout

PARAMETER $= 0$. Note that the computed t values shown in the Minitab and SAS columns are identical. Minitab gives the observed significance level in the column headed p, the SAS printout gives the observed significance level for each T-test in the column titled PR $> |T|$, and SPSS provides it in the column headed SIG T. Note that these observed significance levels have been computed by assuming that the tests are two-tailed. The observed significance levels for one-tailed tests would be equal to half these values.

The Minitab printout gives the value SSE = 7,062 under the Analysis of Variance column headed SS, in the row identified as Error. The value of $s^2 = 196$ is shown in the same row under the column headed MS, and the degrees of freedom DF, 36, appear in the same row. The corresponding values are shown at the top of the SAS printout in the row labeled ERROR, in the columns designated as SUM OF SQUARES, MEAN SQUARE, and DF, respectively. These quantities appear with similar headings in the SPSS printout, except that ERROR is termed RESIDUAL.

The value of R^2, as defined in Section 10.4, is given in the Minitab printout as 93.3% (we defined this quantity as a ratio where $0 \le R^2 \le 1$). It is given on the left of the SAS printout as 0.932624, and it is shown in the SPSS printout as .93262. (Ignore the quantities shown in the Minitab printout as R-sq(adj) and in the SPSS printout as ADJUSTED R SQUARE. These quantities are adjusted for the degrees of freedom associated with the total SS and SSE and are not used or discussed in this text.)

The F statistic for testing the utility of the model (Section 10.4)—that is, testing the null hypothesis that all model parameters (except β_0) equal zero—is shown under the title F VALUE as 166.11 at the top right of the SAS printout. In addition, the SAS printout gives the observed significance level of this F-test under PR > F as 0.0001. This F value 166.10616 is also printed at the left side of the SPSS printout, with the observed significance level given. The F statistic for testing the utility of the model is given in the Minitab printout as 166.11, with a P value of 0.000.

You can also compute the value of the F statistic directly from the mean square entries given in the Analysis of Variance table. Thus, an equivalent form of the F statistic for testing

$$H_0 : \beta_1 = \beta_2 = \cdots = \beta_k = 0$$

is

$$F = \frac{\text{mean square for regression}}{\text{mean square for error (or residuals)}}$$
$$= \frac{\text{mean square for regression}}{s^2}$$

These quantities are given in the Minitab printout under the column marked MS. Thus,

$$F = \frac{32{,}583}{196} = 166$$

a value that agrees (to three significant digits) with the values given in the SAS and SPSS printouts. The logic behind this test and other tests of hypotheses concerning sets of the β parameters is presented in Sections 10.4, 10.5, and 10.7.

We will not comment on the merits or shortcomings of the various packages, because you will have to use the package(s) available at your computer center and become familiar with that output. Most of the computer printouts are similar, and it is relatively easy to learn how to read another output after you have become familiar with one.

10.7 Model Building: Testing Portions of a Model

In Section 10.4, we discussed testing all the parameters in a multiple regression model using the coefficient of determination R^2. Then, in Section 10.5, a T-test for individual model parameters was presented. We now develop a test of sets of β parameters representing a *portion* of the model. In doing so, we also present some techniques that are useful in constructing multiple regression models.

An example will best demonstrate both the need for a test for portions of a regression model and the flexibility of multiple regression models. Suppose a construction firm wishes to compare the performance of its three sales engineers, using the mean profit per sales dollar. The sales engineers bid on jobs in two states, so the true mean profit per sales dollar is to be considered a function of two factors: sales engineer and state. The six means are symbolically represented by μ_{ij}, with the i subscript used for sales engineer ($i = 1, 2, 3$) and the j subscript for state ($j = 1, 2$). The mean values are presented in Table 10.3.

TABLE **10.3**
Mean Profit per Sales Dollar for Six Sales-Engineer/State Combinations

		State	
		S_1	S_2
	E_1	μ_{11}	μ_{12}
Sales Engineer	E_2	μ_{21}	μ_{22}
	E_3	μ_{31}	μ_{32}

Since both sales engineer and state are *qualitative* factors, dummy variables will be used to represent them in the multiple regression model (see Example 10.1). Thus, we define

$$x_1 = \begin{cases} 1 & \text{if state} = S_2 \\ 0 & \text{if state} = S_1 \end{cases}$$

to represent the state effect in the model, and

$$x_2 = \begin{cases} 1 & \text{if sales engineer} = E_2 \\ 0 & \text{if sales engineer} = E_1 \text{ or } E_3 \end{cases}$$

$$x_3 = \begin{cases} 1 & \text{if sales engineer} = E_3 \\ 0 & \text{if sales engineer} = E_1 \text{ or } E_2 \end{cases}$$

to represent the sales engineer effect. Note that two dummy variables are defined to represent the three sales engineers. The reason for this will be apparent when we write the model, which, using $Y = $ profit per sales dollar, is

$$E(Y) = \beta_0 + \underbrace{\beta_1 x_1}_{\text{State}} + \underbrace{\beta_2 x_2 + \beta_3 x_3}_{\substack{\text{Sales} \\ \text{Engineer}}} + \underbrace{\beta_4 x_1 x_2 + \beta_5 x_1 x_3}_{\substack{\text{State} \times \text{Sales} \\ \text{Engineer} \\ \text{Interaction}}}$$

Note that there are six parameters in the model: β_0, β_1, β_2, β_3, β_4, and β_5—one corresponding to each mean in Table 10.3. The correspondence is shown in Table 10.4.

TABLE **10.4**

Correspondence Between Six Sales-Engineer/State Means (μ_{ij}) and Model Parameters β

$$\mu_{11} = E(Y|E_1, S_1) = E(Y|x_1 = x_2 = x_3 = 0) = \beta_0$$
$$\mu_{12} = E(Y|E_1, S_2) = E(Y|x_1 = 1, x_2 = x_3 = 0) = \beta_0 + \beta_1$$
$$\mu_{21} = E(Y|E_2, S_1) = E(Y|x_1 = 0, x_2 = 1, x_3 = 0) = \beta_0 + \beta_2$$
$$\mu_{22} = E(Y|E_2, S_2) = E(Y|x_1 = 1, x_2 = 1, x_3 = 0) = \beta_0 + \beta_1 + \beta_2 + \beta_4$$
$$\mu_{31} = E(Y|E_3, S_1) = E(Y|x_1 = x_2 = 0, x_3 = 1) = \beta_0 + \beta_3$$
$$\mu_{32} = E(Y|E_3, S_2) = E(Y|x_1 = 1, x_2 = 0, x_3 = 1) = \beta_0 + \beta_1 + \beta_3 + \beta_5$$

Other definitions of dummy variables could be used that would generate a different correspondence between the means μ_{ij} and the β parameters of the model.

However, the use of one fewer dummy variable than the number of factor levels (that is, one dummy variable for the two states, and two dummy variables for the three sales engineers) will yield an exact correspondence between the *number* of means and the *number* of model parameters. Fewer dummy variables will result in too few model parameters, and more dummy variables will yield more model parameters than necessary to generate the six mean profit values.

Now suppose the construction firm wants to test whether the relative performance of the three sales engineers is the same in both states, as shown in Figure 10.13. In this case, although the level of profit may shift between states, the shift is the same for all three sales engineers. Therefore, the *relative* position of the engineers' mean profits is the same. This phenomenon is referred to as *adaptivity* of the state and sales-engineer effects and results in a simplification of the model. The interaction terms $\beta_4 x_1 x_2$ and $\beta_5 x_1 x_3$ are now unnecessary, since only β_1 is needed to represent the additive effect of state, as shown in Table 10.5. Note that the difference between the state means for each sales engineer is the same, β_1.

If the sales engineers perform differently in the two states, their mean profits will not maintain the same relative positions. An example of this *nonadditive* relationship

FIGURE **10.13**

Relative Performance of Sales Engineers the Same in Both States: Effects Are Additive

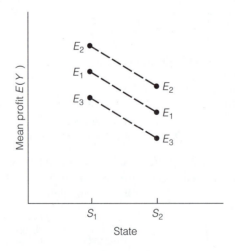

T A B L E 10.5
Representation of Mean Profit in the Additive Model:
$$E(Y) = \beta_0 + \beta_1 x_1 + \beta_2 x_2 + \beta_3 x_3$$

Sales Engineer	State	Mean
E_1	S_1	$\mu_{11} = \beta_0$
E_1	S_2	$\mu_{12} = \beta_0 + \beta_1$
E_2	S_1	$\mu_{21} = \beta_0 + \beta_2$
E_2	S_2	$\mu_{22} = \beta_0 + \beta_1 + \beta_2$
E_3	S_1	$\mu_{31} = \beta_0 + \beta_3$
E_3	S_2	$\mu_{32} = \beta_0 + \beta_1 + \beta_3$

between state and sales engineer is shown in Figure 10.14. The description of this more complex relationship requires all six model parameters:

$$E(Y) = \beta_0 + \beta_1 x_1 + \beta_2 x_2 + \beta_3 x_3 + \beta_4 x_1 x_2 + \beta_5 x_1 x_3$$

Thus, you can see that the difference between an additive and a nonadditive relationship is the interaction terms $\beta_4 x_1 x_2$ and $\beta_5 x_1 x_3$. If β_4 and β_5 are zero, the effects are additive; otherwise, the state/sales-engineer effects are nonadditive.

We will call the interaction model the *complete model*, and the reduced model the *main effects model*. Now suppose we wanted to use some data to test which of these models is more appropriate. This can be done by testing the hypothesis that the β parameters for the interaction terms equal zero:

$$H_0 : \beta_4 = \beta_5 = 0$$
$$H_a : \text{At least one interaction } \beta \text{ parameter differs from zero.}$$

In Section 10.5 we presented the T-test for a single coefficient, and in Section 10.4 we gave the F-test for *all* the β parameters (except β_0) in the model. Now we need a test for *some* of the β parameters in the model. The test procedure is intuitive. First, we use the method of least squares to fit the main effects model and calculate the corresponding sum of squares for error, SSE_1 (the sum of squares of the deviations between observed and predicted y values). Next, we fit the interaction model and

F I G U R E 10.14
Relative Performance of Sales Engineers Not the Same in Both States: Effects Are Nonadditive

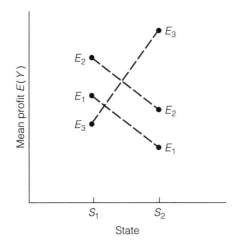

calculate its sum of squares for error, SSE_2. Then we compare SSE_1 to SSE_2 by calculating the difference $SSE_1 - SSE_2$. If the interaction terms contribute to the model, then SSE_2 should be much smaller than SSE_1, and the difference $SSE_1 - SSE_2$ will be large. That is, the larger the difference, the greater the weight of evidence that the variables, sales engineer and state, interact to affect the mean profit per sales dollar in the construction job.

The sum of squares for error will always decrease when new terms are added to the model. The question is whether this decrease is large enough to conclude that it is due to more than just an increase in the number of model terms and to chance. To test the null hypothesis that the interaction terms β_4 and β_5 simultaneously equal zero, we use an F statistic calculated as follows:

$$F = \frac{(SSE_1 - SSE_2)/2}{SSE_2/[n - (5 + 1)]}$$
$$= \frac{\text{drop in SSE/number of } \beta \text{ parameters being tested}}{s^2 \text{ for complete model}}$$

When the assumptions listed in Section 10.3 about the error term ε are satisfied and the β parameters for interaction are all zero (H_0 is true), this F statistic has an F distribution with $\nu_1 = 2$ and $\nu_2 = n - 6$ degrees of freedom. Note that ν_1 is the number of β parameters being tested, and ν_2 is the number of degrees of freedom associated with s^2 in the complete model.

If the interaction terms *do* contribute to the model (H_a is true), we expect the F statistic to be large. Thus, we use a one-tailed test and reject H_0 when F exceeds some critical value F_α, as shown in Figure 10.15.

FIGURE **10.15**
Rejection Region for the F-Test
$H_0 : \beta_4 = \beta_5 = 0$

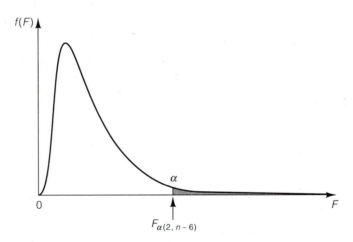

EXAMPLE   **10.3**   The profit per sales dollar Y for the six combinations of sales engineers and states is shown in Table 10.6. Note that the number of construction jobs per combination varies from one for levels (E_1, S_2) to three for levels (E_1, S_1). A total of 12 jobs is sampled.

T A B L E **10.6**
Profit Data for Combinations of
Sales Engineers and States

		State	
		S_1	S_2
Sales Engineer	E_1	$0.065 \\ 0.073 \\ 0.068$	$\$0.036$
	E_2	$0.078 \\ 0.082$	$0.050 \\ 0.043$
	E_3	$0.048 \\ 0.046$	$0.061 \\ 0.062$

a Assume the interaction between E and S is negligible. Fit the model for $E(Y)$ with interaction terms omitted.

b Fit the complete model for $E(Y)$, allowing for the fact that interactions might occur.

c Test the hypothesis that the interaction terms do not contribute to the model.

Solution **a** The computer printout for the main effects model

$$E(Y) = \beta_0 + \overbrace{\beta_1 x_1}^{\substack{S \\ \text{Main} \\ \text{Effect}}} + \overbrace{\beta_2 x_2 + \beta_3 x_3}^{\substack{E \\ \text{Main Effect}}}$$

is given in Figure 10.16. The least-squares prediction equation is

$$\hat{y} = 0.0645 - 0.0158x_1 + 0.0067x_2 - 0.00230x_3$$

F I G U R E **10.16**   Printout for Main Effects Model of Example 10.3

```
SOURCE            DF    SUM OF SQUARES    MEAN SQUARE    F VALUE      PR>F
MODEL              3      0.00085826      0.00028609       1.51     0.2838
ERROR              8      0.00151241      0.00018905
CORRECTED TOTAL   11      0.00237067                     R-SQUARE      STD DEV
                                                        0.362032    0.01374959

                                    T FOR H0:                   STD ERROR OF
PARAMETER      ESTIMATE        PARAMETER = 0    PR>|T|          ESTIMATE
INTERCEPT      0.06445455           8.98        0.0001         0.00718049
X1            -0.01581818          -1.91        0.0928         0.00829131
X2             0.00670455           0.67        0.5190         0.00994093
X3            -0.00229545          -0.23        0.8232         0.00994093
```

b The complete-model printout is given in Figure 10.17. Recall that the complete model is

$$E(Y) = \beta_0 + \beta_1 x_1 + \beta_2 x_2 + \beta_3 x_3 + \beta_4 x_1 x_2 + \beta_5 x_1 x_3$$

The least-squares prediction equation is

$$\hat{y} = 0.0687 - 0.0327 x_1 + 0.0113 x_2 - 0.0217 x_3 - 0.0008 x_1 x_2 + 0.0472 x_1 x_3$$

c Referring to the printouts shown in Figure 10.16 and 10.17, we find the following:

$$\text{Main effects model: SSE}_1 = 0.00151241$$
$$\text{Interaction model: SSE}_2 = 0.00006767$$

The test statistic is

$$F = \frac{(\text{SSE}_1 - \text{SSE}_2)/2}{\text{SSE}_2/(12 - 6)} = \frac{(0.00151241 - 0.00006767)/2}{0.00006767/6}$$

$$= \frac{0.00072237}{0.00001128} = 64.04$$

The critical value of F for $\alpha = 0.05$, $v_1 = 2$, and $v_2 = 6$ is found in Table 7 of the Appendix to be

$$F_{0.05}(2, 6) = 5.14$$

Since the calculated $F = 64.04$ greatly exceeds 5.14, we are quite confident in concluding that the interaction terms contribute to the prediction of Y, profit per sales dollar. They should be retained in the model. ∎

F I G U R E **10.17** Printout for Complete Model (includes interaction) of Example 10.3

SOURCE	DF	SUM OF SQUARES	MEAN SQUARE	F VALUE	PR>F
MODEL	5	0.00230300	0.00046060	40.84	0.0001
ERROR	6	0.00006767	0.00001128		
CORRECTED TOTAL	11	0.00237067		R-SQUARE	STD DEV
				0.971457	0.00335824

PARAMETER	ESTIMATE	T FOR H0: PARAMETER = 0	PR>|T|	STD ERROR OF ESTIMATE
INTERCEPT	0.06866667	35.42	0.0001	0.00193888
X1	-0.03266667	-8.42	0.0002	0.00387776
X2	0.01133333	3.70	0.0101	0.00306564
X3	-0.02166667	-7.07	0.0004	0.00306564
X1*X2	-0.00083333	-0.16	0.8763	0.00512980
X1*X3	0.04716667	9.19	0.0001	0.00512980

The F-test can be used to determine whether *any* set of terms should be included in a model by testing the null hypothesis that the members of a particular set of β parameters simultaneously equal zero. For example, we might want to test to determine whether a set of quadratic terms for quantitative variables or a set of main effect terms for a qualitative variable should be included in a model. The F-test appropriate for testing the null hypothesis that all members of a set of β parameters are equal to zero is summarized below.

***F* - Test for Testing the Null Hypothesis: Set of β Parameters Equal Zero**

Reduced model: $E(Y) = \beta_0 + \beta_1 x_1 + \cdots + \beta_g x_g$

Complete model: $E(Y) = \beta_0 + \beta_1 x_1 + \cdots + \beta_g x_g + \beta_{g+1} x_{g+1} + \cdots + \beta_k x_k$

$H_0 : \beta_{g+1} = \beta_{g+2} = \cdots = \beta_k = 0$

H_a : At least one of the β parameters under test is nonzero.

Test statistic: $F = \dfrac{(SSE_1 - SSE_2)/(k - g)}{SSE_2/[n - (k + 1)]}$

where

$\quad SSE_1$ = sum of squared errors for the reduced model

$\quad SSE_2$ = sum of squared errors for the complete model

$\quad k - g$ = number of β parameters specified in H_0

$\quad k + 1$ = number of β parameters in the complete model (including β_0)

$\quad\quad n$ = total sample size

$\quad$ Rejection region: $F > F_\alpha(v_1, v_2)$

where

$$v_1 = k - g = \text{degrees of freedom for the numerator}$$
$$v_2 = n - (k + 1) = \text{degrees of freedom for the denominator}$$

Exercises

10.13   Suppose you fit the regression model

$$Y = \beta_0 + \beta_1 x_1 + \beta_2 x_2 + \beta_3 x_3 + \beta_4 x_4 + \varepsilon$$

to $n = 25$ data points and you want to test the null hypothesis $\beta_1 = \beta_2 = 0$.

a   Explain how you would find the quantities necessary for the F statistic.

b   How many degrees of freedom would be associated with F?

10.14 An insurance company is experimenting with three different training programs, A, B, and C, for its salespeople. The following main effects model is proposed:

$$E(Y) = \beta_0 + \beta_1 x_1 + \beta_2 x_2 + \beta_3 x_3$$

where

$$Y = \text{monthly sales (in thousands of dollars)}$$
$$x_1 = \text{number of months of experience}$$
$$x_2 = \begin{cases} 1 & \text{if training program B was used} \\ 0 & \text{otherwise} \end{cases}$$
$$x_3 = \begin{cases} 1 & \text{if training program C was used} \\ 0 & \text{otherwise} \end{cases}$$

Training program A is the base level.

a What hypothesis would you test to determine whether the mean monthly sales differ for salespeople trained by the three programs?

b After experimenting with 50 salespeople over a five-year period, the complete model is fit, with the result

$$\hat{y} = 10 + 0.5x_1 + 1.2x_2 - 0.4x_3 \qquad SSE = 140.5$$

Then the reduced model $E(Y) = \beta_0 + \beta_1 x_1$ is fit to the same data, with the result

$$\hat{y} = 11.4 + 0.4x_1 \qquad SSE = 183.2$$

Test the hypothesis you formulated in part (a). Use $\alpha = 0.05$.

10.15 A large research and development company rates the performance of each of the members of its technical staff once a year. Each person is rated on a scale of zero to 100 by his or her immediate supervisor, and this merit rating is used to help determine the size of the person's pay raise for the coming year. The company's personnel department is interested in developing a regression model to help them forecast the merit rating that an applicant for a technical position will receive after he or she has been with the company three years. The company proposes to use the following model to forecast the merit ratings of applicants who have just completed their graduate studies and have no prior related job experience:

$$E(Y) = \beta_0 + \beta_1 x_1 + \beta_2 x_2 + \beta_3 x_1 x_2 + \beta_4 x_1^2 + \beta_5 x_2^2$$

where

$$Y = \text{applicant's merit rating after three years}$$
$$x_1 = \text{applicant's grade-point average (GPA) in graduate school}$$
$$x_2 = \text{applicant's verbal score on the Graduate Record Examination (percentile)}$$

A random sample of $n = 40$ employees who have been with the company more than three years was selected. Each employee's merit rating after three years, his or her graduate-school GPA, and the percentile in which the verbal Graduate Record Exam score fell were recorded. The above model was fit to these data. Below is a portion of the resulting computer printout:

SOURCE	DF	SUM OF SQUARES	MEAN SQUARE
MODEL	5	4911.56	982.31
ERROR	34	1830.44	53.84
TOTAL	39	6742.00	R-SQUARE
			0.729

The reduced model $E(Y) = \beta_0 + \beta_1 x_1 + \beta_2 x_2$ was fit to the same data, and the resulting computer printout is partially reproduced below:

SOURCE	DF	SUM OF SQUARES	MEAN SQUARE
MODEL	2	3544.84	1772.42
ERROR	37	3197.16	86.41
TOTAL	39	6742.00	R-SQUARE
			0.526

a Identify the null and alternative hypotheses for a test to determine whether the complete model contributes information for the prediction of Y.

b Identify the null and alternative hypotheses for a test to determine whether a second-order model contributes more information than a first-order model for the prediction of Y.

c Conduct the hypothesis test you described in part (a). Test using $\alpha = 0.05$. Draw the appropriate conclusions in the context of the problem.

d Conduct the hypothesis test you described in part (b). Test using $\alpha = 0.05$. Draw the appropriate conclusions in the context of the problem.

10.16 In an attempt to reduce the number of work hours lost due to accidents, a company tested each of three safety programs, A, B, and C, at three of the company's nine factories. The proposed complete model is

$$E(Y) = \beta_0 + \beta_1 x_1 + \beta_2 x_2 + \beta_3 x_3$$

where

$$Y = \text{total work hours lost due to accidents for a one-year period}$$
$$\text{beginning six months after the plan is instituted}$$
$$x_1 = \text{total work hours lost due to accidents during the year}$$
$$\text{before the plan was instituted}$$
$$x_2 = \begin{cases} 1 & \text{if program B is in effect} \\ 0 & \text{otherwise} \end{cases}$$
$$x_3 = \begin{cases} 1 & \text{if program C is in effect} \\ 0 & \text{otherwise} \end{cases}$$

After the programs have been in effect for 18 months, the complete model is fit to the $n = 9$ data points, with the result

$$\hat{y} = -2.1 + 0.88 x_1 - 150 x_2 + 35 x_3 \qquad \text{SSE} = 1,527.27$$

Then the reduced model $E(Y) = \beta_0 + \beta_1 x_1$ is fit, with the result

$$\hat{y} = 15.3 + 0.84 x_1 \qquad \text{SSE} = 3,113.14$$

Test to see whether the mean work hours lost differ for the three programs. Use $\alpha = 0.05$.

10.17 The following model was proposed for testing salary discrimination against women in a state university system:

$$E(Y) = \beta_0 + \beta_1 x_1 + \beta_2 x_2 + \beta_3 x_1 x_2 + \beta_4 x_2^2$$

where

$$Y = \text{annual salary (in thousands of dollars)}$$
$$x_1 = \begin{cases} 1 & \text{if female} \\ 0 & \text{if male} \end{cases}$$
$$x_2 = \text{experience (years)}$$

Below is a portion of the computer printout that results from fitting this model to a sample of 200 faculty members in the university system:

```
SOURCE    DF    SUM OF SQUARES    MEAN SQUARE
MODEL      4        2351.70          587.92
ERROR    195         783.90            4.02
TOTAL    199        3135.60        R-SQUARE
                                     0.7500
```

Do these data provide sufficient evidence to support the claim that the mean salary of faculty members is dependent on sex? Use $\alpha = 0.05$.

10.8 Using the Model for Estimation and Prediction

In Section 9.8, we discussed the use of the least-squares line for estimating the mean value of Y, $E(Y)$, for some value of x, say $x = x_p$. We also showed how to use the same fitted model to predict, when $x = x_p$, some value of Y to be observed in the future. Recall that the least-squares line yielded the same value for both the estimate of $E(Y)$ and the prediction of some future value of Y. That is, both were the result of substituting x_p into the prediction equation $\hat{y}_p = \hat{\beta}_0 + \hat{\beta}_1 x_p$ and calculating $\hat{y}_p$. There the equivalence ended. The confidence interval for the mean $E(Y)$ was narrower than the prediction interval for Y, because of the additional uncertainty attributable to the random error ε when predicting some future value of Y.

These same concepts carry over to the multiple regression model. For example, suppose we want to estimate the mean peak power load for a given daily high tempature $x_p = 90°$ F. Assuming the quadratic model represents the true relationship between peak power load and maximum temperature, we want to estimate

$$E(Y) = \beta_0 + \beta_1 x_p + \beta_2 x_p^2$$
$$= \beta_0 + \beta_1(90) + \beta_2(90)^2$$

Substituting into the least-squares prediction equation, the estimate of $E(Y)$ is

$$\hat{y} = \hat{\beta}_0 + \hat{\beta}_1(90) + \hat{\beta}_2(90)^2$$
$$= 1{,}784.188 - 42.3862(90) + 0.27216(90)^2$$
$$= 173.93$$

To form a confidence interval for the mean, we need to know the standard deviation of the sampling distribution for the estimator $\hat{y}$. For multiple regression models, the form of this standard deviation is rather complex. However, most regression computer programs allow us to obtain the confidence intervals for mean values of Y for any given combination of values of the independent variables. This portion of the computer output for the peak power load examples is shown in Figure 10.18. The mean value and corresponding 95% confidence interval for $x = 90$ are shown in the columns labeled ESTIMATED MEAN VALUE, LOWER 95% CL FOR MEAN, and UPPER 95% CL FOR MEAN. Note that

$$\hat{y} = 173.93$$

F I G U R E **10.18**
Printout for Estimated Value and
Corresponding Confidence
Interval for $x = 90$

```
            ESTIMATED        LOWER 95% CL     UPPER 95% CL
    x       MEAN VALUE        FOR MEAN         FOR MEAN
    90   173.92791187      159.14624576    188.70957799
```

which agrees with our earlier calculation. The 95% confidence interval for the true mean of Y is shown to be 159.15 to 188.71 (see Figure 10.19).

If we were interested in predicting the electrical usage for a particular day on which the high temperature is $90°$ F, $\hat{y} = 173.93$ would be used as the predicted value. However, the prediction interval for a particular value of Y will be wider than the confidence interval for the mean value. This is reflected by the printout shown in Figure 10.20, which gives the predicted value of Y and corresponding 95% prediction interval when $x = 90$. The 95% prediction interval for $x = 90$ is 139.91 to 207.94 (see Figure 10.21).

Unfortunately, not all computer packages have the capability to produce confidence intervals for means and prediction intervals for particular Y values. This is a rather serious oversight, since the estimation of mean values and the prediction of particular values represent the culmination of our model-building efforts—using the model to make inferences about the dependent variable Y.

F I G U R E **10.19**
Confidence Interval for Mean
Peak Power Load

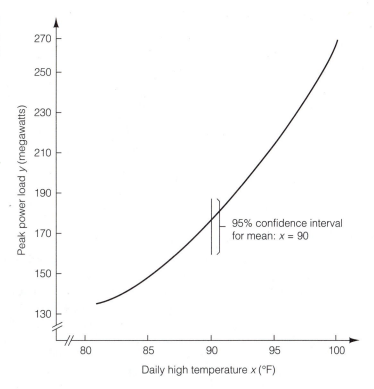

FIGURE **10.20**
Printout for Predicted Value and
Corresponding Prediction Interval
for $x = 90$

x	PREDICTED VALUE	LOWER 95% CL INDIVIDUAL	UPPER 95% CL INDIVIDUAL
90	173.92791187	139.91131797	207.94450577

FIGURE **10.21**
Prediction Interval for Electrical
Usage

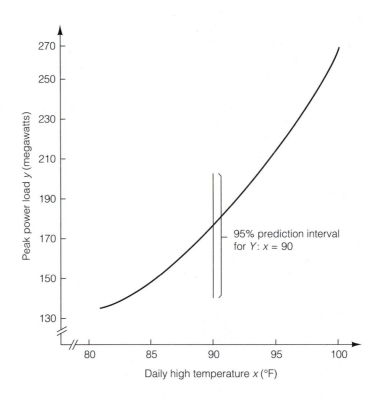

10.9   Multiple Regression: Two Examples

10.9.1   A Response Surface

Many companies manufacture products that are at least partially chemically produced (e.g., steel, paint, gasoline). In many instances, the quality of the finished product is a function of the temperature and pressure at which the chemical reactions take place.

Suppose a manufacturer wanted to model the quality Y of a product as a function of the temperature x_1 and the pressure x_2 at which it is produced. Four inspectors independently assign a quality score between zero and 100 to each product, and then the quality y is calculated by averaging the four scores. An experiment is conducted by varying temperature between 80 and 100°F and varying pressure between 50 and 60 pounds per square inch. The resulting data ($n = 27$) are given in Table 10.7.

Step 1   The first step is to hypothesize a model relating product quality to the temperature and pressure at which the product was manufactured. A model that will allow us

| | x_2 (pounds | | | x_2 (pounds | | | x_2 (pounds | |
x_1 (°F)	per square inch)	y	x_1 (°F)	per square inch)	y	x_1 (°F)	per square inch)	y
80	50	50.8	90	50	63.4	100	50	46.6
80	50	50.7	90	50	61.6	100	50	49.1
80	50	49.4	90	50	63.4	100	50	46.4
80	55	93.7	90	55	93.8	100	55	69.8
80	55	90.9	90	55	92.1	100	55	72.5
80	55	90.9	90	55	97.4	100	55	73.2
80	60	74.5	90	60	70.9	100	60	38.7
80	60	73.0	90	60	68.8	100	60	42.5
80	60	71.2	90	60	71.3	100	60	41.4

T A B L E **10.7**
Temperature, Pressure, and Quality of the Finished Product

to find the setting of temperature and pressure that maximizes quality is the equation for a *paraboloid*. The visualization of the paraboloid appropriate for this application is an inverted bowl-shaped surface, and the corresponding mathematical model for the mean quality at any temperature/pressure setting is

$$E(Y) = \beta_0 + \beta_1 x_1 + \beta_2 x_2 + \beta_3 x_1^2 + \beta_4 x_2^2 + \beta_5 x_1 x_2$$

This model is also referred to as a *complete second-order model* because it contains all first- and second-order terms in x_1 and x_2.[†] Note that a model with only first-order terms (a plane) would have no curvature, so the mean quality could not reach a maximum within the experimental region of temperature/pressure even if the data indicated the probable existence of such a value. The inclusion of second-order terms allows curvature in the three-dimensional response surface traced by mean quality, so a maximum mean quality can be reached if the experimental data indicate that one exists.

Step 2   Next we use the least-squares technique to estimate the model coefficients $\beta_0, \beta_1, \ldots, \beta_5$ of the paraboloid, using the data in Table 10.7. The least-squares model (see Figure 10.22) is

$$\hat{y} = -5,127.90 + 31.10x_1 + 139.75x_2 - 0.133x_1^2 - 1.14x_2^2 - 0.146x_1 x_2$$

A three-dimensional graph of this model is shown in Figure 10.23.

Step 3   The next step is to specify the probability distribution of ε, the random error component. We assume that ε is normally distributed, with a mean of zero and a constant variance σ^2. Furthermore, we assume that the errors are independent. The estimate of the variance σ^2 is given in the printout (Figure 10.22) as

$$s^2 = \text{MSE} = \frac{\text{SSE}}{n - (k+1)} = \frac{59.1784}{27 - (5+1)} = 2.818$$

[†]Recall from analytic geometry that the order of a term is the sum of the exponents of the variables in the term. Thus, $\beta_5 x_1 x_2$ is a second-order term, as is $\beta_3 x_1^2$. The term $\beta_i x_1^2 x_2$ is a third-order term.

F I G U R E **10.22**   Printout for the Product Quality Example

SOURCE	DF	SUM OF SQUARES	MEAN SQUARE	F VALUE	PR>F
MODEL	5	8402.26453714	1680.45290743	596.32	0.0001
ERROR	21	59.17842582	2.81802028		
CORRECTED TOTAL	26	8461.44296296		R-SQUARE	STD DEV
				0.993006	1.67869601

PARAMETER	ESTIMATE	T FOR H0: PARAMETER = 0	PR>T	STD ERROR OF ESTIMATE
INTERCEPT	-5127.89907417	-46.49	0.0001	110.29601483
X1	31.09638889	23.13	0.0001	1.34441322
X2	139.74722222	44.50	0.0001	3.14005411
X1*X1	-0.13338889	-19.46	0.0001	0.00685325
X2*X2	-1.14422222	-41.74	0.0001	0.02741299
X1*X2	-0.14550000	-15.01	0.0001	0.00969196

F I G U R E **10.23**
Plot of Second-Order
Least-Squares Model for the
Product Quality Example

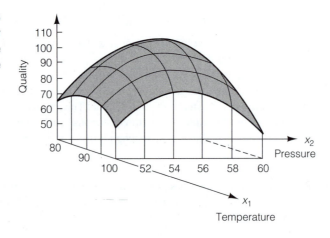

Step 4   Next we want to evaluate the adequacy of the model. First, note that $R^2 = 0.993$. This implies that 99.3% of the variation in y, observed quality ratings for the 27 experiments, is accounted for by the model. The statistical significance of this can be tested:

$$H_0 : \beta_1 = \beta_2 = \beta_3 = \beta_4 = \beta_5 = 0$$

$H_a :$ At least one model coefficient is nonzero.

Test statistic: $F = \dfrac{R^2/k}{(1 - R^2)/[n - (k + 1)]}$

Rejection region: For $\alpha = 0.05$,

$$F > F_\alpha[k, n - (k + 1)] = F_{0.05}(5, 21)$$

that is,

$$F > 2.57$$

The test statistic is given in the printout of Figure 10.22. Since $F = 596.32$ greatly exceeds the tabulated value, we conclude that the model does contribute information about product quality.

Again, we cannot be satisfied with the R^2 and F-test results until we examine residuals for unusual patterns. Figure 10.24 shows the residuals from the second-order model plotted against the predicted values. The pattern appears to be fairly random, but the predicted values do seem to cluster into three groups. Can you think of a reason for this clustering?

Step 5   The culmination of the modeling effort is to use the model for estimation and/or prediction. In this example, suppose the manufacturer is interested in estimating the mean product quality for the setting of temperature and pressure at which the estimated model reaches a maximum. To find this setting, we solve the equations

$$\frac{\partial \hat{y}}{\partial x_1} = \hat{\beta}_1 + 2\hat{\beta}_3 x_1 + \hat{\beta}_5 x_2 = 0$$

$$\frac{\partial \hat{y}}{\partial x_2} = \hat{\beta}_2 + 2\hat{\beta}_4 x_2 + \hat{\beta}_5 x_1 = 0$$

for x_1 and x_2, obtaining $x_1 = 86.25°$ and $x_2 = 55.58$ pounds per square inch. The fact that $\partial^2 \hat{y} / \partial x_1^2 = 2\hat{\beta}_3 < 0$ and $\partial^2 \hat{y} / \partial x_2^2 = 2\hat{\beta}_4 < 0$ ensures that this setting of x_1 and x_2 corresponds to a maximum value of $\hat{y}$.

F I G U R E **10.24**   Residual Plot for the Product Quality Example

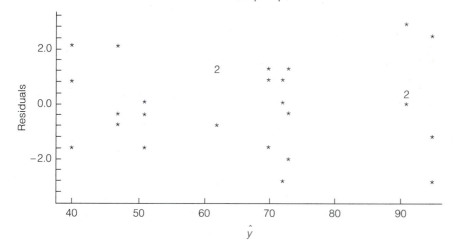

To obtain an estimated mean quality for this temperature/pressure combination, we use the least-squares model:

$$\hat{y} = \hat{\beta}_0 + \hat{\beta}_1(86.25) + \hat{\beta}_2(55.58) + \hat{\beta}_3(86.25)^2 + \hat{\beta}_4(55.58)^2$$
$$+ \hat{\beta}_5(86.25)(55.58)$$

The estimated mean value is given in Figure 10.25, which shows a partial reproduction of the regression printout for this example. The estimated mean quality is 96.9, and a 95% confidence interval is 95.5 to 98.3. Thus, we are confident that the mean quality rating will be between 95.5 and 98.3 when the product is manufactured at 86.25°F and 55.58 pounds per square inch.

F I G U R E **10.25**
Partial Printout for Product
Quality Example

```
                  PREDICTED     LOWER 95% CL   UPPER 95% CL
X1       X2        VALUE         FOR MEAN       FOR MEAN
86.25   55.58   96.87401860   95.46463236    98.28340483
```

10.9.2 Modeling a Time Trend

Global warming is a critical environmental issue. This phenomenon is caused by a buildup of complex gases in the atmosphere, one of which is carbon monoxide (CO_2). The data in Table 10.8 show annual average carbon monoxide concentrations (in ppm) for the Mauna Loa Observatory in Hawaii. Do these data provide evidence of increasing CO_2 amounts? What would be a reasonable model to explain the pattern in CO_2 concentrations over the years?

T A B L E **10.8**
Annual Average CO_2
Concentrations

Year (x)	CO_2 (ppm) (y)	Year (x)	CO_2 (ppm) (y)
1959	316.1	1974	330.4
1960	317.0	1975	331.0
1961	317.7	1976	332.1
1962	318.6	1977	333.6
1963	319.1	1978	335.2
1964	*	1979	336.5
1965	320.4	1980	338.4
1966	321.1	1981	339.5
1967	322.0	1982	340.8
1968	322.8	1983	342.8
1969	324.2	1984	344.3
1970	325.5	1985	345.7
1971	326.5	1986	346.9
1972	327.6	1987	348.6
1973	329.8	1988	351.2

Investigation of the plot of CO_2 concentration against years (Figure 10.26) might suggest a linear trend, and the least-squares fit of a simple linear model produces the following:

```
The regression equation is Y = 243 + 1.20 X1
29 cases used 1 cases contain missing values
```

Predictor	Coef	Stdev	t-ratio	p
Constant	242.887	2.295	105.81	0.000
X1	1.19649	0.03088	38.74	0.000

```
s = 1.433        R-sq = 98.2%
```

This model produces a high value of R^2 and a significant slope, but does the model adequately describe the trends involved? A plot of the residuals (Figure 10.27) suggests that it does not. There is a definite pattern of decreasing residuals followed by increasing residuals.

Perhaps there is curvature here that can be accounted for by fitting an exponential model. Fitting $\ln(y)$ as a function of year produces a slightly better fit:

```
The regression equation is
LN(Y) = 5.54 + 0.00361 X1

29 cases used 1 cases contain missing values
```

Predictor	Coef	Stdev	t-ratio	p
Constant	5.5361	0.00625	885.26	0.000
X1	0.00360512	0.00008414	42.85	0.000

```
s = 0.003905    R-sq = 98.6%
```

F I G U R E **10.26** Plot of CO_2 Concentrations over Time

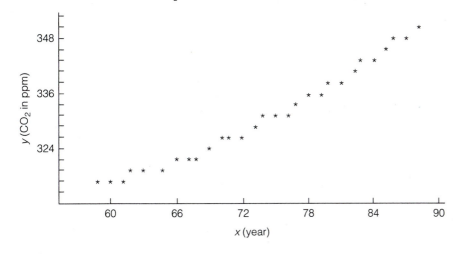

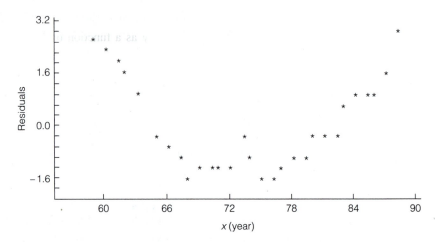

FIGURE **10.27**
Residual Plot for the Simple
Linear Model

However, the residual plot (Figure 10.28) shows no improvement. There is something more complicated going on here!

A more careful look at the residuals in Figure 10.27 might suggest that two different regression lines should be fit to the data, one up to about 1968 and another from 1969 to 1988. This can be accomplished in one operation by defining an indicator variable, x_2, to keep track of the year. One way to do this is to define

$$x_2 = \begin{cases} 0 & \text{if year} \leq 1968 \\ 1 & \text{if year} \geq 1969 \end{cases}$$

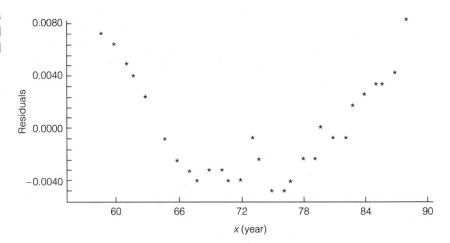

FIGURE **10.28**
Residual Plot for the Exponential
Model

Since slopes will be different for the two lines, an interaction term must be included in the model. Modeling y as a function of x_1, x_2 and $x_1 x_2$ produces the following:

```
The regression equation is
Y = 274 + 0.718 X1 - 45.9 X2 + 0.665 X1X2

29 cases used 1 cases contain missing values

Predictor         Coef           Stdev           t-ratio          p
Constant       273.888           4.059             67.48      0.000
X1             0.71770          0.06390             11.23      0.000
X2             -45.924           4.427            -10.37      0.000
X1X2           0.66531          0.06773              9.82      0.000

s = 0.5794    R-sq = 99.7%    R-sq(adj) = 99.7%

Analysis of Variance

SOURCE            DF            SS            MS             F          P
Regression         3        3130.9        1043.6       3108.61      0.000
Error             25           8.4           0.3
Total             28        3139.3
```

The fit is improved a little (R^2 has increased), and all terms appear to be significant. Moreover, the residual plot (Figure 10.29) shows much smaller residuals, without the definite two-line pattern of Figure 10.27. The stemplot of the residuals in Figure 10.29 suggests that normality may be a reasonable assumption.

Close investigation of the residuals in Figure 10.29 reveals that the two-line model, although better than the other models tried, can still be improved upon. Once we've accounted for major trends, more subtle trends appear. There seems to be a periodic effect in CO_2 concentrations and an increase in later years (after 1978) that is even greater than that shown by the two-line model. The last year (1988) now stands out as producing a large positive residual, perhaps a warning signal for things to come. In summary, there is a significant increase in the slope of CO_2 concentrations over the years that takes effect around 1968, with later years showing an even sharper rate of increase.

As you see from this example, fitting models to data can be a complex task involving a variety of models. Often, there is no single linear model that adequately explains all patterns in the data. Nevertheless, the linear models are very useful for explaining the key features of the data.

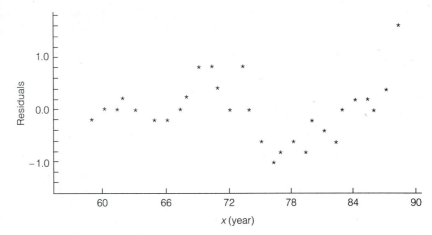

F I G U R E **10.29**
Residual Plot and Stemplot for
the Two-Line Model

Stem-and-leaf of residuals $N = 2$
Leaf unit $= 0.10$ $N^* = 1$

```
   2   -0 | 98
   5   -0 | 766
   7   -0 | 54
   8   -0 | 2
  13   -0 | 11100
  (9)   0 | 000000111
   7    0 | 233
   4    0 |
   4    0 | 7
   3    0 | 88
   1    1 |
   1    1 |
   1    1 | 5
```

10.10   Some Comments on the Assumptions

When we apply a regression analysis to a set of data, we never know for certain that the assumptions of Section 10.3 are satisfied. How far can we deviate from the assumptions and still expect a multiple regression analysis to yield results that will possess the reliability stated in this chapter? How can we detect departures (if they exist) from the assumptions of Section 10.3, and what can we do about them? We provide some partial answers to these questions in this section and direct you to further discussion in succeeding chapters.

Remember (from Sections 10.1 and 10.3) that we model Y as

$$Y = \beta_0 + \beta_1 x_1 + \beta_2 x_2 + \cdots + \beta_k x_k + \varepsilon$$

for a given set of values of $x_1, x_2, \ldots, x_k$, where ε is a random error. The first assumption we made was that the mean value of the random error for any given set of values of $x_1, x_2, \ldots, x_k$ is $E(\varepsilon) = 0$. One consequence of this assumption is that the mean $E(Y)$ for a specific set of values of $x_1, x_2, \ldots, x_k$ is

$$E(Y) = \beta_0 + \beta_1 x_1 + \beta_2 x_2 + \cdots + \beta_k x_k$$

That is,

$$Y = \underbrace{E(Y)}_{\substack{\text{Mean value of } y \\ \text{for specific values} \\ \text{of } x_1, x_2, \cdots, x_k}} + \underbrace{\varepsilon}_{\substack{\text{Random} \\ \text{error}}}$$

The second consequence of the assumption is that the least-squares estimators of the model parameters $\beta_0, \beta_1, \beta_2, \ldots, \beta_k$ will be unbiased regardless of the remaining assumptions made about the random errors and their probability distributions.

The properties of the sampling distributions of the parameter estimators $\hat{\beta}_0, \hat{\beta}_1, \ldots, \hat{\beta}_k$ will depend on the remaining assumptions that we specify concerning the probability distributions of the random errors. You will recall that we assumed that, for any given set of values of $x_1, x_2, \ldots, x_k$, ε has a normal probability distribution with mean equal to zero and variance equal to σ^2. Also, we assumed that the random errors are independent (in a probabilistic sense).

It is unlikely that the assumptions stated above are satisfied exactly for many practical situations. If departures from the assumptions are not too great, experience has shown that a least-squares regression analysis produces estimates—predictions and statistical test results—that possess, for all practical purposes, the properties specified in this chapter. If the observations are likely to be correlated (as in the case of data collected over time), we must check for correlation between the random errors and may have to modify our methodology if correlation is found to exist. A test for correlation of the random errors can be found in W. Mendenhall and J. McClave (*A Second Course in Business Statistics: Regression Analysis*, Dellen, 1981). If the variance of the random error ε changes from one setting of the independent x variables to another, we can sometimes transform the data so that the standard least-squares methodology will be appropriate. Techniques for detecting nonhomogeneous variances of the random errors (a condition called *heteroscedasticity*) and some methods for treating this type of data are also discussed in Mendenhall and McClave, as well as many other texts that focus on regression analysis.

Frequently, the data $(y, x_1, x_2, \ldots, x_k)$ are observational; that is, we simply observe an experimental unit and record values for $y, x_1, x_2, \ldots, x_k$. Do these data violate the assumption that the values $x_1, x_2, \ldots, x_k$ are fixed? For this particular case, if we can assume that $x_1, x_2, \ldots, x_k$ are *measured without error*, the mean value $E(Y)$ can be viewed as a conditional mean. That is, $E(Y)$ gives the mean value of Y, *given* that the x variables assume a specific set of values. With this modification in our thinking, the least-squares regression analysis is applicable to observational data.

Remember that when you perform a regression analysis, the reliability you can place in your inferences is dependent on the assumptions prescribed in Section 10.3 being satisfied. Although the random errors will rarely satisfy these assumptions exactly, the reliability specified by a regression analysis will hold approximately for many types of data encountered in practice.

10.11 Some Pitfalls: Estimability, Multicollinearity, and Extrapolation

There are several problems you should be aware of when constructing a prediction model for some response Y. We discuss a few of the most important in this section.

Problem 1: Parameter Estimability Suppose you want to fit a model relating a firm's monthly profit Y to the advertising expenditure x. We propose the first-order model

$$E(Y) = \beta_0 + \beta_1 x$$

Now suppose we have three months of data, and the firm spent $1,000 on advertising during each month. The data are shown in Figure 10.30. You can see the problem: The parameters of the line cannot be estimated when all the data are concentrated at a single x value. Recall that it takes two points (x values) to fit a straight line. Thus, the parameters are not estimable when only one x value is observed. A similar problem would occur if we attempted to fit the second-order model

$$E(Y) = \beta_0 + \beta_1 x + \beta_2 x^2$$

to a set of data for which only one *or two* different x values were observed (see Figure 10.31). At least three different x values must be observed before a second-order model can be fit to a set of data (that is, before all three parameters are estimable). In general, the number of levels of x must be at least one more than the order of the polynomial in x that you want to fit.

The independent variables will almost always be observed at a sufficient number of levels to permit estimation of the model parameters. However, when the computer program you are using suddenly refuses to fit a model, the problem is probably inestimable parameters.

FIGURE **10.30**
Profit and Advertising
Expenditure Data: Three Months

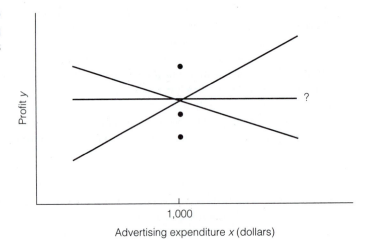

Advertising expenditure x (dollars)

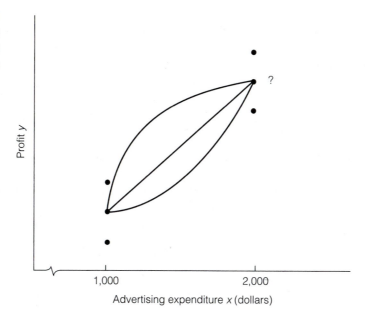

F I G U R E **10.31**
Only Two x Values
Observed—the Second-Order
Model Is Not Estimable

Profit y

1,000   2,000

Advertising expenditure x (dollars)

Problem 2: Multicollinearity   Often two or more of the independent variables used in the model for $E(Y)$ will contribute redundant information because they are correlated with one another. For example, suppose we want to construct a model to predict the gasoline mileage rating of a truck as a function of its load x_1 and the horsepower x_2 of its engine. In general, you would expect heavy loads to require greater horsepower and to result in lower mileage ratings. Thus, although both x_1 and x_2 contribute information for the prediction of mileage rating, some of the information is overlapping because x_1 and x_2 are correlated.

If the model

$$E(Y) = \beta_0 + \beta_1 x_1 + \beta_2 x_2$$

were fit to a set of data, we might find the t values for both $\hat{\beta}_1$ and $\hat{\beta}_2$ (the least-squares estimates) to be nonsignificant. However, the F-test for $H_0 : \beta_1 = \beta_2 = 0$ would probably be highly significant. The tests might seem to be contradictory, but really they are not. The T-tests indicate that the contribution of one variable, say x_1 = load, is not significant after the effect of x_2 = horsepower has been discounted (because x_2 is also in the model). The significant F-test, on the other hand, tells us that at least one of the two variables is making a contribution to the prediction of y (i.e., β_1, β_2, or both differ from zero). In fact, both are probably contributing, but the contribution of one overlaps that of the other. When highly correlated independent variables are present in a regression model, the results can be confusing. The researcher might want to include only one of the variables in the final model.

Problem 3: Prediction Outside the Experimental Region   By the late 1960s, many research economists had developed highly technical models to relate the state of the economy to various economic indices and other independent variables. Many of these models were multiple regression models where, for example, the dependent variable Y might be next year's growth in GNP and the independent variables might include this year's rate of inflation, this year's Consumer Price Index, and other factors. In other words, the model might be constructed to predict the state of next year's economy using this year's knowledge.

Unfortunately, these models were almost unanimously unsuccessful in predicting the recession that occurred in the early 1970s. What went wrong? One of the problems was that the regression models were used to predict Y for values of the independent variables that were outside the region in which the model had been developed. For example, the inflation rate in the late 1960s, when the models were developed, ranged from 6% to 8%. When the double-digit inflation of the early 1970s became a reality, some researchers attempted to use the same models to predict future growth in GNP. As you can see in Figure 10.32, the model can be very accurate for predicting Y when x is in the range of experimentation, but using the model outside that range is a dangerous practice.

F I G U R E **10.32**
Using a Regression Model
Outside the Experimental Region

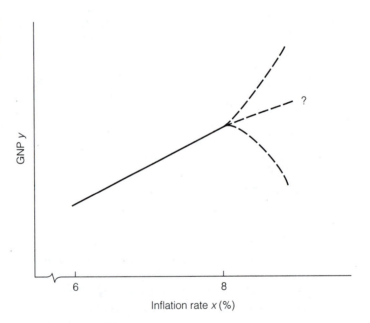

Inflation rate x (%)

10.12 Activities for Students: Fitting Models

How healthy am I? Within our health-conscious society, many people are asking themselves this question. One important measure of health is percent body fat, since too much fat is related to hypertension, diabetes mellitus, and heart disease, as well as to a general lack of mobility. How much is too much? For general health, some experts agree that 10% to 22% body fat is acceptable for men and 20% to 32% is acceptable for women. For athletes, the values should be approximately half those.

A scientific measure of percent body fat begins with a measure of body density. Over the years, various height-weight ratios have been proposed for estimating body density, but all are fairly inaccurate. The most accurate method of measuring body density is hydrostatic (underwater) weighing, but this method is costly in terms of equipment and technicians. A good compromise between height-weight ratios and hydrostatic weighing is a method that predicts body density from a series of skin-fold thickness measurements. Regression techniques are used to relate the skinfold measurements to body densities found by the hydrostatic method. Then the statistical models can be used to predict the body density for a new patient simply as a function of the skinfold measurements.

However, many skinfold measurements can be taken over the human body. Which ones are the best predictors of body density? Scientific studies along these lines have led to the conclusion that skinfold measurements in seven locations are adequate for the task. These seven locations are the chest, axilla (side, under the arm), triceps (front of upper arm), subscapula (below shoulder blade on the back), abdomen, thigh (front), and suprailium (above the pelvis).

Combinations of these seven skinfold measurements, along with age, then form the prediction equations for body density. Since men and women have quite different body compositions, prediction equations must be formulated separately for the two groups. Table 10.9 provides skinfold measurements (in mm) for a sample of 15 women and 14 men. Here, body density was determined by hydrostatic weighing. These data can be used to check the models that have been proposed for predicting body density and to explore other models that might work better. Work through the following steps to see how statistical models help solve this important practical problem related to health.

Sum of Seven   A convenient way to summarize the seven skinfold measurements per person is to simply add them together. Taking this sum as one explanatory variable and age as the other, construct good-fitting models for predicting body density of males and of females. You might begin by looking at scatterplots. The models published in the literature usually have a term that requires squaring the sum of seven. Does this appear to be justified?

Sum of Three   It turns out that the sum of a selected group of three skinfolds often does about as well as the sum of seven in building prediction models for body density. Can you find a sum of three that works well for women? Are the three that work well for men the same as the three for women? The article "Practical Assessment of Body Composition," by Andrew S. Jackson and Michael L. Pollock (*The Physician and Sports Medicine*, vol. 13, no. 5, 1985) suggests that the three skinfolds to use for women are the triceps, thigh, and suprailium, although thigh can be replaced by abdomen with no serious loss in accuracy. For men, the best three are suggested to be the chest, abdomen, and thigh. Does your investigation support these results?

On Your Own   Explore other possible models relating skinfold measurements to body density. See if you can discover other models that fit just as well as those suggested above. Can you find a simpler model that does just as well? Would there be any practical advantage to finding a simpler model?

T A B L E **10.9**   Body Composition: Data Collections from Ultrasound Study

Women

Subscapula	Abdomen	Thigh	Triceps	Chest	Axilla	Suprailiac	Sum	Age	% Fat	Body Density
10.0	19.5	24.5	17.5	14.5	11.0	29.5	126.5	25	19.94	1.05333
7.0	9.5	21.0	15.0	6.5	5.0	5.0	69.0	21	20.038	1.05311
13.0	21.0	17.5	12.5	8.0	13.5	12.5	98.0	20	19.32	10.5472
10.5	14.5	16.0	16.0	11.0	9.5	8.0	85.5	24	18.56	1.05648
13.5	17.5	24.0	15.5	6.5	11.5	16.0	104.5	18	22.08	1.04855
21.0	28.5	29.0	20.0	16.5	21.5	26.5	163.0	23	25.81	1.04034
21.0	34.0	43.5	27.0	16.5	24.0	26.0	192.0	25	30.7804	1.02958
8.0	7.0	17.5	15.0	4.5	9.0	6.0	67.0	19	15.33	1.06376
34.5	42.0	33.0	28.0	26.5	36.0	28.0	228.0	50	44.33	1.00135
12.0	17.0	27.0	18.5	9.5	8.5	9.5	102.0	37	20.74	1.05154
31.0	48.5	47.0	31.5	27.5	25.5	37.5	248.5	67	38.32	1.01367
11.0	11.5	14.0	11.0	7.0	7.0	11.5	73.0	25	9.91	1.07629
11.5	13.5	25.5	14.0	7.0	7.5	13.5	72.5	18	15.81	1.06267
15.0	31.0	20.5	22.0	16.5	17.5	22.0	156.5	42	34.02	1.02268
8.5	14.0	25.5	15.0	7.0	7.0	9.5	86.5	44	23.15	1.04617

Men

Axilla	Subscapula	Chest	Triceps	Abdomen	Suprailiac	Thigh	Sum	Age	% Fat	Body Density
38.5	22.5	23.5	17.0	46.0	51.5	24.0	223	41	31.6	1.02876
15.0	14.0	17.0	15.0	38.0	31.5	19.0	149.5	35	17.99	1.05771
15.5	16.0	16.5	8.5	39.5	30.0	22.0	148.0	63	16.10	1.06201
11.5	11.0	13.5	9.5	31.5	17.0	18.5	112.5	55	8.68	1.07917
17.5	18.0	16.0	8.5	36.0	27.5	18.0	141.5	44	26.87	1.03802
26.0	31.5	23.5	18.0	52.5	39.0	30.0	220.5	27	18.06	1.05756
29.0	44.0	32.0	17.0	51.0	46.5	33.0	252.5	46	28.71	1.03403
19.0	12.0	15.0	12.0	51.5	34.0	34.0	177.5	38	21.71	1.04937
18.0	23.5	19.0	12.0	34.0	23.5	22.0	152.0	43	23.38	1.04567
8.0	11.0	9.5	11.0	25.0	22.5	23.0	110.0	26	9.09	1.07823
7.5	11.5	10.0	9.5	24.0	26.0	12.0	100.5	31	10.77	1.07430
9.0	14.5	5.0	5.0	16.0	16.0	6.5	72.0	24	4.58	1.08892
26.0	25.0	14.5	14.0	51.0	57.5	31.5	219.5	29	21.93	1.04887
9.0	11.0	8.0	5.5	20.0	25.0	7.0	85.5	23	3.82	1.09075

In your study of the data, you might be interested in how percent body fat is determined from the data on body density. For the data in Table 10.9, this determination was made from the equation

$$\% \text{ fat} = [(4.95/\text{body density}) - 4.50](100)$$

10.13 Summary

We have discussed some of the methodology of *multiple regression analysis*, a technique for modeling a dependent variable Y as a function of several independent variables $x_1, x_2, \ldots, x_k$. The steps we follow in constructing and using multiple regression models are much the same as those for the simple straight-line models:

Step 1. The form of the probabilistic model is hypothesized.

Step 2. The model coefficients are estimated by using least squares.

Step 3. The probability distribution of ε is specified, and σ^2 is estimated.

Step 4. The adequacy of the model is checked.

Step 5. If the model is deemed useful, it may be used to make estimates and to predict values of Y to be observed in the future.

We stress that this chapter is not intended to be a complete coverage of multiple regression analysis. Whole texts have been devoted to this topic. However, we have presented the core necessary for a basic understanding of multiple regression.

Supplementary Exercises

10.18 After a regression model is fit to a set of data, a confidence interval for the mean value of Y at a given setting of the independent variables will *always* be narrower than the corresponding prediction interval for a particular value of Y at the same setting of the independent variables. Why?

10.19 Before a particular job is accepted, a computer at a major university estimates the cost of running the job to see if the user's account contains enough money to cover the cost. As part of the job submission, the user must specify estimated values for two variables: central processing unit (CPU) time and number of lines printed. While the CPU time required and the number of lines printed do not account for the complete cost of the run, it is thought that knowledge of their values should allow a good prediction of job cost. The following model is proposed to explain the relationship of CPU time and lines printed to job cost:

$$E(Y) = \beta_0 + \beta_1 x_1 + \beta_2 x_2 + \beta_3 x_1 x_2$$

where

$$Y = \text{job cost (in dollars)}$$
$$x_1 = \text{number of lines printed}$$
$$x_2 = \text{CPU time}$$

Records from 20 previous runs were used to fit this model. The SAS printout is shown.

SOURCE	DF	SUM OF SQUARES	MEAN SQUARE	F VALUE	PR>F
MODEL	3	43.25090461	14.41696820	84.96	0.0001
ERROR	16	2.71515039	0.16969690		
CORRECTED TOTAL	19	45.9660550			

	R-SQUARE	STD DEV
	0.940931	0.41194283

PARAMETER	ESTIMATE	T FOR H0: PARAMETER = 0	PR>T	STD ERROR OF ESTIMATE
INTERCEPT	0.04564705	0.22	0.8318	0.21082636
X1	0.00078505	5.80	0.0001	0.00013537
X2	0.23737262	7.50	0.0001	0.03163301
X1*X2	-0.00003809	-.299	0.0086	0.0001273

X1	X2	PREDICTED VALUE	LOWER 95% CL FOR MEAN	UPPER 95% CL FOR MEAN
2000	42	8.38574865	7.32284845	9.44864885

a Identify the least-squares model that was fit to the data.

b What are the values of SSE and s^2 (estimate of σ^2) for the data?

c Explain what is meant by the statement "This value of SSE [see part (b)] is minimum."

10.20 Refer to Exercise 10.19 and the portion of the SAS printout shown.

a Is there evidence that the model is useful (as a whole) for predicting job cost? Test at $\alpha = 0.05$.

b Is there evidence that the variables x_1 and x_2 interact to affect Y? Test at $\alpha = 0.01$.

c What assumptions are necessary for the tests conducted in parts (a) and (b) to be valid?

10.21 Refer to Exercise 10.19 and the portion of the SAS printout shown. Use a 95% confidence interval to estimate the mean cost of computer jobs that require 42 seconds of CPU time and print 2,000 lines.

10.22 Several states now require all high-school seniors to pass an achievement test before they can graduate. On the test, the seniors must demonstrate their familiarity with basic verbal and mathematical skills. Suppose the educational testing company that creates and administers these exams wants to model the score Y on one of their exams as a function of the student's IQ x_1 and socioeconomic status (SES). The SES is a categorical (or *qualitative*) variable with three levels: low, medium, and high.

$$x_2 = \begin{cases} 1 & \text{if SES is medium} \\ 0 & \text{if SES is low or high} \end{cases}$$

$$x_3 = \begin{cases} 1 & \text{if SES is high} \\ 0 & \text{if SES is low or medium} \end{cases}$$

Data were collected for a random sample of 60 seniors who had taken the test, and the model

$$E(Y) = \beta_0 + \beta_1 x_1 + \beta_2 x_2 + \beta_3 x_3$$

was fit to these data, with the results shown in the following SAS printout.

SOURCE	DF	SUM OF SQUARES	MEAN SQUARE	F VALUE	PR>F
MODEL	3	12268.56439492	4089.52146497	188.33	0.0001
ERROR	56	1216.01893841	21.71462390		STD DEV
CORRECTED TOTAL	59	13484.58333333		R-SQUARE	4.65989527
				0.909822	

PARAMETER	ESTIMATE	T FOR H0: PARAMETER = 0	PR>T	STD ERROR OF ESTIMATE
INTERCEPT	-13.06166081	-3.21	0.0022	4.07101383
X1	0.74193946	17.56	0.0001	0.04224805
X2	18.60320572	12.49	0.0001	1.48895324
X3	13.40965415	8.97	0.0001	1.49417069

a Identify the least-squares equation.

b Interpret the value of R^2 and test to determine whether the data provide sufficient evidence to indicate that this model is useful for predicting achievement test scores.

c Sketch the relationship between predicted achievement test score and IQ for the three levels of SES. [*Note:* Three graphs of $\hat{y}$ versus x_1 must be drawn: the first for the low-SES model $(x_2 = x_3 = 0)$, the second for the medium-SES model $(x_2 = 1, x_3 = 0)$, and the third for the high-SES model $(x_2 = 0, x_3 = 1)$. Note that increase in predicted achievement test score per unit increase in IQ is the same for all three levels of SES; that is, all three lines are parallel.]

10.23 Refer to Exercise 10.22. The same data are now used to fit the model

$$E(Y) = \beta_0 + \beta_1 x_1 + \beta_2 x_2 + \beta_3 x_3 + \beta_4 x_1 x_2 + \beta_5 x_1 x_3$$

Thus, we now add the interaction between IQ and SES to the model. The SAS printout for this model is shown.

SOURCE	DF	SUM OF SQUARES	MEAN SQUARE	F VALUE	PR>F
MODEL	5	12515.10021009	2503.02004202	139.42	0.0001
ERROR	54	969.48312324	17.95339117		
CORRECTED TOTAL	59	13484.58333333		R-SQUARE	STD DEV
				0.928104	4.23714422

PARAMETER	ESTIMATE	T FOR H0: PARAMETER = 0	PR>T	STD ERROR OF ESTIMATE
INTERCEPT	0.60129643	0.11	0.9096	5.26818519
X1	0.59526252	10.70	0.0001	0.05563379
X2	-3.72536406	-0.37	0.7115	10.01967496
X3	-16.23196444	-1.90	0.0631	8.55429931
X1*X2	0.23492147	2.29	0.0260	0.10263908
X1*X3	0.30807756	3.53	0.0009	0.08739554

a Identify the least-squares prediction equation.

b Interpret the value of R^2 and test to determine whether the data provide sufficient evidence to indicate that this model is useful for predicting achievement test scores.

c Sketch the relationship between predicted achievement test score and IQ for the three levels of SES.

d Test to determine whether there is evidence that the mean increase in achievement test score per unit increase in IQ differs for the three levels of SES.

10.24 The EPA wants to model the gas mileage ratings Y of automobiles as a function of their engine size x. A quadratic model

$$Y = \beta_0 + \beta_1 x + \beta_2 x^2$$

is proposed. A sample of 50 engines of varying sizes is selected, and the miles per gallon rating of each is determined. The least-squares model is

$$\hat{y} = 51.3 - 10.1x + 0.15x^2$$

The size x of the engine is measured in hundreds of cubic inches. Also, $s_{\hat{\beta}_2} = 0.0037$ and $R^2 = 0.93$.

a Sketch this model between $x = 1$ and $x = 4$.

b   Is there evidence that the quadratic term in the model is contributing to the prediction of the miles per gallon rating Y? Use $\alpha = 0.05$.

c   Use the model to estimate the mean miles per gallon rating for all cars with 350-cubic-inch engines ($x = 3.5$).

d   Suppose a 95% confidence interval for the quantity estimated in part (c) is (17.2, 18.4). Interpret this interval.

e   Suppose you purchase an automobile with a 350-cubic-inch engine and determine that the miles per gallon rating is 14.7. Is the fact that this value lies outside the confidence interval given in part (d) surprising? Explain.

10.25   To increase the motivation and productivity of workers, an electronics manufacturer decides to experiment with a new pay incentive structure at one of two plants. The experimental plan will be tried at plant A for six months, while workers at plant B will remain on the original pay plan. To evaluate the effectiveness of the new plan, the average assembly time for part of an electronic system was measured for employees at both plants at the beginning and end of the six-month period. Suppose the following model was proposed:

$$Y = \beta_0 + \beta_1 x_1 + \beta_2 x_2 + \varepsilon$$

where

$Y =$ assembly time (hours) at end of six-month period

$x_1 =$ assembly time (hours) at beginning of six-month period

$$x_2 = \begin{cases} 1 & \text{if plant A (dummy variable)} \\ 0 & \text{if plant B} \end{cases}$$

A sample of $n = 42$ observations yielded

$$\hat{y} = 0.11 + 0.98x_1 - 0.53x_2$$

where

$$s_{\hat{\beta}_1} = 0.231 \qquad s_{\hat{\beta}_2} = 0.48$$

Test to see whether, after allowing for the effect of initial assembly time, plant A had a lower mean assembly time than plant B. Use $\alpha = 0.01$. [*Note*: When the (0, 1) coding is used to define a dummy variable, the coefficient of the variable represents the difference between the mean response at the two levels represented by the variable. Thus, the coefficient β_2 is the difference in mean assembly time between plant A and plant B at the end of the six-month period, and $\hat{\beta}_2$ is the sample estimator of that difference.]

10.26   One fact that must be considered in developing a shipping system that is beneficial to both the customer and the seller is time of delivery. A manufacturer of farm equipment can ship its products by either rail or truck. Quadratic models are thought to be adequate in relating time of delivery to distance traveled for both modes of transportation. Consequently, it has been suggested that the following model be fit:

$$E(Y) = \beta_0 + \beta_1 x_1 + \beta_2 x_2 + \beta_3 x_1 x_2 + \beta_4 x_2^2$$

where

$Y =$ shipping time

$$x_1 = \begin{cases} 1 & \text{if rail} \\ 0 & \text{if truck} \end{cases}$$

$x_2 =$ distance to be shipped

a What hypothesis would you test to determine whether the data indicate that the quadratic distance term is useful in the model, that is, whether curvature is present in the relationship between mean delivery time and distance?

b What hypothesis would you test to determine whether there is a difference in mean delivery time by rail and by truck?

10.27 Refer to Exercise 10.26. Suppose the proposed second-order model is fit to a total of 50 observations on delivery time. The sum of squared errors is SSE = 226.12. Then the reduced model

$$E(Y) = \beta_0 + \beta_2 x_2 + \beta_4 x_2^2$$

is fit to the same data, and SSE = 259.34. Test to see whether the data indicate that the mean delivery time differs for rail and truck deliveries.

10.28 *Operations management* is concerned with planning and controlling those organizational functions and systems that produce goods and services (S. Schroeder, *Operations Management: Decision Making in the Operations Function*, McGraw–Hill, 1981). One concern of the operations manager of a production process is the level of productivity of the process. An operations manager at a large manufacturing plant is interested in predicting the level of productivity of assembly line A next year (i.e., the number of units that will be produced by the assembly line next year). To do so, she has decided to use regression analysis to model the level of productivity Y as a function of time x. The number of units produced by assembly line A was determined for each of the past 15 years ($x = 1, 2, \ldots, 15$). The model

$$E(Y) = \beta_0 + \beta_1 x + \beta_2 x^2$$

was fit to these data, using Minitab. The results shown in the printout below were obtained.

```
THE REGRESSION EQUATION IS
Y=-1187-1333 X1-45.6 X2


                                   ST. DEV.    T-RATIO=
          COLUMN    COEFFICIENT    OF COEF.    COEF./S.D.
            --         -1187         446         -2.66
    X1      C2          1333         128         10.38
    X2      C3         -45.59       7.80         -5.84


THE ST.DEV. OF Y ABOUT REGRESSION LINE IS
S=501

WITH (15-3)=12 DEGREES OF FREEDOM
R-SQUARED=97.3 PERCENT
R-SQUARED=96.9 PERCENT. ADJUSTED FOR D.F.

ANALYSIS OF VARIANCE

    DUE TO    DF       SS         MS=SS/DF
REGRESSION     2    110578719     55289359
RESIDUAL      12      3013365       251114
TOTAL    14     113592083
```

a Identify the least-squares prediction equation.

b Find R^2 and interpret its value in the context of this problem.

c   Is there sufficient evidence to indicate that the model is useful for predicting the productivity of assembly line A? Test by using $\alpha = 0.05$.

d   Test the null hypothesis $H_0 : \beta_2 = 0$ against the alternative hypothesis $H_a : \beta_2 \neq 0$ by using $\alpha = 0.05$. Interpret the results of your test in the context of this problem.

e   Which (if any) of the assumptions we make about ε in regression analysis are likely to be violated in this problem? Explain.

10.29   Many companies must accurately estimate their costs before a job is begun in order to acquire a contract and make a profit. For example, a heating and plumbing contractor may base cost estimates for new homes on the total area of the house, the number of baths in the plans, and whether central air conditioning is to be installed.

a   Write a first-order model relating the mean cost of material and labor $E(Y)$ to the area, number of baths, and central air conditioning variables.

b   Write a complete second-order model for the mean cost as a function of the same three variables as in (a).

c   How could you test the research hypothesis that the second-order terms are useful for predicting mean cost?

10.30   Refer to Exercise 10.29. The contractor samples 25 recent jobs and fits both the complete second-order model [part (b)] and the reduced main effects model in [part (a)] so that a test can be conducted to determine whether the additional complexity of the second-order model is necessary. The resulting SSE and R^2 are given in the table.

	SSE	R^2
First-order	8.548	0.950
Second-order	6.133	0.964

a   Is there sufficient evidence to conclude that the second-order terms are important for predicting the mean cost?

b   Suppose the contractor decides to use the main effects model to predict costs. Use the global F-test to determine whether the main effects model is useful for predicting costs.

10.31   A company that services two brands of microcomputers would like to be able to predict the amount of time it takes a service person to perform preventive maintenance on each brand. It believes the following predictive model is appropriate:

$$Y = \beta_0 + \beta_1 x_1 + \beta_2 x_2 + \varepsilon$$

where

Y = maintenance time

$x_1 = \begin{cases} 1 & \text{if brand A} \\ 0 & \text{if brand I} \end{cases}$

x_2 = service person's number of months of experience in preventive maintenance

Ten different service people were randomly selected, and each was randomly assigned to perform preventive maintenance on either a brand A or brand I microcomputer. The data in Table 10.10 were obtained.

a   Fit the model to the data.

b   Investigate whether the overall model is useful. Test by using $\alpha = 0.05$.

c   Find R^2 for the fitted model. Does the value of R^2 support your findings in part (b)? Explain.

d   Find a 90% confidence interval for β_2. Interpret your result in the context of this exercise.

e   Use the fitted model to predict how long it will take a person with six months of experience to service a brand I microcomputer.

TABLE **10.10**

Maintenance Time (hours)	Brand	Experience (months)
2.0	1	2
1.8	1	4
0.8	0	12
1.1	1	12
1.0	0	8
1.5	0	2
1.7	1	6
1.2	0	5
1.4	1	9
1.2	0	7

f   How long would it take the person referred to in part (e) to service ten brand I microcomputers? List any assumptions you made in reaching your prediction.

g   Find a 95% prediction interval for the time required to perform preventive maintenance on a brand A microcomputer by a person with four months of experience.

10.32   Many colleges and universities develop regression models for predicting the grade–point average (GPA) of incoming freshmen. This predicted GPA can then be used to help make admission decisions. Although most models use many independent variables to predict GPA, we will illustrate by choosing two variables:

$$x_1 = \text{verbal score on college entrance examination (percentile)}$$
$$x_2 = \text{mathematics score on college entrance examination (percentile)}$$

The data in Table 10.11 are obtained for a random sample of 40 freshmen at one college.

a   Fit the first-order model (no quadratic and no interaction terms)

$$Y = \beta_0 + \beta_1 x_1 + \beta_2 x_2 + \varepsilon$$

Interpret the value of R^2, and test whether these data indicate that the terms in the model are useful for predicting freshman GPA. Use $\alpha = 0.05$.

b   Sketch the relationship between predicted GPA $\hat{y}$ and verbal score x_1 for the following mathematics scores: $x_2 = 60, 75,$ and 90.

10.33   Refer to Exercise 10.32. Now fit the following second-order model to the data:

$$Y = \beta_0 + \beta_1 x_1 + \beta_2 x_2 + \beta_3 x_1^2 + \beta_4 x_2^2 + \beta_5 x_1 x_2 + \varepsilon$$

a   Interpret the value of R^2, and test whether the data indicate that this model is useful for predicting freshman GPA. Use $\alpha = 0.05$.

b   Sketch the relationship between predicted GPA $\hat{y}$ and the verbal score x_1 for the following mathematics scores: $x_2 = 60, 75,$ and 90. Compare these graphs with those for the first-order model in Exercise 10.32.

c   Test to see whether the interaction term $\beta_5 x_1 x_2$ is important for the prediction of GPA. Use $\alpha = 0.10$. Note that this term permits the distance between three mathematics score curves for GPA versus verbal score to change as the verbal score changes.

10.34   Plastics made under different environmental conditions are known to have differing strengths. A scientist would like to know which combination of temperature and pressure yields a plastic

T A B L E **10.11**

Verbal x_1	Mathematics x_2	GPA Y	Verbal x_1	Mathematics x_2	GPA Y
81	87	3.49	79	75	3.45
68	99	2.89	81	62	2.76
57	86	2.73	50	69	1.90
100	49	1.54	72	70	3.01
54	83	2.56	54	52	1.48
82	86	3.43	65	79	2.98
75	74	3.59	56	78	2.58
58	98	2.86	98	67	2.73
55	54	1.46	97	80	3.27
49	81	2.11	77	90	3.47
64	76	2.69	49	54	1.30
66	59	2.16	39	81	1.22
80	61	2.60	87	69	3.23
100	85	3.30	70	95	3.82
83	76	3.75	57	89	2.93
64	66	2.70	74	67	2.83
83	72	3.15	87	93	3.84
93	54	2.28	90	65	3.01
74	59	2.92	81	76	3.33
51	75	2.48	84	69	3.06

with a high breaking strength. A small preliminary experiment was run at two pressure levels and two temperature levels. The following model was proposed:

$$E(Y) = \beta_0 + \beta_1 x_1 + \beta_2 x_2 + \beta_3 x_1 x_2$$

where

$$Y = \text{breaking strength (pounds)}$$
$$x_1 = \text{temperature } (°F)$$
$$x_2 = \text{pressure (pounds per square inch)}$$

A sample of $n = 16$ observations yielded

$$\hat{y} = 226.8 + 4.9x_1 + 1.2x_2 - 0.7x_1 x_2$$

with

$$s_{\hat{\beta}_1} = 1.11 \qquad s_{\hat{\beta}_2} = 0.27 \qquad s_{\hat{\beta}_3} = 0.34$$

Do the data indicate that there is an interaction between temperature and pressure? Test by using $\alpha = 0.05$.

10.35   Air pollution regulations for power plants are often written so that the maximum amount of pollutant that can be emitted is increased as the plant's output increases. Suppose the data in Table 10.12 are collected over a period of time:

a   Plot the data in a scatterplot.

b   Use the least-squares method to fit a straight line relating sulfur dioxide emission to output. Plot the least-squares line on the scatterplot.

T A B L E **10.12**

Output x (megawatts)	Sulfur Dioxide Emission Y (parts per million)
525	143
452	110
626	173
573	161
422	105
712	240
600	165
555	140
675	210

c Use a computer program to fit a second-order (quadratic) model relating sulfur dioxide emission to output.

d Conduct a test to determine whether the quadratic model provides a better description of the relationship between sulfur dioxide and output than the linear model.

e Use the quadratic model to estimate the mean sulfur dioxide emission when the power output is 500 megawatts. Use a 95% confidence interval.

10.36 The recent expansion of U.S. grain exports has intensified the importance of the linkage between the domestic grain transportation system and international transportation. As a first step in evaluating the economies of this interface, M. Martin and D. Clement (*Transportation Journal*, vol. 22, 1982, pp. 18–26) used multiple regression to estimate ocean transport rates for grain shipped from the Lower Columbia River international ports. These ports include Portland, Oregon; and Vancouver, Longview, and Kalama, Washington. Rates per long ton Y were modeled as a function of the following independent variables:

x_1 = shipment size in long tons

x_2 = distance to destination port in miles

x_3 = bunker fuel price in dollars per barrel

$x_4 = \begin{cases} 1 & \text{if American flagship} \\ 0 & \text{if foreign flagship} \end{cases}$

x_5 = size of the port as measured by the U.S. Defense Mapping Agency's Standards

x_6 = quantity of grain exported from the region during the year of interest

The method of least squares was used to fit the model to 140 observations from the period 1978 through 1980. The following results were obtained (the numbers in parentheses are the t statistics associated with the $\hat{\beta}$ values above them):

$$\hat{Y} = -18.469 - 0.367x_1 + 6.434x_2 - 0.2692x_2^2 + 1.7992x_3 + 50.292x_4 + 2.275x_5 - 0.018x_6$$
$$\quad\quad (-2.76)\quad (-5.62)\quad (3.64)\quad (-2.25)\quad (12.96)\quad (19.14)\quad (1.17)\quad (-2.69)$$
$$R^2 = 0.8979$$
$$F = 130.665$$

a Test $H_0 : \beta_1 = \beta_2 = \beta_3 = \beta_4 = \beta_5 = \beta_6 = \beta_7 = 0$. Use $\alpha = 0.01$. Interpret the results of your test in the context of the problem.

b D. Binkley and G. Harrer (*American Journal of Agricultural Economics*, 1979, pp. 44–51) estimated a similar rate function by using multiple regression, but they used different independent variables. The coefficient of determination for their model was 0.46. Compare the explanatory power of the Binkley and Harrer model with that of Martin and Clement.

   c According to the least-squares model, do freight charges increase with distance? Do they increase at an increasing rate? Explain.

10.37 An economist has proposed the following model to describe the relationship between the number of items produced per day (output) and the number of hours of labor expended per day (input) in a particular production process:

$$Y = \beta_0 + \beta_1 x + \beta_2 x^2 + \varepsilon$$

where

$$Y = \text{number of items produced per day}$$
$$x = \text{number of hours of labor per day}$$

A portion of the computer printout that results from fitting this model to a sample of 25 weeks of production data is shown. Test the hypothesis that, as the amount of input increases, the amount of output also increases, but at a decreasing rate. Do the data provide sufficient evidence to indicate that the *rate* of increase in output per unit increase of input decreases as the input increases? Test by using $\alpha = 0.05$.

```
THE REGRESSION EQUATION IS
Y=-6.17+2.04 X1-0.0323 X2
```

	COLUMN	COEFFICIENT	ST. DEV. OF COEF.	T-RATIO= COEF./S.D.
	--	-6.173	1.666	-3.71
X1	C2	2.036	0.185	11.02
X2	C3	-0.03231	0.00489	-6.60

```
THE ST.DEV. OF Y ABOUT REGRESSION LINE IS
S=1.243
WITH (25-3)=22 DEGREES OF FREEDOM
R-SQUARED=95.5 PERCENT
R-SQUARED=95.1 PERCENT, ADJUSTED FOR D.F.

ANALYSIS OF VARIANCE
```

DUE TO	DF	SS	MS=SS/DF
REGRESSION	2	718.168	359.084
RESIDUAL	22	33.992	1.545
TOTAL	24	752.160	

10.38 An operations engineer is interested in modeling $E(Y)$, the expected length of time per month (in hours) that a machine will be shut down for repairs, as a function of the type of machine (001 or 002) and the age of the machine (in years). He has proposed the following model:

$$E(Y) = \beta_0 + \beta_1 x_1 + \beta_2 x_1^2 + \beta_3 x_2$$

where

$$x_1 = \text{age of machine}$$
$$x_2 = \begin{cases} 1 & \text{if machine type 001} \\ 0 & \text{if machine type 002} \end{cases}$$

Data were obtained on $n = 20$ machine breakdowns and were used to estimate the parameters of the above model. A portion of the regression analysis computer printout is shown below:

SOURCE	DF	SUM OF SQUARES	MEAN SQUARE
MODEL	3	2396.364	798.788
ERROR	16	128.586	8.037
TOTAL	19	2524.950	R-SQUARE
			0.949

The reduced model $E(Y) = \beta_0 + \beta_1 x_1 + \beta_2 x_2$ was fit to the same data. The regression analysis computer printout is partially reproduced below:

SOURCE	DF	SUM OF SQUARES	MEAN SQUARE
MODEL	2	2342.42	1171.21
ERROR	17	182.53	10.74
TOTAL	19	2524.95	R-SQUARE
			0.928

Do these data provide sufficient evidence to conclude that the second-order (x_1^2) term in the model proposed by the operations engineer is necessary? Test by using $\alpha = 0.05$.

10.39 Refer to Exercise 10.38. The data that were used to fit the operations engineer's complete and reduced models are displayed in Table 10.13.

a Use these data to test the null hypothesis that $\beta_1 = \beta_2 = 0$ in the complete model. Test by using $\alpha = 0.10$.

T A B L E **10.13**

Downtime per Month (hours)	Machine Type	Machine Age (years)
10	001	1.0
20	001	2.0
30	001	2.7
40	001	4.1
9	001	1.2
25	001	2.5
19	001	1.9
41	001	5.0
22	001	2.1
12	001	1.1
10	002	2.0
20	002	4.0
30	002	5.0
44	002	8.0
9	002	2.4
25	002	5.1
20	002	3.5
42	002	7.0
20	002	4.0
13	002	2.1

b Carefully interpret the results of the test in the context of the problem.

10.40 Refer to Exercise 10.11, where we presented data on the number of highway deaths and the number of licensed vehicles on the road for the years 1950–1979. We mentioned that the number of deaths Y may also have been affected by the existence of the national 55-mile-per-hour speed limit during the years 1974–1979 (i.e., years 25–30). Define the dummy variable

$$x_2 = \begin{cases} 1 & \text{if 55-mile-per-hour speed limit was in effect} \\ 0 & \text{if not} \end{cases}$$

a Introduce the variable x_2 into the second-order model of Exercise 10.11 to account for the presence or absence of the 55-mile-per-hour speed limit in a given year. Include terms involving the interaction between x_2 and x_1.

b Refer to your model for part (a). Sketch on a single piece of graph paper your visualization of the two response curves, the second-order curves relating Y to x_1 before and after the imposition of the 55-mile-per-hour speed limit.

c Suppose that x_1 and x_2 do not interact. How would that fact affect the graphs of the two response curves of part (b)?

d Suppose that x_1 and x_2 interact. How would this fact affect the graphs of the two response curves?

10.41 In Exercise 10.11, we fit a second-order model to data relating the number Y of U.S. highway deaths per year to the number x_1 of licensed vehicles on the road. In Exercise 10.40, we added a qualitative variable x_2 to account for the presence or absence of the 55-mile-per-hour national speed limit. The accompanying SAS computer printout gives the results of fitting the model

$$E(Y) = \beta_0 + \beta_1 x_1 + \beta_2 x_1^2 + \beta_3 x_2 + \beta_4 x_1 x_2 + \beta_5 x_1^2 x_2$$

to the data. Use this printout and the printout for Exercise 10.11 to determine whether the data provide sufficient evidence to indicate that the qualitative variable (speed limit) contributes information for the prediction of the annual number of highway deaths. Test by using $\alpha = 0.05$. Discuss the practical implications of your test results.

```
DEPENDENT VARIABLE: DEATHS
```

SOURCE	DF	SUM OF SQUARES	MEAN SQUARE	F VALUE
MODEL	5	1428.02852498	285.60570500	41.56
ERROR	24	164.91847502	6.87160313	PR>F
CORRECTED TOTAL	29	1592.94700000		0.0001

R-SQUARE	C.V.	ROOT MSE	DEATHS MEAN
0.896470	5.7752	2.62137428	45.39000000

PARAMETER	ESTIMATE	T FOR H0: PARAMETER = 0	PR>T	STD ERROR OF ESTIMATE
INTERCEPT	16.94054815	2.21	0.0372	7.68054270
VEHICLES	0.34942808	1.89	0.0715	0.18528901
VEHICLES*VEHICLES	-0.00017135	-0.16	0.8724	0.00105538
LIMIT	41.38712279	0.11	0.9155	385.80676597
VEHICLES*LIMIT	-0.75118159	-0.14	0.8878	5.26806364
VEHICLE*VEHICLE* LIMIT	0.00245917	0.14	0.8921	0.01794470

10.42   *Productivity* has been defined as the relationship between inputs and outputs of a productive system. In order to manage a system's productivity, it is necessary to measure it. Productivity is typically measured by dividing a measure of system output by a measure of the inputs to the system. Some examples of productivity measures are sales/salespeople, yards of carpet laid/number of carpet layers, shipments/[(direct labor) + (indirect labor) + (materials)]. Notice that productivity can be improved either by producing greater output with the same inputs or by producing the same output with fewer inputs. In manufacturing operations, productivity ratios like the third example above generally vary with the volume of output produced. The production data in Table 10.14 were collected for a random sample of months for the three regional plants of a particular manufacturing firm. Each plant manufactures the same product.

TABLE **10.14**

North Plant		South Plant		West Plant	
Productivity Ratio	No. of Units Produced	Productivity Ratio	No. of Units Produced	Productivity Ratio	No. of Units Produced
1.30	1,000	1.43	1,015	1.61	501
0.90	400	1.50	925	0.74	140
1.21	650	0.91	150	1.19	303
0.75	200	0.99	222	1.88	930
1.32	850	1.33	545	1.72	776
1.29	600	1.15	402	1.39	400
1.18	756	1.51	709	1.86	810
1.10	500	1.01	176	0.99	220
1.26	925	1.24	392	0.79	160
0.93	300	1.49	699	1.59	626
0.81	258	1.37	800	1.82	640
1.12	590	1.39	660	0.91	190

a   Construct a scatterplot for these data. Plot the North-plant data using dots, the South-plant data using small circles, and the West-plant data using small triangles.

b   Visually fit each plant's response curve to the scatterplot.

c   Based on the results of part (b), propose a second-order regression model that could be used to estimate the relationship between productivity and volume for the three plants.

d   Fit the model you proposed in part (c) to the data.

e   Do the data provide sufficient evidence to conclude that the productivity response curves for the three plants differ? Test by using $\alpha = 0.05$.

f   Do the data provide sufficient evidence to conclude that your second-order model contributes more information for the prediction of productivity than does a first-order model? Test by using $\alpha = 0.05$.

g   Next month, 890 units are scheduled to be produced at the West plant. Use the model you developed in part (d) to predict next month's productivity ratio at the West plant.

10.43   The U.S. Federal Highway Administration publishes data on reaction distance for drivers and braking distance for a typical automobile traveling on dry pavement. Some of the data appear in Table 10.15. Total stopping distance is the sum of reaction distance and braking distance. Find simple linear regression models for estimating each average

a   reaction distance from speed.

b   braking distance from speed.

c   total stopping distance from speed.

Use your answers to fill in a line of the table for an automobile traveling 55 mph.

TABLE **10.15**

Speed (mph)	Reaction Distance (feet)	Braking Distance (feet)
20	22	20
30	33	40
40	44	72
50	55	118
60	66	182
70	77	266

10.44   In building a model to study automobile fuel consumption, H. Biggs and D. Akcelik (*Journal of Transportation Engineering*, vol. 113, no. 1, January 1987, pp. 101–106) began by looking at the relationship between idle fuel consumption and engine capacity. Suppose the data are as in Table 10.16.

TABLE **10.16**

Idle Fuel Consumption (mL/s)	Engine Size (L)
0.18	1.2
0.21	1.2
0.17	1.2
0.31	1.8
0.34	1.8
0.29	1.8
0.42	2.5
0.39	2.5
0.45	2.5
0.52	3.4
0.61	3.4
0.44	3.4
0.62	4.2
0.65	4.2
0.59	4.2

What model do you think best explains the relationship between idle fuel consumption and engine size?

10.45   One important feature of the plates used in high-performance thin-layer chromatography is the water uptake capability. Six plates were modified by increasing the amount of cyanosilane (CN). Water uptake measurements (Rf) were then recorded for three component mixtures (BaP, BaA, Phe) used with each plate. Six measurements were made on each mixture/plate combination. The data are in Table 10.17. Fit a model relating % CN to mean Rf for each compound. (The models could differ.) Do the values of s suggest any complication?

10.46   Low temperatures on winter nights in Florida have caused extensive damage to the state's agriculture. Thus, it is important to be able to predict freezes quite accurately. In an attempt to use satellite data for this purpose, a study was done to compare the surface temperatures detected by the infrared sensors of HCMM and GOES-East satellites. GOES temperature data compared to HCMM sample means are given in the Table 10.18. Find an appropriate model relating the mean HCMM reading to the GOES temperature. What would happen to the R^2 value if the actual HCMM data points were used rather than their means?

T A B L E **10.17**   Rf Values and Standard Deviation for BaP, BaA, and Phe on Modified Plates

Plate No.	Type of Plate	Compound	Mean Rf	s
1	0% CN	BaP	0.17	0.021
		BaA	0.25	0.026
		Phe	0.36	0.033
2	1% CN	BaP	0.21	0.014
		BaA	0.30	0.023
		Phe	0.44	0.046
3	5% CN	BaP	0.23	0.021
		BaA	0.31	0.016
		Phe	0.41	0.025
4	10% CN	BaP	0.36	0.028
		BaA	0.45	0.026
		Phe	0.59	0.027
5	20% CN	BaP	0.46	0.045
		BaA	0.58	0.026
		Phe	0.73	0.027
6	30% CN	BaP	0.46	0.069
		BaA	0.56	0.70
		Phe	0.66	0.095

Source: A. Colsmjo and M. Ericsson, *Journal of High Resolution Chromatography*, vol. 10, April 1987, pp. 177–180.

T A B L E **10.18**

GOES Temp. (K)	HCMM Temp. (K)			
	Mean	S.D.	Range Min.	Max.
275.5	274.5	1.01	272.8	281.2
275.5	275.2	1.25	273.2	281.6
277.5	277.6	2.72	273.6	284.2
276.5	277.3	3.06	273.2	283.9
277.5	277.8	3.53	273.6	283.9
281.0	278.9	0.80	276.9	280.8
277.5	277.9	3.56	273.6	283.5
278.0	279.2	3.49	273.2	283.9
280.0	280.6	3.01	277.3	287.5
285.0	290.9	3.28	281.6	294.2
278.0	276.9	2.07	274.5	283.1
284.5	284.4	2.78	277.3	287.2
284.5	291.5	2.75	282.0	294.2
283.5	285.5	1.87	277.3	287.2
280.0	279.8	2.36	278.9	287.5

Source: E. Chen and H. Allen, *Remote Sensing of the Environment*, vol. 21, 1987, pp. 341–353.

Design of Experiments and the Analysis of Variance

About This Chapter

Throughout this book, we have said much about statistical thinking and statistical tools as essential components of process improvement. Now, virtually all of these ideas and tools are pulled together as we concentrate on perhaps the most important contribution of statistics to quality improvement—the randomized experiment. Notions of data collection, estimation, hypothesis testing, and regression analysis are revisited for the purpose of comparing treatments and making decisions as to which treatment is optimal. Carefully planned and executed experiments are the key to increased knowledge of any process under study and, therefore, to improved quality.

Contents

11.1 Introduction to Designed Experiments

The goal of a statistical investigation is the improvement of a process. That point has been reiterated often throughout this text, and it is the main point to keep in mind as we review and extend the ideas involved in the design of experiments.

A designed experiment is a plan for data collection and analysis directed at making a decision about whether or not one treatment is better than another. We may employ estimation procedures or hypothesis testing procedures in the analysis, but the goal is always a decision about the relative effectiveness of treatments. To review some ideas presented earlier in this book, a designed experiment should follow the scientific method, should follow the model for problem solving given in Chapter 1, and should follow the key elements outlined at the beginning of Chapter 2.

Experimentation is scientific. The scientific method says, basically, that an investigation should begin with a conjecture (question, hypothesis) formulated from prior knowledge, collect new data relevant to the conjecture, and then decide if the conjecture appears to be true. If the conjecture does appear to hold, it may be refined and checked against additional data. If the conjecture does not hold, it may be modified and checked again. Thus, experimentation and the development of knowledge are part of an ongoing process that moves back and forth from conjecture to verification (data) in a continuous cycle.

Experimentation is problem solving. Thus, the procedure of stating the question or problem clearly, collecting and analyzing data, interpreting the data to make decisions, verifying the decisions, and planning the next action provides the pragmatic steps to follow in any experimental investigation. (The reader might want to review the first two sections of Chapter 1 to see how these steps fit together in a real example.)

Experimentation is organized and carried out according to a detailed plan. The key elements listed in Section 2.2 still hold and should be reviewed here. After the question or conjecture has been clearly stated, treatment and control variables are defined, lurking variables may be identified, randomization is employed to assign treatments to experimental units, and unbiased methods of measurement are selected before the data are collected and analyzed.

Let's carry out a hypothetical experiment using the elements outlined above so we can see their relationship to the statistical techniques presented in the remainder of this chapter. Computer chips (semiconductors) are manufactured in a round wafer (about 3 inches in diameter) that holds many chips. The wafers are scored and broken apart to retrieve the chips for actual use in circuits. This process is rather crude and causes much waste. The question for investigation, then, is "How can the breaking process be improved to gain a higher yield of usable chips?" Suppose an engineer suggests a new method, call it method B, to compete with the standard, method A. These two methods form the treatments to be compared in an experiment. (After all, we cannot just take the word of the developer that the new method is better!)

The many possible background (lurking) variables include the day or time the wafer was made, the temperature of the ovens, the chemical composition of the oxides, the work crew who made the wafers, and so on. Initially, the experimenter decides to control none of these variables directly, but rather to attempt to control all of them indirectly by randomization. A number of wafers are selected at random to be broken by method A and others are randomly selected to be broken by method B. Hopefully, the various characteristics of the wafers will be somewhat balanced between the two methods by this randomization; probability theory suggests that such should be the case, especially if a large number of wafers are used in the study. This simple design of an experiment is called a *completely randomized design*. It works well for situations in which direct controls are either not necessary or impossible to arrange.

On thinking more carefully about this scenario, the experimenter decides that the day of the week on which the wafer was made and broken into chips may be important because of differing environmental conditions and different workers. To account for this possibility, the experimental design is changed so that both method A and method B are used on wafers each day of a five-day workweek. Specifically, the methods are used throughout the day on a sample of wafers from each day's production, and the order in which the two methods are used is randomized as well, with the restriction that the same number of observations must be taken on each method each day. The resulting design is called a *randomized block design*. The days form the blocks in which an equal number of measurements is taken on each treatment. The results allow possible treatment differences to show up even if there is great variation in the outcome measures from day to day. In the randomized block design, the differences between days can also be measured, but this difference is not the main focus. The blocks are used as a device to gain more information about the treatment differences.

The number of usable chips per wafer turns out to be much the same from day to day for either treatment, so blocking can be eliminated from future experiments of this type. But another consideration now comes into play. The process under discussion is actually a two-stage process involving scoring the wafer so that the chips are defined and then breaking apart the silicon wafer, hopefully along the score lines. The new method, method B, involves both deeper score lines and a different approach to breaking. Since there are now two types of score lines and two types of breaking, a slightly different arrangement of "treatments" is in order. If each score-line depth/breaking method combination is used, four treatments result (shallow score lines and old breaking method, deep score lines and old breaking method, and so on). Such a design is called a *factorial design* or a factorial arrangement of treatments. Once the four treatments are defined, the measurements on those treatments could actually be taken by using a completely randomized design.

These three designs are the main focus of this chapter; we now proceed with a discussion of how to analyze data from such designed experiments by methods that come under the general topic of *analysis of variance*. These methods extend some of the techniques of Chapters 7 and 8 and make use of regression analysis (Chapters 9 and 10) as a computational tool. As data are presented to you for analysis, pause to review some of the points made above about the design issues so you can judge the value of any decisions that might be made from the data.

11.2 Analysis of Variance for the Completely Randomized Design

The two-sample T-test of Section 8.3 was designed to test the hypothesis that two population means are equal versus the alternative that they differ. Recall that, in order for that test to be used, the experiment must have resulted in two independent random samples, one sample from each population under study. As an illustration, Example 8.10 reported results from a random sample of nine measurements on stamping times for a standard machine and an independently selected random sample of nine measurements from a new machine. This is a simple example of a completely randomized design.

DEFINITION **11.1**

> A **completely randomized design** is a plan for collecting data in which a random sample is selected from each population of interest and the samples are independent. ▪

The stamping-time example consists of only two populations, but Definition 11.1 allows any finite number of populations. For example, suppose mean tensile strengths are to be compared for steel specimens coming from three processes, each involving a different percentage of carbon. If random samples of tensile-strength measurements are taken on specimens from each process and the samples are independent, then the resulting experiment would be a completely randomized design.

The populations of interest in experimental design problems are generally referred to as *treatments*. Thus, the stamping-time problem has two treatments (standard machine and new machine), whereas the tensile-strength problem has three treatments (one for each percentage of carbon in the steel). The notation we will employ for the k-population (or k-treatment) problem is summarized in Table 11.1.

The method we present to analyze data from a completely randomized design (as well as many other designs) is called the *analysis of variance*. The basic inference

TABLE **11.1**
Notation for a Completely
Randomized Design

Populations (treatments)

	1	2	3	$\cdots$	k
Mean	μ_1	μ_2	μ_3	$\cdots$	μ_k
Variance	σ_1^2	σ_2^2	σ_3^2	$\cdots$	σ_k^2

Independent Random Samples

Sample Size	n_1	n_2	n_3	$\cdots$	n_k
Sample Totals	T_1	T_2	T_3	$\cdots$	T_k
Sample Means	$\bar{y}_1$	$\bar{y}_2$	$\bar{y}_3$	$\cdots$	$\bar{y}_k$

Total sample size $= n = n_1 + n_2 + n_3 + \cdots + n_k$
Overall sample total $= \sum y = T_1 + T_2 + \cdots + T_k$
Overall sample mean $= \bar{y} = \sum y / n$
Sum of squares of all n measurements $= \sum y^2$

problem for which analysis of variance provides an answer is the test of the null hypothesis

$$H_0 : \mu_1 = \mu_2 = \cdots = \mu_k$$

versus the alternative hypothesis

$$H_a : \text{At least two treatment means differ.}$$

To see how we might formulate a test statistic for the null hypothesis indicated above, let us once again look at the two-sample problem involving stamping times. Example 8.10 provides the following data:

$$n_1 = 9 \qquad\qquad n_2 = 9$$
$$\bar{y}_1 = 35.22 \text{ seconds} \qquad \bar{y}_2 = 31.56 \text{ seconds}$$
$$(n_1 - 1)s_1^2 = 195.50 \qquad (n_2 - 1)s_2^2 = 160.22$$

If we want to test $H_0 : \mu_1 = \mu_2$ versus $H_a : \mu_1 \neq \mu_2$, the T-test can be used, with the statistic calculated as follows:

$$t = \frac{(\bar{y}_1 - \bar{y}_2)}{s_p \sqrt{\dfrac{1}{n_1} + \dfrac{1}{n_2}}}$$

The s_p here is the square root of the "pooled" s_p^2 given by

$$s_p^2 = \frac{(n_1 - 1)s_1^2 + (n_2 - 1)s_2^2}{n_1 + n_2 - 2}$$

The value of this statistic is calculated in Example 8.10 to be $t = 1.65$. The critical value, with $\alpha = 0.05$ and 16 degrees of freedom, is $t_{0.025} = 2.120$, and hence we cannot reject H_0.

The T-test as presented above cannot be extended to a test of the equality of more than two means. However, if we square the t statistic, we obtain, after some algebraic manipulation,

$$t^2 = \frac{(\bar{y}_1 - \bar{y}_2)^2}{s_p^2 \left(\dfrac{1}{n_1} + \dfrac{1}{n_2} \right)} = \frac{\displaystyle\sum_{i=1}^{2} n_i (\bar{y}_i - \bar{y})^2}{s_p^2}$$

In repeated sampling, this statistic can be shown to have an F distribution with $v_1 = 1$ and $v_2 = n_1 + n_2 - 2$ degrees of freedom in the numerator and denominator, respectively. Thus, we could have tested the above hypothesis by calculating

$$\frac{\displaystyle\sum_{i=1}^{2} n_i (\bar{y}_i - \bar{y})^2}{s_p^2} = \frac{n_1 (\bar{y}_1 - \bar{y})^2 + n_2 (\bar{y}_2 - \bar{y})^2}{s_p^2} = \frac{60.28}{22.23} = 2.71$$

and comparing the value to $F_{0.05}(1, 16) = 4.49$. Again, we would not reject H_0.

This second approach can be generalized to k populations quite easily. In general, the numerator must have a divisor of $(k-1)$, so for testing

$$H_0 : \mu_1 = \mu_2 = \cdots = \mu_k$$

versus

$$H_a : \text{At least two treatment means differ}$$

we use the test statistic

$$F = \frac{\sum_{i=1}^{k} n_i (\bar{y}_i - \bar{y})^2 / (k-1)}{s_p^2}$$

where

$$s_p^2 = \frac{\sum_{i=1}^{k} (n_i - 1) s_i^2}{n - k}$$

By inspecting this ratio, we can see that large deviations among the sample means will cause the numerator to be large and thus may provide evidence for rejection of the null hypothesis. The denominator is an average within-sample variance and is not affected by differences among the sample means.

The usual analysis of variance notation is as follows:

$$\text{SST} = \text{Sum of Squares for Treatments}$$
$$= \sum_{i=1}^{k} n_i (\bar{y}_i - \bar{y})^2$$
$$\text{SSE} = \text{Sum of Squares for Error}$$
$$= s_p^2 (n - k)$$
$$\text{TSS} = \text{Total Sum of Squares}$$
$$= \text{SST} + \text{SSE}$$
$$\text{MST} = \text{Mean Square for Treatments} = \frac{\text{SST}}{k-1}$$

and

$$\text{MSE} = \text{Mean Square for Error} = \frac{\text{SSE}}{n-k} = s_p^2$$

Thus, the F ratio for testing $H_0 : \mu_1 = \mu_2 = \cdots = \mu_k$ is given by

$$F = \frac{\text{MST}}{\text{MSE}}$$

Table 11.2 gives formulas for SST and TSS that minimize computational difficulties but are equivalent to the formulas given above.

As in the two-sample T-test, the population probability distributions must be normal and the population variances must be equal in order for F = MST/MSE to have an F distribution. But the F-test works reasonably well if the populations are

TABLE **11.2**
Calculation Formulas for the Completely Randomized Design

$$TSS = \sum y^2 - \frac{(\sum y)^2}{n}$$

$$SST = \sum_{i=1}^{k} \frac{T_i^2}{n_i} - \frac{(\sum y)^2}{n}$$

$$SSE = TSS - SST$$

only approximately normal (mound-shaped) with similar-size variances. The basic analysis of variance F-test is summarized below.

Test to Compare k Treatment Means for a Completely Randomized Design

$$H_0 : \mu_1 = \mu_2 = \cdots = \mu_k$$

H_a : At least two treatment means differ.

Test statistic: $F = \dfrac{MST}{MSE}$

Assumptions:

1 Each population has a normal probability distribution.
2 The k population variances are equal.

Rejection region : $F > F_\alpha(k - 1, n - k)$

The analysis of variance (ANOVA) summary table is usually written as suggested in Table 11.3.

TABLE **11.3**
ANOVA for Completely Randomized Design

Source	df	SS	MS	F Ratio
Treatments	$k - 1$	SST	MST	MST/MSE
Error	$n - k$	SSE	MSE	
Total	$n - 1$	TSS		

EXAMPLE **11.1**   Over the past ten years, computers and computer training have become integral parts of the curricula of both secondary schools and universities. As a result, younger business professionals tend to be more comfortable with computers than their more senior counterparts.

"The older the person is, the worse it is," said Arnold S. Kahn of the American Psychological Association. "The older they are and the longer they wait before learning, the more dissatisfied they will become. The computer's unrelenting march into offices and factories is often cited as a chief cause of work-related stress. As computers and robots come into the work place, many workers fear they'll never master the new skills required."

Gary W. Dickson (personal communication, 1984), Professor of Management Information Systems at the University of Minnesota, investigated the computer literacy of middle managers with ten years or more of management experience. As part of his study, Dickson designed a questionnaire that he hoped would measure managers' technical knowledge of computers. If the questionnaire was properly designed, the scores received by managers could be used as predictors of their knowledge of computers, with higher scores indicating more knowledge. To check the design of the questionnaire (i.e., its validity), 19 middle managers from the Minneapolis–St. Paul metropolitan area were randomly sampled and asked to complete the questionnaire. Their scores appear in the following table. (The highest possible score on the questionnaire was 169.)

Questionnaire Scores of Middle Managers

Manager I.D.	Level of Technical Expertise	Score
1	A	82
2	A	114
3	A	90
4	A	80
5	B	128
6	B	90
7	C	156
8	A	88
9	A	93
10	B	130
11	A	80
12	A	105
13	B	110
14	B	133
15	C	128
16	B	130
17	B	104
18	C	151
19	C	140

Prior to completing the questionnaire, the managers were asked to describe their knowledge of and experience with computers. This information was used to classify the managers as possessing a low (A), medium (B), or high (C) level of technical computer expertise. These data also appear in the table.

Is there sufficient evidence to conclude that the mean score differs for the three groups of managers?

Solution First, we group the data according to the "treatment" (level of technical expertise) and compute treatment totals.

	Level		
	A	**B**	**C**
	82	128	156
	114	90	128
	90	130	151
	80	110	140
	88	133	
	93	130	
	80	104	
	105		
Totals:	732	825	575

Then, using the calculation formulas for the completely randomized design, we find

$$\text{TSS} = \sum y^2 - \frac{(\sum y)^2}{n} = 250{,}048 - \frac{(2{,}132)^2}{19}$$
$$= 10{,}815.16$$

$$\text{SST} = \sum_{i=1}^{3} \frac{T_i^2}{n_i} - \frac{(\sum y)^2}{n}$$
$$= \frac{(732)^2}{8} + \frac{(825)^2}{7} + \frac{(575)^2}{4} - \frac{(2{,}132)^2}{19}$$
$$= 7{,}633.55$$

and

$$\text{SSE} = \text{TSS} - \text{SST} = 3{,}181.61$$

The test statistic is

$$F = \frac{\text{MST}}{\text{MSE}} = \frac{\text{SST}/(k-1)}{\text{SSE}/(n-k)} = \frac{7{,}633.55/2}{3{,}181.61/16}$$
$$= 19.19$$

Since $F_{0.05}(2, 16) = 3.63$, we can reject the hypothesis that the population mean scores for the three groups of managers are equal.

The ANOVA table follows:

Source	df	SS	MS	F Ratio
Treatments	2	7,633.55	3,816.78	19.19
Error	16	3,181.61	198.85	
Total	18	10,815.16		

Any standard statistical computing software package will produce an analysis of variance table similar to the one given above. The Minitab printout for this problem is shown below.

```
ANALYSIS OF VARIANCE
SOURCE   DF      SS     MS       F       p
FACTOR    2    7634   3817   19.19   0.000
ERROR    16    3182    199
TOTAL    18   10815


LEVEL     N    MEAN   STDEV
A         8   91.50   12.31
B         7  117.86   16.62
C         4  143.75   12.45

POOLED STDEV =   14.10
```

Note that "Treatments" is replaced by "Factor," unless specified otherwise in the computer analysis. The printout also shows each treatment mean and its standard deviation. With Minitab, it is easy to show a dotplot of the original data, as seen below.

Dotplot in Example 11.1

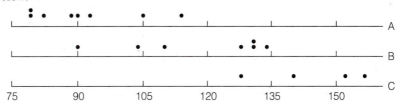

The dotplot shows that the three samples each display about the same amount of variability, but the sample sizes are too small to show normality (or lack of it) in a definitive way. ■

If the equal-variance assumption required by the analysis of variance procedure cannot be met, sometimes a transformation of the data is in order. (We could conduct a test for equality of more than two variances, but we will forgo that in favor of a graphical approach.)

E X A M P L E   **11.2**   Porosity of metal, an important determinant of strength and other properties, can be measured by looking at cross sections of the metal under a microscope. The *pore* (void) is dark, the metal is light, and the boundary between these two phases is often clearly delineated. If a grid is laid on the microscopic field, the number of intersections between the grid lines and the pore boundaries is proportional to the length of that boundary per unit area (and hence is proportional to the pore surface area per unit volume of metal).

Such counts, denoted by N, were made for samples of antimony, Linde copper, and electrolytic copper and are recorded in the table below. (These data came from the Materials Science Department, University of Florida).

	Antimony					Linde Copper					Electrolytic Copper				
	10	9	9	9	10	14	12	15	14	10	42	46	44	39	50
	11	14	11	8	11	17	16	11	13	14	34	42	40	36	37
	7	6	9	7	8	15	11	16	12	6	46	42	43	50	32
	10	12	14	9	8	13	20	17	10	16	41	37	49	28	34
	8	9	8	7	10	10	13	12	17	13	34	38	43	39	42
Mean		9.36					13.48					40.32			
Variance		4.07					9.01					31.58			

Do the mean counts show significant differences?

Solution Obviously, the variation is not constant across the three samples. In fact, the variances are nearly proportional to the means, which indicates that a square root transformation will help stabilize the variances. The boxplots in Figure 11.1 show the original data and the square roots for each metal. Observe that the square root transformation does make the variability about the same for all three samples.

From looking at the data, it is obvious that the electrolytic copper has a much higher mean count; this is borne out by the very high F value (475.16) in the analysis of variance, as shown in the printout below.

```
ANALYSIS OF VARIANCE
SOURCE   DF       SS      MS        F       p
FACTOR    2   153.498   76.749   475.16   0.000
ERROR    72    11.630    0.162
TOTAL    74   165.128


LEVEL    N      MEAN    STDEV
  1     25    3.0430   0.3229
  2     25    3.6479   0.4247
  3     25    6.3347   0.4472
```

What is not so obvious is that the mean count for Linde copper is very significantly higher than the mean count for antimony, as seen in the analysis of variance for only these two sets of samples.

```
ANALYSIS OF VARIANCE
SOURCE   DF       SS      MS       F       p
FACTOR    1     4.573   4.573   32.14   0.000
ERROR    48     6.830   0.142
TOTAL    49    11.403


LEVEL     N      MEAN    STDEV
  1      25    3.0430   0.3229
  2      25    3.6479   0.4247
```

F I G U R E **11.1**   Boxplots for Example 10.2

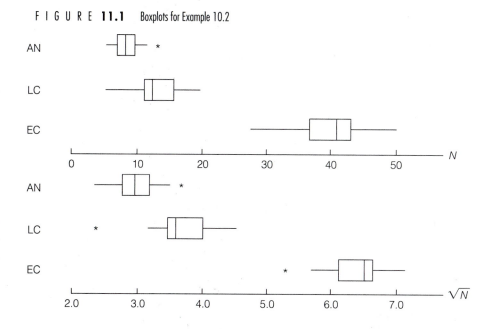

So the mean counts do differ, which is not surprising, but the analysis must be done on transformed data because of the lack of homogeneity of variance.   ▪

We will show in Section 11.3 that the analysis just conducted can be performed using regression models. In Section 11.4, we will show how to compare the pairs of treatment means.

Exercises

11.1   Independent random samples were selected from three normally distributed populations with common (but unknown) variance σ^2. The data are shown below:

Sample 1	Sample 2	Sample 3
3.1	5.4	1.1
4.3	3.6	0.2
1.2	4.0	3.0
	2.9	

a   Compute the appropriate sums of squares and mean squares, and fill in the appropriate entries in the analysis of variance table shown below:

Source	df	SS	MS	*F* Ratio
Treatments				
Error				
Total				

b   Test the hypothesis that the population means are equal (that is, $\mu_1 = \mu_2 = \mu_3$) against the alternative hypothesis that at least one mean is different from the other two. Test by using $\alpha = 0.05$.

11.2   A partially completed ANOVA table for a completely randomized design is shown below:

Source	df	SS	MS	F Ratio
Treatments	4	24.7		
Error				
Total	34	62.4		

a   Complete the ANOVA table.

b   How many treatments are involved in the experiment?

c   Do the data provide sufficient evidence to indicate a difference among the population means? Test by using $\alpha = 0.01$.

11.3   Some varieties of nematodes (round worms that live in the soil and frequently are so small they are invisible to the naked eye) feed upon the roots of lawn grasses and other plants. This pest, which is particularly troublesome in warm climates, can be treated by the application of nematicides. Data collected on the percentage of kill for nematodes for four particular rates of application (dosages given in pounds of active ingredient per acre) are as follows:

Rate of Application

2	3	5	7
86	87	94	90
82	93	99	85
76	89	97	86
		91	

Do the data provide sufficient evidence to indicate a difference in the mean percentage of kill for the four different rates of application of nematicide? Use $\alpha = 0.05$.

11.4   It has been hypothesized that treatment (after casting) of a plastic used in optic lenses will improve wear. Four different treatments are to be tested. To determine whether any differences in mean wear exist among treatments, 28 castings from a single formulation of the plastic were made, and seven castings were randomly assigned to each treatment. Wear was determined by measuring the increase in "haze" after 200 cycles of abrasion (with better wear indicated by small increases).

Treatment

A	B	C	D
9.16	11.95	11.47	11.35
13.29	15.15	9.54	8.73
12.07	14.75	11.26	10.00
11.97	14.79	13.66	9.75
13.31	15.48	11.18	11.71
12.32	13.47	15.05	12.45
11.78	13.06	14.86	12.38

a   Is there evidence of a difference in mean wear among the four treatments? Use $\alpha = 0.05$.

b   Estimate the mean difference in haze increase between treatments B and C, using a 99% confidence interval. [*Hint*: Use the two-sample t statistic.]

c   Find a 90% confidence interval for the mean wear for lenses receiving treatment A. [*Hint*: Use the one-sample t statistic.]

11.5 The application of *management by objectives* (MBO), a method of performance appraisal, was the object of a study by Y. K. Shetty and H. M. Carlisle of Utah State University ("Organizational Correlates of a Management by Objectives Program," *Academy of Management Journal*, 1974, p. 17). The study dealt with the reactions of a university faculty to an MBO program. One hundred nine faculty members were asked to comment on whether they thought the MBO program was successful in improving their performance within their respective departments and the university. Each response was assigned a score from 1 (significant improvement) to 5 (significant decrease). The table shows the sample sizes, sample totals, mean scores, and sums of squares of deviations *within* each sample for samples of scores corresponding to the four academic ranks. Assume that the four samples in the table can be viewed as independent random samples of scores selected from among the four academic ranks.

		Academic Rank		
	Instructor	Assistant Professor	Associate Professor	Professor
Sample Size	15	41	29	24
Sample Total	42.960	145.222	92.249	73.224
Sample Mean	2.864	3.542	3.181	3.051
Within-sample Sum of Squared Deviations	2.0859	14.0186	7.9247	5.6812

a Perform an analysis of variance for the data.

b Arrange the results in an analysis of variance table.

c Do the data provide sufficient evidence to conclude that there is a difference in mean scores among the four academic ranks? Test by using $\alpha = 0.05$.

d Find a 95% confidence interval for the difference in mean scores between instructors and professors.

e Do the data provide sufficient evidence to indicate a difference in mean scores between nontenured faculty members (instructors and assistant professors) and tenured faculty members?

11.6 The concentration of a catalyst used in producing grouted sand is thought to affect its strength. An experiment designed to investigate the effects of three different concentrations of the catalyst utilized five specimens of grout per concentration. The strength of a grouted sand was determined by placing the test specimen in a press and applying pressure until the specimen broke. The pressures required to break the specimens, expressed in pounds per square inch, are shown below.

Concentration of Catalyst

35%	40%	45%
5.9	6.8	9.9
8.1	7.9	9.0
5.6	8.4	8.6
6.3	9.3	7.9
7.7	8.2	8.7

Do the data provide sufficient evidence to indicate a difference in mean strength of the grouted sand among the three concentrations of catalyst? Test by using $\alpha = 0.05$.

11.7 Several companies are experimenting with the concept of paying production workers (generally paid by the hour) on a salary basis. It is believed that absenteeism and tardiness will increase under this plan, yet some companies feel that the working environment and overall productivity will improve. Fifty production workers under the salary plan are monitored at Company A, and fifty workers under the hourly plan are monitored at Company B. The number of work hours missed due to tardiness or absenteeism over a one-year period is recorded for each worker. The results are partially summarized in the table:

Source	df	SS	MS	F Ratio
Company		3,237.2		
Error		16,167.7		
Total	99			

a   Fill in the missing information in the table.

b   Is there evidence at the $\alpha = 0.05$ level of significance that the mean number of hours missed differs for employees of the two companies?

11.8   One of the selling points of golf balls is their durability. An independent testing laboratory is commissioned to compare the durability of three different brands of golf balls. Balls of each type will be put into a machine that hits the balls with the same force that a golfer uses on the course. The number of hits required until the outer covering cracks is recorded for each ball, with the results given in the table below. Ten balls from each manufacturer are randomly selected for testing.

Brand		
A	B	C
310	261	233
235	219	289
279	263	301
306	247	264
237	288	273
284	197	208
259	207	245
273	221	271
219	244	298
301	228	276

Is there evidence that the mean durabilities of the three brands differ? Use $\alpha = 0.05$.

11.9   Eight independent observations on percent copper content were taken on each of four castings of bronze. The sample means for each casting are as follows:

Casting	1	2	3	4
Means	80	81	86	90

SSE = 700

Is there sufficient evidence to say that there are differences among the mean percentages of copper for the four castings? Use $\alpha = 0.01$.

11.10   Three thermometers are used regularly in a certain laboratory. To check the relative accuracies of the thermometers, they are randomly and independently placed in a cell kept at zero degrees Celsius. Each thermometer is placed in the cell four times, with the following results (readings in degrees Celsius):

Thermometer		
1	2	3
0.10	−0.20	0.90
0.90	0.80	0.20
−0.80	−0.30	0.30
−0.20	0.60	−0.30

Are there significant differences among the means for the three thermometers? Use $\alpha = 0.05$.

11.11   Casts of aluminum were subjected to four standard heat treatments and their tensile strengths measured. Five measurements were taken on each treatment, with the following results (in 1,000 psi):

Treatment			
A	**B**	**C**	**D**
35	41	42	31
31	40	49	32
40	43	45	30
36	39	47	32
32	45	48	34

Perform an analysis of variance. Do the mean tensile strengths differ from treatment to treatment, at the 5% significance level?

11.12   How does flextime, which allows workers to set their individual work schedules, affect worker job satisfaction? Researchers recently conducted a study to compare a measure of job satisfaction for workers, using three types of work scheduling: flextime, staggered starting hours, and fixed hours. Workers in each group worked according to their specified work-scheduling system for four months. Although each worker filled out job satisfaction questionnaires both before and after the four-month test period, we will examine only the post–test-period scores. The sample sizes, means, and standard deviations of the scores for the three groups are shown in the table:

	Group		
	Flextime	**Staggered**	**Fixed**
Sample Size	27	59	24
Mean	35.22	31.05	28.71
Standard Deviation	10.22	7.22	9.28

a   Assume that the data were collected according to a completely randomized design. Use the information in the table to calculate the treatment totals and SST.

b   Use the values of the sample standard deviation to calculate the sum of squares of deviations *within* each of the three samples. Then calculate SSE, the sum of these quantities.

c   Construct an analysis of variance table for the data.

d   Do the data provide sufficient evidence to indicate differences in mean job satisfaction scores among the three groups? Test by using $\alpha = 0.05$.

e   Find a 90% confidence interval for the difference in mean job satisfaction scores between workers on flextime and workers on fixed schedules.

f   Do the data provide sufficient evidence to indicate a difference in mean scores between workers on flextime and workers using staggered starting hours? Test by using $\alpha = 0.05$.

11.3   A Linear Model for the Completely Randomized Design

The analysis of variance, as presented in Section 11.2, can be produced through use of the regression techniques given in Chapters 9 and 10. This approach is particularly beneficial if a computer program for multiple regression is available.

Consider the experiment of Example 11.1 involving three treatments. If we let Y denote the response variable for the test score of one manager, we can model Y as

$$Y = \beta_0 + \beta_1 x_1 + \beta_2 x_2 + \varepsilon$$

where

$$x_1 = \begin{cases} 1 & \text{if the response is from treatment B} \\ 0 & \text{otherwise} \end{cases}$$
$$x_2 = \begin{cases} 1 & \text{if the response is from treatment C} \\ 0 & \text{otherwise} \end{cases}$$

and ε is the random error, with $E(\varepsilon) = 0$ and $V(\varepsilon) = \sigma^2$. For a response Y_A from treatment A, $x_1 = 0$ and $x_2 = 0$, and hence

$$E(Y_A) = \beta_0 = \mu_A$$

For a response Y_B from treatment B, $x_1 = 1$ and $x_2 = 0$, so

$$E(Y_B) = \beta_0 + \beta_1 = \mu_B$$

It follows that

$$\beta_1 = \mu_B - \mu_A$$

In like manner,

$$E(Y_C) = \beta_0 + \beta_2 = \mu_C$$

and thus

$$\beta_2 = \mu_C - \mu_A$$

The null hypothesis of interest in the analysis of variance, namely, $H_0 : \mu_A = \mu_B = \mu_C$, is now equivalent to $H_0 : \beta_1 = \beta_2 = 0$.

For fitting the above model by the method of least squares, the data would be arrayed as follows:

	y	x_1	x_2
Treatment A	82	0	0
	114	0	0
	90	0	0
	80	0	0
	88	0	0
	93	0	0
	80	0	0
	105	0	0
Treatment B	128	1	0
	90	1	0
	130	1	0
	110	1	0
	133	1	0
	130	1	0
	104	1	0
Treatment C	156	0	1
	128	0	1
	151	0	1
	140	0	1

The method of least squares will give

$$\hat{\beta}_0 = \bar{y}_A$$
$$\hat{\beta}_1 = \bar{y}_B - \bar{y}_A$$
$$\hat{\beta}_2 = \bar{y}_C - \bar{y}_A$$

The error sum of squares for this model, denoted by SSE_2, turns out to be

$$SSE_2 = TSS - SST = 3,181.61$$

Under $H_0 : \beta_1 = \beta_2 = 0$, the reduced model becomes $Y = \beta_0 + \varepsilon$. The error sum of squares, when this model is fit by the method of least squares, is denoted by SSE_1 and is computed to be

$$SSE_1 = TSS = 10,815.2$$

The F-test discussed in Section 10.6 for testing $H_0 : \beta_1 = \beta_2 = 0$ has the form

$$F = \frac{(SSE_1 - SSE_2)/2}{SSE_2/(n-3)}$$

which is equivalent to

$$F = \frac{[TSS - (TSS - SST)]/2}{SSE_2/(n-3)} = \frac{SST/2}{SSE_2/(n-3)} = \frac{MST}{MSE}$$

Thus, the F-test arising from the regression formulation is equivalent to the analysis of variance F-test given in Section 11.2.

All of the analysis of variance problems discussed in this chapter can be formulated in terms of regression problems, as shown in Sections 11.6 and 11.8; Figure 11.2 shows the Minitab printout for the case described above. Note that SSE_2 is the

F I G U R E 11.2
Linear Model Analysis for
Example 11.1

```
The regression equation is
y=91.5+26.4 x1+52.3 x2
Predictor      Coef    Stdev    t-ratio        p
Constant     91.500    4.986     18.35    0.000
x1           26.357    7.298      3.61    0.002
x2           52.250    8.635      6.05    0.000

s=14.10    R-sq=70.6%    R-sq(adj)=66.9%


Analysis of Variance

SOURCE          DF        SS       MS       F       p
Regression       2    7633.6   3816.8   19.19   0.000
Error           16    3181.6    198.9
Total           18   10815.2
```

error sum of squares for this model, and SSE_1 is the total sum of squares. Also,

$$\hat{\beta}_0 = \bar{y}_A = 91.5$$
$$\hat{\beta}_1 = \bar{y}_B - \bar{y}_A = 117.9 - 91.5 = 26.4$$

and

$$\hat{\beta}_2 = \bar{y}_C - \bar{y}_A = 143.8 - 91.5 = 52.3$$

The regression model approach allows easy access to the residuals for further analysis and checking of assumptions. Figure 11.3a shows the residuals plotted against the normal scores. The strong linear trend in this plot indicates that the normality assumption may be satisfied. The stem-and-leaf plot in Figure 11.3b shows a fairly compact display of residuals, somewhat symmetric about zero. Figure 11.3c shows the residuals for each treatment. Again, they are somewhat symmetric about zero and show about the same amount of variability for each of the three treatments. Thus, the assumptions underlying the analysis of variance (or linear regression) seem to be met satisfactorily.

F I G U R E 11.3
Residual Plots for Example 11.1

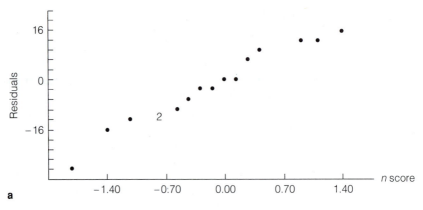

a

Stem-and-leaf of residuals $N = 19$
Leaf unit = 1.0

1	−2	7
1	−2	
2	−1	5
5	−1	311
7	−0	97
(3)	−0	331
9	0	1
8	0	7
7	1	02223
2	1	5
1	2	2

b

F I G U R E **11.3** (Continued)

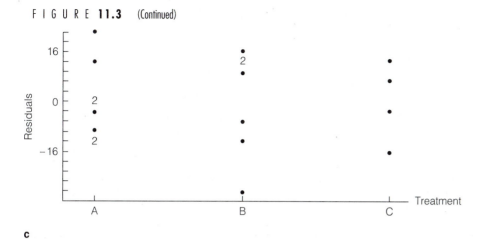

c

Exercises

11.13 Refer to Exercise 11.1. Answer part (b) by writing the appropriate linear model and using regression techniques.

11.14 Refer to Exercise 11.3. Test the assumption of equal percentages of kill for the four rates of application by writing a linear model and using regression techniques.

11.15 Refer to Exercise 11.10. Write a linear model for this experiment. Test for significant differences among the means for the three thermometers by using regression techniques.

11.4 Estimation for the Completely Randomized Design

Confidence intervals for treatment means and differences between treatment means can be produced by the methods introduced in Chapter 8. Recall that we are assuming all treatment populations to be normally distributed with a common variance. Thus, the confidence intervals based on the t distribution can be employed. The common population variance σ^2 is estimated by the pooled sample variance $s^2 = \text{MSE}$.

Since we have k means in an analysis of variance problem, we might want to construct a number of confidence intervals based on the same set of experimental data. For example, we might want to construct confidence intervals for all k means individually or for all possible differences between pairs of means. If such multiple intervals are to be used, we must use extreme care in selecting the confidence co-efficients for the individual intervals so that the *experimentwise* error rate remains small.

For example, suppose we construct two confidence intervals,

$$\overline{Y}_A \pm t_{\alpha/2} \frac{s}{\sqrt{n}} \quad \text{and} \quad \overline{Y}_B \pm t_{\alpha/2} \frac{s}{\sqrt{n}}$$

Now $P(\bar{Y}_A \pm t_{\alpha/2}s/\sqrt{n}$ includes $\mu_A) = 1 - \alpha$, and $P(\bar{Y}_B \pm t_{\alpha/2}s/\sqrt{n}$ includes $\mu_B)$ $= 1 - \alpha$. But

$$P\left(\bar{Y}_A \pm t_{\alpha/2}\frac{s}{\sqrt{n}} \text{ includes } \mu_A \text{ and } \bar{Y}_B \pm t_{\alpha/2}\frac{s}{\sqrt{n}} \text{ includes } \mu_B\right) < 1 - \alpha$$

so the simultaneous coverage probability is *less than* the $(1 - \alpha)$ confidence coefficient that we used on each interval.

If c intervals are to be constructed on one set of experimental data, one method of keeping the simultaneous coverage probability (or confidence coefficient) at a value of *at least* $(1 - \alpha)$ is to make the individual confidence coefficients as close as possible to $1 - (\alpha/c)$. This technique for multiple confidence intervals is outlined in the following summary.

Confidence Intervals for Means in the Completely Randomized Design

Suppose c intervals are to be constructed from one set of data. Single treatment mean μ_i:

$$\bar{y}_i \pm t_{\alpha/2c}\frac{s}{\sqrt{n_i}}$$

Difference between two treatment means $\mu_i - \mu_j$:[†]

$$(\bar{y}_i - \bar{y}_j) \pm t_{\alpha/2c}s\sqrt{\frac{1}{n_i} + \frac{1}{n_j}}$$

Note that $s = \sqrt{\text{MSE}}$ and that all t values depend on $n - k$ degrees of freedom.

EXAMPLE 11.3   Using the data of Example 11.1, construct confidence intervals for all three possible differences between pairs of treatment means so that the simultaneous confidence coefficient is at least 0.95.

Solution   Since there are three intervals to construct ($c = 3$), each interval should be of the form

$$(\bar{y}_i - \bar{y}_j) \pm t_{0.05/2(3)}s\sqrt{\frac{1}{n_i} + \frac{1}{n_j}}$$

Now $t_{0.05/2(3)} = t_{0.05/6}$ is approximately $t_{0.01}$. (This is as close as we can get using Table 5 in the Appendix.) With $n - k = 16$ degrees of freedom, the tabled value is $t_{0.01} = 2.583$.

[†]This is the *Bonferroni* method for conducting multiple comparisons of means.

The three intervals are then constructed as follows:

$$\mu_A - \mu_B : (\bar{y}_A - \bar{y}_B) \pm t_{0.01} s \sqrt{\frac{1}{n_A} + \frac{1}{n_B}} = (91.5 - 117.9) \pm 2.583\sqrt{198.85}\sqrt{\frac{1}{8} + \frac{1}{7}}$$

$$= -26.4 \pm 18.9$$

$$\mu_A - \mu_C : (\bar{y}_A - \bar{y}_C) \pm t_{0.01} s \sqrt{\frac{1}{n_A} + \frac{1}{n_C}} = (91.5 - 143.8) \pm 2.583\sqrt{198.85}\sqrt{\frac{1}{8} + \frac{1}{4}}$$

$$= -52.3 \pm 22.3$$

$$\mu_B - \mu_C : (\bar{y}_B - \bar{y}_C) \pm t_{0.01} s \sqrt{\frac{1}{n_B} + \frac{1}{n_C}} = (117.9 - 143.8) \pm 2.583\sqrt{198.85}\sqrt{\frac{1}{7} + \frac{1}{4}}$$

$$= -25.9 \pm 22.8$$

In interpreting these results, we would be inclined to conclude that all pairs of means differ, because none of the three observed confidence intervals contains zero. Furthermore, inspection of the means leads to the inference that $\mu_C > \mu_B > \mu_A$. The questionnaire appears to be a valid predictor of the level of the managers' technical knowledge of computers. Our combined confidence level for all three confidence intervals is at least 0.95.

If the $t_{0.025}$ value (with 6 degrees of freedom) had been used in place of the $t_{0.01}$ value, all the intervals would have been narrower. However, our combined confidence level would be reduced to something less than 0.95.   ∎

Exercises

11.16   Refer to Exercise 11.1.

a   Find a 90% confidence interval for $(\mu_2 - \mu_3)$. Interpret the interval.

b   What would happen to the width of the confidence interval in part (a) if you quadrupled the number of observations in the two samples?

c   Find a 95% confidence interval for μ_2.

d   Approximately how many observations would be required if you wanted to be able to estimate a population mean correct to within 0.4 with probability equal to 0.95?

11.17   Refer to Exercise 11.2.

a   Suppose $\bar{y}_1 = 3.7$ and $\bar{y}_2 = 4.1$. Do the data provide sufficient evidence to indicate a difference between μ_1 and μ_2? Assume that there are seven observations for each treatment. Test by using $\alpha = 0.10$.

b   Refer to part (a). Find a 90% confidence interval for $(\mu_1 - \mu_2)$.

c   Refer to part (a). Find a 90% confidence interval for μ_1.

11.18   Refer to Exercise 11.3.

a   Estimate the true difference in mean percentage of kill between rate 2 and rate 5. Use a 90% confidence interval.

b   Construct confidence intervals for the six possible differences between treatment means, · with simultaneous confidence coefficient close to 0.90.

11.19   Refer to Exercise 11.6.

a   Find a 95% confidence interval for the difference in mean strength for specimens produced with a 35% concentration of catalyst versus those containing a 45% concentration of catalyst.

b   Construct confidence intervals for the three treatment means, with simultaneous confidence coefficient approximately 0.90.

11.20   Refer to Exercise 11.9. Casting 1 is a standard, and castings 2, 3, and 4 involve slight modifications to the process. Compare the standard with each of the modified processes by producing three confidence intervals with simultaneous confidence coefficient approximately 0.90.

11.5 Analysis of Variance for the Randomized Block Design

In the completely randomized design, only one source of variation, the treatment-to-treatment variation, is specifically considered in the design and analyzed after the data are collected. That is, observations are taken from each of k treatments, and then a test and confidence intervals are constructed to analyze how the treatment means may differ.

Most often, however, the responses of interest, in addition to the treatments under study, are subject to sources of variation. Suppose, for example, that an engineer is studying the gas mileage resulting from four brands of gasoline. If more than one automobile is used in the study, then automobiles would form another important source of variation. To control for this additional variation, it would be crucially important to run each brand of gasoline at least once in each automobile. In this type of experiment, the brands of gasoline are the *treatments* and the automobiles are called the *blocks*. If the order of running the brands in each automobile is randomized, the resulting design is called a *randomized block design*. With the brands denoted by A, B, C, and D and the automobiles by I, II, and III, the structure of the design might look like Figure 11.4. For auto I, brand B was run first, followed in order by A, C, and D.

FIGURE **11.4**
Typical Randomized Block
Design

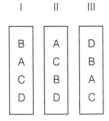

DEFINITION   **11.2**

> A **randomized block design** is a plan for collecting data in which each of k treatments is measured once in each of b blocks. The order of the treatments within the blocks is random. ▪

F I G U R E **11.5**
A Completely Randomized
Design for the Gasoline Mileage
Study

A	B	C	D
—	—	—	—
—	—	—	—
—	—	—	—

The blocking helps the experimenter control a source of variation in responses so that any true treatment differences are more likely to show up in the analysis. Suppose that the gasoline mileage study did, in fact, use more than one automobile, but that we had not blocked on automobiles as indicated above. Assuming that we still take three measurements on each treatment, as indicated in Figure 11.5, two undesirable conditions might result. First, the auto-to-auto variation would add to the variance of the measurements within each sample and hence inflate the MSE. The F-test could then turn out to be nonsignificant even when differences do exist among the true treatment means. Second, all of the A responses could turn out to be from automobile I and all of the B responses from automobile II. If the mean for A and the mean for B are significantly different, we still don't know how to interpret the result. Are the brands really different, or are the automobiles different with respect to gas mileage? A randomized block design will help us avoid both of these difficulties.

After the measurements are completed in a randomized block design, as shown in Figure 11.4, they can be rearranged into a two-way table, as given in Table 11.4.

The total sum of squares (TSS) can now be partitioned into a treatment sum of squares (SST), a block sum of squares (SSB), and an error sum of squares (SSE). The computation formulas are given in Table 11.5.

The test of the null hypothesis that the k treatment means $\mu_1, \mu_2, \ldots, \mu_k$ are equal is, once again, an F-test constructed as the ratio of MST to MSE. In an analogous fashion, we could construct a test of the hypothesis that the block means are equal. The resulting F-test for this case would be the ratio of MSB to MSE.

T A B L E **11.4**
Notation for the Results of the
Randomized Block Experiment

		Treatments				
		1	2	$\cdots$	k	Totals
	1	y_{11}	y_{12}	$\cdots$	y_{1k}	B_1
Blocks	2	y_{21}	y_{22}	$\cdots$	y_{2k}	B_2
	$\vdots$	$\vdots$	$\vdots$	$\cdots$	$\vdots$	$\vdots$
	b	y_{b1}	y_{b2}	$\cdots$	y_{bk}	B_b
Totals		T_1	T_2	$\cdots$	T_k	

Total sample size $= n = bk$
Overall sample total $= \sum y = T_1 + T_2 + \cdots + T_k$
$$= B_1 + B_2 + \cdots + B_b$$
Overall sample mean $= \sum y / n$
Sum of squares of all n measurements $= \sum y^2$
Note that there are k treatments and b blocks.

T A B L E **11.5**
Calculation Formulas for the
Randomized Block Design

$$TSS = \sum y^2 - \frac{(\sum y)^2}{n}$$

$$SST = \frac{1}{b} \sum_{i=1}^{k} T_i^2 - \frac{(\sum y)^2}{n}$$

$$SSB = \frac{1}{k} \sum_{i=1}^{b} B_i^2 - \frac{(\sum y^2)}{n}$$

$$SSE = TSS - SST - SSB$$

$$MST = \frac{SST}{k-1}$$

$$MSB = \frac{SSB}{b-1}$$

$$MSE = \frac{SSE}{n-k-b+1} = \frac{SSE}{(b-1)(k-1)}$$

Test to Compare k Treatment Means for a Randomized Block Design

$H_0: \mu_1 = \mu_2 = \cdots = \mu_k$

H_a: At least two treatment means differ

Test statistic: $F = \dfrac{MST}{MSE}$

Assumptions:

1 Each population (treatment/block combination) has an independent normal probability distribution.

2 The variances of the probability distributions are equal.

Rejection region: $F > F_\alpha[(k-1), (b-1)(k-1)]$

The analysis of variance summary table for a randomized block design is given in Table 11.6.

T A B L E **11.6**
ANOVA for a Randomized Block
Design

Source	df	SS	MS	F Ratio
Treatments	$k-1$	SST	MST	MST/MSE
Blocks	$b-1$	SSB	MSB	MSB/MSE
Errors	$(b-1)(k-1)$	SSE	MSE	
Total	$bk-1$	TSS		

E X A M P L E **11.4** Four chemical treatments for fabric are to be compared with regard to their ability to resist stains. Two different types of fabric are available for the experiment, so it is decided to apply each chemical to a sample of each type of fabric. The result is a randomized block design with four treatments and two blocks. The measurements are as follows:

| | **Block (Fabric)** | | |
	1	2	Totals
1	5	9	14
Treatment 2	3	8	11
(Chemicals) 3	8	13	21
4	4	6	10
Totals	20	36	56

Is there evidence of significant differences among the treatment means? Use $\alpha = 0.05$.

Solution   For these data, the computations (see Table 11.5) are as follows:

$$\text{TSS} = 464 - \frac{(56)^2}{8} = 464 - 392 = 72$$

$$\text{SST} = \frac{1}{2}[(14)^2 + (11)^2 + (21)^2 + (10)^2] - \frac{(56)^2}{8}$$
$$= 429 - 392 = 37$$

$$\text{SSB} = \frac{1}{4}[(20)^2 + (36)^2] - \frac{(56)^2}{8} = 424 - 392 = 32$$

$$\text{SSE} = \text{TSS} - \text{SST} - \text{SSB} = 72 - 37 - 32 = 3$$

The ANOVA summary table then becomes

Source	df	SS	MS	F ratio
Treatments	3	37	$\dfrac{37}{3} = 12.33$	$\dfrac{12.33}{1} = 12.33$
Blocks	1	32	$\dfrac{32}{1} = 32$	$\dfrac{32}{1} = 32$
Errors	3	3	$\dfrac{3}{3} = 1$	
Total	7			

Since $F_{0.05}(3, 3) = 9.28$ and our observed F ratio for treatments is 12.33, we reject $H_0 : \mu_1 = \mu_2 = \mu_3 = \mu_4$ and conclude that there is significant evidence of at least one treatment difference.

We also see in the summary table that the F ratio for blocks is 32. Since $F_{0.05}(1, 3) = 10.13$, we conclude that there is evidence of a difference between the block (fabric) means. That is, the fabrics seem to differ with respect to their ability to resist stains when treated with these chemicals. The decision to "block out" the variability between fabrics appears to have been a good one.

The Minitab printout for this analysis is shown below. Note that the treatment means are provided, along with the standard analysis of variance table. Block means can be easily produced, as well.

```
ANALYSIS OF VARIANCE
SOURCE      DF        SS        MS
TREATMT      3     37.00     12.33
BLOCK        1     32.00     32.00
ERROR        3      3.00      1.00
TOTAL        7     72.00


TREATMT          Mean
        1        7.00
        2        5.50
        3       10.50
        4                  5.00    ▪
```

In the next section, we turn to a brief discussion of how the randomized block experiment can be analyzed by a linear model, and then we consider estimation of means for this design.

Exercises

11.21 A randomized block design was conducted to compare the mean responses for three treatments, A, B, and C, in four blocks. The data are shown below:

		Block		
	1	2	3	4
A	3	6	1	2
Treatment B	5	7	4	6
C	2	3	2	2

a Compute the appropriate sums of squares and mean squares and fill in the entries in the analysis of variance table shown below:

Source	df	SS	MS	F Ratio
Treatment				
Block				
Error				
Total				

b Do the data provide sufficient evidence to indicate a difference among treatment means? Test by using $\alpha = 0.05$.

c Do the data provide sufficient evidence to indicate that blocking was effective in reducing the experimental error? Test by using $\alpha = 0.05$.

11.22   The analysis of variance for a randomized block design produced the ANOVA table entries shown below:

Source	df	SS	MS	F Ratio
Treatment	3	27.1		
Block	5		14.90	
Error		33.4		

Total

a   Complete the ANOVA table.

b   Do the data provide sufficient evidence to indicate a difference among the treatment means? Test by using $\alpha = 0.01$.

c   Do the data provide sufficient evidence to indicate that blocking was a useful design strategy to employ for this experiment? Explain.

11.23   An evaluation of diffusion bonding of zircaloy components is performed. The main objective is to determine which of three elements—nickel, iron, or copper—is the best bonding agent. A series of zircaloy components is bonded, using each of the possible bonding agents. Since there is a great deal of variation among components machined from different ingots, a randomized block design is used, blocking on the ingots. A pair of components from each ingot is bonded together, using each of the three agents, and the pressure (in units of 1,000 pounds per square inch) required to separate the bonded components is measured. The following data are obtained:

| | Bonding Agent | | |
Ingot	Nickel	Iron	Copper
1	67.0	71.9	72.2
2	67.5	68.8	66.4
3	76.0	82.6	74.5
4	72.7	78.1	67.3
5	73.1	74.2	73.2
6	65.8	70.8	68.7
7	75.6	84.9	69.0

Is there evidence of a difference in pressure required to separate the components among the three bonding agents? Use $\alpha = 0.05$.

11.24   A construction firm employs three cost estimators. Usually, only one estimator works on each potential job, but it is advantageous to the company if the estimators are consistent enough so that it does not matter which of the three estimators is assigned to a particular job. To check the consistency of the estimators, several jobs are selected and all three estimators are asked to make estimates. The estimates for each job (in thousands of dollars) by each estimator are given in the table below.

| | Estimator | | |
Job	A	B	C
1	27.3	26.5	28.2
2	66.7	67.3	65.9
3	104.8	102.1	100.8
4	87.6	85.6	86.5
5	54.5	55.6	55.9
6	58.7	59.2	60.1

a Do these estimates provide sufficient evidence that the means for the estimators differ? Use $\alpha = 0.05$.

b Present the complete ANOVA summary table for this experiment.

11.25 A power plant that uses water from the surrounding bay for cooling its condensors is required by the EPA to determine whether discharging its heated water into the bay has a detrimental effect on the flora (plant life) in the water. The EPA requests that the power plant make its investigation at three strategically chosen locations, called *stations*. Stations 1 and 2 are located near the plant's discharge tubes, while Station 3 is located farther out in the bay. During one randomly selected day in each of four months, a diver sent down to each of the stations randomly samples a square-meter area of the bottom and counts the number of blades of the different types of grasses present. The results are as follows for one important grass type:

	Station		
Month	1	2	3
May	28	31	53
June	25	22	61
July	37	30	56
August	20	26	48

a Is there sufficient evidence to indicate that the mean number of blades found per square meter per month differs for the three stations? Use $\alpha = 0.05$.

b Is there sufficient evidence to indicate that the mean number of blades found per square meter differs across the four months? Use $\alpha = 0.05$.

11.26 The Bell Telephone Company's long-distance phone charges may appear to be exorbitant when compared with the charges of some of its competitors, but this is because a comparison of charges between competing companies is often analogous to comparing apples and eggs. Bell's charges for individuals are on a per-call basis. In contrast, its competitors often charge a monthly minimum long-distance fee, reduce the charges as the usage rises, or do both. Shown in the table is a sampling of long-distance charges from Orlando, Florida, to 12 cities for three non-Bell companies offering long-distance service. The data were contained in an advertisement in the *Orlando Sentinel*, 19 March 1984. A note in fine print below the advertisement states that the rates are based on "30 hours of usage" for each of the servicing companies.

From Orlando to	Time	Length of Call (minutes)	Company 1	Company 2	Company 3
New York	Day	2	$0.77	$0.79	$0.66
Chicago	Evening	3	0.69	0.71	0.59
Los Angeles	Day	2	0.87	0.88	0.66
Atlanta	Evening	1	0.22	0.23	0.20
Boston	Day	3	1.15	1.19	0.99
Phoenix	Day	5	1.92	1.98	1.65
West Palm Beach	Evening	2	0.49	0.42	0.40
Miami	Day	3	1.12	1.05	0.99
Denver	Day	10	3.85	3.96	3.30
Houston	Evening	1	0.22	0.23	0.20
Tampa	Day	3	1.06	1.00	0.99
Jacksonville	Day	3	1.06	1.00	0.99

The data in the table are pertinent for companies making phone calls to large cities. Therefore, assume that the cities receiving the calls were randomly selected from among all large cities in the United States.

a   What type of design was used for the data collection?

b   Perform an analysis of variance for the data. Present the results in an ANOVA table.

c   Do the data provide sufficient evidence to indicate differences in mean charges among the three companies? Test by using $\alpha = 0.05$.

d   Company 3 placed the advertisement, so it might be more relevant to compare the charges for companies 1 and 2. Do the data provide sufficient evidence to indicate a difference in mean charges for these two companies? Test by using $\alpha = 0.05$.

11.27   From time to time, one branch office of a company must make shipments to a certain branch office in another state. There are three package delivery services between the two cities where the branch offices are located. Since the price structures for the three delivery services are quite similar, the company wants to compare the delivery times. The company plans to make several different types of shipments to its branch office. To compare the carriers, each shipment will be sent in triplicate, one with each carrier. The results listed in the table are the delivery times in hours.

Shipment	Carrier I	II	III
1	15.2	16.9	17.1
2	14.3	16.4	16.1
3	14.7	15.9	15.7
4	15.1	16.7	17.0
5	14.0	15.6	15.5

Is there evidence of a difference in mean delivery times among the three carriers? Use $\alpha = 0.05$.

11.28   Due to increased energy shortages and costs, utility companies are stressing ways in which home and apartment utility bills can be cut. One utility company reached an agreement with the owner of a new apartment complex to conduct a test of energy-saving plans for apartments. The tests were to be conducted before the apartments were rented. Four apartments were chosen that were identical in size, amount of shade, and direction faced. Four plans were to be tested, one on each apartment. The thermostat was set at 75°F in each apartment, and the monthly utility bill was recorded for each of the three summer months. The results are listed in the table.

Month	Treatment 1	2	3	4
June	$74.44	$68.75	$71.34	$65.47
July	89.96	73.47	83.62	72.33
August	82.00	71.23	79.98	70.87

Treatment 1: no insulation

Treatment 2: insulation in walls and ceilings

Treatment 3: no insulation; awnings for windows

Treatment 4: insulation and awnings for windows

a   Is there evidence that the mean monthly utility bills differ for the four treatments? Use $\alpha = 0.01$.

b   Is there evidence that blocking is important, that is, that the mean utility bills differ for the three months? Use $\alpha = 0.05$.

11.29   A chemist runs an experiment to study the effects of four treatments on the glass transition temperature of a particular polymer compound. Raw material used to make this polymer is bought in small batches. The material is thought to be fairly uniform within a batch but variable between batches. Therefore, each treatment was run on samples from each batch, with the following results (showing temperature in k):

Batch	Treatment I	II	III	IV
1	576	584	562	543
2	515	563	522	536
3	562	555	550	530

a   Do the data provide sufficient evidence to indicate a difference in mean temperature among the four treatments? Use $\alpha = 0.05$.

b   Is there sufficient evidence to indicate a difference in mean temperature among the three batches? Use $\alpha = 0.05$.

c   If the experiment is conducted again in the future, would you recommend any changes in its design?

11.30   The table lists the number of strikes that occurred per year in five U.S. manufacturing industries over the period from 1976 to 1981. Only work stoppages that continued for at least one day and involved six or more workers were counted.

Year	Food and Kindred Products	Primary Metal Industry	Electrical Equipment and Supplies	Fabricated Metal Products	Chemicals and Allied Products
1976	227	197	204	309	129
1977	221	239	199	354	111
1978	171	187	190	360	113
1979	178	202	195	352	143
1980	155	175	140	280	89
1981	109	114	106	203	60

Source: Statistical Abstract of the United States, 1989.

a   Perform an analysis of variance and determine whether there is sufficient evidence to conclude that the mean number of strikes per year differs among the five industries. Test by using $\alpha = 0.05$.

b   Construct the appropriate ANOVA table.

11.6   A Linear Model for the Randomized Block Design

Section 11.3 contains a discussion of the regression approach to the analysis of a completely randomized design. A similar linear model can be written for the randomized block design.

Letting Y denote a response from the randomized block design of Example 11.4, we can write

$$Y = \beta_0 + \beta_1 x_1 + \beta_2 x_2 + \beta_3 x_3 + \beta_4 x_4 + \varepsilon$$

where

$$x_1 = \begin{cases} 1 & \text{if the response is from block 2} \\ 0 & \text{otherwise} \end{cases}$$

$$x_2 = \begin{cases} 1 & \text{if the response is from treatment 2} \\ 0 & \text{otherwise} \end{cases}$$

$$x_3 = \begin{cases} 1 & \text{if the response is from treatment 3} \\ 0 & \text{otherwise} \end{cases}$$

$$x_4 = \begin{cases} 1 & \text{if the response is from treatment 4} \\ 0 & \text{otherwise} \end{cases}$$

and ε is the random error term. The error sum of squares for this model is denoted by SSE_2. Testing the null hypothesis that the four treatment means are equal is now equivalent to testing $H_0 : \beta_2 = \beta_3 = \beta_4 = 0$. The reduced model is then

$$Y = \beta_0 + \beta_1 x_1 + \varepsilon$$

which, when fit by the method of least squares, will produce an error sum of squares denoted by SSE_1.

The F-test for $H_0 : \beta_2 = \beta_3 = \beta_4 = 0$ versus the alternative hypothesis that at least one β_i, $i = 2, 3, 4$, is different from zero then has the form

$$F = \frac{(SSE_1 - SSE_2)(k - 1)}{SSE_2/(b - 1)(k - 1)}$$

$$= \frac{SST/(k - 1)}{SSE_2/(b - 1)(k - 1)} = \frac{MST}{MSE}$$

The test is equivalent to the F-test for equality of treatment means given in Section 11.5.

The regression printouts for the two models (complete and reduced) are shown in Figure 11.6. The sums of squared errors (SSE) for the complete and reduced model are boxed in the figure.

Reduced model: $SSE_1 = 40$

Complete model: $SSE_2 = 3$

Then

$$F = \frac{(40 - 3)/3}{3/3} = 12.33$$

which is exactly the same F value calculated in Example 11.4. The regression approach leads us to the same conclusion as did the ANOVA approach: There is evidence that the mean stain resistances of the four chemicals differ.

The plot of residuals against their normal scores (Figure 11.7) shows a very straight line, indicating that the normality assumption is adequately satisfied for these data.

The randomized block design is only one of many types of block designs. When there are two sources of nuisance variation, it is necessary to block on both sources to eliminate this unwanted variability. Blocking in two directions can be accomplished by using a Latin square design.

To evaluate the toxicity of certain compounds in water, samples must be preserved for long periods of time. Suppose an experiment is being conducted to evaluate the maximum holding time (MHT) for four different preservatives used to treat a mercury-base compound. The MHT is defined as the time that elapses before the solution loses

```
The regression equation is
y=5.00 + 4.00x1-1.50x2+3.50x3-2.00x4


Predictor        Coef      Stdev    t-ratio         p
Constant       5.0000     0.7906      6.32     0.008
x1             4.0000     0.7071      5.66     0.011
x2             -1.500     1.000      -1.50     0.231
x3              3.500     1.000       3.50     0.039
x4             -2.000     1.000      -2.00     0.139


s=1.000    R-sq=95.8%    R-sq(adj)=90.3%


Analysis of Variance


SOURCE          DF         SS        MS        F       p
Regression       4     69.000    17.250    17.25   0.021
Error            3      3.000     1.000
Total            7     72.000
```

a   **Complete Model**

```
The regression equation is
y=5.00+4.00 x1


Predictor        Coef      Stdev    t-ratio         p
Constant        5.000     1.291      3.87   0.008
x1              4.000     1.826      2.19   0.071


s=2.582    R-sq=44.4%     R-sq(adj)=35.2%


Analysis of Variance


SOURCE          DF        SS        MS       F        p
Regression       1    32.000    32.000     4.80    0.071
Error            6    40.000     6.667
Total            7    72.000
```

b   **Reduced Model**

10% of its initial concentration. Both the level of the initial concentration and the analyst who measures the MHT are sources of variation, so an experimental design that blocks on both is necessary for the difference in mean MHT between preservatives to be more accurately estimated. A Latin square design is constructed wherein each preservative is applied exactly *once* to each initial concentration and is analyzed *exactly once* by each analyst. You can see why the design must be "square," since the single application of each treatment level at each block level requires that the number of levels of each block *equal* the number of treatment levels. Thus, to apply

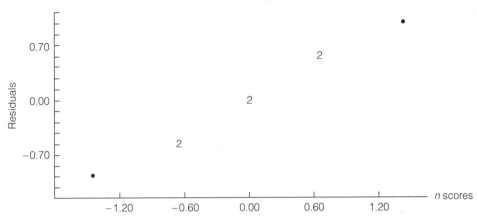

the Latin square design using four different preservatives, we must employ four initial concentrations and four analysts.

The design would be constructed as shown in Figure 11.8, where $P_i =$ preservative i ($i = 1, 2, 3, 4$). Note that each preservative appears exactly once in each row (initial concentration) and in each column (analyst). The resulting design is a 4×4 Latin square. A Latin square design for three treatments will require a 3×3 configuration and, in general, p treatments will require a $p \times p$ array of experimental units. If more observations are desired per treatment, the experimenter would utilize several Latin square configurations in one experiment. In the example above, it would be necessary to run two Latin squares in order to obtain eight observations per treatment.

F I G U R E **11.8**
A Latin Square Design

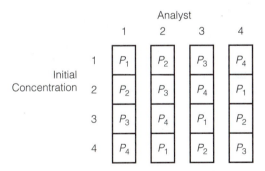

A comparison of mean MHT for any pair of preservatives would eliminate the variability due to initial concentration and to analysts, because each preservative was applied with equal frequency (once) at each level of both blocking factors. Consequently, the block effects would be canceled when mean MHT is compared for any pair of preservatives.

The linear model for the Latin square design is an extension of that for the randomized block design. For the MHT example, the model is

$$Y = \beta_0 + \overbrace{\beta_1 x_1 + \beta_2 x_2 + \beta_3 x_3}^{\text{Initial Concentration}} + \overbrace{\beta_4 x_4 + \beta_5 x_5 + \beta_6 x_6}^{\text{Analyst}}$$
$$+ \underbrace{\beta_7 x_7 + \beta_8 x_8 + \beta_9 x_9}_{\text{Preservative}} + \varepsilon$$

where

$$Y = \text{MHT}$$

$$x_1 = \begin{cases} 1 & \text{if Initial Concentration 1} \\ 0 & \text{otherwise} \end{cases}$$

$$x_2 = \begin{cases} 1 & \text{if Initial Concentration 2} \\ 0 & \text{otherwise} \end{cases}$$

$$x_3 = \begin{cases} 1 & \text{if Initial Concentration 3} \\ 0 & \text{otherwise} \end{cases}$$

$$x_4 = \begin{cases} 1 & \text{if Analyst 1} \\ 0 & \text{otherwise} \end{cases}$$

$$x_5 = \begin{cases} 1 & \text{if Analyst 2} \\ 0 & \text{otherwise} \end{cases}$$

$$x_6 = \begin{cases} 1 & \text{if Analyst 3} \\ 0 & \text{otherwise} \end{cases}$$

$$x_7 = \begin{cases} 1 & \text{if Preservative 1} \\ 0 & \text{otherwise} \end{cases}$$

$$x_8 = \begin{cases} 1 & \text{if Preservative 2} \\ 0 & \text{otherwise} \end{cases}$$

$$x_9 = \begin{cases} 1 & \text{if Preservative 3} \\ 0 & \text{otherwise} \end{cases}$$

Thus, dummy variables are used to represent the block effects as well as the treatments. To test the null hypothesis of no treatment differences, we test $H_0 : \beta_7 = \beta_8 = \beta_9 = 0$ by fitting complete and reduced models, just as we did for the randomized block design. There are equivalent ANOVA formulas (not presented here), similar to those presented in Section 11.5 for the randomized block design, for calculating the F test statistic for treatment (and block) differences.

In summary, there are many types of block designs for comparing treatment means while controlling sources of nuisance variation. As with the randomized block and Latin square designs, other block designs can be analyzed most easily by using the linear model approach, with dummy variables representing both block and treatment effects. Then the complete and reduced model approach is used to test for treatment and block effects on the mean response.

Note that, in order for the F-tests given above to be valid, we must still make the usual assumptions about the independence and normality of population measurements and equal variances across all treatment/block combinations.

Exercises

11.31   Refer to Exercise 11.21. Answer parts (b) and (c) by fitting a linear model to the data and using regression techniques.

11.32   Refer to Exercise 11.25. Answer part (a) by fitting a linear model to the data and using regression techniques.

11.33   Suppose that three automobile engine designs are being compared to determine differences in mean time between breakdowns.

    a   Show how three test automobiles and three different test drivers could be used in a 3×3 Latin square design aimed at comparing the engine designs.

    b   Write the linear model for the design in part (a).

    c   What null hypothesis would you test to determine whether the mean time between breakdowns differs for the engine designs?

11.7 Estimation for the Randomized Block Design

As in the case of the completely randomized design, we might want to estimate the mean for a particular treatment, or the difference between the means of two treatments, after conducting the initial F-test. Since we will generally be interested in c intervals simultaneously, we will want to set the confidence coefficient of each individual interval at $(1 - \alpha/c)$. This will guarantee that the probability of all c intervals simultaneously covering the parameters being estimated is at least $1 - \alpha$. A summary of the estimation procedure is given below.

Confidence Intervals for Means in the Randomized Block Design

Suppose c intervals are to be constructed from one set of data.

$$\text{Single treatment mean } \mu_i: \; \bar{y}_i \pm t_{\alpha/2c} s \sqrt{\frac{1}{b}}$$

Difference between two treatment means $\mu_i - \mu_j :$ [†]

$$(\bar{y}_i - \bar{y}_j) \pm t_{\alpha/2c} s \sqrt{\frac{1}{b} + \frac{1}{b}}$$

Note that $s = \sqrt{\text{MSE}}$ and that all t values depend on $(k - 1)(b - 1)$ degrees of freedom, where k is the number of treatments and b the number of blocks.

E X A M P L E   **11.5**   Refer to the four chemical treatments and two blocks (fabrics) of Example 11.4. It is of interest to estimate simultaneously all possible differences between treatment means. Construct confidence intervals for these differences with a simultaneous confidence coefficient of at least 0.90, approximately.

    Solution   Since there are four treatment means, μ_1, μ_2, μ_3, and μ_4, there will be $c = 6$ differences of the form $\mu_i - \mu_j$. Since the simultaneous confidence coefficient is to be $1 - \alpha = 0.90$, $\alpha/2c$ becomes $0.10/2(6) = 0.008$. The closest tabled t value in Table

---

[†] This is again the *Bonferroni* method for conducting multiple comparisons of means.

5 of the Appendix has a tail area of 0.01, so we will use $t_{0.01}$ as an approximation to $t_{0.008}$.

For the data of Example 11.4, the four sample treatment means are

$$\bar{y}_1 = 7.0, \quad \bar{y}_2 = 5.5, \quad \bar{y}_3 = 10.5, \quad \bar{y}_4 = 5.0$$

Also, $s = \sqrt{\text{MSE}} = \sqrt{1} = 1$ and $b = 2$. Thus, any interval of the form

$$(\bar{y}_i - \bar{y}_j) \pm t_{0.01} s \sqrt{\frac{1}{b} + \frac{1}{b}}$$

will become

$$(\bar{y}_i - \bar{y}_j) \pm (4.541)(1) \sqrt{\frac{1}{2} + \frac{1}{2}}$$

or

$$(\bar{y}_i - \bar{y}_j) \pm 4.541$$

since the degrees of freedom are $(b-1)(k-1) = 3$. The six confidence intervals are as follows:

Parameter	Interval		
$\mu_1 - \mu_2$	$(\bar{y}_1 - \bar{y}_2) \pm 4.541$	or	1.5 ± 4.541
$\mu_1 - \mu_3$	$(\bar{y}_1 - \bar{y}_3) \pm 4.541$	or	-3.5 ± 4.541
$\mu_1 - \mu_4$	$(\bar{y}_1 - \bar{y}_4) \pm 4.541$	or	2.0 ± 4.541
$\mu_2 - \mu_3$	$(\bar{y}_2 - \bar{y}_3) \pm 4.541$	or	-5.0 ± 4.541
$\mu_2 - \mu_4$	$(\bar{y}_2 - \bar{y}_4) \pm 4.541$	or	0.5 ± 4.541
$\mu_3 - \mu_4$	$(\bar{y}_3 - \bar{y}_4) \pm 4.541$	or	5.5 ± 4.541

The sample data would suggest that μ_2 and μ_3 are significantly different and that μ_3 and μ_4 are significantly different, since these intervals do not overlap zero. ■

Exercises

11.34 Refer to Exercise 11.23.

a Form a 95% confidence interval to estimate the true mean difference in pressure between nickel and iron. Interpret this interval.

b Form confidence intervals on the three possible differences between the means for the bonding agents, using a simultaneous confidence coefficient of approximately 0.90.

11.35 Refer to Exercise 11.24. Use a 90% confidence interval to estimate the true difference between the mean responses given by estimators B and C.

11.36 Refer to Exercise 11.25. What are the significant differences among the station means? (Use a simultaneous confidence coefficient of 0.90.)

11.37   Refer to Exercise 11.26. Compare the three telephone companies' mean rates by using a simultaneous confidence coefficient of approximately 0.90.

11.38   Refer to Exercise 11.27.

    a   Use a 99% confidence interval to estimate the difference between the mean delivery times for carriers I and II.

    b   What assumptions are necessary for the validity of the procedures you used in part (a)?

11.8 The Factorial Experiment

The response of interest Y is, in many experimental situations, related to one or more other variables that can be controlled by the experimenter. For example, the yield Y in an experimental study of the process of manufacturing a chemical may be related to the temperature x_1 and pressure x_2 at which the experiment is run. The variables, temperature and pressure, can be controlled by the experimenter. In a study of heat loss through ceilings of houses, the amount of heat loss Y will be related to the thickness of insulation x_1 and the temperature differential x_2 between the inside and outside of the house. Again, the thickness of insulation and the temperature differential can be controlled by the experimenter as he or she designs the study. The strength Y of concrete may be related to the amount of aggregate x_1, the mixing time x_2, and the drying time x_3.

The variables that are thought to affect the response of interest and are controlled by the experimenter are called *factors*. The various settings of these factors in an experiment are called *levels*. A factor/level combination then defines a *treatment*. In the chemical yield example, the temperatures of interest might be 90°, 100°, and 110°C. These are the three experimental levels of the factor "temperature." The pressure settings (levels) of interest might be 400 and 450 psi. Setting temperature at 90°C and pressure at 450 psi would define a particular treatment, and observations on yield could then be obtained for this combination of levels. Note that three levels of temperature and two of pressure would result in $2 \times 3 = 6$ different treatments, if all factor/level combinations are to be used.

An experiment in which treatments are defined by specified factor/level combinations is referred to as a *factorial experiment*. In this section, we assume that r observations on the response of interest are realized from each treatment, with each observation being independently selected. (That is, the factorial experiment is run in a completely randomized design with r observations per treatment.) The objective of the analysis is to decide if, and to what extent, the various factors affect the response variable. Does the yield of chemical increase as temperature increases? Does the strength of concrete decrease as less and less aggregate is used? These are the types of questions we will be able to answer using the results of this section.

11.8.1   Analysis of Variance for the Factorial Experiment

We illustrate the concepts and calculations involved in the analysis of variance for a factorial experiment by considering an example consisting of two factors each at two levels. In the production of a certain industrial chemical, the yield Y depends on the cooking time x_1 and the cooling time x_2. Two cooking times and two cooling times are of interest. Only the relative magnitudes of the levels are important in the analysis of factorial experiments, and thus we can call the two levels of cooking time 0 and 1. That is, x_1 can take on the value 0 or 1. Similarly, we can call the levels of cooling time 0 and 1; that is, x_2 can take on the values 0 or 1. For convenience, we will refer to cooking time as factor A and cooling time as factor B. Schematically, we then have the arrangement of four treatments seen in Figure 11.9, and we will assume that r observations will be available from each.

F I G U R E 11.9
A 2×2 Factorial Experiment
(r observations per treatment)

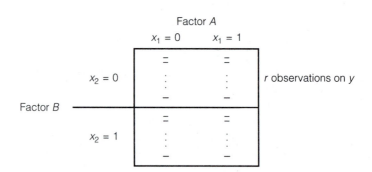

The analysis of this factorial experiment could be accomplished by writing a linear regression model for the response Y as a function of x_1 and x_2 and then employing the theory of Chapters 9 and 10. We will proceed by writing the models for illustrative purposes, but we will then present simple calculation formulas that make the actual fitting of the models by least squares unnecessary.

Since Y depends, supposedly, on x_1 and x_2, we could start with the simple model

$$E(Y) = \beta_0 + \beta_1 x_1 + \beta_2 x_2$$

In Figure 11.10, we see that this model implies that the rate of change on $E(Y)$ as x_1 goes from 0 to 1 is the same (β_1) for each value of x_2. This is referred to as the no-interaction case.

F I G U R E 11.10
$E(Y)$ for a 2×2 Factorial
Experiment with No Interaction

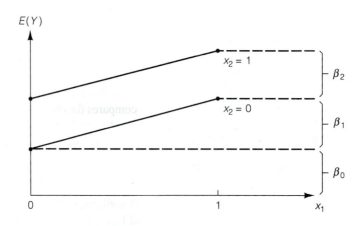

It is quite likely that the rate of change on $E(Y)$ as x_1 goes from 0 to 1 will be different for different values of x_2. To account for this *interaction*, we write the model as

$$E(Y) = \beta_0 + \beta_1 x_1 + \beta_2 x_2 + \beta_3 x_1 x_2$$

$\beta_3 x_1 x_2$ is referred to as the interaction term. Now if $x_2 = 0$, we have

$$E(Y) = \beta_0 + \beta_1 x_1$$

and if $x_2 = 1$, we have

$$E(Y) = (\beta_0 + \beta_2) + (\beta_1 + \beta_3)x_1$$

The slope (rate of change) when $x_2 = 0$ is β_1, but the slope when $x_2 = 1$ changes to $\beta_1 + \beta_3$. Figure 11.11 depicts a possible interaction case.

F I G U R E **11.11**
$E(Y)$ for a 2 × 2 Factorial
Experiment with Interaction

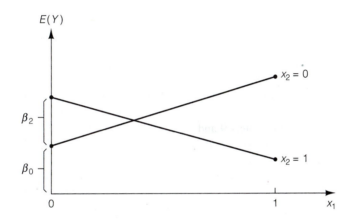

One way to proceed with an analysis of this 2 × 2 factorial experiment is as follows. First test the hypothesis $H_0 : \beta_3 = 0$. If you reject this hypothesis, thus establishing that there is evidence of interaction, *do not* proceed with tests on β_1 and β_2. The significance of β_3 is enough to establish the fact that there are some differences among the treatment means. It may then be best simply to estimate the individual treatment means, or differences between means, by the methods given in Section 11.4.

If the hypothesis $H_0 : \beta_3 = 0$ is not rejected, it is then appropriate to test the hypotheses $H_0 : \beta_2 = 0$ and $H_0 : \beta_1 = 0$. These are often referred to as tests of "main effects." The test of $\beta_2 = 0$ actually compares the observed mean of all observations at $x_2 = 1$ with the corresponding mean at $x_2 = 0$, regardless of the value of x_1. That is, each level of B is averaged over all levels of A, and then the levels of B are compared. This is a reasonable procedure when there is no evidence of interaction, since the change in response from the low to the high level of B is essentially the same for each level of A. Similarly, the test of $\beta_1 = 0$ compares the mean response at the high level of factor A ($x_1 = 1$) with that at the low level of A ($x_1 = 0$), with the means computed over all levels of factor B.

Using the method of fitting complete and reduced models (Section 10.7), we complete the analysis by fitting the models given below and calculating SSE for each:

1 $Y = \beta_0 + \beta_1 x_1 + \beta_2 x_2 + \beta_3 x_1 x_2 + \varepsilon$, SSE_1
2 $Y = \beta_0 + \beta_1 x_1 + \beta_2 x_2 + \varepsilon$, SSE_2
3 $Y = \beta_0 + \beta_1 x_1 + \varepsilon$, SSE_3
4 $Y = \beta_0 + \varepsilon$, SSE_4

Now

$$\text{SSE}_2 - \text{SSE}_1 = \text{SS}(A \times B)$$

the sum of squares for the $A \times B$ interaction. Also,

$$\text{SSE}_3 - \text{SSE}_2 = \text{SS}(B)$$

the sum of squares due to factor B, and

$$\text{SSE}_4 - \text{SSE}_3 = \text{SS}(A)$$

the sum of squares due to factor A. This approach works well if a computer is available for fitting the indicated models.

The sums of squares shown above can be calculated by the direct formulas given in Table 11.7. Table 11.7 shows the general formulas for any two-factor factorial experiment with r observations per cell, assuming that there are a levels of factor A and b levels of factor B. That is, the data must follow the format given in Figure 11.12.

Notice that the treatment sum of squares is now partitioned into a sum of squares for A, a sum of squares for B, and an interaction sum of squares. The degrees of freedom for the latter three sums of squares are $(a-1)$, $(b-1)$, and $(a-1)(b-1)$, respectively. These facts are shown in the analysis of variance table given in Table 11.8.

We illustrate the calculations and F-tests in the following example.

T A B L E **11.7**
Calculation Formulas for a
Two-Factor Factorial Experiment
with r Observations per Cell

$n = abr = $ total number of observations

$$\text{TSS} = \sum y^2 - \frac{(\sum y)^2}{n}$$

$$\text{SST} = \sum_{ij} \frac{T_{ij}^2}{r} - \frac{(\sum y)^2}{n}$$

$$\text{SS}(A) = \sum_{i=1}^{a} \frac{A_i^2}{br} - \frac{(\sum y)^2}{n}$$

$$\text{SS}(B) = \sum_{i=1}^{b} \frac{B_j^2}{ar} - \frac{(\sum y)^2}{n}$$

$$\text{SS}(A \times B) = \text{SST} - \text{SS}(A) - \text{SS}(B)$$

$$\text{SSE} = \text{TSS} - \text{SST}$$

F I G U R E **11.12**
A General Two-Factor Factorial Experiment with r Observation per Cell

		Factor A					
		1	2	3	$\cdots$	a	Totals
	1	— — — T_{11}	— — — T_{12}	— — — T_{13}		— — — T_{1a}	B_1
	2	— — — T_{21}	— — — T_{22}	— — — T_{23}		— — — T_{2a}	B_2
Factor B	3	— — — T_{31}	— — — T_{32}	— — — T_{33}		— — — T_{3a}	B_3
	$\vdots$						
		— — — T_{b1}	— — — T_{b2}	— — — T_{b3}		— — — T_{ba}	B_b
	Totals	A_1	A_2	A_3		A_a	$\sum y$

T_{ij} = Total of observations in row i and column j

T A B L E **11.8** ANOVA Table for a Two-Factor Factorial Experiment in a Completely Randomized Design

Source	df	SS	MS	F ratio
Treatments	$ab-1$	SST	$\text{MST} = \dfrac{\text{SST}}{ab-1}$	
Factor A	$a-1$	SS(A)	$\text{MS}(A) = \dfrac{\text{SS}(A)}{a-1}$	MS(A)/MSE
Factor B	$b-1$	SS(B)	$\text{MS}(B) = \dfrac{\text{SS}(B)}{b-1}$	MS(B)/MSE
$A \times B$ Interaction	$(a-1)(b-1)$	SS($A \times B$)	$\text{MS}(A \times B) = \dfrac{\text{SS}(A+B)}{(a-1)(b-1)}$	MS($A \times B$)/MSE
Error	$ab(r-1)$	SSE	$\text{MSE} = \dfrac{\text{SSE}}{ab(r-1)}$	
Total	$n-1$	TSS		

Note: The assumption is made that all factor combinations have independent normal distributions with equal variances.

E X A M P L E **11.6** For the chemical experiment with two cooking times (factor A) and two cooling times (factor B), the yields are as given below, with $r = 2$ observations per treatment.

$$A$$

	$x_1 = 0$	$x_1 = 1$	
$x_2 = 0$	9 8 $\overline{17}$	5 6 $\overline{11}$	28
$x_2 = 1$	8 7 $\overline{15}$	3 4 $\overline{7}$	22
	32	18	50

B

Perform an analysis of variance.

Solution   Using the calculation formulas of Table 11.7, with $a = 2$, $b = 2$, and $r = 2$, we have

$$\text{TSS} = 9^2 + 8^2 + \cdots + 3^2 + 4^2 - \frac{(50)^2}{8} = 31.5$$

$$\text{SST} = \frac{1}{2}[(17)^2 + (11)^2 + (15)^2 + (7)^2] - \frac{(50)^2}{8} = 29.5$$

$$\text{SS}(A) = \frac{1}{4}[(32)^2 + (18)^2] - \frac{(50)^2}{8} = 24.5$$

$$\text{SS}(B) = \frac{1}{4}[(28)^2 + (22)^2] - \frac{(50)^2}{8} = 4.5$$

$$\text{SSE}(A \times B) = \text{SST} - \text{SS}(A) - \text{SS}(B) = 0.5$$

and

$$\text{SSE} = \text{TSS} - \text{SST} = 2.0$$

The analysis of variance, using the format of Table 11.8, proceeds as follows:

Source	df	SS	MS	F Ratio
Treatments	3	29.5		
A	1	24.5	24.5	$\dfrac{24.5}{0.5} = 49.0$
B	1	4.5	4.5	$\dfrac{4.5}{0.5} = 9.0$
$A \times B$	1	0.5	0.5	$\dfrac{0.5}{0.5} = 1.0$
Error	4	2.0	0.5	
Total	7	31.5		

Since $F_{0.05}(1, 4) = 7.71$, the interaction effect is not significant. We can then proceed with "main effect" tests for factors A and B. The F ratio of 49.0 for factor A is highly significant. Thus, there is a significant difference between the mean response at $x_1 = 0$ and that at $x_1 = 1$. Looking at the data, we see that the yield falls off as cooking time goes from the low level to the high level.

The F ratio of 9.0 for factor B is also significant. The mean yield also seems to decrease as cooling time is changed from the low to the high level.   ▪

In Example 11.7, we show an analysis in which the interaction term is highly significant.

E X A M P L E   **11.7**   Suppose a chemical experiment like the one of Example 11.6 (two factors each at two levels) gave the following results:

$$A$$

	$x_1 = 0$	$x_1 = 1$	
$x_2 = 0$	9 8	5 6	28
$x_2 = 1$	3 4	8 7	22
	24	26	50

B is on the left, labeling the rows $x_2 = 0$ and $x_2 = 1$.

Perform an analysis of variance. If the interaction is significant, construct confidence intervals for the six possible differences between treatment means.

Solution   The data involve the same responses as in Example 11.6, with the observations in the lower left and lower right cells interchanged. Thus, we still have

$$\text{TSS} = 9^2 + 8^2 + \cdots + 8^2 + 7^2 - \frac{(50)^2}{8} = 31.5$$

$$\text{SST} = \frac{1}{2}[(17)^2 + (11)^2 + (7)^2 + (15)^2] - \frac{(50)^2}{8} = 29.5$$

and

$$\text{SSE} = \text{TSS} - \text{SST} = 2.0$$

Now

$$\text{SS}(A) = \frac{1}{4}[(24)^2 + (26)^2] - \frac{(50)^2}{8} = 0.5$$

$$\text{SS}(B) = \frac{1}{4}[(28)^2 + (22)^2] - \frac{(50)^2}{8} = 4.5$$

and

$$\text{SSE}(A \times B) = \text{SST} - \text{SS}(A) - \text{SS}(B) = 24.5$$

The analysis of variance table is as follows:

Source	df	SS	MS	F Ratio
Treatments	3	29.5		
A	1	0.5	0.5	
B	1	4.5	4.5	
$A \times B$	1	24.5	24.5	49.0
Error	4	2.0	0.5	
Total	7	31.5		

At the 5% level, the interaction term is highly significant. Hence, we will not make any "main effect" tests but instead will place confidence intervals on all possible differences between treatment means. For convenience, we will identify the $(x_1 = 0, x_2 = 0)$ combination as treatment 1 with mean μ_1, $(x_1 = 1, x_2 = 0)$ as treatment 2 with mean μ_2, $(x_1 = 0, x_2 = 1)$ as treatment 3 with mean μ_3, and $(x_1 = 1, x_2 = 1)$ as treatment 4 with mean μ_4. To form a confidence interval on $\mu_i - \mu_j$, we follow the format of Section 11.4, which gives the interval in the form

$$(\bar{y}_i - \bar{y}_j) \pm t_{\alpha/2c} s \sqrt{\frac{1}{n_i} + \frac{1}{n_j}}$$

In this problem, $c = 6$ and $n_i = n_j = 2$. Also, $s = \sqrt{\text{MSE}}$ and is based on 4 degrees of freedom. If we settle for $1 - \alpha = 0.90$, then $\alpha/2c = 0.10/12 \approx 0.01$. From Table 5 in the Appendix, $t_{0.01} = 3.747$, with 4 degrees of freedom. Thus, all six intervals will be of the form

$$(\bar{y}_i - \bar{y}_j) \pm 3.747\sqrt{0.5}\sqrt{\frac{1}{2} + \frac{1}{2}}$$

or

$$(\bar{y}_i - \bar{y}_j) \pm 2.65$$

The sample means are given by

$$\bar{y}_1 = 8.5, \quad \bar{y}_2 = 5.5, \quad \bar{y}_3 = 3.5, \quad \bar{y}_4 = 7.5$$

and the confidence intervals would then be

$$(\bar{y}_1 - \bar{y}_2) \pm 2.65 \quad \text{or} \quad 3.0 \pm 2.65^*$$
$$(\bar{y}_1 - \bar{y}_3) \pm 2.65 \quad \text{or} \quad 5.0 \pm 2.65^*$$
$$(\bar{y}_1 - \bar{y}_4) \pm 2.65 \quad \text{or} \quad 1.0 \pm 2.65$$
$$(\bar{y}_2 - \bar{y}_3) \pm 2.65 \quad \text{or} \quad 2.0 \pm 2.65$$
$$(\bar{y}_2 - \bar{y}_4) \pm 2.65 \quad \text{or} \quad -2.0 \pm 2.65$$
$$(\bar{y}_3 - \bar{y}_4) \pm 2.65 \quad \text{or} \quad -4.0 \pm 2.65^*$$

The significant differences are then between μ_1 and μ_2, μ_1 and μ_3, and μ_3 and μ_4 (see the intervals marked with an asterisk). In practical terms, the yield is reduced significantly as x_1 goes from 0 to 1 and x_2 remains at 0. Also, the yield is increased as x_1 goes from 0 to 1 and x_2 remains at 1. The yield decreases as x_2 goes from 0 to 1 with x_1 at 0, but there is no significant change in mean yield between x_2 at 0 and x_2 at 1 with x_1 at 1. Since μ_1 and μ_4 do not appear to differ, the best choices for maximizing yield appear to be either $(x_1 = 0, x_2 = 0)$ or $(x_1 = 1, x_2 = 1)$.   ∎

11.8.2   Fitting Higher-Order Models

Suppose the chemical example under discussion had three cooking times (levels of factor A) and two cooling times (levels of factor B). If the three cooking times were equally spaced (such as 20 minutes, 30 minutes, and 40 minutes), then the levels of A could be coded as $x_1 = -1$, $x_1 = 0$, and $x_1 = 1$. Again, only the *relative* magnitudes of the levels are important in the analysis. Since there are now three levels of factor A, a quadratic term in x_1 can be added to the model. Thus, the complete model would have the form

$$E(Y) = \beta_0 + \beta_1 x_1 + \beta_2 x_1^2 + \beta_3 x_2 + \beta_4 x_1 x_2 + \beta_5 x_1^2 x_2$$

Note that if $x_2 = 0$, we simply have a quadratic model in x_1, given by

$$E(Y) = \beta_0 + \beta_1 x_1 + \beta_2 x_1^2$$

If $x_2 = 1$, we have another quadratic model in x_1, but the coefficients have changed:

$$E(Y) = (\beta_0 + \beta_3) + (\beta_1 + \beta_4)x_1 + (\beta_2 + \beta_5)x_1^2$$

If $\beta_4 = \beta_5 = 0$, the two curves will have the same shape, but if either β_4 or β_5 differs from zero, the curves will differ in shape. Thus, β_4 and β_5 are both components of interaction. (See Figure 11.13.)

F I G U R E **11.13** $E(Y)$ for a 3×2 Factorial Experiment

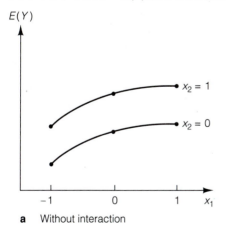

a   Without interaction

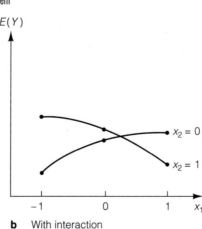

b   With interaction

In the analysis, we first test $H_0 : \beta_4 = \beta_5 = 0$ (no interaction). If we reject this hypothesis, then we do no further F-tests, but we may compare treatment means. If we do not reject the hypothesis of no interaction, we proceed with tests of $H_0 : \beta_3 = 0$ (no main effect for factor B) and $H_0 : \beta_1 = \beta_2 = 0$ (no main effect for factor A). The test of $H_0 : \beta_3 = 0$ merely compares the mean responses at the high and low levels of factor B. The test of $H_0 : \beta_1 = \beta_2 = 0$ looks for both linear and quadratic trends among the three means for the levels of factor A, averaged across the levels of factor B.

The analysis could be conducted by fitting complete and reduced models and using the results of Section 10.7, or by using the computational formulas of Table 11.7. We illustrate both approaches with the following example.

EXAMPLE 11.8 In the manufacture of a certain beverage, an important measurement is the percentage of impurities present in the final product. The following data show the percentage of impurities present in samples taken from products manufactured at three different temperatures (factor A) and two sterilization times (factor B). The three levels of A were actually 75°, 100°, and 125°C. The two levels of B were actually 15 minutes and 20 minutes. The data are as follows:

		A: 75°C	A: 100°C	A: 125°C	
B	15 min.	14.05 14.93 28.98	10.55 9.48 20.03	7.55 6.59 14.14	63.15
B	20 min.	16.56 15.85 32.41	13.63 11.75 25.38	9.23 8.78 18.01	75.80
		61.39	45.41	32.15	138.95

a Perform an analysis of variance using the formulas of Table 11.7.

b Perform an analysis of variance using the regression approach.

Solution **a** Using the formulas of Table 11.7, with $a = 3$, $b = 2$, and $r = 2$, we have

$$\text{TSS} = (14.05)^2 + \cdots + (8.78)^2 - \frac{(138.95)^2}{12} = 124.56$$

$$\text{SST} = \frac{1}{2}[(28.98)^2 + (20.03)^2 + \cdots + (18.01)^2] - \frac{(138.95)^2}{12} = 121.02$$

$$\text{SS}(A) = \frac{1}{4}[(61.39)^2 + (45.41)^2 + (32.15)^2] - \frac{(138.95)^2}{12} = 107.18$$

$$\text{SS}(B) = \frac{1}{6}[(63.15)^2 + (75.80)^2] - \frac{(138.95)^2}{12} = 13.34$$

$$\text{SS}(A \times B) = \text{SST} - \text{SS}(A) - \text{SS}(B) = 0.50$$

and

$$SSE = TSS - SST = 3.54$$

The analysis of variance table then has the following form:

Source	df	SS	MS	F Ratio
Treatments	5	121.02		
A	2	107.18	53.59	90.83
B	1	13.34	13.34	22.59
$A \times B$	2	0.50	0.25	0.42
Error	6	3.54	0.59	
Total	11	124.56		

Since $F_{0.05}(2, 5) = 5.14$ and $F_{0.05}(1, 6) = 5.99$, it is clear that the interaction is not significant and that the main effects for both factor A and factor B are highly significant. The means for factor A tend to decrease as the temperature goes from low to high, and this decreasing trend is of approximately the same degree for both the low and high levels of factor B.

b   For the regression approach, we begin by fitting the complete model, with interaction, as shown in the following printout. [These models use a $(-1, 0, +1)$ coding for A and a $(-1, +1)$ coding for B.]

```
The regression equation is
y=11.4-3.66x1+1.34x2+0.340x1**2+0.055x1x2-0.425x1**2x2

Predictor      Coef     Stdev    t-ratio        p
Constant    11.3525    0.3841      29.56    0.000
x1          -3.6550    0.2716     -13.46    0.000
x2           1.3375    0.3841       3.48    0.013
x1**2        0.3400    0.4704       0.72    0.497
x1x2         0.0550    0.2716       0.20    0.846
x1**2x2     -0.4250    0.4704      -0.90    0.401

s=0.7682    R-sq=97.2%    R-sq(adj)=94.8%

Analysis of Variance

SOURCE         DF          SS        MS        F        p
Regression      5     121.022    24.204    41.01    0.000
Error           6       3.541     0.590
Total          11     124.562
```

Thus, $SSE_1 = 3.541$.

To test for significant interaction ($H_0 : \beta_4 = \beta_5 = 0$), we fit the reduced model as follows:

```
The regression equation is
y=11.4-3.66x1+1.05x2+0.340x1**2
```

```
Predictor      Coef      Stdev     t-ratio          p
Constant    11.3525     0.3556       31.92      0.000
x1          -3.6550     0.2515      -14.54      0.000
x2           1.0542     0.2053        5.13      0.000
x1**2        0.3400     0.4355        0.78      0.457
```

```
s=0.7112    R-sq=96.8%    R-sq(adj)=95.5%
```

```
Analysis of Variance
```

```
SOURCE        DF         SS         MS         F          p
Regression     3    120.516     40.172     79.41      0.000
Error          8      4.047      0.506
Total         11    124.563
```

We now have $\text{SSE}_2 = 4.047$, and the F statistic becomes

$$\frac{(4.047 - 3.541)/2}{3.541/6} = \frac{0.25}{0.59} = 0.42$$

the same as the analysis of variance F ratio for interaction.

　　To test for significant main effect for factor B, we must reduce the model even further by letting the coefficient of x_2 equal zero.

```
The regression equation is
y=11.4-3.66x1+0.340x1**2
```

```
Predictor      Coef      Stdev     t-ratio          p
Constant    11.3525     0.6949       16.34      0.000
x1          -3.6550     0.4913       -7.44      0.000
x1**2        0.3400     0.8510        0.40      0.699
```

```
s=1.390    R-sq=86.0%    R-sq(adj)=82.9%
```

```
Analysis of Variance
```

```
SOURCE        DF         SS         MS         F          p
Regression     2    107.180     53.590     27.75      0.000
Error          9     17.382      1.931
Total         11    124.563
```

Now

$$\text{SSE}_3 = 17.382$$

and

$$\text{SS}(B) = \text{SSE}_3 - \text{SSE}_2$$
$$= 17.382 - 4.047 = 13.335$$

Also,

$$SSE_4 = SST = 124.563$$

so that

$$SS(A) = SSE_4 - SSE_3$$
$$= 124.563 - 17.382 = 107.181$$

Note that $SS(B)$ and $SS(A)$ as calculated from the regression models agree with the calculations from the analysis of variance table.

Since the regression model allows easy checking of residuals, we plot these residuals (from the main effects model, with no interaction) against their normal scores, as shown in Figure 11.14. The nearly straight-line plot indicates that the normality assumption is reasonable. The residuals are also plotted against the three levels of A, the two levels of B and the predicted values ($\hat{y}$) in Figure 11.15. Note that variation is greatest at the middle level of A and the high level of B, which might suggest that the process is more difficult to control at those levels. The plot of residuals against predicted values shows a fairly random scatter, as should be produced by a model that fits well. ■

F I G U R E **11.14**   Normal Scores Plot of Residuals from Main Effects Model

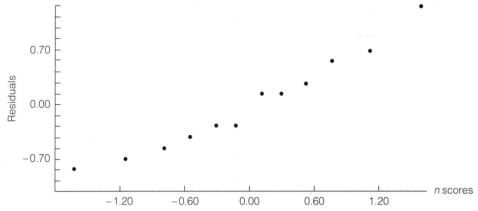

Two-level factorial experiments (2×2 or, in general, $k \times 2$ if there are k factors of interest) are often used as screening experiments in preliminary investigations. If a process depends on many factors, the two-level experiments are run to determine which factors seem to be affecting the response most dramatically. A higher-level experiment can be run subsequently to find optimal settings for the levels of those factors. Since two-level factorial experiments are so common, we present more details related to their analysis, focusing on graphical techniques.

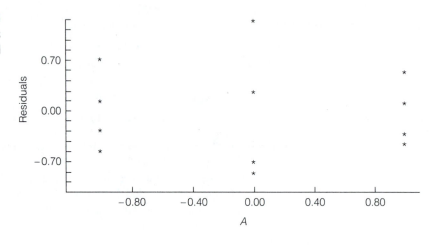

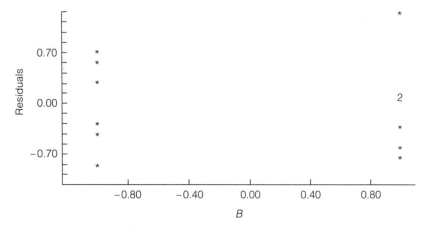

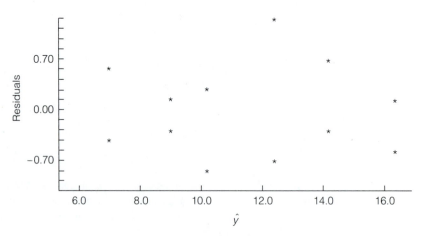

11.8.3   Graphic Methods of Analysis

Going back to the data of Example 11.6 for a 2×2 factorial experiment, we first observe that the numerical values of x_1 and x_2 have no effect on the analysis. Thus, we will simply refer to the low levels as $-$ and the high levels as $+$. We then have two observations on yield at $A^- B^-$ (low cooking time, low cooling time), two at $A^- B^+$, and so on. A convenient way to display the data is on back-to-back dotplots, as in Figure 11.16. We can immediately see a large shift from high to low values as we move from A^- to A^+, and a less dramatic shift downward as we move from B^- to B^+.

FIGURE **11.16**
Dotplots for Factors A and B

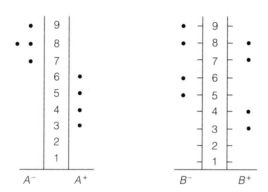

In a similar way, interaction terms can be plotted. The upper left and lower right cells of the data table make positive ($+$) contributions to the interaction, and the other two cells make negative ($-$) contributions. Thus, the interaction dotplot looks like Figure 11.17. Although the $(AB)^+$ values are more spread out, there is little shift of location as we move from $(AB)^-$ to $(AB)^+$. Thus, there is little interaction effect.

FIGURE **11.17**
Dotplot for Interaction

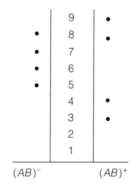

The shifts seen in Figures 11.16 and 11.17 can be quantified more precisely in terms of averages. The average of all eight yields is 6.25, while the average of A^+ yields is 4.5 and the average of A^- yields is 8.0. Similar averages can be calculated for B^+ and B^-. Figure 11.18 shows the overall average as the center line and the factor-level (cell) averages plotted relative to this center. This is a convenient way to show all averages on one plot.

F I G U R E **11.18**

Plot of Factor-Level Means

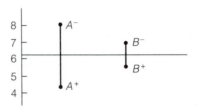

Notice that the distance between A^+ and the overall mean is the same as the distance between A^- and the overall mean, with a similar result holding for factor B. This distance,

$$\frac{1}{2}(A^+ \text{ average} - A^- \text{ average}) = -1.75$$

is called the *main effect* due to A. It tells us that, on the average, the high cooking time lowers the yield by 1.75 units compared to the experimentwide average yield. Similarly, the main effect for B, -0.75, tells us that the high cooling time lowers the average yield by 0.75 units.

The *interaction effect* is measured by

$$\frac{1}{2}[(AB)^+ \text{ average} - (AB)^- \text{ average}] = -0.25$$

and turns out to be small compared to the main effects in this case.

A simple and elegant graphical interpretation of main effects and interaction is developed in Figure 11.19. First we plot the A^- and A^+ averages and connect them with a dashed line. Since we can think of A^- as being at -1 and A^+ as being at $+1$ on the horizontal axis, these points are two units apart. This makes the slope of the dashed line equal to the main effect for A, -1.75 in this case.

We then plot the $A^- B^-$ and $A^+ B^-$ averages and connect them with the line labeled B^-. Repeating the procedure for $A^- B^+$ and $A^+ B^+$ gives the B^+ line. The slope of the B^+ line is

$$\frac{1}{2}(A^+ B^+ \text{ average} - A^- B^+ \text{ average}) = \frac{1}{2}(3.5 - 7.5) = -2.0$$

The difference between the slopes of the dashed line (main effect for A) and the B^+ line is

$$[-2.0 - (-1.75)] = -0.25$$

which is the interaction effect. Whereas a main effect measures the difference between an overall average and a factor-level average, an interaction effect measures the difference between the slope of a main effect line and the slope for that factor plotted at a fixed level of a second factor. To see what happens to these lines for a situation with a large interaction, make a similar plot for the data in Example 11.7.

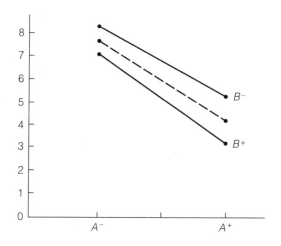

F I G U R E **11.19**
Plot of Main Effect and
Interaction Lines

EXAMPLE   **11.9**   An experiment was conducted to compare the strengths of two brands of facial tissues, the famous brands X and Y. The two brands (two levels of factor A) were each tested under both dry and damp conditions (the two levels of factor B). Thus, the four treatments tested are

$A^- B^-$	(brand X, dry)
$A^+ B^-$	(brand Y, dry)
$A^- B^+$	(brand X, damp)
$A^+ B^+$	(brand Y, damp)

Four tissues were tested under each treatment. The strength measurement was produced as follows. A tissue was stretched over the opening of a plastic cup and held by a rubber band. Then a marble was dropped onto the stretched tissue. The lowest height (in inches) from which a marble that went through the tissue was dropped is the recorded strength measurement. The data are shown in the accompanying table. (One of the authors actually conducted the experiment!)

		A	
		−	+
B	−	14, 14 12, 11	10, 11 11, 12
	+	4, 4 3, 5	5, 5 6, 6

The 14 in the upper left corner means that a tissue of brand X that was dry broke first from a marble dropped from 14 inches above its surface. Analyze the data for significant main effects and interaction.

Solution   Since these are two-digit measurements, stem-and-leaf plots can be used in place of dotplots. The main effect and interaction plots are shown below.

$$
\begin{array}{cc}
\begin{array}{c|c|c}
 & 1 & \\
4421 & 1 & 0112 \\
5 & 0 & 5566 \\
443 & 0 &
\end{array} &
\begin{array}{c|c|c}
 & 1 & \\
44221110 & 1 & \\
 & 0 & 55566 \\
 & 0 & 344
\end{array} &
\begin{array}{c|c|c}
 & 1 & \\
2110 & 1 & 1244 \\
5 & 0 & 5566 \\
443 & 0 &
\end{array} \\
A^{-} \qquad A^{+} & B^{-} \qquad B^{+} & (AB)^{-} \qquad (AB)^{+}
\end{array}
$$

These plots show little shift as we move from A^{-} to A^{+}, although A^{-} values are more spread out than A^{+} values. A dramatic shift is seen as we move from B^{-} to B^{+}, indicating a large main effect for moisture content. There is some slight shift to larger values as we move from $(AB)^{-}$ to $(AB)^{+}$, signaling a possible interaction.

The main effect for A (brand) is

$$
\frac{1}{2}(A^{+} \text{ average} - A^{-}\text{average}) = \frac{1}{2}(8.250 - 8.375) = -0.0625
$$

which indicates little difference in strength between brands.

The main effect for B (dry versus damp) is

$$
\frac{1}{2}(B^{+} \text{ average} - B^{-}\text{average}) = \frac{1}{2}(4.750 - 11.875) = -3.5625
$$

a rather large value in comparison to the sizes of the measurements. It should not be surprising that damp tissues have less strength than dry tissues. We do not have to experiment very carefully to see this result!

The interaction effect is

$$
\frac{1}{2}[(AB)^{+} \text{ average} -(AB)^{-} \text{ average}] = \frac{1}{2}(9.125 - 7.500) = 0.8125
$$

This is much larger than the main effect for A and indicates that the interaction is obscuring the effect of brands. A more careful look at the data suggests that brand X tissues tend to be stronger than brand Y tissues when dry, but brand Y tissues are slightly stronger when damp. When do you want tissues to be strong? Do you think you could state a preference for either brand?

The main effect plot here could be misleading because of the interaction, so we show only the interaction plot in Figure 11.20. Note that the two solid lines have obviously different slopes, indicating some interaction.

F I G U R E **11.20**
Interaction Plot for Example
11.9

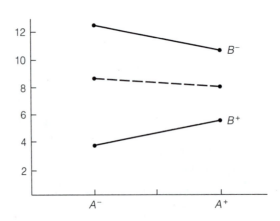

The analysis of variance shown below for these data shows a large interaction mean square and a very large mean square due to factor B, just as we expected. Both of these effects are highly significant.

```
ANALYSIS OF VARIANCE

SOURCE              DF              SS              MS
A                    1           0.063           0.063
B                    1         203.062         203.062
INTERACTION          1          10.563          10.563
ERROR               12          11.750           0.979
TOTAL               15         225.437    ■
```

The analyses of factorial experiments with more than two factors proceed along similar lines. However, we will not present those results here. The interested reader should consult a text on experimental design.

Exercises

11.39 In pressure sintering of alumina, two important variables controlled by the experimenter are the pressure and time of sintering. An experiment involving two pressures and two times, with three specimens tested on each pressure/time combination, showed the following densities (in g/cc):

		Pressure	
		100 psi	200 psi
Time	10 min.	3.6, 3.5, 3.3	3.8, 3.9, 3.8
	20 min.	3.4, 3.7, 3.7	4.1, 3.9, 4.2

a   Perform an analysis of variance, constructing appropriate tests for interaction and main effects. Use $\alpha = 0.05$.

b   Estimate the difference between the true densities for specimens sintered for 20 minutes and those sintered for 10 minutes. Use a 95% confidence coefficient.

11.40 The yield percentage Y of a chemical process depends on the temperature at which the process is run and the length of time the process is active. For two levels of temperature and two lengths of time, the yields were as follows (with two observations per treatment):

		Temperature	
		Low	High
Time	Low	24 25	28 30
	High	26 28	23 22

a Perform an analysis of variance, testing first for interaction. Use $\alpha = 0.05$.

b If interaction is significant, construct confidence intervals on the six possible differences between treatment means, using a simultaneous confidence coefficient of approximately 0.90.

11.41 The yield percentage Y of a certain precipitate depends on the concentration of the reactant and the rate of addition of diammonium hydrogen phosphate. Experiments were run at three different concentration levels and three addition rates. The yield percentages, with two observations per treatment, were as follows:

<table>
<tr><td></td><td></td><td colspan="3" align="center">Concentration of Reactant</td></tr>
<tr><td></td><td></td><td align="center">−1</td><td align="center">0</td><td align="center">1</td></tr>
<tr><td rowspan="6">Addition
Rate</td><td rowspan="2">−1</td><td align="center">90.1</td><td align="center">92.4</td><td align="center">96.4</td></tr>
<tr><td align="center">90.3</td><td align="center">91.8</td><td align="center">96.8</td></tr>
<tr><td rowspan="2">0</td><td align="center">91.2</td><td align="center">94.3</td><td align="center">98.2</td></tr>
<tr><td align="center">92.3</td><td align="center">93.9</td><td align="center">97.6</td></tr>
<tr><td rowspan="2">1</td><td align="center">92.4</td><td align="center">96.1</td><td align="center">99.0</td></tr>
<tr><td align="center">92.5</td><td align="center">95.8</td><td align="center">98.9</td></tr>
</table>

a Perform an analysis of variance, constructing all appropriate tests at the 5% significance level.

b Estimate the average yield percentage, in a 95% confidence interval, for the treatment at the middle level of both concentration and addition rate.

11.42 How do women compare with men in their ability to perform laborious tasks that require strength? Some information on this question is provided in a study by M. D. Phillips and R. L. Pepper of the fire-fighting ability of men and women ("Shipboard Fire-Fighting Performance of Females and Males," *Human Factors*, 24, 1982). Phillips and Pepper conducted a 2×2 factorial experiment to investigate the effect of the factor sex (male or female) and the factor weight (light or heavy) on the length of time required for a person to perform a particular fire-fighting task. Eight persons were selected for each of the $2 \times 2 = 4$ sex/weight categories of the 2×2 factorial experiment, and the length of time needed to complete the task was recorded for each of the 32 persons. The means and standard deviations of the four samples are shown in the table.

	Light		Heavy	
	Mean	Standard Deviation	Mean	Standard Deviation
Female	18.30	6.81	14.50	2.93
Male	13.00	5.04	12.25	5.70

a Calculate the total of the $n = 8$ time measurements for each of the four categories of the 2×2 factorial experiment.

b Use the result of part (a) to calculate the sums of squares for sex, weight, and the sex/weight interaction.

c Calculate each sample variance. Then calculate the sum of squares of deviations *within* each sample for each of the four samples.

d   Calculate SSE. [*Hint*: SSE is the pooled sum of squares of the deviations calculated in part (c).]

e   Now that you know SS(Sex), SS(Weight), SS(Sex $\times$ Weight), and SSE, find SS(Total).

f   Summarize the calculations in an analysis of variance table.

g   Explain the practical significance of the presence (or absence) of sex/weight interaction. Do the data provide evidence of a sex/weight interaction?

h   Do the data provide sufficient evidence to indicate a difference in time required to complete the task between light men and light women? Test by using $\alpha = 0.05$.

i   Do the data provide sufficient evidence to indicate a difference in time to complete the task between heavy men and heavy women? Test by using $\alpha = 0.05$.

11.43   Refer to Exercise 11.42. Phillips and Pepper (1982) give data on another 2×2 factorial experiment utilizing 20 males and 20 females. The experiment involved the same treatments, with ten persons assigned to each sex/weight category. The response measured for each person was the pulling force the person was able to exert on the starter cord of a P-250 fire pump. The means and standard deviations of the four samples (corresponding to the $2 \times 2 = 4$ categories of the experiment) are shown in the table.

	Light Mean	Light Standard Deviation	Heavy Mean	Heavy Standard Deviation
Female	46.26	14.23	62.72	13.97
Male	88.07	8.32	86.29	12.45

a   Use the procedures outlined in Exercise 11.42 to perform an analysis of variance for the experiment. Display your results in an ANOVA table.

b   Explain the practical significance of the presence (or absence) of sex/weight interaction. Do the data provide sufficient evidence of a sex/weight interaction?

c   Do the data provide sufficient evidence to indicate a difference in force exerted between light men and light women? Test by using $\alpha = 0.05$.

d   Do the data provide sufficient evidence to indicate a difference in force exerted between heavy men and heavy women? Test by using $\alpha = 0.05$.

11.44   In analyzing coal samples for ash content, two types of crucibles and three temperatures were used in a complete factorial arrangement, with the following results:

		Crucible Steel		Crucible Silica	
	825	8.7	7.2	9.3	9.1
Temperature	875	9.4	9.6	9.7	9.8
	925	10.1	10.2	10.4	10.7

(Two independent observations were taken on each treatment.) Note that "crucible" is not a quantitative factor, but it can still be considered to have two levels even though the levels cannot be ordered as to high and low.

Perform an analysis of variance, conducting appropriate tests at the 5% significance level. Does there appear to be a difference between the two types of crucible with respect to average ash content?

11.45 Four different types of heads were tested on each of two sealing machines. Four independent measurements of strain were then made on each head/machine combination. The data are as follows:

	Head 1	2	3	4
Machine A	3	2	6	3
	0	1	5	0
	4	1	8	1
	1	3	8	1
Machine B	6	7	2	4
	8	6	0	7
	6	3	1	6
	5	4	2	7

a Perform an analysis of variance, making all appropriate tests at the 5% level of significance.

b If you were to use machine A, which head type would you recommend for maximum strain resistance in the seal? Why?

11.46 An experiment was conducted to determine the effects of two alloying elements (carbon and manganese) on the ductility of specimens of metal. Two specimens from each treatment were measured, and the data (in work required to break a specimen of standard dimension) are as follows:

		Carbon 0.2%	0.5%
Manganese	0.5%	34.5	38.2
		37.5	39.4
	1.0%	36.4	42.8
		37.1	43.4

a Perform an analysis of variance, conducting all appropriate tests at the 5% significance level.

b Which treatment would you recommend to maximize average breaking strength? Why?

11.9 Activities for Students: Designing an Experiment

After studying 11 chapters of statistics, it is time to put your knowledge into practice! This is an open-ended activity that allows each student to design and analyze his or her own experiment. Before we offer suggestions on some areas for possible investigation, let's review the essential components of a designed experiment as they were first introduced in Chapter 2.

1 Clearly define the *question* to be investigated.

2 Identify the key variables to be used as *treatments*.

3 Identify the other important variables that can be *controlled*.

4 Identify important background (lurking) variables that cannot be controlled but should be balanced by *randomization*.

5 Randomly assign treatments to the *experimental units*.

6 Decide on a *method of measurement* that will minimize *measurement bias*.

7 Organize the *data collection* and *data analysis*.

8 Plan a careful and thorough *data analysis*.

9 Write *conclusions* in light of the original question.

10 Plan a *follow-up* study to answer the question more completely or to answer the next logical question on the issue at hand.

At this point, we know that steps 2 through 5 may involve a completely randomized design or a randomized block design (or a more complex design not covered in this text) and that treatments may arise from combinations of controllable factors set in a factorial arrangement. Keep these considerations in mind as you think about designing a study of your own.

The best ideas for designed experiments originate in your own field of interest. So begin by thinking carefully about questions in your field that could possibly be answered by an experiment that you can design and execute. In planning your project, go through the ten steps listed above and decide how you plan to handle each one.

If no ideas seem readily available, here are a few suggestions that have worked for others:

1 How does exercise affect pulse rate, for those who exercise regularly and those who do not?

2 Which paper airplane design stays afloat the longest or can be controlled to land closest to a target?

3 Which Lego car design is best for speed? for distance?

4 What food mixture (cookies, cake, salad) seems to produce the best taste?

5 What is my optimum travel path to school in order to minimize time en route?

6 Should I change my bowling stance (or tennis grip or golf swing) to improve my performance?

If ideas are still needed, read the article by W. G. Hunter on "Some Ideas About Teaching Design of Experiments, with 2^5 Examples of Experiments Conducted by Students," *The American Statistician*, vol. 31, no. 1, 1977. This article contains an abundance of ideas on experiments actually conducted by college students. In any event, have fun!

11.10 Summary

Before measurements are actually made in any experimental investigation, careful attention should be paid to the *design* of the experiment. If the experimental objective is the comparison of k treatment means and no other major source of variation is present, then a *completely randomized design* will be adequate. If a second source of variation is present, it can often be controlled through the use of a *randomized block design*. These two designs serve as an introduction to the topic of design of experiments, but several more complex designs, such as the *Latin square*, can be considered.

The analysis of variance for either design considered in this chapter can be carried out either through a linear model (regression) approach or through the use of direct calculation formulas. The reader would benefit by trying both approaches on some examples to gain familiarity with the concepts and techniques.

The *factorial experiment* arises in situations in which treatments are defined by various combinations of factor levels. Notice that the factorial arrangement defines the treatments of interest but is *not* in itself a design. Factorial experiments can be run in completely randomized, randomized block, or other designs, but we considered their use only in the completely randomized design in this chapter. Many more topics dealing with factorial experiments can be found in texts that focus on experimental design and analysis of variance.

Supplementary Exercises

11.47   Three methods have been devised to reduce the time spent in transferring materials from one location to another. With no previous information available on the effectiveness of these three approaches, a study is performed. Each approach is tried several times, and the amount of time to completion (in hours) is recorded in the table.

Method		
A	B	C
8.2	7.9	7.1
7.1	8.1	7.4
7.8	8.3	6.9
8.9	8.5	6.8
8.8	7.6	
	8.5	

a What type of experimental design was used?

b Is there evidence that the mean time to completion of the task differs for the three methods? Use $\alpha = 0.01$.

c Form a 95% confidence interval for the mean time to completion for method B.

11.48 One important consideration in determining which location is best for a new retail business is the amount of traffic that passes the location each business day. Counters are placed at each of four locations on the five weekdays, and the number of cars passing each location is recorded in the following table:

	Location			
Day	I	II	III	IV
1	453	482	444	395
2	500	605	505	490
3	392	400	383	390
4	441	450	429	405
5	427	431	440	430

a What type of design does this represent?

b Is there evidence of a difference in the mean number of cars per day at the four locations?

c Estimate the difference between the mean numbers of cars that pass locations I and III each weekday.

11.49 Mileage tests were performed to compare three different brands of regular gas. Four different automobiles were used in the experiment, and each brand of gas was used in each car until the mileage was determined. The results, in miles per gallon, are shown in the table.

	Automobile			
Brand	1	2	3	4
A	20.2	18.7	19.7	17.9
B	19.7	19.0	20.3	19.0
C	18.3	18.5	17.9	21.1

a Is there evidence of a difference in the mean mileage rating among the three brands of gasoline? Use $\alpha = 0.05$.

b Construct the ANOVA summary table for this experiment.

c Is there evidence of a difference in the mean mileage for the four models? That is, is blocking important in this type of experiment? Use $\alpha = 0.05$.

d Form a 99% confidence interval for the difference between the mileage ratings of brands B and C.

e Form confidence intervals for the three possible differences between the means for the brands, with a simultaneous confidence coefficient of approximately 0.90.

11.50   The table shows the partially completed analysis of variance for a two-factor factorial experiment.

Source	df	SS	MS	F Ratio
A	3	2.6		
B	5	9.2		
AB			3.1	
Error		18.7		
Total	47			

a   Complete the analysis of variance table.

b   Give the number of levels for each factor and the number of observations per factor-level combination.

c   Do the data provide sufficient evidence to indicate an interaction between the factors? Test by using $\alpha = 0.05$.

d   State the practical implications of your test results in part (c).

11.51   The data shown in the table are for a 4×3 factorial experiment with two observations per factor-level combination.

Level of B

		1	2	3
	1	2	5	1
		4	6	3
Level of A	2	5	2	10
		4	2	9
	3	7	1	5
		10	0	3
	4	8	12	7
		7	11	4

a   Perform an analysis of variance for the data and display the results in an analysis of variance table.

b   Do the data provide sufficient information to indicate an interaction between the factors? Test by using $\alpha = 0.05$.

c   Suppose the objective of the experiment is to select the factor-level combination with the largest mean. Based on the data and using a simultaneous confidence coefficient of approximately 0.90, which pairs of means appear to differ?

11.52   The set of activities and decisions through which a firm moves from its initial awareness of an innovative industrial procedure to its final adoption or rejection of the innovation is referred to as the *industrial adoption process*. The process can be described as having five stages: (1) awareness, (2) interest (additional information requested), (3) evaluation (advantages and disadvantages compared), (4) trial (innovation tested), and (5) adoption. As part of a study of the industrial adoption process, S. Ozanne and R. Churchill (1971) hypothesized that firms use a greater number of informational inputs to the process (e.g., visits by salespersons) in the later stages than in the earlier stages. In particular, they tested the hypotheses that a greater number of informational inputs are used in the interest stage than in the awareness stage and that a greater number are used in the evaluation stage than in the interest stage. Ozanne and Churchill collected the information given in the table on the number of informational inputs used by a sample of 37 industrial firms that had recently adopted a particular new automatic machine tool.

	Number of Information Sources Used				Number of Information Sources Used		
Company	Awareness Stage	Interest Stage	Evaluation Stage	Company	Awareness Stage	Interest Stage	Evaluation Stage
1	2	2	3	20	1	1	1
2	1	1	2	21	1	2	1
3	3	2	3	22	1	4	3
4	2	1	2	23	2	3	3
5	3	2	4	24	1	1	1
6	3	4	6	25	3	1	4
7	1	1	2	26	1	1	5
8	1	3	4	27	1	3	1
9	2	3	3	28	1	1	1
10	3	2	4	29	3	2	2
11	1	2	4	30	4	2	3
12	3	3	3	31	1	2	3
13	4	4	4	32	1	3	2
14	4	2	7	33	1	2	2
15	3	2	2	34	2	3	2
16	4	5	4	35	1	2	4
17	2	1	1	36	2	2	3
18	1	1	3	37	2	3	2
19	1	3	2				

Source: S. Ozanne and R. Churchill, *Marketing Models, Behavioral Science Applications*, Intex Educational Publishers, 1971, pp. 249–265.

a Do the data provide sufficient evidence to indicate that differences exist in the mean number of informational inputs of the three stages of the industrial adoption process studied by Ozanne and Churchill? Test by using $\alpha = 0.05$.

b Find a 90% confidence interval for the difference in the mean number of informational inputs between the evaluation and awareness stages.

11.53 England has experimented with different 40-hour workweeks to maximize production and minimize expenses. A factory tested a 5-day week (8 hours per day), a 4-day week (10 hours per day), and a $3\frac{1}{3}$-day week (12 hours per day), with the weekly production results shown in the table (in thousands of dollars worth of items produced).

8-Hour Day	10-Hour Day	12-Hour Day
87	75	95
96	82	76
75	90	87
90	80	82
72	73	65
86		

a What type of experimental design was employed here?

b Construct an ANOVA summary table for this experiment.

c Is there evidence of a difference in the mean productivities for the three lengths of work-days?

d Form a 90% confidence interval for the mean weekly productivity when 12-hour workdays are used.

11.54 To compare the preferences of technicians for three brands of calculators, each of three technicians was required to perform an identical series of calculations on each of the three calculators, A, B, and C. To avoid the possibility of fatigue, a suitable time period separated each set of calculations, and the calculators were used in random order by each technician. A preference rating, based on a 0–100 scale, was recorded for each machine/technician combination. These data are shown in the table:

Technician	Calculator Brand		
	A	B	C
1	85	90	95
2	70	70	75
3	65	60	80

a Do the data provide sufficient evidence to indicate a difference in technician preference among the three brands? Use $\alpha = 0.05$.

b Why did the experimenter have each technician test all three calculators? Why not randomly assign three different technicians to each calculator?

11.55 Sixteen workers were randomly selected to participate in an experiment to determine the effects of work scheduling and method of payment on attitude toward the job. Two types of scheduling were employed, the standard 8 to 5 workday and a modification whereby the worker was permitted to start the day at either 7 A.M. or 8 A.M. and to vary the starting time as desired; in addition, the worker was allowed to choose, on a daily basis, either a $\frac{1}{2}$-hour or 1-hour lunch period. The two methods of payment were a standard hourly rate and a reduced hourly rate with an added piece rate based on the worker's production. Four workers were randomly assigned to each of the four scheduling/payment combinations, and each completed an attitude test after one month on the job. The test scores are shown in the table.

		Payment	
		Hourly Rate	Hourly and Piece Rate
Scheduling	8 to 5	54, 68 55, 63	89, 75 71, 83
	Worker-modified Schedule	79, 65 62, 74	83, 94 91, 86

a Construct an analysis of variance table for the data.

b Do the data provide sufficient information to indicate a factor interaction? Test by using $\alpha = 0.05$. Explain the practical implications of your test.

c Do the data indicate that any of the scheduling/payment combinations produce a mean attitude score that is clearly higher than the other three? Test by using a simultaneous Type I error rate of approximately $\alpha = 0.05$, and interpret your results.

11.56 An experiment was conducted to compare the yields of orange juice for six different juice extractors. Because of a possibility of a variation in the amount of juice per orange from one truckload of oranges to another, equal weights of oranges from a single truckload were assigned to each extractor, and this process was repeated for 15 loads. The amount of juice recorded for each extractor and each truckload produced the following sums of squares:

Source	df	SS	MS	F Ratio
Extractor		84.71		
Truckload		159.29		
Error		94.33		
Total		339.33		

a Complete the ANOVA table.

b Do the data provide sufficient evidence to indicate a difference in mean amount of juice extracted by the six extractors? Use $\alpha = 0.05$.

11.57 A farmer wants to determine the effect of five different concentrations of lime on the pH (acidity) of the soil on a farm. Fifteen soil samples are to be used in the experiment, five from each of the three different locations. The five soil samples from each location are then randomly assigned to the five concentrations of lime, and one week after the lime is applied the pH of the soil is measured. The data are shown in the following table:

	Lime Concentration				
Location	0	1	2	3	4
I	3.2	3.6	3.9	4.0	4.1
II	3.6	3.7	4.2	4.3	4.3
III	3.5	2.9	4.0	3.9	4.2

a What type of experimental design was used here?

b Do the data provide sufficient evidence to indicate that the five concentrations of lime have different mean soil pH levels? Use $\alpha = 0.05$.

c Is there evidence of a difference in soil pH levels among locations? Test by using $\alpha = 0.05$.

11.58 In the hope of attracting more riders, a city transit company plans to have express bus service from a suburban terminal to the downtown business district. These buses should save travel time. The city decides to perform a study of the effect of four different plans (such as a special bus lane and traffic signal progression) on the travel time for the buses. Travel times (in minutes) are measured for several weekdays during a morning rush-hour trip while each plan is in effect. The results are recorded in the table.

	Plan		
1	2	3	4
27	25	34	30
25	28	29	33
29	30	32	31
26	27	31	
	24	36	

a What type of experimental design was employed?

b Is there evidence of a difference in the mean travel times for the four plans? Use $\alpha = 0.01$.

c Form a 95% confidence interval for the difference between plan 1 (express lane) and plan 3 (a control: no special travel arrangements).

11.59   Five sheets of writing paper are randomly selected from each of three batches produced by a certain company. A measure of brightness is obtained for each sheet, with the following results:

	Batch	
1	2	3
28	34	27
32	36	25
25	32	29
27	38	31
26	39	21

a   Do the mean brightness measurements seem to differ among the three batches? Use $\alpha = 0.05$.

b   If you were to select the batch with the largest mean brightness, which would you select? Why?

11.60   In order to provide its clients with comparative information on two large suburban residential communities, a realtor wants to know the average home value in each community. Eight homes are selected at random within each community and are appraised by the realtor. The appraisals are given in the table (in thousands of dollars). Can you conclude that the average home value is different in the two communities? You have three ways of analyzing this problem [parts (a) through (c)].

Community	
A	B
43.5	73.5
49.5	62.0
38.0	47.5
66.5	36.5
57.5	44.5
32.0	56.0
67.5	68.0
71.5	63.5

a   Use the two-sample t statistic to test $H_0 : \mu_A = \mu_B$.

b   Consider the regression model

$$y = \beta_0 + \beta_1 x + \varepsilon$$

where

$$x = \begin{cases} 1 & \text{if community } B \\ 0 & \text{if community } A \end{cases}$$

$$y = \text{appraised price}$$

Since $\beta_1 = \mu_B - \mu_A$, testing $H_0 : \beta_1 = 0$ is equivalent to testing $H_0 : \mu_A = \mu_B$. Use the partial reproduction of the regression printout shown here to test $H_0 : \beta_1 = 0$. Use $\alpha = 0.05$.

```
SOURCE              df    SUM OF SQUARES    MEAN SQUARE    F VALUE      PR>F
MODEL                1       40.64062500    40.64062500       0.21    0.6501
ERROR               14     2648.71875000   189.19419643
CORRECTED TOTAL     15     2689.35937500                   R-SQUARE    ROOT MSE
                                                           0.015112  13.75478813


                              T FOR H0:     PR>|T|   STD ERROR OF
PARAMETER      ESTIMATE     PARAMETER=0               ESTIMATE
INTERCEPT    53.25000000        10.95      0.0001    4.86305198
X             3.18750000         0.46      0.6501    6.87739406
```

   c Use the ANOVA method to test $H_0 : \mu_A = \mu_B$. Use $\alpha = 0.05$.

   d Using the results of the three tests in parts (a) through (c), verify that the tests are equivalent (for this special case, $k = 2$) for the completely randomized design in terms of the test-statistic value and rejection region. For the three methods used, what are the advantages and disadvantages (limitations) of using each method in analyzing results for this type of experimental design?

11.61 Drying stresses in wood produce acoustic emissions (AE), and the rate of acoustic emission can be monitored as a check on drying conditions. The data shown below are the results of measuring peak values of acoustic emission over 10-second intervals for wood specimens with controlled diameter, thickness, temperature, and relative humidity (RH). Which factors appear to affect the AE rates?

Test No.	Diameter (mm)	Thickness (mm)	Dry-Bulb Temp. (°C)	RH (%)	Peak Values of AE Event Count Rates (1/10 s)
1	180	20	40	40	780
2	180	20	50	50	695
3	180	20	60	30	1,140
4	180	25	40	40	270
5	180	25	50	50	340
6	180	25	60	30	490
7	180	30	40	50	160
8	180	30	50	30	270
9	180	30	60	40	185
10	125	20	40	50	620
11	125	20	50	30	1,395
12	125	20	60	40	1,470
13	125	25	40	30	645
14	125	25	50	40	1,020
15	125	25	60	50	755
16	125	30	40	30	545
17	125	30	50	40	635
18	125	30	60	50	335

Source: M. Naguchi, S. Kitayama, K. Satoyoshi, and J. Umetsu, *Forest Products Journal,* 37(1), 1987, pp. 28–34.

11.62 In a study of wood-laminating adhesives by R. Kreibich and R. Hemingway (*Forest Products Journal*, 37(2), 1987, pp. 43–46), tannin is added to the usual phenol-resorcinol-formaldehyde (PRF) resin and tested for strength. For four mixtures of resin and tannin and two types of test (test A, dry shear test; test B, two-hour boil test), the resulting shear strengths of the wood specimens (in psi) were as follows:

		Test	
		A	B
	50/50	1,960	1,460
		2,267	140
PRF/Tannin	40/60	1,450	1,620
Ratio		1,683	2.287
	30/70	2,080	760
		1,750	130
	23/77	1,493	650
		1,850	90

What can you conclude about the effects of the tannin mixtures on the strength of the laminate?

11.63 Acid precipitation studies in southeastern Arizona involved sampling rain from two sites (Bisbee and the Tucson Research Ranch of the National Audubon Society) and sampling air from smelter plumes. The data for five elements, in terms of their ratios to copper, are given in the table below. One objective of the study is to see if the plume samples differ from the rain samples at the other sites. What can you say about this aspect of the study?

Summary Statistics for Metals in Precipitation and Plume Samples

Sample	Statistic	Mass Ratio (g/g)				
		As/Cu	Cd/Cu	Pb/Cu	Sb/Cu	Zn/Cu
Bisbee	Mean	0.46	0.068	1.03	0.073	1.64
(8/84–10/84)	S.E.	0.15	0.017	0.16	0.019	0.31
	Min.	0.14	0.029	0.31	0.043	0.58
	Max.	1.33	0.19	0.26	0.108	3.79
	n	7	10	13	3	13
TRR	Mean	0.48	0.087	0.94	0.078	4.2
(8/84–10/84)	S.E.	0.12	0.024	0.17	0.018	1.1
	Min.	0.13	0.019	0.26	0.037	0.8
	Max.	1.3	0.30	2.1	0.16	12.2
	n	11	11	11	7	11
TRR	Mean	0.56	0.074	0.82	0.10	9.7
(8/84–9/85)	S.E.	0.067	0.011	0.07	0.016	1.6
	Min.	0.086	0.014	0.08	0.020	0.8
	Max.	1.33	0.30	2.1	0.17	55
	n	31	31	49	11	49
Smelter	Mean	0.72	0.077	0.90	0.054	1.7
Plumes	S.E.	0.37	0.022	0.23	0.034	0.67
	Min.	0.20	0.01	0.44	0.014	0.11
	Max.	2.2	0.15	1.6	0.19	3.8
	n	5	5	4	5	5

Source: C. Blanchard and M. Stromberg, *Atmospheric Environment*, vol. 21, no. 11, 1987, pp. 2375–2381.

11.64   Lumber moisture content (MC) must be monitored during drying in order to maximize the lumber quality and minimize the drying costs. Four types of moisture meters were studied on three levels of moisture content in lumber by T. Breiner, D. Arganbright, and W. Pong (*Forest Products Journal*, 37(4), 1987, pp. 9–16). Of the four meters used, two were hand-held (resistance and power-loss types) and two were in-line (high- and low-frequency). From the data given below, can you determine whether any significant differences exist among the meters for the various MC levels?

Variation in Average MC Between Three Repeat Runs (CV = coefficient of variation = $SD/\bar{x}$)

	Resistance	Power-Loss	High-Frequency	Low-Frequency
20% MC Level	$\bar{X} = 17.8$ $SD = 0.3$ $CV = 1.7\%$	$\bar{X} = 14.2$ $SD = 0.2$ $CV = 1.4\%$	$\bar{X} = 16.8$ $SD = 0.1$ $CV = 0.06\%$	$\bar{X} = 19.4$ $SD = 0.1$ $CV = 0.3\%$
15% MC Level	$\bar{X} = 13.5$ $SD = 0.2$ $CV = 1.1\%$	$\bar{X} = 11.9$ $SD = 0.1$ $CV = 0.9\%$	$\bar{X} = 13.7$ $SD = 0.1$ $CV = 0.8\%$	$\bar{X} = 16.9$ $SD = 0.1$ $CV = 0.5\%$
10% MC Level	$\bar{X} = 11.4$ $SD = 0.1$ $CV = 0.9\%$	$\bar{X} = 10.1$ $SD = 0.1$ $CV = 1.3\%$	$\bar{X} = 11.3$ $SD = 0.1$ $CV = 0.6\%$	$\bar{X} = 15.1$ $SD = 0.1$ $CV = 0.5\%$

11.65   Mouthpiece forces produced while playing the trumpet were studied by J. Barbenel, P. Kerry, and J. Davies (*Journal of Biomechanics*, vol. 21, no. 5, 1988, pp. 417–424). Twenty-five high-proficiency players and 35 medium-proficiency players produced results summarized in the following table:

Results of Maximum Force Tests: Mean Value (interquartile range)

Test	High-Proficiency Popular	High-Proficiency Classical	Medium-Proficiency Popular	Medium-Proficiency Classical
C major scale	72 (53–83)	63 (51–72)	51 (43–59)	44 (38–57)
G_5 at 95 dB				
Maximum force	85 (63–97)	73 (59–83)	54 (46–66)	46 (28–56)
Normal force	24 (19–28)	27 (20–35)	28 (22–35)	26 (18–29)
Minimum force	22 (18–30)	23 (16–27)	24 (20–28)	23 (15–30)

Given only mean values and interquartile ranges, how would you describe the differences, if any, between high-proficiency and medium-proficiency players playing a C major scale? Do there appear to be differences in mean forces between popular and classical play of C major scales? Describe the main differences, if any, among maximum, normal, and minimum forces measured during the playing of G_5 at 95 dB.

11.66   Aerosol particles in the atmosphere influence the transmission of solar radiation by absorption and scattering. Absorption causes direct heating of the atmosphere, and scattering causes some radiation to be reflected to space and diffuses radiation reaching the ground. Thus, accurate measurements of absorption and scattering coefficients are quite important. The data given below come from a study by C. Kilsby and M. Smith (*Atmospheric Environment*, vol. 21, no. 10, 1987, pp. 2233–2246). Estimate the difference between aircraft and ground mean scattering coefficients (with a 95% confidence coefficient), assuming the samples are of size 3. Repeat, assuming the samples are of size 20. Perform similar estimations for the absorption coefficients.

Comparison of Ground and Aircraft Measurements Around Noon on 27 July 1984 (standard deviations given in parentheses)

Instrument	Quantity	Aircraft Measurement	Ground Measurement	Ratio Aircraft/Ground
Nephelometer	Scattering coefficient at 0.55 μm (10^{-4}m)	1.2($\pm$0.2)	0.8($\pm$0.2)	1.5($\pm$0.5)
Filter sample	Absorption coefficient at 0.55 μm (10^{-6}/m)	15.5($\pm$2.5)	8.3($\pm$1.3)	1.9($\pm$0.6)
FSSP	Particle concentration $0.25 < r < 3.75\mu$m (cm^{-3})	3.5($\pm$1.0)	2.6($\pm$1.0)	1.3($\pm$0.8)

11.67 The water quality beneath five retention/recharge basins for an urban area was investigated by the EPA. Data on salinity of the water for samples taken at varying depths from each basin are shown in the accompanying table. How would you suggest comparing the mean salinity for the five basins, in light of the fact that measurement depths are not the same? Carry out a statistical analysis of these means.

		Salinity		
Basin Name	Depth of Sample, m	Mean	Standard Deviation	No. of Observations
F	2.74	0.123	0.063	10
	8.53	0.099	0.016	3
	15.8	0.112	0.035	15
	20.7	0.153	0.075	18
G	2.74	0.119	NC	1
	14.0	0.541	NC	1
	23.8	0.744	0.090	16
M	3.64	0.294	NC	1
	11.3	0.150	0.035	7
	20.7	0.148	NC	1
	26.2	0.109	0.034	12
EE	3.66	0.085	0.042	13
	7.01	0.158	0.054	12
	10.1	0.175	0.035	14
MM	3.66	0.050	0.012	5
	8.53	0.054	0.015	8
	17.7	0.061	0.020	8

Source: H. Nightingale, *Water Resources Bulletin*, vol. 23, no. 2, 1987, pp. 197–205.

11.68 How do back and arm strengths of men and women change as the strength test moves from static to dynamic situations? This question was investigated by S. Kumar, D. Chaffin, and M. Redfern (*Journal of Biomechanics*, vol. 21, no. 1, 1988, pp. 35–44). The dynamic conditions required movement at 20, 60, and 100 cm/sec. Ten male and ten female young adult volunteers without a history of back pain participated in the study. From the data given below, analyze the differences between male and female strengths, the differences between back and arm strengths, and the pattern of strengths as one moves from a static strength test through three levels of dynamic strength tests.

	Task	Mean Peak Strength
Males		
Back	Static	726
	Slow	672
	Medium	639
	Fast	597
Arm	Static	521
	Slow	399
	Medium	332
	Fast	275
Females		
Back	Static	503
	Slow	487
	Medium	432
	Fast	436
Arm	Static	296
	Slow	266
	Medium	221
	Fast	192

11.69 Instantaneous flood discharge rates (in m³/sec) were recorded over many years for 11 rivers in Connecticut. The sample means and standard deviations are reported below. Noting that the data are highly variable, how would you compare the mean discharge rates for these rivers? Perform an analysis to locate significant differences among these means, if possible.

River	Location	Drainage Area (sq. km)	Record	Years of Record	$\bar{x}$	s_x
Scantic	Broad Brook	178	1930–1982	53	42.0	51.3
Burlington	Burlington	92	1937–1982	46	9.7	7.3
Connecticut	Hartford	3,500	1905–1982	78	980.1	761.5
North Branch Park	Hartford	125	1956–1982	27	45.6	50.4
Salmon	East Hampton	575	1930–1981	52	107.4	93.6
Eightmile	North Plain	320	1938–1982	45	30.4	26.0
East Branch Eightmile	North Lyme	197	1937–1982	46	24.9	23.3
Quinnipiac	Wallingford	205	1932–1982	51	62.3	37.9
Blackberry	Canaan	215	1943–1982	30	57.8	67.8
Tenmile	Gaylordsville	430	1921–1982	52	103.5	81.3
Still	Lanesville	495	1935–1982	48	45.7	37.9

Source: D. Jain and V. Singh, *Water Resources Bulletin*, vol. 23, no. 1, 1987, pp. 59–71.

11.70 Does channelization have an effect on stream flow? This question is to be answered from data on annual stream flow and annual precipitation for two watersheds in Florida. (See A.

Shirmohammadi, J. Sheridan, and W. Knisel, *Water Resources Bulletin*, vol. 23, no. 1, 1987, pp. 103–111.) The data are summarized in the table below. Does there seem to be a significant difference in mean stream flow before and after channelization, even with differences in precipitation present?

Watershed	Area (km^2)	Status	Record Period	Annual Precipitation (mm)		Annual Stream Flow (mm)	
				Mean	S.D.	Mean	S.D.
W3	40.7	Before channelization	1956–1963	1,259	230.6	315.7	200.9
		After channelization	1966–1981	1,205	166.0	279.5	177.7
W2	255.6	Before channelization	1956–1961	1,308	272.9	442.6	256.3
		After channelization	1969–1972	1,301	216.1	396.6	216.6

Appendix

T A B L E 1 Random Numbers

Row \ Column	1	2	3	4	5	6	7	8	9	10	11	12	13	14
1	10480	15011	01536	02011	81647	91646	69179	14194	62590	36207	20969	99570	91291	90700
2	22368	46753	25595	85393	30995	89198	27982	53402	93965	34095	52666	19174	39615	99505
3	24130	48360	22527	97265	76393	64809	15179	24830	49340	32081	30680	19655	63348	58629
4	42167	93093	06243	61680	07856	16376	39440	53537	71341	57004	00849	74917	97758	16379
5	37570	39975	81837	16656	06121	91782	60468	81305	49684	60672	14110	06927	01263	54613
6	77912	06907	11008	42751	27756	53498	18602	70659	90655	15053	21916	81825	44394	42880
7	99562	72905	56420	69994	98872	31016	71194	18738	44013	48840	63213	21069	10634	12952
8	96301	91977	05463	07972	18876	20922	94595	56869	69014	60045	18425	84903	42508	32307
9	89579	14342	63661	10281	17453	18103	57740	84378	25331	12566	58678	44947	05585	56941
10	85475	36875	53342	53988	53060	59533	38867	62300	08158	17983	16439	11458	18593	64952
11	28918	69578	88231	33276	70997	79936	56865	05859	90106	31595	01547	85590	91610	78188
12	63553	40961	48235	03427	49626	69445	18663	72695	52180	20847	12234	90511	33703	90322
13	09429	93969	52636	92737	88974	33488	36320	17617	30015	08272	84115	27156	30613	74952
14	10365	61129	87529	85689	48237	52267	67689	93394	01511	26358	85104	20285	29975	89868
15	07119	97336	71048	08178	77233	13916	47564	81056	97735	85977	29372	74461	28551	90707
16	51085	12765	51821	51259	77452	16308	60756	92144	49442	53900	70960	63990	75601	40719
17	02368	21382	52404	60268	89368	19885	55322	44819	01188	65255	64835	44919	05944	55157
18	01011	54092	33362	94904	31273	04146	18594	29852	71585	85030	51132	01915	92747	64951
19	52162	53916	46369	58586	23216	14513	83149	98736	23495	64350	94738	17752	35156	35749
20	07056	97628	33787	09998	42698	06691	76988	13602	51851	46104	88916	19509	25625	58104
21	48663	91245	85828	14346	09172	30168	90229	04734	59193	22178	30421	61666	99904	32812
22	54164	58492	22421	74103	47070	25306	76468	26384	58151	06646	21524	15227	96909	44592
23	32639	32363	05597	24200	13363	38005	94342	28728	35806	06912	17012	64161	18296	22851
24	29334	27001	87637	87308	58731	00256	45834	15398	46557	41135	10367	07684	36188	18510
25	02488	33062	28834	07351	19731	92420	60952	61280	50001	67658	32586	86679	50720	94953
26	81525	72295	04839	96423	24878	82651	66566	14778	76797	14780	13300	87074	79666	95725
27	29676	20591	68086	26432	46901	20849	89768	81536	86645	12659	92259	57102	80428	25280
28	00742	57392	39064	66432	84673	40027	32832	61362	98947	96067	64760	64584	96096	98253
29	05366	04213	25669	26422	44407	44048	37937	63904	45766	66134	75470	66520	34693	90449
30	91921	26418	64117	94305	26766	25940	39972	22209	71500	64568	91402	42416	07844	69618
31	00582	04711	87917	77341	42206	35126	74087	99547	81817	42607	43808	76655	62028	76630
32	00725	69884	62797	56170	86324	88072	76222	36086	84637	93161	76038	65855	77919	88006
33	69011	65795	95876	55293	18988	27354	26575	08625	40801	59920	29841	80150	12777	48501
34	25976	57948	29888	88604	67917	48708	18912	82271	65424	69774	33611	54262	85963	03547

Row \ Column	1	2	3	4	5	6	7	8	9	10	11	12	13	14
35	09763	83473	73577	12908	30883	18317	28290	35797	05998	41688	34952	37888	38917	88050
36	91576	42595	27958	30134	04024	86385	29880	99730	55536	84855	29080	09250	79656	73211
37	17955	56349	90999	49127	20044	59931	06115	20542	18059	02008	73708	83517	36103	42791
38	46503	18584	18845	49618	02304	51038	20655	58727	28168	15475	56942	53389	20562	87338
39	92157	89634	94824	78171	84610	82834	09922	25417	44137	48413	25555	21246	35509	20468
40	14577	62765	35605	81263	39667	47358	56873	56307	61607	49518	89656	20103	77490	18062
41	98427	07523	33362	64270	01638	92477	66969	98420	04880	45585	46565	04102	46880	45709
42	34914	63976	88720	82765	34476	17032	87589	40836	32427	70002	70663	88863	77775	69348
43	70060	28277	39475	46473	23219	53416	94970	25832	69975	94884	19661	72828	00102	66794
44	53976	54914	06990	67245	68350	82948	11398	42878	80287	88267	47363	46634	06541	97809
45	76072	29515	40980	07391	58745	25774	22987	80059	39911	96189	41151	14222	60697	59583
46	90725	52210	83974	29992	65831	38857	50490	83765	55657	14361	31720	57375	56228	41546
47	64364	67412	33339	31926	14883	24413	59744	92351	97473	89286	35931	04110	23726	51900
48	08962	00358	31662	25388	61642	34072	81249	35648	56891	69352	48373	45578	78547	81788
49	95012	68379	93526	70765	10592	04542	76463	54328	02349	17247	28865	14777	62730	92277
50	15664	10493	20492	38391	91132	21999	59516	81652	27195	48223	46751	22923	32261	85653
51	16408	81899	04153	53381	79401	21438	83035	92350	36693	31238	59649	91754	72772	02338
52	18629	81953	05520	91962	04739	13092	97662	24822	94730	06496	35090	04822	86774	98289
53	73115	35101	47498	87637	99016	71060	88824	71013	18735	20286	23153	72924	35165	43040
54	57491	16703	23167	49323	45021	33132	12544	41035	80780	45393	44812	12515	98931	91202
55	30405	83946	23792	14422	15059	45799	22716	19792	09983	74353	68668	30429	70735	25499
56	16631	35006	85900	98275	32388	52390	16815	69298	82732	38480	73817	32523	41961	44437
57	96773	20206	42559	78985	05300	22164	24369	54224	35083	19687	11052	91491	60383	19746
58	38935	64202	14349	82674	66523	44133	00697	35552	35970	19124	63318	29686	03387	59846
59	31624	76384	17403	53363	44167	64486	64758	75366	76554	31601	12614	33072	60332	92325
60	78919	19474	23632	27889	47914	02584	37680	20801	72152	39339	34806	08930	85001	87820
61	03931	33309	57047	74211	63445	17361	62825	39908	05607	91284	68833	25570	38818	46920
62	74426	33278	43972	10119	89917	15665	52872	73823	73144	88662	88970	74492	51805	99378
63	09066	00903	20795	95452	92648	45454	09552	88815	16553	51125	79375	97596	16296	66092

64	42238	12426	87025	14267	20979	04508	64535	86064	31355	29472	47689	05974	52468	16834
65	16153	08002	26504	41744	81959	65642	74240	00033	56302	67107	77510	70625	28725	34191
66	21457	40742	29820	96783	29400	21840	15035	33310	34537	06116	95240	15957	16572	06004
67	21581	57802	02050	89728	17937	37621	47075	97403	42080	48626	68995	43805	33386	21597
68	55612	78095	83197	33732	05810	24813	86902	16489	60397	03264	88525	42786	05269	92532
69	44657	66999	99324	51281	84463	60563	79312	68876	93454	25471	93911	25650	12682	73572
70	91340	84979	46949	81973	37949	61023	43997	80644	15263	43942	89203	71795	99533	50501
71	91227	21199	31935	27022	84067	05462	35216	29891	14486	68607	41867	14951	91696	85065
72	50001	38140	66321	19924	72163	09538	12151	91903	06878	18749	34405	56087	82790	70925
73	65390	05224	72958	28609	81406	39147	25549	42627	48542	45233	57202	94617	23772	07896
74	27504	96131	83944	41575	10573	08619	64482	36152	73923	05184	94142	25299	84387	34925
75	31769	94851	39117	89632	00959	16487	65536	39782	49071	17095	02330	74301	00275	48280
76	11508	70225	51111	38351	19444	66499	71945	13442	05422	78675	84081	66938	93654	59894
77	37449	30362	06694	54690	04052	53115	62757	78662	95348	11163	81651	50245	34971	52924
78	46515	70331	85922	38329	57015	15765	97161	45349	17869	61796	66345	81073	49106	79860
79	30986	81223	42416	58353	21532	30502	32305	05174	86482	07901	54339	58861	74818	46942
80	63798	64995	46583	09785	44160	78128	83991	92520	42865	83531	30377	35909	81250	54238
81	82486	84846	99254	67632	43218	50076	21361	51202	64816	88124	41870	52689	51275	83556
82	21885	32906	92431	09060	64297	51674	64126	26123	62570	05155	59194	52799	28225	85762
83	60336	98782	07408	53458	13564	59089	26445	85205	29789	41001	12535	12133	14645	23541
84	43937	46891	24010	25560	86355	33941	25786	71899	54990	15475	95434	98227	21824	19585
85	97656	63175	89303	16275	07100	92063	21942	47348	18611	20203	18534	03862	78095	50136
86	03299	01221	05418	38982	55758	92237	26759	21216	86367	98442	08303	56613	91511	75928
87	79626	06486	03574	17668	07785	76020	79924	83325	25651	88428	85076	72811	22717	50585
88	85636	68335	47539	03129	65651	11977	02510	99447	26113	68645	34327	15152	55230	93448
89	18039	14367	61337	06177	12143	46609	32989	64708	74014	00533	35398	58408	13261	47908
90	08362	15656	60627	36478	65648	16764	53412	07832	09013	41574	17639	82163	60859	75567
91	79556	29068	04142	16268	15387	12856	66227	22478	38358	73373	88732	09443	82558	05250
92	92608	82674	27072	32534	17075	27698	98204	11951	63863	34648	88022	56148	34925	57031
93	23982	25835	40055	67006	12293	02753	14827	35071	23235	99704	37543	11601	35503	85171
94	09915	96306	05908	97901	28395	14186	00821	70426	80703	75647	76310	88717	37890	40129
95	59037	33300	26695	62247	69927	76123	50842	86654	43834	70959	79725	93872	28117	19233
96	42488	78077	69882	61657	34136	79180	97526	04098	43092	73571	80799	76536	71255	64239
97	46764	86273	63003	93017	31204	36692	40202	57306	35275	55543	53203	18098	47625	88684
98	03237	45430	55417	63282	90816	17349	88298	36600	90183	78406	06216	95787	42579	90730
99	86591	81482	52667	61582	14972	90053	89534	49199	76036	43716	97548	04379	46370	28672
100	38534	01715	94964	87288	65680	39560	39772	86537	12918	62738	19636	51132	25739	56947

Source: Abridged from W. H. Beyer, ed., *CRC Standard Mathematical Tables,* 24th ed. (Cleveland, The Chemical Rubber Company), 1976. Reproduced by permission of the publisher.

T A B L E **2** Binomial Probabilities

Tabulated values are $\sum_{x=0}^{k} p(x)$. (Computations are rounded at the third decimal place.)

(a) $n = 5$

k \ p	0.01	0.05	0.10	0.20	0.30	0.40	0.50	0.60	0.70	0.80	0.90	0.95	0.99
0	0.951	0.774	0.590	0.328	0.168	0.078	0.031	0.010	0.002	0.000	0.000	0.000	0.000
1	0.999	0.977	0.919	0.737	0.528	0.337	0.188	0.087	0.031	0.007	0.000	0.000	0.000
2	1.000	0.999	0.991	0.942	0.837	0.683	0.500	0.317	0.163	0.058	0.009	0.001	0.000
3	1.000	1.000	1.000	0.993	0.969	0.913	0.812	0.663	0.472	0.263	0.081	0.023	0.001
4	1.000	1.000	1.000	1.000	0.998	0.990	0.969	0.922	0.832	0.672	0.410	0.226	0.049

(b) $n = 10$

k \ p	0.01	0.05	0.10	0.20	0.30	0.40	0.50	0.60	0.70	0.80	0.90	0.95	0.99
0	0.904	0.599	0.349	0.107	0.028	0.006	0.001	0.000	0.000	0.000	0.000	0.000	0.000
1	0.996	0.914	0.736	0.376	0.149	0.046	0.011	0.002	0.000	0.000	0.000	0.000	0.000
2	1.000	0.988	0.930	0.678	0.383	0.167	0.055	0.012	0.002	0.000	0.000	0.000	0.000
3	1.000	0.999	0.987	0.879	0.650	0.382	0.172	0.055	0.011	0.001	0.000	0.000	0.000
4	1.000	1.000	0.998	0.967	0.850	0.633	0.377	0.166	0.047	0.006	0.000	0.000	0.000
5	1.000	1.000	1.000	0.994	0.953	0.834	0.623	0.367	0.150	0.033	0.002	0.000	0.000
6	1.000	1.000	1.000	0.999	0.989	0.945	0.828	0.618	0.350	0.121	0.013	0.001	0.000
7	1.000	1.000	1.000	1.000	0.998	0.988	0.945	0.833	0.617	0.322	0.070	0.012	0.000
8	1.000	1.000	1.000	1.000	1.000	0.998	0.989	0.954	0.851	0.624	0.264	0.086	0.004
9	1.000	1.000	1.000	1.000	1.000	1.000	0.999	0.994	0.972	0.893	0.651	0.401	0.096

(c) $n = 15$

k \ p	0.01	0.05	0.10	0.20	0.30	0.40	0.50	0.60	0.70	0.80	0.90	0.95	0.99
0	0.860	0.463	0.206	0.035	0.005	0.000	0.000	0.000	0.000	0.000	0.000	0.000	0.000
1	0.990	0.829	0.549	0.167	0.035	0.005	0.000	0.000	0.000	0.000	0.000	0.000	0.000
2	1.000	0.964	0.816	0.398	0.127	0.027	0.004	0.000	0.000	0.000	0.000	0.000	0.000
3	1.000	0.995	0.944	0.648	0.297	0.091	0.018	0.002	0.000	0.000	0.000	0.000	0.000
4	1.000	0.999	0.987	0.836	0.515	0.217	0.059	0.009	0.001	0.000	0.000	0.000	0.000
5	1.000	1.000	0.998	0.939	0.722	0.403	0.151	0.034	0.004	0.000	0.000	0.000	0.000
6	1.000	1.000	1.000	0.982	0.869	0.610	0.304	0.095	0.015	0.001	0.000	0.000	0.000
7	1.000	1.000	1.000	0.996	0.950	0.787	0.500	0.213	0.050	0.004	0.000	0.000	0.000
8	1.000	1.000	1.000	0.999	0.985	0.905	0.696	0.390	0.131	0.018	0.000	0.000	0.000
9	1.000	1.000	1.000	1.000	0.996	0.966	0.849	0.597	0.278	0.061	0.002	0.000	0.000
10	1.000	1.000	1.000	1.000	0.999	0.991	0.941	0.783	0.485	0.164	0.013	0.001	0.000
11	1.000	1.000	1.000	1.000	1.000	0.998	0.982	0.909	0.703	0.352	0.056	0.005	0.000
12	1.000	1.000	1.000	1.000	1.000	1.000	0.996	0.973	0.873	0.602	0.184	0.036	0.000
13	1.000	1.000	1.000	1.000	1.000	1.000	1.000	0.995	0.965	0.833	0.451	0.171	0.010
14	1.000	1.000	1.000	1.000	1.000	1.000	1.000	1.000	0.995	0.965	0.794	0.537	0.140

T A B L E **2** (Continued)

(d) $n = 20$

k	0.01	0.05	0.10	0.20	0.30	0.40	0.50	0.60	0.70	0.80	0.90	0.95	0.99
0	0.818	0.358	0.122	0.002	0.001	0.000	0.000	0.000	0.000	0.000	0.000	0.000	0.000
1	0.983	0.736	0.392	0.069	0.008	0.001	0.000	0.000	0.000	0.000	0.000	0.000	0.000
2	0.999	0.925	0.677	0.206	0.035	0.004	0.000	0.000	0.000	0.000	0.000	0.000	0.000
3	1.000	0.984	0.867	0.411	0.107	0.016	0.001	0.000	0.000	0.000	0.000	0.000	0.000
4	1.000	0.997	0.957	0.630	0.238	0.051	0.006	0.000	0.000	0.000	0.000	0.000	0.000
5	1.000	1.000	0.989	0.804	0.416	0.126	0.021	0.002	0.000	0.000	0.000	0.000	0.000
6	1.000	1.000	0.998	0.913	0.608	0.250	0.058	0.006	0.000	0.000	0.000	0.000	0.000
7	1.000	1.000	1.000	0.968	0.772	0.416	0.132	0.021	0.001	0.000	0.000	0.000	0.000
8	1.000	1.000	1.000	0.990	0.887	0.596	0.252	0.057	0.005	0.000	0.000	0.000	0.000
9	1.000	1.000	1.000	0.997	0.952	0.755	0.412	0.128	0.017	0.001	0.000	0.000	0.000
10	1.000	1.000	1.000	0.999	0.983	0.872	0.588	0.245	0.048	0.003	0.000	0.000	0.000
11	1.000	1.000	1.000	1.000	0.995	0.943	0.748	0.404	0.113	0.010	0.000	0.000	0.000
12	1.000	1.000	1.000	1.000	0.999	0.979	0.868	0.584	0.228	0.032	0.000	0.000	0.000
13	1.000	1.000	1.000	1.000	1.000	0.994	0.942	0.750	0.392	0.087	0.002	0.000	0.000
14	1.000	1.000	1.000	1.000	1.000	0.998	0.979	0.874	0.584	0.196	0.011	0.000	0.000
15	1.000	1.000	1.000	1.000	1.000	1.000	0.994	0.949	0.762	0.370	0.043	0.003	0.000
16	1.000	1.000	1.000	1.000	1.000	1.000	0.999	0.984	0.893	0.589	0.133	0.016	0.000
17	1.000	1.000	1.000	1.000	1.000	1.000	1.000	0.996	0.965	0.794	0.323	0.075	0.001
18	1.000	1.000	1.000	1.000	1.000	1.000	1.000	0.999	0.992	0.931	0.608	0.264	0.017
19	1.000	1.000	1.000	1.000	1.000	1.000	1.000	1.000	0.999	0.988	0.878	0.642	0.182

(e) $n = 25$

k	0.01	0.05	0.10	0.20	0.30	0.40	0.50	0.60	0.70	0.80	0.90	0.95	0.99
0	0.778	0.277	0.072	0.004	0.000	0.000	0.000	0.000	0.000	0.000	0.000	0.000	0.000
1	0.974	0.642	0.271	0.027	0.002	0.000	0.000	0.000	0.000	0.000	0.000	0.000	0.000
2	0.998	0.873	0.537	0.098	0.009	0.000	0.000	0.000	0.000	0.000	0.000	0.000	0.000
3	1.000	0.966	0.764	0.234	0.033	0.002	0.000	0.000	0.000	0.000	0.000	0.000	0.000
4	1.000	0.993	0.902	0.421	0.090	0.009	0.000	0.000	0.000	0.000	0.000	0.000	0.000
5	1.000	0.999	0.967	0.617	0.193	0.029	0.002	0.000	0.000	0.000	0.000	0.000	0.000
6	1.000	1.000	0.991	0.780	0.341	0.074	0.007	0.000	0.000	0.000	0.000	0.000	0.000
7	1.000	1.000	0.998	0.891	0.512	0.154	0.022	0.001	0.000	0.000	0.000	0.000	0.000
8	1.000	1.000	1.000	0.953	0.677	0.274	0.054	0.004	0.000	0.000	0.000	0.000	0.000
9	1.000	1.000	1.000	0.983	0.811	0.425	0.115	0.013	0.000	0.000	0.000	0.000	0.000
10	1.000	1.000	1.000	0.994	0.902	0.586	0.212	0.034	0.002	0.000	0.000	0.000	0.000
11	1.000	1.000	1.000	0.998	0.956	0.732	0.345	0.078	0.006	0.000	0.000	0.000	0.000
12	1.000	1.000	1.000	1.000	0.983	0.846	0.500	0.154	0.017	0.000	0.000	0.000	0.000
13	1.000	1.000	1.000	1.000	0.994	0.922	0.655	0.268	0.044	0.002	0.000	0.000	0.000
14	1.000	1.000	1.000	1.000	0.998	0.966	0.788	0.414	0.098	0.006	0.000	0.000	0.000
15	1.000	1.000	1.000	1.000	1.000	0.987	0.885	0.575	0.189	0.017	0.000	0.000	0.000
16	1.000	1.000	1.000	1.000	1.000	0.996	0.946	0.726	0.323	0.047	0.000	0.000	0.000
17	1.000	1.000	1.000	1.000	1.000	0.999	0.978	0.846	0.488	0.109	0.002	0.000	0.000
18	1.000	1.000	1.000	1.000	1.000	1.000	0.993	0.926	0.659	0.220	0.009	0.000	0.000
19	1.000	1.000	1.000	1.000	1.000	1.000	0.998	0.971	0.807	0.383	0.033	0.001	0.000
20	1.000	1.000	1.000	1.000	1.000	1.000	1.000	0.991	0.910	0.579	0.098	0.007	0.000
21	1.000	1.000	1.000	1.000	1.000	1.000	1.000	0.998	0.967	0.766	0.236	0.034	0.000
22	1.000	1.000	1.000	1.000	1.000	1.000	1.000	1.000	0.991	0.902	0.463	0.127	0.002
23	1.000	1.000	1.000	1.000	1.000	1.000	1.000	1.000	0.998	0.973	0.729	0.358	0.026
24	1.000	1.000	1.000	1.000	1.000	1.000	1.000	1.000	1.000	0.996	0.928	0.723	0.222

T A B L E **3** Poisson Distribution Function

$$F(x, \lambda) = \sum_{k=0}^{x} e^{-\lambda} \frac{\lambda^k}{k!}$$

λ \ x	0	1	2	3	4	5	6	7	8	9
0.02	0.980	1.000								
0.04	0.961	0.999	1.000							
0.06	0.942	0.998	1.000							
0.08	0.923	0.997	1.000							
0.10	0.905	0.995	1.000							
0.15	0.861	0.990	0.999	1.000						
0.20	0.819	0.982	0.999	1.000						
0.25	0.779	0.974	0.998	1.000						
0.30	0.741	0.963	0.996	1.000						
0.35	0.705	0.951	0.994	1.000						
0.40	0.670	0.938	0.992	0.999	1.000					
0.45	0.638	0.925	0.989	0.999	1.000					
0.50	0.607	0.910	0.986	0.998	1.000					
0.55	0.577	0.894	0.982	0.998	1.000					
0.60	0.549	0.878	0.977	0.997	1.000					
0.65	0.522	0.861	0.972	0.996	0.999	1.000				
0.70	0.497	0.844	0.966	0.994	0.999	1.000				
0.75	0.475	0.827	0.959	0.993	0.999	1.000				
0.80	0.449	0.809	0.953	0.991	0.999	1.000				
0.85	0.427	0.791	0.945	0.989	0.998	1.000				
0.90	0.407	0.772	0.937	0.987	0.998	1.000				
0.95	0.387	0.754	0.929	0.981	0.997	1.000				
1.00	0.368	0.736	0.920	0.981	0.996	0.999	1.000			
1.1	0.333	0.699	0.900	0.974	0.995	0.999	1.000			
1.2	0.301	0.663	0.879	0.966	0.992	0.998	1.000			
1.3	0.273	0.627	0.857	0.957	0.989	0.998	1.000			
1.4	0.247	0.592	0.833	0.946	0.986	0.997	0.999	1.000		
1.5	0.223	0.558	0.809	0.934	0.981	0.996	0.999	1.000		
1.6	0.202	0.525	0.783	0.921	0.976	0.994	0.999	1.000		
1.7	0.183	0.493	0.757	0.907	0.970	0.992	0.998	1.000		
1.8	0.165	0.463	0.731	0.891	0.964	0.990	0.997	0.999	1.000	
1.9	0.150	0.434	0.704	0.875	0.956	0.987	0.997	0.999	1.000	
2.0	0.135	0.406	0.677	0.857	0.947	0.983	0.995	0.999	1.000	

Source: Reprinted by permission from E. C. Molina, *Poisson's Exponential Binomial Limit* (Princeton, N.J., D. Van Nostrand Company, Inc.), 1947.

T A B L E **3** (Continued)

λ \ x	0	1	2	3	4	5	6	7	8	9
2.2	0.111	0.355	0.623	0.819	0.928	0.975	0.993	0.998	1.000	
2.4	0.091	0.308	0.570	0.779	0.904	0.964	0.988	0.997	0.999	1.000
2.6	0.074	0.267	0.518	0.736	0.877	0.951	0.983	0.995	0.999	1.000
2.8	0.061	0.231	0.469	0.692	0.848	0.935	0.976	0.992	0.998	0.999
3.0	0.050	0.199	0.423	0.647	0.815	0.916	0.966	0.988	0.996	0.999
3.2	0.041	0.171	0.380	0.603	0.781	0.895	0.955	0.983	0.994	0.998
3.4	0.033	0.147	0.340	0.558	0.744	0.871	0.942	0.977	0.992	0.997
3.6	0.027	0.126	0.303	0.515	0.706	0.844	0.927	0.969	0.988	0.996
3.8	0.022	0.107	0.269	0.473	0.668	0.816	0.909	0.960	0.984	0.994
4.0	0.018	0.092	0.238	0.433	0.629	0.785	0.889	0.949	0.979	0.992
4.2	0.015	0.078	0.210	0.395	0.590	0.753	0.867	0.936	0.972	0.989
4.4	0.012	0.066	0.185	0.359	0.551	0.720	0.844	0.921	0.964	0.985
4.6	0.010	0.056	0.163	0.326	0.513	0.686	0.818	0.905	0.955	0.980
4.8	0.008	0.048	0.143	0.294	0.476	0.651	0.791	0.887	0.944	0.975
5.0	0.007	0.040	0.125	0.265	0.440	0.616	0.762	0.867	0.932	0.968
5.2	0.006	0.034	0.109	0.238	0.406	0.581	0.732	0.845	0.918	0.960
5.4	0.005	0.029	0.095	0.213	0.373	0.546	0.702	0.822	0.903	0.951
5.6	0.004	0.024	0.082	0.191	0.342	0.512	0.670	0.797	0.886	0.941
5.8	0.003	0.021	0.072	0.170	0.313	0.478	0.638	0.771	0.867	0.929
6.0	0.002	0.017	0.062	0.151	0.285	0.446	0.606	0.744	0.847	0.916

λ	10	11	12	13	14	15	16
2.8	1.000						
3.0	1.000						
3.2	1.000						
3.4	0.999	1.000					
3.6	0.999	1.000					
3.8	0.998	0.999	1.000				
4.0	0.997	0.999	1.000				
4.2	0.996	0.999	1.000				
4.4	0.994	0.998	0.999	1.000			
4.6	0.992	0.997	0.999	1.000			
4.8	0.990	0.996	0.999	1.000			
5.0	0.986	0.995	0.998	0.999	1.000		
5.2	0.982	0.993	0.997	0.999	1.000		
5.4	0.977	0.990	0.996	0.999	1.000		
5.6	0.972	0.988	0.995	0.998	0.999	1.000	
5.8	0.965	0.984	0.993	0.997	0.999	1.000	
6.0	0.957	0.980	0.991	0.996	0.999	0.999	1.000

T A B L E **3** (Continued)

λ \ x	0	1	2	3	4	5	6	7	8	9
6.2	0.002	0.015	0.054	0.134	0.259	0.414	0.574	0.716	0.826	0.902
6.4	0.002	0.012	0.046	0.119	0.235	0.384	0.542	0.687	0.803	0.886
6.6	0.001	0.010	0.040	0.105	0.213	0.355	0.511	0.658	0.780	0.869
6.8	0.001	0.009	0.034	0.093	0.192	0.327	0.480	0.628	0.755	0.850
7.0	0.001	0.007	0.030	0.082	0.173	0.301	0.450	0.599	0.729	0.830
7.2	0.001	0.006	0.025	0.072	0.156	0.276	0.420	0.569	0.703	0.810
7.4	0.001	0.005	0.022	0.063	0.140	0.253	0.392	0.539	0.676	0.788
7.6	0.001	0.004	0.019	0.055	0.125	0.231	0.365	0.510	0.648	0.765
7.8	0.000	0.004	0.016	0.048	0.112	0.210	0.338	0.481	0.620	0.741
8.0	0.000	0.003	0.014	0.042	0.100	0.191	0.313	0.453	0.593	0.717
8.5	0.000	0.002	0.009	0.030	0.074	0.150	0.256	0.386	0.523	0.653
9.0	0.000	0.001	0.006	0.021	0.055	0.116	0.207	0.324	0.456	0.587
9.5	0.000	0.001	0.004	0.015	0.040	0.089	0.165	0.269	0.392	0.522
10.0	0.000	0.000	0.003	0.010	0.029	0.067	0.130	0.220	0.333	0.458

λ	10	11	12	13	14	15	16	17	18	19
6.2	0.949	0.975	0.989	0.995	0.998	0.999	1.000			
6.4	0.939	0.969	0.986	0.994	0.997	0.999	1.000			
6.6	0.927	0.963	0.982	0.992	0.997	0.999	0.999	1.000		
6.8	0.915	0.955	0.978	0.990	0.996	0.998	0.999	1.000		
7.0	0.901	0.947	0.973	0.987	0.994	0.998	0.999	1.000		
7.2	0.887	0.937	0.967	0.984	0.993	0.997	0.999	0.999	1.000	
7.4	0.871	0.926	0.961	0.980	0.991	0.996	0.998	0.999	1.000	
7.6	0.854	0.915	0.954	0.976	0.989	0.995	0.998	0.999	1.000	
7.8	0.835	0.902	0.945	0.971	0.986	0.993	0.997	0.999	1.000	
8.0	0.816	0.888	0.936	0.966	0.983	0.992	0.996	0.998	0.999	1.000
8.5	0.763	0.849	0.909	0.949	0.973	0.986	0.993	0.997	0.999	0.999
9.0	0.706	0.803	0.876	0.926	0.959	0.978	0.989	0.995	0.998	0.999
9.5	0.645	0.752	0.836	0.898	0.940	0.967	0.982	0.991	0.996	0.998
10.0	0.583	0.697	0.792	0.864	0.917	0.951	0.973	0.986	0.993	0.997

λ	20	21	22							
8.5	1.000									
9.0	1.000									
9.5	0.999	1.000								
10.0	0.998	0.999	1.000							

T A B L E **3**   (Continued)

λ \ x	0	1	2	3	4	5	6	7	8	9
10.5	0.000	0.000	0.002	0.007	0.021	0.050	0.102	0.179	0.279	0.397
11.0	0.000	0.000	0.001	0.005	0.015	0.038	0.079	0.143	0.232	0.341
11.5	0.000	0.000	0.001	0.003	0.011	0.028	0.060	0.114	0.191	0.289
12.0	0.000	0.000	0.001	0.002	0.008	0.020	0.046	0.090	0.155	0.242
12.5	0.000	0.000	0.000	0.002	0.005	0.015	0.035	0.070	0.125	0.201
13.0	0.000	0.000	0.000	0.001	0.004	0.011	0.026	0.054	0.100	0.166
13.5	0.000	0.000	0.000	0.001	0.003	0.008	0.019	0.041	0.079	0.135
14.0	0.000	0.000	0.000	0.000	0.002	0.006	0.014	0.032	0.062	0.109
14.5	0.000	0.000	0.000	0.000	0.001	0.004	0.010	0.024	0.048	0.088
15.0	0.000	0.000	0.000	0.000	0.001	0.003	0.008	0.018	0.037	0.070

	10	11	12	13	14	15	16	17	18	19
10.5	0.521	0.639	0.742	0.825	0.888	0.932	0.960	0.978	0.988	0.994
11.0	0.460	0.579	0.689	0.781	0.854	0.907	0.944	0.968	0.982	0.991
11.5	0.402	0.520	0.633	0.733	0.815	0.878	0.924	0.954	0.974	0.986
12.0	0.347	0.462	0.576	0.682	0.772	0.844	0.899	0.937	0.963	0.979
12.5	0.297	0.406	0.519	0.628	0.725	0.806	0.869	0.916	0.948	0.969
13.0	0.252	0.353	0.463	0.573	0.675	0.764	0.835	0.890	0.930	0.957
13.5	0.211	0.304	0.409	0.518	0.623	0.718	0.798	0.861	0.908	0.942
14.0	0.176	0.260	0.358	0.464	0.570	0.669	0.756	0.827	0.883	0.923
14.5	0.145	0.220	0.311	0.413	0.518	0.619	0.711	0.790	0.853	0.901
15.0	0.118	0.185	0.268	0.363	0.466	0.568	0.664	0.749	0.819	0.875

	20	21	22	23	24	25	26	27	28	29
10.5	0.997	0.999	0.999	1.000						
11.0	0.995	0.998	0.999	1.000						
11.5	0.992	0.996	0.998	0.999	1.000					
12.0	0.988	0.994	0.997	0.999	0.999	1.000				
12.5	0.983	0.991	0.995	0.998	0.999	0.999	1.000			
13.0	0.975	0.986	0.992	0.996	0.998	0.999	1.000			
13.5	0.965	0.980	0.989	0.994	0.997	0.998	0.999	1.000		
14.0	0.952	0.971	0.983	0.991	0.995	0.997	0.999	0.999	1.000	
14.5	0.936	0.960	0.976	0.986	0.992	0.996	0.998	0.999	0.999	1.000
15.0	0.917	0.947	0.967	0.981	0.989	0.994	0.997	0.998	0.999	1.000

T A B L E **4**   Normal Curve Areas

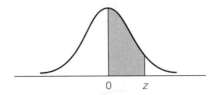

z	0.00	0.01	0.02	0.03	0.04	0.05	0.06	0.07	0.08	0.09
0.0	0.0000	0.0040	0.0080	0.0120	0.0160	0.0199	0.0239	0.0279	0.0319	0.0359
0.1	0.0398	0.0438	0.0478	0.0517	0.0557	0.0596	0.0636	0.0675	0.0714	0.0753
0.2	0.0793	0.0832	0.0871	0.0910	0.0948	0.0987	0.1026	0.1064	0.1103	0.1141
0.3	0.1179	0.1217	0.1255	0.1293	0.1331	0.1368	0.1406	0.1443	0.1480	0.1517
0.4	0.1554	0.1591	0.1628	0.1664	0.1700	0.1736	0.1772	0.1808	0.1844	0.1879
0.5	0.1915	0.1950	0.1985	0.2019	0.2054	0.2088	0.2123	0.2157	0.2190	0.2224
0.6	0.2257	0.2291	0.2324	0.2357	0.2389	0.2422	0.2454	0.2486	0.2517	0.2549
0.7	0.2580	0.2611	0.2642	0.2673	0.2704	0.2734	0.2764	0.2794	0.2823	0.2852
0.8	0.2881	0.2910	0.2939	0.2967	0.2995	0.3023	0.3051	0.3078	0.3106	0.3133
0.9	0.3159	0.3186	0.3212	0.3238	0.3264	0.3289	0.3315	0.3340	0.3365	0.3389
1.0	0.3413	0.3438	0.3461	0.3485	0.3508	0.3531	0.3554	0.3577	0.3599	0.3621
1.1	0.3643	0.3665	0.3686	0.3708	0.3729	0.3749	0.3770	0.3790	0.3810	0.3830
1.2	0.3849	0.3869	0.3888	0.3907	0.3925	0.3944	0.3962	0.3980	0.3997	0.4015
1.3	0.4032	0.4049	0.4066	0.4082	0.4099	0.4115	0.4131	0.4147	0.4162	0.4177
1.4	0.4192	0.4207	0.4222	0.4236	0.4251	0.4265	0.4279	0.4292	0.4306	0.4319
1.5	0.4332	0.4345	0.4357	0.4370	0.4382	0.4394	0.4406	0.4418	0.4429	0.4441
1.6	0.4452	0.4463	0.4474	0.4484	0.4495	0.4505	0.4515	0.4525	0.4535	0.4545
1.7	0.4554	0.4564	0.4573	0.4582	0.4591	0.4599	0.4608	0.4616	0.4625	0.4633
1.8	0.4641	0.4649	0.4656	0.4664	0.4671	0.4678	0.4686	0.4693	0.4699	0.4706
1.9	0.4713	0.4719	0.4726	0.4732	0.4738	0.4744	0.4750	0.4756	0.4761	0.4767
2.0	0.4772	0.4778	0.4783	0.4788	0.4793	0.4798	0.4803	0.4808	0.4812	0.4817
2.1	0.4821	0.4826	0.4830	0.4834	0.4838	0.4842	0.4846	0.4850	0.4854	0.4857
2.2	0.4861	0.4864	0.4868	0.4871	0.4875	0.4878	0.4881	0.4884	0.4887	0.4890
2.3	0.4893	0.4896	0.4898	0.4901	0.4904	0.4906	0.4909	0.4911	0.4913	0.4916
2.4	0.4918	0.4920	0.4922	0.4925	0.4927	0.4929	0.4931	0.4932	0.4934	0.4936
2.5	0.4938	0.4940	0.4941	0.4943	0.4945	0.4946	0.4948	0.4949	0.4951	0.4952
2.6	0.4953	0.4955	0.4956	0.4957	0.4959	0.4960	0.4961	0.4962	0.4963	0.4964
2.7	0.4965	0.4966	0.4967	0.4968	0.4969	0.4970	0.4971	0.4972	0.4973	0.4974
2.8	0.4974	0.4975	0.4976	0.4977	0.4977	0.4978	0.4979	0.4979	0.4980	0.4981
2.9	0.4981	0.4982	0.4982	0.4983	0.4984	0.4984	0.4985	0.4985	0.4986	0.4986
3.0	0.4987	0.4987	0.4987	0.4988	0.4988	0.4989	0.4989	0.4989	0.4990	0.4990

Source: Abridged from Table I of A. Hald. *Statistical Tables and Formulas* (New York: John Wiley & Sons, Inc.), 1952. Reproduced by permission of A. Hald and the publisher.

T A B L E **5**   Critical Values of *t*

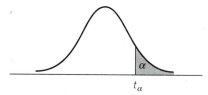

t_α

Degrees of Freedom	$t_{0.100}$	$t_{0.050}$	$t_{0.025}$	$t_{0.010}$	$t_{0.005}$
1	3.078	6.314	12.706	31.821	63.657
2	1.886	2.920	4.303	6.695	9.925
3	1.638	2.353	3.182	4.541	5.841
4	1.533	2.132	2.776	3.747	4.604
5	1.476	2.015	2.571	3.365	4.032
6	1.440	1.943	2.447	3.143	3.707
7	1.415	1.895	2.365	2.998	3.499
8	1.397	1.860	2.306	2.896	3.355
9	1.383	1.833	2.262	2.821	3.250
10	1.372	1.812	2.228	2.764	3.169
11	1.363	1.796	2.201	2.718	3.106
12	1.356	1.782	2.179	2.681	3.055
13	1.350	1.771	2.160	2.650	3.012
14	1.345	1.761	2.145	2.624	2.977
15	1.341	1.753	2.131	2.602	2.947
16	1.337	1.746	2.120	2.583	2.921
17	1.333	1.740	2.110	2.567	2.898
18	1.330	1.734	2.101	2.552	2.878
19	1.328	1.729	2.093	2.539	2.861
20	1.325	1.725	2.086	2.528	2.845
21	1.323	1.721	2.080	2.518	2.831
22	1.321	1.717	2.074	2.508	2.819
23	1.319	1.714	2.069	2.500	2.807
24	1.318	1.711	2.064	2.492	2.797
25	1.316	1.708	2.060	2.485	2.787
26	1.315	1.706	2.056	2.479	2.779
27	1.314	1.703	2.052	2.473	2.771
28	1.313	1.701	2.048	2.467	2.763
29	1.311	1.699	2.045	2.462	2.756
∞	1.282	1.645	1.960	2.326	2.576

Source: From M. Merrington, "Table of Percentage Points of the *t*-Distribution," *Biometrika*, 32, 1941, p. 300. Reproduced by permission of the *Biometrika* Trustees.

T A B L E 6 Critical Values of χ^2

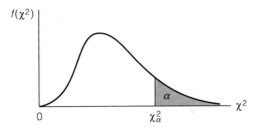

Degrees of Freedom	$\chi^2_{0.995}$	$\chi^2_{0.990}$	$\chi^2_{0.975}$	$\chi^2_{0.950}$	$\chi^2_{0.900}$
1	0.0000393	0.0001571	0.0009821	0.0039321	0.0157908
2	0.0100251	0.0201007	0.0506356	0.102587	0.210720
3	0.0717212	0.114832	0.215795	0.351846	0.584375
4	0.206990	0.297110	0.484419	0.710721	1.063623
5	0.411740	0.554300	0.831211	1.145476	1.61031
6	0.675727	0.872085	1.237347	1.63539	2.20413
7	0.989265	1.239043	1.68987	2.16735	2.83311
8	1.344419	1.646482	2.17973	2.73264	3.48954
9	1.734926	2.087912	2.70039	3.32511	4.16816
10	2.15585	2.55821	3.24697	3.94030	4.86518
11	2.60321	3.05347	3.81575	4.57481	5.57779
12	3.07382	3.57056	4.40379	5.22603	6.30380
13	3.56503	4.10691	5.00874	5.89186	7.04150
14	4.07468	4.66043	5.62872	6.57063	7.78953
15	4.60094	5.22935	6.26214	7.26094	8.54675
16	5.14224	5.81221	6.90766	7.96164	9.31223
17	5.69724	6.40776	7.56418	8.67176	10.0852
18	6.26481	7.01491	8.23075	9.39046	10.8649
19	6.84398	7.63273	8.90655	10.1170	11.6509
20	7.43386	8.26040	9.59083	10.8508	12.4426
21	8.03366	8.89720	10.28293	11.5913	13.2396
22	8.64272	9.54249	10.9823	12.3380	14.0415
23	9.26042	10.19567	11.6885	13.0905	14.8479
24	9.88623	10.8564	12.4011	13.8484	15.6587
25	10.5197	11.5240	13.1197	14.6114	16.4734
26	11.1603	12.1981	13.8439	15.3791	17.2919
27	11.8076	12.8786	14.5733	16.1513	18.1138
28	12.4613	13.5648	15.3079	16.9279	18.9392
29	13.1211	14.2565	16.0471	17.7083	19.7677
30	13.7867	14.9535	16.7908	18.4926	20.5992
40	20.7065	22.1643	24.4331	26.5093	29.0505
50	27.9907	29.7067	32.3574	34.7642	37.6886
60	35.5346	37.4848	40.4817	43.1879	46.4589
70	43.2752	45.4418	48.7576	51.7393	55.3290
80	51.1720	53.5400	57.1532	60.3915	64.2778
90	59.1963	61.7541	65.6466	69.1260	73.2912
100	67.3276	70.0648	74.2219	77.9295	82.3581

Source: From C. M. Thompson, "Tables of the Percentage Points of the χ^2 Distribution," *Biometrika,* 32, 1941, pp. 188–189. Reproduced by permission of the *Biometrika* Trustees.

T A B L E **6**   (Continued)

relevant probability

Degrees of Freedom	$\chi^2_{0.100}$	$\chi^2_{0.050}$	$\chi^2_{0.025}$	$\chi^2_{0.010}$	$\chi^2_{0.005}$
1	2.70554	3.84146	5.02389	6.63490	7.87944
2	4.60517	5.99147	7.37776	9.21034	10.5966
3	6.25139	7.81473	9.34840	11.3449	12.8381
4	7.77944	9.48773	11.1433	13.2767	14.8602
5	9.23635	11.0705	12.8325	15.0863	16.7496
6	10.6446	12.5916	14.4494	16.8119	18.5476
7	12.0170	14.0671	16.0128	18.4753	20.2777
8	13.3616	15.5073	17.5346	20.0902	21.9550
9	14.6837	16.9190	19.0228	21.6660	23.5893
10	15.9871	18.3070	20.4831	23.2093	25.1882
11	17.2750	19.6751	21.9200	24.7250	26.7569
12	18.5494	21.0261	23.3367	26.2170	28.2995
13	19.8119	22.3621	24.7356	27.6883	29.8194
14	21.0642	23.6848	26.1190	29.1413	31.3193
15	22.3072	24.9958	27.4884	30.5779	32.8013
16	23.5418	26.2962	28.8454	31.9999	34.2672
17	24.7690	27.5871	30.1910	33.4087	35.7185
18	25.9894	28.8693	31.5264	34.8053	37.1564
19	27.2036	30.1435	32.8523	36.1908	38.5822
20	28.4120	31.4104	34.1696	37.5662	39.9968
21	29.6151	32.6705	35.4789	38.9321	41.4010
22	30.8133	33.9244	36.7807	40.2894	42.7956
23	32.0069	35.1725	38.0757	41.6384	44.1813
24	33.1963	36.4151	39.3641	42.9798	45.5585
25	34.3816	37.6525	40.6465	44.3141	46.9278
26	35.5631	38.8852	41.9232	45.6417	48.2899
27	36.7412	40.1133	43.1944	46.9630	49.6449
28	37.9159	41.3372	44.4607	48.2782	50.9933
29	39.0875	42.5569	45.7222	49.5879	52.3356
30	40.2560	43.7729	46.9792	50.8922	53.6720
40	51.8050	55.7585	59.3417	63.6907	66.7659
50	63.1671	67.5048	71.4202	76.1539	79.4900
60	74.3970	79.0819	83.2976	88.3794	91.9517
70	85.5271	90.5312	95.0231	100.425	104.215
80	96.5782	101.879	106.629	112.329	116.321
90	107.565	113.145	118.136	124.116	128.299
100	118.498	124.342	129.561	135.807	140.169

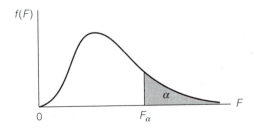

ν_2 \ ν_1	Numerator Degrees of Freedom								
	1	2	3	4	5	6	7	8	9
1	161.4	199.5	215.7	224.6	230.2	234.0	236.8	238.9	240.5
2	18.51	19.00	19.16	19.25	19.30	19.33	19.35	19.37	19.38
3	10.13	9.55	9.28	9.12	9.01	8.94	8.89	8.85	8.81
4	7.71	6.94	6.59	6.39	6.26	6.16	6.09	6.04	6.00
5	6.61	5.79	5.41	5.19	5.05	4.95	4.88	4.82	4.77
6	5.99	5.14	4.76	4.53	4.39	4.28	4.21	4.15	4.10
7	5.59	4.74	4.35	4.12	3.97	3.87	3.79	3.73	3.68
8	5.32	4.46	4.07	3.84	3.69	3.58	3.50	3.44	3.39
9	5.12	4.26	3.86	3.63	3.48	3.37	3.29	3.23	3.18
10	4.96	4.10	3.71	3.48	3.33	3.22	3.14	3.07	3.02
11	4.84	3.98	3.59	3.36	3.20	3.09	3.01	2.95	2.90
12	4.75	3.89	3.49	3.26	3.11	3.00	2.91	2.85	2.80
13	4.67	3.81	3.41	3.18	3.03	2.92	2.83	2.77	2.71
14	4.60	3.74	3.34	3.11	2.96	2.85	2.76	2.70	2.65
15	4.54	3.68	3.29	3.06	2.90	2.79	2.71	2.64	2.59
16	4.49	3.63	3.24	3.01	2.85	2.74	2.66	2.59	2.54
17	4.45	3.59	3.20	2.96	2.81	2.70	2.61	2.55	2.49
18	4.41	3.55	3.16	2.93	2.77	2.66	2.58	2.51	2.46
19	4.38	3.52	3.13	2.90	2.74	2.63	2.54	2.48	2.42
20	4.35	3.49	3.10	2.87	2.71	2.60	2.51	2.45	2.39
21	4.32	3.47	3.07	2.84	2.68	2.57	2.49	2.42	2.37
22	4.30	3.44	3.05	2.82	2.66	2.55	2.46	2.40	2.34
23	4.28	3.42	3.03	2.80	2.64	2.53	2.44	2.37	2.32
24	4.26	3.40	3.01	2.78	2.62	2.51	2.42	2.36	2.30
25	4.24	3.39	2.99	2.76	2.60	2.49	2.40	2.34	2.28
26	4.23	3.37	2.98	2.74	2.59	2.47	2.39	2.32	2.27
27	4.21	3.35	2.96	2.73	2.57	2.46	2.37	2.31	2.25
28	4.20	3.34	2.95	2.71	2.56	2.45	2.36	2.29	2.24
29	4.18	3.33	2.93	2.70	2.55	2.43	2.35	2.28	2.22
30	4.17	3.32	2.92	2.69	2.53	2.42	2.33	2.27	2.21
40	4.08	3.23	2.84	2.61	2.45	2.34	2.25	2.18	2.12
60	4.00	3.15	2.76	2.53	2.37	2.25	2.17	2.10	2.04
120	3.92	3.07	2.68	2.45	2.29	2.17	2.09	2.02	1.96
∞	3.84	3.00	2.60	2.37	2.21	2.10	2.01	1.94	1.88

Denominator Degrees of Freedom

Source: From M. Merrington and C. M. Thompson, "Tables of Percentage Points of the Inverted Beta (F)-Distribution," *Biometrika,* 33, 1943, pp. 73–88. Reproduced by permission of the *Biometrika* Trustees.

T A B L E **7** (Continued)

ν_2 \ ν_1	Numerator Degrees of Freedom									
	10	12	15	20	24	30	40	60	120	∞
1	241.9	243.9	245.9	248.0	249.1	250.1	251.1	252.2	253.3	254.3
2	19.40	19.41	19.43	19.45	19.45	19.46	19.47	19.48	19.49	19.50
3	8.79	8.74	8.70	8.66	8.64	8.62	8.59	8.57	8.55	8.53
4	5.96	5.91	5.86	5.80	5.77	5.75	5.72	5.69	5.66	5.63
5	4.74	4.68	4.62	4.56	4.53	4.50	4.46	4.43	4.40	4.36
6	4.06	4.00	3.94	3.87	3.84	3.81	3.77	3.74	3.70	3.67
7	3.64	3.57	3.51	3.44	3.41	3.38	3.34	3.30	3.27	3.23
8	3.35	3.28	3.22	3.15	3.12	3.08	3.04	3.01	2.97	2.93
9	3.14	3.07	3.01	2.94	2.90	2.86	2.83	2.79	2.75	2.71
10	2.98	2.91	2.85	2.77	2.74	2.70	2.66	2.62	2.58	2.54
11	2.85	2.79	2.72	2.65	2.61	2.57	2.53	2.49	2.45	2.40
12	2.75	2.69	2.62	2.54	2.51	2.47	2.43	2.38	2.34	2.30
13	2.67	2.60	2.53	2.46	2.42	2.38	2.34	2.30	2.25	2.21
14	2.60	2.53	2.46	2.39	2.35	2.31	2.27	2.22	2.18	2.13
15	2.54	2.48	2.40	2.33	2.29	2.25	2.20	2.16	2.11	2.07
16	2.49	2.42	2.35	2.28	2.24	2.19	2.15	2.11	2.06	2.01
17	2.45	2.38	2.31	2.23	2.19	2.15	2.10	2.06	2.01	1.96
18	2.41	2.34	2.27	2.19	2.15	2.11	2.06	2.02	1.97	1.92
19	2.38	2.31	2.23	2.16	2.11	2.07	2.03	1.98	1.93	1.88
20	2.35	2.28	2.20	2.12	2.08	2.04	1.99	1.95	1.90	1.84
21	2.32	2.25	2.18	2.10	2.05	2.01	1.96	1.92	1.87	1.81
22	2.30	2.23	2.15	2.07	2.03	1.98	1.94	1.89	1.84	1.78
23	2.27	2.20	2.13	2.05	2.01	1.96	1.91	1.86	1.81	1.76
24	2.25	2.18	2.11	2.03	1.98	1.94	1.89	1.84	1.79	1.73
25	2.24	2.16	2.09	2.01	1.96	1.92	1.87	1.82	1.77	1.71
26	2.22	2.15	2.07	1.99	1.95	1.90	1.85	1.80	1.75	1.69
27	2.20	2.13	2.06	1.97	1.93	1.88	1.84	1.79	1.73	1.67
28	2.19	2.12	2.04	1.96	1.91	1.87	1.82	1.77	1.71	1.65
29	2.18	2.10	2.03	1.94	1.90	1.85	1.81	1.75	1.70	1.64
30	2.16	2.09	2.01	1.93	1.89	1.84	1.79	1.74	1.68	1.62
40	2.08	2.00	1.92	1.84	1.79	1.74	1.69	1.64	1.58	1.51
60	1.99	1.92	1.84	1.75	1.70	1.65	1.59	1.53	1.47	1.39
120	1.91	1.83	1.75	1.66	1.61	1.55	1.50	1.43	1.35	1.25
∞	1.83	1.75	1.67	1.57	1.52	1.46	1.39	1.32	1.22	1.00

Denominator Degrees of Freedom

TABLE **8**   Percentage Points of the F Distribution, $\alpha = 0.01$

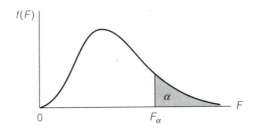

ν_2	Numerator Degrees of Freedom ν_1								
	1	2	3	4	5	6	7	8	9
1	4,052	4,999.5	5,403	5,625	5,764	5,859	5,928	5,982	6,022
2	98.50	99.00	99.17	99.25	99.30	99.33	99.36	99.37	99.39
3	34.12	30.82	29.46	28.71	28.24	27.91	27.67	27.49	27.35
4	21.20	18.00	16.69	15.98	15.52	15.21	14.98	14.80	14.66
5	16.26	13.27	12.06	11.39	10.97	10.67	10.46	10.29	10.16
6	13.75	10.92	9.78	9.15	8.75	8.47	8.26	8.10	7.98
7	12.25	9.55	8.45	7.85	7.46	7.19	6.99	6.84	6.72
8	11.26	8.65	7.59	7.01	6.63	6.37	6.18	6.03	5.91
9	10.56	8.02	6.99	6.42	6.06	5.80	5.61	5.47	5.35
10	10.04	7.56	6.55	5.99	5.64	5.39	5.20	5.06	4.94
11	9.65	7.21	6.22	5.67	5.32	5.07	4.89	4.74	4.63
12	9.33	6.93	5.95	5.41	5.06	4.82	4.64	4.50	4.39
13	9.07	6.70	5.74	5.21	4.86	4.62	4.44	4.30	4.19
14	8.86	6.51	5.56	5.04	4.69	4.46	4.28	4.14	4.03
15	8.68	6.36	5.42	4.89	4.56	4.32	4.14	4.00	3.89
16	8.53	6.23	5.29	4.77	4.44	4.20	4.03	3.89	3.78
17	8.40	6.11	5.18	4.67	4.34	4.10	3.93	3.79	3.68
18	8.29	6.01	5.09	4.58	4.25	4.01	3.84	3.71	3.60
19	8.18	5.93	5.01	4.50	4.17	3.94	3.77	3.63	3.52
20	8.10	5.85	4.94	4.43	4.10	3.87	3.70	3.56	3.46
21	8.02	5.78	4.87	4.37	4.04	3.81	3.64	3.51	3.40
22	7.95	5.72	4.82	4.31	3.99	3.76	3.59	3.45	3.35
23	7.88	5.66	4.76	4.26	3.94	3.71	3.54	3.41	3.30
24	7.82	5.61	4.72	4.22	3.90	3.67	3.50	3.36	3.26
25	7.77	5.57	4.68	4.18	3.85	3.63	3.46	3.32	3.22
26	7.72	5.53	4.64	4.14	3.82	3.59	3.42	3.29	3.18
27	7.68	5.49	4.60	4.11	3.78	3.56	3.39	3.26	3.15
28	7.64	5.45	4.57	4.07	3.75	3.53	3.36	3.23	3.12
29	7.60	5.42	4.54	4.04	3.73	3.50	3.33	3.20	3.09
30	7.56	5.39	4.51	4.02	3.70	3.47	3.30	3.17	3.07
40	7.31	5.18	4.31	3.83	3.51	3.29	3.12	2.99	2.89
60	7.08	4.98	4.13	3.65	3.34	3.12	2.95	2.82	2.72
120	6.85	4.79	3.95	3.48	3.17	2.96	2.79	2.66	2.56
∞	6.63	4.61	3.78	3.32	3.02	2.80	2.64	2.51	2.41

Source: From M. Merrington and C. M. Thompson, "Tables of Percentage Points of the Inverted Beta (F)-Distribution," *Biometrika,* 33, 1943, pp. 73–88. Reproduced by permission of the *Biometrika* Trustees.

T A B L E **8** (Continued)

ν_2 \ ν_1	Numerator Degrees of Freedom									
	10	12	15	20	24	30	40	60	120	∞
1	6,056	6,106	6,157	6,209	6,235	6,261	6,287	6,313	6,339	6,366
2	99.40	99.42	99.43	99.45	99.46	99.47	99.47	99.48	99.49	99.50
3	27.23	27.05	26.87	26.69	26.60	26.50	26.41	26.32	26.22	26.13
4	14.55	14.37	14.20	14.02	13.93	13.84	13.75	13.65	13.56	13.46
5	10.05	9.89	9.72	9.55	9.47	9.38	9.29	9.20	9.11	9.02
6	7.87	7.72	7.56	7.40	7.31	7.23	7.14	7.06	6.97	6.88
7	6.62	6.47	6.31	6.16	6.07	5.99	5.91	5.82	5.74	5.65
8	5.81	5.67	5.52	5.36	5.28	5.20	5.12	5.03	4.95	4.86
9	5.26	5.11	4.96	4.81	4.73	4.65	4.57	4.48	4.40	4.31
10	4.85	4.71	4.56	4.41	4.33	4.25	4.17	4.08	4.00	3.91
11	4.54	4.40	4.25	4.10	4.02	3.94	3.86	3.78	3.69	3.60
12	4.30	4.16	4.01	3.86	3.78	3.70	3.62	3.54	3.45	3.36
13	4.10	3.96	3.82	3.66	3.59	3.51	3.43	3.34	3.25	3.17
14	3.94	3.80	3.66	3.51	3.43	3.35	3.27	3.18	3.09	3.00
15	3.80	3.67	3.52	3.37	3.29	3.21	3.13	3.05	2.96	2.87
16	3.69	3.55	3.41	3.26	3.18	3.10	3.02	2.93	2.84	2.75
17	3.59	3.46	3.31	3.16	3.08	3.00	2.92	2.83	2.75	2.65
18	3.51	3.37	3.23	3.08	3.00	2.92	2.84	2.75	2.66	2.57
19	3.43	3.30	3.15	3.00	2.92	2.84	2.76	2.67	2.58	2.49
20	3.37	3.23	3.09	2.94	2.86	2.78	2.69	2.61	2.52	2.42
21	3.31	3.17	3.03	2.88	2.80	2.72	2.64	2.55	2.46	2.36
22	3.26	3.12	2.98	2.83	2.75	2.67	2.58	2.50	2.40	2.31
23	3.21	3.07	2.93	2.78	2.70	2.62	2.54	2.45	2.35	2.26
24	3.17	3.03	2.89	2.74	2.66	2.58	2.49	2.40	2.31	2.21
25	3.13	2.99	2.85	2.70	2.62	2.54	2.45	2.36	2.27	2.17
26	3.09	2.96	2.81	2.66	2.58	2.50	2.42	2.33	2.23	2.13
27	3.06	2.93	2.78	2.63	2.55	2.47	2.38	2.29	2.20	2.10
28	3.03	2.90	2.75	2.60	2.52	2.44	2.35	2.26	2.17	2.06
29	3.00	2.87	2.73	2.57	2.49	2.41	2.33	2.23	2.14	2.03
30	2.98	2.84	2.70	2.55	2.47	2.39	2.30	2.21	2.11	2.01
40	2.80	2.66	2.52	2.37	2.29	2.20	2.11	2.02	1.92	1.80
60	2.63	2.50	2.35	2.20	2.12	2.03	1.94	1.84	1.73	1.60
120	2.47	2.34	2.19	2.03	1.95	1.86	1.76	1.66	1.53	1.38
∞	2.32	2.18	2.04	1.88	1.79	1.70	1.59	1.47	1.32	1.00

Denominator Degrees of Freedom

T A B L E **9** Factors for Computing Control Chart Lines

Observations in Sample, n	Chart for Averages — Factors for Control Limits			Chart for Standard Deviations — Factors for Central Line		Chart for Standard Deviations — Factors for Control Limits				Chart for Ranges — Factors for Central Line		Chart for Ranges — Factors for Control Limits				
	A_f	A_2	A_3	c_4	$1/c_4$	B_3	B_4	B_5	B_6	d_2	$1/d_2$	d_3	D_1	D_2	D_3	D_4
2	2.121	1.880	2.659	0.7979	1.2533	0	3.267	0	2.606	1.128	0.8862	0.853	0	3.686	0	3.267
3	1.732	1.023	1.954	0.8862	1.1284	0	2.568	0	2.276	1.693	0.5908	0.888	0	4.358	0	2.575
4	1.500	0.729	1.628	0.9213	1.0854	0	2.266	0	2.088	2.059	0.4857	0.880	0	4.698	0	2.282
5	1.342	0.577	1.427	0.9400	1.0638	0	2.089	0	1.964	2.326	0.4299	0.864	0	4.918	0	2.114
6	1.225	0.483	1.287	0.9515	1.0510	0.030	1.970	0.029	1.874	2.534	0.3946	0.848	0	5.079	0	2.004
7	1.134	0.419	1.182	0.9594	1.0424	0.118	1.882	0.113	1.806	2.704	0.3698	0.833	0.205	5.204	0.076	1.924
8	1.061	0.373	1.099	0.9650	1.0363	0.185	1.815	0.179	1.751	2.847	0.3512	0.820	0.388	5.307	0.136	1.864
9	1.000	0.337	1.032	0.9693	1.0317	0.239	1.761	0.232	1.707	2.970	0.3367	0.808	0.547	5.393	0.184	1.816
10	0.949	0.308	0.975	0.9727	1.0281	0.284	1.716	0.276	1.669	3.078	0.3249	0.797	0.686	5.469	0.223	1.777
11	0.905	0.285	0.927	0.9754	1.0253	0.321	1.679	0.313	1.637	3.173	0.3152	0.787	0.811	5.535	0.256	1.744
12	0.866	0.266	0.886	0.9776	1.0230	0.354	1.646	0.346	1.610	3.258	0.3069	0.778	0.923	5.594	0.283	1.717
13	0.832	0.249	0.850	0.9794	1.0210	0.382	1.618	0.374	1.585	3.336	0.2998	0.770	1.025	5.647	0.307	1.693
14	0.802	0.235	0.817	0.9810	1.0194	0.406	1.594	0.399	1.563	3.407	0.2935	0.763	1.118	5.696	0.328	1.672
15	0.755	0.223	0.789	0.9823	1.0180	0.428	1.572	0.421	1.544	3.472	0.2880	0.756	1.203	5.740	0.347	1.653
16	0.750	0.212	0.763	0.9835	1.0168	0.448	1.552	0.440	1.526	3.532	0.2831	0.750	1.282	5.782	0.363	1.637
17	0.728	0.203	0.739	0.9845	1.0157	0.466	1.534	0.458	1.511	3.588	0.2787	0.744	1.356	5.820	0.378	1.622
18	0.707	0.194	0.718	0.9854	1.0148	0.482	1.518	0.475	1.496	3.640	0.2747	0.739	1.424	5.856	0.391	1.609
19	0.688	0.187	0.698	0.9862	1.0140	0.497	1.503	0.490	1.483	3.689	0.2711	0.733	1.489	5.889	0.404	1.596
20	0.671	0.180	0.680	0.9869	1.0132	0.510	1.490	0.504	1.470	3.735	0.2677	0.729	1.549	5.921	0.415	1.585
21	0.655	0.173	0.663	0.9876	1.0126	0.523	1.477	0.516	1.459	3.778	0.2647	0.724	1.606	5.951	0.425	1.575
22	0.640	0.167	0.647	0.9882	1.0120	0.534	1.466	0.528	1.448	3.819	0.2618	0.720	1.660	5.979	0.435	1.565
23	0.626	0.162	0.633	0.9887	1.0114	0.545	1.455	0.539	1.438	3.858	0.2592	0.716	1.711	6.006	0.443	1.557
24	0.612	0.157	0.619	0.9892	1.0109	0.555	1.445	0.549	1.429	3.895	0.2567	0.712	1.759	6.032	0.452	1.548
25	0.600	0.135	0.606	0.9896	1.0105	0.565	1.435	0.559	1.420	3.931	0.2544	0.708	1.805	6.056	0.459	1.541
Over 25	$3/\sqrt{n}$	$\cdots$	a	b	c	d	e	f	g	$\cdots$	$\cdots$	$\cdots$	$\cdots$	$\cdots$	$\cdots$	$\cdots$

Source: "Manual on Presentation of Data and Control Chart Analysis," 6th ed., ASTM, Philadelphia, Pa, p. 91.
Note: Values of all factors in this table were recomputed in 1987 by A. T. A. Holden of the Rochester Institute of Technology. The computed values for d_2 and d_3 as tabulated agree with appropriately rounded values from H. L. Harter, in *Order Statistics and Their Use in Testing and Estimation*, vol. 1, 1969, p. 376.

$$a = 3/\sqrt{n - 0.5}, \quad b = (4n - 4)/(4n - 3), \quad c = (4n - 3)/(4n - 4), \quad d = 1 - 3/\sqrt{2n - 2.5}, \quad e = 1 + 3/\sqrt{2n - 2.5},$$

$$f = (4n - 4)/(4n - 3) - 3/\sqrt{2n - 1.5}, \quad g = (4n - 4)/(4n - 3) + 3/\sqrt{2n - 1.5}$$

T A B L E **10**   Sample Size Code Letters: MIL-STD-105D

Lot or Batch Size	Special Inspection Levels				General Inspection Levels		
	S-1	S-2	S-3	S-4	I	II	III
2–8	A	A	A	A	A	A	B
9–15	A	A	A	A	A	B	C
16–25	A	A	B	B	B	C	D
26–50	A	B	B	C	C	D	E
51–90	B	B	C	C	C	E	F
91–150	B	B	C	D	D	F	G
151–280	B	C	D	E	E	G	H
281–500	B	C	D	E	F	H	J
501–1,200	C	C	E	F	G	J	K
1,201–3,200	C	D	E	G	H	K	L
3,201–10,000	C	D	F	G	J	L	M
10,001–35,000	C	D	F	H	K	M	N
35,001–150,000	D	E	G	J	L	N	P
150,001–500,000	D	E	G	J	M	P	Q
500,001 and over	D	E	H	K	N	Q	R

T A B L E   11   Master Table for Normal Inspection (Single Sampling): MIL-STD-105D

Acceptable Quality Levels (normal inspection)

Sample Size Code Letter	Sample Size	0.010	0.015	0.025	0.040	0.065	0.10	0.15	0.25	0.40	0.65	1.0	1.5	2.5	4.0	6.5	10	15	25	40	65	100	150	250	400	650	1.000
A	2	↓	↓	↓	↓	↓	↓	↓	↓	↓	↓	↓	↓	↓	↓	↓	↓	0 1	1 2	2 3	3 4	5 6	7 8	10 11	14 15	21 22	30 31
B	3	↓	↓	↓	↓	↓	↓	↓	↓	↓	↓	↓	↓	↓	↓	↓	0 1	1 2	2 3	3 4	5 6	7 8	10 11	14 15	21 22	30 31	44 45
C	5	↓	↓	↓	↓	↓	↓	↓	↓	↓	↓	↓	↓	↓	↓	0 1	1 2	2 3	3 4	5 6	7 8	10 11	14 15	21 22	30 31	44 45	↑
D	8	↓	↓	↓	↓	↓	↓	↓	↓	↓	↓	↓	↓	↓	0 1	1 2	2 3	3 4	5 6	7 8	10 11	14 15	21 22	30 31	44 45	↑	↑
E	13	↓	↓	↓	↓	↓	↓	↓	↓	↓	↓	↓	↓	0 1	1 2	2 3	3 4	5 6	7 8	10 11	14 15	21 22	30 31	44 45	↑	↑	↑
F	20	↓	↓	↓	↓	↓	↓	↓	↓	↓	↓	↓	0 1	1 2	2 3	3 4	5 6	7 8	10 11	14 15	21 22	30 31	44 45	↑	↑	↑	↑
G	32	↓	↓	↓	↓	↓	↓	↓	↓	↓	↓	0 1	1 2	2 3	3 4	5 6	7 8	10 11	14 15	21 22	30 31	44 45	↑	↑	↑	↑	↑
H	50	↓	↓	↓	↓	↓	↓	↓	↓	↓	0 1	1 2	2 3	3 4	5 6	7 8	10 11	14 15	21 22	30 31	44 45	↑	↑	↑	↑	↑	↑
J	80	↓	↓	↓	↓	↓	↓	↓	↓	0 1	1 2	2 3	3 4	5 6	7 8	10 11	14 15	21 22	30 31	44 45	↑	↑	↑	↑	↑	↑	↑
K	125	↓	↓	↓	↓	↓	↓	↓	0 1	1 2	2 3	3 4	5 6	7 8	10 11	14 15	21 22	30 31	44 45	↑	↑	↑	↑	↑	↑	↑	↑
L	200	↓	↓	↓	↓	↓	↓	0 1	1 2	2 3	3 4	5 6	7 8	10 11	14 15	21 22	30 31	44 45	↑	↑	↑	↑	↑	↑	↑	↑	↑
M	315	↓	↓	↓	↓	↓	0 1	1 2	2 3	3 4	5 6	7 8	10 11	14 15	21 22	30 31	44 45	↑	↑	↑	↑	↑	↑	↑	↑	↑	↑
N	500	↓	↓	↓	↓	0 1	1 2	2 3	3 4	5 6	7 8	10 11	14 15	21 22	30 31	44 45	↑	↑	↑	↑	↑	↑	↑	↑	↑	↑	↑
P	800	↓	↓	↓	0 1	1 2	2 3	3 4	5 6	7 8	10 11	14 15	21 22	30 31	44 45	↑	↑	↑	↑	↑	↑	↑	↑	↑	↑	↑	↑
Q	1,250	↓	↓	0 1	1 2	2 3	3 4	5 6	7 8	10 11	14 15	21 22	30 31	44 45	↑	↑	↑	↑	↑	↑	↑	↑	↑	↑	↑	↑	↑
R	2,000	↓	0 1	1 2	2 3	3 4	5 6	7 8	10 11	14 15	21 22	30 31	44 45	↑	↑	↑	↑	↑	↑	↑	↑	↑	↑	↑	↑	↑	↑

(Each AQL column shows the pair Ac Re.)

↓ = use first sampling plan below arrow. If sample size equals or exceeds lot or batch size, do 100% inspection.

↑ = use first sampling plan above arrow.

Ac = acceptance number.

Re = rejection number.

T A B L E **12** Master Table for Tightened Inspection (Single Sampling): MIL-STD-105D

Values shown as **Ac Re** (Ac = acceptance number, Re = rejection number). ↓ = use first sampling plan below arrow. ↑ = use first sampling plan above arrow.

Sample Size Code Letter	Sample Size	0.010	0.015	0.025	0.040	0.065	0.10	0.15	0.25	0.40	0.65	1.0	1.5	2.5	4.0	6.5	10	15	25	40	65	100	150	250	400	650	1,000
A	2	↓	↓	↓	↓	↓	↓	↓	↓	↓	↓	↓	↓	↓	↓	↓	↓	↓	0 1	1 2	2 3	3 4	5 6	8 9	12 13	18 19	27 28
B	3	↓	↓	↓	↓	↓	↓	↓	↓	↓	↓	↓	↓	↓	↓	↓	↓	0 1	1 2	2 3	3 4	5 6	8 9	12 13	18 19	27 28	41 42
C	5	↓	↓	↓	↓	↓	↓	↓	↓	↓	↓	↓	↓	↓	↓	↓	0 1	1 2	2 3	3 4	5 6	8 9	12 13	18 19	27 28	41 42	↑
D	8	↓	↓	↓	↓	↓	↓	↓	↓	↓	↓	↓	↓	↓	↓	0 1	1 2	2 3	3 4	5 6	8 9	12 13	18 19	27 28	41 42	↑	↑
E	13	↓	↓	↓	↓	↓	↓	↓	↓	↓	↓	↓	↓	↓	0 1	1 2	2 3	3 4	5 6	8 9	12 13	18 19	27 28	41 42	↑	↑	↑
F	20	↓	↓	↓	↓	↓	↓	↓	↓	↓	↓	↓	↓	0 1	1 2	2 3	3 4	5 6	8 9	12 13	18 19	27 28	41 42	↑	↑	↑	↑
G	32	↓	↓	↓	↓	↓	↓	↓	↓	↓	↓	↓	0 1	1 2	2 3	3 4	5 6	8 9	12 13	18 19	27 28	41 42	↑	↑	↑	↑	↑
H	50	↓	↓	↓	↓	↓	↓	↓	↓	↓	↓	0 1	1 2	2 3	3 4	5 6	8 9	12 13	18 19	27 28	41 42	↑	↑	↑	↑	↑	↑
J	80	↓	↓	↓	↓	↓	↓	↓	↓	↓	0 1	1 2	2 3	3 4	5 6	8 9	12 13	18 19	27 28	41 42	↑	↑	↑	↑	↑	↑	↑
K	125	↓	↓	↓	↓	↓	↓	↓	↓	0 1	1 2	2 3	3 4	5 6	8 9	12 13	18 19	27 28	41 42	↑	↑	↑	↑	↑	↑	↑	↑
L	200	↓	↓	↓	↓	↓	↓	↓	0 1	1 2	2 3	3 4	5 6	8 9	12 13	18 19	27 28	41 42	↑	↑	↑	↑	↑	↑	↑	↑	↑
M	315	↓	↓	↓	↓	↓	↓	0 1	1 2	2 3	3 4	5 6	8 9	12 13	18 19	27 28	41 42	↑	↑	↑	↑	↑	↑	↑	↑	↑	↑
N	500	↓	↓	↓	↓	↓	0 1	1 2	2 3	3 4	5 6	8 9	12 13	18 19	27 28	41 42	↑	↑	↑	↑	↑	↑	↑	↑	↑	↑	↑
P	800	↓	↓	↓	↓	0 1	1 2	2 3	3 4	5 6	8 9	12 13	18 19	27 28	41 42	↑	↑	↑	↑	↑	↑	↑	↑	↑	↑	↑	↑
Q	1,250	↓	↓	↓	0 1	1 2	2 3	3 4	5 6	8 9	12 13	18 19	27 28	41 42	↑	↑	↑	↑	↑	↑	↑	↑	↑	↑	↑	↑	↑
R	2,000	↓	↓	0 1	1 2	2 3	3 4	5 6	8 9	12 13	18 19	27 28	41 42	↑	↑	↑	↑	↑	↑	↑	↑	↑	↑	↑	↑	↑	↑
S	3,150	↓	0 1	1 2	2 3	3 4	5 6	8 9	12 13	18 19	27 28	41 42	↑	↑	↑	↑	↑	↑	↑	↑	↑	↑	↑	↑	↑	↑	↑

Acceptable Quality Levels (tightened inspection)

↓ = use first sampling plan below arrow. If sample size equals or exceeds lot or batch size, do 100% inspection.

↑ = use first sampling plan above arrow.

Ac = acceptance number.

Re = rejection number.

T A B L E **13** Master Table for Reduced Inspection (Single Sampling): MIL-STD-105D

Acceptable Quality Levels (reduced inspection)*

(Each cell gives Ac Re. ↓ = use first sampling plan below arrow. ↑ = use first sampling plan above arrow.)

Sample Size Code Letter	Sample Size	0.010	0.015	0.025	0.040	0.065	0.10	0.15	0.25	0.40	0.65	1.0	1.5	2.5	4.0	6.5	10	15	25	40	65	100	150	250	400	650	1.000
A	2	↓	↓	↓	↓	↓	↓	↓	↓	↓	↓	↓	↓	↓	↓	0 1	↑	↑	1 2	2 3	3 4	5 6	7 8	10 11	14 15	21 22	30 31
B	2	↓	↓	↓	↓	↓	↓	↓	↓	↓	↓	↓	↓	↓	0 1	↑	↑	0 2	1 3	2 4	3 5	5 6	7 8	10 11	14 15	21 22	30 31
C	2	↓	↓	↓	↓	↓	↓	↓	↓	↓	↓	↓	↓	0 1	↑	↑	0 2	1 3	1 4	2 5	3 6	5 8	7 10	10 13	14 17	21 24	↑
D	3	↓	↓	↓	↓	↓	↓	↓	↓	↓	↓	↓	0 1	↑	↑	0 2	1 3	1 4	2 5	3 6	5 8	7 10	10 13	14 17	21 24	↑	↑
E	5	↓	↓	↓	↓	↓	↓	↓	↓	↓	↓	0 1	↑	↑	0 2	1 3	1 4	2 5	3 6	5 8	7 10	10 13	14 17	21 24	↑	↑	↑
F	8	↓	↓	↓	↓	↓	↓	↓	↓	↓	0 1	↑	↑	0 2	1 3	1 4	2 5	3 6	5 8	7 10	10 13	14 17	21 24	↑	↑	↑	↑
G	13	↓	↓	↓	↓	↓	↓	↓	↓	0 1	↑	↑	0 2	1 3	1 4	2 5	3 6	5 8	7 10	10 13	14 17	21 24	↑	↑	↑	↑	↑
H	20	↓	↓	↓	↓	↓	↓	↓	0 1	↑	↑	0 2	1 3	1 4	2 5	3 6	5 8	7 10	10 13	14 17	21 24	↑	↑	↑	↑	↑	↑
J	32	↓	↓	↓	↓	↓	↓	0 1	↑	↑	0 2	1 3	1 4	2 5	3 6	5 8	7 10	10 13	14 17	21 24	↑	↑	↑	↑	↑	↑	↑
K	50	↓	↓	↓	↓	↓	0 1	↑	↑	0 2	1 3	1 4	2 5	3 6	5 8	7 10	10 13	14 17	21 24	↑	↑	↑	↑	↑	↑	↑	↑
L	80	↓	↓	↓	↓	0 1	↑	↑	0 2	1 3	1 4	2 5	3 6	5 8	7 10	10 13	14 17	21 24	↑	↑	↑	↑	↑	↑	↑	↑	↑
M	125	↓	↓	↓	0 1	↑	↑	0 2	1 3	1 4	2 5	3 6	5 8	7 10	10 13	14 17	21 24	↑	↑	↑	↑	↑	↑	↑	↑	↑	↑
N	200	↓	↓	0 1	↑	↑	0 2	1 3	1 4	2 5	3 6	5 8	7 10	10 13	14 17	21 24	↑	↑	↑	↑	↑	↑	↑	↑	↑	↑	↑
P	315	↓	0 1	↑	↑	0 2	1 3	1 4	2 5	3 6	5 8	7 10	10 13	14 17	21 24	↑	↑	↑	↑	↑	↑	↑	↑	↑	↑	↑	↑
Q	500	0 1	↑	↑	0 2	1 3	1 4	2 5	3 6	5 8	7 10	10 13	14 17	21 24	↑	↑	↑	↑	↑	↑	↑	↑	↑	↑	↑	↑	↑
R	800	↑	↑	0 2	1 3	1 4	2 5	3 6	5 8	7 10	10 13	14 17	21 24	↑	↑	↑	↑	↑	↑	↑	↑	↑	↑	↑	↑	↑	↑

↓ = use first sampling plan below arrow. If sample size equals or exceeds lot or batch size, do 100% inspection.

↑ = use first sampling plan above arrow.

Ac = acceptance number.

Re = rejection number.

*If the acceptance number has been exceeded but the rejection number has not been reached, accept the lot but reinstate normal inspection.

T A B L E **14**   MIL-STD-414

(a) AQL Conversion Table

For Specified AQL Values Falling Within These Ranges	Use This AQL Value
to 0.049	0.04
0.050–0.069	0.065
0.070–0.109	0.10
0.110–0.164	0.15
0.165–0.279	0.25
0.280–0.439	0.40
0.440–0.699	0.65
0.700–1.09	1.0
1.10–1.64	1.5
1.65–2.79	2.5
2.80–4.39	4.0
4.40–6.99	6.5
7.00–10.9	10.0
11.00–16.4	15.0

(b) Sample Size Code Letters*

Lot Size	Inspection Levels				
	I	II	III	IV	V
3–8	B	B	B	B	C
9–15	B	B	B	B	D
16–25	B	B	B	C	E
26–40	B	B	B	D	F
41–65	B	B	C	E	G
66–110	B	B	D	F	H
111–180	B	C	E	G	I
181–300	B	D	F	H	J
301–500	C	E	G	I	K
501–800	D	F	H	J	L
801–1,300	E	G	I	K	L
1,301–3,200	F	H	J	L	M
3,201–8,000	G	I	L	M	N
8,001–22,000	H	J	M	N	O
22,001–110,000	I	K	N	O	P
110,001–550,000	I	K	O	P	Q
550,001 and over	I	K	P	Q	Q

*Sample size code letters given in body of table are applicable when the indicated inspection levels are to be used.

T A B L E **15**

MIL-STD-414 (standard deviation method) Master Table for Normal and Tightened Inspection for Plans Based on Variability Unknown (single specification limit—form 1)

Sample Size Code Letter	Sample Size	Acceptable Quality Levels (normal inspection)													
		0.04 k	0.065 k	0.10 k	0.15 k	0.25 k	0.40 k	0.65 k	1.00 k	1.50 k	2.50 k	4.00 k	6.50 k	10.00 k	15.00 k
B	3	↓	↓	↓	↓	↓	↓	↓	↓	↓	1.12	0.958	0.765	0.566	0.341
C	4	↓	↓	↓	↓	↓	↓	↓	1.45	1.34	1.17	1.01	0.814	0.617	0.393
D	5	↓	↓	↓	↓	↓	↓	1.65	1.53	1.40	1.24	1.07	0.874	0.675	0.455
E	7	↓	↓	↓	↓	2.00	1.88	1.75	1.62	1.50	1.33	1.15	0.955	0.755	0.536
F	10	↓	↓	↓	2.24	2.11	1.98	1.84	1.72	1.58	1.41	1.23	1.03	0.828	0.611
G	15	2.64	2.53	2.42	2.32	2.20	2.06	1.91	1.79	1.65	1.47	1.30	1.09	0.886	0.664
H	20	2.69	2.58	2.47	2.36	2.24	2.11	1.96	1.82	1.69	1.51	1.33	1.12	0.917	0.695
I	25	2.72	2.61	2.50	2.40	2.26	2.14	1.98	1.85	1.72	1.53	1.35	1.14	0.936	0.712
J	30	2.73	2.61	2.51	2.41	2.28	2.15	2.00	1.86	1.73	1.55	1.36	1.15	0.946	0.723
K	35	2.77	2.65	2.54	2.45	2.31	2.18	2.03	1.89	1.76	1.57	1.39	1.18	0.969	0.745
L	40	2.77	2.66	2.55	2.44	2.31	2.18	2.03	1.89	1.76	1.58	1.39	1.18	0.971	0.746
M	50	2.83	2.71	2.60	2.50	2.35	2.22	2.08	1.93	1.80	1.61	1.42	1.21	1.00	0.774
N	75	2.90	2.77	2.66	2.55	2.41	2.27	2.12	1.98	1.84	1.65	1.46	1.24	1.03	0.804
O	100	2.92	2.80	2.69	2.58	2.43	2.29	2.14	2.00	1.86	1.67	1.48	1.26	1.05	0.819
P	150	2.96	2.84	2.73	2.61	2.47	2.33	2.18	2.03	1.89	1.70	1.51	1.29	1.07	0.841
Q	200	2.97	2.85	2.73	2.62	2.47	2.33	2.18	2.04	1.89	1.70	1.51	1.29	1.07	0.845
		0.065	0.10	0.15	0.25	0.40	0.65	1.00	1.50	2.50	4.00	6.50	10.00	15.00	

Acceptable Quality Levels (tightened inspection)

Note: All AQL values are in percent defective.

↓ = use first sampling plan below arrow, that is, both sample size and k value. When sample size equals or exceeds lot size, every item in the lot must be inspected.

TABLE **16** Factors for Two-Sided Tolerance Limits

δ n	1 − α = 0.95			1 − α = 0.99		
	0.90	0.95	0.99	0.90	0.95	0.99
2	32.019	37.674	48.430	160.193	188.491	242.300
3	8.380	9.916	12.861	18.930	22.401	29.055
4	5.369	6.370	8.299	9.398	11.150	14.527
5	4.275	5.079	6.634	6.612	7.855	10.260
6	3.712	4.414	5.775	5.337	6.345	8.301
7	3.369	4.007	5.248	4.613	5.488	7.187
8	3.136	3.732	4.891	4.147	4.936	6.468
9	2.967	3.532	4.631	3.822	4.550	5.966
10	2.839	3.379	4.433	3.582	4.265	5.594
11	2.737	3.259	4.277	3.397	4.045	5.308
12	2.655	3.162	4.150	3.250	3.870	5.079
13	2.587	3.081	4.044	3.130	3.727	4.893
14	2.529	3.012	3.955	3.029	3.608	4.737
15	2.480	2.954	3.878	2.945	3.507	4.605
16	2.437	2.903	3.812	2.872	3.421	4.492
17	2.400	2.858	3.754	2.808	3.345	4.393
18	2.366	2.819	3.702	2.753	3.279	4.307
19	2.337	2.784	3.656	2.703	3.221	4.230
20	2.310	2.752	3.615	2.659	3.168	4.161
25	2.208	2.631	3.457	2.494	2.972	3.904
30	2.140	2.549	3.350	2.385	2.841	3.733
35	2.090	2.490	3.272	2.306	2.748	3.611
40	2.052	2.445	3.213	2.247	2.677	3.518
45	2.021	2.408	3.165	2.200	2.621	3.444
50	1.996	2.379	3.126	2.162	2.576	3.385
55	1.976	2.354	3.094	2.130	2.538	3.335
60	1.958	2.333	3.066	2.103	2.506	3.293
65	1.943	2.315	3.042	2.080	2.478	3.257
70	1.929	2.299	3.021	2.060	2.454	3.225
75	1.917	2.285	3.002	2.042	2.433	3.197
80	1.907	2.272	2.986	2.026	2.414	3.173
85	1.897	2.261	2.971	2.012	2.397	3.150
90	1.889	2.251	2.958	1.999	2.382	3.130
95	1.881	2.241	2.945	1.987	2.368	3.112
100	1.874	2.233	2.934	1.977	2.355	3.096
150	1.825	2.175	2.859	1.905	2.270	2.983
200	1.798	2.143	2.816	1.865	2.222	2.921
250	1.780	2.121	2.788	1.839	2.191	2.880
300	1.767	2.106	2.767	1.820	2.169	2.850
400	1.749	2.084	2.739	1.794	2.138	2.809
500	1.737	2.070	2.721	1.777	2.117	2.783
600	1.729	2.060	2.707	1.764	2.102	2.763
700	1.722	2.052	2.697	1.755	2.091	2.748
800	1.717	2.046	2.688	1.747	2.082	2.736
900	1.712	2.040	2.682	1.741	2.075	2.726
1,000	1.709	2.036	2.676	1.736	2.068	2.718
∞	1.645	1.960	2.576	1.645	1.960	2.576

Source: From C. Eisenart, M. W. Hastay, and W. A. Wallis. *Techniques of Statistical Analysis*, McGraw-Hill Book Company, Inc., 1947. Reproduced with permission of McGraw-Hill.

References

American Society for Testing and Materials (1992). *Manual on Presentation of Data and Control Chart Analysis,* 6th ed., ASTM, Philadelphia, Pa.

Bowker, A. H., and G. J. Lieberman (1972). *Engineering Statistics,* 2nd ed., Prentice-Hall, Inc., Englewood Cliffs, N.J.

Ford (1987). *Continuing Process Control and Process Capability Improvement,* Ford Motor Company, Dearborn, Mich.

Grant, E. L., and R. S. Leavenworth (1980). *Statistical Quality Control,* 5th ed., McGraw-Hill Book Co., New York.

Guttman, I.; S. S. Wilks; and J. S. Hunter (1971). *Introductory Engineering Statistics,* 2nd ed., John Wiley & Sons, New York.

Koopmans, L. H. (1987). *An Introduction to Contemporary Statistical Methods,* PWS-KENT Publishing Co., Boston.

Kume, H. (1992). *Statistical Methods for Quality Improvement,* The Association for Overseas Technical Scholarship (AOTS), Tokyo.

Lapin, L. L. (1990). *Probability and Statistics for Modern Engineering,* 2nd ed., PWS-KENT Publishing Co., Boston.

McClave, J. T., and F. H. Dietrich (1985). *Statistics,* 3rd ed., Dellen Publishing Co., San Francisco.

Mendenhall, W.; D. D. Wackerly; and R. L. Scheaffer (1990). *Mathematical Statistics with Applications,* 4th ed., PWS-KENT Publishing Co., Boston.

Miller, I., and J. E. Freund (1977). *Probability and Statistics for Engineers,* 2nd ed., Prentice-Hall Inc., Englewood Cliffs, N.J.

Ott, L. (1988). *An Introduction to Statistical Methods and Data Analysis,* 3rd ed., PWS-KENT Publishing Co., Boston.

Ross, S. (1984). *A First Course in Probability,* 2nd ed., Macmillan Publishing Co., New York.

Ryan, T. P. (1989). *Statistical Methods for Quality Improvement,* John Wiley & Sons, New York.

Stephens, M. A. (1974). "EDF Statistics for Goodness of Fit and Some Comparisons." *Journal of Am. Stat. Assn.,* vol. 69, no. 347, pp. 730–737.

Thompson, J. R., and J. Koronacki (1993). *Statistical Process Control for Quality Improvement,* Chapman and Hall, New York.

Tukey, J. W. (1977) *Exploratory Data Analysis,* Addison-Wesley Publishing Co.

Wadsworth, H. M.; K. S. Stephens; and A. B. Godfrey (1986). *Modern Methods for Quality Control and Improvement,* John Wiley & Sons, New York.

Walpole, R. E., and R. H. Myers (1985). *Probability and Statistics for Engineers and Scientists,* 3rd ed., Macmillan Publishing Co., New York.

Answers to Selected Exercises

Chapter 1

1.4

Pre-eruption		Posteruption
9	0.6	
8 5	0.7	
2	0.8	
	0.9	
	1.0	
8	1.1	
6	1.2	2 4 9
	1.3	
	1.4	0 5 6
6	1.5	
	1.6	3
	1.7	9
	1.8	
	1.9	0

Leaf unit $= 0.01$

1.6 **a** 64%; 45%

b 20–29; 30–39

1.9 **b** $\bar{x} = 10.35$, $s = 49.00$

d No. The extreme data point at $+205.2$ inflates the mean and standard deviation.

1.11 Yes

1.13 **c** $\bar{x}_1 = 4.37$, $s_1 = 0.34$; $\bar{x}_2 = 1.12$, $s_2 = 0.82$

1.15 $\bar{x}_1 = 83.4$, $s_1 = 89.20$; $\bar{x}_2 = 35.8$, $s_2 = 39.16$

1.17 SAT: $\bar{x} = 474.30$, $s = 6.96$; ACT: $\bar{x} = 17.75$, $s = 0.84$

f 18.7

1.18 1980: $\bar{x} = 22.23$, $s = 2.08$; 1990: $\bar{x} = 129.64$, $s = 5.61$

b Empirical rule better for 1980

1.19 $\bar{x} = 0.344$, $s = 0.022$

c Yes

d Yes

Chapter 2

2.2 **a** The order given is correct.
b All sequences have the same probability.

2.6 **a** 921,000
b A share represents fewer households.

2.8 **a** Experiment

2.9 **a** 33%
b Observational study

2.10 **b**

	White	Black	Asian	Native American	
Engineer	2.446	0.043	0.151	0.011	2.7 million
Computer	6.180	0.259	0.462	0.007	7.0 million

2.11 **b** 13 **d** 9

2.13 **a** *JD, JS, DM, DN, MN, JM, JN, DS, MS, SN*
b 7 **c** 6 **d** $\overline{A}$
e $\overline{A} = \{MS, MN, SN\}$
$AB = B = \{JM, JS, JN, DM, DS, DN\}$
$A \cup B = A = \{JD, JM, JS, JN, DM, DS, DN\}$
$\overline{AB} = \{JD, MS, MN, SN\}$

2.16 **a** *L, R, S* **b** $P(L) = P(R) = P(S) = \frac{1}{3}$ **c** $\frac{2}{3}$

2.17 **b** 0.9 **d** 0.1
f 0.1

2.19 **b** 0.05 **d** 0.74

2.21 **b** $\frac{5}{9}$

2.23 **a** (I, I), (I, II), (I, III), (II, I), (II, II), (II, III), (III, I), (III, II), (III, III)
b $\frac{1}{3}$ **c** $\frac{5}{9}$

2.25 **a**   42   **b**   21

2.27 5,040

2.30 **a** $\frac{3}{5}$ **b** $\frac{2}{5}$ **c** $\frac{3}{10}$

2.31 **b** $\frac{9}{64}$ **d** $\frac{36}{64}$

2.33 **a**   24   **b** $\frac{1}{2}$

2.36 **a** $\frac{2}{3}$ **b** $\frac{1}{9}$ **c** $\frac{1}{3}$

2.37 **b** $\frac{32,949}{64,053} = 0.5144$ **d** $\frac{7,067}{16,065} = 0.4399$

2.39 **a**   0.10   **b**   0.03   **c**   0.06   **d**   0.018

2.41 **a** $\frac{306}{380} = 0.8053$ **b** $\frac{378}{380} = 0.9947$
        c $\frac{306}{378} = 0.8096$

2.43 **a**   0.7225   **b**   0.19

2.45 0.81, 0.99   **2.46**   0.40

2.47 0.8235   **2.48**   0.0833

2.49 **c**   34,600; 9,700
        d   No

2.52

Cholesterol	Odds Ratio
≤ 159	0.236
160–209	0.309
210–259	0.608
≥ 260	0.596

2.53 Odds ratios: **a**   1.18   **b**   0.68; 0

2.55 **a**   HHHH, HHHT, HHTH, HHTT, HTHH, HTHT, HTTH, HTTT, THHH, THHT, THTH, THTT, TTHH, TTHT, TTTH, TTTT
        b $A = \{HHHT, HHTH, HTHH, THHH\}$ **c** $\frac{1}{4}$

2.57 **a** 0.07, 0.32, 0.29   **b** 0.23, 0.43, 0.39
c 0.59, 0.75, 0.65

2.60 **b** 0.18   **d** $\frac{18}{57} = 0.3158$

2.61 120   **2.63** 720   **2.65** 40,320

2.67 **a** 0.216   **b** 0.936   **c** 0.648

2.69 $\frac{1}{16}$ **2.70** 0.5952

2.71 **a** $20(\frac{1}{64}) = \frac{5}{16}$ **b** $27(\frac{1}{2})^{10}$

2.73 Not independent

2.75 $\frac{1}{16} = 0.0625$ **2.78** Not independent

2.79 $\frac{1}{2}$

2.83 **a** 0.00892   **b** 0.9890

2.87 0.5073

Chapter 3

3.1

x	0	1	2	3
$p(x)$	$\frac{1}{30}$	$\frac{9}{30}$	$\frac{15}{30}$	$\frac{5}{30}$

3.3

x	0	1	2	3
$p(x)$	0.2585	0.4419	0.2518	0.0478

Independence; no; not unusual

3.5 **a**

x	0	1	2	3	4
$p(x)$	0.0053	0.0575	0.2331	0.4201	0.2840

b 0.9947

3.7

x	0	1	2	3
$p(x)$	$\frac{8}{27}$	$\frac{12}{27}$	$\frac{6}{27}$	$\frac{1}{27}$

y	0	1	2	3
$p(y)$	$\frac{2,744}{3,375}$	$\frac{588}{3,375}$	$\frac{42}{3,375}$	$\frac{1}{3,375}$

$x + y$	0	1	2	3	4	5	6
$p(x + y)$	0.2409	0.41297	0.26179	0.07445	0.00935	0.00053	0.00001

3.9 **a**

x	0	1	2
$p(x)$	$\frac{1}{6}$	$\frac{4}{6}$	$\frac{1}{6}$

b

x	0	1
$p(x)$	$\frac{1}{2}$	$\frac{1}{2}$

c

x	0
$p(x)$	1

3.11 **a** $0, \frac{2}{3}$ **b** $0, \frac{12}{5}$ **c** Box II

3.13 $\mu = 37; \sigma = 11.2$ approximately

3.15 0.4; 0.44; 0.6633

3.17 **a** (1,4702, 6.5298)   **b** Yes

3.19 **a** (84.1886, 115.8114)   **b** No

3.21 **a** 0.1536   **b** 0.1808   **c** 0.9728
d 0.8   **e** 0.64

3.23 **a** 0.537   **b.** 0.098

3.25 **a** 16   **b** 3.2

3.27   8   **3.29**   0.5931

3.31   a   0.99   b   0.9999

3.33   a   0.1536   b   0.9728

3.35   3.96   **3.37**   840

3.39   a   0.648   b   1

3.41   a   0.04374   b   0.99144

3.43   a $\frac{10}{9}, \frac{10}{81}$ b $\frac{10}{3}, \frac{30}{81}$

3.45   150; 4,500; no

3.47   a   0.128   b   0.03072

3.49   a   0.06561   b $\frac{40}{9}, \frac{40}{81}$

3.51   a $\frac{3}{16}$ b $\frac{3}{16}$ c $\frac{1}{8}$ d $\frac{1}{2}$

3.53   a   0.0183   b   0.908   c   0.997

3.55   a   0.467   b   0.188

3.57   a   0.8187   b   0.5488

3.59   a   0.140   b   0.042   c   0.997

3.61   a $128e^{-16}$ b $128e^{-16}$

3.63   320; 56.5685

3.65 λ^2

3.67   a   0.001   b   0.000017

3.69 $\frac{1}{42}$

3.71   a $\frac{4}{5}$ b $\frac{1}{5}$

3.73   a

y	0	1	2
$p(y)$	$\frac{7}{15}$	$\frac{7}{15}$	$\frac{1}{15}$

   b

y	0	1	2	3
$p(y)$	$\frac{1}{6}$	$\frac{1}{2}$	$\frac{3}{10}$	$\frac{1}{30}$

3.75 $\frac{3}{14}$

3.77 **a** $\frac{9}{14}$ **b** $\frac{13}{14}$ **c** 1

3.79 **a** 1   **b** 1   **c** $\frac{18}{19}$ **d** $\frac{49}{57}$ **e** $\frac{728}{969}$

3.81 $\frac{10}{21}$

3.83 $E(Y) = np;\ V(Y) = np(1 - p)$

3.89 0.32805; 0.99999

3.91 **a** 1   **b** 0.5905   **c** 0.1681
  d 0.03125   **e** 0

3.93 **a** $n = 25,\ a = 5$ **b** $n = 25,\ a = 5$

3.95 **a** 0.758   **b** 12; 12   **c** (5.0718, 18.9282)

3.97 **a** 5   **b** 0.007   **c** 0.384

3.99 0.993

3.103 0.18522

3.105 0.01536; 0.0256

3.107 $E(Y) = 900;\ V(Y) = 90;\ K = 2,\ (881.026, 918.974)$

3.109 **a** $p(y) = \binom{4}{y}\left(\frac{1}{3}\right)^y\left(\frac{2}{3}\right)^{4-y}$ **b** $\frac{1}{9}$ **c** $\frac{4}{3}$ **d** $\frac{8}{9}$

3.111 **a** 0.1192   **b** 0.117; yes

3.113 3

3.115 No

Chapter 4

4.1 **a** Discrete **b** Discrete

c Discrete **d** Continuous

4.3 **a** $\frac{27}{32}$ **b** 4

4.5 **b** $F(x) = 0, \quad x \le 5$

$$= \frac{(x - 7)^3}{8} + 1, \quad 5 \le x \le 7$$

$$= 1, \quad x > 7$$

c $\frac{7}{8}$ **d** $\frac{37}{56}$

4.7 **a** $\frac{3}{4}$ **b** $\frac{4}{5}$ **c** 1

d $F(x) = 0, x < 0$

$$= x^2, 0 \le x \le 1$$

$$= 1, x > 1$$

Yes

4.9 $60; \frac{1}{3}$

4.11 4

4.13 **a** 5.5; 0.15 **b** $k = 2, (4.7254, 6.2746)$

c About 58% of the time

4.15 **a** $F(x) = 0, x < a$

$$= \frac{x - a}{b - a}, a \le x \le b$$

$$= 1, x > b$$

b $\frac{b - c}{b - a}$ **c** $\frac{b - d}{b - c}$

4.17 **a** $\frac{1}{20}$ **b** $\frac{1}{20}$ **c** $\frac{1}{2}$

4.19 $\frac{3}{4}$

4.21 **a** $\frac{1}{8}$ **b** $\frac{1}{8}$ **c** $\frac{1}{4}$

4.23 **a** $\frac{2}{7}$ **b** $\frac{3}{200}$; $\frac{49}{120,000}$

4.25 $\frac{1}{6}$

4.27 **a** $\frac{1}{2}$ **b** $\frac{1}{4}$

4.29 **a** 60; $\frac{100}{3}$ **b** 4

4.31 **a** $e^{-5/4}$ **b** $e^{-5/6} - e^{-5/4}$

4.33 **a** e^{-2} **b** 460.52cfs

4.35 **a** $1,100$; $2,920,000$ **b** No; $P(C > 2,000)$
                                                  $= e^{-1.94}$

4.37 **a** e^{-1} **b** $e^{-1/2}$

4.39 **a** $e^{-1/2}$ **b** $e^{-1/4}$ **c** No

4.41 **a** $e^{-5/4}$ **b** $(1 - e^{-5/4})^2$

4.43 **a** $1 - e^{-5/11}$ **b** $e^{-15/11} - e^{-30/11}$
    c 50.66 minutes

4.45 **a** 3.2; 6.4 **b** $(0, 8.26)$

4.47 **a** 276; $47,664$ **b** $(0, 930.963)$

4.49 **a** 20; 200 **b** 10; 50

4.51 **a** 240; 189.74 **b** $(0, 809.21)$

4.53 9.6; 30.72; $f(y) = \dfrac{y^2 e^{-y/3.2}}{\Gamma(3)(3.2)^3}$, $y > 0$
                                 $= 0$, $y \le 0$

4.55   **a**   0.3849   **b**   0.3159   **c**   0.3227
      d   0.1586   **e**   0.0366

4.57   0.0062

4.59   .0730

4.61   **a**   0.9544   **b**   0.8297

4.63   0.6170; 0.3174

4.65   **a**   0.0062   **b**   225.6

4.67   13.67

4.69   **a**   Yes   **b**   Yes   **c**   No; no   **d**   4

4.73   **a**   60   **b** $\frac{4}{7}$; $\frac{3}{98}$

4.75   **a** $\frac{52}{3}$; $\frac{1{,}348}{45}$ **b**   (6.39, 28.28)

4.77 $\frac{2}{3}$; 240°

4.79   **a** $\frac{1}{2}$; $\frac{1}{28}$ **b** $\frac{1}{2}$; $\frac{1}{20}$ **c** $\frac{1}{2}$; $\frac{1}{12}$
      d   Case (a)

4.81   **a** $1 - e^{-1}$ **b** $\sqrt{\pi}$

4.83   0.6576

4.85   0.06573

4.87   **a** $e^{-2.5}$ **b**   0.01855
      c $\dfrac{\sqrt{10\pi}}{2}$; $10(1 - \frac{\pi}{4})$

4.89   42.9193

4.91 $R(t) = e^{-t_2/\theta}$; $t_2 > 0$

4.93 a

4.97 $e^{t^2/2}$

4.99 a $\frac{1}{2}$ b $F(y) = 0, y < 0$
$$= \frac{y^2}{4}, 0 \leq y \leq 2$$
$$= 1, y > 2$$

d $\frac{3}{4}$ e $\frac{3}{4}$

4.101 a $\frac{6}{5}$

b
$$0, y \leq -1$$
$$F(y) = 0.2(y + 1), -1 < y \leq 0$$
$$= 0.2 + 0.2y + \tfrac{3}{5}y^2, 0 < y \leq 1$$
$$= 1, y > 1$$

d 0; 0.2; 1   e $\frac{1}{4}$ f $\frac{2}{5}; \frac{41}{150}$

4.103 15.87%

4.105 0.073

4.107 $\frac{c}{4}, \frac{c(6 - c)}{16}$ b $\frac{c}{(2 - t)^2}$ c 4

4.109 $(0.0062)^3$

4.111 a $1 - e^{-6}$ b $f(y) = \dfrac{y^3 e^{-y/40}}{\Gamma(4)(10)^4}, y > 0$
$$= 0, y \leq 0$$

4.113 $1 - e^{-4}$

4.115 $53.58

4.117 0.736

4.119 a Wiebull with $\gamma = 2, \theta = \dfrac{1}{\lambda\pi}$ b $\dfrac{1}{2\sqrt{\lambda}}$

4.121 a $e^{11}(10^{-2}g); (e^{38} - e^{22})(10^{-4}g^2)$
b (0, 3,570,244.56)   c 0.8023

Chapter 5

5.1 **a**

x_2 \ x_1	0	1	2
0	$\frac{1}{9}$	$\frac{2}{9}$	$\frac{1}{9}$
1	$\frac{2}{9}$	$\frac{2}{9}$	0
2	$\frac{1}{9}$	0	0

b

x_1	0	1	2
$p(x_1)$	$\frac{4}{9}$	$\frac{4}{9}$	$\frac{1}{9}$

c $\frac{1}{2}$

5.3 **a**

	X_1	
	0	1
0	.063	.078
1	.101	.056
X_2 2	.163	.065
3	.169	.055
4	.193	.057

b

X_1	
0	1
.45	.55
.64	.36
.71	.29
.75	.25
.77	.23

c

0	.092	.250
1	.146	.179
X_2 2	.236	.210
3	.245	.178
4	.280	.185

5.5 **a** $f(x_1) = 1, 0 \le x_1 \le 1$
   $= 0,$ otherwise

   b $\frac{1}{2}$ **c** Yes

5.7 **a** $\frac{7}{8}$ **b** $\frac{1}{2}$ **c** $\frac{2}{3}$

5.9 **a** $\frac{21}{64}$ **b** $\frac{1}{3}$ **c** No

5.11 **a** Yes   **b** $\dfrac{3e^{-1}}{2}$

5.13 $\frac{11}{36}$

5.15 $\frac{23}{144}$

5.17 **a** 0.2; 0.16; 0.2; 0.16

   b -0.04 **c** 0.4; 0.24

5.19 **a** $\frac{2}{3}; \frac{1}{18}$ **b** $(\frac{1}{3}, 1)$

5.21 **a** 0.0972 **b** $\frac{1}{2}$ **c** e^{-1}

    d $f_1(y_1) = y_1 e^{-y_1}, 0 \le y_1 < \infty$
             $= 0$, otherwise
       $f_2(y_2) = e^{-y_2}, 0 \le y_2 < \infty$
             $= 0$, otherwise

5.23 **a** 1 **b** 1 **c** e^{-2}

5.27 66,960

5.29 0.07776; independence

5.31 **a** $\frac{4}{27}$ **b** $\frac{1}{27}$ **c** $\frac{4}{9}$

5.33 2.5; 4.875

5.35 **a** 0.08575 **b** 0.7627 **c** 40; 24

5.37 0.05213

5.39 **a** 0.1587 **b** 0.3085

5.41 **a** 4 **b** $f_1(x_1) = 2x_1, 0 \le x_1 \le 1$
                     $f_2(x_2) = 2x_2, 0 \le x_2 \le 1$

    c $F(x_1, x_2) = 0, x_1, x_2 < 0$
               $= x_1^2 x_2^2, 0 \le x_1, x_2 \le 1$
               $= x_1^2, 0 \le x_1 \le 1, x_2 > 1$
               $= x_2^2, 0 \le x_2 \le 1, x_1 > 1$
               $= 1, x_1, x_2 > 1$

    d $\frac{9}{64}$ **e** $\frac{1}{4}$

5.43 **a**

x_1	x_2	$p(x_1, x_2)$
0	0	0
0	1	$\frac{1}{28}$
0	2	$\frac{1}{14}$
0	3	$\frac{1}{84}$
1	0	$\frac{1}{21}$
1	1	$\frac{2}{7}$
1	2	$\frac{1}{7}$
2	0	$\frac{1}{7}$
2	1	$\frac{3}{14}$
3	0	$\frac{1}{21}$

b

	0	1	2	3	$p(x_1)$
0	0	$\frac{1}{28}$	$\frac{1}{14}$	$\frac{1}{84}$	$\frac{10}{84}$
1	$\frac{1}{21}$	$\frac{2}{7}$	$\frac{1}{7}$	0	$\frac{10}{21}$
2	$\frac{1}{7}$	$\frac{3}{14}$	0	0	$\frac{5}{14}$
3	$\frac{1}{21}$	0	0	0	$\frac{1}{21}$
$p(x_2)$	$\frac{5}{21}$	$\frac{15}{28}$	$\frac{3}{14}$	$\frac{1}{84}$	

c $\frac{9}{16}$

5.45 a $\dfrac{2x_1}{1 - x_2^2}, \ 0 \le x_2 \le x_1 \le 1$

b $\dfrac{1}{x_1}, \ 0 \le x_2 \le x_1 \le 1$

d $\frac{5}{12}$

5.47 a $f(x_1) = \begin{cases} \dfrac{2}{\pi}\sqrt{1 - x_1^2}, & -1 \le x_1 \le 1 \\ 0, & \text{otherwise} \end{cases}$

b $\frac{1}{2}$

5.49 a 1

b $f(x_1) = \begin{cases} \dfrac{x_1}{2}, & 0 \le x_1 \le 2 \\ 0, & \text{otherwise} \end{cases}$

$f(x_2) = \begin{cases} 2(1 - x_2), & 0 \le x_2 \le 1 \\ 0, & \text{otherwise} \end{cases}$

c $f(x_1 | x_2) = \begin{cases} \dfrac{1}{2(1 - x_2)}, & 0 \le 2x_2 \le x_1 \le 2 \\ 0, & \text{otherwise} \end{cases}$

d $f(x_2|x_1) = \begin{cases} \dfrac{2}{x_1}, & 0 \le 2x_2 \le x_1 \le 2 \\ 0, & \text{otherwise} \end{cases}$

e $\frac{1}{2}$ **f** $\frac{8}{9}$

5.51 **a** $\frac{2}{3}$ **b** $\frac{1}{2}; \frac{1}{18}$ **c** 0

5.53 **a** -0.3333 **b** $\frac{7}{3}; \frac{7}{18}$ **c** $\frac{7}{3}; \frac{7}{18}$

5.55 **a** 2 **b** $\frac{2}{3}$

5.57 $\left(\frac{1}{2}\right)^{(x+1)}$, $x = 0, 1, 2, \ldots$

5.59 42; 26; no

5.61 **a** 1 **b** 1

5.63 300

5.65 **a** $(p_1 e^{t_1} + p_2 e^{t_2} + p_3 e^{t_3})^n$

    b $-np_1 p_2$

Chapter 6

6.1 **b** $\bar{x} = 1.6633$, $s^2 = .228097$

    c $(0.7081, 2.6185)$

6.3 **b** $\bar{x} = 3.5$, $s^2 = 3.0172$

6.5 $\bar{x}_1 = 102.78$, $s_1 = 45.92$
$\bar{x}_2 = 85.62$, $s_2 = 18.67$
$\bar{x}_3 = 89.44$, $s_3 = 10.97$
$\bar{x}_4 = 102.54$, $s_4 = 47.34$

6.9 0.6826   **6.11** 0.9090

6.13 a 0.5328   b 0.9772

   c Independent random sample

6.15 0.1587

6.17 a 0.9938   b 110

6.19 0.0013   **6.21** 0.9876

6.23 0.0062

6.25 a 0.5930   b 0.0559   c 0.0017

6.27 0.1292

6.29 0.1539   **6.31** 0.0023

6.33 0.0043

6.35 10.148   **6.37** 29

6.39 No; $P(S^2 > 0.065) \approx 0.1$

6.41 a $0.025 < P(S^2 > 80) < .05$

   b $P(S^2 < 20) < 0.005$

   c (21.13, 78.87)

6.43 $a = 119.80658, b = 356.9625$

6.45 a (14.7756, 15.281)

   b No; (14.7613, 15.2595)

6.47 (14.5162, 15.5404); no

6.49 a (0, 0.1443)   b No; (0, 0.1377)

6.51 a (0, 0.1906)   b No

6.53 (0, 11.8015)

6.57 C_{pk} will increase; proportion will decrease.

6.61 $6 \cdot 10^{-7}$

6.65 0.9484

6.67 0.0274

6.69 183.07

Chapter 7

7.1 a $\hat{\theta}_1, \hat{\theta}_2, \hat{\theta}_3, \hat{\theta}_4,$ b $\hat{\theta}_4$

7.3 a $\overline{X}$ c $\frac{1}{n}\sum_{i=1}^{n} X_i^2 + 3\overline{X}$

7.5 $\frac{1}{12n} + \frac{1}{4}$

7.7 $\overline{X} - 1.645\, S\sqrt{\frac{n-1}{2}}\,\Gamma(\frac{n-1}{2})/\Gamma(\frac{n}{2})$

7.9 (205.81, 214.19)

7.11 (17.98, 18.82)   **7.13** 139

7.15 (0.0563, 0.1837)

7.17 (0.2636, 0.5364)

7.19 (0.658, 0.842)

7.21 (173.79, 186.21)

7.23 (9.52, 10.08)

7.25 (0.15, 0.53)

7.27 (475.25, 478.75)

7.29 (399.25, 700.75)

7.31 No

7.35 193

7.37 (0.2521, 0.5879); yes

7.39 (−0.1193, 0.3574)

7.41 (−1.5198, 12.3865), (0.158, 12.201)

7.43 (−80.61, −31.873)

7.45 (0.928, 9.87)

7.47 (−0.9659, 0.0087)

7.49 (−244.9688, 178.5243); no

7.51 (−0.0265, 0.0465). The methods are not significantly different using a 95% confidence interval.

7.53 (28.72, 29.28)

7.55 (0, 15.938)

7.57 (2.917, 5.583)

7.59 a (10.0625, 10.1375)

   b (10.0605, 10.1395)

   c (10.0413, 10.1587)

7.61 (1.04126, 1.1587)

7.63 a (1.04378, 1.1562)

   b (1.04606, 1.1539); no

7.65 a $n = 77$ b $n = 18$

   c $n = 93$ d $n = 22$

7.67 a $\bar{X} \pm Z_{\alpha/2}\sqrt{\bar{X}/n}$ b (3.608, 4.392)

7.69 $\hat{\beta} = \bar{X}/\alpha$ **7.71** $\hat{p} = \frac{1}{x}$

7.73 $60\hat{p} + 20\hat{p}(1 + 19\hat{p})$

7.75 $\hat{\lambda} = \dfrac{Y + 1}{2}$

7.77   (1.926, 2.274)

7.79   (59.75, 70.25)

7.81   **a**   (2.0988, 2.6712)   **b**   (3.0017, 3.8503)

7.83   **a**   (0.1664, 1.1724)   **b**   (0.6910, 6.9879)

7.85   (0.000263, 0.001855)

7.87 $m = \dfrac{n\sigma_1}{\sigma_1 + \sigma_2}$

7.89 $2\bar{y} + \bar{x} \pm t_{.025}\sqrt{\hat{\sigma}^2(\frac{4}{n} + \frac{3}{m})}$

   $\hat{\sigma}^2 = \dfrac{(n-1)S_y^2 + (\frac{m-1}{3})S_x^2}{n + m - 2}$

7.91 $\hat{R} = \dfrac{X}{n - X}$

Chapter 8

8.1 $Z = -4.216$; reject

8.3   70

8.5   0.1151

8.7 $Z = -1.22$; no. P-value $= 0.2224$

8.9 $Z = 0.527$; no

8.11 $t = -2.52$; no

8.13 $t = -7.47$; yes

8.15 $t = -0.3945$; no

8.17 $Z = 1.77$; no

8.19 $Z = -3.16$; yes

8.21 $t = -0.753$; no

8.23 $\chi^2 = 20.16$; no

8.25 $\chi^2 = 78.71$; yes

8.27 $Z = 3.648$; yes

8.29 $t = 0.1024$; no

8.31 $t = 5.94$; yes

8.33 $F = 1.0936$; do not reject

8.35 $t = 1.214$; no

8.37 $t = 7.016$; yes

8.39 $t = 1.109$; no

8.41 $t = -0.254$; no

8.43 $t = -0.349$; no

8.45 $F = 1.0678$; no

8.47 $\chi^2 = 1.23$; do not reject

8.49 $\chi^2 = 1.671$; do not reject

8.51 $\chi^2 = 0.542$; no

8.53 **a** $\chi^2 = 13.709$; no   **b** $Z = 2.0494$; yes

8.55 **a** $\chi^2 = 2.0533$; do not reject
b $Z = .4899$; do not reject

8.57 $\chi^2 = 2.48$; no

8.59 $\chi^2 = 9.62$; no

8.61 $\chi^2 = 20.20$; yes

8.63 $\chi^2 = 3.24$; do not reject

8.65 $\chi^2 = 55.70063$; reject

8.67 Modified $D = .6679$; do not reject

8.69 Modified $D = 3.8916$; reject

8.71 Modified $D = 0.2132$; no

8.75 **a** Sample size: 35; accept if $\dfrac{20 - \bar{X}}{S} \geq 1.57$

b Sample size: 35; accept if $\dfrac{20 - \bar{X}}{S} \geq 1.76$

8.77 **a** Sample size: 20; accept if $\dfrac{\bar{X} - L}{S} \geq 0.917$

b Sample size: 20; accept if $\dfrac{\bar{X} - L}{S} \geq 1.12$

8.79 $t = -13.44$; no

8.81 $Z = -2.108$; no

8.83 $t = -0.5392$; no

8.85 $\chi^2 = 21.875$; yes

8.87 $Z = -4.66$; yes

8.89 $Z = 1.17$; no

8.91 $F = 12.25$; no

Chapter 9

9.1   **a** $\hat{\beta}_0 = 0$, $\hat{\beta}_1 = \frac{6}{7}$

9.3 $\hat{y}_i = -0.06 + (2.78)x_i$

9.5 $\hat{y}_i = 480.2 - 3.75x_i$

9.7 $\hat{y}_i = 10.20833 + 1.925x_i$

9.9   8.1: 1.14286, .28571; 8.2: 1.6, .53333;
8.3: 0.199, .066333; 8.4: 0.039469, .0098673;
8.5: 114.87443, 19.14574; 8.6: 88.61638, 14.76940;
8.7: 241.79167, 24.17917; 8.8: 87.78550, 10.97319

9.11 $t = 6.7082$; yes

9.13   (2.26, 3.30)

9.15 $t = -17.9417$; yes

9.17   (5.75, 13.5)

9.19   **a** $\hat{y}_i = 36.68 - 0.333x_i$

      b   1.34687, 0.16836

      c $t = -7.377$; yes

9.21   c $r_1 = .96528$, $r_2 = .99605$ d   Yes

9.23   a $\hat{y}_i = -235.116 + 1.27x_i$
      b $t = 18.299$; yes   c $r^2 = .97667$

9.25   a $(0.908, 2.521)$ b $(0.0256, 3.403)$

9.27 $(1.337, 2.033)$

9.29 $(187.169, 210.569)$

9.31 $(4.991, 21.107)$

9.33   a $\hat{y}_i = 9.5538 + 2.6708x_i$ b $t = 8.475$; yes
      c $(16.911, 19.290)$ d $(17.140, 17.992)$

9.35   a 0.91548 b 0.83810 c $t = 5.0877$; yes

9.37 $r^2 = 0.59210$

9.39   a $\hat{y}_i = -1{,}094.28686 + 181.1026x_i$
      b $(141.936, 220.270)$

9.41   a $\hat{y}_i = -13.49037 - 0.052829x_i$
      b $t = -6.836$; reject
      c 0.85382 d $(0.5987, 1)$

9.43   a   Conoco, reject; DuPont, reject; Mobil, reject; Seagram, reject
      b 0.076 c   Conoco

9.45   a $\hat{y}_i = 44.17197 - 0.0254777x_i$
      c $t = -0.0294$; do not reject   d   No
      e 0.000288209

9.47 **a** $r = 0.89140$, $r^2 = 0.79460$

b $t = 4.817$; yes

9.49 $\ln(y) = -64.3 + 0.037x$, where $x = $ year. Growth is faster through 1878 and then is slower than average from model.

9.51 $y = 910 - 0.455x$

9.53 Both show significant increases using current dollars. Women show significant increase while men do not using constant dollars.

9.55 $y = -0.33 + 12.32x$; increase

Chapter 10

10.1 **a** Yes   **b** $F = 38.84$; reject

10.3 $F = 1.056$; do not reject

10.5 **a** $\hat{y}_i = 20.0911 - 0.6705x_i + 0.009535x_i^2$

c $t = 1.507$; no

d $\hat{y}_i = 19.2791 - 0.4449x_i$

e $(-0.5177, -0.3722)$

10.7 $t = 3.039$; reject

10.11 **a** $y = \beta_0 + \beta_1 x_1 + \beta_2 x_1^2 + \epsilon$

b $F = 44.50$; reject   **c** $p = 0.0001$

d $t = -4.683$; yes   **e** $p = 0.0001$

10.13 **b** $\nu_1 = 2, \nu_2 = 20$

10.15 **a** $H_0: \beta_1 = \cdots = \beta_5 = 0; H_a$: at least one of the β_is is nonzero

b $H_0: \beta_3 = \beta_4 = \beta_5 = 0; H_a$: at least one of $\beta_3, \beta_4, \beta_5$ is nonzero

c $F = 18.29$; reject   **d** $F = 8.46$; reject

10.17 $F = 1.409$; do not reject

10.19 **a** $\hat{y} = .04565 + .000785x_1 + .23737x_2 - .0000381x_1x_2$

b SSE $= 2.71515039, s^2 = 0.1696969$

c SSE is minimized by the choice of $\hat{\beta}_0, \hat{\beta}_1, \hat{\beta}_2,$ and $\hat{\beta}_3$.

10.21 ($7.32, $9.45)

10.23 **a** $\hat{y} = 0.60130 + 0.59526x_1 - 3.72536x_2 - 16.23196x_3 + 0.23492x_1x_2 + 0.30808x_1x_3$

b $R^2 = 0.928104, F = 139.42$; reject

d $F = 6.866$; reject

10.25 $t = -1.104$; do not reject

10.27 $F = 3.306$; reject

10.29 **a** $y = \beta_0 + \beta_1x_1 + \beta_2x_2 + \beta_3x_3 + \epsilon; x_1 =$ area, $x_2 =$ number of baths, $x_3 = 1$ for central air, $x_3 = 0$ otherwise

b $y = \beta_0 + \beta_1x_1 + \beta_2x_2 + \beta_3x_3 + \beta_4x_1x_2 + \beta_5x_1x_3 + \beta_6x_2x_3 + \beta_7x_1^2 + \beta_8x_2^2 + \epsilon$

c $H_0: \beta_4 = \cdots = \beta_8 = 0; H_a$: At least one $\beta_i, i = 4, \cdots, 8$ is nonzero.

10.31 **a** $\hat{y} = 1.67939 + 0.44414x_1 - 0.079322x_2$

b $F = 150.62$; reject   **c** $R^2 = 0.97729$; yes

d $(-0.0907, -0.0680)$ **e** 1.2036 hours

f 12.036 hours

10.33 **a** $R^2 = 0.936572$, $F = 100.41$; reject

      c $t = 1.67$; do not reject

10.35 **b** $\hat{y} = -93.12768 + 0.44458x$

      c $\hat{y} = 204.46031 - 0.63800x + 0.00095925x^2$

      d $t = 3.64$; reject

10.37 $t = -6.6$; reject   **10.38** $F = 6.712$; yes

10.41 $F = 9.987$; reject

10.43 **a** $y = 1.1x$

      b $y = 36.2 - 2.34x + 0.08x^2$

      c $y = 36.2 - 1.24x + 0.08x^2$

10.45 B_aA: $y = 0.286 + 1.114x$

      B_aP: $y = 0.202 + 1.024x$

      Phe: $y = 0.411 + 1.10x$

Chapter 11

11.1 **a**

Source	df	SS	MS	F
Treatment	2	11.0752	5.5376	3.151
Error	7	12.3008	1.7573	
Total	9	23.3760		

      b $F = 3.151$; do not reject

11.3 $F = 8.634$; yes

11.5 **b**

Source	df	SS	MS	F
Treatment	3	6.8160	2.2720	8.0283
Error	105	29.715	0.2830	
Total	108	36.531		

    c $F = 8.0283$; yes

    d $(-0.5302, 0.1562)$ **e**   No

11.7 **a**

Source	df	SS	MS	F
Treatment	1	3,237.2	3,237.2	19.622
Error	98	16,167.7	164.9765	
Total	99	19,404.9		

    b $F = 19.622$; yes

11.9 $F = 6.907$; yes

11.11 $F = 29.915$; yes

11.13 $\hat{y} = 1.4333 + 1.4333x_1 + 2.5417x_2$;

$x_1 = \begin{cases} 1 & \text{Sample 1} \\ 0 & \text{otherwise} \end{cases}$, $\quad x_2 = \begin{cases} 1 & \text{Sample 2} \\ 0 & \text{otherwise} \end{cases}$;

$F = 3.15$; do not reject

11.15 $\hat{y} = 0.275 - 0.275x_1 - 0.05x_2$;

$x_1 = \begin{cases} 1 & \text{Thermometer 1} \\ 0 & \text{otherwise} \end{cases}$,

$x_2 = \begin{cases} 1 & \text{Thermometer 2} \\ 0 & \text{otherwise} \end{cases}$;

$F = 0.24$; do not reject

11.17   **a** $t = -0.6676$; no

    b $(-1.3857, 0.5857)$

    c $(3.0030, 4.3970)$

11.19 **a** $(-3.3795, -0.8205)$

 b $i = 1, j = 2 \ (5.7231, 7.7169)$
$i = 1, j = 3 \ (7.1231, 9.1169)$
$i = 2, j = 3 \ (7.8231, 9.8169)$

11.21 **a**

Source	df	SS	MS	F
Treatment	2	23.1667	11.5833	12.636
Block	3	14.25	4.75	5.181
Error	6	5.5	0.9167	
Total	11	42.9167		

 b $F = 12.636$; yes   **c** $F = 5.181$; yes

11.23 $F = 6.3588$; yes

11.24 **a** $F = 0.3381$; no

 b

Source	df	SS	MS	F
Treatment	2	0.9412	0.4706	0.3381
Block	5	10,239.9695	2,047.9939	1,471.3844
Error	10	13.9188	1.3919	
Total	17	10,254.8294		

11.25 **a** $F = 39.4626$; yes   **b** $F = 1.9251$; no

11.27 $F = 83.8239$; yes

11.29 **a** $F = 2.320$; no   **b** $F = 4.6761$; no
  c Do not block

11.31 $y = \beta_0 + \beta_1 x_1 + \beta_2 x_2 + \beta_3 x_3 + \beta_4 x_4 + \beta_5 x_5 + \epsilon$;

$x_1 = \begin{cases} 1 & \text{Block 1} \\ 0 & \text{otherwise} \end{cases}$, $x_2 = \begin{cases} 1 & \text{Block 2} \\ 0 & \text{otherwise} \end{cases}$,

$x_3 = \begin{cases} 1 & \text{Block 3} \\ 0 & \text{otherwise} \end{cases}$,

$x_4 = \begin{cases} 1 & \text{Treatment A} \\ 0 & \text{otherwise} \end{cases}$, $x_5 = \begin{cases} 1 & \text{Treatment B} \\ 0 & \text{otherwise} \end{cases}$

11.33 **a**

	Auto		
	1	2	3
Driver 1	D_1	D_2	D_3
Driver 2	D_3	D_1	D_2
Driver 3	D_2	D_3	D_1

b $y = \beta_0 + \beta_1 x_1 + \beta_2 x_2 + \beta_3 x_3 + \beta_4 x_4 + \beta_5 x_5 + \beta_6 x_6 + \epsilon$;

$x_1 = \begin{cases} 1 & \text{Auto 1} \\ 0 & \text{otherwise} \end{cases}$, $x_2 = \begin{cases} 1 & \text{Auto 2} \\ 0 & \text{otherwise} \end{cases}$,

$x_3 = \begin{cases} 1 & \text{Driver 1} \\ 0 & \text{otherwise} \end{cases}$,

$x_4 = \begin{cases} 1 & \text{Driver 2} \\ 0 & \text{otherwise} \end{cases}$, $x_5 = \begin{cases} 1 & \text{Design 1} \\ 0 & \text{otherwise} \end{cases}$,

$x_6 = \begin{cases} 1 & \text{Design 2} \\ 0 & \text{otherwise} \end{cases}$

c $H_0: \beta_5 = \beta_6 = 0$

11.35 $(-1.418, 1.051)$

11.37 $(-0.0943, 0.0910)$, $(0.0574, 0.2426)$, $(0.0591, 0.2443)$

11.39 **a**

Source	F	
A	26.0417	Significant
B	5.0417	Not significant
$A \times B$	0.3750	Not significant

   b $(-0.3716, 0.00495)$

11.41 **a**

Source	F	
A	458.8507	Significant
B	89.8981	Significant
$A \times B$	3.1915	Not significant

   b $(93.416, 94.584)$

11.43 **a b**

Source	df	SS	MS	F	
A	1	538.756	538.756	3.4653	Not significant
B	1	10,686.361	10,686.361	68.7360	Significant
$A \times B$	1	831.744	831.744	5.3499	Significant
Error	36	5,596.9083	155.4697		
Total	39				

   c Yes   **d** Yes

11.45 **a**

Source	F	
A	0.4764	Not significant
B	11.4503	Significant
$A \times B$	23.4712	Significant

b Head 3

11.47 **a** Completely randomized design

b $F = 7.0245$; yes   **c** (7.701, 8.599)

11.49 **a** $F = 0.1925$; no

b

Source	df	SS	MS	F
Treatment	2	0.63166	0.31583	0.1925
Block	3	0.85583	0.28528	0.1739
Error	6	9.84167	1.64028	
Total	11	11.32916		

c $F = 0.1739$; no   **d** (−2.807, 3.907)

e (−2.865, 2.115), (−2.315, 2.665), (−1.940, 3.040)

11.51 **a**

Source	df	SS	MS	F
A	3	74.3333	24.7778	16.5185
B	2	4.0833	2.0417	1.3613
$A \times B$	6	168.9167	28.1527	18.7685
Error	12	18.0	1.5	
Total	23	265.3333		

b $F = 18.7685$; yes

11.53 **a** Completely randomized design

b

Source	df	SS	MS	F
Treatment	2	57.6042	28.8021	0.3375
Error	13	1,109.3333	85.3333	
Total	15	1,166.9374		

  c $F = 0.3375$; no   **d** (73.684, 88.316)

11.55 **a**

Source	df	SS	MS	F
A	1	1,444	1,444	29.4694
B	1	361	361	7.3673
$A \times B$	1	1	1	.0204
Error	12	588	49	
Total	15	2,394		

  b $F = .0204$; no

  c $(-50.854, -25.146)$, $(-31.854, -6.146)$. Use hourly and piece rates and worker-modified schedule.

11.57 **a** Randomized block design

  b $F = 10.4673$; yes   **c** $F = 3.5036$; no

11.59 **a** $F = 12.6139$; yes   **b** Batch 2

11.61 Thickness and diameter affect AE rates significantly.

11.63 The plume samples differ from the rain samples on the elements As, Sb, and Zn.

11.66 0.40 ± 0.50 ($n = 3$) (not significant)
0.40 ± 0.09 ($n = 20$)
7.2 ± 6.21 ($n = 3$)
7.2 ± 1.17 ($n = 20$)